THE STARS AND GALAXIES

FIFTH EDITION

21ST CENTURY
ASTRONOMY

THE STARS AND GALAXIES

FIFTH EDITION

21ST CENTURY
ASTRONOMY

LAURA KAY • Barnard College

STACY PALEN • Weber State University

GEORGE BLUMENTHAL • University of California—Santa Cruz

W. W. NORTON & COMPANY

NEW YORK • LONDON

W. W. Norton & Company has been independent since its founding in 1923, when William Warder Norton and Mary D. Herter Norton first published lectures delivered at the People's Institute, the adult education division of New York City's Cooper Union. The firm soon expanded its program beyond the Institute, publishing books by celebrated academics from America and abroad. By midcentury, the two major pillars of Norton's publishing program—trade books and college texts—were firmly established. In the 1950s, the Norton family transferred control of the company to its employees, and today—with a staff of four hundred and a comparable number of trade, college, and professional titles published each year—W. W. Norton & Company stands as the largest and oldest publishing house owned wholly by its employees.

Editor: Erik Fahlgren
Project Editor: Diane Cipollone
Developmental Editor: Becky Kohn
Assistant Editor: Arielle Holstein
Manuscript Editor: Christopher Curioli
Managing Editor, College: Marian Johnson
Managing Editor, College Digital Media: Kim Yi
Production Manager: Andy Ensor
Media Editor: Rob Bellinger
Associate Media Editor: Julia Sammaritano
Media Project Editor: Danielle Belfiore
Media Editorial Assistants: Ruth Bolster and Liz Vogt
Marketing Manager: Stacy Loyal
Design Director: Rubina Yeh
Designer: Anna Reich
Photo Editor: Trish Marx
Permissions Manager: Megan Jackson
Permissions Assistant: Elizabeth Trammell
Composition: Graphic World
Manufacturing: Transcontinental

The Library of Congress cataloged another edition as follows:

Kay, Laura.
    21st century astronomy. — Fifth edition / Laura Kay, Barnard College, Stacy Palen, Weber State University, George Blumenthal, University of California-Santa Cruz.
        pages cm
    Previous edition: 21st century astronomy (New York : W.W. Norton & Company, 2013).
    Includes index.
    ISBN 978-0-393-93899-9 (pbk.)
    1. Astronomy—Textbooks. I. Palen, Stacy. II. Blumenthal, George (George Ray) III. Title. IV. Title: Twenty-first century astronomy.
    QB45.2.A14 2016
    520—dc23
                                2015023646
ISBN 978-0-393-60336-1
W. W. Norton & Company, Inc., 500 Fifth Avenue, New York, NY 10110-0017
wwnorton.com
W. W. Norton & Company Ltd., Castle House, 75/76 Wells Street, London W1T 3QT
1 2 3 4 5 6 7 8 9 0

Laura Kay thanks her wife, M.P.M. She dedicates this book to her late uncle, Lee Jacobi, for an early introduction to physics, and to her late colleagues at Barnard College, Tally Kampen and Sally Chapman.

Stacy Palen thanks her husband, John Armstrong, for his patient support during this project.

George Blumenthal gratefully thanks his wife, Kelly Weisberg, and his children, Aaron and Sarah Blumenthal, for their support during this project. He also wants to thank Professor Robert Greenler for stimulating his interest in all things related to physics.

Brief Contents

CHAPTERS 8–12 ARE NOT INCLUDED IN THIS EDITION.

Contents

PART I   Introduction to Astronomy

Chapter 6   The Tools of the Astronomer   142

PART II   The Solar System

Chapter 7   The Birth and Evolution of Planetary Systems   172

PART III   Stars and Stellar Evolution

PART IV   Galaxies, the Universe, and Cosmology

Working It Out

CHAPTERS 8–12 ARE NOT INCLUDED IN THIS EDITION.

AstroTours

CHAPTERS 8–12 ARE NOT INCLUDED IN THIS EDITION.

AstroTour animations are available from the free Student Site at the Digital Landing Page, and they are also integrated into assignable SmartWork exercises. Offline versions of the animations for classroom presentation are available from the Instructor's Resource USB Drive. digital.wwnorton.com/astro5.

Astronomy in Action Videos

CHAPTERS 8–12 ARE NOT INCLUDED IN THIS EDITION.

digital.wwnorton.com/astro5

Nebraska Simulations

CHAPTERS 8–12 ARE NOT INCLUDED IN THIS EDITION.

digital.wwnorton.com/astro5

Preface

Dear Student

Why is it a good idea to take a science course, and in particular, why is astronomy a course worth taking? Many people choose to learn about astronomy because they are curious about the universe. Your instructor likely has two basic goals in mind for you as you take this course. The first is to understand some basic physical concepts and how they apply to the universe around us. The second is to think like a scientist and learn to use the scientific method not only to answer questions in this course but also to make decisions in your life. We have written the fifth edition of *21st Century Astronomy* with these two goals in mind.

Throughout this book, we emphasize not only the content of astronomy (for example, the differences among the planets, the formation of chemical elements) but also *how* we know what we know. The scientific method is a valuable tool that you can carry with you and use for the rest of your life. One way we highlight the process of science is the **Process of Science Figures**. In each chapter, we have chosen one discovery and provided a visual representation illustrating the discovery or a principle of the process of science. In these figures, we try to illustrate that science is not a tidy process, and that discoveries are sometimes made by different groups, sometimes by accident, but always because people are trying to answer a question and show why or how we think something is the way it is.

The most effective way to learn something is to "do" it. Whether playing an instrument or a sport or becoming a good cook, reading "how" can only take you so far. The same is true of learning astronomy. We have written this book to help you "do" as you learn. We have created several tools in every chapter to make reading a more active process. At the beginning of each chapter, we have provided a set of Learning Goals to guide you as you read. There is a lot of information in every chapter, and the Learning Goals should help you focus on the most important points. We present a big-picture question in association with the chapter-opening figure at the beginning of each chapter. For each of these, we have tried to pose a question that is not only relevant to its chapter but also something you may have wondered about. We hope that these questions, plus the photographs that accompany them, capture your attention as well as your imagination.

In addition, there are **Check Your Understanding** questions at the end of each chapter section. These questions are designed to be answered quickly if you have understood the previous section. The answers are provided in the back of the book so you can check your answer and decide if further review is necessary.

As a citizen of the world, you make judgments about science, distinguishing between good science and pseudoscience. You use these judgments to make decisions in the grocery store, pharmacy, car dealership, and voting booth. You may base these decisions on the presentation of information you receive through the media, which is very different from the presentation in class. One important skill is the ability to recognize what is credible and to question what is not. To help you

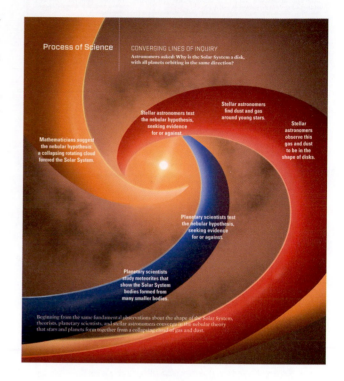

Process of Science

CONVERGING LINES OF INQUIRY
Astronomers asked: Why is the Solar System a disk, with all planets orbiting in the same direction?

Mathematicians suggest the nebular hypothesis: a collapsing rotating cloud formed the Solar System.

Stellar astronomers test the nebular hypothesis, seeking evidence for or against.

Stellar astronomers find dust and gas around young stars.

Stellar astronomers observe this gas and dust to be in the shape of disks.

Planetary scientists test the nebular hypothesis, seeking evidence for or against.

Planetary scientists study meteorites that show the Solar System bodies formed from many smaller bodies.

Beginning from the same fundamental observations about the shape of the Solar System, theorists, planetary scientists, and stellar astronomers converge in the nebular theory that stars and planets form together from a collapsing cloud of gas and dust.

CHECK YOUR UNDERSTANDING 7.4

Suppose that astronomers found a rocky, terrestrial planet beyond the orbit of Neptune. What is the most likely explanation for its origin? (a) It formed close to the Sun and migrated outward. (b) It formed in that location and was not disturbed by migration. (c) It formed later in the Sun's history than other planets. (d) It is a captured planet that formed around another star.

READING ASTRONOMY NEWS

ARTICLES QUESTIONS

A system with five planets was observed by NASA's Kepler space telescope.

Earth-Size Planet Found in the "Habitable Zone" of Another Star

By Science@NASA

Using NASA's Kepler space telescope, astronomers have discovered the first Earth-size planet orbiting in the "habitable zone" of another star (see Figure 7.23). The planet, named "Kepler-186f," orbits an M dwarf, or red dwarf, a class of stars that makes up 70 percent of the stars in the Milky Way Galaxy. The discovery of Kepler-186f confirms that planets the size of Earth exist in the habitable zone of stars other than our Sun.

The "habitable zone" is defined as the range of distances from a star where liquid water might pool on the surface of an orbiting planet. While planets have previously been found in the habitable zone, the previous finds are all at least 40 percent larger in size than Earth, and understanding their makeup is challenging. Kepler-186f is more reminiscent of Earth.

Kepler-186f orbits its parent M dwarf star once every 130 days and receives one-third the energy that Earth gets from the Sun, placing it nearer the outer edge of the habitable zone. On the surface of Kepler-186f, the brightness of its star at high noon is only as bright as our Sun appears to us about an hour before sunset.

"M dwarfs are the most numerous stars," said Elisa Quintana, research scientist at the SETI Institute at NASA's Ames Research Center in Moffett Field, California, and lead author of the paper published today in the journal *Science*. "The first signs of other life in the galaxy may well come from planets orbiting an M dwarf."

However, "being in the habitable zone does not mean we know this planet is habitable," cautions Thomas Barclay, a research scientist at the Bay Area Environmental Research Institute at Ames, and coauthor of the paper. "The temperature on the planet is strongly dependent on what kind of atmosphere the planet has. Kepler-186f can be thought of as an Earth-cousin rather than an Earth-twin. It has many properties that resemble Earth."

Kepler-186f resides in the Kepler-186 system, about 500 light-years from Earth in the constellation Cygnus. The system is also home to four companion planets: Kepler-186b, Kepler-186c, Kepler-186d, and Kepler-186e, whiz around their sun every four, seven, 13, and 22 days, respectively, making them too hot for life as we know it. These four inner planets all measure less than 1.5 times the size of Earth.

Although the size of Kepler-186f is known, its mass and composition are not. Previous research, however, suggests that a planet the size of Kepler-186f is likely to be rocky.

"The discovery of Kepler-186f is a significant step toward finding worlds like our planet Earth," said Paul Hertz, NASA's Astrophysics Division director at the agency's headquarters in Washington.

The next steps in the search for distant life include looking for true Earth-twins—Earth-size planets orbiting within the habitable zone of a Sun-like star—and measuring their chemical compositions. The Kepler space telescope, which simultaneously and continuously measured the brightness of more than 150,000 stars, is NASA's first mission capable of detecting Earth-size planets around stars like our Sun.

Looking ahead, Hertz said, "future NASA missions, like the Transiting Exoplanet Survey Satellite and the James Webb Space Telescope, will discover the nearest rocky exoplanets and determine their composition and atmospheric conditions, continuing humankind's quest to find truly Earth-like worlds."

ARTICLES QUESTIONS

1. This NASA press release was picked up by business and international news feeds. Why do you think coverage of this discovery was so widespread?
2. The planet is closer to its star than Earth is to the Sun yet receives much less energy. What does that imply about the temperature of the star?
3. Why is the mass of this planet not yet known? What method will be used to find its mass?
4. How will astronomers estimate the planet's composition?
5. Why is this planet called a "cousin" of Earth?

hone this skill, we have provided **Reading Astronomy News** sections at the end of every chapter. These features include a news article with questions to help you make sense of how science is presented to you. It is important that you learn to be critical of the information you receive, and these features will help you do that.

While we know a lot about the universe, science is an ongoing process, and we continue to search for new answers. To give you a glimpse of what we don't know, we provide an **Unanswered Questions** feature near the end of each chapter. Most of these questions represent topics that scientists are currently studying.

? UNANSWERED QUESTIONS

- How typical is the Solar System? Only within the past few years have astronomers found other systems containing four or more planets, and so far the observed distributions of large and small planets in these multiplanet systems have looked different from those of the Solar System. Computer simulations of planetary system formation suggest that a system with an orbital stability and a planetary distribution like those of the Solar System may develop only rarely. Improved supercomputers can run more complex simulations, which can be compared with the observations to understand better how solar systems are configured.
- How Earth-like must a planet be before scientists declare it to be "another Earth"? An editorial in the science journal *Nature* cautioned that scientists should define "Earth-like" in advance—before multiple discoveries of planets "similar" to Earth are announced and a media frenzy ensues. Must a planet be of similar size and mass, be located in the habitable zone, and have spectroscopic evidence of liquid water before we call it "Earth 2.0"?

The language of science is mathematics, and it can be as challenging to learn as any other language. The choice to use mathematics as the language of science is not arbitrary; nature "speaks" math. To learn about nature, you will need to speak its language. We don't want the language of math to obscure the concepts, so we have placed this book's mathematics in **Working It Out** boxes to make it clear when we are beginning and ending a mathematical argument, so that you can spend time with the concepts in the chapter text and then revisit the mathematics of the concept to study the formal language of the argument. You will learn to work with data and identify when data aren't quite right. We want you to be comfortable reading, hearing, and speaking the language of science, and we will provide you with tools to make it easier.

7.3 Working It Out Estimating the Radius of an Extrasolar Planet

The masses of extrasolar planets can often be estimated using Kepler's laws and the conservation of angular momentum. When planets are detected by the transit method, astronomers can estimate the radius of an extrasolar planet. In this method, astronomers look for planets that eclipse their stars and observe how much the star's light decreases during this eclipse (see Figure 7.19). In the Solar System when Venus or Mercury transits the Sun, a black circular disk is visible on the face of the circular Sun. During the transit, the amount of light from the transited star is reduced by the area of the circular disk of the planet divided by the area of the circular disk of the star:

$$\text{Percentage reduction in light} = \frac{\text{Area of disk of planet}}{\text{Area of disk of star}}$$

$$= \frac{\pi R^2_{\rm planet}}{\pi R^2_{\rm star}} = \frac{R^2_{\rm planet}}{R^2_{\rm star}}$$

Then, to solve for the radius of the planet, astronomers need an estimate of the radius of the star and a measurement of the percentage reduction in light during the transit. The radius of a star is estimated from the surface temperature and the luminosity of the star.

Let's consider an example. Kepler-11 is a system of at least six planets that transit a star. The radius of the star, $R_{\rm star}$, is estimated to be 1.1 times the radius of the Sun, or $1.1 \times (7.0 \times 10^5\ \text{km}) = 7.7 \times 10^5\ \text{km}$. The light from planet Kepler-11c is observed to decrease by 0.077 percent, or 0.00077 (see Figure 7.19). What is Kepler-11c's size?

$$0.00077 = \frac{R^2_{\rm Kepler\text{-}11c}}{R^2_{\rm star}} = \frac{R^2_{\rm Kepler\text{-}11c}}{(7.7 \times 10^5\ \text{km})^2}$$

$$R^2_{\rm Kepler\text{-}11c} = 4.5 \times 10^8\ \text{km}^2$$

$$R_{\rm Kepler\text{-}11c} = 2.1 \times 10^4\ \text{km}$$

Dividing Kepler-11c's radius by the radius of Earth (6,400 km) shows that the planet Kepler-11c has a radius of 3.3 $R_{\rm Earth}$.

Origins: The Death of the Dinosaurs 227

Origins

The Death of the Dinosaurs

When large impacts happen on Earth, they can have far-reaching consequences for Earth's climate and for terrestrial life. One of the biggest and most significant impacts happened at the end of the Cretaceous Period, which lasted from 146 million years ago to 65 million years ago. At the end of the Cretaceous Period, more than 50 percent of all living species, including the dinosaurs, became extinct. This mass extinction is marked in Earth's fossil record by the Cretaceous-Tertiary boundary, or *K-T boundary* (the *K* comes from *Kreide*, German for "Cretaceous"). Fossils of dinosaurs and other now-extinct life-forms are found in older layers below the K-T boundary. Fossils in the newer rocks above the K-T boundary lack more than half of all previous species but contain a record of many other newly evolving species. Big winners in the new order were the mammals—distant ancestors of humans—that moved into ecological niches vacated by extinct species.

How do scientists know that an impact was involved? The K-T boundary is marked in the fossil record in many areas by a layer of clay. Studies at more than 100 locations around the world have found that this layer contains large amounts of the element iridium, as well as traces of soot. Iridium is very rare in Earth's crust but is common in meteorites. The soot at the K-T boundary possibly indicates that widespread fires burned the world over. The thickness of the layer of clay at the K-T boundary and the concentration of iridium increases toward what is today the Yucatán Peninsula in Mexico. Although the original crater has largely been erased by erosion, geophysical surveys and rocks from drill holes in this area show a deeply deformed subsurface rock structure, similar to that seen at known impact sites. These results provide compelling evidence that 65 million years ago, an asteroid about 10 km in diameter struck the area, throwing great clouds of red-hot dust and other debris into the atmosphere (**Figure 8.30**) and possibly igniting a worldwide conflagration. The energy of the impact is estimated to have been more than that released by 5 billion nuclear bombs.

An impact of this energy clearly would have had a devastating effect on terrestrial life. In addition to the possible firestorm ignited by the impact, computer models suggest there would have been earthquakes and tsunamis. Dust from the collision and soot from the firestorms thrown into Earth's upper atmosphere would have remained there for years, blocking out sunlight and plunging Earth into decades of a cold and dark "impact winter." Recent measurements of ancient microbes in ocean sediments suggest that Earth may have cooled by 7°C. The firestorms, temperature changes, and decreased food supplies could have led to a mass starvation that would have been especially hard on large animals such as the dinosaurs.

Not all paleontologists believe that this mass extinction was the result of an impact; some think volcanic activity was important as well. However, the evidence is compelling that a great impact did occur at the end of the Cretaceous Period. Life on our planet has had its course altered by sudden and cataclysmic events when asteroids and comets have slammed into Earth. It seems very possible that we owe our existence to the luck of our remote ancestors—small rodent-like mammals—that could live amid the destruction after such an impact 65 million years ago.

Figure 8.30 This artist's rendition depicts an asteroid or comet, perhaps 10 km across, striking Earth 65 million years ago in what is now the Yucatán Peninsula in Mexico. The lasting effects of the impact might have killed off most forms of terrestrial life, including the dinosaurs.

Each chapter concludes with an **Origins** section, which relates material or subjects found in the chapter to questions about the origin of the universe and the origin of life in the universe and on Earth. Astrobiologists have made much progress in recent years on understanding how conditions in the universe may have helped or hindered the origin of life, and in each Origins we explore an example from its chapter that relates to how the universe and life formed and evolved.

At the end of each chapter, we have provided several types of questions, problems, and activities for you to practice your skills. The Test Your Understanding questions focus on more detailed facts and concepts from the chapter. Thinking about the Concepts questions ask you to synthesize information and explain the "how" or "why" of a situation. Applying the Concepts problems give you a chance to practice the quantitative skills you learned in the chapter and to work through a situation mathematically. The **Using the Web** questions and **Explorations** represent other opportunities to "learn by doing." Using the Web sends you to websites of space missions, observatories, experiments, or archives to access recent observations, results, or press releases. Other sites are for "citizen science" projects in which you can contribute to the analysis of new data.

Explorations show you how to use the concepts and skills you learned in an interactive way. Most of the book's Explorations ask you to use animations and simulations on the Student Site, while the others are hands-on, paper-and-pencil activities that use everyday objects such as ice cubes or balloons.

The resources outside of the book (at the Student Site) can help you understand and visualize many of the physical concepts described in the book. **AstroTours** and **Nebraska Simulations** are represented by icons in the margins of the book. There is also a series of short **Astronomy in Action** videos that are represented by icons in the margins and available at the Student Site. These videos feature one of the authors (and several students) demonstrating physical concepts at work. Your instructor might assign these videos to you or you might choose to watch them on your own to create a better picture of each concept in your mind.

Astronomy gives you a sense of perspective that no other field of study offers. The universe is vast, fascinating, and beautiful, filled with a wealth of objects that, surprisingly, can be understood using only a handful of principles. By the end of this book, you will have gained a sense of your place in the universe.

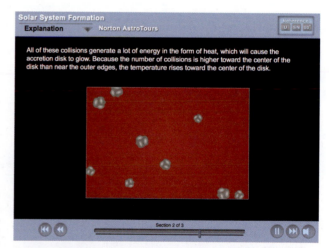

 AstroTour

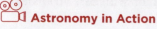 **Astronomy in Action**

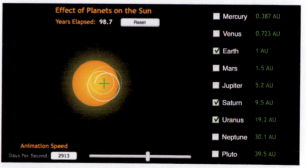

 **Nebraska Simulation**

Dear Instructor

We wrote this book with a few overarching goals: to inspire students, to make the material interactive, and to create a useful and flexible tool that can support multiple learning styles.

As scientists and as teachers, we are passionate about the work we do. We hope to share that passion with students and inspire them to engage in science on their own. Through our own experience, familiarity with education research, and surveys of instructors, we have come to know a great deal about how students learn and what goals teachers have for their students. We have explicitly addressed many of these goals and learning styles in this book, sometimes in large, immediately visible ways such as the inclusion of features but also through less obvious efforts such as questions and problems that relate astronomical concepts to everyday situations or a fresh approach to organizing material.

For example, many teachers state that they would like their students to become "educated scientific consumers" and "critical thinkers" or that their students should "be able to read a news story about science and understand its significance." We have specifically addressed these goals in our Reading Astronomy News feature, which presents a news article and a series of questions that guide a student's critical thinking about the article, the data presented, and the sources.

In nearly every chapter, we have Visual Analogy figures that compare astronomy concepts to everyday events or objects. Through these analogies, we strive to make the material more interesting, relevant, and memorable.

Education research shows that the most effective way to learn is by doing. Exploration activities at the end of each chapter are hands-on, asking students to take the concepts they've learned in the chapter and apply them as they interact with animations and simulations on the Student Site or work through pencil-and-paper activities. Many of these Explorations incorporate everyday objects and can be used either in your classroom or as activities at home. The Using the Web problems direct students to "citizen science" projects, where they can contribute to the analysis of new astronomical data. Other problems send students to websites of space missions, observatories, collaborative projects, and catalogs to access the most current observations, results, and news releases. These Web problems can be used for homework, lab exercises, recitations, or "writing across the curriculum" projects.

We also believe students should be exposed to the more formal language of science—mathematics. We have placed the math in Working It Out boxes, so it does not interrupt the flow of the text or get in the way of students' understanding of conceptual material. But we've gone further by beginning with fundamental ideas in early Working It Out boxes and slowly building in complexity through the book. We've also worked to remove some of the stumbling blocks that affect student confidence by providing calculator hints, references to earlier Working It Out boxes, and detailed, fully worked examples. Many chapters include problems on reading and interpreting graphs. Appendix 1, "Mathematical Tools," has also been reorganized and expanded.

Discussion of basic physics is contained in Part I to accommodate courses that use the *Solar System* or *Stars and Galaxies* volumes. A "just-in-time" approach to introducing the physics is still possible by bringing in material from Chapters 2–6 as needed. For example, the sections on tidal forces in Chapter 4 can be taught along with the moons of the Solar System in Part II, or with mass transfer in binary stars in Part III, or with galaxy interactions in Part IV. Spectral lines in Chapter 5 can be taught with planetary atmospheres in Part II or with stellar spectral types in Part III, and so on.

In our overall organization, we have made several efforts to encourage students to engage with the material and build confidence in their scientific skills as they proceed through the book. For planets, stars and galaxies, we have organized the material to cover the general case first and then delve into more details with specific examples. Thus, you will find "planetary systems" before our own Solar System, "stars" before the Sun, and "galaxies" before the Milky Way. This allows us to avoid frustrating students by making assumptions about what they know about stars or galaxies or forward-referencing to basic definitions and overarching concepts. This organization also implicitly helps students understand their place in the universe: our galaxy and our star are each one of many. They are specific examples of a physical universe in which the same laws apply everywhere. Planets have been organized comparatively to emphasize that science is a process of studying individual examples that lead to collective conclusions. All of these organizational choices were made with the student perspective in mind and a clear sense of the logical hierarchy of the material.

Even our layout has been designed to maximize student engagement—one wide text column is interrupted as seldom as possible. Material from the earlier edition's Connections boxes has been streamlined and incorporated into the text.

We have continued to respond to commentary from you, our colleagues. We have reorganized the material in the first half of Part IV to reflect user feedback. We begin in Chapter 19 by introducing galaxies as a whole and our measurements of them, including recession velocities. Then we address the Milky Way in Chapter 20—a specific example of a galaxy that we can discuss in detail. This follows the repeating motif of moving from the general to the specific that exists throughout the text and gives students a basic grounding in the concepts of spiral galaxies, supermassive black holes, and dark matter before they need to apply those concepts to the specific example of our own galaxy. Chapter 21, "The Expanding Universe," covers the cosmological principle, the Hubble expansion, and the observational evidence for the Big Bang.

We revised each chapter, streamlining some topics, and updating the science to reflect the progress in the field. When appropriate, we have updated the Origins sections, which often illustrate how astrobiologists and other scientists approach the study of a scientific question from the chapter related to the origin of the universe and of life. We have enhanced the material on exoplanets and incorporated material about exoplanets into other chapters when appropriate. We include new images of Mars, Ceres, Comet 67P/Churyumov-Gerasimenko, and Pluto. We note the discovery of our new home supercluster, Laniakea. We've updated the cosmology sections on the highest-redshift objects and the first stars and galaxies.

Many professors find themselves under pressure from accrediting bodies or internal assessment offices to assess their courses in terms of learning goals. To help you with this, we've revised each chapter's Learning Goals and organized the end-of-chapter Summary to correspond to the chapter's Learning Goals. In SmartWork, questions and problems are tagged and can be sorted by Learning Goal. SmartWork contains more than 2,000 questions and problems that are tied directly to this text, including the Check Your Understanding questions and versions of the Reading Astronomy News and Exploration questions. Any of these could be used as a reading quiz to be completed before class or as homework. Every question in SmartWork has hints and answer-specific feedback so that students are coached to work toward the correct answer. An instructor can easily modify any of the provided questions, answers, and feedback or can create his or her own questions.

We've also created a series of 23 videos explaining and demonstrating concepts from the text, accompanied by questions integrated into SmartWork. You might assign these videos prior to lecture—either as part of a flipped modality or as a "reading quiz." In either case, you can use the diagnostic feedback from the questions in SmartWork to tailor your in-class discussions. Or you might show them in class, to stimulate discussion. Or you might simply use them as a jumping-off point—to get ideas for activities to do with your own students.

We continue to look for better ways to engage students, so please let us know how these features work for your students.

Ancillaries for Students
digital.wwnorton.com/astro5

SmartWork Online Homework

Steven Desch, Guilford Technical Community College
Violet Mager, Penn State Wilkes-Barre
Todd Young, Wayne State College

More than 2,000 questions support *21st Century Astronomy, Fifth Edition*—all with answer-specific feedback, hints, and ebook links. Questions include Summary Self-Tests, versions of the Explorations (based on AstroTours and the University of Nebraska simulations), and new Reading Astronomy News questions. Also new to this edition, Astronomy in Action video questions focus on getting students to come to class prepared and on overcoming common misconceptions. Rounding out the SmartWork course, Process of Science Guided Inquiry Assignments help students apply the scientific method to important questions in astronomy, challenging them to think like scientists.

Student Site

W. W. Norton's free and open student website features the following:

- Thirty AstroTour animations. These animations, some of which are interactive, use art from the text to help students visualize important physical and astronomical concepts. All are now tablet-compatible.

- Nebraska Simulations (sometimes called applets or NAAPs, for Nebraska Astronomy Applet Programs). These simulations allow students to manipulate variables and see how physical systems work.

- Twenty-three Astronomy in Action videos that feature author Stacy Palen demonstrating the most important concepts in a visual, easy to understand, and memorable way.

Learning Astronomy by Doing Astronomy: Collaborative Lecture Activities

Stacy Palen, Weber State University
Ana Larson, University of Washington

Students learn best by doing. Devising, writing, testing, and revising suitable in-class activities that use real astronomical data, illuminate astronomical concepts, and pose probing questions that ask students to confront misconceptions can be challenging and time consuming. In this workbook, the authors draw on their experience teaching thousands of students in many different types of courses (large in-class, small in-class, hybrid, online, flipped, and so forth) to bring 30 field-tested activities that can be used in any classroom today. The activities have been designed to require no special software, materials, or equipment and to take no more than 50 minutes to do.

Starry Night Planetarium Software (College Version) and Workbook

Steven Desch, Guilford Technical Community College
Michael Marks, Bristol Community College

Starry Night is a realistic, user-friendly planetarium simulation program designed to allow students in urban areas to perform observational activities on a computer screen. Norton's unique accompanying workbook offers observation assignments that guide students' virtual explorations and help them apply what they've learned from the text reading assignments.

For Instructors

Instructor's Manual

Ben Sugerman, Goucher College

This resource includes brief chapter overviews; suggested discussion points; notes on the AstroTour animations, Nebraska Simulations, and Astronomy in Action videos contained on the Instructor Resource USB Drive (described later); and worked solutions to all end-of-chapter questions and problems, including answers to all Reading Astronomy News and Check Your Understanding questions found in the textbook.

PowerPoint Lecture Slides

Jack Hughes, Rutgers University
Jack Brockway, Radford University

These ready-made lecture slides integrate selected textbook art, all Check Your Understanding and Working It Out questions from the text, and links to the AstroTour animations. Designed with accompanying lecture outlines, these lecture slides are fully editable and are available in Microsoft PowerPoint format.

Test Bank

Joshua Thomas, Clarkson University
Parviz Ghavamian, Towson University
Adriana Durbala, University of Wisconsin–Stevens Point

The Test Bank has been revised using Bloom's Taxonomy and provides a quality bank of more than 2,400 multiple-choice and short-answer questions. Each chapter of the Test Bank consists of six question levels classified according to Bloom's Taxonomy:

Remembering

Understanding

Applying

Analyzing

Evaluating

Creating

Questions are further classified by section and difficulty, making it easy to construct tests and quizzes that are meaningful and diagnostic. The Test Bank assesses a common set of Learning Objectives consistent with the textbook and SmartWork online homework.

Norton Instructor's Resource Site

This Web resource contains the following resources to download:

- Test Bank, available in ExamView, Word RTF, and PDF formats
- Instructor's Manual in PDF format
- Lecture PowerPoint slides with lecture notes
- All art and tables in JPEG and PPT formats
- Starry Night College, W. W. Norton Edition, Instructor's Manual
- AstroTour animations
- Selected Nebraska Simulations
- Coursepacks, available in BlackBoard, Angel, Desire2Learn, and Moodle formats

Coursepacks

Norton's Coursepacks, available for use in various Learning Management Systems (LMSs), feature all Test Bank questions, links to the AstroTours and Nebraska Simulations, worksheets based on the Explorations and Astronomy in Action videos, and automatically graded versions of the end-of-chapter Test Your Understanding multiple-choice questions. Coursepacks are available in BlackBoard, Canvas, Desire2Learn, and Moodle formats.

Instructor Resource USB Drive

This USB drive contains all instructor resources found on the Instructor's Resource Site, including offline versions of the Astronomy in Action videos, AstroTour animations, and Nebraska Simulations.

Acknowledgments

The authors would like to acknowledge the extraordinary efforts of the staff at W. W. Norton: Arielle Holstein, who kept things flowing smoothly; Diane Cipollone, who shepherded manuscript through the redesigned layout; Jane Miller and Trish Marx for managing the numerous photos; and the copy editor, Christopher Curioli, who made sure that all the grammar and punctuation survived the multiple rounds of the editing process. We would especially like to thank Becky Kohn for the developmental editing process, and our editor Erik Fahlgren for his degree of commitment to the project.

Andy Ensor managed the production. Hope Miller Goodell was the design director. Rob Bellinger and Julia Sammaritano worked on the media and supplements, and Stacy Loyal will help get this book into the hands of people who can use it.

We gratefully acknowledge the contributions of the authors who worked on previous editions of *21st Century Astronomy*: Dave Burstein, Ron Greeley, Jeff Hester, Brad Smith, Howard Voss, and Gary Wegner, with special thanks to Dave for starting the project, to Jeff for leading the original authors through the first edition, and to Brad for leading the second and third editions.

Laura Kay
Stacy Palen
George Blumenthal

And we would like to thank the reviewers, whose input at every stage improved the book:

Reviewers for the Fifth Edition

James Applegate, Columbia University
Matthew Bailey, University of Nevada–Reno
Fabien Baron, Georgia State University
Bob Becker, University of California–Davis
Miles Blanton, Bowling Green State University
Jean Brodie, University of California–Santa Cruz
Gerald Cecil, University of North Carolina at Chapel Hill
Damian Christian, California State University–Northridge
Micol Christopher, Mt. San Antonio College
Bethany Cobb, George Washington University
Kate Dellenbusch, Bowling Green State University
Karna Desai, Indiana University–Bloomington
Matthias Dietrich, The Ohio State University
Yuri Efremenko, University of Tennessee
Robert Egler, North Carolina State University
Jason Ferguson, Wichita State University
Jay Gallagher, University of Wisconsin–Madison
Richard Gelderman, Western Kentucky University
Douglas Gobeille, University of Rhode Island
Greg Gowens, University of West Georgia
Aaron Grocholski, Louisiana State University
Peter Hahn, George Washington University

Javier Hasbun, University of West Georgia
Lynnette Hoerner, Red Rocks Community College
Michael Hood, Mt. San Antonio College
Douglas Ingram, Texas Christian University
Bill Keel, University of Alabama
Charles Kerton, Iowa State University
David Kirkby, University of California–Irvine
Mark Kruse, Duke University
Silas Laycock, University of Massachusetts–Lowell
Alexandre Lazarian, University of Wisconsin–Madison
Lauren Likkel, University of Wisconsin–Eau Claire
Catherine Lovekin, Mount Allison University
Lori Lubin, University of California–Davis
Loris Magnani, University of Georgia
Joseph McMullin, Pima Community College
David Menke, Pima Community College
Bahram Mobasher, University of California–Riverside
Edward M. Murphy, University of Virginia
Robert Mutel, University of Iowa
Harold Nations, College of Southern Nevada–Charleston
Richard Nolthenius, Cabrillo College
Chris Packham, University of Texas at San Antonio

Jay Pehl, Indiana University–Purdue University Indianapolis
Michael Reid, University of Toronto
Edward Rhoads, Indiana University–Purdue University Indianapolis
Russell Robb, University of Victoria
James Roberts, University of North Texas
Lindsay Rocks, Front Range Community College
Eric Schlegel, University of Texas at San Antonio
Ohad Shemmer, University of North Texas
Allyn Smith, Austin Peay State University
Inseok Song, University of Georgia
James Sowell, Georgia Institute of Technology
Adriane Steinacker, University of California–Santa Cruz
Gregg Stiesberg, Ithaca College
Michael Strauss, Princeton University
Edmund Sutton, University of Illinois at Urbana-Champaign
Glenn Tiede, Bowling Green State University
Christy Tremonti, University of Wisconsin–Madison
Chandra Vanajakshi, College of San Mateo
Frederick Walter, Stony Brook University
Kevin Williams, Buffalo State
Kurtis Williams, Texas A&M University–Commerce
David Wittman, University of California–Davis
Brian Woodahl, Indiana University–Purdue University Indianapolis
Garett Yoder, Eastern Kentucky University

Reviewers of Previous Editions

Scott Atkins, University of South Dakota
Timothy Barker, Wheaton College
Peter A. Becker, George Mason University
Timothy C. Beers, Michigan State University
David Bennum, University of Nevada–Reno
Edwin Bergin, University of Pittsburgh
William Blass, University of Tennessee
Steve Bloom, Hampden Sydney College
Daniel Boice, University of Texas at San Antonio
Bram Boroson, Clayton State University
David Branning, Trinity College
Julie Bray-Ali, Mt. San Antonio College
Jack Brockway, Radford University
Suzanne Bushnell, McNeese State University
Paul Butterworth, George Washington University
Juan E. Cabanela, Minnesota State University–Moorhead
Amy Campbell, Louisiana State University
Michael Carini, West Kentucky University
Gerald Cecil, University of North Carolina at Chapel Hill
Supriya Chakrabarti, Boston University
Damian Christian, California State University–Northridge

Micol Christopher, Mt. San Antonio College
Robert Cicerone, Bridgewater State College
David Cinabro, Wayne State University
Judith Cohen, California Institute of Technology
Eric M. Collins, California State University–Northridge
John Cowan, University of Oklahoma–Norman
Debashis Dasgupta, University of Wisconsin–Milwaukee
Kate Dellenbusch, Bowling Green State University
Robert Dick, Carleton University
Matthias Dietrich, The Ohio State University
Gregory Dolise, Harrisburg Area Community College
Yuri Efremenko, University of Tennessee
Tom English, Guilford Technical Community College
David Ennis, The Ohio State University
Jason Ferguson, Wichita State University
John Finley, Purdue University
Matthew Francis, Lambuth University
Kevin Gannon, College of Saint Rose
Todd Gary, O'More College of Design
Christopher Gay, Santa Fe College
Parviz Ghavamian, Towson University
Martha Gilmore, Wesleyan University
Greg Gowens, University of West Georgia
Bill Gutsch, St. Peter's College
Karl Haisch, Utah Valley University
Javier Hasbun, University of West Georgia
Charles Hawkins, Northern Kentucky University
Sebastian Heinz, University of Wisconsin–Madison
Barry Hillard, Baldwin Wallace College
Paul Hintzen, California State University–Long Beach
Paul Hodge, University of Washington
William A. Hollerman, University of Louisiana at Lafayette
Hal Hollingsworth, Florida International University
Olencka Hubickyj-Cabot, San Jose State University
Kevin M. Huffenberger, University of Miami
James Imamura, University of Oregon
Douglas Ingram, Texas Christian University
Adam Johnston, Weber State University
Steven Kawaler, Iowa State University
Bill Keel, University of Alabama
Charles Kerton, Iowa State University
Monika Kress, San Jose State University
Jessica Lair, Eastern Kentucky University
Alex Lazarian, University of Wisconsin–Madison
Kevin Lee, University of Nebraska–Lincoln
Matthew Lister, Purdue University
M. A. K. Lodhi, Texas Tech University
Leslie Looney, University of Illinois at Urbana–Champaign
Jack MacConnell, Case Western Reserve University
Kevin Mackay, University of South Florida
Dale Mais, Indiana University–South Bend

Michael Marks, Bristol Community College
Norm Markworth, Stephen F. Austin State University
Kevin Marshall, Bucknell University
Stephan Martin, Bristol Community College
Justin Mason, Ivy Tech Community College
Amanda Maxham, University of Nevada–Las Vegas
Chris McCarthy, San Francisco State University
Ben McGimsey, Georgia State University
Charles McGruder, West Kentucky University
Janet E. McLarty-Schroeder, Cerritos College
Stanimir Metchev, Stony Brook University
Chris Mihos, Case Western University
Milan Mijic, California State University–Los Angeles
J. Scott Miller, University of Louisville
Scott Miller, Sam Houston State University
Kent Montgomery, Texas A&M University–Commerce
Andrew Morrison, Illinois Wesleyan University
Edward M. Murphy, University of Virginia
Kentaro Nagamine, University of Nevada–Las Vegas
Ylva Pihlström, University of New Mexico
Jascha Polet, California State Polytechnic University
Dora Preminger, California State University–Northridge
Daniel Proga, University of Nevada–Las Vegas
Laurie Reed, Saginaw Valley State University
Judit Györgyey Ries, University of Texas
Allen Rogel, Bowling Green State University
Kenneth Rumstay, Valdosta State University
Masao Sako, University of Pennsylvania
Samir Salim, Indiana University–Bloomington
Ata Sarajedini, University of Florida

Eric Schlegel, University of Texas at San Antonio
Paul Schmidtke, Arizona State University
Ann Schmiedekamp, Pennsylvania State University
Jonathan Secaur, Kent State University
Ohad Shemmer, University of North Texas
Caroline Simpson, Florida International University
Paul P. Sipiera, William Rainey Harper College
Ian Skilling, University of Pittsburgh
Tammy Smecker-Hane, University of California–Irvine
Allyn Smith, Austin Peay State University
Jason Smolinski, State University of New York at Oneonta
Roger Stanley, San Antonio College
Ben Sugerman, Goucher College
Neal Sumerlin, Lynchburg College
Angelle Tanner, Mississippi State University
Christopher Taylor, California State University–Sacramento
Donald Terndrup, The Ohio State University
Todd Thompson, The Ohio State University
Glenn Tiede, Bowling Green State University
Frances Timmes, Arizona State University
Trina Van Ausdal, Salt Lake Community College
Walter Van Hamme, Florida International University
Karen Vanlandingham, West Chester University
Nilakshi Veerabathina, University of Texas at Arlington
Paul Voytas, Wittenberg University
Ezekiel Walker, University of North Texas
James Webb, Florida International University
Paul Wiita, Georgia State University
Richard Williamon, Emory University
David Wittman, University of California–Davis

About the Authors

Laura Kay is a professor of Physics and Astronomy at Barnard College, where she has taught since 1991. She received a BS degree in physics and an AB degree in feminist studies from Stanford University, and MS and PhD degrees in astronomy and astrophysics from the University of California–Santa Cruz. As a graduate student she spent 13 months at the Amundsen Scott station at the South Pole in Antarctica. She studies active galactic nuclei, using ground-based and space telescopes. She teaches courses in astronomy, astrobiology, women and science, and polar exploration. At Barnard she has served as chair of the Physics & Astronomy Department, chair of the Women's Studies Department, chair of Faculty Governance, and interim associate dean for Curriculum and Governance.

Stacy Palen is an award-winning professor in the physics department and the director of the Ott Planetarium at Weber State University. She received her BS in physics from Rutgers University and her PhD in physics from the University of Iowa. As a lecturer and postdoc at the University of Washington, she taught Introductory Astronomy more than 20 times over 4 years. Since joining Weber State, she has been very active in science outreach activities ranging from star parties to running the state Science Olympiad. Stacy does research in formal and informal astronomy education and the death of Sun-like stars. She spends much of her time thinking, teaching, and writing about the applications of science in everyday life. She then puts that science to use on her small farm in Ogden, Utah.

George Blumenthal is chancellor at the University of California–Santa Cruz, where he has been a professor of astronomy and astrophysics since 1972. He received his BS degree from the University of Wisconsin–Milwaukee and his PhD in physics from the University of California–San Diego. As a theoretical astrophysicist, George's research encompasses several broad areas, including the nature of the dark matter that constitutes most of the mass in the universe, the origin of galaxies and other large structures in the universe, the earliest moments in the universe, astrophysical radiation processes, and the structure of active galactic nuclei such as quasars. Besides teaching and conducting research, he has served as Chair of the UC–Santa Cruz Astronomy and Astrophysics Department, has chaired the Academic Senate for both the UC–Santa Cruz campus and the entire University of California system, and has served as the faculty representative to the UC Board of Regents.

THE STARS AND GALAXIES

FIFTH EDITION

21ST CENTURY
ASTRONOMY

1 Thinking Like an Astronomer

This is a fascinating time to be studying this most ancient of the sciences. Loosely translated, the word **astronomy** means "patterns among the stars." But modern astronomy—the astronomy we will talk about in this book—is about far more than looking at the sky and cataloging the visible stars. The contents of the universe, the origin and fate of the universe, and the nature of space and time have become the subjects of rigorous scientific investigation. Humans have long speculated about our *origins*. How and when did the Sun, Earth, and Moon form? Are other galaxies, stars, planets, and moons similar to our own? The answers that scientists are finding to these questions are changing not only our view of the cosmos but also our view of ourselves.

LEARNING GOALS

In this chapter, we will begin the study of astronomy by exploring our place in the universe and the methods of science. By the conclusion of this chapter, you should be able to:

LG 1 Describe the size and age of the universe and Earth's place in it.

LG 2 Use the scientific method to study the universe.

LG 3 Demonstrate how scientists use mathematics, including graphs, to find patterns in nature.

The first view of Earth seen from deep space. In December 1968, *Apollo 8* astronauts photographed Earth above the Moon's limb. ▶ ▶ ▶

What is your
cosmic
address?

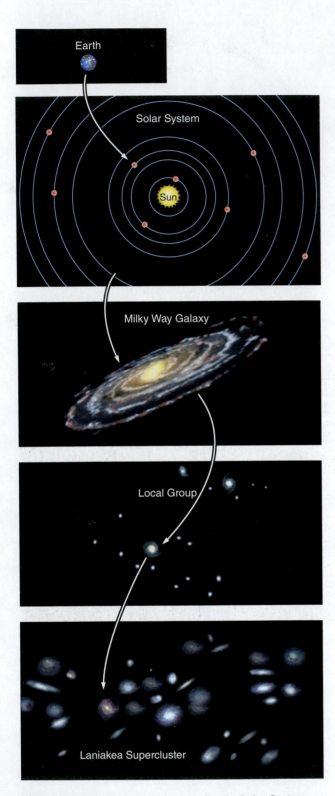

Figure 1.1 Our cosmic address is Earth, Solar System, Milky Way Galaxy, Local Group, Laniakea Supercluster. We live on Earth, a planet orbiting the Sun in our Solar System, which is a star in the Milky Way Galaxy. The Milky Way is a large galaxy within the Local Group of galaxies, which in turn is located in the Laniakea Supercluster.

1.1 Earth Occupies a Small Place in the Universe

Astronomers contemplate our place in the universe by studying Earth's position in space and time. Locating Earth in the larger universe is the first step in learning the science of astronomy. In this section, you will get a feel for the neighborhood in which Earth is located. You will also begin to explore the scale of the universe in space and time.

Our Place in the Universe

Most people receive their postal mail at an address—building number, street, city, state, and country. We can expand our view to include the enormously vast universe we live in. What is our "cosmic address"? We reside on a planet called Earth, which is orbiting under the influence of gravity around a star called the Sun. The **Sun** is a typical, middle-aged star and seems extraordinary only because of its importance to us within our own **Solar System**. Our Solar System consists of eight planets: Mercury, Venus, Earth, Mars, Jupiter, Saturn, Uranus, and Neptune. It also contains many smaller bodies, such as dwarf planets, asteroids, and comets. All of these objects are bound to the Sun by gravity.

The Sun is located about halfway out from the center of the **Milky Way Galaxy**, a flattened collection of stars, gas, and dust. Our Sun is just one among several hundred billion stars scattered throughout our galaxy, and many of these stars are themselves surrounded by planets.

The Milky Way is a member of a collection of a few dozen galaxies called the **Local Group**. Most galaxies in this group are much smaller than the Milky Way. Looking farther outward, the Local Group is part of a vastly larger collection of thousands of galaxies—a **supercluster**—called the Laniakea Supercluster. There are millions of superclusters in the observable universe.

We can now define our cosmic address—Earth, Solar System, Milky Way Galaxy, Local Group, Laniakea Supercluster—as illustrated in **Figure 1.1**. Yet even this address is not complete, as the Laniakea Supercluster encompasses only the *local universe*. The part of the universe that we can see—the *observable universe*—extends to 50 times the size of Laniakea in every direction. Within this volume, there are about as many galaxies as there are stars in the Milky Way—several hundred billion. The universe is not only much larger than the local universe but also contains much more than the observed planets, stars, and galaxies. Up to 95 percent of the mass of the universe is made up of matter that does not interact with light, known as *dark matter*, and a form of energy that permeates all of space, known as *dark energy*. Neither of these is well understood, and they are among the many exciting areas of research in astronomy.

The Scale of the Universe

As you saw in Figure 1.1, the size of the universe completely dwarfs our human experience. We can start by comparing astronomical sizes and distances to something more familiar. For example, the diameter of our Moon is about equal to the distance between the offices of the first two authors of this book, in New York, New York, and Ogden, Utah (**Figure 1.2a**). The distance from Earth to the Moon is about 100 times the Moon's diameter, and the planet Saturn with its majestic rings would fill much of that distance (Figure 1.2b). The distance from Earth to the Sun is about 400 times the Earth–Moon distance, and the distance to the planet Neptune is about 30 times the Earth–Sun distance.

But as we move out from the Solar System to the stars, the distances become so enormous that they are difficult to comprehend. The nearest star is about 9,000 times farther away from the Sun than the Sun's distance to the planet Neptune. The diameter of our Milky Way Galaxy is 30,000 times the distance to that nearest star. The Andromeda Galaxy, the nearest similar large galaxy to the Milky Way, is about 600,000 times farther away than that nearest star. The diameter of the Local Group of galaxies is about 4 times the distance to Andromeda, and the diameter of the recently identified Laniakea Supercluster, which includes the Local Group and many other galaxy groups, is 50 times larger than the Local Group. As noted earlier, this is just one of millions of superclusters in the observable universe.

To get a better sense of these distances, imagine a model in which the objects and distances in the universe are 1 billion times smaller than they really are. In this model, Earth is about the size of a marble or a peanut M&M (about 1.3 centimeters, or half an inch), the Moon is 38 centimeters (cm) away, and the Sun is 150 meters away. Neptune is 4.5 kilometers (km) from the Sun, and the nearest star to the Sun is about 40,000 km away (or about the length of the circumference of the real Earth). The model Milky Way Galaxy would fill the Solar System nearly to the orbit of Saturn. The distance between the model Milky Way and Andromeda galaxies would fill the Solar System 20 times farther, out beyond humanity's most distant space probe. The model Laniakea Supercluster would fill the Solar System and go about one-eighth of the way to the nearest star.

When thinking about the distances in the universe, it can be helpful to discuss the time it takes to travel to various places. If someone asks you how far it is to the nearest city, you might say 100 km or you might say 1 hour. In either case, you will have given that person an idea of how far the city is. In astronomy, the speed of a car on the highway is far too slow to be useful. Instead, we use the fastest speed in the universe—the speed of light. Light travels at 300,000 kilometers per second (km/s). Light can circle Earth, a distance of 40,000 km, in just under $\frac{1}{7}$ of a second. So we say that the circumference of Earth is $\frac{1}{7}$ of a light-second. Even relatively small distances in astronomy are so vast that they are measured in units of **light-years (ly)**: the distance light travels in 1 year, about 9.5 trillion km, or 6 trillion miles.

Because light takes time to reach us, we see astronomical objects as they were in the past: the extent back in time depends on the object's distance from us. Because light takes $1\frac{1}{4}$ seconds to reach us from the Moon, we see the Moon as it was $1\frac{1}{4}$ seconds ago. Because light takes $8\frac{1}{3}$ minutes to reach us from the Sun, we see the Sun as it was $8\frac{1}{3}$ minutes ago. We see the nearest star as it was more than 4 years ago and objects across the Milky Way as they were tens of thousands of years ago. The light from the Virgo Cluster of galaxies has been traveling 50 million years to reach us. The light from the most distant observable objects has been traveling for almost the age of the universe—nearly 13.8 billion

(a)

(b)

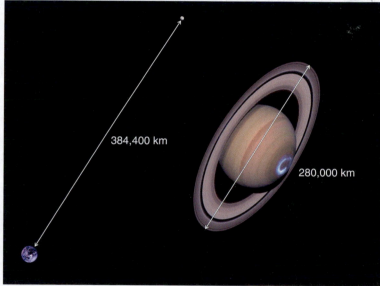

Figure 1.2 (a) The diameter of the Moon is about the same as the distance between New York, New York, and Ogden, Utah. (b) The size of Saturn, including the rings, is about 70 percent of the distance between Earth and the Moon.

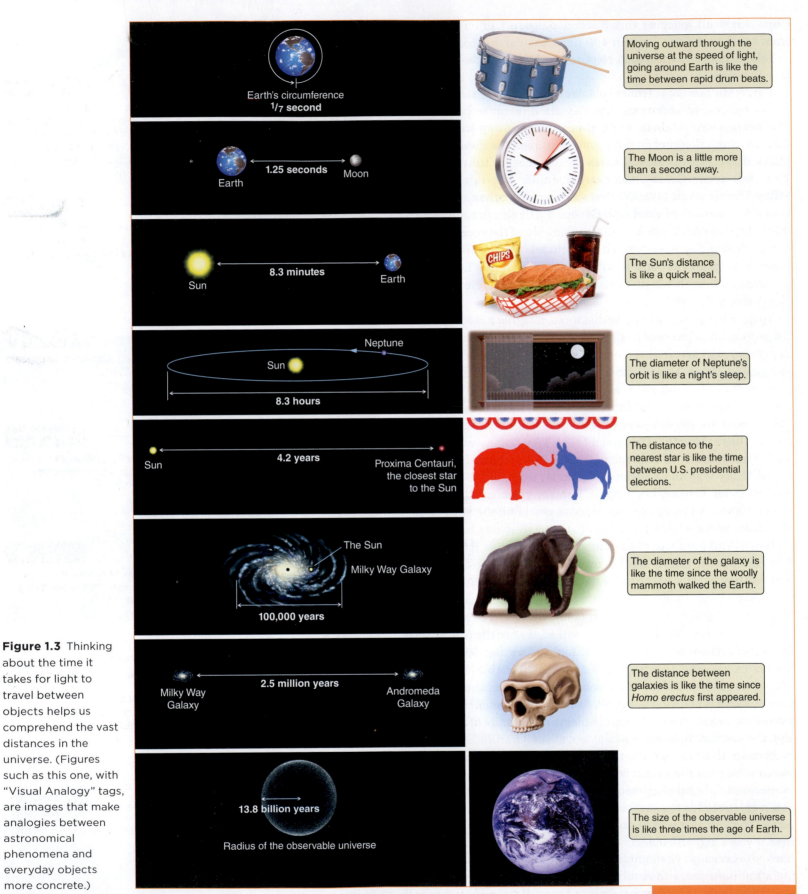

Figure 1.3 Thinking about the time it takes for light to travel between objects helps us comprehend the vast distances in the universe. (Figures such as this one, with "Visual Analogy" tags, are images that make analogies between astronomical phenomena and everyday objects more concrete.)

Earth's circumference
1/7 second

Moving outward through the universe at the speed of light, going around Earth is like the time between rapid drum beats.

Earth — 1.25 seconds → Moon

The Moon is a little more than a second away.

Sun ← 8.3 minutes → Earth

The Sun's distance is like a quick meal.

Neptune
Sun
8.3 hours

The diameter of Neptune's orbit is like a night's sleep.

Sun ← 4.2 years → Proxima Centauri, the closest star to the Sun

The distance to the nearest star is like the time between U.S. presidential elections.

The Sun
Milky Way Galaxy
100,000 years

The diameter of the galaxy is like the time since the woolly mammoth walked the Earth.

Milky Way Galaxy ← 2.5 million years → Andromeda Galaxy

The distance between galaxies is like the time since *Homo erectus* first appeared.

13.8 billion years
Radius of the observable universe

The size of the observable universe is like three times the age of Earth.

VISUAL ANALOGY

years. **Figure 1.3** begins with Earth and progresses outward to the observable universe.

The vast distances from Earth to other objects in the universe tell us that we occupy a very small part of the space in the universe and a very small part of time. Earth and the Solar System are only about one-third the age of the universe. Animals have existed on Earth for even less time. Imagine the age of the universe and the important events in it as if they took place within a single day, as illustrated in **Figure 1.4**. In this timeline, the Big Bang begins the cosmic day at midnight, and the original light chemical elements are created within the first 2 seconds. The first stars and galaxies appear within the first 10 minutes. Our Solar System formed from recycled gas and dust left over from previous generations of stars, at about 4 P.M. on this cosmic clock. The first bacterial life appears on Earth at 5:20 P.M., the first animals at 11:20 P.M., and modern humans at 11:59:59.8 P.M.—with only a fifth of a second to go in this cosmic day. We humans appeared quite recently in the history of the universe.

CHECK YOUR UNDERSTANDING 1.1

Rank the following in order of size: (a) a light-minute, (b) a light-year, (c) a light-hour, (d) the radius of Earth, (e) the distance from Earth to the Sun, (f) the radius of the Solar System.

1.2 Science Is a Way of Viewing the Universe

Humans have long paid attention to the sky and the stars and developed the dynamic science of astronomy. New discoveries happen frequently, and ideas about the universe are evolving rapidly. To view the universe through the eyes of an astronomer, you will need to understand how science itself works. Throughout this book, we will emphasize not only scientific discoveries but also the process of science. In this section, we will examine the scientific method.

The Scientific Method

The **scientific method** is a systematic way of testing new ideas or explanations. Often, scientists begin with a fact—an observation or a measurement. For example, you might observe that the weather changes in a predictable way each year and wonder why that happens. You then create a **hypothesis**, a testable explanation of the observation: "I think that it is cold in the winter and warm in the summer because Earth is closer to the Sun in the summer." You and your colleagues come up with a test: if it is cold in the winter and warm in the summer because Earth is closer to the Sun in the summer, then it will be cold in the winter everywhere on the planet—Australia should have winter at the same time of year as the United States. This test can be used to check your hypothesis. You travel from the United States to Australia in January and find that it is summer in Australia. Your hypothesis has just been proved incorrect, so we say that it has been **falsified**. This is different than the meaning in common usage, where one might think of "falsified" evidence as having been manipulated to misrepresent the truth. There are two important elements of your test that all scientific tests share. Your observation is reproducible: anyone who goes to Australia will find the same result.

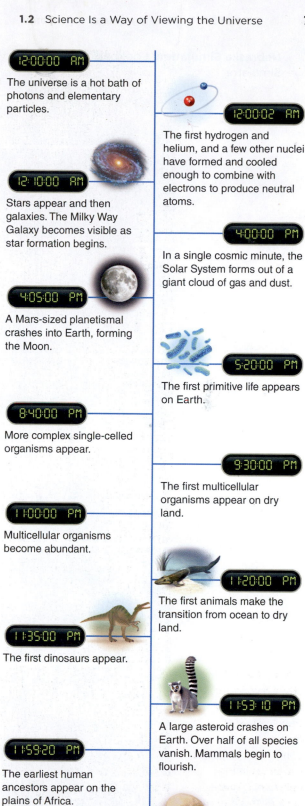

12:00:00 AM The universe is a hot bath of photons and elementary particles.

12:00:02 AM The first hydrogen and helium, and a few other nuclei have formed and cooled enough to combine with electrons to produce neutral atoms.

12:10:00 AM Stars appear and then galaxies. The Milky Way Galaxy becomes visible as star formation begins.

4:00:00 PM In a single cosmic minute, the Solar System forms out of a giant cloud of gas and dust.

4:05:00 PM A Mars-sized planetismal crashes into Earth, forming the Moon.

5:20:00 PM The first primitive life appears on Earth.

8:40:00 PM More complex single-celled organisms appear.

9:30:00 PM The first multicellular organisms appear on dry land.

11:00:00 PM Multicellular organisms become abundant.

11:20:00 PM The first animals make the transition from ocean to dry land.

11:35:00 PM The first dinosaurs appear.

11:53:10 PM A large asteroid crashes on Earth. Over half of all species vanish. Mammals begin to flourish.

11:59:20 PM The earliest human ancestors appear on the plains of Africa.

11:59:58.5 PM *Homo sapiens* first appears.

11:59:59.8 PM Modern Humans

Figure 1.4 This cosmic timeline presents the history of the universe as a 24-hour day.

Nebraska Simulation: Lookback Time
Simulator

Your result is also repeatable: if you conducted a similar test next year or the year after, you would get the same result. Because you have falsified your hypothesis, you must revise or completely change it to be consistent with the new data.

Any idea that is not testable—that is not falsifiable—must be accepted or rejected based on intuition alone, so it is not a scientific idea. A falsifiable hypothesis or idea does not have to be testable using current technology, but we must be able to imagine an experiment or observation that could prove the idea wrong if we could carry it out. As continuing tests support a hypothesis by failing to disprove it, scientists come to accept the hypothesis as a theory. A **theory** is a well-developed idea or group of ideas that is tied to known physical laws and makes testable predictions. As in the previous paragraph, the scientific meaning is different than the meaning in common usage. In everyday language, theory may mean a guess: "Do you have a theory about who did it?" In everyday language, a theory can be something we don't take too seriously. "After all," people say, "it's only a theory."

In stark contrast, scientists use the word *theory* to mean a carefully constructed proposition that takes into account every piece of relevant data as well as our entire understanding of how the world works. A theory has been used to make testable predictions, and all of those predictions have come true. Every attempt to prove it false has failed. A classic example is Einstein's theory of relativity, which we cover in some depth in Chapter 18. For more than a century, scientists have tested the predictions of the theory of relativity and have not been able to falsify it. Even after 100 years of verification, if a prediction of the theory of relativity failed tomorrow, the theory would require revision or replacement. As Einstein himself noted, a theory that fails only one test is proved false. In this sense, all scientific knowledge is subject to challenge.

In the loosely defined hierarchy of scientific knowledge, an *idea* is a notion about how something might be. Moving up the hierarchy we come to a *fact*, which is an observation or measurement. For example, the measured value of Earth's radius is a fact. A *hypothesis* is an idea that leads to testable predictions. A hypothesis may be the forerunner of a scientific theory, or it may be based on an existing theory, or both. At the top of the hierarchy is a *theory*: an idea that has been examined carefully, is consistent with all existing theoretical and observational knowledge, and makes testable predictions. Ultimately, the success of the predictions is the deciding factor between competing theories. A scientific *law* is a series of observations that leads to an ability to make predictions but has no underlying explanation of why the phenomenon occurs. So we might have a "law of daytime" that says the Sun rises and sets once each day. We could have a "theory of daytime" that says the Sun rises and sets once each day because Earth spins on its axis. Scientists themselves can be sloppy about the way they use these words, and you will sometimes see them used differently than we have defined them here.

As shown in the **Process of Science Figure**, the scientific method follows a specific sequence. Scientists begin with an observation or idea, followed by careful analysis, followed by a hypothesis, followed by prediction, followed by further observations or experiments to test the prediction. A hypothesis may lead to a scientific theory, or it may be based on an existing theory, or both. Ultimately, the basis for deciding among competing theories is the success of their predictions. Scientists can use theories to take their knowledge a step further by building theoretical models. A **theoretical model** is a detailed description of the properties of a particular object or system in terms of known physical laws or theories, which are used to connect the theoretical model to the behavior of a complex system.

The construction of new theories is often guided by scientific **principles**, which are general ideas or a sense about the universe that will guide the

THE SCIENTIFIC METHOD

The scientific method is a formal procedure used to test the validity of scientific hypotheses and theories.

Start with an observation or idea.

Suggest a hypothesis.

Make a prediction.

If the test or observation does not support the hypothesis, make more observations, revise the hypothesis, or choose a new one.

Perform a test, experiment, or additional observation and analyze the data.

If the test provides evidence to support the hypothesis, make additional predictions and test them.

An idea or observation leads to a falsifiable hypothesis that is either accepted as a tested theory or rejected on the basis of observational or experimental tests of its predictions. The blue loop goes on indefinitely as scientists continue to test the theory.

construction of new theories. Two principles used in the study of astronomy are the *cosmological principle* and *Occam's razor*.

The **cosmological principle** is the testable assumption that the physical laws that apply here and now also apply everywhere and at all times. This principle also encompasses the assumption that there are no special locations or directions in the universe. By extension, the cosmological principle asserts that matter and energy obey the same physical laws throughout space and time as they do today on Earth. This principle means that the same physical laws that we observe and apply in laboratories on Earth can be used to understand what goes on in the centers of stars or in the hearts of distant galaxies. Each new theory that comes from applying the cosmological principle to observations of the universe around us adds to our confidence in the validity of this principle, which we will discuss in more detail in Chapter 21.

Occam's razor states that when we are faced with two hypotheses that explain all the observations equally well, we should use the one that requires the fewest assumptions until we have evidence to the contrary. For example, a hypothesis that atoms are constructed differently in the Andromeda Galaxy and in the Milky Way Galaxy would violate the cosmological principle. This hypothesis would also require a large number of assumptions to explain how atoms in the Andromeda Galaxy are constructed differently and yet still appear to behave in the same way as atoms in the Milky Way. For example, suppose you assume that the center of an atom in Andromeda is negatively charged, opposite to that in the Milky Way, where the center of an atom is positively charged. However, that assumption would require you to make an assumption about the location of the boundary between Andromeda-like matter and Milky Way–like matter and about why atoms on the boundary between the two regions did not destroy each other. You would also need an assumption about why atoms in the two regions are constructed so differently. If reasonable experimental evidence ever challenges the validity of the cosmological principle, scientists will construct a new description of the universe that takes this new data into account. Until then, the cosmological principle is the hypothesis that has the fewest assumptions, satisfying Occam's razor. To date, the cosmological principle has repeatedly been tested and remains valid.

In many sciences, researchers can conduct controlled experiments to test different hypotheses. This experimental method is often not available to astronomers: we cannot change the tilt of Earth or the temperature of a star to see what happens. Instead, astronomers work from observations or existing models. They make multiple observations using various methods and create mathematical and physical models based on established science to explain the observations.

For example, when astronomers first discovered planets orbiting other stars, these new *extrasolar planets* were most often giant gaseous planets (similar to Jupiter) in short-period orbits very close to their star. These planets are the easiest to discover because they have a strong pull on their star, as we will discuss in Chapter 7. However, their proximity to their star is completely unlike the situation in our own Solar System, where the giant planets are far from the Sun. These discoveries challenged existing ideas about how our Solar System formed. As different observers using multiple telescopes found more and more of these planets, astronomers realized they needed new explanations of planet formation to explain how such large planets could wind up so close to their star. Astronomers could not build different Solar Systems—but they could create computer simulations of planetary systems using the known laws of physics. When they did, they found that planets can migrate, moving to orbits closer to or farther from their star. Planetary scientists search for evidence to test this idea in our own Solar

System, where this may have occurred early in the life of the Sun. In this example, the observations occurred before the theory was constructed.

Alternatively, an astronomer might make predictions from an existing successful mathematical or physical model and then conduct observations and analyze data to test the predictions. One example is the discovery of black holes. In the late 18th century, two scientists hypothesized the existence of "dark stars": massive objects having such strong gravity that light could not escape. At that time, the scientists did not have a way to test this hypothesis. More than 100 years later, in the early 20th century, Karl Schwarzschild studied Einstein's relativity equations and calculated that these collapsed dark stars would be very small, with a radius of only a few kilometers. Fifty years later, these objects were named *black holes*. There was still no evidence of their existence until the 1970s and 1980s, when the new technology of space-based X-ray telescopes made possible the observations needed to test the hypothesis. Einstein's existing theory made the discovery of black holes possible.

The scientific method provides the rules for testing whether an idea is false, but it offers no insight into where the idea came from in the first place or how an experiment was designed. Scientists discussing their work use words such as *insight*, *intuition*, and *creativity*. Scientists speak of a beautiful theory in the same way that an artist speaks of a beautiful painting or a musician speaks of a beautiful performance. Science has an aesthetic that is as human and as profound as any found in the arts.

Yet science is not the same as art or music in one important respect. Whereas art and music are judged by a human jury alone, the validity of a scientific hypothesis or theory is subject to the natural world. Nature alone provides the final decisions about which theories can be kept and which theories must be discarded. It does not matter what we want to be true. In the history of science, many a beautiful and beloved theory has been abandoned.

Scientific Revolutions

Scientific inquiry is necessarily dynamic. Scientists do not have all the answers and must constantly refine their ideas in response to new data and new insights. The vulnerability of knowledge may seem like a weakness. "Gee, you really don't know anything," the cynical person might say. But this vulnerability is actually the greatest strength of the scientific process: it means that science self-corrects. Incorrect ideas are eventually overturned by new information. In science, even our most cherished ideas about the nature of the physical world remain fair game, subject to challenge by new evidence. Many of history's best scientists earned their status by falsifying a universally accepted idea. This is a powerful motivation for scientists to challenge old ideas constantly—to formulate and test new explanations for their observations.

For example, Sir Isaac Newton developed classical physics in the 17th century to explain motion, forces, and gravity. Newtonian physics (discussed in detail in Chapters 3 and 4) withstood the scrutiny of scientists for more than 200 years. Yet during the late 19th and early 20th centuries, a series of scientific revolutions completely changed our understanding of the nature of reality. The work of Albert Einstein (**Figure 1.5**) is representative of these scientific revolutions. Einstein's special and general theories of relativity replaced Newton's mechanics. Einstein did not prove Newton wrong but rather showed that Newton's theories were a special case of a far more general and powerful set of physical laws. Einstein's new ideas unified the concepts of mass and energy and destroyed the conventional notion of space and time as separate concepts.

Figure 1.5 Albert Einstein is perhaps the most famous scientist of the 20th century, and he was *Time* magazine's selection for Person of the Century. Einstein helped to usher in two different scientific revolutions.

Throughout this text, you will encounter many other discoveries that forced scientists to abandon accepted theories. Einstein himself never embraced the view of the world offered by *quantum mechanics*—a second revolution he helped start. Yet quantum mechanics, a statistical description of the behavior of particles smaller than atoms, has held up for more than 100 years. In science, all authorities are subject to challenge, even Einstein.

Science is a way of thinking about the world. It is a search for the relationships that make our world what it is. Scientific inquiry assumes that nature operates by consistent, explicable, inviolate rules. Scientific knowledge is an accumulated collection of ideas about how the universe works, yet scientists are always aware that what is known today may be superseded tomorrow. A scientist assumes that there is an order in the universe and that the human mind is capable of grasping the essence of the rules underlying that order. Scientists build on these assumptions to make predictions and then test those predictions, finding the underlying rules that allow humanity to solve problems, invent new technologies, or find a new appreciation for the natural world. In the final analysis, science has found such a central place in our civilization because *science works*.

CHECK YOUR UNDERSTANDING 1.2

The scientific method is a process by which scientists: (a) prove theories to be known facts; (b) gain confidence in theories by failing to prove them wrong; (c) show all theories to be wrong; (d) survey what the majority of people think about a theory.

1.3 Astronomers Use Mathematics to Find Patterns

Scientific thinking allows scientists to make predictions. Once a pattern has been observed, for example the daily rising and setting of the Sun, scientists can predict what will happen next.

Imagine that the patterns in your life became disrupted, so that the world became entirely unpredictable. For example, what would life be like if you could not predict whether an object you dropped would fall up or down? Or what if one morning the Sun rises in the east and sets in the west, and the next day it rises in the west and sets in the east? In fact, objects do fall toward the ground. The Sun rises, sets, and then rises again at predictable times and in predictable locations. Spring turns into summer, summer turns into autumn, autumn turns into

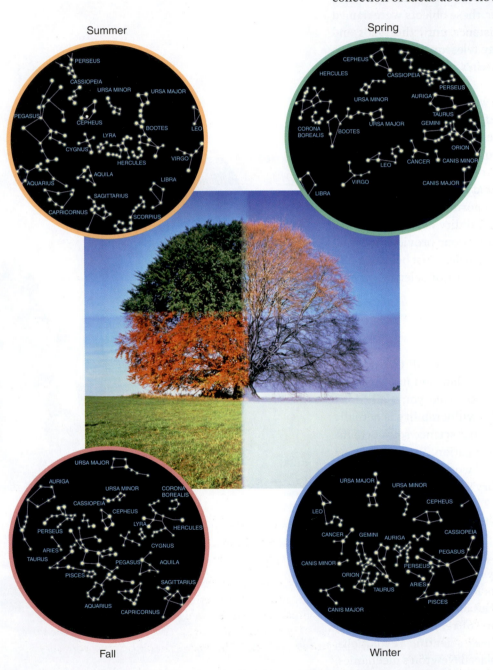

Figure 1.6 Since ancient times, people recognized that patterns in the sky change with the seasons. These and other patterns shape our lives. These star maps show the sky in the Northern Hemisphere during each season.

winter, and winter turns into spring. The rhythms of nature produce patterns in our lives, and these patterns give us clues about the nature of the physical world. Astronomers identify and characterize these patterns and use them to understand the world around us. Some of the most easily identified patterns in nature are those we see in the sky. What in the sky will look different or the same a week from now? A month from now? A year from now? As you can see in **Figure 1.6**, patterns in the sky mark the changing of the seasons. Because planting and harvesting times are determined by the seasons, it is no surprise that astronomy—which studies these patterns that are so important to agriculture—is the oldest of all sciences. We will see many other examples of patterns in the sky in the next chapter.

Astronomers use mathematics to analyze patterns and to communicate complex material compactly and accurately. Because the study of patterns in nature is so important to science, it should come as no surprise that mathematics is the language of science. Many people find mathematics to be a major obstacle that prevents them from appreciating the beauty and elegance of the world as seen through the eyes of a scientist. It is our goal in this book to explain any necessary math in everyday language. We will describe what equations mean and help you use them in a way that allows you to connect scientific concepts to the world. Your responsibility is to accept the challenge and make an honest effort to understand the material. **Working It Out 1.1** and **Working It Out 1.2** in this chapter review some basics of mathematical tools and graphs. At the back of the book, Appendix 1 explains some essential mathematical tools, and Appendix 2 contains physical constants of nature. Other appendixes contain data tables with key information about planets, moons, and stars.

CHECK YOUR UNDERSTANDING 1.3

When you see a pattern in nature, it is usually evidence of: (a) a theory being displayed; (b) a breakdown of random clustering; (c) an underlying physical law.

1.1 Working It Out Mathematical Tools

Mathematics provides scientists many of the tools that they need to understand the patterns they observe and to communicate that understanding to others. Following are a few tools that will be useful in our study of astronomy:

Scientific notation. Scientific notation is how we handle numbers of vastly different sizes. Writing out 7,540,000,000,000,000,000,000 in standard notation is very inefficient. Scientific notation uses the first few digits (the *significant* ones) and counts the number of decimal places to create the condensed form 7.54×10^{21}. Similarly, rather than writing out 0.000000000005, we write 5×10^{-12}. The exponent on the 10 is positive or negative depending on the direction that the decimal point was moved. For example, the average distance to the Sun is 149,600,000 km, but astronomers usually express it as 1.496×10^8 km.

Ratios. Ratios are a useful way to compare things. A star may be "10 times as massive as the Sun" or "10,000 times as luminous as the Sun." These expressions are ratios.

Proportionality. Often, understanding a concept amounts to understanding the *sense* of the relationships that it predicts or describes. "If you have twice as far to go, it will take you twice as long to get here." "If you have half as much money, you will be able to buy only half as much gas." These are examples of proportionality.

Appendix 1 has a more detailed explanation of mathematical tools used in this book.

1.2 Working It Out Reading a Graph

Scientists often convey complex information and mathematical patterns in graphical form. Reading graphs is a skill that is important not only in astronomy but also in life. Economists, social and political scientists, mortgage brokers, financial analysts, retirement planners, doctors, and scientists all use graphs to evaluate and communicate important information.

Graphs typically have two axes: a horizontal axis (the *x*-axis) and a vertical axis (the *y*-axis). Typically, the *x*-axis shows an independent variable, which is the one you might have control over in an experiment. The *y*-axis shows the dependent variable, which—in many experiments—is the variable a researcher is studying.

Graphs can take different shapes. Suppose we plot the distance a car travels over a period of time, as shown in **Figure 1.7a**. In a linear graph, each interval on an axis represents the same-sized step. Each step on the horizontal axis of the graph in Figure 1.7a represents 5 minutes. Each step on the vertical axis represents a distance of 5 km traveled by the car. Data are plotted on the graph, with one dot for each observation; for example, the distance the car has traveled after 20 minutes is 20 km.

Drawing a line through these data indicates the trend of the data. To understand what the trend means, scientists often find the slope of the line, which is the relationship of the line's rise along the *y*-axis to its movement along the *x*-axis. To find the slope, we look at the change between two points on the vertical axis divided by the change between two points on the horizontal axis; for example, finding the slope of the line gives

$$\text{Slope} = \frac{\text{Change in vertical axis}}{\text{Change in horizontal axis}}$$

$$= \frac{(15 - 10)\,\text{km}}{(15 - 10)\,\text{min}}$$

$$= 1\,\text{km/min}$$

In this case, the trend tells us that the car is traveling at 1 kilometer per minute (km/min), or 60 kilometers per hour (km/h). The slope of a line often contains extra information that is useful.

Many observations of natural processes do not result in a straight line on a graph. An example of this is an exponential process. Think about what happens when you catch a cold. When you get up in the morning at 7:00 A.M. you feel fine. At 9:00 A.M. you feel a little tired. By 11:00 A.M. you have a bit of a sore throat or a sniffle and think, "I wonder if I'm getting sick," and by 1:00 P.M. you have a runny nose and congestion and fever and chills. This is an exponential process, because the virus that has infected you reproduces exponentially.

For the sake of this discussion, suppose the virus produces one copy of itself each time it invades a cell. (In fact, viruses produce between 1,000 and 10,000 copies each time they invade a cell, so the exponential curve is actually much steeper.) One virus infects a cell and multiplies, so now there are two viruses—the original and a copy. These viruses invade two new cells, and each one produces a copy. Now there are four viruses. After the next cell invasion, there are eight. Then 16, 32, 64, 128, 256, 512, 1,024, 2,048, and so on. This behavior is plotted in Figure 1.7b.

It can be difficult to see what's happening in the early stages of an exponential curve, because the later numbers are so much larger than the earlier ones. For this reason, we sometimes plot this type of data *logarithmically*, by putting the logarithm (the power of 10) of the data on the vertical axis, as shown in Figure 1.7c. Now each step on the axis represents 10 times as many viruses as the previous step. Even though we draw all the steps the same size on the page, they represent different-sized steps in the data—the number of viruses, for example. We often use this technique in astronomy because it has a second, related advantage: very large variations in the data can easily fit on the same graph.

Each time you see a graph, you should first understand the axes—what data are plotted on this graph? Then you should check whether the axes are linear or logarithmic. Finally, you can look at the actual data or lines in the graph to understand how the system behaves.

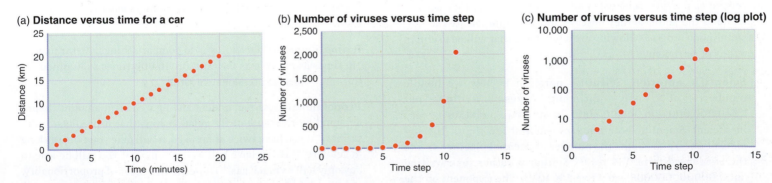

(a) **Distance versus time for a car**

(b) **Number of viruses versus time step**

(c) **Number of viruses versus time step (log plot)**

Figure 1.7 Graphs like these show relationships between quantities. (a) The time and distance traveled. (b, c) These graphs show the relationship between time and the number of viruses.

Origins

An Introduction

How and when did the universe begin? What combination of events led to the existence of humans as sentient beings living on a small rocky planet orbiting a typical middle-aged star? Are there others like us scattered throughout the galaxy?

Earlier in the chapter we mentioned the theme of origins. Throughout this book, you will see that this theme involves much more than how humans came to be on Earth. In these Origins sections, which conclude each chapter, we'll look into the origin of the universe and the origin of life on Earth. We will also examine the possibilities of life elsewhere in the Solar System and beyond—a subject called **astrobiology**. Our origins theme will include the discovery of planets around other stars and how they compare with the planets of our own Solar System.

Later in the book, we will present observational evidence that supports the **Big Bang** theory, which states that the universe started expanding from an infinitesimal size about 13.8 billion years ago. Only the lightest chemical elements were found in substantial amounts in the early universe: hydrogen and helium, and tiny amounts of lithium, and beryllium. However, we live on a planet with a central core consisting mostly of very heavy elements such as iron and nickel, surrounded by outer layers made up of rocks containing large amounts of silicon and various other elements, all heavier than the original elements. The human body contains carbon, nitrogen, oxygen, calcium, phosphorus, and a host of other chemical elements—except for hydrogen itself, all are heavier than hydrogen and helium. If these heavier elements that make up Earth and our bodies were not present in the early universe, where did they come from?

The answer to this question lies within the stars (**Figure 1.8**). In the core of a star, light elements, such as hydrogen, combine to form more massive atoms, which eventually leads to atoms such as carbon. When a star nears the end of its life, it often loses much of its material—including some of the new atoms formed in its interior—by blasting it back into interstellar space. This material combines with material lost from other stars—some of which produced even more massive atoms as they exploded—to form large clouds of dust and gas. Those clouds go on to make new stars and planets, similar to our Sun and Solar System. Prior "generations" of stars supplied the building blocks for the chemical processes that we see in the universe, including life. The atoms that make up much of what we see were formed in the cores of stars. The phrase "We are stardust" is not just poetry. We are actually made of recycled stardust.

The study of origins also provides examples of the process of science. Many of the physical processes in chemistry, geology, planetary science, physics, and astronomy that are seen on Earth or in the Solar System are observed across the galaxy and throughout the universe. But as of this writing, the only biology we know about is that existing on Earth. Thus, at this point in human history, much of what scientists can say about the origin of life on Earth and the possibility of life elsewhere is reasoned extrapolation and educated speculation. In these Origins sections, we'll address some of these hypotheses and try to be clear about which are speculative and which have been tested.

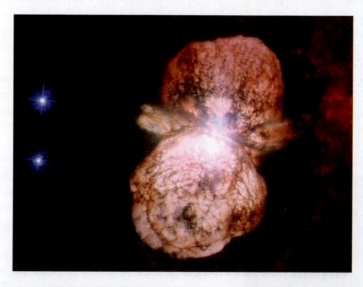

Figure 1.8 You and everything around you are composed of atoms forged in the interiors of stars that lived and died before the Sun and Earth were formed. The supermassive star Eta Carinae, shown here, is currently ejecting a cloud of chemically enriched material just as earlier generations of stars once did to enrich the gas that would become our Solar System.

This article illustrates how direct and indirect observations both contribute to new discoveries.

Probe Detects Southern Sea Under Ice on Saturnian Moon Enceladus

By **ALAN BOYLE, NBC News**

NASA's *Cassini* orbiter has detected the faint signature of a hidden southern ocean beneath the ice of the Saturnian moon Enceladus, confirming past suspicions and sparking fresh speculation about extraterrestrial marine life.

"It makes the interior of Enceladus a very attractive potential place to look for life," said Jonathan Lunine, a planetary scientist at Cornell University and a member of the science team reporting the discovery in this week's issue of the journal *Science*.

Astrobiologists have had Enceladus on their list since 2006, when *Cassini* detected geysers of water spewing up from fissures in the southern hemisphere (**Figure 1.9**). However, it took much more subtle observations to confirm the source.

"It was not a surprise to find a water reservoir . . . but the mass and geometry of this reservoir were unknown," Luciano Iess of Sapienza University of Rome, the *Science* study's lead author, told reporters during a teleconference.

Iess and his colleagues say the reservoir is a sea of liquid water, buried under 19 to 25 miles (30 to 40 kilometers) of ice. The sea is at least 6 miles (10 kilometers) deep and extends at least halfway up from the south pole toward the equator in every direction.

"This means that it is as large, or larger, than Lake Superior," said Caltech's David Stevenson, another coauthor of the *Science* study.

How scientists know

It took masterful feats of observation and calculation to figure all that out.

Astronomers began by measuring slight variations in *Cassini*'s velocity as it sped past the 310-mile-wide (500-kilometer-wide) moon on three occasions between 2010 and 2012. Those changes amounted to mere millimeters per second, and could be detected only by analyzing the Doppler shifts in the radio transmissions from the spacecraft. (A classic example of Doppler shift is the rise and fall in the pitch of a train's whistle as it zooms past you.)

The velocity variations were caused by anomalies in Enceladus's gravitational field; that is, regions of the moon that had more or less mass than average. Astronomers had already known about a huge depression in Enceladus's southern hemisphere, so they expected the mass concentration to be less in the south. But after taking everything they knew into account, researchers determined that the concentration was more massive than it should have been.

The best way to explain the extra mass was to assume the existence of a sea in the south, lying between Enceladus's rocky core and icy shell (**Figure 1.10**). Liquid water is denser than water ice, as illustrated by the ice cubes floating in a glass of water.

Planetary physicist William McKinnon of Washington University in St. Louis told *Science* that the interpretation made sense. "You could create a model without water, but people wouldn't find it satisfying," he said.

Looking for life

The gravity measurements mesh nicely with the presence of those water geysers spewing out from Enceladus's southern fissures, which

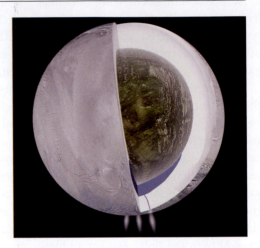

Figure 1.10 An illustration of one model of the interior of Enceladus, showing the rocky core and a southern sea, with water making its way up through cracks in the moon's icy shell and spewing out as jets of water vapor and ice.

are also known as "tiger stripes." Enceladus's core undergoes tidal flexing as it circles Saturn, and that flexing is thought to generate heat that's concentrated at the poles.

Astronomers suggest that there's enough heat at the south pole to melt the ice and push seawater up to the surface through the fissures.

That scenario is exciting for astrobiologists, because it means the sea could be in contact with organic-rich silicate material at the bottom, at the right temperature for sustaining life.

Earlier observations from *Cassini* have shown that the water in Enceladus's geysers contains salts as well as organic molecules such as methane and ethane. However, the spacecraft's instruments aren't designed to detect the heavier organic molecules that would constitute evidence for life, Lunine said.

The easiest way to check for life would be to send a probe with the right kind of instruments through Enceladus's geysers to look for the right chemicals.

Enceladus isn't the only game in town when it comes to the search for life, however.

Figure 1.9 A backlit view of Enceladus from the *Cassini* orbiter shows illuminated jets of water spewing out from surface fissures.

Scientists say that Europa, one of Jupiter's moons, also appears to have an ice-covered sea. Last December, researchers reported evidence that Europa is also spewing geysers of water into space. Such findings have led NASA to seek $15 million to start work on a mission to Europa.

The Europa mission alone would require years to plan and at least $1 billion in funding. Right now, Enceladus is a lower priority for future exploration, and it's not clear when the moon's southern ocean will get a closer look. But the latest findings suggest that places like Enceladus and Europa (and perhaps Ganymede and Callisto, two more ice-covered moons of Jupiter) could represent astrobiological frontiers at least as promising as Mars.

"I look at this as a cornucopia of habitable environments in the outer solar system," Lunine said.

ARTICLES   **QUESTIONS**

1. How large is Enceladus? How does this compare with the size of our Moon or with the distance between Chicago and St. Louis or between Santa Cruz and Los Angeles? How large is the ocean?
2. Was the original discovery of geysers on Enceladus in 2006 from an observation, a hypothesis, or a model?
3. How did scientists make this new discovery? Did they directly observe the water? If not, how did they conclude that it exists?
4. Why is it important to have multiple observations leading to a conclusion?
5. Why are scientists excited about the discovery of water on another world?

Summary

Astronomy seeks answers to many compelling questions about the universe. It uses all available tools to follow the scientific method. The process of science is based on objective reality, physical evidence, and testable hypotheses. Scientists continually strive to improve their understanding of the natural world and must be willing to challenge accepted truths as new information becomes available. We are a product of the universe: the very atoms we're made of were formed in stars that died long before the Sun and Earth were formed.

LG 1 **Describe the size and age of the universe and Earth's place in it.** We reside on a planet orbiting a star at the center of a solar system in a vast galaxy that is one of many galaxies in the universe. We occupy a very tiny part of the universe in space and time.

LG 2 **Use the scientific method to study the universe.** The scientific method is an approach to learning about the physical universe. It includes observations, forming hypotheses, making predictions to enable the testing and refining of those hypotheses, and repeated testing of theories. All scientific knowledge is provisional. Like art, literature, and music, science is a creative human activity; it is also a remarkably powerful, successful, and aesthetically beautiful way of viewing the world.

LG 3 **Demonstrate how scientists use mathematics, including graphs, to find patterns in nature.** Mathematics provides many of the tools that astronomers need to understand the patterns we see and to communicate that understanding to others.

? UNANSWERED QUESTIONS

- What makes up the universe? We have listed planets, stars, and galaxies as making up the cosmos, but astronomers now have evidence that 95 percent of the universe is composed of dark matter and dark energy, which we do not understand. Scientists are using the largest telescopes and particle colliders on Earth, as well as telescopes and experiments in space, to explore what makes up dark matter and what constitutes dark energy.

- Does life as we know it exist elsewhere in the universe? At the time of this writing, there is no scientific evidence that life exists on any other planet. Our universe is enormously large and has existed for a great length of time. What if life is too far away or existed too long ago for us ever to "meet"?

Questions and Problems

Test Your Understanding

1. Rank the following in order of increasing size.
   - a. Local Group
   - b. Milky Way
   - c. Solar System
   - d. universe
   - e. Sun
   - f. Earth
   - g. Laniakea Supercluster
   - h. Virgo Supercluster

2. If an event were to take place on the Sun, how long would it take for the light it generates to reach us?
   - a. 8 minutes
   - b. 11 hours
   - c. 1 second
   - d. 1 day
   - e. It would reach us instantaneously.

3. *Understanding* in science means that
   - a. we have accumulated lots of facts.
   - b. we are able to connect facts through an underlying idea.
   - c. we are able to predict events on the basis of accumulated facts.
   - d. we are able to predict events on the basis of an underlying idea.

4. The cosmological principle states that
   - a. on a large scale, the universe is the same everywhere at a given time.
   - b. the universe is the same at all times.
   - c. our location is special.
   - d. all of the above

5. The Sun is part of
   - a. the Solar System.
   - b. the Milky Way Galaxy.
   - c. the universe.
   - d. all of the above

6. A light-year is a measure of
   - a. distance.
   - b. time.
   - c. speed.
   - d. mass.

7. Occam's razor states that
   - a. the universe is expanding in all directions.
   - b. the laws of nature are the same everywhere in the universe.
   - c. if two hypotheses fit the facts equally well, the simpler one is the more likely to apply.
   - d. patterns in nature are really manifestations of random occurrences.

8. The circumference of Earth is $\frac{1}{7}$ of a light-second. Therefore,
   - a. if you were traveling at the speed of light, you would travel around Earth 7 times in 1 second.
   - b. light travels a distance equal to Earth's circumference in $\frac{1}{7}$ of a second.
   - c. neither a nor b
   - d. both a and b

9. According to the graphs in Figures 1.7b and c, by how much did the number of viruses increase in four time steps?
   - a. It doubled.
   - b. It tripled.
   - c. It quadrupled.
   - d. It went up more than 10 times.

10. Any explanation of a phenomenon that includes a supernatural influence is not scientific because
    - a. it does not have a hypothesis.
    - b. it is wrong.
    - c. people who believe in the supernatural are not credible.
    - d. science is the study of the natural world.

11. "All scientific knowledge is provisional." In this context, *provisional* means
    - a. "wrong."
    - b. "relative."
    - c. "temporary."
    - d. "incomplete."

12. When we observe a star that is 10 light-years away, we are seeing that star
    - a. as it is today.
    - b. as it was 10 days ago.
    - c. as it was 10 years ago.
    - d. as it was 20 years ago.

13. Which of the following was *not* made in the Big Bang?
    - a. hydrogen
    - b. lithium
    - c. beryllium
    - d. carbon

14. "We are stardust" means that
    - a. Earth exists because of the collision of two stars.
    - b. the atoms in our bodies have passed through (and in many cases formed in) stars.
    - c. Earth is primarily formed of material that used to be in the Sun.
    - d. Earth and the other planets will eventually form a star.

15. The following astronomical events led to the formation of you. Place them in order of their occurrence over astronomical time.
    - a. Stars die and distribute heavy elements into the space between the stars.
    - b. Hydrogen and helium are made in the Big Bang.
    - c. Enriched dust and gas gather into clouds in interstellar space.
    - d. Stars are born and process light elements into heavier ones.
    - e. The Sun and planets form from a cloud of interstellar dust and gas.

Thinking about the Concepts

16. Suppose you lived on the planet named "Tau Ceti e" that orbits Tau Ceti, a nearby star in our galaxy. How would you write your cosmic address?

17. Imagine yourself living on a planet orbiting a star in a very distant galaxy. What does the cosmological principle tell you about the physical laws at this distant location?

18. If the Sun suddenly exploded, how soon after the explosion would we know about it?

19. If a star exploded in the Andromeda Galaxy, how long would it take that information to reach Earth?

20. Give an example of a scientific theory that has been superseded by a newer theory. As scientists developed this new theory, where on the Process of Science Figure did a change occur so that the old theory became invalid and the new theory was accepted?

21. Some people have proposed the theory that Earth was visited by extraterrestrials (aliens) in the remote past. Can you think of any tests that could support or refute that theory? Is it falsifiable? Would you regard this proposal as science or pseudoscience?

22. What does the word *falsifiable* mean? Give an example of an idea that is not falsifiable. Give an example of an idea that is falsifiable.

23. Explain how the word *theory* is used differently in the context of science than in common everyday language.

24. What is the difference between a *hypothesis* and a *theory* in science?

25. Suppose the tabloid newspaper at your local supermarket claimed that children born under a full Moon become better students than children born at other times.
    a. Is this theory falsifiable?
    b. If so, how could it be tested?

26. A textbook published in 1945 stated that light takes 800,000 years to reach Earth from the Andromeda Galaxy. In this book, we assert that it takes 2,500,000 years. What does this difference tell you about a scientific "fact" and how our knowledge evolves with time?

27. Astrology makes testable predictions. For example, it predicts that the horoscope for your star sign on any day should fit you better than horoscopes for other star signs. Read the daily horoscopes for all of the astrological signs in a newspaper or online. How many of them might fit the day you had yesterday? Repeat the experiment every day for a week and keep a record of which horoscopes fit your day each day. Was your horoscope sign consistently the best description of your experiences?

28. A scientist on television states that it is a known fact that life does not exist beyond Earth. Would you consider this scientist reputable? Explain your answer.

29. Some astrologers use elaborate mathematical formulas and procedures to predict the future. Does this show that astrology is a science? Why or why not?

30. Why can it be said that we are made of stardust? Explain why this statement is true.

Applying the Concepts

31. Review Working It Out 1.1. Convert the following numbers to scientific notation:
    a. 7,000,000,000
    b. 0.00346
    c. 1,238

32. Review Working It Out 1.1. Convert the following numbers to standard notation:
    a. 5.34×10^8
    b. 4.1×10^3
    c. 6.24×10^{-5}

33. If a car is traveling at 35 km/h, how far does it travel in
    a. 1 hour?
    b. half an hour?
    c. 1 minute?

34. Review Appendix 1.7. The surface area of a sphere is proportional to the square of its radius. How many times larger is the surface area if the radius is
    a. doubled?
    b. tripled?
    c. halved (divided by 2)?
    d. divided by 3?

35. The average distance from Earth to the Moon is 384,400 km. How many days would it take you, traveling at 800 km/h—the typical speed of jet aircraft—to reach the Moon?

36. The average distance from Earth to the Moon is 384,400 km. In the late 1960s, astronauts reached the Moon in about 3 days. How fast (on average) must they have been traveling (in kilometers per hour) to cover this distance in this time? Compare this speed to the speed of a jet aircraft (800 km/h).

37. (a) If it takes about 8 minutes for light to travel from the Sun to Earth, and Neptune is 30 times farther from Earth than the Sun is, how long does it take light to reach Earth from Neptune? (b) Radio waves travel at the speed of light. What does this fact imply about the problems you would have if you tried to conduct a two-way conversation between Earth and a spacecraft orbiting Neptune?

38. The distance from Earth to Mars varies from 56 million km to 400 million km. How long does it take a radio signal traveling at the speed of light to reach a spacecraft on Mars when Mars is closest and when Mars is farthest away?

39. The surface area of a sphere is proportional to the square of its radius. The radius of the Moon is only about one-quarter that of Earth. How does the surface area of the Moon compare with that of Earth?

40. A remote Web page may sometimes reach your computer by going through a satellite orbiting approximately 3.6×10^4 km above Earth's surface. What is the minimum delay, in seconds, that the Web page takes to show up on your computer?

41. Imagine that you have become a biologist, studying rats in Indonesia. Most of the time, Indonesian rats maintain a constant population. Every half century, however, these rats suddenly begin to multiply exponentially! Then the population crashes back to the constant level. Sketch a graph that shows the rat population over two of these episodes.

42. New York is 2,444 miles from Los Angeles. What is that distance in car-hours? In car-days? (Assume a travel speed of 70 mph.)

43. Some theorize that a tray of hot water will freeze more quickly than a tray of cold water when both are placed in a freezer.
    a. Does this theory make sense to you?
    b. Is the theory falsifiable?
    c. Do the experiment yourself. Note the results. Was your intuition borne out?

44. A pizzeria offers a 9-inch-diameter pizza for $12 and an 18-inch-diameter pizza for $24. Are both offerings equally economical? If not, which is the better deal? Explain your reasoning.

45. The circumference of a circle is given by $C = 2\pi r$, where r is the radius of the circle.
    a. Calculate the approximate circumference of Earth's orbit around the Sun, assuming that the orbit is a circle with a radius of 1.5×10^8 km.
    b. Noting that there are 8,766 hours in a year, how fast, in kilometers per hour, does Earth move in its orbit?
    c. How far along in its orbit does Earth move in 1 day?

USING THE WEB

46. Go to the interactive "Scale of the Universe" Web page at the Astronomy Picture of the Day website (http://apod.nasa.gov/apod/ap140112.html). Start at 10^0 (human size) and scale upward; clicking on an object gives you its exact size. What astronomical bodies are about the size of the United States? What objects are about the size of Earth? What stars are larger than the distance from Earth to the Sun? How many light-days is the distance of *Voyager 1* to the Earth? What objects are about the size of the distance from the Sun to the nearest star? How much larger is the Milky Way than the size of the Solar System? How much larger is the Local Group

than the Milky Way? How much larger is the observable universe than the Local Group?

47. a. For a video representation of the scale of the universe, view the short video *The Known Universe* at the Hayden Planetarium website (http://www.haydenplanetarium.org/universe), which takes the viewer on a journey from the Himalayan mountains to the most distant galaxies. How far have broadcast radio programs from Earth traveled? Is the Sun a particularly luminous star compared to others? Do you think the video is effective for showing the size and scale of the universe?
    b. A similar film produced in 1996 in IMAX format, *Cosmic Voyage*, can be found online at http://topdocumentaryfilms.com/cosmic-voyage. Watch the "powers of ten" zoom out to the cosmos, starting at the 7-minute mark, for about 5 minutes. Do the "powers of ten" circles add to your understanding of the size and scale of the universe? (The original film *Powers of Ten* can be viewed online at http://apod.nasa.gov/apod/ap150324.html, but notably it extends a few powers of ten less than the newer film.)

48. Go to the Astronomy Picture of the Day (APOD) app or website (http://apod.nasa.gov/apod) and click on "Archive" to look at the recent pictures and videos. Submissions to this website come from all around the world. Pick one and read the explanation. Was the image or video taken from Earth or from space? Is it a combination of several images? Does it show Earth, our Solar System, objects in our Milky Way Galaxy, more distant galaxies, or something else? Is the explanation understandable to someone who has not studied astronomy? Do you think this website promotes a general interest in astronomy?

49. Throughout this book, we will examine how discoveries in astronomy and space are covered in the media. Go to your favorite news website (or to one assigned by your instructor) and find a recent article about astronomy or space. Does this website have a separate section for science? Is the article you selected based on a press release, on interviews with scientists, or on an article in a scientific journal? Use Google News or the equivalent to see how widespread the coverage of this story is. Have many newspapers carried it? Has it been picked up internationally? Has it been discussed in blogs? Do you think this story was interesting enough to be covered?

50. Go to a blog about astronomy or space. Is the blogger a scientist, a science writer, a student, or an enthusiastic amateur astronomer? What is the current topic of interest? Is it controversial? Are readers making many comments? Is this blog something you would want to read again?

smartwork

If your instructor assigns homework in SmartWork, access your assignments at digital.wwnorton.com/astro5.

EXPLORATION

Logic is fundamental to the study of science and to scientific thinking. A logical fallacy is an error in reasoning, which good scientific thinking avoids. For example, "because Einstein said so" is not an adequate argument. No matter how famous the scientist is (even if he is Einstein), he or she must still supply a logical argument and evidence to support a claim. Anyone who claims that something must be true because Einstein said it has committed the logical fallacy known as an *appeal to authority*. There are many types of logical fallacies, but a few of them crop up often enough in discussions about science that you should be aware of them.

Ad hominem. In an ad hominem fallacy, you attack the person who is making the argument, instead of the argument itself. Here is an extreme example of an ad hominem argument: "A famous politician says Earth is warming. But I think this politician is an idiot. So Earth can't be warming."

Appeal to belief. This fallacy has the general pattern "Most people believe X is true; therefore X is true." For example, "Most people believe Earth orbits the Sun. Therefore, Earth orbits the Sun." Note that even if the conclusion is correct, you may still have committed a logical fallacy in your argument.

Begging the question. In this fallacy, also known as circular reasoning, you assume the claim is true and then use this assumption as evidence to prove the claim is true. For example, "I am trustworthy; therefore I must be telling the truth." No real evidence is presented for the conclusion.

Biased sample. If a sample drawn from a smaller pool has a bias, then conclusions about the sample cannot be applied to a larger pool. For example, imagine you poll students at your university and find that 30 percent of them visit the library one or more times per week. Then you conclude that 30 percent of Americans visit the library one or more times per week. You have committed the biased sample fallacy, because university students are not a representative sample of the American public.

Post hoc ergo propter hoc. *Post hoc ergo propter hoc* is Latin for "after this, therefore because of this." Just because one thing follows another doesn't mean that one caused the other. For example, "There was an eclipse and then the king died. Therefore, the eclipse killed the king." This fallacy is often connected to related inverse reasoning: "If we can prevent an eclipse, the king won't die."

Slippery slope. In this fallacy, you claim that a chain reaction of events will take place, inevitably leading to a conclusion that no one could want. For example, "If I don't get A's in all of my classes, I will not ever be able to get into graduate school, and then I won't ever be able to get a good job, and then I will be living in a van down by the river until I'm old and starve to death." None of these steps actually follows inevitably from the one before.

Following are some examples of logical fallacies. Identify the type of fallacy represented. Each of the fallacies we just discussed is represented once.

1 You get a chain email threatening terrible consequences if you break the chain. You move it to your spam box. Later that day you get in a car accident. The following morning, you retrieve the chain email and send it along.

_____

_____

2 If I get question 1 on the assignment wrong, then I'll get question 2 wrong as well, and before you know it, I will never catch up in the class.

_____

_____

3 All my friends love the band Degenerate Electrons. Therefore, all people my age love this band.

_____

_____

4 Eighty percent of Americans believe in the tooth fairy. Therefore, the tooth fairy exists.

_____

_____

5 My professor says that the universe is expanding. But my professor is a geek, and I don't like geeks. So the universe can't be expanding.

_____

_____

6 When applying for a job, you use a friend as a reference. Your prospective employer asks you how she can be sure your friend is trustworthy, and you say, "I can vouch for him."

_____

_____

2 Patterns in the Sky—Motions of Earth and the Moon

Ancient peoples learned that they could use the patterns they observed in the sky to predict the changing length of day, the change of seasons, and the changes in the appearance of the Moon. Some people understood these patterns well enough to create complicated calendars and predict rare eclipses. But now we can see these patterns with the perspective of centuries of modern science, and we can explain these changes as a consequence of the motions of Earth and the Moon. Discovering what causes these patterns has shown us the way outward into the universe.

LEARNING GOALS

In this chapter, we will examine the patterns in the sky and on Earth and the underlying motions that cause these patterns. By the conclusion of this chapter, you should be able to:

LG 1 Describe how Earth's rotation about its axis and revolution around the Sun affect our perception of celestial motions as seen from different places on Earth.

LG 2 Explain why there are different seasons throughout the year.

LG 3 Describe the factors that create the phases of the Moon.

LG 4 Sketch the alignment of Earth, the Moon, and the Sun during eclipses of the Sun and the Moon

The changing seasons bring noticeable variations to the landscape. ▶ ▶ ▶

What causes
the seasons?

(a)

(b)

(c)

Figure 2.1 (a) One of the suspected uses of Stonehenge 4,000 years ago was to keep track of celestial events. (b) The Mayan El Caracol at Chichén Itzá in Mexico (906 CE) is believed to have been designed to align with the planet Venus. (c) The Beijing Ancient Observatory in China (1442 CE) includes ancient astronomical instruments as well as some brought by European Jesuits in the 17th and 18th centuries.

▶❚❚ **AstroTour:** The Celestial Sphere and the Ecliptic

2.1 Earth Spins on Its Axis

Ancient humans may not have known that they were "stardust," but they did sense that there was a connection between their lives on Earth and the sky above. Before our modern technological civilization, people's lives were more attuned to the ebb and flow of nature, which includes the patterns in the sky. By watching the repeating patterns of the Sun, Moon, and stars in the sky, people found that they could predict when the seasons would change and when the rains would come and the crops would grow. Ancient astronomers with knowledge of the sky—priests and priestesses, natural philosophers and explorers—all had knowledge of the world that others did not, and knowledge of the world was power. Some of these early observations and ideas about these patterns live on today in the names of stars and the apparent grouping of stars we call **constellations**, in calendars based on the Moon and Sun that are still in use by many cultures, and in the astronomical names of the days of the week.

From Mesopotamia to Africa, from Europe to Asia, from the Americas to the British Isles, the archaeological record holds evidence that ancient cultures built structures that were sometimes used to study astronomical positions and events. **Figure 2.1** shows some examples of these. Pre-telescopic astronomical observatories from the 8th through 17th centuries were used to study the sky for time-keeping and navigation. Many of these structures and observatories are now national historical or UNESCO World Heritage sites.

The Celestial Sphere

Long before Christopher Columbus sailed, Aristotle and other Greek philosophers knew that Earth is a sphere. However, because Earth seems stationary, they did not realize that the changes they observed in the sky from day to day and year to year are caused by Earth's motions. As we will see in this subsection, Earth's rotation on its axis determines the passage of day and night, which dictates the rhythm of life on Earth.

One reason the ancients did not suspect that Earth rotates was that they could not perceive Earth's spinning motion. As Earth rotates about its axis, its surface is moving quite fast—about 1,674 kilometers per hour (km/h) at the equator. We do not "feel" that motion any more than we would feel the motion of a car with a perfectly smooth ride cruising down a straight highway. Nor do we feel the direction of Earth's spin, although the hourly motion of the Sun, Moon, and stars across the sky reveals it. Earth's **North Pole** is at the north end of Earth's rotation axis. Imagine you are in space far above Earth's North Pole. From there you would see Earth complete a counterclockwise rotation, once each 24-hour period, as shown in **Figure 2.2**. As the rotating Earth carries an observer on the surface from west to east, objects in the sky appear to move in the other direction, from east to west. As seen from Earth's surface, the path each object takes across the sky is called its **apparent daily motion**.

To help visualize the apparent daily motions of the Sun and stars, it is useful to think of the sky as a huge sphere with the stars painted on its surface and Earth at its center. From ancient Greek times to the Renaissance, most people believed this to be a true representation of the heavens. Astronomers call this imaginary sphere, shown in **Figure 2.3**, the **celestial sphere**. The celestial sphere is a useful concept because it is easy to visualize, but never forget that it is, in fact, imaginary.

Each point on the celestial sphere indicates a direction in space. Directly above Earth's North Pole is the **north celestial pole (NCP)**. Directly above Earth's **South Pole**, which is at the south end of Earth's rotation axis, is the **south celestial pole (SCP)**. Directly above Earth's **equator** is the **celestial equator**, an imaginary circle that divides the sky into a northern half and a southern half. Just as the north celestial pole is the projection of the direction of Earth's North Pole into the sky, the celestial equator is the projection of the plane of Earth's equator into the sky. Just as Earth's North Pole is 90° away from Earth's equator, the north celestial pole is 90° away from the celestial equator. If you are in the Northern Hemisphere and you point one arm toward the celestial equator and one arm toward the north celestial pole, your arms will always form a right angle, so the north celestial pole is 90° away from the celestial equator. If you are in the Southern Hemisphere, the same holds true there: the angle between the celestial equator and the south celestial pole is always 90° as well.

Between the celestial poles and the equator, objects have positions on the celestial sphere with coordinates analogous to latitude and longitude on Earth. **Latitude** is an indication of distance north or south from Earth's equator. On the celestial sphere, **declination** similarly indicates the distance of an object north or south of the celestial equator (from 0° to ±90°). On Earth, **longitude** measures how far east or west you are from the Royal Observatory in Greenwich, England. **Right ascension** on the celestial sphere is similar to longitude on Earth and measures the angular distance of a celestial body eastward along the celestial equator from the point where the Sun's path crosses the celestial equator from south to north. These coordinates are used to locate objects in the sky quickly. The **ecliptic** is the path of the Sun in the sky throughout the year. Detailed descriptions and illustrations of latitude and longitude, and of celestial coordinates used with the celestial sphere, can be found in Appendix 7.

The **zenith** is the point in the sky directly above you wherever you are, as shown in **Figure 2.3a**. You can find the **horizon** by standing up and pointing your right hand at the zenith and your left hand straight out from your side. Turn in a complete circle. Your left hand has traced out the entire horizon. You can divide the sky into an east half and a west half with a line that runs from the horizon at due north through the zenith to the horizon at due south. This imaginary north–south line is called the **meridian**, shown as a dashed line in Figure 2.3a. Figure 2.3b shows these locations on the celestial sphere. The meridian line continues around the far side of the celestial sphere, through the **nadir** (the point directly below you), and back to the starting point due north.

Take a moment to visualize all these locations in space. To see how to use the celestial sphere, consider the Sun at noon and at midnight. Local noon occurs when the Sun crosses the meridian at your location. This is the highest point above the horizon that the Sun will reach on any given day. The highest point is almost never the zenith. You have to be in a specific place on a specific day for the Sun to be directly over your head at noon, for example, at a latitude 23.5° north of the

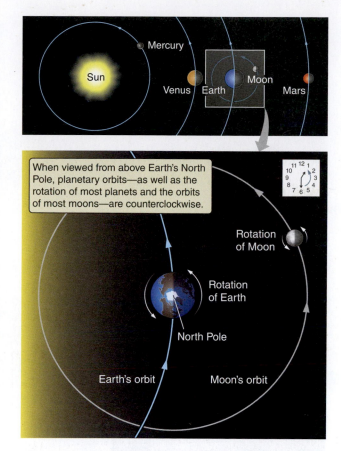

Figure 2.2 Motions in the Solar System, as viewed from above Earth's North Pole. (Not drawn to scale.)

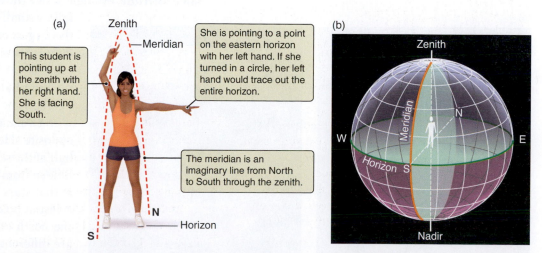

Figure 2.3 (a) The meridian is a line on the celestial sphere that runs from north to south, dividing the sky into an east half and a west half. (b) At any location on Earth, the sky is divided into an east half and a west half by an imaginary meridian projected onto the celestial sphere.

Astronomy in Action: Vocabulary of the Celestial Sphere

Nebraska Simulations: Celestial and Horizon Systems Comparison; Rotating Sky Explorer

▶❚ **AstroTour:** The View from the Poles

Figure 2.4 As viewed from (a) Earth's North Pole, (b) stars move throughout the night in counter-clockwise, circular paths about the zenith. (c) The same half of the sky is always visible from the North Pole.

(a)

North celestial pole (NCP)

North Pole

This disk represents the horizon, the boundary between the part of the sky you can see and the part that is blocked from view by Earth.

Equator

South Pole

From the North Pole looking directly overhead, the **north celestial pole (NCP)** is at the zenith.

(b)

North celestial pole at the zenith

As Earth rotates, the stars appear to move in a counterclockwise direction around the **NCP**.

(c)

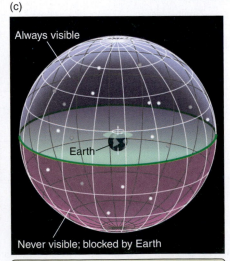

Always visible

Earth

Never visible; blocked by Earth

From the North Pole, you always see the same half of the sky.

equator on June 20. Local midnight occurs when the Sun is precisely opposite from its position at local noon. From our perspective on Earth, the celestial sphere appears to rotate, carrying the Sun across the sky to its highest point at noon, over toward the west to set in the evening. In *reality*, the Sun remains in the same place in space through the entire 24-hour period, and Earth rotates so that any given location on Earth faces a different direction at every moment. When it is noon where you live, Earth has rotated so that you face most directly toward the Sun. Half a day later, at midnight, your location on Earth has rotated to face most directly away from the Sun.

The View from the Poles

The apparent daily motions of the stars and the Sun depend on where you live. For example, the apparent daily motions of celestial objects in a northern place such as Alaska are quite different from the apparent daily motions seen from a tropical island such as Hawaii. To understand why your location matters, let's examine the view of the stars from the poles—and then use these to guide our thinking about the view of the stars from other latitudes.

Imagine that you are standing on the North Pole watching the sky as in **Figure 2.4a**. At the North Pole, the north celestial pole is directly overhead at the zenith. Ignore the Sun for the moment and pretend that you can always see stars in the sky. You are standing where Earth's axis of rotation intersects its surface, which is like standing at the center of a rotating carousel. As Earth rotates, the spot directly above you remains fixed over your head while everything else in the sky appears to revolve in a counterclockwise direction around this spot. Figure 2.4b depicts this overhead view.

No matter where you are on Earth, you can see only half of the sky at any one time. The horizon is the boundary between the part of the sky you can see and the other half of the sky that is blocked by Earth. Except at the poles, the visible half of the sky changes constantly as Earth rotates, because the zenith points to different locations in the sky as Earth carries you around. In contrast, if you are standing at the North Pole, the zenith is always in the same location in space, so the objects visible from the North Pole follow circular paths that always have the same **altitude**, or angle above the horizon. Objects close to the zenith appear to follow small circles, while objects near the horizon follow the largest circles (Figure 2.4b). The view from the North Pole is special because from there, nothing rises or sets each day as Earth turns: from there you will always see the *same* half of the celestial sphere (Figure 2.4c).

The view from Earth's South Pole is much the same—with two major differences. First, the South Pole is on the opposite side of Earth from the North Pole, so the visible half of the sky at the South Pole is precisely the half that is hidden from view at the North Pole. The second difference is that stars appear to move clockwise around the south celestial pole rather than counterclockwise as they do at the north celestial pole. To visualize why these motions are different, stand up and spin around from right to left. As you look at the ceiling, things appear to move in a counterclockwise direction, but as you look at the floor, they appear to be moving clockwise.

CHECK YOUR UNDERSTANDING 2.1
No matter where you are on Earth, stars appear to rotate about a point called the:
(a) zenith; (b) celestial pole; (c) nadir; (d) meridian.

The View Away from the Poles

Suppose that you leave the North Pole to travel south to lower latitudes. Imagine a line from the center of Earth to your location on the surface of the planet, as in **Figure 2.5**. Now imagine a second line from the center of Earth to the point on the equator closest to you. The angle between these two lines is your latitude. At the North Pole, for example, these two imaginary lines form a 90° angle. At the equator, they form a 0° angle. So the latitude of the North Pole is 90° north, and the latitude of the equator is 0°. The South Pole is at latitude 90° south.

Your latitude determines the part of the sky that you can see throughout the year. As you move south from the North Pole, your zenith moves away from the north celestial pole, and so the horizon moves as well. At the North Pole, the horizon makes a 90° angle with the north celestial pole, which is at the zenith. At a latitude of 60° north, as in Figure 2.5, your horizon is tilted 60° from the north celestial pole. The angle between your horizon and the north celestial pole is equal to your latitude no matter where you are on Earth. The situation is the same in the Southern Hemisphere—your latitude is the altitude of the south celestial pole. At the equator, at a latitude of 0°, the north and south celestial poles would be at the northern and southern horizons, respectively.

One way to solidify your understanding of the view of the sky at different latitudes is to draw pictures like the one in Figure 2.5. If you can draw a picture like this for any latitude—filling in the values for each angle in the drawing and imagining what the sky looks like from that location—then you will be well on your way to developing a working knowledge of the appearance of the sky. That knowledge will prove useful later, when we discuss a variety of phenomena, such as the changing of the seasons.

Motions of the Stars and the Celestial Poles

Figure 2.6 shows two time-lapse views of the sky from different latitudes. The apparent motions of the stars about the celestial poles also differs from latitude to latitude. The visible part of the sky constantly changes, as stars rise and set with Earth's rotation. From this perspective the horizon appears fixed, while the stars appear to move. From these different latitudes, if we focus our attention on the north celestial pole, we see much the same thing we saw from Earth's North Pole. The north celestial pole remains fixed in the sky, and all of the stars appear to move throughout the night in counterclockwise, circular paths around that point. But because the north celestial pole is no longer directly overhead as it was at the North Pole, the apparent circular paths of the stars are now tipped relative to the horizon. (More correctly, your horizon is now tipped relative to the apparent circular paths of the stars.)

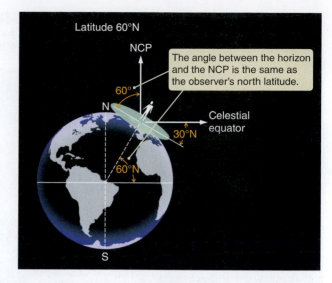

Figure 2.5 Your perspective on the sky depends on your location on Earth. The locations of the celestial poles and the celestial equator in an observer's sky depend on the observer's latitude. In this case, an observer at latitude 60° north sees the north celestial pole at an altitude of 60° above the northern horizon and the celestial equator 30° above the southern horizon.

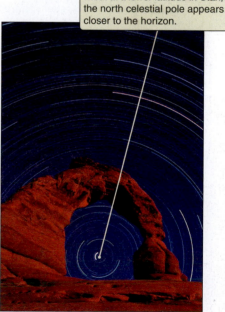

Figure 2.6 Time exposures of the sky showing the apparent motions of stars through the night. Note the difference in the circumpolar portion of the sky as seen from the two different northern latitudes.

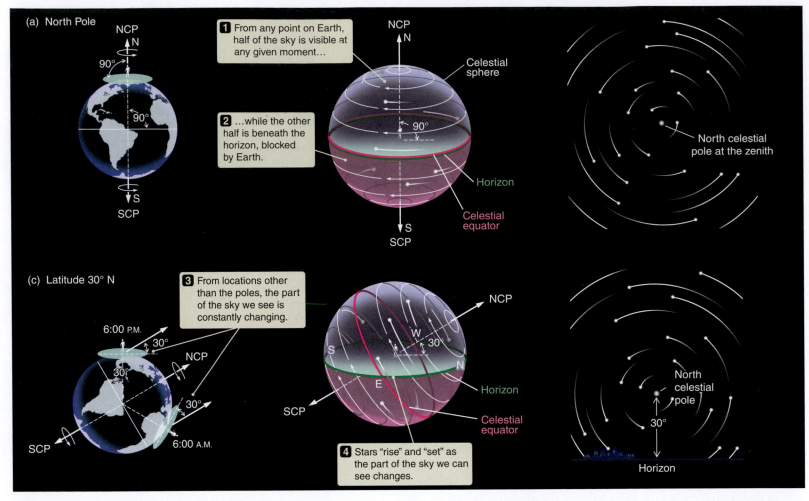

(a) North Pole

1 From any point on Earth, half of the sky is visible at any given moment…

2 …while the other half is beneath the horizon, blocked by Earth.

NCP
N
Celestial sphere
90°
Horizon
Celestial equator
S
SCP

North celestial pole at the zenith

(c) Latitude 30° N

3 From locations other than the poles, the part of the sky we see is constantly changing.

6:00 P.M.
30°
NCP
30°
30°
SCP
6:00 A.M.

NCP
W
30°
S
E
N
Horizon
SCP
Celestial equator

4 Stars "rise" and "set" as the part of the sky we can see changes.

North celestial pole
30°
Horizon

Figure 2.7 The celestial sphere is shown here as viewed by observers at four different latitudes. At all locations other than the poles, stars rise and set as the part of the celestial sphere that we see changes during the day.

From the vantage point of an observer in the Northern Hemisphere, stars near the north celestial pole never dip below the horizon and thus stay visible all night. Recall from Figure 2.5 that the latitude is equal to the altitude of the north celestial pole. Stars closer to the north celestial pole than this angle never dip below the horizon as they complete their apparent paths around the pole. These stars are called **circumpolar** ("around the pole") stars. Another group of stars, near the south celestial pole, never rise above the horizon in the Northern Hemisphere. Stars between those that never rise and those in the circumpolar region can be seen for *only part* of each 24-hour day. These stars appear to rise and set as Earth turns. The only place on Earth where you can see the entire sky over the course of 24 hours is the equator. From the equator, the north and south celestial poles sit on the northern and southern horizons, respectively, and all of the stars move through the sky each 24-hour day. (Even though the Sun lights the sky for roughly half of this time, the stars are still there.)

Figure 2.7 shows the orientation of the sky as seen by observers at four different latitudes. For an observer at the North Pole (Figure 2.7a), the celestial equator lies exactly along the horizon. The north celestial pole is at the zenith, and the

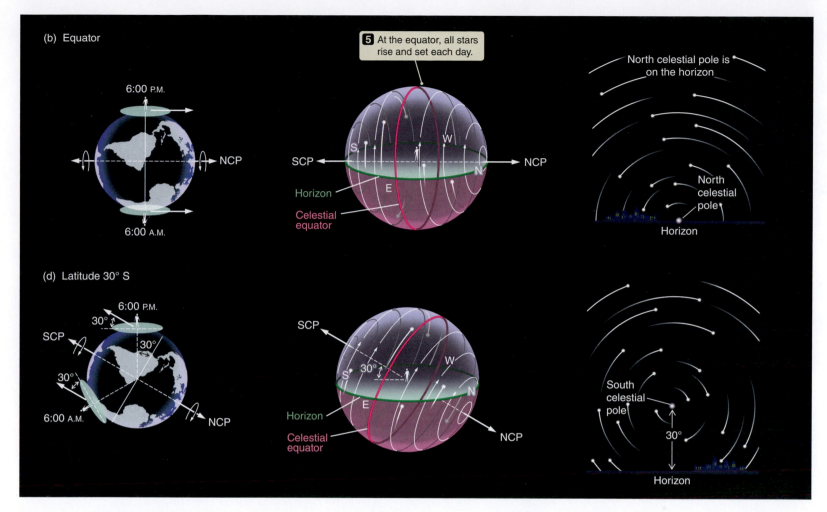

Figure 2.7 Continued.

southern half of the sky is never visible. Stars neither rise nor set; their paths form circles parallel to the horizon. For an observer at the equator (Figure 2.7b), the celestial poles are both at the horizon, and all the stars are visible in a 24-hour period, rising straight up and setting straight down each day.

At other latitudes, the celestial equator intersects the horizon due east and due west. Therefore, a star on the celestial equator rises due east and sets due west. Stars located north of the celestial equator rise north of east and set north of west. Stars located south of the celestial equator rise south of east and set south of west.

From everywhere else on Earth (except at the poles), half of the celestial equator is always visible above the horizon. Therefore, any object located on the celestial equator is visible half of the time—above the horizon for 12 hours each day. Objects that are not on the celestial equator are above the horizon for differing amounts of time. Figures 2.7c and d show that stars in the observer's hemisphere are visible for more than half the day because more than half of each star's path in the sky is above the horizon. In contrast, stars in the opposite hemisphere are visible for less than half the day because less than half of each star's path in the sky is above the horizon.

Nebraska Simulations: Meridional Altitude Simulator

Nebraska Simulation: Declination Ranges Simulator

Nebraska Simulations: Big Dipper Clock

For example, as seen from the Northern Hemisphere, stars north of the celestial equator remain above the horizon for more than 12 hours each day. The farther north the star is, the longer it stays up. Circumpolar stars are the extreme example of this phenomenon; they are always above the horizon. In contrast, stars south of the celestial equator are above the horizon for less than 12 hours each day. The farther south a star is, the less time it is visible. For an observer in the Northern Hemisphere, stars located close to the south celestial pole never rise above the horizon.

Using the Stars for Navigation Since ancient times, travelers have used the stars for navigation. They would find the north or south celestial poles by recognizing the stars that surround them. In the Northern Hemisphere, a moderately bright star happens to be located close to the north celestial pole (**Figure 2.8a**). This star is called Polaris, the "North Star." The altitude of Polaris in the sky is nearly equal to your latitude. If you are in Phoenix, Arizona, for example (latitude 33.5° north), the north celestial pole has an altitude of 33.5°. In Fairbanks, Alaska (latitude 64.6° north), the north celestial pole sits much higher, with an altitude of 64.6°. Similarly in the Southern Hemisphere, the constellation Crux (commonly called the Southern Cross) points to a star near the south celestial pole (Figure 2.8b). A navigator who has located a pole star can identify north and south, and therefore east, west, and her latitude. This enables the navigator to determine which direction to travel. The location of the north celestial pole in the sky was used to measure the size of Earth, as described in **Working It Out 2.1**. Determining your longitude by astronomical methods is much more complicated because of Earth's rotation. Longitude cannot be determined from astronomical observation alone.

(a)

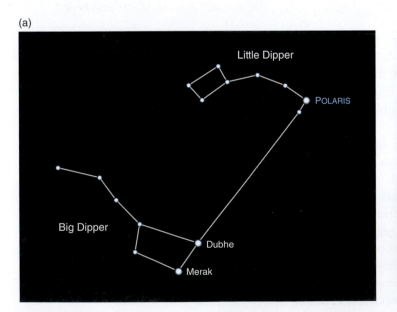

(b)

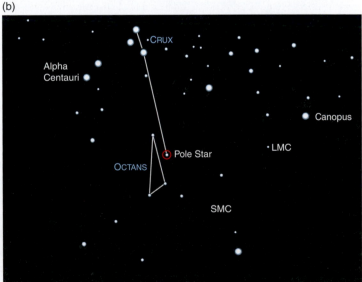

Figure 2.8 Groups of stars near the pole stars in the sky can be used to locate a pole star. (a) Two bright "pointer stars" stars in the cup of the Big Dipper point toward Polaris, the "North Star." (b) In the Southern Hemisphere, the constellation Crux and two of its bright "pointer stars" can be used to locate the relatively faint southern pole star.

2.1 Working It Out   How to Estimate the Size of Earth

We can use the location of the north celestial pole in the sky to estimate the size of Earth. Suppose we start out in Phoenix, Arizona, and we observe the north celestial pole to be 33.5° above the horizon. If we head north, by the time we reach the Grand Canyon, about 290 kilometers (km) from Phoenix, we notice that the north celestial pole has risen to about 36° above the horizon. This difference between 33.5° and 36° (2.5°) is 1/144 of the way around a circle. (A circle is 360°, and 2.5°/360° = 1/144.)

This means that we must have traveled 1/144 of the way around the circumference, C, of Earth by traveling the 290 km between Phoenix and the Grand Canyon. In other words,

$$\frac{1}{144} \times C = 290 \text{ km}$$

Rearranging the expression, the circumference of Earth is given by

$$C = 144 \times 290 \text{ km} \approx 42{,}000 \text{ km}$$

The actual circumference of Earth is just over 40,000 km, so our simple calculation was close. The circumference of a circle is equal to 2π multiplied by its radius. So, the radius of Earth is given by

$$\text{Radius} = \frac{C}{2\pi} = \frac{40{,}000 \text{ km}}{2\pi} = 6{,}400 \text{ km}$$

It was in much this same way that the Greek astronomer Eratosthenes (276–194 BCE) made the first accurate measurements of the size of Earth in about 230 BCE. As illustrated in **Figure 2.9**, Eratosthenes used the distance between his home city of Alexandria and the city of Syene (currently Aswân, in Egypt), which was 5,000 "stadia." He noticed that on the first day of summer in Syene, the sunlight reflected directly off the water in a deep well, so the Sun must have been nearly at the zenith. By measuring the shadow of the Sun from an upright stick in Alexandria, he saw that the Sun was about 7.2° south of the zenith on the same date. Assuming Earth was spherical and Syene was directly south of Alexandria, he determined the

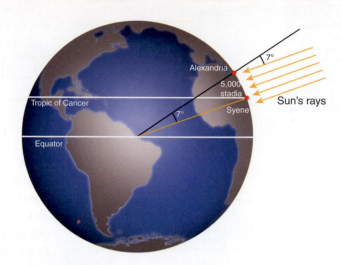

Figure 2.9 Eratosthenes estimated the size of Earth using observations and basic calculations.

distance between the two cities to be 7.2° divided by 360, or 1/50 of the circumference of Earth.

It was difficult to estimate distances accurately in those days, and although historians know Eratosthenes concluded that the cities were 5,000 stadia apart, they are still not at all sure of the value of his stadion unit. If the stadion was 185 meters, then Eratosthenes would have worked the math in a similar way:

$$\frac{1}{50} \times C = 5{,}000 \text{ stadia} \times 185 \text{ meters/stadion}$$

$$= 925{,}000 \text{ meters} = 925 \text{ km}$$

Eratosthenes would have found the circumference of Earth to be

$$C = 50 \times 925 \text{ km} = 46{,}250 \text{ km}$$

only about 16 percent higher than the modern value.

Relative Motions and Frame of Reference

Why don't we feel the motion of Earth as it spins on its axis and moves through space in its orbit around the Sun? Astronomers use the concept of a **frame of reference**, which is a coordinate system within which an observer measures positions and motions. The difference in motion between two individual frames of reference is called the **relative motion**. For example, imagine that you are riding in a car traveling down a straight section of highway at a constant speed. If you are not looking out the window or feeling road vibrations, there is no experiment you can easily do to tell the difference between riding in a car down a straight section of highway at constant speed and sitting in the car while it is parked in

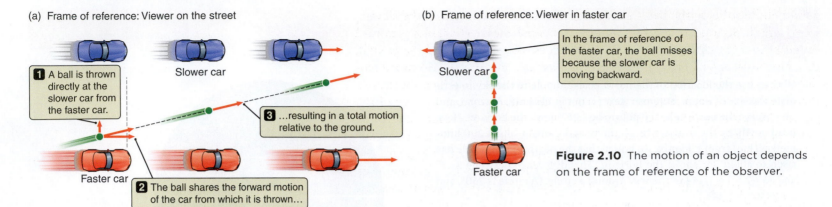

(a) Frame of reference: Viewer on the street

1 A ball is thrown directly at the slower car from the faster car.

Slower car

3 ...resulting in a total motion relative to the ground.

Faster car

2 The ball shares the forward motion of the car from which it is thrown...

(b) Frame of reference: Viewer in faster car

In the frame of reference of the faster car, the ball misses because the slower car is moving backward.

Slower car

Faster car

Figure 2.10 The motion of an object depends on the frame of reference of the observer.

(c) Frame of reference: Viewer in slower car

In the frame of reference of the slower car, the ball misses because the ball and the faster car are moving forward.

Slower car

Faster car

your driveway. Because everything in the car is moving together, the relative motions between objects in the car are all that can be measured, and they are all zero, no motion is observed. Similarly, the resulting relative motions between objects that are near each other on Earth are zero. This is why we do not notice Earth's motion.

Now imagine that two cars are driving down the road at different speeds, as shown in **Figure 2.10a**. Ignoring for the moment any real-world complications, like wind resistance, if you were to throw a ball from the faster-moving car directly out the side window at the slower-moving car as the two cars passed, you would miss. The ball shares the forward motion of the faster car, so the ball outruns the forward motion of the slower car. As shown in Figure 2.10b, from your perspective in the faster car, the slower car lagged behind the ball. From the slower car's perspective, represented in Figure 2.10c, your car and the ball sped on ahead.

Although we cannot feel Earth's rotation, it influences things as diverse as the motion of weather patterns on Earth and how an artillery gunner must aim at a distant target. An object on the surface of Earth moves in a circle each day around Earth's rotation axis. This circle is larger for objects near Earth's equator and smaller for objects closer to one of Earth's poles. But all objects must complete their circular motion in exactly 1 day. As you can see in **Figure 2.11**, the surface of Earth moves faster at the equator than at higher latitudes. An object closer to the equator has farther to go each day than does an object nearer a pole. Therefore, the object nearer the equator must be moving faster than the object at higher latitude. If an object starts out at one latitude and then moves to another, its apparent motion over the surface of Earth is influenced by this difference in speed.

Now consider two locations at different latitudes. Imagine that a cannonball is launched directly north from a point in the Northern Hemisphere, as shown in Figure 2.11a. Because the cannon is located nearer to the equator than its target is, the cannonball is moving toward the east faster than its target. Even though the cannonball is fired toward the north, it shares the eastward velocity of the cannon itself. This means that the cannonball is *also* moving toward the east faster than

The ground near the equator is like the faster car in Figure 2.10. The ground at higher latitudes is like the slower car.

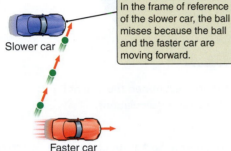

Slower car

Faster car

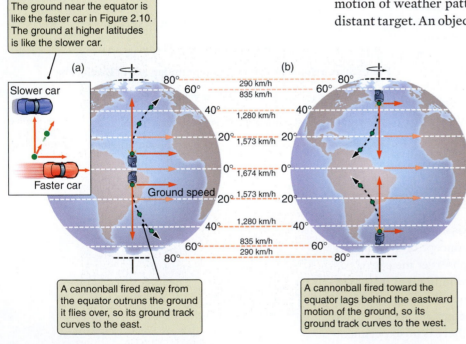

(a)

(b)

80°	
60°	290 km/h
	835 km/h
40°	1,280 km/h
20°	1,573 km/h
0°	1,674 km/h
20°	1,573 km/h
40°	1,280 km/h
60°	835 km/h
	290 km/h
80°	

Ground speed

A cannonball fired away from the equator outruns the ground it flies over, so its ground track curves to the east.

A cannonball fired toward the equator lags behind the eastward motion of the ground, so its ground track curves to the west.

Figure 2.11 The Coriolis effect causes objects to be deflected as they move across the surface of Earth. The green dashed line shows the path of the cannonball as measured by a local observer.

its target. Recall how the ball thrown from the faster car outpaced the slower-moving car in Figure 2.10. Similarly, as the cannonball flies north, it moves toward the east faster than the ground underneath it does. To an observer on the ground, the cannonball appears to curve toward the east as it outruns the eastward motion of the ground it is crossing. The farther north the cannonball flies, the greater the difference between its eastward velocity and the eastward velocity of the ground. Thus, the cannonball follows a path that appears to curve more and more to the east the farther north it goes. If you are located in the Northern Hemisphere and fire a cannonball *south* toward the equator, as shown in Figure 2.11b, the opposite effect will occur. Now the cannon is moving toward the east more slowly than its target. As the cannonball flies toward the south, its eastward motion lags behind that of the ground underneath it, and the cannonball appears to curve toward the west.

This curving motion of objects from the difference in Earth's rotation speeds at different latitudes is called the **Coriolis effect**. In the Northern Hemisphere, the Coriolis effect causes a cannonball fired north to drift to the east as seen from the surface of Earth. In other words, the cannonball appears to curve to the right. A cannonball fired south appears to curve to the west, which also gives it the appearance of curving to the right. In the Northern Hemisphere, the Coriolis effect seems to deflect things to the *right*. If you think through this example for the Southern Hemisphere, you will see that south of the equator, the Coriolis effect seems to deflect things to the *left*. In between, at the equator itself, the Coriolis effect vanishes.

On Earth the effect is enough to deflect a fly ball hit north or south into deep left field in a stadium in the northern United States by about a half a centimeter. At some time or other, the Coriolis effect from the rotation of Earth has probably determined the outcome of a ball game.

CHECK YOUR UNDERSTANDING 2.2

If the star Polaris has an altitude of 35°, then we know that: (a) our longitude is 55° east; (b) our latitude is 55° north; (c) our longitude is 35° west; (d) our latitude is 35° north.

2.2 Revolution about the Sun Leads to Changes during the Year

Earth orbits (or **revolves**) around the Sun in the same direction that Earth spins about its axis—counterclockwise as viewed from above Earth's North Pole (see **Figure 2.12**). A **year** is the time it takes for Earth to complete one revolution around the Sun. The motion of Earth around the Sun is responsible for many of the patterns of change we see in the sky and on Earth, including changes in the stars we see overhead. Because of this motion, the stars in the night sky change throughout the year, and Earth experiences seasons.

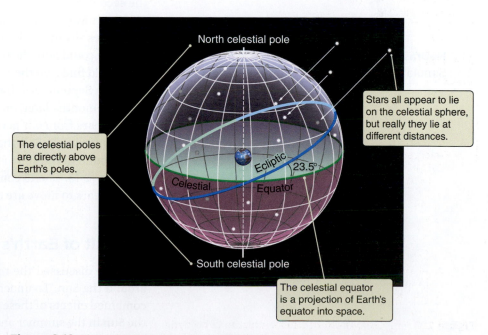

Figure 2.12 The celestial sphere is a useful fiction for thinking about the appearance and apparent motion of the stars in the sky.

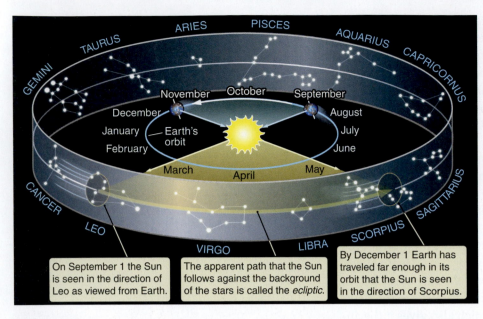

On September 1 the Sun is seen in the direction of Leo as viewed from Earth.

The apparent path that the Sun follows against the background of the stars is called the *ecliptic*.

By December 1 Earth has traveled far enough in its orbit that the Sun is seen in the direction of Scorpius.

Figure 2.13 As Earth orbits around the Sun, the Sun's apparent position against the background of stars changes. The imaginary circle traced by the annual path of the Sun is called the ecliptic. The blue band shows 12 of the zodiacal constellations along the ecliptic.

Nebraska Simulations: Ecliptic (Zodiac) Simulator

Constellations and the Zodiac

As shown in **Figure 2.13**, as Earth orbits the Sun, our view of the night sky changes. Six months from now, Earth will be on the other side of the Sun. The stars that are overhead at midnight 6 months from now are those that are near overhead at noon today. In order to follow the patterns of the Sun and the stars, early humans grouped together stars that formed recognizable patterns, called **constellations**. But people from different cultures saw different patterns and projected ideas from their own cultures onto what they saw in the sky. Constellations named for winged horses, dragons, and other imaginary images, and the stories that go with them, are creations of the human imagination. If you look at the sky, no obvious pictures of these images emerge. Instead, there is only the random pattern of stars—about 5,000 of them visible to the naked eye— spread out across the sky.

Modern constellations visible from the Northern Hemisphere draw heavily from the list of constellations compiled 2,000 years ago by the Alexandrian astronomer Ptolemy. Modern constellation names in the southern sky come from European explorers visiting the Southern Hemisphere during the 17th and 18th centuries. Today, astronomers use an officially sanctioned set of 88 constellations as a kind of road map of the sky. Every star in the sky lies within the borders of a single constellation, and the names of constellations are used in naming the stars that lie within their boundaries. For example, Sirius, the brightest star in the sky, lies within the boundaries of the constellation Canis Major (meaning "big dog"). Following the Greek alphabet, the brightest star in a constellation is called *alpha*, the second brightest is called *beta*, and so forth. The official name of Sirius is Alpha Canis Majoris, indicating that it is the brightest star in Canis Major. Appendix 6 provides sky maps showing the constellations.

If you could note the position of the Sun relative to the stars each day for a year, you would find that the Sun traces out a great circle against the background of the stars. On September 1, the Sun appears to be in the direction of the constellation Leo. Six months later, on March 1, Earth is on the other side of the Sun, and the Sun appears from our perspective on Earth to be in the direction of the constellation Aquarius. Recall that the apparent path that the Sun follows against the background of the stars is called the ecliptic and is illustrated as the yellow band in Figure 2.13. The 13 constellations that lie along the ecliptic through which the Sun appears to move are called the constellations of the **zodiac**.

The Tilt of Earth's Axis and the Seasons

We have discussed the rotation of Earth on its axis and the revolution of Earth around the Sun. To understand why the seasons change, we need to consider the combined effects of these two motions. Many people believe that Earth is closer to the Sun in the summer and farther away in the winter, and this change in distance causes the seasons. Can this hypothesis be falsified? We can make a prediction that if the distance from Earth to the Sun caused the seasons, then all of Earth should experience summer at the same time of year. But the United States experiences

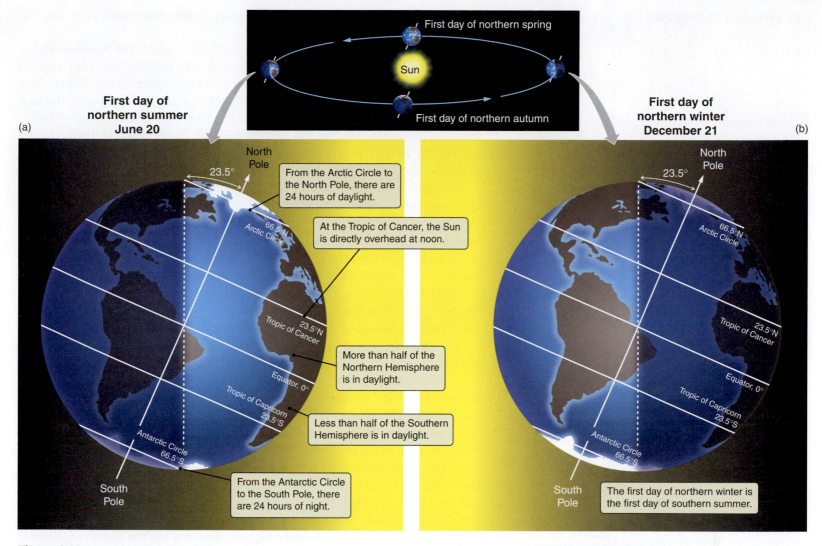

Figure 2.14 (a) On the first day of the northern summer (around June 20, the summer solstice), the northern end of Earth's axis is tilted most nearly toward the Sun, while the Southern Hemisphere is tipped away. (b) Six months later, on the first day of the northern winter (around December 21, the winter solstice), the situation is reversed. Seasons are opposite in the two hemispheres.

summer in June, while Australia experiences summer in December. In modern times, we can directly measure the distance, and we find that Earth is actually closest to the Sun at the beginning of January. We have just falsified this hypothesis, and we need to look for another one that explains all of the available facts.

We observe that as Earth orbits the Sun, the Sun appears to move along the ecliptic, which is tilted 23.5° with respect to the celestial equator. This occurs because Earth's axis of rotation is tilted by 23.5° from the perpendicular to Earth's orbital plane. We notice that during the summer, the days are longer than in winter, and the Sun is higher in the sky as it crosses the meridian in summer than in winter.

Figure 2.14 shows that as Earth moves around the Sun, its axis always points towards Polaris, in the same direction in space. During its orbit, sometimes Earth is on one side of the Sun, and sometimes on the other side. Therefore, sometimes Earth's North Pole is tilted more toward the Sun, and other times the South Pole is tilted more towards the Sun. When Earth's North Pole is tilted toward the Sun, an observer on Earth views the Sun north of the celestial equator; for observers in the Northern Hemisphere, the Sun is above the horizon more than 12 hours each day, thus the days are longer than 12 hours. Six months later, when Earth's North

▶‖ **AstroTour:** The Earth Spins and Revolves

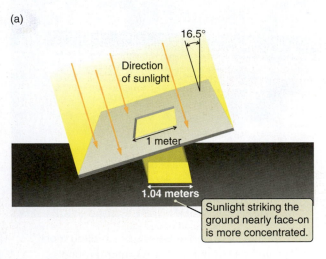

(a)

16.5°

Direction of sunlight

1 meter

1.04 meters

Sunlight striking the ground nearly face-on is more concentrated.

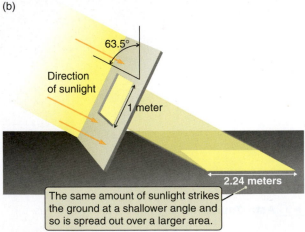

(b)

63.5°

Direction of sunlight

1 meter

2.24 meters

The same amount of sunlight strikes the ground at a shallower angle and so is spread out over a larger area.

Figure 2.15 Local noon at latitude 40° north. (a) On the first day of northern summer, sunlight strikes the ground almost face-on. (b) On the first day of northern winter, sunlight strikes the ground more obliquely, and less than half as much sunlight falls on each square meter of ground each second.

Pole is tilted away from the Sun, an observer in the same place views the Sun south of the celestial equator.

In the preceding paragraph, we were careful to specify the *Northern* Hemisphere because seasons are opposite in the Southern Hemisphere. Look again at Figure 2.14. Around June 20, while the Northern Hemisphere is enjoying the long days and short nights of summer, Earth's South Pole is tilted away from the Sun. It is winter in the Southern Hemisphere; less than half of the Southern Hemisphere is illuminated by the Sun, and the days are shorter than 12 hours. On December 21, Earth's South Pole is tilted toward the Sun. It is summer in the Southern Hemisphere; the days are long and the nights are short there.

To understand how the combination of Earth's axial tilt and its path around the Sun creates seasons, consider a limiting case. If Earth's spin axis were exactly perpendicular to the plane of Earth's orbit (the **ecliptic plane**), then the Sun would always be on the celestial equator. At every latitude, the Sun would follow the same path through the sky every day, rising due east each morning and setting due west each evening. The Sun would be above the horizon exactly half the time, and days and nights would always be exactly 12 hours long everywhere on Earth. In short, if Earth's spin axis were exactly perpendicular to the plane of Earth's orbit, each day would be just like the last, and there would be no seasons.

The differing length of days through the year is part of the explanation for seasonal temperature changes, but it is not the whole story. Another important effect relates to the angle at which the Sun's rays strike Earth. The Sun is higher in the sky during the summer than it is during the winter, and sunlight strikes the ground *more directly* during the summer than during the winter. To see why this is important, study **Figure 2.15**. During the summer, Earth's surface is more nearly face-on to the incoming sunlight. More energy falls on each square meter of ground each second; the light is concentrated and bright. During the winter, the surface of Earth is more tilted with respect to the sunlight, so the light is more diffuse. Less energy falls on each square meter of the ground each second. This is the main reason why it is hotter in the summer and colder in the winter. As you can see in the **Process of Science Figure**, determining the causes of seasonal change requires accounting for all the known facts.

We can compare the average temperatures found at different latitudes on Earth to see the effect of the height of the Sun in the sky. Near the equator, the Sun passes high overhead every day, regardless of the season. As a result, the average temperatures are warm throughout the year. At high latitudes, however, the Sun is *never* high in the sky, and the average temperatures can be cold and harsh even during the summer. In between, at latitude 40° north, which stretches across the United States from northern California to New Jersey, more than twice as much solar energy falls on each square meter of ground per second at noon on June 20 as falls there at noon on December 21. These two effects—the directness of sunlight and the differing length of the night—mean that the Sun heats a hemisphere more during summer than winter.

The Solstices and the Equinoxes

Four days during Earth's orbit mark unique moments in the year. The day when the Sun is highest in the sky as it crosses the meridian—the line from due north to due south that passes overhead—is called the **summer solstice**. On this day, the Sun rises farthest north of east and sets farthest north of west. This occurs each year near June 20, the first day of summer in the Northern Hemisphere. This orientation of Earth and Sun is shown in Figure 2.14a.

Process of Science

THEORIES MUST FIT ALL THE KNOWN FACTS

Many people misunderstand the phenomenon of changing seasons because they do not account for all the relevant facts.

Take 1

The Hypothesis

We have seasons because Earth is closer to the Sun in summer and farther away in winter.

The Test

If this is true, both the Northern and Southern hemispheres would have summer in July.
The Northern and Southern hemispheres experience opposite seasons.

The Conclusion

The hypothesis is falsified.

Take 2

The Hypothesis

We have seasons because the tilt of Earth's axis causes one hemisphere to be significantly closer to the Sun than the other.

The Test

If this is true, the distances must be very different to cause such a large effect. Earth is tiny compared to its distance from the Sun: the difference in distance between hemispheres is less than 0.004 percent of the distance from the Sun.

The Conclusion

The hypothesis is falsified.

Take 3

The Hypothesis

We have seasons because Earth's tilt changes the distribution of energy—one hemisphere receives more light than the other.

The Test

If this is true, the amount of sunlight striking the ground in the summer should be more than in the winter, and the days should be longer in summer.

The Conclusion

Seasons are caused primarily by a change in illumination due to Earth's tilt. During winter, less energy falls on each square meter of ground per second.

New information often challenges misconceptions.

(a)

Motion of Earth around the Sun

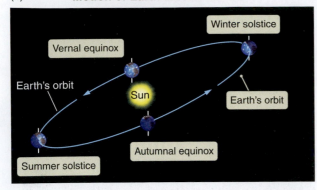

(b) **Apparent motion of the Sun seen from Earth**

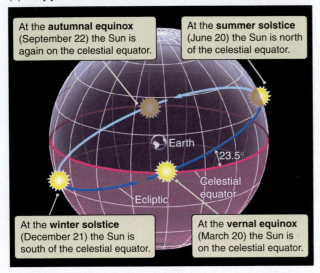

Figure 2.16 The motion of Earth around the Sun is shown from the frame of reference of (a) the Sun and (b) Earth.

📹 **Astronomy in Action:** The Earth-Moon-Sun System

Six months after the summer solstice, the North Pole is tilted away from the Sun. This day is the **winter solstice** in the Northern Hemisphere, shown in Figure 2.14b. The winter solstice occurs each year about December 21, the shortest day of the year and the first day of winter in the Northern Hemisphere. Almost all cultural traditions in the Northern Hemisphere include a major celebration of some sort in late December. These winter festivals celebrate the return of the source of Earth's light and warmth. The days have stopped growing shorter and are beginning to get longer, and spring will come again.

Between the two solstices, there are two days when the ecliptic crosses the celestial equator. On these days, the Sun lies directly above Earth's equator. We call these days *equinoxes*, which means "equal night," because the entire Earth experiences 12 hours of daylight and 12 hours of darkness. Halfway between summer solstice and winter solstice, the **autumnal equinox** marks the beginning of fall in the Northern Hemisphere; it occurs around September 22. Halfway between winter solstice and summer solstice, the **vernal equinox** marks the beginning of spring in the Northern Hemisphere; it occurs around March 20.

Figure 2.16 shows the solstices and equinoxes from two perspectives. Figure 2.16a shows Earth in orbit around a stationary Sun, and Figure 2.16b shows the Sun's apparent motion along the celestial sphere, which is how it appears to observers on Earth. In both cases, we are looking at the plane of Earth's orbit from the side, so that it is shown in perspective and looks quite flattened. We have also tilted the images so that the North Pole of Earth points straight up. The equinoxes correspond to the points in the sky where the celestial equator meets the ecliptic. Practice shifting between these two perspectives. You will know that you understand these differing perspectives when you are able to look at a position in either panel and predict the corresponding positions of the Sun and Earth in the other panel.

Just as it takes time for a pot of water on a stove to heat up when the burner is turned up and time for the pot to cool off when the burner is turned down, it takes time for Earth to respond to changes in heating from the Sun. The hottest months of northern summer are usually July and August, which come after the summer solstice, when the days are growing shorter. Similarly, the coldest months of northern winter are usually January and February, which occur after the winter solstice, when the days are growing longer. Temperature changes on Earth lag behind changes in the amount of heating we receive from the Sun.

This picture of the seasons must be modified somewhat near Earth's poles. At latitudes north of 66.5° north and south of 66.5° south, the Sun is circumpolar for

Figure 2.17 This composite photo shows the midnight Sun, which can be seen in latitudes above 66.5° north (or south). In the 360-degree panoramic view, the Sun moves 15° each hour.

a part of the year surrounding the first day of summer. These lines of latitude are the **Arctic Circle** and the **Antarctic Circle** (see Figure 2.14). When the Sun is circumpolar, it is above the horizon 24 hours a day, earning the polar regions the nickname "land of the midnight Sun" (**Figure 2.17**). There is an equally long period surrounding the first day of winter when the Sun never rises and the nights are 24 hours long. The Sun never rises high in the Arctic or Antarctic sky, so the sunlight is never very direct. Even with the long days at the height of summer, the Arctic and Antarctic regions remain relatively cool.

In contrast, on the equator, *all* stars are above the horizon 12 hours a day, and the Sun is no exception. On the equator, days and nights are 12 hours long throughout the year. The Sun passes directly overhead on the first day of spring and the first day of autumn because these are the days when the Sun is on the celestial equator. Sunlight is most direct, perpendicular to the ground, at the equator on these days. On the summer solstice, the Sun is at its northernmost point along the ecliptic. On this day, and on the winter solstice, the Sun is farthest from the zenith at noon, and therefore sunlight is least direct at the equator.

As shown in Figure 2.14, latitude 23.5° north is called the Tropic of Cancer, and latitude 23.5° south is called the Tropic of Capricorn. The band between these two latitudes is called the **Tropics**. If you live in the tropics—in Rio de Janeiro or Honolulu, for example—the Sun will be directly overhead at noon twice during the year.

Precession of the Equinoxes

When the Alexandrian astronomer Ptolemy and his associates were formalizing their knowledge of the positions and motions of objects in the sky 2,000 years ago, the Sun appeared in the constellation Cancer on the first day of northern summer and in the constellation Capricornus on the first day of northern winter. Today, the Sun is in Taurus on the first day of northern summer and in Sagittarius on the first day of northern winter. Why have the constellations in which solstices appear changed? There are actually two motions associated with Earth and its axis: Earth spins on its axis, but its axis also wobbles like the axis of a spinning top (**Figure 2.18**). The wobble is very slow: it takes about 26,000 years for the north celestial pole to complete one trip around a large circle centered on the north ecliptic pole. Currently, Polaris is the star we see near the north celestial pole. However, if you could travel several thousand years into the past or future, you would find that the point about which the northern sky appears to rotate is no longer near Polaris, but instead the stars rotate about another point on the path shown in Figure 2.18b. This figure shows the path of the north celestial pole through the sky during one cycle of this wobble.

The celestial equator is perpendicular to Earth's axis. Therefore, as Earth's axis wobbles, the celestial equator must also wobble. As the celestial equator wobbles, the locations where it crosses the ecliptic—the equinoxes—change as well. During each 26,000-year wobble of Earth's axis, the locations of the equinoxes make one complete circuit around the celestial equator. This change of the position of the equinox, due to the wobble of Earth's axis, is called the **precession of the equinoxes**.

CHECK YOUR UNDERSTANDING 2.3

If Earth's axis were tilted by 45°, instead of its actual tilt, how would the seasons be different than they are currently? (a) The seasons would remain the same. (b) Summers would be colder. (c) Winters would be shorter. (d) Winters would be colder.

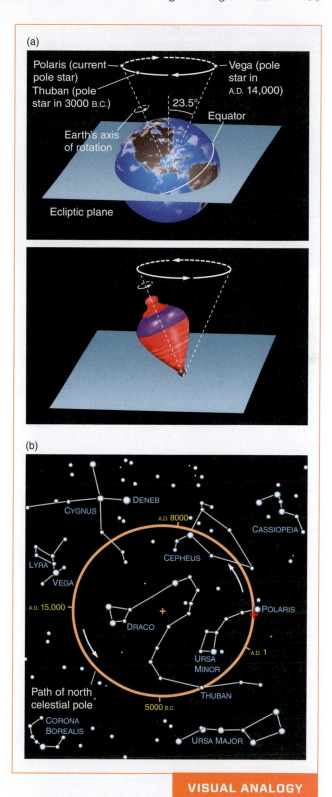

VISUAL ANALOGY

Figure 2.18 (a) Earth's axis of rotation changes orientation in the same way that the axis of a spinning top changes orientation. (b) This precession causes the projection of Earth's rotation axis to move in a circle, centered on the north ecliptic pole (orange cross in the center). The red cross marks the projection of Earth's axis on the sky in the early 21st century.

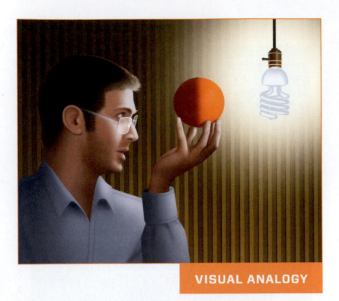

VISUAL ANALOGY

Figure 2.19 An orange and a lamp can help you visualize the changing phases of the Moon.

Astronomy in Action: Phases of the Moon

AstroTour: The Moon's Orbit: Eclipses and Phases

2.3 The Moon's Appearance Changes as It Orbits Earth

The most prominent object in our sky after the Sun is the Moon. Just as Earth orbits around the Sun, the Moon orbits around Earth once every 27.3 days. In this section, we will discuss the phases of the Moon as seen from Earth.

The Changing Phases of the Moon

The Moon and its changing aspects have long fascinated humans. We speak of the "man in the Moon," the "harvest Moon," and sometimes a "blue Moon." In mythology, the Moon was the Roman goddess Diana, the Greek goddess Artemis, and the Inuit god Igaluk. The Moon has been the frequent subject of mythology, art, literature, and music.

Unlike the Sun, the Moon has no light source of its own; it shines by reflected sunlight. As the Moon orbits Earth, our view of the illuminated portion of the Moon is constantly changing. These different appearances of the Moon are called **phases** of the Moon. As the Moon orbits Earth, our view of the illuminated portion of the Moon is constantly changing. During a new Moon, when the Moon is between Earth and the Sun, the side facing away from us is illuminated, and during a full Moon, when the Earth is between the Sun and the Moon, the side facing toward us is illuminated. The rest of the time, only part of the illuminated portion can be seen from Earth. Sometimes the Moon appears as a circular disk in the sky. Other times it is nothing more than a thin sliver or its face appears dark.

To help you visualize the changing phases of the Moon, use an orange, a lamp, and your head. Your head is Earth, the orange is the Moon, and the lamp is the Sun (**Figure 2.19**). Turn off all the other lights in the room, and step back as far from the lamp as you can. Hold up the orange slightly above your head so that it is illuminated from one side by the lamp. Move the orange clockwise around your head and watch how the appearance of the orange changes. When you are between the orange and the lamp, the face of the orange that is toward you is fully illuminated. The orange appears to be a bright, circular disk. As the orange moves around its circle, you will see a progression of lighted shapes, depending on how much of the bright side and how much of the dark side of the orange you can see. This progression of shapes exactly mimics the changing phases of the Moon.

Figure 2.20 shows the changing phases of the Moon. The **new Moon** occurs when the Moon is between Earth and the Sun. The far side is illuminated, but the near side is in darkness and we cannot see it. The new Moon appears close to the Sun in the sky, so it is up in the daytime with the Sun: it rises in the east at sunrise, crosses the meridian near noon, and sets in the west near sunset. A new Moon is never above the horizon in the nighttime sky.

A few days after a new Moon, as the Moon orbits Earth, a sliver of its illuminated half, called a **waxing crescent Moon**, becomes visible. *Waxing* here means "growing in size and brilliance"; the name refers to the fact that the Moon appears to be "filling out" from night to night at this time. From our perspective, the Moon has also moved away from the Sun in the sky. Because the Moon travels around Earth in the same direction in which Earth rotates, we now see the Moon trailing the Sun, so it is east of the Sun in the sky. A waxing crescent Moon is

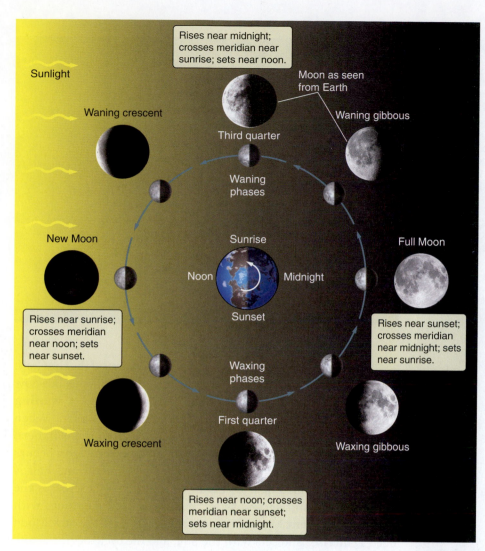

Sunlight

Waning crescent

Rises near midnight; crosses meridian near sunrise; sets near noon.

Moon as seen from Earth

Third quarter

Waning gibbous

Waning phases

New Moon

Sunrise

Noon   Midnight

Sunset

Full Moon

Rises near sunrise; crosses meridian near noon; sets near sunset.

Rises near sunset; crosses meridian near midnight; sets near sunrise.

Waxing phases

Waxing crescent

First quarter

Waxing gibbous

Rises near noon; crosses meridian near sunset; sets near midnight.

Figure 2.20 The inner circle of images (connected by blue arrows) shows the Moon as it orbits Earth, as seen by an observer far above Earth's North Pole. The Sun is on the left. The outer ring of images shows the corresponding phases of the Moon as seen from Earth.

visible in the western sky in the evening, near the setting Sun but remaining above the horizon after the Sun sets. The "horns" of the crescent always point directly away from the Sun.

As the Moon moves farther along in its orbit, and the angle between the Sun and Moon grows, more and more of its near side becomes illuminated. About a week after the new Moon, half of the near side of the Moon is illuminated and half is in darkness. This phase is called a **first quarter Moon** because the Moon has moved a quarter of the way around Earth and has completed the first quarter of its cycle from new Moon to new Moon. A look at Figure 2.20 shows that the first quarter Moon rises at noon, crosses the meridian at sunset, and sets at midnight.

As the Moon moves beyond first quarter, more than half of its near side is illuminated. This phase is called a **waxing gibbous Moon**, from the Latin *gibbus*, meaning, "hump." The waxing gibbous Moon continues nightly to "grow" until finally we see the entire near side of the Moon illuminated—a **full Moon**. Earth is now between the Sun and the Moon, which appear opposite each other in the sky when viewed from Earth. The full Moon rises as the Sun sets, crosses the meridian at midnight, and sets in the morning as the Sun rises.

(a)   (b)

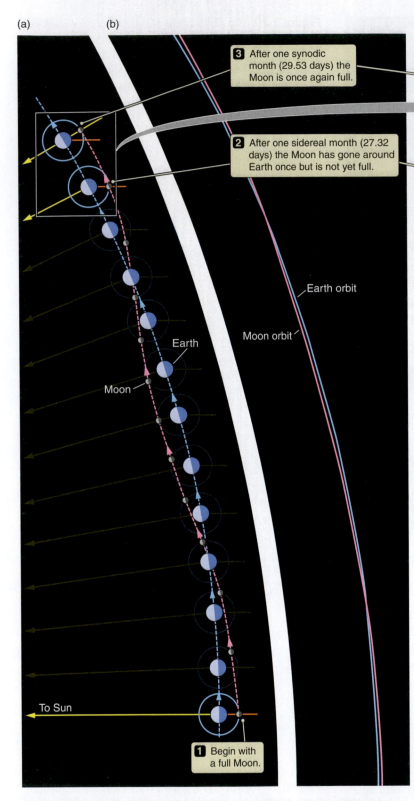

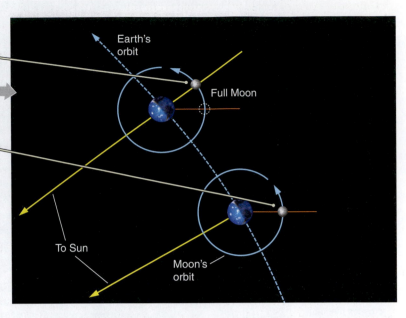

3 After one synodic month (29.53 days) the Moon is once again full.

2 After one sidereal month (27.32 days) the Moon has gone around Earth once but is not yet full.

Earth orbit

Moon orbit

Earth

Moon

To Sun

1 Begin with a full Moon.

Earth's orbit

Full Moon

To Sun

Moon's orbit

Figure 2.21 (a) The Moon completes one sidereal orbit in 27.32 days, but the synodic period (the period between phases seen from Earth) from one full Moon to the next is 29.53 days. The horizontal orange line to the right of the Moon indicates a fixed direction in space. (b) The orbits of Earth and the Moon are shown here to scale although the sizes of Earth and the Moon are not.

The second half of the Moon's orbit is the reverse of the first half. The Moon continues in its orbit, again appearing gibbous but now becoming smaller each night. This phase is called a **waning gibbous Moon** because *waning* means "becoming smaller." When the Moon is waning, the left side—as viewed from the Northern Hemisphere—appears illuminated. A **third quarter Moon** occurs when half of the near side is illuminated by sunlight and half is in darkness. A third quarter Moon rises at midnight, crosses the meridian near sunrise, and sets at noon. The cycle continues with a **waning crescent Moon** in the morning sky, until the new Moon once again rises and sets with the Sun, and the cycle begins again. Notice that when the Moon is farther from the Sun than Earth is, it is in gibbous (or full) phases. When the Moon is closer to the Sun than Earth is, it is in crescent (or new) phases.

You can always tell a waxing Moon from a waning Moon because the side that is illuminated is always the side facing the Sun. When the Moon is waxing, it appears in the evening sky, so its western side is illuminated. This is the right side as viewed from the Northern Hemisphere. Conversely, when the Moon is waning, the eastern side appears bright. This is the left side as viewed from the Northern Hemisphere.

Figure 2.21 illustrates two types of lunar periods, the first one based on the Moon's orbit in space, and the second based on the alignment of the moving Moon, Earth, and Sun. The Moon completes one orbit around Earth in 27.32 days: this **sidereal period** is how long it takes to return to the same location in its orbit. However, because of the changing relationships among Earth, the Moon, and the Sun due to Earth's orbital motion, it takes 29.53 days to go from one full Moon to the next. This is known as its **synodic period** and is the basis for our "month" because it is what we can easily observe from Earth.

Do not try to memorize all possible combinations of where the Moon is in the sky at what phase and at what time of day. Instead, work on understanding the motion and phases of the Moon, and then use your understanding to figure out the specifics of any given case. To study the phases of the Moon, draw a picture like Figure 2.20, and use it to follow the Moon around its orbit. From your drawing, figure out what phase you would see and where it would appear in the sky at a given time of day. You might also try the simulations described in "Exploration: The Phases of the Moon" at the end of the chapter.

The Visible Face of the Moon

Although the Moon's illumination varies at different parts of its orbit, one aspect of the Moon's appearance that does not change is the face of the Moon that we see. If we were to go outside next week or next month, or 20 years from now, or 20,000 centuries from now, we would still see the same side of the Moon that we see tonight. This happens because the Moon rotates on its axis exactly once for each revolution that it makes around Earth.

Imagine walking around the Washington Monument while keeping your face toward the monument at all times. By the time you complete one circle around the monument, your head has turned completely around once. When you were south of the monument, you were facing north; when you were east of the monument, you were facing west; and so on. But someone looking at you from the monument would always see your face. The Moon does exactly the same thing, rotating on its axis once per revolution around Earth, always keeping the same face toward Earth (**Figure 2.22**). When an object's revolution and rotation are synchronized (or in sync) with each other, its called **synchronous rotation**. We will see other examples of this in our Solar System.

The Moon's *far side*, facing away from Earth, is often called the "dark side of the Moon." In fact, there is no side of the Moon that is always dark. At any given time, half of the Moon is in sunlight and half is in darkness—just as at any given time, half of Earth is in sunlight and half is in darkness. The side of the Moon that faces away from Earth, the "far side," spends just as much time in sunlight as the side of the Moon that faces toward Earth does.

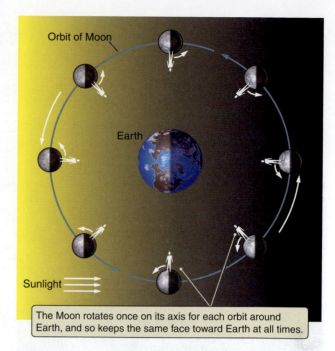

The Moon rotates once on its axis for each orbit around Earth, and so keeps the same face toward Earth at all times.

Figure 2.22 The Moon rotates once on its axis for each orbit around Earth—an effect called synchronous rotation. In this illustration, the Sun is far to the left of the Earth-Moon system.

CHECK YOUR UNDERSTANDING 2.4

You see the Moon rising just as the Sun is setting. What phase is the Moon in?

 Nebraska Simulations: Lunar Phase Simulator

2.4 Calendars Are Based on the Day, Month, and Year

Archeologists tell us that the development of agriculture was crucial for the rise of human civilization, and keeping track of the seasons and best times of the year to plant and harvest was critical to successful farming. Records going back to the dawn of humanity suggest that people kept track of time by following the patterns in the sky, especially those of the Sun, the Moon and the stars. Some anthropologists have speculated that notches on fragments of bone found in southern France represent a 33,000-year-old lunar calendar. In this section, we will examine some different calendars.

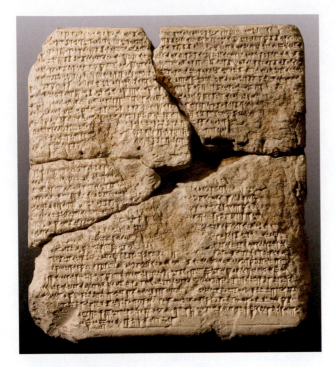

Figure 2.23 The ancient Egyptian calendar used a system of 12 months, plus festival days.

Lunar and Lunisolar Calendars

As civilizations developed around the globe, different cultures tried to solve the "problem" of the calendar. The rates of rotation of Earth and revolutions of the Moon around Earth and Earth around the Sun are not even multiples of each other. A lunar cycle (full Moon to full Moon) is 29.5 days, and a solar cycle—the time it takes for the Sun to appear to move from and return to its highest possible point in the sky at noon on the summer solstice—is 365.24 days. One solar cycle has 12.38 lunar cycles. These fractions of days and months are what make calendars complicated.

Some of the oldest known calendars come from the Egyptians, the Babylonians, and the Chinese. The ancient Egyptians used a system of 12 months of 30 days each—which added up to 360 days—and then added five "festival days" to the end of the year (**Figure 2.23**). Without leap years, this 365-day year led to a drift of the seasons, so an extra month was added when necessary. When we consider how we celebrate the days between the modern December holidays and the New Year, an end-of-year calendar break for festivals seems like a good solution!

The Babylonians started the 24-hour day and 7-day week—7 for the Sun, the Moon, and the 5 planets visible with the naked eye. They created the first *lunisolar* calendar, in which a month began with the first sighting of the lunar crescent, and a 13th month was added when needed to catch up to the solar year. As the Babylonians developed mathematics, they discovered that 235 lunar months equals 19 solar years (and 6,940 days). Then they created a calendar cycle that consisted of 19 years, in which 12 of the years have 12 months and 7 of the years have 13 months, and then the cycle repeats. The ancient Hebrew calendar adopted this cycle, and the Jewish calendar still uses it today. This type of calendar keeps holidays in the same season from year to year, even though the dates are different.

The ancient Chinese calendar dating back several thousand years occasionally added a 13th month. By about 500 BCE, the Chinese were using a year of 365.25 days and a system similar to the Babylonians of adding a 13th month into some years. A few cultures used stellar calendars, for example following the position of a bright star like Sirius or certain prominent groups of stars in their sky, such as the Pleiades or the Big Dipper, to mark out a year.

The Islamic calendar is a purely lunar calendar, with no 13th lunar month added in. Their 12 months of 29 or 30 days each add up to 354 days—11.24 days short of the solar year. For this reason, the Islamic New Year and all other holidays drift earlier in each successive solar year. In the Islamic calendar, a holiday may fall in the winter in some years, and then a few years later it will have moved back to autumn.

The Modern Civil Calendar

The international civil calendar used today is a solar calendar known as the **Gregorian calendar**. It is based on the **tropical year**, which measures the 365.242 solar days from one vernal equinox to the next. A **solar day** is the 24-hour period of Earth's rotation that brings the Sun back to the same local meridian. This is in contrast to the **sidereal day**, which is the time it takes for Earth to make one rotation and face the exact same star on the meridian. The sidereal day is about 23 hours 56 minutes and differs from the solar day because of Earth's motion around the Sun.

The Gregorian calendar includes a system of **leap years**—years in which a 29th day is added to the month of February—decreed by Julius Caesar in 45 BCE to make up for the extra fraction of a day. Leap years prevent the seasons from slowly sliding through the year to become increasingly out of sync with the months, so we don't end up experiencing winter in December one year and in August other years.

The Gregorian calendar is named for Pope Gregory XIII. He was concerned that the Easter holiday, which falls on the first Sunday after the first full Moon following March 21, was drifting away from the vernal (spring) Equinox. Julius Caesar's rule of one leap year every 4 years resulted in an average year of 365.25 days, but the actual year is 365.242 days. This difference of 0.008 day is about 11.5 minutes per year, or 3 days every 400 years, and by Gregory's time it had caused the date of the vernal equinox to drift in the Julian calendar by about 10 days. So in 1582, Pope Gregory decreed that 10 days would be deleted from the calendar to move the vernal equinox back to March 21. To make this work out better in the future, he declared that only century years divisible by 400 are leap years, thereby deleting 3 leap years (and the 3 days) every 400 years. Catholic countries followed this system immediately, but Protestant countries did not adopt it until the 1700s. Eastern Orthodox countries, including Russia, did not switch from the Julian to the Gregorian calendar until the 1900s. One slight further revision—making years divisible by 4,000 into common 365-day years—has been proposed so the modern Gregorian calendar will slip by about only 1 day in 20,000 years.

Despite international adoption of the Gregorian calendar, billions of people still celebrate holidays and festivals according to a lunisolar or lunar calendar. Chinese New Year, Passover, Easter, Ramadan, Rosh Hashanah, and Diwali, among others, have dates that change from one year to the next because they are based on lunar months from lunisolar or lunar calendars. The astronomy of people from long ago is still in use today.

 Nebraska Simulations: Synodic Lag

CHECK YOUR UNDERSTANDING 2.5

Suppose that the astronomical cycles were even multiples of each other, so that a month was precisely 30 days, and a year was precisely 12 months. How would this change the dates of "wandering" holidays such as Chinese New Year or Ramadan?

2.5 Eclipses Result from the Alignment of Earth, Moon, and the Sun

For ancient peoples attuned to the patterns of the sky, it must have been terrifying to look up to see the Sun being eaten away as if by a giant dragon or the full Moon turning the color of blood. An **eclipse** is the total or partial obscuration of one celestial body, or the light from that body, by another celestial body. Archaeological evidence suggests that ancient peoples put great effort into trying to figure out the pattern of eclipses and thereby bring them into the orderly scheme of the heavens. Stonehenge, pictured in Figure 2.1a, may have enabled its builders to predict when eclipses might occur. Ancient Chinese, Babylonian, and Greek astronomers had figured out that eclipses occur in cycles, and they were able to use their knowledge to make predictions about when and where eclipses would occur. In this section, we will describe the different types of eclipses and their frequency.

Figure 2.24 Different parts of the Sun are blocked at different places within the Moon's shadow. An observer on Earth in the umbra (point A) sees a total solar eclipse, observers in the penumbra (points B and C) see a partially eclipsed Sun, and observers at point D see an annular solar eclipse.

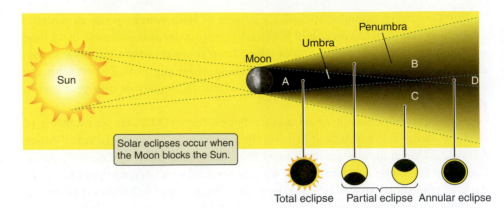

Solar eclipses occur when the Moon blocks the Sun.

Total eclipse   Partial eclipse   Annular eclipse

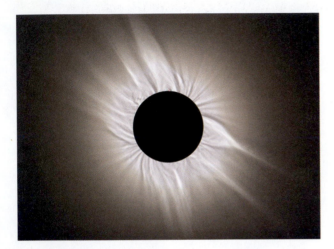

Figure 2.25 The full spectacle of a total eclipse of the Sun.

Solar Eclipses

A **solar eclipse** occurs when the Moon passes between Earth and the Sun; observers on Earth in the shadow of the Moon will see the eclipse. There are three different types of solar eclipses: *total*, *partial*, and *annular*. Consider the structure of the shadow of the Sun cast by a round object such as the Moon, as shown in **Figure 2.24**. An observer at point A would be unable to see any part of the surface of the Sun. This darkest, inner part of the shadow is called the **umbra**. If a location on Earth passes through the Moon's umbra, the Sun's light is totally blocked by the Moon, and a **total solar eclipse** will be observed (**Figures 2.25 and 2.26a**). At points B and C in Figure 2.24, an observer can see one side of the disk of the Sun but not the other. This outer region, which is only partially in shadow, is the **penumbra**. If a location on the surface of Earth passes through the Moon's penumbra, viewers at that location will observe a **partial solar eclipse**, in which the disk of the Moon blocks the light from a portion of the Sun's disk.

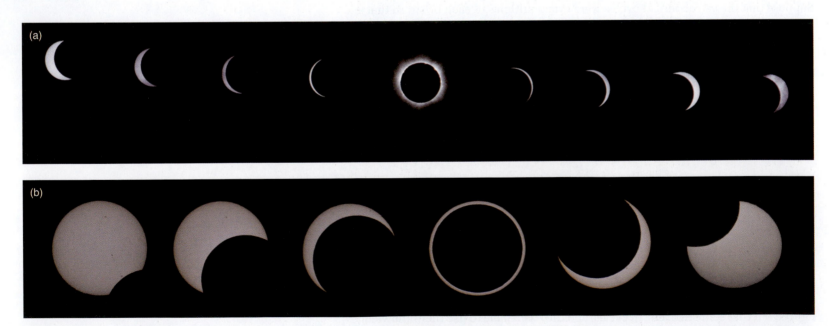

Figure 2.26 Time sequences of images of the Sun taken (a) during a total solar eclipse and (b) during the annular solar eclipse of May 20, 2012. The Sun set during the ending phases.

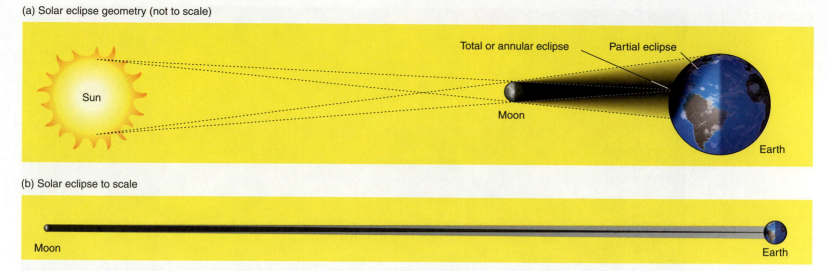

(a) Solar eclipse geometry (not to scale)

Sun

Total or annular eclipse   Partial eclipse

Moon

Earth

(b) Solar eclipse to scale

Moon

Earth

Figure 2.27 (a, b) A solar eclipse occurs when the shadow of the Moon falls on the surface of Earth. Note that (b) is drawn to proper scale.

In the third type of solar eclipse, called an **annular solar eclipse**, the Sun appears as a bright ring surrounding the dark disk of the Moon (Figure 2.26b). An observer at point D in Figure 2.24 is far enough from the Moon that the Moon's apparent size in the sky is smaller than the Sun's. The apparent size of an object in the sky depends on the object's actual size and its distance from us. The Sun is about 400 times the diameter of the Moon, and the distance between the Sun and Earth is about 400 times more than the distance between the Moon and Earth. As a result, the Moon and Sun have almost exactly the same apparent size in the sky. Another factor is that the Moon's orbit is not a perfect circle. When the Moon and Earth are a bit closer together than average, the Moon appears slightly larger in the sky than the Sun. An eclipse occurring at that time will be total for some observers. When the Moon and Earth are farther apart than average, the Moon appears smaller than the Sun, so eclipses occurring during this time will be annular for some observers. Among all solar eclipses, one-third are total at some location on the surface of Earth, one-third are annular, and one-third are seen only as a partial eclipse.

Figure 2.27 shows the geometry of a solar eclipse when the Moon's shadow falls on the surface of Earth. Figures like this usually show Earth and the Moon much closer together than they really are. The page is too small to draw them correctly and still see the critical details. The relative sizes and distances between Earth and the Moon are roughly equivalent to the difference between a basketball and a tennis ball placed 7 meters apart. Figure 2.27b shows the geometry of a solar eclipse with Earth, the Moon, and the separation between them drawn to the correct scale. Compare this drawing to Figure 2.27a and you will understand why drawings of Earth and the Moon are rarely drawn to the correct scale. If the Sun were drawn to scale in Figure 2.27a, it would be bigger than your head and located almost 64 meters off the left side of the page.

From any particular location, you are more likely to observe a partial solar eclipse than a total solar eclipse. Where the Moon's penumbra touches Earth, it has a diameter of almost 7,000 km—large enough to cover a substantial fraction of Earth. Thus, a partial solar eclipse is often visible from many locations on

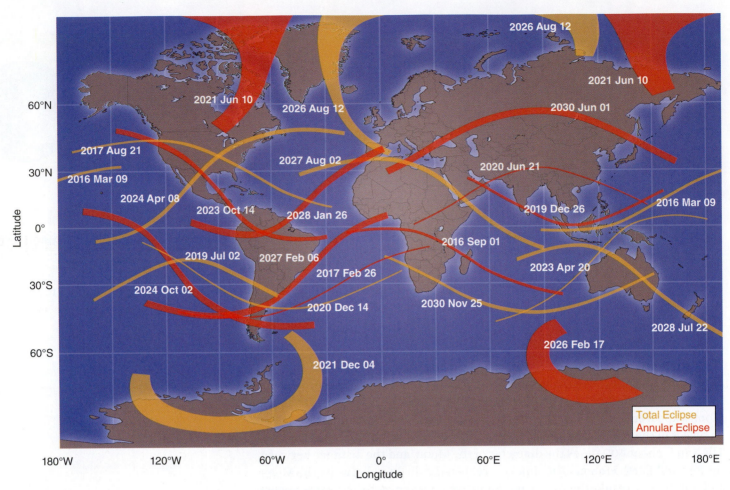

Figure 2.28 The paths of total solar eclipses through 2020. Solar eclipses occurring in Earth's polar regions cover more territory because the Moon's shadow hits the ground obliquely.

Earth. In contrast, the path along which a total solar eclipse can be seen, shown in **Figure 2.28**, covers only a tiny fraction of Earth's surface. Even when the distance between Earth and the Moon is at a minimum, the umbra is only 270 km wide at the surface of Earth. As the Moon moves along in its orbit, this tiny shadow sweeps across the face of Earth at speeds of a few thousand kilometers per hour. Additionally, the Moon's shadow falls on the curved surface of Earth, causing the region shaded by the Moon during a solar eclipse to be elongated by differing amounts. The curvature can even cause an eclipse that started out as annular to become total.

The result is that a total solar eclipse can never last longer than 7½ minutes and is usually significantly shorter. Even so, it is one of the most amazing sights in nature. People all over the world flock to the most remote corners of Earth to witness the fleeting spectacle of the bright disk of the Sun blotted out of the daytime sky. Perhaps you saw some of the annular eclipse that was visible from much of the United States in May 2012. The first total solar eclipse since 1979 that will be visible in the continental United States will take place in August 2017 (followed by another in 2024). Annular eclipses will be visible from parts of the United States in 2021 and 2023. Viewing a solar eclipse should be on your lifetime to-do list!

(a) Lunar eclipse geometry (not to scale)

(b) Lunar eclipse to scale

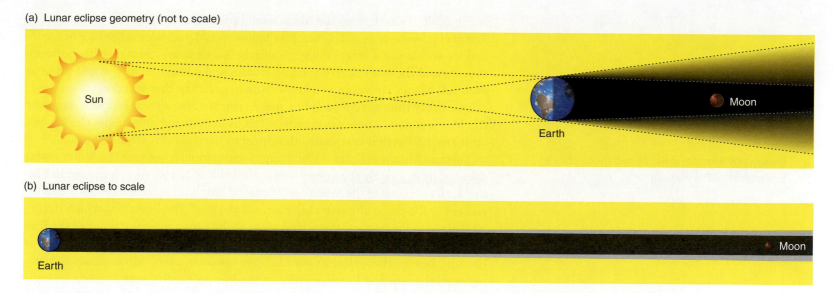

Figure 2.29 (a, b) A lunar eclipse occurs when the Moon passes through Earth's shadow. Note that (b) is drawn to proper scale.

Lunar Eclipses

Lunar eclipses occur when the Moon moves through the shadow of Earth. The geometry of a lunar eclipse is shown in **Figure 2.29a** and is drawn to scale in Figure 2.29b. Here Earth is between the Sun and the Moon. Because Earth is much larger than the Moon, the dark umbra of Earth's shadow at the distance of the Moon is more than 2½ times the diameter of the Moon. A **total lunar eclipse**, when the Moon is entirely within Earth's shadow, lasts as long as 1 hour 40 minutes. In a total lunar eclipse, the Moon often appears red (**Figure 2.30a**). This "blood-red Moon," as it has been called in literature and poetry, occurs because the Moon is being illuminated by red light from the Sun that is bent as it travels through Earth's atmosphere and hits the Moon. Other colors of light are absorbed or scattered away from the Moon by Earth's atmosphere and therefore do not illuminate it.

A **penumbral lunar eclipse** occurs when the Moon passes through the penumbra of Earth's shadow; these are noticeable only from a very dark location or when the Moon passes within about 1,000 km of the umbra. If Earth's shadow incompletely covers the Moon, some of the disk of the Moon remains bright and some of it is in shadow. This is called a **partial lunar eclipse**. Figure 2.30b shows a composite of images taken at different times during a partial lunar eclipse. In the center image, the Moon is nearly completely eclipsed by Earth's shadow.

Many more people have observed a total lunar eclipse than have observed a total solar eclipse. To see a total solar eclipse, you must be located within that very narrow band of the Moon's shadow as it moves across Earth's surface. In contrast, when the Moon is immersed in Earth's shadow, anyone located in the hemisphere of Earth that is facing the Moon can see it. As a result, from any location, total eclipses of the Moon are relatively common, and you may have seen at least one.

Frequency of Eclipse Seasons

How did some people in ancient cultures successfully predict eclipses? From their understanding of lunar and solar cycles for making calendars, they were able to compute cycles of eclipses. Imagine Earth, the Moon, and the Sun all sitting on the same flat tabletop. If the Moon's orbit were in exactly the same plane

(a)

(b)

Figure 2.30 (a) During a total lunar eclipse, the Moon often appears blood red. (b) A time-lapse series of photographs of a partial lunar eclipse clearly shows Earth's shadow. Note the size of Earth's shadow compared to the size of the Moon.

Nebraska Simulations: Moon Inclinations; Eclipse Shadow Simulator

as the orbit of Earth, then the Moon would pass directly between Earth and the Sun at every new Moon. The Moon's shadow would pass across the face of Earth, and we would see a solar eclipse. Similarly, Earth would pass directly between the Sun and the Moon every synodic month, and a lunar eclipse would occur at each full Moon. However, you know from experience that you don't see a lunar eclipse every time the Moon is full, nor do you observe a solar eclipse every time the Moon is new. These observations tell us something about how the Moon's orbit around Earth is oriented with respect to Earth's orbit around the Sun.

Solar and lunar eclipses do not happen every month because the Moon's orbit does not lie in exactly the same plane as the orbit of Earth. As you can see in **Figure 2.31**, the plane of the Moon's orbit around Earth is inclined by about 5.2° with respect to the plane of Earth's orbit around the Sun. The line along which the orbital plane of the Sun and the orbital plane of the Moon intersect is called the **line of nodes**. For part of the year, the line of nodes points in the general direction of the Sun. During these times, called **eclipse seasons**, a new Moon passes directly between the Sun and Earth, casting its shadow on Earth's surface and causing a solar eclipse. Similarly, a full Moon occurring during an eclipse season passes through Earth's shadow, and a lunar eclipse results. An eclipse season lasts only 38 days. That's how long the Sun is close enough to the line of nodes for eclipses to occur. Most of the time the line of nodes points farther away from the Sun, and Earth, Moon, and Sun cannot line up closely enough for an eclipse to occur. At these times, a solar eclipse cannot take place because the shadow of a new Moon passes "above" or "below" Earth. Similarly, no lunar eclipse can occur because a full Moon passes "above" or "below" the shadow of Earth.

If the plane of the Moon's orbit always had the same orientation, then eclipse seasons would occur twice a year, as suggested by Figure 2.31. In actuality, eclipse

Figure 2.31 Eclipses are possible only when the Sun, Moon, and Earth lie along (or very close to) an imaginary line known as the line of nodes. When the Sun does not lie along the line of nodes, Earth passes under or over the shadow of a new Moon, and a full Moon passes under or over the shadow of Earth.

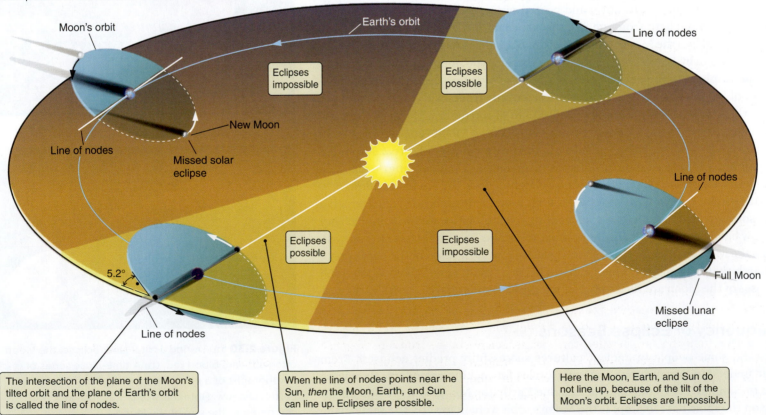

seasons occur about every 5 months 20 days. The roughly 10-day difference is due to the fact that the plane of the Moon's orbit slowly wobbles, much like the wobble of a spinning plate balanced on the end of a circus performer's stick. As it does so, the line of nodes changes direction. This wobble rotates in the direction opposite the direction of the Moon's motion in its orbit. That is, the line of nodes moves clockwise as viewed from above Earth's orbital plane. One wobble of the Moon's orbit takes 18.6 years, so we say that the line of nodes regresses at a rate of 360° every 18.6 years, or 19.4° per year. This amounts to about a 20-day regression each year. If January 1 marks the middle of an eclipse season, the next eclipse season will be centered around June 20, and the one after that around December 10.

Nebraska Simulations: Eclipse Table

CHECK YOUR UNDERSTANDING 2.6

If the Moon were in its same orbital plane but twice as far from Earth, which of the following would happen? (a) The Moon would not go through phases. (b) Total eclipses of the Sun would not be possible. (c) Annular eclipses of the Sun would not be possible. (d) Total eclipses of the Moon would not be possible.

Origins

The Obliquity of Earth

The various motions of Earth give rise to the most basic of patterns faced by life on Earth. Earth's rotation is responsible for the cycle of night and day. Earth's axial tilt and its passage around the Sun bring the change of the seasons. As life evolved on Earth, it had to adapt to these patterns.

The range of climate on Earth based on distance from the equator likely has contributed to the broad diversity of life of our planet. Earth's biodiversity includes life that adapted to the long, cold polar nights and life in equatorial latitudes that adapted to much higher temperatures. Earth's life adapted to seasonal patterns in rain and drought, leading to acquired seasonal patterns of migration and reproduction. If Earth had no axial tilt, the poles would continually be in winter and probably too cold for humans. Midlatitudes would not have the cool winters that are needed for many food crops. Latitudes near the equator would be even consistently warmer than they are now.

Life might have been affected by periodic changes in Earth's axial tilt. If Earth's tilt were larger than 23.5°, the seasonal variation would be even stronger. If the tilt were smaller, the seasonal variation would be weaker. Chinese, Indian, Greek, and Arabic records going back 3,000 years indicate that the tilt was estimated in ancient times by measuring the length of the shadow from a vertical pole on the day of the solstice. We now know that for the past few million years, Earth's axial tilt actually varied from 22.1° to 24.5° over a 41,000-year cycle. The Moon's gravity is responsible for maintaining the tilt within this small range over the past half-billion years—about the time since animal life greatly diversified on Earth. Currently, the tilt is about midway between the two extremes and getting smaller. It will reach its minimum value of 22.1° in about 10,000 years. Scientists are studying whether this variation in tilt correlates with periods of temperature change on Earth, especially the times of ice ages.

Nebraska Simulation: Obliquity Simulator

READING ASTRONOMY NEWS

The first total solar eclipse in the continental United States in decades will take place in August 2017. In this story, one town claims that it will be the best place to see the eclipse.

Thousands Expected in Hopkinsville for 2017 Solar Eclipse

By **ADAM GHASSEMI**

HOPKINSVILLE, Kentucky—Hopkinsville has a population of a little more than 30,000 but in 2017 it's expected to more than double for just a few days.

"It's going to be a really big deal," said Cheryl Cook with the Hopkinsville-Christian County Convention and Visitors Bureau who started planning seven years ago for the biggest event they've ever seen. "It could cause traffic jams from here to Nashville," she said.

Visitors are coming from places as far away as Germany, Australia, and Japan to see a total eclipse for 2:40 of darkness.

It's so rare, a map of the projected path has been hanging in the office of Austin Peay Astronomy Professor Spencer Buckner for 15 years.

"The best place on Earth that will have the longest period of totality is actually just north and west of Hopkinsville," Buckner said Friday. "I don't know of another one that will happen in the next few hundred years."

City leaders want to make sure they don't miss any details, like visitors have protective eyewear to see the eclipse, or cutting electricity off in certain parts of the city. That way when things do go dark, lights won't automatically come on and disrupt the view.

"We're hoping for a bright, sunny, warm day. No clouds in the sky," Cook said.

Scientists are planning how they'll capture and study the eclipse along its cross-country path from Oregon to South Carolina, while "eclipse chasers" have their trips to Hopkinsville already confirmed.

"Over 400 rooms booked," said Chairman of the Hopkinsville-Christian County Convention and Visitors Bureau Board Jeff Smith. "I've never seen anything like this."

When the Moon perfectly aligns with the Sun there won't be any better place in the world to see it on August 21, 2017.

"The world's coming to Hopkinsville," Cook went on to say.

Spectators will be able to see the eclipse from a number of places across Southern Kentucky and Middle Tennessee, but people just 70 miles from Hopkinsville in Nashville will get 44 fewer seconds to witness it.

ARTICLES   QUESTIONS

1. Why are people excited about this solar eclipse?
2. How long is this eclipse as seen from Hopkinsville? Explain why people in Nashville observe a much shorter eclipse.
3. Would you predict that there will be a lunar eclipse in summer 2017? If so, what are the possible dates?
4. Look at the map on the NASA eclipse page: http://eclipse.gsfc.nasa.gov/SEgoogle/SEgoogle2001/SE2017Aug21Tgoogle.html. Will you be able to see the eclipse from your school?
5. The claim in this story was challenged by people who argued that Hopkinsville may have the shadow of the Moon passing closest to Earth's center, but a town in Illinois will have 0.1 seconds more of totality. Will most people care about this 0.1-second difference in totality? What other factors will likely affect where people go to see this eclipse?

"Thousands expected in Hopkinsville for 2017 solar eclipse," by Adam Ghassemi. NewsChannel 5 Network, January 17, 2014. © WTVF-TV. Reprinted by permission of NewsChannel 5 Network, LLC.

Summary

The motions of Earth and the Moon are responsible for many of the repeating patterns that can be observed in the sky. Calendars keep track of time using these patterns. Earth's rotation about its axis causes daily patterns of rising and setting. Earth's revolution around the Sun causes yearly patterns of the stars in the sky and the passage of the seasons. The tilt of Earth on its axis changes both the length of daytime and the intensity of sunlight, causing the seasons. The Moon's revolution around Earth causes the month-long pattern of the phases of the Moon. Occasionally, alignments of Earth, the Moon, and the Sun cause eclipses. The tilt of Earth's axis causes the variation in climate. This tilt varies slightly over tens of thousands of years. Life on Earth adapted to these seasonal variations.

LG 1 **Describe how Earth's rotation about its axis and revolution around the Sun affect our perception of celestial motions as seen from different places on Earth.** The daily rotation of Earth on its axis causes the apparent daily motion of the Sun, Moon, and stars. Our location on Earth and Earth's location in its orbit around the Sun determine which stars we see at night. You can determine your latitude from the altitude of the pole star. When observing objects in our sky, we need to consider the relative motion of Earth and other objects. The ecliptic is the path that the Sun appears to take through the stars.

LG 2 **Explain why there are different seasons throughout the year.** A year is the time it takes for Earth to complete one revolution around the Sun. Constellations are patterns of stars that reappear in the same place in the sky at the same time of night each year. The tilt of Earth's axis determines the seasons by changing the angle at which sunlight strikes the surface of Earth in different locations. The changing angle of sunlight and the differing length of the day cause the seasonal variations on Earth. The changing seasons are marked by equinoxes and solstices.

LG 3 **Describe the factors that create the phases of the Moon.** The relative locations of the Sun, Earth, and Moon determine the phases of the Moon. The Moon takes one sidereal month to complete one revolution around Earth and one synodic month to go through a cycle of phases. The Moon's motion around Earth causes it to be illuminated differently at different times. When the Moon is farther from the Sun than Earth is, it is in gibbous phases. When the Moon is closer to the Sun than Earth is, it is in crescent phases.

LG 4 **Sketch the alignment of Earth, the Moon, and the Sun during eclipses of the Sun and the Moon.** A solar eclipse occurs when the new Moon is in the plane of Earth and the Sun and the shadow of the Moon falls on Earth. A lunar eclipse occurs when the full Moon is in the plane of Earth and the Sun and the shadow of Earth falls on the Moon. Twice a year, at new or at full Moon, the Moon is exactly in line between Earth and the Sun. At these times, eclipses occur.

? UNANSWERED QUESTION

- How long will Earth continue to have total solar eclipses? These occur because the Moon and the Sun are coincidentally the same size in our sky, but will that always be the case? The observed size of an object in the sky depends on its actual diameter and its distance from us. One or both of these can change. The Moon is slowly moving away from Earth by about 4 meters per century. Over time, the Moon will appear smaller in the sky, and it won't be able to cover the full disk of the Sun. While we can measure the current rate of the Moon's movement away from Earth, we are less certain of how this rate may change with time. A lesser and more uncertain effect comes from the Sun—which will continue to brighten slowly, as it has throughout its history. With this brightening, the actual diameter of the Sun may slightly increase, and it will appear larger in our sky. A more distant Moon and a larger Sun will eventually result in an end to total eclipses on Earth.

Questions and Problems

Test Your Understanding

1. Constellations are groups of stars that
   a. are close to each other in space.
   b. are bound to each other by gravity.
   c. are close to each other in Earth's sky.
   d. all have the same composition.

2. Where on Earth can you stand and, over the course of a year, see the entire sky?
   a. only at the North Pole
   b. at either pole
   c. at the equator
   d. anywhere

3. Day and night are caused by
   a. the tilt of Earth on its axis.
   b. the rotation of Earth on its axis.
   c. the revolution of Earth around the Sun.
   d. the revolution of the Sun around Earth.

4. Polaris, the North Star, is unique because
   a. it is the brightest star in the night sky.
   b. it is the only star in the sky that doesn't move throughout the night.
   c. it is always located at the zenith, for any observer.
   d. it has a longer path above the horizon than any other star.

5. There is an angle between the ecliptic and the celestial equator because
   a. Earth's axis is tilted with respect to its orbit.
   b. Earth's orbit is tilted with respect to the orbits of other planets.
   c. the Sun follows a rising and falling path through space.
   d. the Sun's orbit is tilted with respect to Earth's.

6. The tilt of Earth's axis causes the seasons because
   a. one hemisphere of Earth is closer to the Sun in summer.
   b. the days are longer in summer.
   c. the rays of light strike the ground more directly in summer.
   d. both a and b
   e. both b and c

7. Which is *not* true on the vernal and autumnal equinoxes?
   a. Every place on Earth has 12 hours of daylight and 12 hours of darkness.
   b. The Sun rises due east and sets due west.
   c. The Sun is located on the celestial equator.
   d. The motion of the stars in the sky is different than on other days.

8. We always see the same side of the Moon because
   a. the Moon does not rotate on its axis.
   b. the Moon rotates on its axis once for each revolution around Earth.
   c. when the other side of the Moon is facing Earth, it is unlit.
   d. when the other side of the Moon is facing Earth, it is on the opposite side of Earth.

9. You see the Moon on the meridian at sunrise. The phase of the Moon is
   a. waxing gibbous.
   b. full.
   c. first quarter.
   d. third quarter.

10. A lunar eclipse occurs when _____ shadow falls on
    _____ .
    a. Earth's; the Moon
    b. the Moon's; Earth
    c. the Sun's; the Moon
    d. the Sun's; Earth

11. Different cultures created different calendars because
    a. they had measured different lengths of the day, month, and year.
    b. they used different definitions of the day, month, and year.
    c. the number of days in a month and the number of days and months in a year are not integers.
    d. calendars are completely arbitrary.

12. Which stars we see at night depends on
    a. our location on Earth.
    b. Earth's location in its orbit.
    c. the time of the observation.
    d. all of the above

13. On the summer solstice in June, the Sun will be directly above _____ and all locations north of _____ will experience daylight all day.
    a. the Tropic of Cancer; the Antarctic Circle
    b. the Tropic of Capricorn; the Arctic Circle
    c. the Tropic of Cancer; the Arctic Circle
    d. the Tropic of Capricorn; the Antarctic Circle

14. The Sun, Moon, and stars
    a. appear to move each day because the celestial sphere rotates about Earth.
    b. change their relative positions over time.
    c. rise north or south of west and set north or south of east, depending on their location on the celestial sphere.
    d. always remain in the same position relative to each other.

15. You see the first quarter Moon on the meridian. Where is the Sun?
    a. on the western horizon
    b. on the eastern horizon
    c. below the horizon
    d. on the meridian.

Thinking about the Concepts

16. Polaris was used for navigation by seafarers such as Columbus as they sailed from Europe to North America. When Magellan sailed the South Seas, he could not use Polaris for navigation. Explain why.

17. If you were standing at Earth's North Pole, where would you see the north celestial pole relative to your zenith?

18. Observers in the Northern Hemisphere see the zodiacal constellation Gemini in the winter. Why do they not see it in the summer?

19. Imagine that you are flying along in a jetliner.
    a. Describe ways to tell that you are moving.
    b. If you look down at a building, which way is it moving relative to you?

20. Astronomers are sometimes asked to serve as expert witnesses in court cases. Suppose you are called in as an expert witness, and the defendant states that he could not see the pedestrian because the full Moon was casting long shadows across the street at midnight. Is this claim credible? Why or why not?

21. Imagine that one person was developing a theory of seasons as described in the three "takes" in the Process of Science Figure in this chapter. Compare this process with the flowchart of the Process of Science Figure in Chapter 1. Describe what the development of this theory would look like on that flowchart.

22. Why is the winter solstice *not* the coldest time of year?

23. Earth spins on its axis and wobbles like a top.
    a. How long does it take to complete one spin?
    b. How long does it take to complete one wobble?

24. What is the approximate time of day when you see the full Moon near the meridian? At what time is the first quarter (waxing) Moon on the eastern horizon? Use a sketch to help explain your answers.

25. Assume that the Moon's orbit is circular. Imagine that you are standing on the side of the Moon that faces Earth.
    a. How would Earth appear to move in the sky as the Moon made one revolution around Earth?
    b. How would the "phases of Earth" appear to you compared to the phases of the Moon as seen from Earth?

26. If people on Earth were observing a lunar eclipse, what would you see from the Moon?

27. From any given location, why are you more likely to witness a partial eclipse of the Sun than a total eclipse?

28. Why do we not see a lunar eclipse each time the Moon is full or witness a solar eclipse each time the Moon is new?

29. How would Earth's temperature variation be different if it was tilted 90° on its axis (like the planet Uranus)?

30. Explain how a cyclic change in Earth's tilt could affect its seasonal temperatures.

Applying the Concepts

31. Earth is spinning along at 1,674 km/h at the equator. Use this fact, along with the length of the day, to calculate Earth's equatorial diameter.

32. Determine the latitude where you live. Draw and label a diagram showing that your latitude is the same as (a) the altitude of the north celestial pole and (b) the angle (along the meridian) between the celestial equator and your local zenith. What is the altitude of the Sun at noon as seen from your home at the times of the winter solstice and the summer solstice?

33. Using a protractor, you estimate an angle of 40° between your zenith and Polaris. What is the altitude of Polaris? What is your latitude? Are you in the continental United States or Canada?

34. The southernmost star in a group of stars known as the Southern Cross lies approximately 65° south of the celestial equator. What is the farthest-north latitude for which the entire Southern Cross is visible? Can it be seen in any U.S. states? If so, which ones?

35. Imagine that you are standing on the South Pole at the time of the southern summer solstice.
    a. How far above the horizon will the Sun be at noon?
    b. How far above (or below) the horizon will the Sun be at midnight?

36. Suppose the tilt of Earth's equator relative to its orbit were 10° instead of 23.5°. At what latitudes would the Arctic and Antarctic circles and the Tropics of Cancer and Capricorn be located?

37. The Moon's orbit is tilted by about 5° relative to Earth's orbit around the Sun. What is the highest altitude in the sky that the Moon can reach, as seen in Philadelphia (latitude 40° north)?

38. Suppose you would like to witness the midnight Sun (when the Sun appears just above the northern horizon at midnight), but you don't want to travel any farther north than necessary.
    a. How far north (that is, to which latitude) would you have to go?
    b. At what time of year would you make this trip?

39. If, as some historians believe, the Egyptian stadion was about 157.5 meters, then what would Eratosthenes have computed for the size of Earth? How close is this to the modern value?

40. a. The vernal equinox is now in the zodiacal constellation Pisces. The precession of Earth's axis will eventually cause the vernal equinox to move into Aquarius. How long, on average, does the vernal equinox spend in each of the 12 zodiacal constellations?
    b. Stonehenge was erected roughly 4,000 years ago. Referring to the zodiacal constellations shown in Figure 2.14, identify the constellation in which these ancient builders saw the vernal equinox.

41. Referring to Figure 2.19, estimate when Vega, the fifth-brightest star in our sky (excluding the Sun), will once again be the northern pole star.

42. The apparent diameter of the Moon in the sky is approximately $\frac{1}{2}°$. How long does it take the Moon to move 360°? About how long does it take the Moon to move a distance equal to its own diameter across the sky?

43. The apparent size of an object in the sky is proportional to its actual diameter divided by its distance. The Moon has a radius of 1,737 km, with an average distance of 3.780×10^5 km from Earth. The Sun has a radius of 696,000 km, with an average distance of 1.496×10^8 km from Earth. Show that the apparent sizes of the Moon and Sun in our sky are approximately the same.

44. Earth has an average radius of 6,371 km. If you were standing on the Moon, how much larger would Earth appear in the lunar sky than the Moon appears in our sky?

45. How would the length of the eclipse season change if the plane of the Moon's orbit were inclined less than its current 5.2° to the plane of Earth's orbit? Explain your answer.

to the next? Bring up the "Duration of Days/Darkness Table for One Year" page for your location. Are the days getting longer or shorter? When do the shortest and the longest days occur? Look up a location in the opposite hemisphere (Northern or Southern). When are the days shortest and longest?

47. Go to the "Earth and Moon Viewer" website (http://fourmilab. ch/earthview). Under "Viewing the Earth," click on "latitude, longitude and altitude" and enter your approximate latitude and longitude, and 40,000 for altitude; then select "View Earth." Are you in daytime or nighttime? Now play with the locations; keep the same latitude but change to the opposite hemisphere (Northern or Southern). Is it still night or day? Go back to your latitude, and this time enter 180° minus your longitude, and change from west to east, or from east to west, so that you are looking at the opposite side of Earth. Is it night or day there? What do you see at the North Pole (latitude 90° north) and the South Pole (latitude 90° south)? At the bottom of your screen you can play with the time. Move back 12 hours. What do you observe at your location and at the poles?

48. Go to the U.S. Naval Observatory website (USNO "Data Services," at http://aa.usno.navy.mil/data). Look up the Moon data for the current day. When will it rise and set? What is the phase? How will it change over the next 4 weeks. Enter one day at a time or look at the yearlong tables for moonrise and moonset and for the dates of primary phases. What time of day does a third quarter Moon rise? When (and in what phases) can you see the Moon in the daytime?

49. Using the times of moonrise and moonset that you located in question 48, make a plan to observe the Moon directly at least once a day for a week. Take a picture of the Moon (or make a sketch) every day. How is the brightness of the Moon changing? If it's daytime, how far is the Moon from the Sun in the sky? If it's nighttime, are the stars that are near the Moon in the sky the same every night?

50. Go to the "NASA Eclipse" website (http://eclipse.gsfc.nasa. gov/eclipse.html). When is the next lunar eclipse? Will it be visible at your location if the skies are clear? Is it a total or partial eclipse? When is the next solar eclipse? Will it be visible at your location? Compare the fraction of Earth that the solar eclipse will affect with the fraction for the lunar eclipse. Why are lunar eclipses visible in so many more locations?

USING THE WEB

46. Go to the U.S. Naval Observatory website (USNO "Data Services," at http://aa.usno.navy.mil/data). Look up the times for sunrise and sunset for your location for the current week. (You can change the dates one at a time or bring up a table for the entire month.) How are the times changing from one day

smartwork

If your instructor assigns homework in SmartWork, access your assignments at digital.wwnorton.com/astro5.

EXPLORATION

The Phasesof the Moon

Visit the Student Site at the Digital Landing Page, and open the Lunar Phase Simulator applet in Chapter 2.

Study the diagrams shown in the simulator. The largest window shows a view of the Earth-Moon system as seen from above Earth's North Pole. The Sun is far off the screen to the left. An observer stands on Earth. The small window at upper right shows the appearance of the Moon as seen from the Northern Hemisphere. The small window at lower right shows the observer's location, with the Sun and Moon pictured as flat disks in the sky.

1 Given the relative positions of the observer and the Sun, approximately what time is it for this observer: 6:00 A.M., noon, 6:00 P.M., or midnight?

2 Where is the Moon in the observer's sky: on the eastern horizon, on the western horizon, below the horizon, or crossing the meridian?

3 What is the phase of the Moon?

4 Imagine yourself on the Moon in the image shown in the larger window. If you looked toward Earth, what phase of Earth would you see?

Now select "start animation." Allow the animation to run until the Moon is 20 percent illuminated (as shown in the upper small window on the right).

5 In which direction does the Moon orbit Earth: clockwise or counterclockwise?

6 For observers in the Northern Hemisphere, which side of the Moon is illuminated first after a new Moon: right or left?

7 If you observe a crescent Moon with the horns of the crescent pointing right, is the Moon waxing or waning?

Grab the Moon with your mouse, and drag it to first quarter. Drag the observer so that her local time is approximately midnight.

8 Where is the first quarter Moon in the observer's sky: on the eastern horizon, on the western horizon, below the horizon, or crossing the meridian?

These three things are related: the time, the phase of the Moon, and the Moon's location in the sky.

9 Arrange the observer and the Moon so that the Moon is full and crossing the meridian. What time is it for the observer: 6:00 A.M., noon, 6:00 P.M., or midnight?

10 Arrange the observer and the Moon so that the Moon is in third quarter and the time for the observer is approximately noon (the Sun is on the meridian). Where is the Moon in the observer's sky: on the eastern horizon, on the western horizon, below the horizon, or crossing the meridian?

11 Arrange the observer and the Moon so that it is approximately 6:00 P.M. (sunset) for the observer, and the Moon is just rising on the eastern horizon. What is the phase of the Moon?

There are many other combinations of the time, phase of the Moon, and Moon's location to play with. Challenge yourself to be able to set up any two of the three and find the third. When you can do this without the simulator, just by making the picture in your head, you will really understand the phases of the Moon.

3 Motion of Astronomical Bodies

The birth of modern astronomy dates back to the time when astronomers and mathematicians discovered regular patterns in the motions of the planets. A successful theory of how Earth moved and how it fit in with its neighbors in the Solar System was the first step toward understanding Earth's place in the universe.

LEARNING GOALS

In this chapter, we will examine how astronomers came to understand that Earth and other planets orbit the Sun. By the conclusion of this chapter, you should be able to:

LG 1 Describe and contrast the geocentric and heliocentric models of the Solar System.

LG 2 Use Kepler's laws to describe the motion of objects in the Solar System.

LG 3 Explain how Galileo's astronomical discoveries provided empirical evidence for the heliocentric model.

LG 4 Describe the work of Galileo and Newton, which led them to discover the physical laws that govern the motion of all objects.

The shadows of a few of the Galilean moons of Jupiter fall on the planet. ▶ ▶ ▶

Why doesn't
Jupiter orbit
Earth?

3.1 The Motions of Planets in the Sky

Nebraska Simulation: Ptolemaic Orbit of Mars

Figure 3.1 In the Ptolemaic view of the heavens, Earth is at the center, orbited by the Moon, Mercury, Venus, the Sun, Mars, Jupiter, and Saturn.

When people in ancient times looked up at the sky, they saw that the Sun, Moon, and stars rose in the east and set in the west and appeared to be moving around Earth. The ancient peoples were aware of five planets (*planet* means "wandering star") because they moved in a generally eastward direction from one night to the next among the stars, whose positions were fixed on the celestial sphere. But they did not know that Earth was similar to these planets. A successful theory of how Earth and the planets move and how Earth fits in with its neighbors in the Solar System was the first step to understanding Earth's place in the universe. The history of the progression of ideas—from Earth at the center of all things to Earth as a tiny, insignificant rock—is a prime example of the self-correcting nature of science.

The Geocentric Model

As you saw in the previous chapter, hard evidence of the motions of Earth was remarkably difficult to come by. Largely because the motion of Earth through space cannot be felt, early astronomers developed a **geocentric** (Earth-centered) model of the Solar System to explain what they observed in the sky. When people looked up at the sky, the Sun, Moon, planets, and stars appeared to be moving around Earth. Before the 17th century, most educated people believed that the Sun, the Moon, and the known planets all moved in circles around a stationary Earth. **Figure 3.1** illustrates Ptolemy's geocentric model.

However, the geocentric model did not account for all observations. Ancient astronomers knew that the planets would occasionally do something unusual. Most of the time, the planets have an eastward **prograde motion**, in which each night they move a little eastward compared to the background stars. But sometimes the ancients observed apparent **retrograde motion**, in which the planets appear to move westward for a period of time before resuming their normal eastward travel. This retrograde motion is shown for Mars in **Figure 3.2**.

This odd behavior of the five "naked eye" planets—Mercury, Venus, Mars, Jupiter, and Saturn—created a puzzling problem for the geocentric model as it was summarized in 150 CE by the Alexandrian astronomer Ptolemy (Claudius Ptolemaeus, 90–168 CE). Some astronomers and philosophers in ancient times, for example Aristarchus of Samos (310–230 BCE), hypothesized that the Sun might be the center of the Solar System, but they did not have the tools to test the hypothesis or the mathematical insight to formulate a more complete and testable model. But most other astronomers of the time were skeptical, because they thought that if Earth moved around the Sun, they should feel Earth's motion. Therefore they preferred the geocentric model, in which the Sun, Moon, and planets all moved in perfect circles around a stationary Earth, with the "fixed stars" being located somewhere way out beyond the planets.

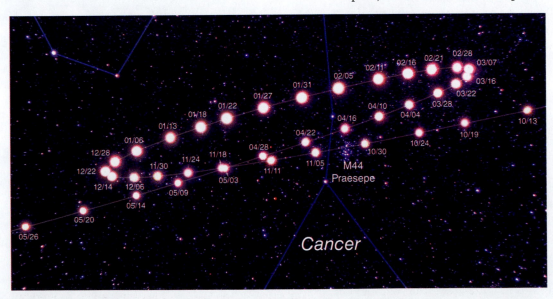

Figure 3.2 This time-lapse photographic series shows Mars as it moves in apparent retrograde motion.

Because the geocentric model in its simplest form failed to explain retrograde motion of the planets, Ptolemy added an embellishment called an *epicycle*, a small circle superimposed on each planet's larger circle, as illustrated in **Figure 3.3**. In this model, as the planet travels along its larger circle around Earth, it is also moving along a smaller circle. When its motion along the smaller circle was in a direction opposite to that of the forward motion of the larger circle, its forward motion would be reversed. Ptolemy's model had many of these epicycles, but they made the model *work*, in the sense that this model was reasonably successful at predicting the positions of planets in the sky. For nearly 1,500 years, Ptolemy's model of the heavens was the accepted paradigm in the Western world.

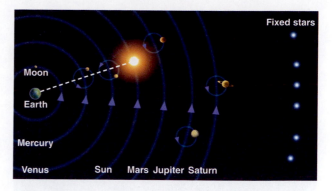

Figure 3.3 To reconcile retrograde motion with the geocentric model of the Solar System, additional loops called epicycles were added to each planet's circular orbit around Earth.

CHECK YOUR UNDERSTANDING 3.1

How did the ancients know the planets were different from the stars?

Copernicus Proposes a Heliocentric Model

Nicolaus Copernicus (1473–1543—**Figure 3.4**) is famous for placing the Sun rather than Earth at the center of the Solar System. He was not the first person to consider the idea that Earth orbited the Sun, and probably he had read the ideas of ancient Greek and medieval Arab astronomers who considered putting the Sun at the center of the Solar System. However, he was the first to develop a comprehensive mathematical model that could be tested by later astronomers. This work was the beginning of the Copernican Revolution. Through the work of 16th- and 17th-century scientists such as Tycho Brahe, Galileo Galilei, Johannes Kepler, and Isaac Newton, the **heliocentric** (Sun-centered) theory of the Solar System became one of the best-corroborated theories in all of science.

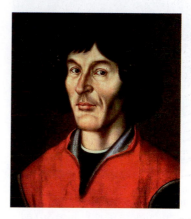

Figure 3.4 Nicolaus Copernicus rejected the ancient Greek model of an Earth-centered universe and replaced it with a model that centered on the Sun.

Nebraska Simulation: Retrograde Motion

Copernicus was multilingual and highly educated: he studied philosophy, canon (Catholic) law, medicine, economics, mathematics, and astronomy in his native Poland and in Italy. Copernicus conducted astronomical observations from a small tower, and sometime around 1514 he started writing about heliocentricity. Eighteen years later, he completed his manuscript. He did not publish the book because he knew his ideas would be controversial: philosophical and religious views of the time held that humanity and thus Earth must be the center of the universe. Late in his life, Copernicus was finally persuaded to publish his ideas, and his great work *De revolutionibus orbium coelestium* ("On the Revolutions of the Heavenly Spheres") appeared in 1543, the year of his death. This work pointed the way toward the modern cosmological principle introduced in Chapter 1—the idea that our location in the universe is not special.

Figure 3.5 shows Copernicus's model with the planets orbiting around the Sun. This model explained the observed motions of Earth, the Moon, and the planets, including retrograde motion, much more simply than the geocentric model did. Think about when you were in a car or train and you passed a slower-moving car or train, and it seemed as if the other vehicle was moving backward. It can be hard to tell which vehicle is moving and in which direction without an external frame of reference. Copernicus provided that frame of reference for the Sun and its planets.

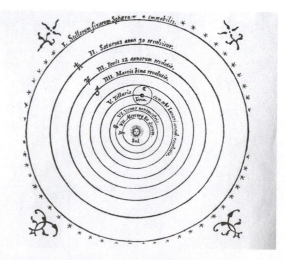

Figure 3.5 This illustration shows the Copernican heliocentric view of the Solar System (II–VII) and the fixed stars (I). The Sun is at the center and is orbited by Mercury, Venus, Earth, Mars, Jupiter, and Saturn. The Moon orbits Earth.

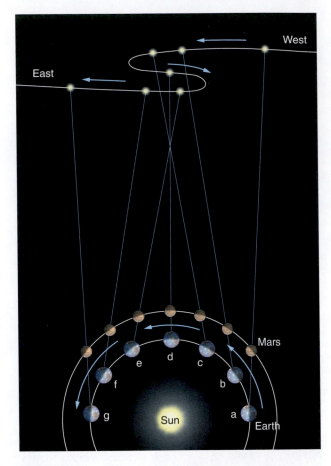

Figure 3.6 The Copernican model explains the apparent retrograde motion of Mars (see Figure 3.1) as seen in Earth's sky when Earth passes Mars in its orbit. The figure is not to scale.

In the Copernican model, the planets farther from the Sun undergo apparent retrograde motion when Earth overtakes them in their orbits. **Figure 3.6** illustrates this for Mars. Conversely, the planets closer to the Sun than Earth is— Mercury and Venus—move in apparent retrograde motion when overtaking Earth. Except for the Sun, all Solar System objects exhibit apparent retrograde motion. The magnitude of the effect diminishes with increasing distance from Earth. Retrograde motion is an illusion caused by the relative motion between Earth and the other planets.

Copernicus still conceived of the planets as moving in perfectly circular orbits, and as a result he needed to use some epicycles to match the observations. His model made testable predictions of the location of each planet on a given night, which were at least as accurate, but not more accurate, than those of the geocentric model. But this heliocentric model was, overall, simpler than the geocentric model and became the basis for further refinements in understanding how Earth moved. As copies of *De Revolutionibus* and Copernicus's ideas slowly spread across Europe, other scientists were excited by the heliocentric model, and a scientific revolution began.

Scaling the Solar System

Copernicus's model assumed that the planets traveled around the Sun in circular orbits with constant speeds. From his observations he deduced the correct order of the planets and concluded that planets closer to the Sun traveled faster than planets farther from the Sun. He also realized he needed to consider two categories of planets: **inferior** planets are closer to the Sun than Earth is; and **superior** planets are farther from the Sun than Earth is.

In Copernicus's model, periodically Earth, another planet, and the Sun line up in space to form either a straight line or a right triangle. As shown in **Figure 3.7a**, when a superior planet is in line with the Sun and Earth but on the other side of the Sun from Earth, we call the configuration a *conjunction*. A superior planet in conjunction will rise and set in the sky with the Sun. Note that when a superior planet is in conjunction, it is the farthest away from Earth that it gets and

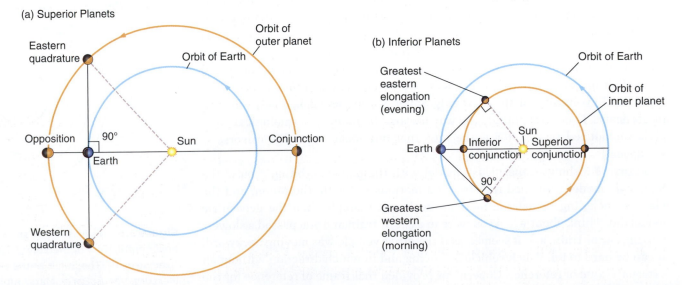

Figure 3.7 The diagrams show planetary configurations for (a) superior (outer) planets and (b) inferior (inner) planets.

therefore is at its faintest. You won't see the planet at all exactly at conjunction, because it's on the far side of the Sun.

In contrast, when a superior planet is in line with the Sun and Earth on the same side of the Sun as Earth, we call the configuration an *opposition*. At opposition, the superior planet is "opposite" the Sun in the sky: like a full Moon, it rises when the Sun sets and sets when the Sun rises. When the superior planet is in opposition, it is the closest it gets to Earth and thus is at its brightest; therefore, opposition is the best time to observe the planet in the sky. Opposition is also the time when the planet exhibits retrograde motion, because that is exactly when Earth is overtaking the planet in its orbit. *Quadrature* is when Earth, the Sun, and a superior planet form a right triangle in space.

For an inferior planet, the configurations are slightly different (Figure 3.7b). If the inferior planet is between Earth and the Sun, we call the configuration an *inferior conjunction*: this is when the planet is closest to Earth. If the inferior planet is on the other side of the Sun from Earth, we call the configuration a *superior conjunction*: this is when the planet is farthest from Earth. When the inferior planet forms a right triangle with Earth and the Sun—and thus is the farthest it gets from the Sun in the sky—we call the configuration the *greatest elongation*. When you are standing on Earth and looking toward the inner planets, you always see them close to the Sun in the sky, so you see Mercury and Venus only within a few hours of sunrise or sunset. The best time to observe these planets is at greatest elongation because they will have the greatest separation from the Sun in the sky.

Copernicus used the geometry of these alignments along with his observations of the positions of the planets in the sky, including their altitudes and the times they rose and set, to estimate the planet–Sun distances in terms of the Earth–Sun distance. He realized there were two types of orbital periods. Similar to the terms used for lunar orbits, a planet's *sidereal period* is how long it takes the planet to make one orbit around the Sun with respect to the stars and return to the same point in space. A planet's *synodic period* is how long it takes the planet to return to the same configuration with the Sun and Earth, such as inferior conjunction to inferior conjunction or opposition to opposition. The synodic period is what can be observed from Earth; for example, from opposition to opposition.

As shown in **Figure 3.8a**, Earth and the superior planet are in opposition at point A. Superior planets move around the Sun more slowly than Earth does, so Earth will complete one orbit around the Sun and then catch up to the superior planet to form the next opposition at point B. In Figure 3.8b, Earth and the inferior planet are in inferior conjunction at point A. An inferior planet moves around the Sun faster than Earth does, so it completes one sidereal period and then must continue in its orbit to catch up to Earth for the next inferior conjunction at point B. The numerical details are shown in **Working It Out 3.1**. **Table 3.1** shows that these relative distances calculated by Copernicus are remarkably close to distances obtained by modern methods. Copernicus's model not only predicted planetary positions in the sky but also could be used to compare the distances between the planets and the Sun accurately and thus to set the scale of the Solar System.

CHECK YOUR UNDERSTANDING 3.2

The planet Uranus will be observed in retrograde motion when: (a) Uranus is closest to the Sun; (b) Uranus is farthest from the Sun; (c) Earth overtakes Uranus in its orbit; (d) Uranus overtakes Earth in its orbit.

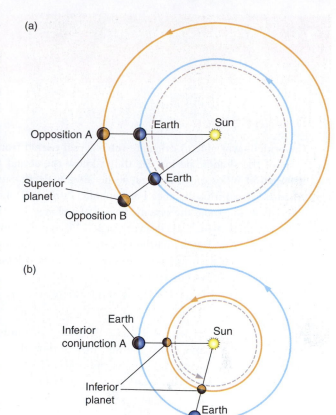

Figure 3.8 The synodic periods of planets indicate how long it takes them to return to the same configuration with Earth and the Sun. (a) Earth completes one orbit around the Sun first and then catches up to the superior planet. (b) Inferior planets complete a full orbit around the Sun first and then catch up to Earth.

 **Nebraska Simulation:** Planetary Configurations Simulator

	TABLE 3.1	**Copernicus's Scale of the Solar System**	
Planet		Copernicus's Value (AU)	Modern Value (AU)
Mercury		0.38	0.39
Venus		0.72	0.72
Earth		1.00	1.00
Mars		1.52	1.52
Jupiter		5.22	5.20
Saturn		9.17	9.58

AU = astronomical unit.

Orbital Periods

Copernicus was able to calculate the sidereal period from observations of the synodic period. He didn't know the actual Earth–Sun distance in miles or kilometers, so he let the Earth–Sun distance equal 1. We call this distance 1 astronomical unit (AU). Let P be the sidereal period and S the synodic period of a planet. E is the sidereal period of Earth, which equals 1 year, or 365.25 days. By thinking about the distance that Earth and the planet move in one synodic period, and noting that an inferior planet orbits the Sun in less time than Earth does, it can be shown that

$$\frac{1}{P} = \frac{1}{E} + \frac{1}{S}$$

for an inferior planet, with P, E, and S all in the same units of days or years. Similarly, Earth orbits the Sun in less time than a superior planet does, so the planet has traveled only part of its orbit around the Sun after 1 Earth year. The equation for a superior planet is

$$\frac{1}{P} = \frac{1}{E} - \frac{1}{S}$$

Let's consider Saturn as an example. The time that passes between one opposition—the date of maximum brightness—and the next shows that Saturn's synodic period is 378 days, or $378 \div 365.25 = 1.035$ years. Then to compute its sidereal period, P of Saturn in years, we use $S = 1.035$ and $E = 1$:

$$\frac{1}{P} = \frac{1}{1\,\text{yr}} - \frac{1}{1.035\,\text{yr}}$$

$$= 1 - 0.966 = 0.034\,\text{yr}^{-1}$$

and thus

$$P = \frac{1}{0.034\,\text{yr}^{-1}} = 29.4\,\text{yr}$$

The sidereal period of Saturn is 29.4 years, meaning that it takes Saturn 29.4 years to travel around the Sun and return to where it started in space.

 Nebraska Simulation: Synodic Period Calculator

Scaling the Solar System

Copernicus used the configurations of the planets shown in Figure 3.7 along with the sidereal periods of the planets to compute the relative distances of the planets. For the superior planets, he measured the fraction of the circular orbit that the planet completed in the time between opposition and quadrature, and then used trigonometry to solve for the planet–Sun distance in astronomical units (see Figure 3.7a). For the inferior planets, he had a right triangle at the point of greatest elongation, and then he used right-triangle trigonometry to solve for the planet–Sun distance in astronomical units (see Figure 3.7b). Copernicus's values are impressively similar to modern values (see Table 3.1). Copernicus still did not know the actual value of the astronomical unit in miles, but he was able to compute accurately the relative distances of the planets from the Sun for the first time.

3.2 Kepler's Laws Describe Planetary Motion

Copernicus did not understand *why* the planets move about the Sun, but he realized that his heliocentric picture provided a way to compute the relative distances of the planets. His theory is an example of **empirical science**, which seeks to describe patterns in nature with as much accuracy as possible. Copernicus's work was revolutionary because he was able to challenge the accepted geocentric model and propose that Earth is one planet among many. His conclusions paved the way for other great empiricists, including Tycho Brahe and Johannes Kepler.

Tycho Brahe's Observations

Tycho Brahe (1546–1601—**Figure 3.9**) was a Danish astronomer of noble birth who entered university at age 13 to study philosophy and law. After seeing a partial solar eclipse in 1560, Tycho (conventionally referred to by his first name) became interested in astronomy. A few years later, he observed Jupiter and Saturn near each other in the sky, but not in the exact positions predicted by the astro-

Figure 3.9 Tycho Brahe, known commonly as Tycho, was one of the greatest astronomical observers before the invention of the telescope.

nomical tables based on Ptolemy's model. Tycho gave up studying law and devoted himself to making better tables of the positions of the planets in the sky.

The king of Denmark granted Tycho the island of Hven, located between Sweden and Denmark, to build an observatory. Tycho designed and built new instruments, operated a printing press, and taught students and others how to conduct observations. With the assistance of his sister Sophie, Tycho carefully measured the precise positions of planets in the sky over several decades, developing the most comprehensive set of planetary data available at that time. He created his own geo-heliocentric model, shown in **Figure 3.10**. In Tycho's model, the planets orbit the Sun, and the Sun and planets orbit Earth. This model gained limited acceptance among people who preferred to keep Earth at the center for philosophical or religious reasons. Tycho lost his financial support when the king died, and in 1600 he relocated to Prague.

Kepler's Laws

In 1600, Tycho hired a more mathematically inclined astronomer, Johannes Kepler (1571–1630—**Figure 3.11**), as his assistant. Kepler, who had studied the ideas of Copernicus, was responsible for the next major step toward understanding the motions of the planets. Upon Tycho's death, Kepler inherited the records of his observations. Working first with Tycho's observations of Mars, Kepler deduced three empirical rules, now generally referred to as **Kepler's laws**, which accurately describe the motions of the planets. These laws are empirical: they use prior data to make predictions about future behavior but do not include an underlying theory of why the objects behave as they do.

Kepler's First Law When Kepler compared Tycho's extensive planetary observations with predictions from Copernicus's heliocentric model, he expected the data to confirm circular orbits for planets orbiting the Sun. Instead he found disagreements between his predictions and Tycho's observations. He was not the first to notice such discrepancies. Rather than discarding Copernicus's model, Kepler made some revisions.

Kepler discovered that if he replaced Copernicus's circular orbits with *elliptical* orbits, he could predict the positions of planets for any day, and found that his predictions fit Tycho's observations almost perfectly. An **ellipse** is a shape that looks like an elongated circle. It is symmetric from right to left and from top to bottom. As shown in **Figure 3.12a**, you can draw an ellipse by attaching the two

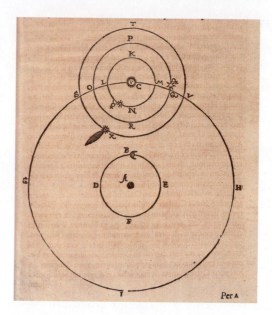

Figure 3.10 This reproduction shows Tycho's geo-heliocentric model with the Moon and Sun orbiting Earth, and the other planets orbiting the Sun.

Figure 3.11 Johannes Kepler explained the motions of the planets with three empirically based laws.

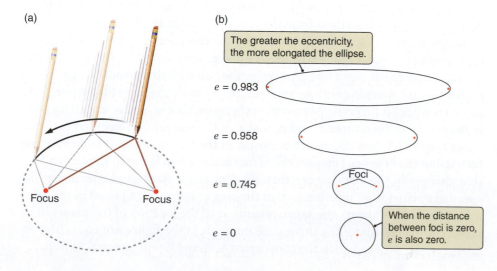

(a)

Focus   Focus

(b)

The greater the eccentricity, the more elongated the ellipse.

e = 0.983

e = 0.958

e = 0.745   Foci

e = 0   When the distance between foci is zero, e is also zero.

▶❚❙ **AstroTour:** Kepler's Laws

🖐 **Nebraska Simulation:** Eccentricity Demonstrator

Figure 3.12 (a) We can draw an ellipse by attaching a length of string to a piece of paper at two points (called foci) and then pulling the string around as shown. (b) Ellipses range from circles to elongated eccentric shapes. e = eccentricity.

Kepler's First Law

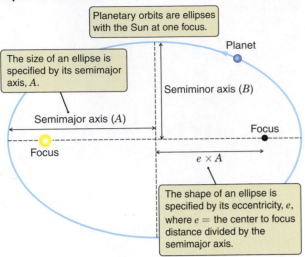

> Planetary orbits are ellipses with the Sun at one focus.

> The size of an ellipse is specified by its semimajor axis, A.

Semiminor axis (B)

Semimajor axis (A)

Planet

Focus

Focus

$e \times A$

> The shape of an ellipse is specified by its eccentricity, e, where e = the center to focus distance divided by the semimajor axis.

Figure 3.13 Planets move on elliptical orbits with the Sun at one focus. The eccentricity is given by the center-to-focus distance divided by the semimajor axis.

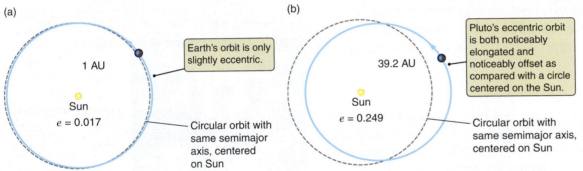

(a)

1 AU

Sun

$e = 0.017$

> Earth's orbit is only slightly eccentric.

Circular orbit with same semimajor axis, centered on Sun

(b)

39.2 AU

Sun

$e = 0.249$

> Pluto's eccentric orbit is both noticeably elongated and noticeably offset as compared with a circle centered on the Sun.

Circular orbit with same semimajor axis, centered on Sun

Figure 3.14 The orbits of (a) Earth and (b) Pluto compared with circles around the Sun. e = eccentricity.

Kepler's Second Law

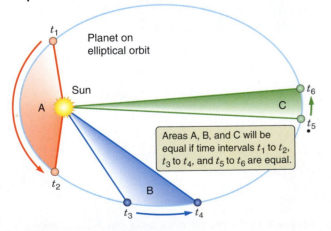

t_1

Planet on elliptical orbit

Sun

A

C

t_6

t_5

> Areas A, B, and C will be equal if time intervals t_1 to t_2, t_3 to t_4, and t_5 to t_6 are equal.

t_2

B

t_3

t_4

Figure 3.15 An imaginary line between a planet and the Sun sweeps out an area as the planet orbits. If the three intervals of time shown are equal, then the three areas A, B, and C will be the same.

ends of a piece of string to a piece of paper, stretching the string tight with the tip of a pencil, and then drawing around those two points while keeping the string tight. Each of the points at which the string is attached is a **focus** (plural: *foci*) of the ellipse. In Figure 3.12b, you can see that as the two foci become closer together, the ellipse becomes more circular. As the two foci move farther apart, the ellipse becomes more and more elongated. The **eccentricity (*e*)** of an ellipse measures this elongation; it is determined by the separation between the two foci divided by the length of the long axis. A circle has an eccentricity of 0 because the two foci coincide at the center of the circle. The more elongated the ellipse becomes, the closer its eccentricity gets to 1.

Kepler's first law of planetary motion states that the orbit of each planet is an ellipse with the Sun located at one focus (see the **Process of Science Figure**). There is nothing but empty space at the other focus. **Figure 3.13** illustrates Kepler's first law and shows how the features of an ellipse match observed planetary motions. The dashed lines in Figure 3.13 represent the two main axes of the ellipse. Half of the length of the long axis (the major axis) of the ellipse is called the **semimajor axis**, often denoted by the letter A. The semimajor axis of an orbit is the average distance between one focus and the ellipse itself. The average distance between the Sun and a planet equals the semimajor axis of the planet's orbit.

The eccentricities of planetary orbits vary widely, but most planetary objects in our Solar System have nearly circular orbits. As shown in **Figure 3.14a**, Earth's orbit is very nearly a circle centered on the Sun; with an eccentricity of 0.017, the distance variation is small. In contrast, dwarf planet Pluto's orbit, as shown in Figure 3.14b, has an eccentricity of 0.249. The orbit is noticeably elongated, with the Sun offset from center.

Kepler's Second Law From a close analysis of Tycho's observational data of changes in the positions of the planets, Kepler found that a planet moves fastest when it is closest to the Sun and slowest when it is farthest from the Sun. For example, we now measure Earth's average speed in its orbit around the Sun at 29.8 kilometers per second (km/s). When Earth is closest to the Sun, it travels at 30.3 km/s. When it is farthest from the Sun, it travels at 29.3 km/s.

Kepler found an elegant way to describe the changing speed of a planet in its orbit around the Sun. **Figure 3.15** shows a planet at six different points in its orbit (t_1 to t_6). Imagine a straight line connecting the Sun with this planet. We can think of this line as "sweeping out" an area as it moves with the planet from one point to another. Area A (in orange) is swept out between times t_1 and t_2, area B (in blue) is swept out between times t_3 and t_4, and area C (in green) is swept out between times t_5 and t_6. When the planet is closest to the Sun (area A), it is moving the fastest, but the distance between the planet and the Sun is small. Kepler realized that changes in the distance between the Sun and a planet and changes in the speed of a planet work together so that the area swept out by a planet in the same amount of time is always the same, regardless of the location of the planet in its orbit. This means that if the three time intervals in the figure are equal (that is, $t_1 \rightarrow t_2 = t_3 \rightarrow t_4 = t_5 \rightarrow t_6$), then the three areas A, B, and C will be equal as well.

Process of Science

THEORIES ARE FALSIFIABLE

Early astronomers studied the motions of the planets but did not understand why they behave as they do.

The Big Idea: Copernicus proposes that planets move in circular heliocentric orbits, with epicycles.

The Observation: Tycho observes and collects lots of data about planet positions.

The Prediction: Kepler uses Copernicus's model to predict where planets should be.

The Test: Kepler compares predictions with data—they disagree. Copernicus's idea is falsified!

The New Big Idea: Planet orbits are not circular. They are elliptical.

Model gains acceptance with a physical understanding of Newton's laws.

Scientific theories must be, in principle, falsifiable. Disproving an old theory always leads to deeper understanding.

Kepler's second law, also called Kepler's **law of equal areas**, states that the imaginary line connecting a planet to the Sun sweeps out equal areas in equal times, regardless of where the planet is in its orbit. This law applies to only one planet at a time. The area swept out by Earth in a given time is always the same. Likewise, the area swept out by Mars in a given time is always the same. But the area swept out by Earth and the area swept out by Mars in a given time are *not* the same. This law can be used to find the speed of a planet anywhere in its orbit.

CHECK YOUR UNDERSTANDING 3.3

Kepler's second law says that if a planet is in an elliptical orbit around a star, then the planet moves fastest when the planet is: (a) farthest from the star; (b) closest to the star; (c) located at one of the foci; (d) closest to another planet.

Kepler's Third Law Kepler looked for patterns in the orbital periods of the planets. He found that compared to planets closer to the Sun, planets farther from the Sun travel on longer orbits and they move more slowly in their orbits around the Sun. Kepler discovered a mathematical relationship between the sidereal period of a planet's orbit—how many years it takes to go around the Sun and return to the same position in space—and its average distance from the Sun in astronomical units. **Kepler's third law** states that in these units, the square of the sidereal period (P) of a planet's orbit is equal to the cube of the semimajor axis (A) of the planet's orbit.

Table 3.2 shows the periods and semimajor axes of the orbits of the eight classical planets and three of the dwarf planets, along with the values of the ratio P^2 divided by A^3. These data are also plotted in **Figure 3.16**. Kepler referred to this relationship as his **harmonic law** or, more poetically, as the "Harmony of the Worlds." Kepler's third law is explored further in **Working It Out 3.2**. Kepler's laws enhanced the heliocentric mathematical model of Copernicus and led to its greater acceptance.

 **Nebraska Simulation:** Planetary Orbit Simulator

3.2 | Working It Out Kepler's Third Law

Kepler's third law states that the square of the period of a planet's orbit, measured in years P_{years}, is equal to the cube of the semimajor axis of the planet's orbit, measured in astronomical units A_{AU}. Translated into math, the law says

$$(P_{years})^2 = (A_{AU})^3$$

Kepler used units based on Earth—astronomical units and years—as a matter of convenience. If other units were used, then P^2 would still be proportional to A^3, but the constant of proportionality would not be 1.

As an example of using this law, suppose that you want to know the average radius of Neptune's orbit in astronomical units. First you need to find out how long Neptune's period is in Earth years, which

you can determine by observing the synodic period and then computing its sidereal period from that (using Working It Out 3.1). Neptune's sidereal period is 165 years. Plugging this number into Kepler's third law gives this result:

$$(P_{years})^2 = (165)^2 = 27{,}225 = (A_{AU})^3$$

To solve this equation, you must first square 165 to get 27,225 and then take its cube root (see Appendix 1 for calculator hints). Then

$$A_{AU} = \sqrt[3]{27{,}225} = 30.1$$

The semimajor axis of Neptune's orbit, that is, the average distance between Neptune and the Sun, is 30.1 AU.

| TABLE 3.2 | Kepler's Third Law: $P^2 = A^3$ | | |

Planet	Period P (years)	Semimajor Axis A (AU)	$\dfrac{P^2}{A^3}$
Mercury	0.241	0.387	$\dfrac{0.241^2}{0.387^3} = 1.00$
Venus	0.615	0.723	$\dfrac{0.615^2}{0.723^3} = 1.00$
Earth	1.000	1.000	$\dfrac{1.000^2}{1.000^3} = 1.00$
Mars	1.881	1.524	$\dfrac{1.881^2}{1.524^3} = 1.00$
Ceres	4.599	2.765	$\dfrac{4.559^2}{2.765^3} = 1.00$
Jupiter	11.86	5.204	$\dfrac{11.86^2}{5.204^3} = 1.00$
Saturn	29.46	9.582	$\dfrac{29.46^2}{9.582^3} = 0.99*$
Uranus	84.01	19.201	$\dfrac{84.01^2}{19.201^3} = 1.00$
Neptune	164.79	30.047	$\dfrac{164.79^2}{30.047^3} = 1.00$
Pluto	247.68	39.236	$\dfrac{247.68^2}{39.236^3} = 1.02*$
Eris	557.00	67.696	$\dfrac{557.00^2}{67.696^3} = 1.00$

*Slight perturbations from the gravity of other planets are the reason that these ratios are not exactly 1.00.

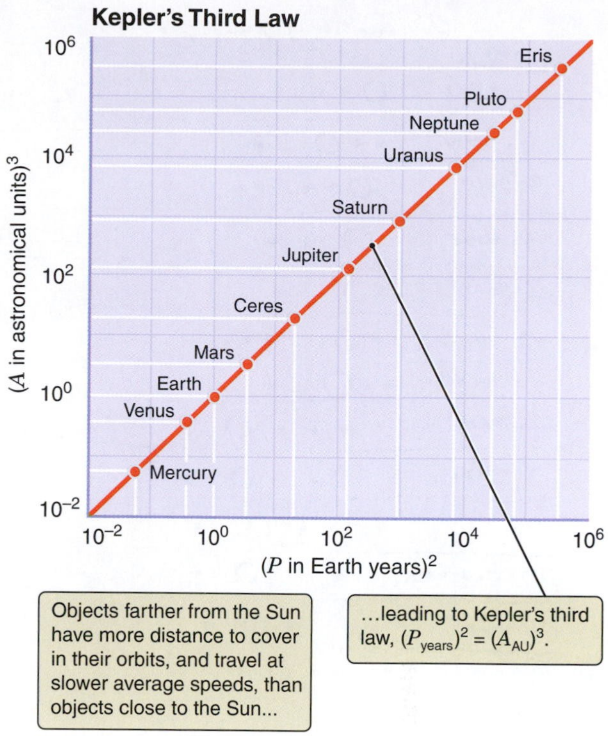

Kepler's Third Law

Objects farther from the Sun have more distance to cover in their orbits, and travel at slower average speeds, than objects close to the Sun...

...leading to Kepler's third law, $(P_{years})^2 = (A_{AU})^3$.

Figure 3.16 A plot of A^3 versus P^2 for objects in our Solar System shows that they obey Kepler's third law. (Note that by plotting powers of 10 on each axis, we are able to fit both large and small values on the same plot. We will do this frequently.)

CHECK YOUR UNDERSTANDING 3.4

Place the following in order from largest to smallest semimajor axis: (a) a planet with a period of 84 Earth days; (b) a planet with a period of 1 Earth year; (c) a planet with a period of 2 Earth years; (d) a planet with a period of 0.5 Earth year.

3.3 Galileo's Observations Supported the Heliocentric Model

Galileo Galilei (1564–1642—**Figure 3.17**)—one of the heroes of astronomy—was the first to use a telescope to conduct and report on significant discoveries about astronomical objects. Galileo's telescopes were relatively small, yet sufficient for him to observe spots on the Sun, the uneven surface and craters of the Moon, and the large number of stars in the band of light in the sky called the Milky Way.

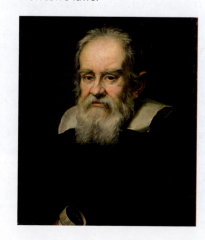

Figure 3.17 Galileo Galilei laid the physical framework for Newton's laws.

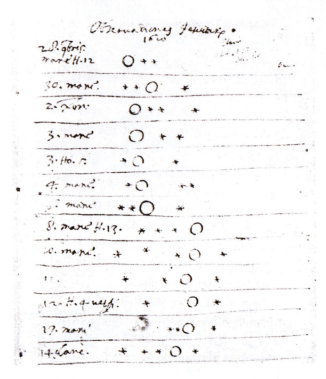

Figure 3.18 This page from Galileo's notebook shows his observations of the four largest moons of Jupiter.

Nebraska Simulations: Phases of Venus; Ptolemaic Phases of Venus

▶Ⅱ **AstroTour:** Velocity, Acceleration, Inertia

Figure 3.19 Modern photographs of the phases of Venus show that when we see Venus more illuminated, it also appears smaller, implying that Venus is farther away at that time.

Galileo's Observations

Galileo provided the first observational evidence that some objects in the sky do not orbit Earth. When Galileo turned his telescope to the planet Jupiter, he observed four "stars" in a line near the planet. Over time he observed that the objects changed position from night to night (**Figure 3.18**). Galileo hypothesized that these objects were moons in orbit around Jupiter. These four moons are the largest of Jupiter's many moons and are still called the Galilean moons. Galileo also estimated the relative distance of each moon from Jupiter and the periods of their orbits, and he was able to show that they followed Kepler's third law.

Galileo also observed that the planet Venus went through phases like the Moon. He noticed that the phases of Venus were correlated with the size of the image of Venus in his telescope. In a geocentric model in which Venus orbits Earth like the Moon does, the apparent size of Venus would be constant, like we see for the Moon. In the heliocentric model, the Earth–Venus distance varies, and the size of Venus changes accordingly. When Venus is in its gibbous to full phases, it is farther away, on the other side of the Sun than the side Earth is on, and it is smaller in the sky. When Venus is in its crescent to new phases, it is closer, on the same side of the Sun as the side Earth is on, and it is larger in the sky (**Figure 3.19**). The observations of Jupiter's moons and the phases of Venus in particular convinced Galileo that Copernicus was correct in placing the Sun at the center of the Solar System.

In addition to his astronomical observations, Galileo did important work on the motion of objects. Unlike natural philosophers, who thought about objects in motion but did not actually experiment with them, Galileo conducted experiments with falling and rolling objects. As with his telescopes, Galileo improved or developed new technology to enable him to conduct these experiments. For example, by carefully rolling balls down an inclined plane and by dropping various objects from a height, he found that the distance traveled by a falling object is proportional to the square of the time it has been falling. If he simultaneously dropped two objects of different masses, they reached the ground at the same time, demonstrating that all objects falling to Earth accelerate at the same rate, independent of their mass.

Galileo's observations and experiments with many types of moving objects, such as carts and balls, led him to disagree with the Greek philosophers about when and why objects continue to move or come to rest. Prior to Galileo, it was thought that the natural state of an object was to be at rest. But Galileo found that the natural state of an object is to keep doing what it was doing until a force acts on it. That is, an object in motion continues moving along a straight line with a constant speed until a force acts on it to change its state of motion. This idea of *inertia*, which was later adopted by Newton as his first law of motion, has implications for not only the motion of carts and balls but also the orbits of planets.

Dialogue Concerning the Two Chief World Systems

Much has been written about the considerable danger Galileo faced because of his work. His later life was consumed by conflict with the Catholic Church over his support of the Copernican system. In 1632, Galileo published his best-selling book, *Dialogo sopra i due massimi sistemi del mondo* ("Dialogue Concerning the

Two Chief World Systems"). The *Dialogo* presents a brilliant philosopher named Salviati as the champion of the Copernican heliocentric view of the universe. The defender of an Earth-centered universe, Simplicio—who uses arguments made by the classical Greek philosophers and the pope—sounds silly and ignorant.

Galileo, a religious man who had two daughters in a convent, thought he had the tacit approval of the Catholic Church for his book. But when he placed a number of the pope's geocentric arguments in the unflattering mouth of Simplicio, the perceived attack on the pope attracted the attention of the church. Galileo was put on trial for heresy, sentenced to prison, and eventually placed under house arrest. To escape a harsher sentence, Galileo was forced publicly to recant his belief in the Copernican theory that Earth moves around the Sun. According to one story, as he left the courtroom, Galileo stamped his foot on the ground and muttered, "And yet it moves!"

The *Dialogo* was placed on the pope's Index of Prohibited Works, along with Copernicus's *De Revolutionibus*, but it traveled across Europe, was translated into other languages, and was read by other scientists. (Two centuries later, in 1835, the church finally removed the uncensored version of the *Dialogo* from its prohibited list.) Galileo spent his final years compiling his research on inertia and other ideas into the book *Discourses and Mathematical Demonstrations Relating to Two New Sciences*, which was published in 1638 in Holland, outside the jurisdiction of the Catholic Church.

CHECK YOUR UNDERSTANDING 3.5

Which of Galileo's astronomical observations were best explained by a heliocentric model? (a) sunspots; (b) craters on the Moon; (c) moons of Jupiter; (d) phases of Venus

---

3.4 Newton's Three Laws Help to Explain the Motion of Celestial Bodies

Empirical laws, like Kepler's laws, describe what happens, but they do not explain why. Kepler described the orbits of planets as ellipses, but he did not explain why they should be so. To take that next step in the scientific process, scientists use basic physical principles and the tools of mathematics to derive the empirically determined laws. Alternatively, a scientist might start with physical laws and predict relationships, which are then verified or falsified by experiment and observation. If these predictions are verified by experiment and observation, the scientist may have determined something fundamental about how the universe works.

Sir Isaac Newton (1642–1727—**Figure 3.20**) took this next step in explaining the nature of motion. Newton was a student of mathematics at Cambridge University when it closed down because of the Great Plague and students were sent home to the safer countryside. Over the next 2 years, he continued to study on his own, and at the age of 23 he invented calculus, which would become crucial to his development of the physics of motion. The German mathematician Gottfried Leibniz independently developed calculus around the same time.

Building on the work of Kepler, Galileo, and others, Newton proposed three physical laws that govern the motions of all objects in the sky and on Earth. In this section, we will examine these three laws, which are essential to an understanding of the motions of the planets and all other celestial bodies.

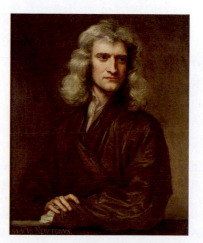

Figure 3.20 Sir Isaac Newton formulated three laws of motion.

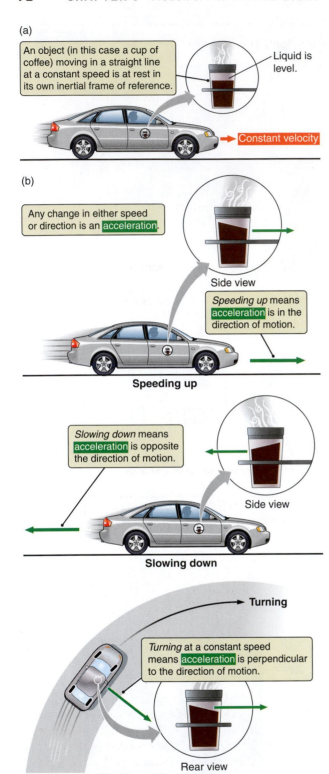

(a) An object (in this case a cup of coffee) moving in a straight line at a constant speed is at rest in its own inertial frame of reference.

Liquid is level.

Constant velocity

(b) Any change in either speed or direction is an acceleration.

Side view

Speeding up means acceleration is in the direction of motion.

Speeding up

Slowing down means acceleration is opposite the direction of motion.

Side view

Slowing down

Turning

Turning at a constant speed means acceleration is perpendicular to the direction of motion.

Rear view

Figure 3.21 (a) An object moving in a straight line at a constant speed is at rest in its own inertial frame of reference. (b) Any change in the velocity of an object is an acceleration. When you are driving, for example, any time your speed changes or you follow a curve in the road, you are experiencing an acceleration. (Throughout the text, velocity arrows will be shown as red, and acceleration arrows will be shown as green.)

Newton's First Law: Objects at Rest Stay at Rest; Objects in Motion Stay in Motion

A **force** is a push or a pull on an object. It is possible for two or more forces to oppose one another in such a way that they are perfectly balanced and cancel out. For example, gravity pulls down on you as you sit in your chair. But the chair pushes up on you with an exactly equal and opposite force. So you remain motionless. Forces that cancel out have no effect on an object's motion. When forces add together to produce an effect, we often use the term *net force*, or sometimes just *force*.

Imagine that you are driving a car, and your phone is on the seat next to you. A rabbit runs across the road in front of you, and you press the brakes hard. You feel the seat belt tighten to restrain you. At the same time, your phone flies off the seat and hits the dashboard. You have just experienced what Newton describes in his first law of motion. **Inertia** is the tendency of an object to maintain its state—either of motion or of rest—until it is pushed or pulled by a net force. In the case of the stopping car, you did not hit the dashboard because the force of the seat belt slowed you down. The phone did hit the dashboard because no such force acted upon it.

Newton's first law of motion describes inertia and states that an object in motion tends to stay in motion, in the same direction, until a net force acts upon it; and an object at rest tends to stay at rest until a net force acts upon it. Galileo's law of inertia became the cornerstone of physics as Newton's first law.

Recall from Section 2.1 of Chapter 2 the concept of a frame of reference. Within a frame of reference, only the relative motions between objects have any meaning. Without external clues, you cannot tell the difference between sitting still and traveling at constant speed in a straight line. For example, if you close your eyes while riding in the passenger seat of a quiet car on a smooth road, you would feel as though you were sitting still. Returning to the earlier example, your phone was "at rest" beside you on the front seat of your car, but a person standing by the side of the road would see the phone moving past at the same speed as the car. People in a car approaching you would see the phone moving quite fast—at the speed they are traveling plus the speed you are traveling! All of these perspectives are equally valid, and all of these speeds of the phone are correct when measured in the appropriate reference frame.

A reference frame moving in a straight line at a constant speed is an **inertial frame of reference**. Any inertial frame of reference is as good as another. As illustrated in **Figure 3.21a**, in the inertial frame of reference of a cup of coffee, the cup is at rest in its own frame even if the car is moving quickly down the road.

Newton's Second Law: Motion Is Changed by Forces

What if a net force does act? In the earlier example, you were traveling in the car, and your motion was slowed when the force of the seat belt acted upon you. Forces change an object's motion—by changing either the speed or the direction. This reflects **Newton's second law of motion**: if a net force acts on an object, then the object's motion changes.

As an example of changes in an object's motion, think about a car. When you are in the driver's seat of a car, you have a number of controls, including a gas pedal and a brake pedal. You use these to make the car speed up or slow down. A *change in speed* is one way the car's motion can change. But you also have the steering wheel in your hands. When you are moving down the road and you turn the wheel, your speed does not necessarily change, but the direction of your motion does. A *change in direction* is also a kind of change in motion.

Together, the combined speed and direction of an object's motion are called the object's **velocity**. "Traveling at 50 kilometers per hour (km/h)" indicates speed; "traveling north at 50 km/h" indicates velocity. The rate at which the velocity of an object changes is called **acceleration**. Acceleration tells you how rapidly a change in velocity happens. For example, if you go from 0 to 100 km/h in 4 seconds, you feel a strong push from the seat back as it shoves your body forward, causing you to accelerate along with the car. However, if you take 2 minutes to go from 0 to 100 km/h, the acceleration is so slight that you hardly notice it.

Partly because the gas pedal on a car is often called the accelerator, some people think *acceleration* always means that an object is speeding up. But we need to stress that, as used in physics, any change in speed or direction is an acceleration. Figure 3.21b illustrates the point by showing what happens to the coffee in a coffee cup as the car speeds up, slows down, or turns. Slamming on your brakes and going from 100 to 0 km/h in 4 seconds is just as much acceleration as going from 0 to 100 km/h in 4 seconds. Similarly, the acceleration you experience as you go through a fast, tight turn at a constant speed is every bit as real as the acceleration you feel when you slam your foot on the gas pedal or the brake pedal. Speeding up, slowing down, turning left, turning right—if you are not moving in a straight line at a constant speed, you are experiencing an acceleration.

Newton's second law of motion says that a net force causes acceleration. The acceleration an object experiences depends on two things. First, as shown in **Figure 3.22**, the acceleration depends on the strength of the net force acting on the object to change its motion. If the forces acting on the object do *not* add up to zero, then there is a net force and the object accelerates (Figure 3.22a). The stronger the net force, the greater the acceleration. Push on something twice as hard and it experiences twice as much acceleration (Figure 3.22b). Push on something 3 times as hard and its acceleration is 3 times as great. The resulting change in motion occurs in the direction the net force points. Push an object away from you, and it will accelerate away from you.

The acceleration that an object experiences also depends on its inertia. You can push some objects easily, for example, an empty box from a refrigerator delivery. However, the actual refrigerator, even though it is about the same size, is not easily shoved around. The greater the mass, the greater the inertia, and the *less* acceleration will occur in response to the same net force, as shown in Figure 3.22c. This relationship among acceleration, force, and mass is expressed mathematically in **Working It Out 3.3**.

Newton's Third Law: Whatever Is Pushed, Pushes Back

Imagine that you are standing on a skateboard and pushing yourself along with your foot. Each shove of your foot against the ground sends you faster along your way. But why does this happen? Your muscles flex, and your foot exerts a force on the ground. (Earth itself does not noticeably accelerate, because its great mass gives it great inertia.) Yet this does not explain why *you* experience an acceleration. The fact that you accelerate means that as you push on the ground, the ground must be pushing back on you.

Part of Newton's genius was his ability to see patterns in such everyday events. Newton realized that *every* time one object exerts a force on another, the second

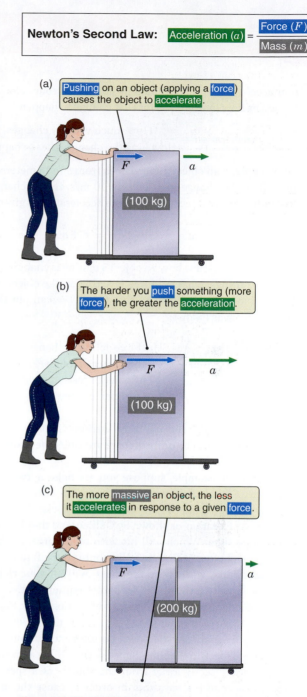

Newton's Second Law: Acceleration $(a) = \dfrac{\text{Force }(F)}{\text{Mass }(m)}$

(a) Pushing on an object (applying a force) causes the object to accelerate.

F a (100 kg)

(b) The harder you push something (more force), the greater the acceleration.

F a (100 kg)

(c) The more massive an object, the less it accelerates in response to a given force.

F a (200 kg)

Figure 3.22 Newton's second law of motion says that the acceleration experienced by an object is determined by the force acting on the object divided by the object's mass. (Throughout the text, force arrows will be shown as blue.)

Astronomy in Action: Velocity, Force, and Acceleration

AstroTour: Velocity, Acceleration, Inertia

3.3 Working It Out Using Newton's Laws

Your acceleration is determined by how much your velocity changes, divided by how long it takes for that change to happen:

$$\text{Acceleration} = \frac{\text{How much velocity changes}}{\text{How long the change takes to happen}}$$

For example, if an object's speed goes from 5 to 15 meters per second (m/s), then the change in velocity is 10 m/s. If that change happens over the course of 2 seconds, then the acceleration is given by

$$a = \frac{15\,\text{m/s} - 5\,\text{m/s}}{2\,\text{s}} = 5\,\text{m/s}^2$$

If we want to know how an object's motion is changing, we need to know two things: what net force is acting on the object, and what is the resistance of the object to that force? We can put the idea into equation form as follows:

$$\begin{pmatrix} \text{The} \\ \text{acceleration} \\ \text{experienced} \\ \text{by an object} \end{pmatrix} = \frac{\begin{array}{c}\text{The force acting to change} \\ \text{the object's motion}\end{array}}{\begin{array}{c}\text{The object's resistance} \\ \text{to that change}\end{array}} = \frac{\text{Force}}{\text{Mass}}$$

Newton's second law above is often written as Force = mass × acceleration, or $F = ma$. The units of force are called **newtons (N)**, so that $1\,\text{N} = 1\,\text{kg m/s}^2$.

As a simple example, suppose you are holding two blocks of the same size. The block in your right hand has twice the mass of the block in your left hand. When you drop the blocks, they both fall with the same acceleration, as shown by Galileo, and they hit your two feet at the same time. Which will hit with more force: the block falling onto your right foot or the one falling onto your left foot? The block in your right hand, with twice the mass, will hit your right foot with twice the force that the other block hits your left foot.

To see how Newton's three laws of motion work together, study **Figure 3.23**. An astronaut is adrift in space, motionless with respect to the nearby space shuttle. With no tether to pull on, how can the astronaut get back to the ship? Suppose the 100-kg astronaut throws a 1-kg wrench directly away from the shuttle at a speed of 10 m/s. Newton's second law says that in order to cause the motion of the wrench to change, the astronaut has to apply a force to it in the direction away from the shuttle. Newton's third law says that the wrench must therefore push back on the astronaut with as much force but in the opposite direction. The force of the wrench on the astronaut causes the astronaut to begin drifting toward the shuttle. How fast will the astronaut move? Turn to Newton's second law again. Because the astronaut has more mass, she will accelerate less than the wrench will. A force that causes the 1-kg wrench to accelerate to 10 m/s will

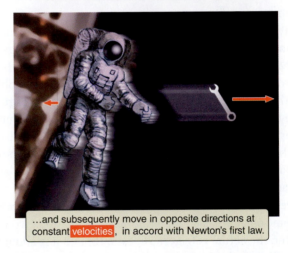

An astronaut adrift in space pushes on a wrench, which, according to Newton's third law, pushes back on the astronaut.

Space shuttle

While in contact with each other, the wrench and the astronaut experience accelerations proportional to the inverse of their masses…

…and subsequently move in opposite directions at constant velocities, in accord with Newton's first law.

Figure 3.23 According to Newton's laws, if an astronaut adrift in space throws a wrench, the two will move in opposite directions. Their speeds will depend on their masses: the same force will produce a smaller acceleration of a more massive object than of a less massive object. (Acceleration and velocity arrows are not drawn to scale.)

have much less effect on the 100-kg astronaut. Because acceleration equals force divided by mass, the 100-kg astronaut will experience only 1/100 as much acceleration as the 1-kg wrench. The astronaut will drift toward the shuttle, but only at the leisurely rate of 1/100 × 10 m/s, or 0.1 m/s.

object exerts a matching force on the first. That second force is exactly as strong as the first force but is in exactly the opposite direction. When you are accelerating yourself on the skateboard, you push backward on Earth, and Earth pushes you forward. As shown in **Figure 3.24**, a woman pulling a load on a cart pulls on the rope, and the rope pulls back. A car tire pushes back on the road, and the road pushes forward on the tire. Earth pulls on the Moon, and the Moon pulls on Earth. A rocket engine pushes hot gases out of its nozzle, and those hot gases push back on the rocket, propelling it into space.

All of these force pairs are examples of **Newton's third law of motion**, which says that forces always come in pairs, and the forces of a pair are always equal in strength but opposite in direction. The forces in these pairs always act on two different objects. Your weight pushes down on the floor, and the floor pushes back up on your feet with the same amount of force. For every force there is *always* an equal force in the opposite direction.

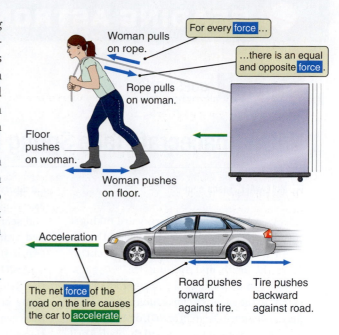

Figure 3.24 Newton's third law states that for every force there is always an equal and opposite force. These opposing forces always act on the two different objects in the same pair.

CHECK YOUR UNDERSTANDING 3.6

Imagine a planet moving in a perfectly circular orbit around the Sun. Is this planet experiencing acceleration? (a) Yes, because it is changing its speed all the time. (b) Yes, because it is changing its direction of motion all the time. (c) No, because its speed is not changing all the time. (d) No, because planets do not experience accelerations.

Origins

Planets and Orbits

In addition to the planets in our own Solar System, several thousand planets have been detected orbiting stars other than our own. Their orbits can be calculated and understood by applying the same three Kepler's laws that we have discussed here. Astrobiologists think that the orbit of a planet around its star affects its chances of developing life.

Consider the average distance of the planet from its star. You might intuitively guess that a planet close to its star will receive more energy from its star than that received by a planet far from its star. If Earth were closer to the Sun, it would be hotter throughout the year—perhaps so hot that water would evaporate and not exist as a liquid. If Earth were farther from the Sun, it would be colder and perhaps all water would freeze. We know that liquid water was a crucial element for the formation of life

on Earth. So some astronomers look for planets at a distance from their star such that liquid water can exist (this distance will vary depending on the temperature and size of the star).

We might also think about the eccentricity of a planet's orbit: recall from Figure 3.14a that Earth's orbit differs from a circle by less than 2 percent. Thus, Earth's distance from the Sun does not vary much throughout the year; and as we saw in Chapter 2, seasonal variation on Earth is caused by the tilt of Earth's axis, not by the slight changes in its distance from the Sun. However, if we look at our neighboring planet Mars, which has about the same axial tilt as Earth, we see a greater seasonal variation because of its more eccentric orbit. The distance between Mars and the Sun varies from 1.38 AU (207 million km) to 1.67 AU (249 million km)—an

eccentricity of 9 percent. As a result, the seasons on Mars are not equal. They are shorter when Mars is closer to the Sun and moving faster and longer when Mars is farther from the Sun and moving slower. The inequality of the seasons on Mars has an effect on the overall stability of its temperature and climate. When we look at planets orbiting other stars, we see that many have orbital eccentricities even higher than that of Mars and therefore large variations in temperature.

Earth is at the right distance from the Sun to have temperatures that permit water to be liquid, and its orbital eccentricity is low enough that the average planetary temperature does not change much over the course of its orbit. These orbital characteristics have contributed to making the conditions on Earth suitable for the development of life.

ARTICLES   QUESTIONS

NASA Spacecraft Take Spring Break at Mars

By **MIKE WALL, Space.com**

NASA's robotic Mars explorers are taking a cosmic break for the next few weeks, thanks to an unfavorable planetary alignment of Mars, the Earth, and the Sun.

Mission controllers won't send any commands to the agency's *Opportunity* rover, *Mars Reconnaissance Orbiter* (*MRO*), or *Mars Odyssey* orbiter from today (April 9) through April 26. The blackout is even longer for NASA's car-size *Curiosity* rover, which is slated to go solo from April 4 through May 1.

The cause of the communications moratorium is a phenomenon called a Mars solar conjunction, during which the Sun comes between Earth and the Red Planet (**Figure 3.25**). Our star can disrupt and degrade interplanetary signals in this formation, so mission teams won't be taking any chances.

"Receiving a partial command could confuse the spacecraft, putting them in grave danger," NASA officials explain in a video posted last month by the agency's Jet Propulsion Laboratory (JPL) in Pasadena, California.

Opportunity and *Curiosity* will continue performing stationary science work, using commands already beamed to the rovers.

Curiosity will focus on gathering weather data, assessing the martian radiation environment, and searching for signs of subsurface water and hydrated minerals, officials said Monday (April 8).

MRO and *Odyssey* will also keep studying the Red Planet from above, and they'll continue to serve as communications links between the rovers and Earth. The conjunction will also affect the European Space Agency's *Mars Express* orbiter, officials have said.

Odyssey will send rover data home as usual during conjunction, though the orbiter may have to relay information multiple times due to dropouts. *MRO*, on the other hand, entered record-only mode on April 4. The spacecraft will probably have about 52 gigabits of data to relay when it's ready to start transmitting again on May 1, *MRO* officials have said.

Mars solar conjunctions occur every 26 months, so NASA's Red Planet veterans have dealt with them before. This is the fifth conjunction for *Opportunity*, in fact, and the sixth for *Odyssey*, which began orbiting Mars in 2001.

But it'll be the first for *Curiosity*, which touched down on August 5, kicking off a two-year surface mission to determine if the Red

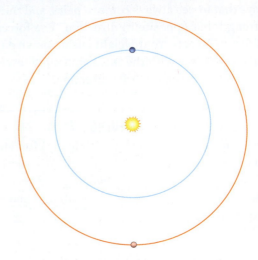

Figure 3.25 Every 26 months, Mars and Earth are on opposite sides of the Sun, making communication between the two planets impossible.

Planet has ever been capable of supporting microbial life.

"The biggest difference for this 2013 conjunction is having *Curiosity* on Mars," *Odyssey* mission manager Chris Potts, of NASA's Jet Propulsion Laboratory in Pasadena, California, said in a statement last month.

ARTICLES   QUESTIONS

1. How often do these Mars solar conjunctions occur? Does this interval correspond to the sidereal period or the synodic period of Mars?
2. Give two reasons that conjunction is a bad time to view Mars in the sky.
3. Make a sketch of the Mars solar conjunction, showing Mars, Earth, and the Sun. Using the orbital periods in Table 3.2, add in the positions of Mars and Earth at the next Mars conjunction. Will it take place at the same location in space?
4. View this short video from NASA: http://www.jpl.nasa.gov/video/?id=1204. Do you think it explains the concepts well to someone who is not reading an astronomy textbook?
5. If people were on a mission on Mars, a loss of contact would be very troubling. How might NASA plan to avoid losing contact with astronauts on Mars during conjunctions?

Summary

Early astronomers hypothesized that Earth was stationary at the center of the Solar System. Later astronomers realized that a Sun-centered Solar System was much simpler and could explain the observations. Planets, like Jupiter, orbit the Sun, not Earth. Kepler's laws describe the elliptical orbits of planets around the Sun, including details about how fast a planet travels at various points in its orbit. These laws helped Newton to advance science by developing his laws of motion, which govern the motion of all objects (not just orbiting ones). Orbital semimajor axis, eccentricity, and stability may affect a planet's suitability to foster life.

LG 1 **Describe and contrast the geocentric and heliocentric models of the Solar System.** Earth's motion is hard to detect, so prior to the Copernican Revolution, most people accepted a geocentric model of the Solar System, in which all objects orbit around Earth. In particular, apparent retrograde motion of the planets was difficult to understand in this model. Copernicus created the first comprehensive mathematical model of the Solar System with the Sun at the center, called a heliocentric model. His model explained apparent retrograde motion as a visual illusion seen when an inner planet passes an outer planet in their orbits.

LG 2 **Use Kepler's laws to describe the motion of objects in the Solar System.** Using Tycho's observational data, Kepler developed empirical rules to describe the motions of the planets. Kepler's three laws state that (1) planets move in elliptical orbits around the Sun, (2) planets move fastest when closest to the Sun and slowest when farthest from the Sun, so that the planets sweep out equal areas in equal times, and (3) the orbital period of a planet squared equals the semimajor axis of its orbit cubed, or $P^2 = A^3$.

LG 3 **Explain how Galileo's astronomical discoveries provided empirical evidence for the heliocentric model.** Galileo invented astronomical telescopes and used them to observe moons in orbit around Jupiter. He also saw Venus going through phases like the Moon, but changing its apparent size in each phase. These astronomical observations were difficult to explain with a geocentric model.

LG 4 **Describe the work of Galileo and Newton, which led them to discover the physical laws that govern the motion of all objects.** Galileo studied the physics of falling objects and discovered the principle of inertia. Newton's laws state that (1) objects do not change their motion unless they experience a net force, (2) Force = mass × acceleration, and (3) "every force has an equal and opposite force." Net forces cause accelerations; that is, changes in motion. Inertia resists changes in motion.

? UNANSWERED QUESTIONS

- Would the history of scientific discoveries in physics and astronomy have been different if the Catholic Church had not prosecuted Galileo? Galileo wrote the *Dialogo* after being ordered by the Catholic Church in 1616 not to "hold or defend" the idea that Earth moves and the Sun is still. And he wrote his equally famous *Discorsi e Dimostrazioni Matematiche* (often shortened in English to "Two New Sciences") while under house arrest after his trial. However undeterred Galileo appeared to be, the effects of the decrees, prohibitions, and prosecutions might have dissuaded other scientists in Catholic countries from pursuing this type of work. Indeed, after Galileo's experiences, the center of the scientific revolution moved north to Protestant Europe.

- What percentage of planets are in unstable orbits? In younger planetary systems, planets might migrate in their orbits because of the presence of other massive planets nearby. We will see in Chapter 7 that Uranus and Neptune might have migrated in this way. Some planets have been discovered moving through the galaxy without any obvious orbit around a star—and therefore are not in the stable orbits we see in our own Solar System.

Questions and Problems

Test Your Understanding

1. An *empirical science* is one that is based on
   a. hypothesis.
   b. calculus.
   c. computer models.
   d. observed data.

2. When Earth catches up to a slower-moving outer planet and passes it in its orbit, the planet
   a. exhibits retrograde motion.
   b. slows down because it feels Earth's gravitational pull.
   c. decreases in brightness as it passes through Earth's shadow.
   d. moves into a more elliptical orbit.

3. Copernicus's model of the Solar System was superior to Ptolemy's because
   a. it had a mathematical basis that could be used to predict the positions of planets.
   b. it was much more accurate.
   c. it did not require any epicycles.
   d. it fit the telescopic data better.

4. An inferior planet is one that is
   a. smaller than Earth.
   b. larger than Earth.
   c. closer to the Sun than Earth is.
   d. farther from the Sun than Earth is.

5. The time it takes for a planet to come back to the same position relative to the Sun is called its _____ period.
   a. synodic
   b. sidereal
   c. heliocentric
   d. geocentric

6. Suppose a planet is discovered orbiting a star in a highly elliptical orbit. While the planet is close to the star it moves _____, but while it is far away it moves _____.
   a. faster; slower
   b. slower; faster
   c. retrograde; prograde
   d. prograde; retrograde

7. If a superior planet is observed from Earth to have a synodic period of 1.2 years, what is its sidereal period?
   a. 0.54 years
   b. 1.8 years
   c. 4.0 years
   d. 6.0 years

8. A net force must be acting when an object
   a. accelerates.
   b. changes direction but not speed.
   c. changes speed but not direction.
   d. all of the above

9. For Earth, $P^2/A^3 = 1.0$ (in appropriate units). Suppose a new dwarf planet is discovered that is 14 times as far from the Sun as Earth is. For this planet,
   a. $P^2/A^3 = 1.0$.
   b. $P^2/A^3 > 1.0$.
   c. $P^2/A^3 < 1.0$.
   d. one can't know the value of P^2/A^3 without more information.

10. Galileo observed that Venus had phases that correlated with its size in his telescope. From this information, you may conclude that Venus
    a. is the center of the Solar System.
    b. orbits the Sun.
    c. orbits Earth.
    d. orbits the Moon.

11. Kepler's second law says that
    a. planetary orbits are ellipses with the Sun at one focus.
    b. the square of a planet's orbital period equals the cube of its semimajor axis.
    c. net forces cause changes in motion.
    d. planets move fastest when they are closest to the Sun.

12. Suppose you read in the newspaper that a new planet has been found. Its average speed in orbit is 33 km/s. When it is closest to its star it moves at 31 km/s, and when it is farthest from its star it moves at 35 km/s. This story is in error because
    a. the average speed is far too fast.
    b. Kepler's third law says the planet has to sweep out equal areas in equal times, so the speed of the planet cannot change.
    c. Kepler's second law says the planet must move fastest when it is closest, not when it is farthest away.
    d. using these numbers, the square of the orbital period will not be equal to the cube of the semimajor axis.

13. Galileo observed that Jupiter has moons. From this information, you may conclude that
    a. Jupiter is the center of the Solar System.
    b. Jupiter orbits the Sun.
    c. Jupiter orbits Earth.
    d. some things do not orbit Earth.

14. If you start from rest and accelerate at 10 mph/s and end up traveling at 60 mph, how long did it take?
    a. 1 second
    b. 6 seconds
    c. 60 seconds
    d. 0.6 seconds

15. Planets with high eccentricity may be unlikely candidates for life because
    a. the speed varies too much.
    b. the period varies too much.
    c. the temperature varies too much.
    d. the orbit varies too much.

Thinking about the Concepts

16. Study Figure 3.1. During normal motion, does Mars move toward the east or west? Which direction does it travel when moving retrogradely? For how many days did Mars move retrogradely? If one of the martian missions were photographing *Earth* in the sky during these days, what would it have observed?

17. Copernicus and Kepler engaged in what is called empirical science. What do we mean by *empirical*?

18. Explain why the synodic period of Saturn is very close to 1 Earth year (a sketch may help).

19. Make a sketch of Earth, Venus, and the Sun in a geocentric model and in a heliocentric model. Label the Sun and Earth, and then show the changing phases of Venus over the course of one orbit of Venus. What would we observe in each model? Why was the invention of the telescope necessary to distinguish between these models?

20. Experiment with falling objects as Galileo did. Drop pairs of objects of different masses—do they reach the ground at the same time? Do they hit the ground with the same force? Does this work with a sheet of paper or a tissue—why or why not?

21. The speed of a planet in its orbit varies in its journey around the Sun. At what point in its orbit is the planet moving the fastest? At what point is it moving the slowest?

22. The orbit of the Moon around Earth also is elliptical, with an eccentricity of 0.05. How does this compare with the eccentricity of Earth's orbit? How do these elliptical orbits explain the types of solar eclipses discussed in Chapter 2?

23. Galileo came up with the concept of inertia. What do we mean by *inertia*?

24. If Kepler had lived on Mars, would he have deduced the same empirical laws for the motion of the planets? Explain.

25. What is the difference between speed and velocity? between velocity and acceleration?

26. When involved in an automobile collision, a person not wearing a seat belt will move through the car and often strike the windshield directly. Which of Newton's laws explains why the person continues forward, even though the car stopped?

27. When riding in a car, we can sense changes in speed or direction through the forces that the car applies on us. Do we wear seat belts in cars and airplanes to protect us from speed or from acceleration? Explain your answer.

28. An astronaut standing on Earth can easily lift a wrench having a mass of 1 kg, but not a scientific instrument with a mass of 100 kg. In the International Space Station, she is quite capable of manipulating both, although the scientific instrument responds much more slowly than the wrench. Explain why.

29. The Process of Science Figure illustrates that scientific ideas are always open to challenge. Construct an argument that this constant process of challenging and falsifying ideas is a strength of science, rather than a weakness.

30. How might you expect conditions on Earth to be different if the eccentricity of its orbit was 0.17 instead of 0.017?

Applying the Concepts

31. Study the graph in Figure 3.16. Is this graph linear or logarithmic? From the data on the graph, find the approximate semimajor axis and period of Saturn. Show your work.

32. Study Figure 3.19, which shows that the apparent size of Venus changes as it goes through phases. Approximately how many times larger is Venus in the sky at the tiniest crescent than at the gibbous phase shown? Therefore, approximately how many times closer is Venus to us at the phase of that tiniest crescent than at the gibbous phase?

33. Suppose a new dwarf planet is discovered orbiting the Sun with a semimajor axis of 50 AU. What would be the orbital period of this new dwarf planet?

34. Planet Neptune's orbital period is 164.8 years. What is the semimajor axis of its orbit? How much time passes between oppositions of Neptune?

35. Dwarf planet Ceres is located at 2.77 AU from the Sun. Its synodic period is 1.278 years.
    a. Use Working It Out 3.1 to find the sidereal period in years.
    b. Use Kepler's law to find the sidereal period in years.
    c. Compare your results for (a) and (b).

36. Suppose you read online that "experts have discovered a new planet with a distance from the Sun of 2 AU and a period of 3 years." Use Kepler's third law to argue that this is impossible.

37. Show, as Galileo did, that Kepler's third law applies to the four moons of Jupiter that he discovered by calculating P^2 divided by A^3 for each moon. (Data on the moons can be found in Appendix 4.)

38. In a period of 3 months, a planet travels 30,000 km with an average speed of 3.8 m/s. Some time later, the same planet travels 65,000 km in 3 months. How fast is the planet traveling at this later time? During which period is the planet closer to the Sun?

39. If you were on Mars, how often would you see retrograde motion of *Earth* in the martian night sky? (You can view a simulation at http://mars.jpl.nasa.gov/allaboutmars/nightsky/retrograde/.)

40. The elliptical orbit of a comet recently visited by a spacecraft is 1.24 AU from the Sun at its closest approach and 5.68 AU from the Sun at its farthest.
    a. Sketch the orbit of the comet. When is it moving fastest? When is it moving slowest?
    b. What is the semimajor axis of its orbit? How long does it take to go around the Sun?
    c. What is the distance from the Sun to the "center" of the ellipse? What is the eccentricity of the comet's orbit?

41. You are driving down a straight road at a speed of 90 km/h, and you see another car approaching you at a speed of 110 km/h along the road.
    a. Relative to your own frame of reference, how fast is the other car approaching you?
    b. Relative to the other driver's frame of reference, how fast are you approaching the other driver's car?

42. During the latter half of the 19th century, a few astronomers thought there might be a planet circling the Sun inside Mercury's orbit. They even gave it a name: Vulcan. We now know that Vulcan does not exist. If a planet with an orbit one-fourth the size of Mercury's actually existed, what would be its orbital period relative to that of Mercury?

43. Suppose you are pushing a small refrigerator of mass 50 kg on wheels. You push with a force of 100 N.
    a. What is the refrigerator's acceleration?
    b. Assume the refrigerator starts at rest. How long will the refrigerator accelerate at this rate before it gets away from you (that is, before it is moving faster than you can run—of the order 10 m/s)?

44. If a 100-kg astronaut pushes on a 5,000-kg satellite and the satellite experiences an acceleration of 0.1 m/s², what is the acceleration experienced by the astronaut in the opposite direction?

45. Sketch the orbit of Mars using the information provided in the "Origins: Planets and Orbits" section of the chapter for the closest and farthest distances of Mars from the Sun.
    a. What is its major axis? What is its semimajor axis?
    b. What is the distance from the "center" of the orbit to the Sun? Compute the eccentricity of the orbit. Compare this with the eccentricity of Earth's orbit.

USING THE WEB

46. Go to the Web page "This Week's Sky at a Glance" (http://skyandtelescope.com/observing/ataglance) at the *Sky & Telescope* magazine website. Which planets are visible in your sky this week? Why are Mercury and Venus visible only in the morning before sunrise or in the evening just after sunset? Before telescopes, how did people know the planets were different from the stars?

47. Look up the dates for the next opposition of Mars, Jupiter, or Saturn. One source is the NASA "Sky Events Calendar" at http://eclipse.gsfc.nasa.gov/SKYCAL/SKYCAL.html. Check only the "Planet Events" box in "Section 2: Sky Events"; and in Section 3, generate a Sky Events Calendar for the year. If you are coming up on an opposition, take pictures of the planet over the next few weeks. Can you see its position move in retrograde motion with respect to the background stars?

48. Refer to the Web page from question 47 to find the current observational positions of all the planets.
    a. Which ones are in or near to conjunction, opposition, or greatest elongation?
    b. Which are visible in the morning sky? in the evening sky?
    c. To connect when we see the planets on Earth with the physical alignments of the planets with Earth and the Sun in space, sketch the Solar System with Earth, Sun, and planets as it looks today from "above." Check your result using NASA's "Solar System Simulator" (http://space.jpl.nasa.gov): set it for "Show Me Solar System" as seen from above, and look at the field of view of 2°, 20°, and 45° to see the inner and then the outer planets. Does the simulator agree with your sketch?

49. Go to the Museo Galileo website and view the exhibit on Galileo's telescope (http://www.museogalileo.it/en/explore/exhibitions/pastexhibitions/galileostelescope.html). What did his telescope look like? What other instruments did he use? From the museum page you can link to short videos (in English) on his science and his trial (http://catalogue.museogalileo.it/index/VideoIndexByThematicArea.html#s7). For example, click on "Galileo's micrometer": How did he measure the separation of the moons from Jupiter? How did this measurement allow him to show that the moons obeyed Kepler's law? Why is Galileo often considered the first modern scientist? Why is his middle finger on display in the museum?

50. Go to the online "Extrasolar Planets Encyclopedia" (http://exoplanet.eu/catalog).
    a. Find a planet with an orbital period similar to that of Earth. What is the semimajor axis of its orbit? If it is very different from 1 AU, then the mass of the star is different from that of the Sun. Click on the star name in the first column to see the star's mass. What is the orbital eccentricity?
    b. Click on "Planet" to sort by name, and select a star with multiple planets. Verify that Kepler's third law applies by showing that the value of P^2/A^3 is about the same for each of the planets of this star. How eccentric are the orbits of the multiple planets?

smartwork

If your instructor assigns homework in SmartWork, access your assignments at digital.wwnorton.com/astro5.

EXPLORATION

digital.wwnorton.com/astro5

In this Exploration, we will examine how Kepler's laws apply to the orbit of Mercury. Visit the Student Site at the Digital Landing page, and open the Planetary Orbit Simulator applet. This simulator animates the orbits of the planets, enabling you to control the simulation speed, as well as a number of other parameters. Here we focus on exploring the orbit of Mercury, but you may wish to spend some time examining the orbits of other planets as well.

Kepler's First Law

To begin exploring the simulation, in the "Orbit Settings" panel, use the drop-down menu next to "set parameters for" to select "Mercury" and then click "OK." Click the "Kepler's 1st Law" tab at the bottom of the control panel. Use the radio buttons to select "show empty focus" and "show center."

1 How would you describe the shape of Mercury's orbit?

_____

_____

_____

Deselect "show empty focus" and "show center," and select "show semiminor axis" and "show semimajor axis." Under "Visualization Options," select "show grid."

2 Use the grid markings to estimate the ratio of the semiminor axis to the semimajor axis.

_____

_____

_____

3 Calculate the eccentricity of Mercury's orbit from this ratio using $e = [1 - (\text{Ratio})^2]^{1/2}$.

_____

_____

_____

Kepler's Second Law

Click on "reset" near the top of the control panel, set parameters for Mercury, and click "OK." Then click on the "Kepler's 2nd Law" tab at the bottom of the control panel. Slide the "adjust size" slider to the right, until the fractional sweep size is $\frac{1}{8}$.

Click on "start sweeping." The planet moves around its orbit, and the simulation fills in area until one-eighth of the ellipse is filled. Click on "start sweeping" again as the planet arrives at the rightmost point in its orbit (that is, at the point in its orbit farthest from the Sun). You

may need to slow the animation rate using the slider under "Animation Controls." Click on "show grid" under the visualization options. (If the moving planet annoys you, you can pause the animation.) One easy way to estimate an area is to count the number of squares.

4 Count the number of squares in the yellow area and in the red area. You will need to decide what to do with fractional squares. Are the areas the same? Should they be?

_____

_____

Kepler's Third Law

Click on "reset" near the top of the control panel, set parameters for Mercury, and then click on the "Kepler's 3rd Law" tab at the bottom of the control panel. Select "show solar system orbits" in the "Visualization Options" panel. Study the graph. Use the eccentricity slider to change the eccentricity of the simulated planet. Make the eccentricity first smaller and then larger.

5 Did anything in the graph change?

_____

_____

6 What do your observations of the graph tell you about the dependence of the period on the eccentricity?

_____

_____

Set parameters back to those for Mercury. Now use the semimajor axis slider to change the semimajor axis of the simulated planet.

7 What happens to the period when you make the semimajor axis smaller?

_____

_____

8 What happens when you make it larger?

_____

_____

9 What do these results tell you about the dependence of the period on the semimajor axis?

_____

_____

4 Gravity and Orbits

In this chapter, we explore the physical laws that explain the regular patterns in the motions of the planets. Because the Sun is far more massive than all the other parts of the Solar System combined, its gravity shapes the motions of every object in its vicinity, from the almost circular orbits of some planets to the extremely elongated orbits of comets.

LEARNING GOALS

By the conclusion of this chapter, you should be able to:

LG 1 Explain the elements of Newton's universal law of gravitation.

LG 2 Use the laws of motion and gravitation to explain planetary orbits.

LG 3 Explain how tidal forces from the Sun and Moon create Earth's tides.

LG 4 Describe the effects of tidal forces on solid bodies.

The International Space Station in orbit around Earth. ▶▶▶

What keeps a space station in orbit?

4.1 Gravity Is a Force between Any Two Objects Due to Their Masses

In Chapter 3, we explored Kepler's work on the movement of the planets around the Sun and Newton's laws of motion. In this chapter, we build on those concepts to look at Newton's universal law of gravitation. Although some of the properties of gravity were observed before Newton, his work connected the everyday phenomenon of falling objects to the motion of the planets around the Sun. Newton's theory of gravity combined Kepler's empirical laws and Newton's own laws of motion.

Gravity, Mass, and Weight

Many forces that we see in everyday life involve direct contact between objects. The cue ball on a pool table slams into the eight ball, knocking it into the pocket. The shoe of the child pushing a scooter presses directly against the surface of the pavement. When there is physical contact between two objects, the source of the forces between them is easy to see.

If you drop a ball, it falls toward the ground. The ball picks up speed as it falls, accelerating downward toward Earth. Newton's second law says that where there is acceleration, there is force. But where is the force that causes the ball to accelerate? The ball falling toward Earth is an example of a different kind of force, one that acts at a distance across the intervening void of space. The ball falling toward Earth is accelerating in response to the force of gravity. **Gravity**, one of the fundamental forces of nature, is the mutually attractive force between objects with mass.

Recall from Chapter 3 that Galileo discovered that all freely falling objects accelerate toward Earth at the same rate, regardless of their mass. Drop a marble and a book at the same time and from the same height, and they will hit the ground together. Note that air resistance becomes a factor at higher speeds, but it is negligible for dense, slow objects. The acceleration of falling objects due to gravity near the surface of Earth, also measured experimentally by Galileo, is usually written as g (lowercase) and has an average value across the surface of Earth of 9.8 meters per second squared (m/s^2). The value of g varies slightly across Earth's surface, ranging from 9.78 m/s^2 at the equator to 9.83 m/s^2 at the poles. This variation exists because Earth is not a perfect sphere: its rotation makes it flatter at the poles, so the radius of Earth is smaller at the poles.

After working out the laws governing the motion of objects, Newton realized that if all objects, no matter what their mass, fall with the same acceleration, then the gravitational *force* on an object must be determined by the object's *mass*.

Recall Newton's second law from Chapter 3: acceleration equals force divided by mass, or $a = F/m$. The acceleration due to gravity can be the same for all objects only if the value of the force divided by the mass is the same for all objects. In other words, an object twice as massive has double the gravitational force acting on it; an object 3 times as massive has triple the gravitational force acting on it, and so on.

The gravitational force acting on an object attracted by a planet is called the object's **weight**. On the surface of Earth, weight equals mass multiplied by the acceleration of gravity at Earth's surface, g. In common language, we often use weight and mass interchangeably. To be more scientifically precise, astronomers use *mass* to refer to the amount of stuff in an object and *weight* to refer to the force exerted on that object by the planet's gravitational pull. Your mass is the same no matter what planet or moon you are on, but your weight will be different:

$$F_{weight} = m \times g$$

where F_{weight} is an object's weight in newtons (N), the metric unit of force; m is the object's mass in kilograms (kg); and g is Earth's constant for acceleration due to gravity, 9.8 m/s². On Earth, an object with a *mass* of 1 kg has a *weight* of 9.8 N. As illustrated in **Figure 4.1**, on the Moon the acceleration due to gravity is about 6 times lower at 1.6 m/s², so the 1-kg mass would have a weight of 1.6 N. On the Moon, your weight would be about one-sixth of your weight on Earth.

Newton's Law of Gravity

As Newton told the story, he saw an apple fall from a tree to the ground, and he reasoned that if gravity is a force that depends on mass, then there should be a gravitational force between *any* two masses, including between a falling apple and Earth. This great insight came from applying his third law of motion to gravity. Recall that Newton's third law states that for every force there is an equal and opposite force. Therefore, if Earth exerts a force of 9.8 N on a 1-kg mass sitting on its surface, then that 1-kg mass must also exert a force of 9.8 N on Earth. Drop a 7-kg bowling ball and it falls toward Earth, but at the same time Earth falls toward the 7-kg bowling ball. The reason we do not notice the motion of Earth is that Earth is very massive, so it has a lot of resistance to changes in its motion. In the time it takes a 7-kg bowling ball to fall to the ground from a height of 1 kilometer (km), Earth has "fallen" toward the bowling ball by only a tiny fraction of the size of an atom.

Newton reasoned that if doubling the mass of any object doubles the gravitational force between the object and Earth, then doubling the mass of Earth ought to do the same thing. In short, the gravitational force between Earth and an object must be equal to the product of the two masses multiplied by *something*:

Gravitational force = Something × Mass of Earth × Mass of object

If the mass of the object is 2 times greater, then the force of gravity will be 2 times greater. Likewise, if the mass of Earth happened to be 3 times what it is, the force of gravity would also have to be 3 times greater. If both the mass of Earth and the mass of the object were greater by these amounts, the gravitational force would increase by a factor of 2 × 3, or 6 times. Because objects fall toward the center of Earth, we know that gravity is an attractive force acting along a line between the two masses.

If gravity is a force that depends on mass, then there should be a gravitational force between *any* two masses. Suppose we have two masses—call them mass 1 and mass 2, or m_1 and m_2 for short. The gravitational force between them is *something* multiplied by the product of the masses:

Gravitational force between two objects = Something × m_1 × m_2

We have gotten this far just by combining Galileo's observations of falling objects with (1) Newton's laws of motion and (2) Newton's belief that Earth is a

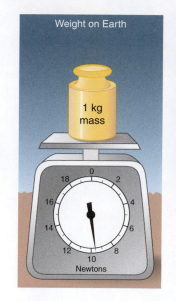
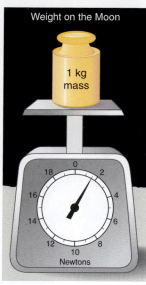

Figure 4.1 On the Moon, a mass of 1 kg has $\frac{1}{6}$ the weight (displayed in newtons) that it has on Earth.

mass just like any other mass. But what about that "something" in the previous expression?

Kepler had already thought about this question. He reasoned that because the Sun is the focal point for planetary orbits, the Sun must exert an influence over the motions of the planets. Kepler speculated that this influence must grow weaker with distance from the Sun, because the planets closer to the Sun moved much faster than the farther ones. Kepler did not know about forces or inertia or gravity as the cause of celestial motion, but he thought that geometry alone suggested how this solar "influence" might change for planets progressively farther from the Sun.

To see why the influence must become weaker, imagine you have a certain amount of paint to spread over the surface of a sphere. If the sphere is small, you will get a thick coat of paint. But if the sphere is larger, the paint has to spread farther, and you will get a thinner coat. The surface area of a sphere depends on the square of the sphere's radius. Double the radius of a sphere, and the sphere's surface area becomes 4 times what it was. If you paint this new, larger sphere, the paint must cover 4 times as much area, and the thickness of the paint will be only a fourth of what it was on the smaller sphere. Triple the radius of the sphere: the sphere's surface will be 9 times larger and the coat of paint will be only one-ninth as thick.

Kepler reasoned that as the influence of the Sun extended farther and farther into space, it would have to spread out to cover the surface of a larger and larger imaginary sphere centered on the Sun. The influence of the Sun should diminish with the square of the distance from the Sun—a relationship known as an **inverse square law**.

Kepler had an interesting idea, but not a scientific hypothesis with testable predictions. He lacked an explanation for how the Sun influences the planets and the mathematical tools to calculate how an object would move under such an influence. Newton had both. If gravity is a force between *any* two objects, then there should be a gravitational force between the Sun and each of the planets. If this gravitational force were the same as Kepler's "influence," then gravity might behave according to an inverse square law.

Newton's expression for gravity came to look like this:

$$\text{Gravitational force between two objects} = \text{Something} \times \frac{m_1 \times m_2}{(\text{Distance between objects})^2}$$

There is still a "something" left in this expression, and that something is a constant of proportionality. This constant determines the strength of gravity between objects, and it is the same for all pairs of objects. Newton named it the **universal gravitational constant**, written as G (uppercase). Newton estimated this gravitational constant G by using Galileo's measurement of g, estimates of Earth's radius, and a guess at the mass of Earth by assuming it had about the same density as typical rocks. It was not until many years later that the actual value of G was first measured. Today the value of G is accepted as 6.67×10^{-11} N m^2/kg^2 or its equivalents: 6.67×10^{-11} m^3/(kg s^2) or 6.67×10^{-20} km^3/(kg s^2).

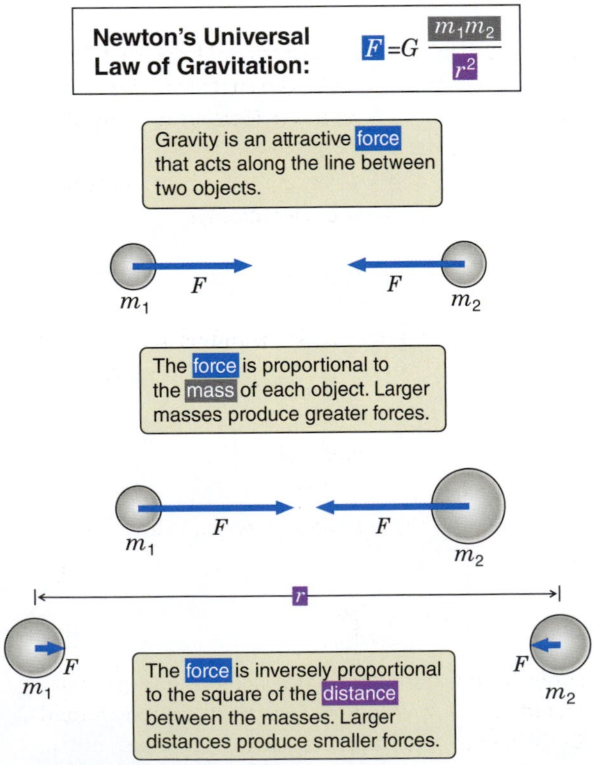

A Universal Law of Gravitation

Newton's **universal law of gravitation**, illustrated in **Figure 4.2**, states that gravity is a force between any two objects having mass and has the following properties:

1. It is an attractive force acting along a straight line between the two objects.

Newton's Universal Law of Gravitation: $F = G \dfrac{m_1 m_2}{r^2}$

Gravity is an attractive force that acts along the line between two objects.

m_1 F F m_2

The force is proportional to the mass of each object. Larger masses produce greater forces.

m_1 F F m_2

r

F m_1 The force is inversely proportional to the square of the distance between the masses. Larger distances produce smaller forces. F m_2

Figure 4.2 Gravity is an attractive force between two objects. The force of gravity depends on the masses of the objects, m_1 and m_2, and the distance, r, between them.

2. It is proportional to the mass of one object (m_1) multiplied by the mass of the other object (m_2). If we double m_1, then the force (F) increases by a factor of 2. Likewise, if we double m_2, F increases by a factor of 2.

3. It is inversely proportional to the square of the distance r between the centers of the two objects: As seen in **Figure 4.3**, if we double r, F decreases by a factor of 4. If we triple r, F falls by a factor of 9.

Written as a mathematical formula, the universal law of gravitation states

$$F_{grav} = G \times \frac{m_1 \times m_2}{r^2}$$

where F_{grav} is the force of gravity between two objects, m_1 and m_2 are the masses of objects 1 and 2, r is the distance between the centers of mass of the two objects, and G is the universal gravitational constant. The relationship between the force of gravity and the masses and separation distance between two objects is further explored in **Working It Out 4.1**.

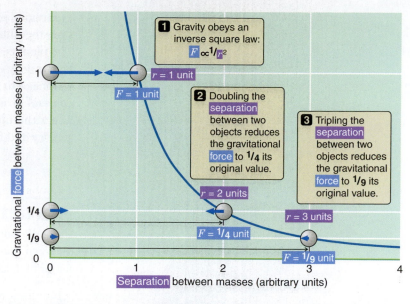

Figure 4.3 As two objects move apart, the gravitational force between them decreases by the inverse square of the distance between them.

4.1 **Working It Out** Playing with Newton's Laws of Motion and Gravitation

For any two objects, the force of gravity is directly proportional to the masses and inversely proportional to the *square* of the distance between them. Let's look at a few examples of how to use this equation:

Changing the Distance

How would the gravitational force between Earth and the Moon change if the distance between them were doubled? In this example, the masses of the Sun and Moon stay the same, and r becomes $2r$. We can calculate how the force changes by writing the equation for distance r and again for distance $2r$, and then taking a ratio to compare them:

$$\frac{F_{grav\ at\ distance\ 2r}}{F_{grav\ at\ distance\ r}} = \frac{G \times \dfrac{M_{Earth}M_{Moon}}{(2r)^2}}{G \times \dfrac{M_{Earth}M_{Moon}}{r^2}}$$

We can cancel out the constant G and the masses of Earth and the Moon, which do not change. Then you need to multiply both the numerator and denominator of the fraction by r^2, and remember that both the 2 and the r get squared in $(2r)^2 = 4r^2$. The equation becomes

$$\frac{F_{grav\ at\ distance\ 2r}}{F_{grav\ at\ distance\ r}} = \frac{\cancel{G} \times \cancel{M_{Earth}} \times \cancel{M_{Moon}}}{\cancel{G} \times \cancel{M_{Earth}} \times \cancel{M_{Moon}}} \times \frac{r^2}{(2r)^2}$$

$$= \frac{\cancel{r^2}}{4\cancel{r^2}} = \frac{1}{4}$$

Doubling the distance reduced the force by a factor of 4; that is, the force is $\frac{1}{4}$ as strong.

 Nebraska Simulation: Gravity Algebra

Gravitational Acceleration

There are two ways to think about the gravitational force that Earth exerts on an object with mass m located on the surface of Earth. Recall Newton's second law of motion: $F = m \times a$. Here we are considering the gravitational force and the acceleration due to gravity, or $F_{grav} = m \times g$. The other way to think about the force is from the perspective of the universal law of gravitation, which says

$$F_{grav} = G \times \frac{M_{Earth} \times m}{R_{Earth}^2}$$

Here, M_{Earth} is the mass of Earth, and R_{Earth} is the radius of Earth. The two expressions describing this force must be equal to each other. Therefore,

$$\cancel{m} \times g = G \times \frac{M_{Earth} \times \cancel{m}}{R_{Earth}^2}$$

The mass m is on both sides of the equation, so we can cancel it out. The equation then becomes

$$g = G \times \frac{M_{Earth}}{R_{Earth}^2}$$

The expression shows that the gravitational acceleration (g) experienced by an object of mass m on the surface of Earth is determined by the mass of Earth and by the radius of Earth. The mass of the object itself (m) appears nowhere in this expression, so changing m has no effect on the gravitational acceleration of an object on Earth.

Gravity pulls you toward the center of Earth. Gravity holds the planets and stars together and keeps the thin blanket of air we breathe close to Earth's surface. The planets, including Earth, orbit around the Sun, and gravity holds them in orbit. Gravity caused a vast interstellar cloud of gas and dust to collapse 4.5 billion years ago to form the Sun, Earth, and the rest of the Solar System. Gravity binds colossal groups of stars into galaxies. Gravity shapes space and time, and it can affect the ultimate fate of the universe. We will return often to the concept of gravity and find it central to an understanding of the universe.

(a)

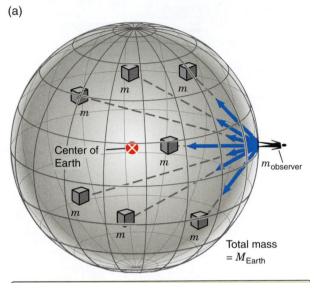

An object on the surface of a spherical mass (such as Earth) feels a gravitational attraction toward each small part of the sphere.

(b)

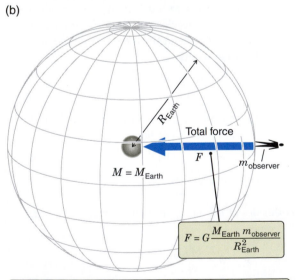

$$F = G \frac{M_{Earth}\, m_{observer}}{R^2_{Earth}}$$

The net force is the same as if we scooped up the mass of the entire sphere and concentrated it at a point at the center.

Figure 4.4 Outside a sphere, the net gravitational force due to a spherical mass is the same as the gravitational force from the same mass concentrated at a point at the center of the sphere.

CHECK YOUR UNDERSTANDING 4.1

If the distance between Earth and the Sun were cut in half, the gravitational force between these two objects would: (a) decrease by a factor of 4; (b) decrease by a factor of 2; (c) increase by a factor of 2; (d) increase by a factor of 4.

Gravity Differs from Place to Place within an Object

You can think of Earth as a collection of small masses, each of which feels a gravitational attraction toward every other small part of Earth. The mutual gravitational attraction that occurs among all parts of the same object is called self-gravity.

As you sit reading this book, you are exerting a gravitational attraction on every other fragment of Earth, and every other fragment of Earth is exerting a gravitational attraction on you. Your gravitational interaction is strongest with the parts of Earth closest to you. The parts of Earth that are on the other side of our planet are much farther from you, so their pull on you is correspondingly less.

The net effect of all these forces is to pull you (or any other object) toward the center of Earth. If you drop a hammer, it falls directly toward the ground. Because Earth is nearly spherical, for every piece of Earth pulling you toward your right, a corresponding piece of Earth is pulling you toward your left with just as much force. For every piece of Earth pulling you forward, a corresponding piece of Earth is pulling you backward. As you can see in **Figure 4.4a**, because Earth is almost spherically symmetric, all of these "sideways" forces cancel out, leaving behind an overall force that points toward Earth's center.

Some parts of Earth are closer to you and others are farther away, but there is an average distance between you and all of the small fragments of Earth that are pulling on you. This average distance is the distance between you and the center of Earth. As illustrated in Figure 4.4b, the overall pull that you feel is the same as it would be if all the mass of Earth was concentrated at a single point located at the very center of the planet.

This relationship is true for any spherically symmetric object. Outside the object, the gravity from such an object behaves as if all the mass of that object were concentrated at a point at its center. This relationship will be important in many applications. For example, when you estimate your weight on another planet, you are calculating the force of gravity between you and the planet. The "distance" in the gravitational equation will be the distance between you and the center of the planet, which is just the radius of the planet.

CHECK YOUR UNDERSTANDING 4.2

If Earth shrank to a smaller radius but kept the same mass, would the gravitational force between Earth and the Moon become: (a) smaller; (b) larger; or (c) stay the same? Would everyone's weight on Earth: (a) increase; (b) decrease; or (c) stay the same?

4.2 An Orbit Is One Body "Falling around" Another

Kepler's laws on the motions of planets enabled astronomers to predict the positions of the planets accurately, but these laws did not explain why the planets behave as they do. Newton's work explained why planets orbit the Sun.

Newton used his laws of motion and his proposed law of gravity to predict the paths of planetary orbits. His calculations showed that these orbits should be ellipses with the Sun at one focus, that planets should move faster when closer to the Sun, and that the square of the period of a planet's orbit should vary as the cube of the semimajor axis of that elliptical orbit. Newton's universal law of gravitation *predicted* that planets should orbit the Sun in just the way that Kepler's empirical laws described. By explaining Kepler's laws, Newton found important corroboration for his law of gravitation.

Gravity and Orbits

Newton's laws tell us how forces change an object's motion and how objects interact with each other through gravity. To know where an object will be at any given time, we carefully have to "add up" the object's motion over time. Newton invented calculus to do this, but we will aim just for a conceptual understanding.

In Newton's time, the closest thing to making a heavy object fly was shooting cannonballs out of a cannon, so he used cannonballs in "thought experiments" about planetary motions. If one drops a cannonball, it falls directly to the ground, like any other mass does. However, as you can see in **Figure 4.5a**, a cannonball fired out of a cannon that is level with the ground behaves differently. The cannonball still falls to the ground in the same amount of time as it does when it is dropped, but while falling it also travels over the ground, following a curved path that carries it a horizontal distance before it finally lands. As shown in Figure 4.5b, the faster the cannonball moves when it is fired from the cannon, the farther it will go before it hits the ground.

In the real world this experiment reaches a natural limit. To travel through air, the cannonball must push the air out of its way—an effect normally referred to as *air resistance*—which slows it down. But because this is only a thought experiment, we can ignore such real-world complications. Instead imagine that, having inertia, the cannonball continues along its course until it runs into something. The faster the cannonball moves when it is fired, the farther it goes before hitting the ground. If the cannonball flies far enough, Earth's surface curves out from under it, as shown in Figure 4.5c. As illustrated in Figure 4.5d, eventually a point is reached where the cannonball is flying so fast that the surface of Earth curves away from the cannonball at exactly the same rate at which the cannonball is falling toward Earth. When this occurs, the cannonball, which always falls

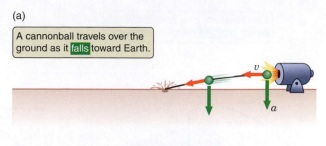
(a) A cannonball travels over the ground as it falls toward Earth.

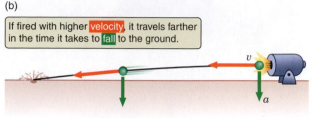
(b) If fired with higher velocity, it travels farther in the time it takes to fall to the ground.

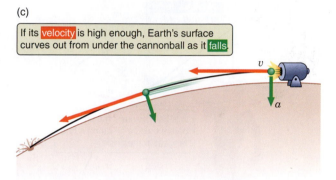
(c) If its velocity is high enough, Earth's surface curves out from under the cannonball as it falls.

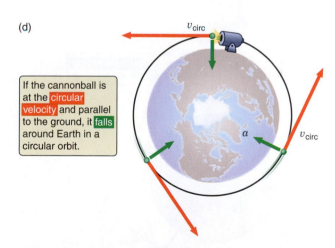
(d) If the cannonball is at the circular velocity and parallel to the ground, it falls around Earth in a circular orbit.

Figure 4.5 Newton realized that a cannonball fired at the right speed would fall around Earth in a circle. Velocity (*v*) is indicated by a red arrow and acceleration (*a*) by a green arrow.

▶❚❚ **AstroTour:** Newton's Laws and Universal Gravitation

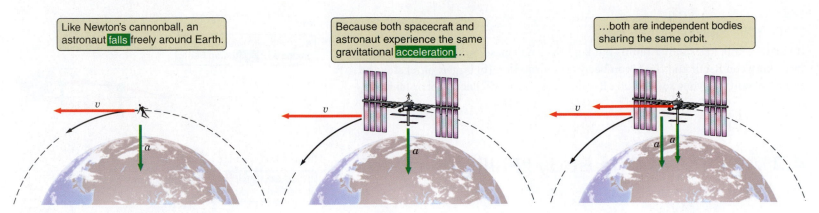

Figure 4.6 A "weightless" astronaut has not escaped Earth's gravity. Rather, an astronaut and a spacecraft share the same orbit as they fall around Earth together.

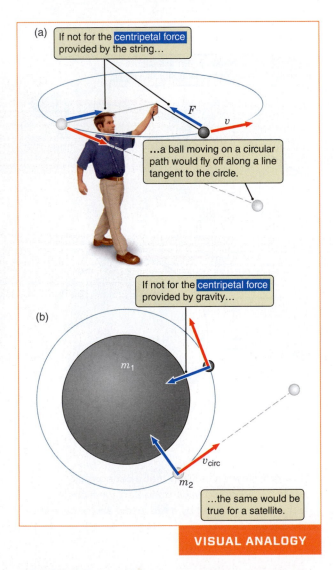

Figure 4.7 (a) A string provides the centripetal force that keeps a ball moving in a circle. (We are ignoring the smaller force of gravity that also acts on the ball.) (b) Similarly, gravity provides the centripetal force that holds a satellite in a circular orbit.

toward the center of Earth, is, in a sense, falling around the world. An **orbit** is the path of one object that freely falls around another.

Why do astronauts appear to float freely about the cabin of a spacecraft? It is not because they have escaped Earth's gravity; it is Earth's gravity that holds them in their orbit. Instead the answer lies in Galileo's early observation that all objects fall with the same acceleration, regardless of their mass. The astronauts and the spacecraft are both in orbit around Earth, moving in the same direction, at the same speed, and experiencing the same gravitational acceleration, so they fall around Earth together. **Figure 4.6** demonstrates this point. The astronaut is orbiting Earth just as the spacecraft is orbiting Earth. On the surface of Earth, your body tries to fall toward the center of Earth, but the ground gets in the way. You experience your weight when you are standing on Earth because the ground pushes on you to oppose the force of gravity, which pulls you downward. In the spacecraft, however, nothing interrupts the astronaut's fall, because the spacecraft is falling around Earth in just the same orbit. The astronaut is in **free fall**, falling freely in Earth's gravity. The **Process of Science Figure** illustrates the universality of Newton's law of gravitation.

What Velocity Is Needed to Reach Orbit?

How fast must Newton's cannonball be fired for it to fall around the world? The cannonball would be in **uniform circular motion**, which means it moves along a circular path at constant speed. This type of motion is discussed in more depth in Appendix 8. Another example of uniform circular motion is a ball whirling around your head on a string, illustrated in **Figure 4.7a**. If you let go of the string, the ball will fly off in a straight line in whatever direction it is traveling at the time, just as Newton's first law predicts for an object in motion. The string prevents the ball from flying off by constantly changing the direction the ball is traveling. The central force of the string on the ball is called a **centripetal force**: a force toward the center of the circle. Using a more massive ball, speeding up its motion, or making the string shorter so that the turn is tighter all increase the force needed to keep a ball moving in a circle.

In the case of Newton's cannonball (or a satellite), there is no string to hold the ball in its circular motion. Instead, the force is provided by gravity, as illustrated in Figure 4.7b. The force of gravity must be just enough to keep the satellite moving on its circular path. Because this force has a specific strength, it follows that

Process of Science

UNIVERSALITY

The laws of physics are the same everywhere and at all times. The principle underlies our understanding of the natural world.

Galileo determined that all objects have the same gravitational acceleration.

Newton's law of universal gravitation extended this observation to the Solar System.

Apollo 15 **commander David Scott tested the law with a feather and a hammer on the Moon. With no air resistance, the feather and the hammer fell at the same rate.**

The same physical laws apply to falling objects, to planets orbiting the Sun, to stars orbiting within the galaxy, and to galaxies orbiting each other.

the satellite must be moving at a particular speed around the circle, which we call its **circular velocity** (v_{circ}). If the satellite were moving at any other velocity, it would not be moving in a circular orbit. Remember the cannonball: if it is moving too slowly, it will drop below the circular path and hit the ground. Similarly, if the cannonball is moving too fast, its motion will carry it above the circular orbit. Only a cannonball moving at just the right velocity—the circular velocity—will fall around Earth on a circular path (see Figure 4.5d).

Newton's thought experiment became a reality in 1957, when the Soviet Union launched the first human-made object to orbit Earth. They used a rocket to lift Sputnik 1, an object about the size of a basketball, high enough above Earth's upper atmosphere that air resistance wasn't an issue. Sputnik 1 was given a high enough speed that it fell around Earth, just as Newton's imaginary cannonball.

When one object is falling around a much more massive object, we say that the less massive object is a **satellite** of the more massive object. Planets are satellites of the Sun, and moons are natural satellites of planets. Newton's imaginary cannonball and Sputnik 1 were satellites (*sputnik* means "satellite" in Russian). A spacecraft orbiting Earth and the astronauts inside of it are independent satellites of Earth that conveniently happen to share the same orbit.

The Shape of Orbits

Some Earth satellites travel a circular path at constant speed. Just like the ball on the string, satellites traveling at the circular velocity remain the same distance from Earth at all times, neither speeding up nor slowing down in orbit. But what if the satellite then fired its rockets and started traveling *faster* than the circular velocity? The pull of Earth is as strong as ever, but because the satellite has greater speed, its path is not bent by Earth's gravity sharply enough to hold it in a circle. So the satellite begins to climb above a circular orbit.

As the distance between the satellite and Earth increases, the satellite slows down. Think about a ball thrown upwards into the air, illustrated in **Figure 4.8a**. As the ball climbs higher, the pull of Earth's gravity opposes its motion, slowing the ball down. The ball climbs more and more slowly until its vertical motion

Nebraska Simulation: Earth Orbit Plot

▶❚❚ AstroTour: Elliptical Orbit

Figure 4.8 (a) A ball thrown into the air slows as it climbs away from Earth and then speeds up as it heads back toward Earth. (b) A planet on an elliptical orbit around the Sun does the same thing. (Although no planet has an orbit as eccentric as the one shown here, the orbits of comets can be far more eccentric.)

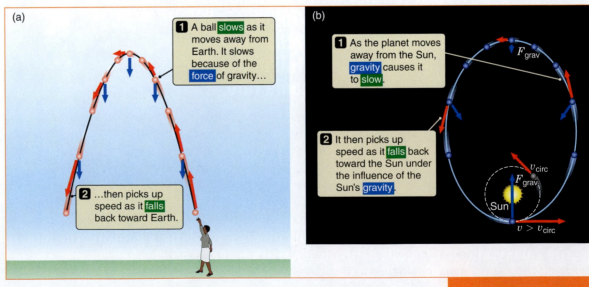

(a)
1 A ball slows as it moves away from Earth. It slows because of the force of gravity…

2 …then picks up speed as it falls back toward Earth.

(b)
1 As the planet moves away from the Sun, gravity causes it to slow.

2 It then picks up speed as it falls back toward the Sun under the influence of the Sun's gravity.

F_{grav}

v_{circ}

F_{grav}

Sun

$v > v_{circ}$

VISUAL ANALOGY

stops for an instant and then is reversed; the ball then begins to fall back toward Earth, speeding up along the way. A satellite does exactly the same thing. As the satellite climbs above a circular orbit and begins to move away from Earth, Earth's gravity opposes the satellite's outward motion, slowing the satellite down. The farther the satellite is from Earth, the more slowly the satellite moves—just like the ball thrown into the air. Also just like the ball, the satellite reaches a maximum height on its curving path and then begins falling back toward Earth, while Earth's gravity speeds it up as it gets closer and closer to Earth. The satellite's orbit has changed from circular to elliptical.

Any object in an elliptical orbit, including a planet orbiting the Sun, will therefore move faster when it is closer to the object it is orbiting due to gravity. Recall from Chapter 3 that Kepler's second law says that a planet moves fastest when it is closest to the Sun and slowest when it is farthest from the Sun. As shown in Figure 4.8b, planets lose speed as they pull away from the Sun and then gain that speed back as they fall inward toward the Sun.

Newton's laws do more than explain Kepler's laws: they predict different types of orbits beyond Kepler's empirical experience. **Figure 4.9a** shows a series of satellite orbits, each with the same point of closest approach to Earth but with different velocities at that point, as indicated in Figure 4.9b. The greater the speed a satellite has at its closest approach to Earth, the farther the satellite is able to pull away from Earth, and the more eccentric its orbit becomes. As long as it remains elliptical, no matter how eccentric, the orbit is called a **bound orbit** because the satellite is gravitationally bound to the object it is orbiting.

In this sequence of faster and faster satellites there comes a point at which the satellite is moving so fast that gravity is unable to reverse its outward motion, so the object travels away from Earth, never to return. The lowest speed at which an object can permanently leave the gravitational grasp of another mass is called the **escape velocity**, v_{esc}. Once a satellite's velocity at closest approach equals or exceeds v_{esc}, and it is no longer gravitationally bound to the object it was orbiting, we say it is in an **unbound orbit**.

As Figure 4.9 shows, an object with a velocity *less* than the escape velocity (v_{esc}) will be on an elliptically shaped orbit and will follow the same path over and over again. Unbound orbits do not close like an ellipse (see Figure 4.9a). An object such as a comet on an unbound orbit makes only a single pass around the Sun and then continues away from the Sun into deep space, never to return. Circular velocity and escape velocity are further explored in **Working It Out 4.2**.

Measuring Mass Using Newton's Version of Kepler's Law

Newton's calculations opened up an entirely new way of investigating the universe. He showed that the same physical laws that describe the flight of a cannonball on Earth—or the legendary apple falling from a tree—also describe the motions of the planets through the heavens. His laws of motion and gravitation predict all three of Kepler's empirical laws of planetary motion. Newton's version of Kepler's laws can be used to measure the mass of the Sun from the orbit of Earth, as seen in **Working It Out 4.3**.

When a much lower mass object such as Earth is orbiting a much more massive object such as the Sun, the Sun's gravity has a strong influence on Earth, but Earth's gravity has little effect on the Sun. Therefore, it is a good approximation to say that the Sun remains motionless as Earth orbits around it.

(a) Representative orbits

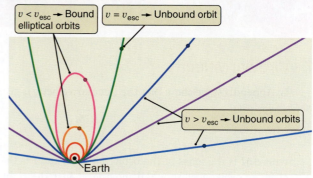

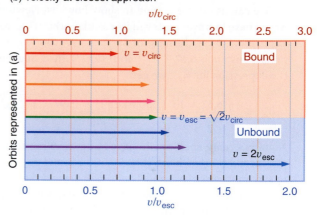

(b) Velocity at closest approach

Figure 4.9 (a) A range of different orbits that share the same point of closest approach but differ in velocity at that point. (b) Closest-approach velocities for the orbits in (a). An object's velocity at closest approach determines the orbit shape and whether the orbit is bound or unbound. v_{circ} = circular velocity; v_{esc} = escape velocity.

4.2 Working It Out Circular Velocity and Escape Velocity

Circular Velocity

In Appendix 7, we show that the circular velocity is given by

$$v_{circ} = \sqrt{\frac{GM}{r}}$$

where M is the mass of the orbited object, and r is the radius of the circular orbit. A cannonball moving at just the right velocity—the circular velocity—will fall around Earth on a circular path.

We can use this equation to show how fast Newton's cannonball would have to travel to stay in its circular orbit. The average radius of Earth is 6,370 km, the mass of Earth is 5.97×10^{24} kg, and the gravitational constant is 6.67×10^{-20} km³/(kg s²). Inserting these values into the expression for v_{circ}, we get

$$v_{circ} = \sqrt{\frac{[6.67 \times 10^{-20}\,\text{km}^3/(\text{kg s}^2)] \times (5.97 \times 10^{24}\,\text{kg})}{6,370\,\text{km}}} = 7.9\,\text{km/s}$$

Newton's cannonball would have to be traveling about 8 kilometers per second (km/s)—more than 28,000 kilometers per hour (km/h)—to stay in its circular orbit. That's well beyond the reach of a typical cannon, but rockets routinely attain these speeds.

Now let's compare this speed with that needed to launch a satellite from the Moon into orbit just above the lunar surface. The radius of the Moon is 1,740 km, and its mass is 7.35×10^{22} kg. These values give the following circular velocity:

$$v_{circ} = \sqrt{\frac{[6.67 \times 10^{-20}\,\text{km}^3/(\text{kg s}^2)] \times (7.35 \times 10^{22}\,\text{kg})}{1,740\,\text{km}}} = 1.7\,\text{km/s}$$

The velocity needed to launch a satellite into a low circular orbit is considerably lower on the Moon than on Earth.

Escape Velocity

Sending a spacecraft to another planet requires launching it with a velocity greater than Earth's escape velocity. The escape velocity is a factor of $\sqrt{2}$, or approximately 1.41, times the circular velocity. This relation can be expressed as

$$v_{esc} = \sqrt{2} \times v_{circ} = \sqrt{\frac{2GM}{R}}$$

Using the numbers in the above example, we can calculate the escape velocity from the surface of Earth:

$$v_{esc} = \sqrt{2} \times v_{circ} = 1.41 \times 7.9\,\text{km/s} = 11.2\,\text{km/s}$$

To leave Earth, a rocket must have a speed of 11.2 km/s, or 40,300 km/h.

As with weight, the escape velocity from other astronomical objects will be different than the escape velocity from Earth. Ida is a small asteroid orbiting the Sun between the orbits of Mars and Jupiter. Ida has an average radius of 15.7 km and a mass of 4.2×10^{16} kg. Therefore,

$$v_{esc} = \sqrt{\frac{2 \times [6.67 \times 10^{-20}\,\text{km}^3/(\text{kg s}^2)] \times (4.2 \times 10^{16}\,\text{kg})}{15.7\,\text{km}}}$$

$$v_{esc} = 0.019\,\text{km/s} = 68\,\text{km/h}$$

A baseball thrown at about 130 km/h would easily escape from Ida's surface and fly off into interplanetary space.

However, later we will see that sometimes the two objects are closer to having the same mass; for example, dwarf planet Pluto and its moon Charon, or a large planet and a star, or two stars. In these examples, both objects experience significant accelerations in response to their mutual gravitational attraction. The two objects are both orbiting about a common point located between them, called the **center of mass**, so we now must think of them as falling around each other. Each mass is moving on its own elliptical orbit around the two objects' mutual center of mass. From measuring the size and period of any orbit, we can calculate the *sum* of the masses of the orbiting objects. Almost all knowledge about the masses of astronomical objects comes directly from the application of Newton's version of Kepler's third law.

Astronomy in Action: Center of Mass

CHECK YOUR UNDERSTANDING 4.3

If we wanted to increase the Hubble Space Telescope's altitude above Earth and keep it in a stable orbit, we also would need to: (a) increase its orbital speed; (b) increase its weight; (c) decrease its weight; (d) decrease its orbital speed.

Newton's Version of Kepler's Third Law

The time it takes for a planet to complete one orbit around the Sun equals the distance traveled divided by the planet's speed. For simplicity, let's assume the orbit is circular. Thus, the time it takes an object to make one trip around the Sun is the circumference of the circle ($2\pi r$) divided by the object's speed (v). The speed of the planet must be equal to the circular velocity discussed in Working It Out 4.2. Putting these relationships together, we have

$$\text{Orbital period } (P) = \frac{\text{Circumference of orbit}}{\text{Circular velocity}} = \frac{2\pi r}{\sqrt{\dfrac{GM_{Sun}}{r}}}$$

Squaring both sides of the equation gives

$$P^2 = \frac{4\pi^2 r^2}{\dfrac{GM_{Sun}}{r}} = \frac{4\pi^2}{GM_{Sun}} \times r^3$$

The square of the period of an orbit is equal to a constant ($4\pi^2/GM_{Sun}$) multiplied by the cube of the radius of the orbit. This is Kepler's third law ($P^2 = constants \times A^3$) applied to circular orbits. When Kepler used Earth units of years and astronomical units for the planets, he was taking a ratio of their periods and orbital radii with those for Earth, so the constants cancelled out. Using calculus, Newton was similarly able to derive Kepler's third law for elliptical orbits with semimajor axis A instead of radius r.

Mass of the Sun

If we can measure the size and period of any orbit, then we can use Newton's universal law of gravitation to calculate the mass of the object being orbited. To do so, we rearrange Newton's form of Kepler's third law above to read

$$M = \frac{4\pi^2}{G} \times \frac{A^3}{P^2}$$

Everything on the right side of this equation is either a constant (4, π, and G) or a quantity that we can measure (the semimajor axis A and period P of an orbit). The left side of the equation is the mass of the object at the focus of the ellipse. For example, we can find the mass of the Sun by noting the period and semimajor axis of the orbit of a planet around the Sun. Let's use the numbers for Earth. Whenever we have an equation with G, it is best to put everything else into the same units as G (km, kg, s). So first we must compute the number of seconds in 1 year: $P = 1$ yr $= 365.24$ days/yr $\times$ 24 h/day $\times$ 60 min/h $\times$ 60 s/min $= 3.16 \times 10^7$ s. The semimajor axis $A = 1$ AU $= 1.5 \times 10^8$ km. Then the mass of the Sun can be computed:

$$M_{Sun} = \frac{4\pi^2}{G} \times \frac{A^3}{P^2} = \frac{4\pi^2}{6.67 \times 10^{-20}\,km^3/(kg\,s^2)} \times \frac{(1.5 \times 10^8\,km)^3}{(3.16 \times 10^7\,s)^2}$$

$$M_{Sun} = 2.00 \times 10^{30}\,kg$$

We could have used the period and semimajor axis of any Solar System planet to get the same result.

4.3 Tidal Forces Are Caused by Gravity

The rise and the fall of the oceans are called Earth's **tides**. Coastal dwellers long ago noted that the strength of the tides varies with the phase of the Moon. Tides are strongest during a new or a full Moon and are weakest during first quarter or third quarter Moon. In this section, we see how tides result from differences between the strength of the gravitational pull of the Moon and Sun on one part of Earth in comparison to their pull on other parts of Earth.

 **Astronomy in Action:** Tides

Tides and the Moon

Figure 4.4 demonstrated that each small part of an object feels a gravitational attraction toward every other small part of the object, and this self-gravity differs from place to place. In addition, each small part of an object feels a gravitational attraction toward every other mass in the universe, and these external forces differ from place to place within the object as well.

AstroTour: Tides and the Moon

   The Moon's gravity pulls on Earth as if the mass of the Moon is concentrated at the Moon's center. The side of Earth that faces the Moon is closer to the Moon than is the rest of Earth, so it feels a stronger-than-average gravitational

 Nebraska Simulation: Tidal Bulge Simulation

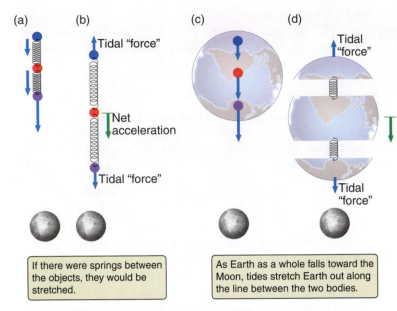

(a)

(b)

Tidal "force"

Net acceleration

Tidal "force"

(c)

(d)

Tidal "force"

Tidal "force"

If there were springs between the objects, they would be stretched.

As Earth as a whole falls toward the Moon, tides stretch Earth out along the line between the two bodies.

Figure 4.10 (a) Imagine three objects connected by springs. (b) The springs are stretched as if there were forces pulling outward on each end of the chain. (c) Similarly, three locations on Earth experience different gravitational attractions toward the Moon. (d) The difference in the Moon's gravitational attraction across Earth causes of Earth's tides.

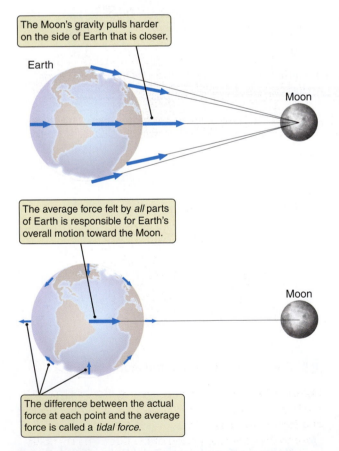

The Moon's gravity pulls harder on the side of Earth that is closer.

Earth

Moon

The average force felt by *all* parts of Earth is responsible for Earth's overall motion toward the Moon.

Moon

The difference between the actual force at each point and the average force is called a *tidal force.*

Figure 4.11 Tidal forces stretch Earth along the line between Earth and the Moon but compress Earth perpendicular to this line.

attraction toward the Moon. In contrast, the side of Earth facing away from the Moon is farther than average from the Moon, so it feels a weaker-than-average attraction toward the Moon. The pull of the Moon on the near side of Earth is about 7 percent greater than its pull on the far side of Earth.

To understand the consequence of this variation in the pull of the Moon, imagine three rocks being pulled by gravity toward the Moon. A rock closer to the Moon feels a stronger force than a rock farther from the Moon. Now suppose the three rocks are connected by springs (**Figure 4.10a**). As the rocks are pulled toward the Moon, the purple rock pulls away from the red rock, and the red rock pulls away from the blue rock. Therefore, the differences in the gravitational forces they feel will stretch *both* of the springs (Figure 4.10b). Now instead of springs, imagine that the rocks are at different places on Earth (Figure 4.10c). On the side of Earth away from the Moon, the force is smaller (as indicated by the shorter arrow), so that part gets left behind (Figure 4.10d). These differences in the Moon's gravitational attraction on different parts of Earth are called **tidal forces**.

Figure 4.11 shows how Earth is stretched, causing a tidal bulge. The Moon is not pushing the far side of Earth away; rather, it simply is not pulling on the far side of Earth as hard as it is pulling on the planet as a whole. The far side of Earth is "left behind" as the rest of the planet is pulled more strongly toward the Moon. Figure 4.11 shows that there is also a net force squeezing inward on Earth in the direction perpendicular to the line between Earth and the Moon. Together, the stretching by tidal forces along the line between Earth and the Moon and the squeezing by tidal forces perpendicular to this line distort the shape of Earth like a rubber ball caught in the middle of a tug-of-war.

If the surface of Earth was perfectly smooth and covered with a uniform ocean and Earth did not rotate, then the Moon would pull our oceans into an elongated **tidal bulge** like that shown in **Figure 4.12a**. The water would be at its deepest on the side toward the Moon and on the side away from the Moon and at its shallowest midway between. However, Earth is *not* covered with perfectly uniform oceans, and Earth does rotate. As any point on Earth rotates through the ocean's tidal bulges, that point experiences the ebb and flow of the tides. In addition, friction between the spinning Earth and its tidal bulge drags the oceanic tidal bulge around in the direction of Earth's rotation, as illustrated in Figure 4.12b.

Follow along in Figure 4.12c as you imagine riding on Earth throughout the course of a day. You begin as the rotating Earth carries you through the tidal bulge on the Moonward side of the planet. Because Earth's rotation drags the tidal bulge, the Moon is not exactly overhead but is instead high in the western sky. When you are at the high point in the tidal bulge, the ocean around you has risen higher than average—called a *high tide*. About $6\frac{1}{4}$ hours later, somewhat after the Moon has settled beneath the western horizon, the rotation of Earth carries you through a point where the ocean is lower than average—called a *low tide*. If you wait another $6\frac{1}{4}$ hours, it is again high tide. You are now passing through the region where ocean water is "left behind" (relative to Earth as a whole) in the tidal bulge on the side of Earth that is away from the Moon. The Moon, which is responsible for the tides you see, is itself at that time hidden from view on the far side of Earth. About $6\frac{1}{4}$ hours later, sometime after the Moon has risen above the

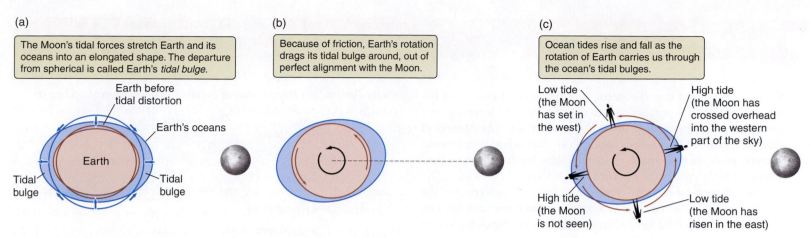

Figure 4.12 (a) Tidal forces pull Earth and its oceans into a tidal bulge. (b) Earth's rotation pulls its tidal bulge slightly out of alignment with the Moon. (c) As Earth's rotation carries us through these bulges, we experience the ocean tides. The magnitude of the tides has been exaggerated in these diagrams for clarity. In these figures, the observer is looking down from above Earth's North Pole. Sizes and distances are not to scale.

eastern horizon, it is low tide. About 25 hours after you started this journey—the amount of time the Moon takes to return to the same point in the sky from which it started—you again pass through the tidal bulge on the near side of the planet. This is the age-old pattern by which mariners have lived their lives for millennia: the twice-daily coming and going of high tide, shifting through the day in lock-step with the passing of the Moon.

Local geography, such as the shapes of Earth's shorelines and ocean basins, complicate the simple picture of tides. In addition, there are oceanwide oscillations similar to water sloshing around in a basin. As they respond to the tidal forces from the Moon, Earth's oceans flow around the various landmasses that break up the water covering our planet. Some places, like the Mediterranean Sea and the Baltic Sea, are protected from tides by their relatively small sizes and the narrow passages connecting these bodies of water with the larger ocean. In other places, the shape of the land funnels the tidal surge from a large region of ocean into a relatively small area, concentrating its effect, as at the Bay of Fundy (**Figure 4.13**). Tidal effects from the Sun and Moon are very slight even in the Great Lakes, where the water is more affected by local weather and geography.

Solar Tides

The tides resulting from the pull of the Moon are called **lunar tides**. The Sun also influences Earth's tides. The gravitational pull of the Sun causes Earth to stretch along a line pointing approximately in the direction of the Sun. The side of Earth closer to the Sun is pulled toward the Sun more strongly than is the side of Earth away from the Sun, just as the side of Earth closest to the Moon is pulled more strongly toward the Moon. Tides on Earth due to differences in the gravitational pull of the Sun are called **solar tides**. Although the absolute strength of the Sun's pull on Earth is nearly 200 times greater than the strength of the Moon's pull on Earth, the Sun's gravitational attraction does not change by much from one side of Earth to the other, because the Sun is much farther away than the Moon. As a result, solar tides are only about half as strong as lunar tides (**Working It Out 4.4**).

(a)

(b)

Figure 4.13 The world's most extreme tides are found in the Bay of Fundy in eastern Canada. Water rocks back and forth in this bay with a period of about 13 hours, close to the 12.5-hour period of the tides. The shape of the basin amplifies the tides so that difference in water depth between low tide (a) and high tide (b) is extreme; typically about 14.5 meters and as much as 16.6 meters.

4.4 | Working It Out | Tidal Forces

Earlier you learned that the strength of the gravitational force between two bodies is proportional to their masses and inversely proportional to the square of the distance between them. The strength of tidal forces caused by one body acting on another is also proportional to the mass of the body that is raising the tides, but it is inversely proportional to the *cube* of the distance between them.

The equation for the tidal force comes from the difference of the gravitational force on one side of a body compared with the force on the other side. The tidal force acting on Earth by the Moon is given by

$$F_{tidal}(Moon) = \frac{2GM_{Earth}M_{Moon}R_{Earth}}{d^3_{Earth\text{-}Moon}}$$

where R_{Earth} is Earth's radius and $d_{Earth\text{-}Moon}$ is the distance between Earth and the Moon.

As an example, let's compare the tidal force acting on Earth by the Moon with the tidal force acting on Earth by the Sun. F_{tidal} from the Moon is given in the preceding equation; F_{tidal} from the Sun is given by

$$F_{tidal}(Sun) = \frac{2GM_{Earth}M_{Sun}R_{Earth}}{d^3_{Earth\text{-}Sun}}$$

We know the Moon is much closer to Earth than the Sun is, but the Sun is much more massive than the Moon. To compare the lunar and solar tides, we can take a ratio of the tidal forces and proceed in a similar way to our comparison of gravitational forces in Working It Out 4.1:

$$\frac{F_{tidal}(Moon)}{F_{tidal}(Sun)} = \frac{\dfrac{2GM_{Earth}M_{Moon}R_{Earth}}{d^3_{Earth\text{-}Moon}}}{\dfrac{2GM_{Earth}M_{Sun}R_{Earth}}{d^3_{Earth\text{-}Sun}}}$$

Canceling out the constant G and the terms common in both equations (M_{Earth} and R_{Earth}) gives

$$\frac{F_{tidal}(Moon)}{F_{tidal}(Sun)} = \frac{M_{Moon}}{M_{Sun}} \times \frac{d^3_{Earth\text{-}Sun}}{d^3_{Earth\text{-}Moon}}$$

$$\frac{F_{tidal}(Moon)}{F_{tidal}(Sun)} = \frac{M_{Moon}}{M_{Sun}} \times \left(\frac{d_{Earth\text{-}Sun}}{d_{Earth\text{-}Moon}}\right)^3$$

Using the values from Appendix 4, $M_{Moon} = 7.35 \times 10^{22}$ kg, $M_{Sun} = 2 \times 10^{30}$ kg, $d_{Earth\text{-}Moon} = 384{,}400$ km, and $d_{Earth\text{-}Sun} = 1.5 \times 10^8$ km gives

$$\frac{F_{tidal}(Moon)}{F_{tidal}(Sun)} = \frac{7.35 \times 10^{22}\,\text{kg}}{2 \times 10^{30}\,\text{kg}} \times \left(\frac{1.5 \times 10^8\,\text{km}}{384{,}400\,\text{km}}\right)^3 = 2.2$$

So the tidal force from the Moon is 2.2 times stronger than the tidal force from the Sun, which is why we often hear that tides are caused by the Moon. But the Sun is an important factor, too, and that's why the tides change depending on the alignment of the Moon and the Sun with Earth.

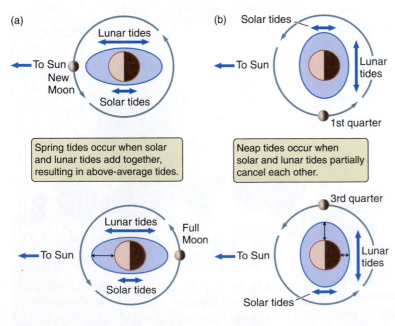

Figure 4.14 Solar tides are about half as strong as lunar tides. The interactions of solar and lunar tides result in either (a) spring tides when they are added together or (b) neap tides when they partially cancel each other.

Solar and lunar tides interact. As shown in **Figure 4.14a**, when the Moon and the Sun are lined up with Earth, at either new or full Moon, the lunar and solar tides on Earth overlap. This creates more extreme tides ranging from extra-high high tides to extra-low low tides. The extreme tides near the new or full Moon are called **spring tides**—not because of the season, but because the water appears to spring out of the sea. Conversely—illustrated in Figure 4.14b—when the Moon, Earth, and Sun make a right angle, at the Moon's first and third quarters, the lunar and solar tidal forces stretch Earth in different directions, creating less extreme tides known as **neap tides**. The word *neap* is derived from the Saxon word *neafte*, which means "scarcity": at these times of the month, shellfish and other food gathered in the tidal region are less accessible because the low tide is higher than at other times. Neap tides are only about half as strong as average tides and only a third as strong as spring tides.

CHECK YOUR UNDERSTANDING 4.4

Rank in order of the strongest tides: (a) new Moon in July; (b) first quarter Moon in July; (c) full Moon in January; (d) third quarter Moon in January.

4.4 Tidal Forces Affect Solid Bodies

In the previous section, we focused on the movements of the liquid of Earth's oceans in response to the tidal forces from the Moon and Sun. But these tidal forces also affect the solid body of Earth. As Earth rotates through its tidal bulge, the solid body of the planet is constantly being deformed by tidal forces. Earth is somewhat elastic (like a rubber ball), and tidal stresses cause a vertical displacement of about 30 centimeters (cm) between high tide and low tide, or roughly a third of the displacement of the oceans. It takes energy to deform the shape of a solid object. (If you want a practical demonstration of this fact, hold a rubber ball in your hand and squeeze and release it a few dozen times.) This energy from the deformation is converted into thermal energy by friction in Earth's interior. This friction opposes and takes energy from the rotation of Earth, causing Earth to gradually slow. Earth's internal friction adds to the slowing caused by friction between Earth and its oceans as the planet rotates through the tidal bulge of the oceans. As a result, Earth's days are currently lengthening by about 1.7 milliseconds (ms) every century. This sounds small but it adds up: when dinosaurs ruled, the day was closer to 23 hours long, and 200 million years into the future, the day will be close to 25 hours.

Tidal Locking

Other solid bodies besides Earth experience tidal forces. For example, the Moon has no bodies of liquid to make tidal forces obvious, but its shape is distorted in the same manner as Earth. Because of Earth's much greater mass and the Moon's smaller radius, the tidal effects of Earth on the Moon are about 20 times as great as the tidal effects of the Moon on Earth. Given that the average tidal deformation of Earth is about 30 cm, the average tidal deformation of the Moon should be about 6 meters. However, what we actually observe on the Moon is a tidal bulge of about 20 meters. This unexpectedly large displacement exists because the Moon's tidal bulge was "frozen" into its relatively rigid crust at an earlier time, when the Moon was closer to Earth and tidal forces were much stronger than they are today. Planetary scientists sometimes call this deformation the Moon's *fossil tidal bulge*.

Recall from Chapter 2 that the Moon's rotation period exactly equals its orbital period. This synchronous rotation of the Moon is a result of **tidal locking**. Early in its history, the period of the Moon's rotation was almost certainly different from its orbital period. As the Moon rotated through its extreme tidal bulge, however, friction within the Moon's crust was tremendous, rapidly slowing the Moon's rotation. After a fairly short time, the period of the Moon's rotation equaled the period of its orbit. When its orbital and rotation periods became equalized, the Moon no longer rotated with respect to its tidal bulge. Instead, the Moon and its tidal bulge rotated *together*, in lockstep with the Moon's orbit around Earth. As illustrated in **Figure 4.15**, this scenario continues today as the tidally distorted Moon orbits Earth, always keeping the same face and the long axis of its tidal bulge toward Earth.

Tidal forces affect not only the rotations of the Moon and Earth, but also their orbits. Because of its tidal bulge, Earth is not a perfectly spherical body. Therefore, the material in Earth's tidal bulge on the side nearer the Moon pulls on the Moon more strongly than does material in the tidal bulge on the back side of

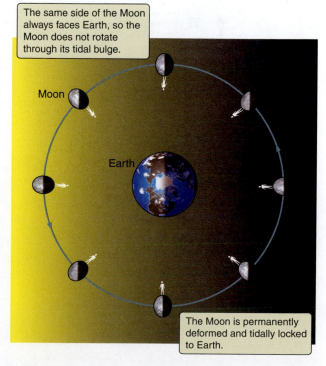

Figure 4.15 Tidal forces due to Earth's gravity lock the Moon's rotation to its orbital period.

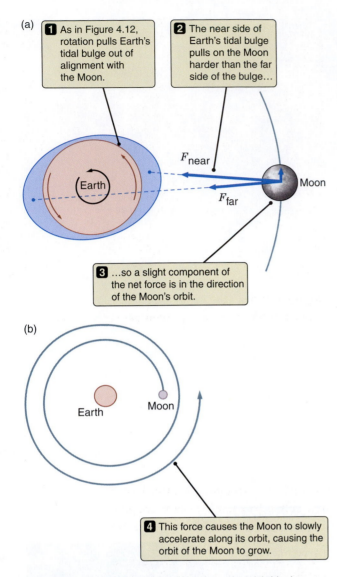

(a)

1 As in Figure 4.12, rotation pulls Earth's tidal bulge out of alignment with the Moon.

2 The near side of Earth's tidal bulge pulls on the Moon harder than the far side of the bulge…

Earth

F_{near}

F_{far}

Moon

3 …so a slight component of the net force is in the direction of the Moon's orbit.

(b)

Earth

Moon

4 This force causes the Moon to slowly accelerate along its orbit, causing the orbit of the Moon to grow.

Figure 4.16 Interaction between Earth's tidal bulge and the Moon causes the Moon to accelerate in its orbit and the Moon's orbit to grow.

Earth. Because the tidal bulge on the Moonward side of Earth "leads" the Moon somewhat, as shown in **Figure 4.16**, the gravitational attraction of the bulge causes the Moon to accelerate slightly along the direction of its orbit around Earth. The rotation of Earth is dragging the Moon along with it. The acceleration of the Moon in the direction of its orbital motion causes the orbit of the Moon to grow larger. At present, the Moon is drifting away from Earth at a rate of 3.83 cm per year.

As the Moon grows more distant, the length of the lunar month increases by about 0.014 second each century. If this increase in the radius of the Moon's orbit were to continue long enough (about 50 billion years), Earth would become tidally locked to the Moon, just as the Moon is now tidally locked to Earth. At that point, the period of rotation of Earth, the period of rotation of the Moon, and the orbital period of the Moon would all be exactly the same—about 47 of our current days—and the Moon would be about 43 percent farther from Earth than it is today. However, this situation will never actually occur—at least not before the Sun itself has burned out.

The effects of tidal forces can be seen throughout the Solar System. Most of the moons in the Solar System are tidally locked to their parent planets, and in the case of dwarf planet Pluto and its largest moon, Charon, each is tidally locked to the other.

Tidal locking is only one way that orbits and rotations can be coupled together. Tidal forces have coupled the planet Mercury's rotation to its very elliptical orbit around the Sun. Yet unlike the Moon with its synchronous rotation, Mercury spins on its axis three times for every two trips around the Sun. The period of Mercury's orbit—87.97 Earth days—is exactly $1\frac{1}{2}$ times the 58.64 days that it takes Mercury to spin once on its axis. When Mercury comes to the point in its orbit that is closest to the Sun, one hemisphere faces the Sun, and then in the next orbit the other hemisphere faces the Sun.

Tidal Forces on Many Scales

We normally think of the effects of tidal forces as small compared to the force of gravity holding an object together, yet tidal effects can be extremely destructive. Consider for a moment the fate of a small moon, asteroid, or comet that wanders too close to a massive planet such as Jupiter or Saturn. All objects in the Solar System larger than about a kilometer in diameter are held together by their self-gravity. However, the self-gravity of a small object such as an asteroid, a comet, or a small moon is feeble. In contrast, the tidal forces close to a massive object such as Jupiter can be very strong. If the tidal forces trying to tear an object apart become greater than the self-gravity trying to hold the object together, the object will break into pieces.

The **Roche limit** is the distance at which a planet's tidal forces are greater than the self-gravity of a smaller object—such as a moon, asteroid, or comet—causing the object to break apart. For a smaller object having the same density as the planet, the Roche limit is about 2.45 times the planet's radius. Such an object bound together solely by its own gravity can remain intact when it is outside a planet's Roche limit, but not when it is inside the limit. Objects such as the International Space Station and other Earth satellites are not torn apart, even though they orbit well within Earth's Roche limit, because chemical bonds hold them together, not just self-gravity.

We have concentrated on the role that tidal forces play on Earth and the Moon. We find tidal forces throughout the Solar System and the universe. Any time two objects of significant size or two collections of objects interact gravitationally, the gravitational forces will differ from one place to another within the objects, giving rise to tidal effects. Tidal disruption of small bodies is the source of the particles that make up the rings of the giant planets. Tidal interactions can cause material from one star in a binary pair to be pulled onto the other star. Tidal effects can strip stars from clusters consisting of thousands of stars. Galaxies can pass close enough together to strongly interact gravitationally. When this happens, as in **Figure 4.17**, tidal effects can grossly distort both galaxies taking part in the interaction. Tidal forces even play a role in shaping huge collections of galaxies—the largest known structures in the universe.

Figure 4.17 The tidal "tails" seen here are characteristic of tidal interactions between galaxies.

CHECK YOUR UNDERSTANDING 4.5

The Moon always keeps the same face toward Earth because of: (a) tidal locking; (b) tidal forces from the Sun; (c) tidal forces from Earth; (d) tidal forces from the Earth and Sun.

Origins

Tidal Forces and Life

In this chapter, we noted that Earth's rotation is slowing down as the Moon slowly moves away into a larger orbit. In the distant past the Moon was closer and Earth rotated faster, so tides would have been stronger and the interval between high tides would have been shorter. The tides are also affected by the configuration of the continents and oceans on Earth, so it is not known precisely how much faster Earth rotated billions of years ago. But the stronger and more frequent tides would have provided additional energy to the oceans of the young Earth.

Scientists debate whether life on Earth originated deep in the ocean, on the surface of the ocean, or on land (see Chapter 24). The tides shaped the regions in the margins between land and ocean, such as tide pools and coastal flats. Some think that these border regions, which alternate between wet and dry with the tides, could have been places where concentrations of biochemicals periodically became high enough for more complex reactions to take place. These complex reactions were important to the earliest life. Later, these border regions may have been important as advanced life moved from the sea to the land.

Elsewhere in the Solar System, the giant planets Jupiter and Saturn are far from the Sun, and thus very cold. Jupiter and Saturn each have many moons, and the closest moons would experience strong tidal forces from their respective planet. As you saw in Reading Astronomy News in Chapter 1, several of these moons are thought to have a liquid ocean underneath an icy surface. Tidal forces from Jupiter or Saturn provide the heat to keep the water in a liquid state. Astrobiologists think that these subsurface liquid oceans are perhaps the most probable location for life elsewhere in the Solar System.

We have seen that on Earth, the tidal forces from the Sun are about half as strong as those from the Moon. A planet with a closer orbit would experience much stronger tidal forces from its star. In Chapter 7, you will see that many of the planets detected outside of the Solar System have orbits very close to their stars. These planets experience strong tidal forces, and they might be tidally locked so that they have synchronous rotation as the Moon does around Earth, with one side of the planet always facing the star and one side facing away. How might life on Earth have evolved differently if half of the planet was in perpetual night and half in perpetual day?

ARTICLES   QUESTIONS

Astronomers and physicists have wondered if the value of Newton's gravitational constant G has changed over time. This press release reports on one experiment to measure the value of G.

Exploding Stars Prove Newton's Law of Gravity Unchanged over Cosmic Time

By **Swinburne Media Centre**

Australian astronomers have combined all observations of supernovae ever made to determine that the strength of gravity has remained unchanged over the last 9 billion years.

Newton's gravitational constant, known as *G*, describes the attractive force between two objects, together with the separation between them and their masses. It had previously been suggested that *G* could have been slowly changing over the 13.8 billion years since the Big Bang.

If *G* had been decreasing over time, for example, this would mean the Earth's distance to the Sun was slightly smaller in the past, meaning that we would experience longer seasons now compared to much earlier points in the Earth's history.

But researchers at Swinburne University of Technology in Melbourne have now analysed the light given off by 580 supernova explosions in the nearby and far universe and have shown that the strength of gravity has not changed.

"Looking back in cosmic time to find out how the laws of physics may have changed is not new," said Professor Jeremy Mould. "But supernova cosmology now allows us to do this with gravity."

A Type Ia supernova marks the violent death of a star called a white dwarf, which is as massive as our Sun but packed into a ball the size of our Earth.

Telescopes can detect the light from this explosion and use its brightness as a "standard candle" to measure distances in the universe, a tool that helped Australian astronomer Professor Brian Schmidt in his 2011 Nobel Prize–winning work discovering the mysterious force Dark Energy.

Professor Mould and his PhD student Syed Uddin at the Swinburne Centre for Astrophysics and Supercomputing and the ARC Centre of Excellence for All-sky Astrophysics (CAASTRO) assumed that these supernova explosions happen when a white dwarf reaches a critical mass or after colliding with other stars to "tip it over the edge."

"This critical mass depends on Newton's gravitational constant *G* and allows us to monitor it over billions of years of cosmic time—instead of only decades as was the case in previous studies," Professor Mould said.

Despite these vastly different time spans, their results agree with findings from the Lunar Laser Ranging Experiment that has been measuring the distance between the Earth and the Moon since NASA's *Apollo* missions in the 1960s and has been able to monitor possible variations in *G* at very high precision.

"Our cosmological analysis complements experimental efforts to describe and constrain the laws of physics in a new way and over cosmic time," Mr Uddin said.

In their current publication, the Swinburne researchers were able to set an upper limit on the change in Newton's gravitational constant of 1 part in 10 billion per year over the past 9 billion years.

This research was published in *Publications of the Astronomical Society of Australia*.

ARTICLES   QUESTIONS

1. Explain how observing very distant supernova means observing back in time.
2. Why are physicists still measuring *G*, centuries after Newton described this constant?
3. What quantities that you studied in the chapter would change if the value of *G* changed over time?
4. Why is it important to the process of science that this result agrees with a result from a totally different experiment?
5. If the distance to the Sun were slightly smaller in the past, would solar tides have been weaker or stronger? Would you expect this to have been as important to tides on Earth as changes in the distance of the Moon?

Summary

Objects stay in orbit because of gravity. Newton's laws of motion and his proposed law of gravity predict the paths of planetary orbits and explain Kepler's laws. Newton's calculations showed that these orbits should be ellipses with the Sun at one focus, that planets should move faster when closer to the Sun, and that the square of the period of a planet's orbit should vary as the cube of the semimajor axis of that elliptical orbit. Newton also showed mathematically Galileo's conclusion that falling objects have an accelerated motion independent of their mass. Tidal forces provide energy to Earth's oceans. Tide pools on Earth may have been a site of early biochemical reactions. Some moons in the outer Solar System might have liquid water because of tidal heating from their respective planet.

LG 1 **Explain the elements of Newton's universal law of gravitation.** Gravity is a force between any two objects due to their masses. As one of the fundamental forces of nature, gravity binds the universe together. The force of gravity is proportional to the mass of each object and inversely proportional to the square of the distance between them.

LG 2 **Use the concepts of motion and gravitation to explain planetary orbits.** An orbit is one body "falling around" another. Planets orbit the Sun in elliptical orbits. All objects affected by gravity move either on bound elliptical orbits or unbound paths. Orbits are ultimately given their shape by the gravitational attraction of the objects involved, which in turn is a reflection of the masses of these objects.

LG 3 **Explain how tidal forces from the Sun and Moon create Earth's tides.** Tides on Earth are the result of differences between how hard the Moon and Sun pull on one part of Earth in comparison with their pull on other parts of Earth. The primary cause of tides is the Moon, which stretches out Earth. The tides are the strongest when the Sun, Moon, and Earth are aligned. As Earth rotates, tides rise and fall twice each day.

LG 4 **Describe the effects of tidal forces on solid bodies.** Tidal forces lock the Moon's rotation to its orbit around Earth. Tidal forces can break up an object if its gets too close to a more massive object. Tidal forces are observed throughout the universe, in planets and moons, pairs of stars, and interacting galaxies.

? UNANSWERED QUESTION

- What range of gravities will support human life? Humans have evolved to live on Earth's surface, but what happens when humans go elsewhere? What are the limits for our hearts, lungs, eyes, and bones? At the higher end of human tolerance, fighter pilots have been trained to experience about 10 times the normal surface gravity on Earth for very short periods of time (too long and they black out). Astronauts who spend several months in near-weightless conditions experience medical problems such as bone loss. On the Moon or Mars, humans will weigh much less than on Earth. Numerous science fiction tales have been written about what happens to children born on a space station or on another planet or moon with low surface gravity: would their hearts and bodies ever be able to adjust to the higher surface gravity of Earth or must they stay in space forever?

Questions and Problems

Test Your Understanding

1. In Newton's universal law of gravitation, the force is
   a. proportional to both masses.
   b. proportional to the radius.
   c. proportional to the radius squared.
   d. inversely proportional to the orbiting mass.

2. Rank the following objects in order of their circular velocities, from smallest to largest.
   a. a 5-kg object orbiting Earth halfway to the Moon
   b. a 10-kg object orbiting Earth just above Earth's surface
   c. a 15-kg object orbiting Earth at the same distance as the Moon
   d. a 20-kg object orbiting Earth one-quarter of the way to the Moon

3. An object in a(n) _____ orbit in the Solar System will remain in its orbit forever. An object in a(n) _____ orbit will escape from the Solar System.
   a. unbound; bound
   b. circular; elliptical
   c. bound; unbound
   d. elliptical; circular

4. Compared to your mass on Earth, on the Moon your mass would be
   a. lower because the Moon is smaller than Earth.
   b. lower because the Moon has less mass than Earth.
   c. higher because of the combination of the Moon's mass and size.
   d. the same, mass doesn't change.

5. If you went to Mars, your weight would be
   a. higher because you are closer to the center of the planet.
   b. lower because Mars has two small moons instead of one big moon, so there's less tidal force.
   c. lower because Mars has lower mass and a smaller radius that together produce a lower gravitational force.
   d. the same as on Earth.

6. Venus has about 80 percent of Earth's mass and about 95 percent of Earth's radius. Your weight on Venus will be
   a. 20 percent more than on Earth.
   b. 20 percent less than on Earth.
   c. 10 percent more than on Earth.
   d. 10 percent less than on Earth.

7. The connection between gravity and orbits enables astronomers to measure the _____ of stars and planets.
   a. distances
   b. sizes
   c. masses
   d. compositions

8. If the Moon had twice the mass that it does, how would the strength of lunar tides change?
   a. The highs would be higher, and the lows would be lower.
   b. Both the highs and the lows would be higher.
   c. The highs would be lower, and the lows would be higher.
   d. Nothing would change.

9. If Earth had half of its current radius, how would the strength of lunar tides change?
   a. The highs would be higher, and the lows would be lower.
   b. Both the highs and the lows would be higher.
   c. The highs would be lower, and the lows would be higher.
   d. Both the highs and the lows would be lower.

10. If the Moon were 2 times closer to Earth than it is now, the gravitational force between Earth and the Moon would be
    a. 2 times stronger.
    b. 4 times stronger.
    c. 8 times stronger.
    d. 16 times stronger.

11. If the Moon were 2 times closer to Earth than it is now, the tides would be
    a. 2 times stronger.
    b. 4 times stronger.
    c. 8 times stronger.
    d. 16 times stronger.

12. If two objects are tidally locked to each other,
    a. the tides always stay on the same place on each object.
    b. the objects always remain in the same place in each other's sky.
    c. the objects are falling together.
    d. both a and b

13. Spring tides occur only when
    a. the Sun is near the vernal equinox.
    b. the Moon's phase is new or full.
    c. the Moon's phase is first quarter or third quarter.
    d. it is either spring or fall.

14. If an object crosses from farther to closer than the Roche limit, it
    a. can no longer be seen.
    b. begins to accelerate very quickly.
    c. slows down.
    d. may be torn apart.

15. Self-gravity is
    a. the gravitational pull of a person.
    b. the force that holds objects like people and lamps together.
    c. the gravitational interaction of all the parts of a body.
    d. the force that holds objects on Earth.

Thinking about the Concepts

16. Both Kepler's laws and Newton's laws tell us something about the motion of the planets, but there is a fundamental difference between them. What is that difference?

17. Explain the difference between circular velocity and escape velocity. Which of these must be larger? Why?

18. Explain the difference between weight and mass.

19. Weight on Earth is proportional to mass. On the Moon, too, weight is proportional to mass, but the constant of proportionality is different on the Moon than it is on Earth. Why? Explain why this difference does not violate the universality of physical law, described in the Process of Science Figure.

20. Two comets are leaving the vicinity of the Sun, one traveling in an elliptical orbit and the other in a unbound orbit. What can you say about the future of these two comets? Would you expect either of them to return eventually?

21. What is the advantage of launching satellites from spaceports located near the equator? Would you expect satellites to be launched to the east or to the west? Why?

22. Explain how to use celestial orbits to estimate an object's mass. What are the observational quantities you need to make this mass estimation?

23. Suppose astronomers discovered an object approaching the Sun in an unbound orbit. What would that say about the origin of the object?

24. What determines the strength of gravity at various radii between Earth's center and its surface?

25. The best time to dig for clams along the seashore is when the ocean tide is at its lowest. What phases of the Moon would be best for clam digging? What would be the best times of day during those phases?

26. The Moon is on the meridian at your seaside home, but your tide calendar does not show that it is high tide. What might explain this apparent discrepancy?

27. We may have an intuitive feeling for why lunar tides raise sea level on the side of Earth facing the Moon, but why is sea level also raised on the side facing away from the Moon?

28. Tides raise and lower the level of Earth's oceans. Can they do the same for Earth's landmasses? Explain your answer.

29. Lunar tides raise the ocean surface less than 1 meter. How can tides as large as 5–10 meters occur?

30. Most commercial satellites are well inside the Roche limit as they orbit Earth. Why are they not torn apart?

Applying the Concepts

31. Mars has about one-tenth the mass of Earth and about half of Earth's radius. What is the value of gravitational acceleration on the surface of Mars compared to that on Earth? Estimate your mass and weight on Mars compared with your mass and weight on Earth. Do Hollywood movies showing people on Mars accurately portray this difference in weight?

32. Earth speeds along at 29.8 km/s in its orbit. Neptune's nearly circular orbit has a radius of 4.5×10^9 km, and the planet takes 164.8 years to make one trip around the Sun. Calculate the speed at which Neptune moves along in its orbit.

33. Venus's circular velocity is 35.03 km/s, and its orbital radius is 1.082×10^8 km. Use this information to calculate the mass of the Sun.

34. At the surface of Earth, the escape velocity is 11.2 km/s. What would be the escape velocity at the surface of a very small asteroid having a radius 10^{-4} that of Earth and a mass 10^{-12} that of Earth?

35. How long does it take Newton's cannonball, moving at 7.9 km/s just above Earth's surface, to complete one orbit around Earth?

36. When a spacecraft is sent to Mars, it is first launched into an Earth orbit with circular velocity.
    a. Describe the shape of this orbit.
    b. What minimum velocity must we give the spacecraft to send it on its way to Mars?

37. Earth's average radius is 6,370 km and its mass is 5.97×10^{24} kg. Show that the acceleration of gravity at the surface of Earth is 9.81 m/s².

38. Using 6,370 km for Earth's radius, compare the gravitational force acting on a NASA rocket when it is sitting on its launch-pad with the gravitational force acting on it when it is orbiting 350 km above Earth's surface.

39. The International Space Station travels on a nearly circular orbit 350 km above Earth's surface. What is its orbital speed?

40. Rearrange the terms in the last equation in Working It Out 4.1 to calculate the mass of Earth, using the measured values of g, G, and R_{Earth}.

41. As described in Working It Out 4.4, tidal force is proportional to the masses of the two objects and is inversely proportional to the cube of the distance between them. Some astrologers claim that your destiny is determined by the "influence" of the planets that are rising above the horizon at the moment of your birth. Compare the tidal force of Jupiter (mass 1.9×10^{27} kg; distance 7.8×10^8 km) with that of the doctor in attendance at your birth (mass 80 kg, distance 1 meter).

42. The asteroid Ida (mass 4.2×10^{16} kg) is attended by a tiny asteroidal moon, Dactyl, which orbits Ida at an average distance of 90 km. Neglecting the mass of the tiny moon, what is Dactyl's orbital period in hours?

43. Suppose you go skydiving.
    a. Just as you fall out of the airplane, what is your gravitational acceleration?
    b. Would this acceleration be bigger, smaller, or the same if you were strapped to a flight instructor, and so had twice the mass?
    c. Just as you fall out of the airplane, what is the gravitational force on you? (Assume your mass is 70 kg.)
    d. Would the gravitational force be bigger, smaller, or the same if you were strapped to a flight instructor, and so had twice the mass?

44. Assume that a planet just like Earth is orbiting the bright star Vega at a distance of 1 astronomical unit (AU). The mass of Vega is twice that of the Sun.
    a. How long in Earth years will it take to complete one orbit around Vega?
    b. How fast is the Earth-like planet traveling in its orbit around Vega?

45. Suppose in the past the Moon was 80 percent of the distance from Earth that it is now. Calculate how much stronger the lunar tides would have been. How would the neap and spring tides be different from now?

46. Go to NASA's "Apollo 15 Hammer-Feather Drop" Web page (http://nssdc.gsfc.nasa.gov/planetary/lunar/apollo_15_feather_drop.html) and watch the video from *Apollo 15* of astronaut David Scott dropping the hammer and falcon feather on the Moon. (You might find a better version on YouTube.) What did this experiment show? What would happen if you tried this on Earth with a feather and a hammer? Would it work? Suppose instead you dropped the hammer and a big nail. How would they fall? How does the acceleration of falling objects on the Moon compare to the acceleration of falling objects on Earth?

47. Go to the Exploratorium's "Your Weight on Other Worlds" Web page (http://exploratorium.edu/ronh/weight), which will calculate your weight on other planets and moons in our Solar System. On which objects would your weight be higher than it is on Earth? What difficulties would human bodies have in a higher-gravity environment? For example, would it be easy to get up out of bed and walk? What are the possible short-term and long-term effects of lower gravity on the human body? Can you think of some types of life on Earth that might adapt well to a different gravity?

48. a. Watch the first 12 minutes of the episode "Gravity" in the *Universe* series (http://www.history.com/shows/the-universe/videos/the-universe-gravity) to see several illustrated examples of gravity. Why does the tennis ball appear to float when dropped at the top of the roller coaster? How was Newton able to imagine a satellite orbiting Earth centuries before it was possible? Why is it the speed of the cannonball that determines whether it goes into orbit? What was the technical difficulty in launching a satellite?

b. In the same video as in part (a), watch the trip on the zero-G plane, at 23–29 minutes. How does the plane simulate zero-G? Why does it last for only 20–30 seconds? How is this similar to the roller coaster in part (a)?

49. Go to a website that will show you the times for high and low tides; for example, http://saltwatertides.com. Pick a location and bring up the tide table for today and the next 14 days. Why are there two high tides and two low tides every day? What is the difference in the height of the water between high and low tides? In the last few columns of the table, the times of moonrise and moonset are indicated, as well as the percent of lunar illumination. Does the time of the high tide lead or follow the highest position of the Moon in the sky? Compare with Figure 4.12c: what phases of the Moon have the greatest differences in the height of high and low tides?

50. Figure 4.17 shows two galaxies pulling at each other, most likely after they have already collided. Go to http://www.cita.utoronto.ca/~dubinski/nbody/ and scroll down to "Movie 2" to see a simulation of this interaction. Galaxies are not solid objects: they contain stars and gas and dust and a lot of empty space. Explain how these tidal tails can result from this type of interaction.

smartwork

If your instructor assigns homework in SmartWork, access your assignments at digital.wwnorton.com/astro5.

EXPLORATION

digital.wwnorton.com/astro5

In the Exploration of Chapter 3, we used the Planetary Orbit Simulator to explore Kepler's laws for Mercury. Now that we know how Newton's laws explain why Kepler's laws describe orbits, we will revisit the simulator to explore the Newtonian features of Mercury's orbit. Visit the Student Site at the Digital Landing Page, and open the Planetary Orbit Simulator applet.

Acceleration

To begin exploring the simulation, set parameters for "Mercury" in the "Orbit Settings" panel and then click "OK." Click the "Newtonian Features" tab at the bottom of the control panel. Select "show solar system orbits" and "show grid" under "Visualization Options." Change the animation rate to 0.01, and select the "start animation" button.

Examine the graph at the bottom of the panel.

1 Where is Mercury in its orbit when the acceleration is smallest?

_____

_____

2 Where is Mercury in its orbit when the acceleration is largest?

_____

_____

3 What are the values of the largest and smallest accelerations?

_____

_____

In the "Newtonian Features" graph, mark the boxes for vector and line that correspond to the acceleration. Specifying these parameters will insert an arrow that shows the direction of the acceleration and a line that extends the arrow.

4 To what Solar System object does the arrow point?

_____

_____

5 In what direction is the force on the planet?

_____

_____

Velocity

Examine the graph at the bottom of the panel again.

6 Where is Mercury in its orbit when the velocity is smallest?

_____

_____

7 Where is Mercury in its orbit when the velocity is largest?

_____

8 What are the values of the largest and smallest velocities?

_____

Add the velocity vector and line to the simulation by clicking on the boxes in the graph window. Study the resulting arrows carefully.

9 Are the velocity and the acceleration always perpendicular (is the angle between them always 90°)?

_____

_____

10 If the orbit were a perfect circle, what would be the angle between the velocity and the acceleration?

_____

_____

Hypothetical Planet

In the "Orbit Settings" panel, change the semimajor axis to 0.8 AU.

11 How does this imaginary planet's orbital period now compare to Mercury's?

_____

_____

Now change the semimajor axis to 0.1 AU.

12 How does this planet's orbital period now compare to Mercury's?

_____

_____

13 Summarize your observations of the relationship between the speed of an orbiting object and the semimajor axis.

_____

_____

5 Light

Our knowledge of the universe beyond Earth comes from light emitted, absorbed, or reflected by astronomical objects. Light carries information about the temperature, composition, and speed of the objects. Light also tells us about the nature of the material that the light passed through on its way to Earth. Yet light plays a far larger role in astronomy than that of being a messenger. Light is one of the primary means by which energy is transported throughout the universe. Stars, planets, and vast clouds of gas and dust filling the space between the stars heat up as they absorb light and cool off as they emit light. Light carries energy generated in the heart of a star outward through the star and off into space. Light transports energy from the Sun outward through the Solar System, heating the planets; and light carries energy away from each planet, allowing each one to cool. The balance between these two processes establishes each planet's temperature and therefore a planet's possible suitability for life.

LEARNING GOALS

An astronomer must try to understand the universe by the light and other particles that reach Earth from distant objects. By the conclusion of this chapter, you should be able to:

LG 1   Describe the wave and particle properties of light, and describe the electromagnetic spectrum.

LG 2   Describe how to measure the chemical composition of distant objects using the unique spectral lines of different types of atoms.

LG 3   Describe the Doppler effect and how it can be used to measure the motion of distant objects.

LG 4   Explain how the spectrum of light that an object emits depends on its temperature.

LG 5   Differentiate luminosity from brightness, and illustrate how distance affects each.

The visible part of the electromagnetic spectrum is laid out in the colors of this rainbow. ▶ ▶ ▶

What is light?

5.1 Light Brings Us the News of the Universe

Since the earliest investigations of light, there has been disagreement over the question of whether light is composed of particles or if it is a disturbance that travels from one point to another, called a **wave**. Scientists have since come to understand that light sometimes acts like a wave and at other times acts like a particle. We will begin this section with a discussion of how fast light travels. We will then discuss its wavelike properties. After that we will look at how light behaves as a particle.

The Speed of Light

In the 1670s, Danish astronomer Ole Rømer (1644–1710) studied the movement of the moons of Jupiter, measuring the times when each moon disappeared behind the planet. To his amazement, the observed times did not follow the regular schedule that he predicted from Kepler's laws. Sometimes the moons disappeared behind Jupiter sooner than expected, and at other times they disappeared behind Jupiter later than expected. Rømer realized that the difference depended on where Earth was in its orbit. If he began tracking the moons when Earth was closest to Jupiter, by the time Earth was farthest from Jupiter the moons were almost 17 minutes "late." When Earth was once again closest to Jupiter, the moons again passed behind Jupiter at the predicted times.

Rømer correctly concluded that his observations were not a failure of Kepler's laws. Instead, he was seeing the first clear evidence that light travels at a finite speed. As shown in **Figure 5.1**, the moons appeared "late" when Earth was farther from Jupiter because of the time needed for light to travel the extra distance between the two planets. Over the course of Earth's yearly trip around the Sun, the distance between Earth and Jupiter changes by 2 astronomical units (AU). The speed of light equals this distance divided by Rømer's 16.7-minute delay, or about 3×10^5 kilometers per second (km/s). The value that Rømer actually announced in 1676 was a bit on the low side—2.25×10^5 km/s—because the size of

Figure 5.1 Danish astronomer Ole Rømer realized that apparent differences between the predicted and observed orbital motions of Jupiter's moons depend on the distance between Earth and Jupiter. He used these observations to measure the speed of light. (The superscript letters in the expression "$16^m 40^s$" stand for minutes and seconds of time, respectively.)

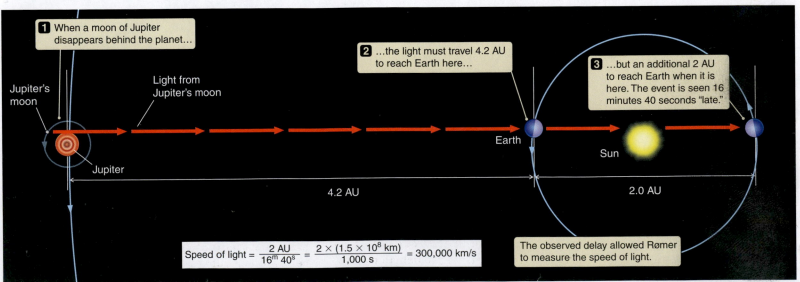

1 When a moon of Jupiter disappears behind the planet…

2 …the light must travel 4.2 AU to reach Earth here…

3 …but an additional 2 AU to reach Earth when it is here. The event is seen 16 minutes 40 seconds "late."

Jupiter's moon

Light from Jupiter's moon

Jupiter

Earth

Sun

4.2 AU

2.0 AU

$$\text{Speed of light} = \frac{2\ \text{AU}}{16^m\ 40^s} = \frac{2 \times (1.5 \times 10^8\ \text{km})}{1{,}000\ \text{s}} = 300{,}000\ \text{km/s}$$

The observed delay allowed Rømer to measure the speed of light.

Earth's orbit was not well known. Modern measurements of the speed of light give a value of 2.99792458×10^5 km/s in a **vacuum** (a region of space devoid of matter). The speed of light in a vacuum is one of nature's fundamental constants, usually written as c (lowercase). The speed of light through any medium, such as air or glass, is always less than c.

The International Space Station moves around Earth at a speed of about 28,000 kilometers per hour (km/h), taking 91 minutes to complete one orbit. Light travels almost 40,000 times faster than this and can circle Earth in only $\frac{1}{7}$ of a second. Because light is so fast, the travel time of light is a convenient way of expressing cosmic distances. It takes light $1\frac{1}{4}$ seconds to travel between Earth and the Moon, so we say that the Moon is $1\frac{1}{4}$ light-*seconds* from Earth. The Sun is $8\frac{1}{3}$ light-*minutes* away, and the next-nearest star is $4\frac{1}{3}$ light-*years* distant. Thus, a *light-year* is defined as the distance traveled by light in 1 year, or about 9.5 trillion km. Although it is sometimes misused as a measure of time, the light-year is a measure of distance.

As light travels at this high speed, it carries energy from place to place. **Energy** is the ability to do work, and it comes in many forms. **Kinetic energy** is the energy of moving objects. **Thermal energy** is closely related to kinetic energy and is the sum of all the random motion of atoms, molecules, and particles, by which we measure their temperature. For example, when light from the Sun strikes the pavement, the pavement heats up. That energy was carried from the Sun to the pavement by light. Rømer knew how long it took for light to travel a given distance, but it would take more than 200 years for physicists to figure out what light actually is.

Light as an Electromagnetic Wave

In the late 19th century, the Scottish physicist James Clerk Maxwell (1831–1879) introduced the concept that electricity and magnetism are two components of the same physical phenomenon. An **electric force** is the push or pull between electrically charged particles that make up atoms, such as protons and electrons, arising from their electric charges. Particles with opposite charges attract, and those with like charges repel. A **magnetic force** is a force between electrically charged particles arising from their motion.

To describe these electric and magnetic forces, Maxwell considered what happens when charged particles move in *electric fields* and *magnetic fields*. An **electric field** is a measure of the electric force on a charge at any point in space. Similarly, a **magnetic field** is a measure of the magnetic force acting on a small magnet at any point in space.

Maxwell summarized the behavior of electric fields and magnetic fields in four elegant equations. Among other things, these equations say that a changing electric field causes a magnetic field, and that a changing magnetic field causes an electric field. A change in the motion of a charged particle causes a changing electric field, which causes a changing magnetic field, which causes a changing electric field, and so on. You can see this interaction in **Figure 5.2**. Once the process starts, a self-sustaining procession of oscillating electric and magnetic fields moves out in all directions through space. In other words, an accelerating charged particle gives rise to an **electromagnetic wave**. These electromagnetic waves, and the accelerating charges that generate them, are the sources of electromagnetic radiation. Maxwell's equations also predict the speed at which an electromagnetic wave should travel, which agrees with the measured speed of light (c).

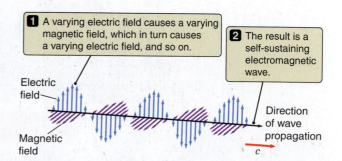

1 A varying electric field causes a varying magnetic field, which in turn causes a varying electric field, and so on.

2 The result is a self-sustaining electromagnetic wave.

Electric field

Magnetic field

Direction of wave propagation

c

Figure 5.2 An electromagnetic wave consists of oscillating electric and magnetic fields that are perpendicular both to each other and to the direction in which the wave travels.

Maxwell's wave description of light also gives us an idea of how light originates and how it interacts with matter. When a drop of water falls from the faucet into a sink full of water, it causes a disturbance, or wave, like the one shown in **Figure 5.3a**. The wave moves outward as a ripple on the surface of the water. As shown in Figure 5.3b, electromagnetic waves resulting from periodic changes in the strength of the electric and magnetic fields move out through space away from their source in much the same way. However, the ripples in the sink are distortions of the water's surface, and they require a **medium**: a substance to travel through. Light waves move through empty space—what we call a vacuum—in the absence of a medium.

Now imagine that a soap bubble is floating in the sink, illustrated in **Figure 5.4a**. The bubble remains stationary until the ripple from the dripping faucet reaches it. As the ripple passes by, the rising and falling water causes the bubble to rise and fall. This can only happen if the wave is carrying energy—a conserved quantity that gives objects and particles the ability to do work. Light waves similarly carry energy through space and cause electrically charged particles to vibrate, as in Figure 5.4b.

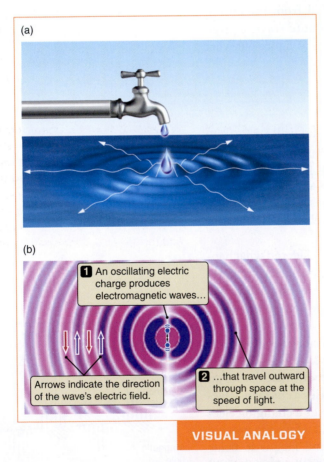

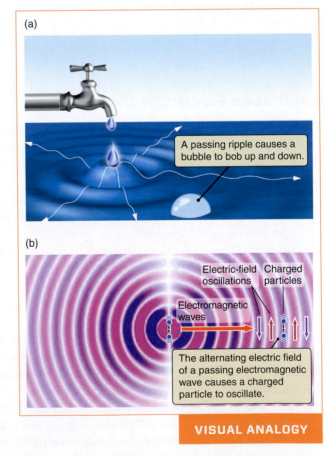

Figure 5.3 (a) A drop falling into water generates waves that move outward across the water's surface. (b) In similar fashion, an oscillating (accelerated) electric charge generates electromagnetic waves that move away at the speed of light.

Figure 5.4 (a) When waves moving across the surface of water reach a bubble, they cause the bubble to bob up and down. (b) Similarly, a passing electromagnetic wave causes an electric charge to oscillate in response to the wave.

Characterizing Waves

In this book you will learn about several kinds of waves, including electromagnetic waves crossing the vast expanse of the universe and earthquakes traveling through Earth. Waves are generally characterized by four quantities: *amplitude*, *speed*, *frequency*, and *wavelength*. Each of these quantities is illustrated in **Figure 5.5**. The **amplitude** of a wave is the height of the wave above the undisturbed position (Figure 5.5a). For water waves, the amplitude is how far the water is lifted up by the wave. In the case of light, the amplitude of a light wave is related to the brightness of the light. A water wave travels at a particular speed, v (Figure 5.5b), through the water. The water itself doesn't travel; it just moves up and down at the same location. For waves like those in water, this speed is variable and depends on the density of the substance the wave moves through, among other things. Light, in contrast, always moves through a vacuum at the same speed, $c \approx 300{,}000$ km/s.

The distance from one crest of a wave to the next is the **wavelength**, usually denoted by the Greek letter lambda, λ (Figure 5.5c). The number of wave crests passing a point in space each second is called the wave's **frequency**, f. The unit of frequency is cycles per second, which is called **hertz** (abbreviated **Hz**) after the 19th century physicist Heinrich Hertz (1857–1894), who was the first to experimentally confirm Maxwell's predictions about electromagnetic radiation. Panels (c) and (d) of Figure 5.5 show that waves with longer wavelengths have lower frequencies, and waves with shorter wavelengths have higher frequencies. Higher-frequency waves carry more energy. Think about standing on an ocean beach: if the ocean waves are more frequent, they will be more energetic.

Waves travel a distance of one wavelength each cycle, so the speed of a wave can be found by multiplying the frequency and the wavelength. Translating this idea into math, we have $v = \lambda f$. The speed of light in a vacuum is always c, so once the wavelength of a wave of light is known, its frequency is known, and vice versa. Because light travels at constant speed, its wavelength and frequency are inversely proportional to each other: if the wavelength increases, the frequency decreases. A tremendous amount of information can be carried by waves; for example, complex and beautiful music, which travels by sound waves. As you continue your study of the universe, time and time again you will find that the information you receive, whether about the interior of Earth or about a distant star or galaxy, rides in on a wave.

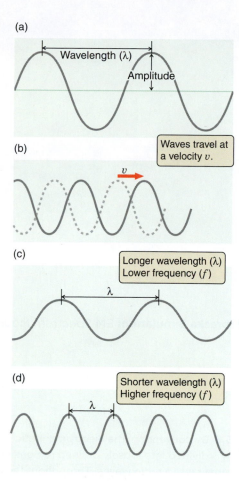

Figure 5.5 A wave is characterized by the distance from one peak to the next (wavelength, λ), the frequency of the peaks (f), the maximum deviations from the medium's undisturbed state (amplitude), and the speed (v) at which the wave pattern travels from one place to another. In an electromagnetic wave, the amplitude is the maximum strength of the electric field, and the speed of light is written as c.

The Electromagnetic Spectrum

Most light signals are made up of many wavelengths. You have almost certainly seen a rainbow, spread out across the sky, as in the chapter opening figure. A rainbow is created when white light interacts with water droplets and is spread out into its component colors. Light spread out by wavelength is called a **spectrum**. At the long-wavelength (and therefore low-frequency) end of the visible spectrum is red light. At the other end is violet light. A commonly used unit for the wavelength of visible light is the **nanometer**, abbreviated **nm**. A nanometer is one-billionth (10^{-9}) of a meter. Human eyes can see light between violet (about 380 nm) and red (750 nm). Stretched out between the two, in a rainbow, is the rest of the visible spectrum.

The light-sensitive cells in our eyes respond to visible light. But this is only a small sample of the range of possible wavelengths for electromagnetic radiation.

Light can have wavelengths that are much shorter or much longer than your eyes can perceive. The whole range of different wavelengths of light is collectively called the **electromagnetic spectrum**, illustrated in **Figure 5.6**. Most of the electromagnetic spectrum, and therefore most of the information in the universe, is invisible to the human eye. To detect light outside the visible, we must use specialized detectors of various kinds, as we will discuss in Chapter 6.

Refer to Figure 5.6 as we take a tour of the electromagnetic spectrum, beginning with the shortest wavelengths and working our way to the longest ones. The very shortest wavelengths of light are called **gamma rays**, or sometimes gamma radiation. Because this light has the shortest wavelengths, it has the highest frequency and the highest energy, so it penetrates matter easily. Wavelengths between 0.1 nm and 40 nm are called **X-rays**. You have probably encountered X-rays at the dentist or in an emergency room—X-ray light has enough energy to penetrate through skin and muscle but is stopped by denser bone. **Ultraviolet (UV) radiation** has wavelengths between 40 and about 380 nm—longer than X-rays but shorter than visible light. You are familiar with this type of light from sunburns: UV light has enough energy to penetrate into your skin, but not much farther.

Infrared (IR) radiation has longer wavelengths than the reddest wavelengths in the visible range. You are familiar with a small wavelength range of this kind of radiation because you often feel it as heat. When you hold your hand next to a hot stove, some of the heat you feel is carried to your hand by infrared radiation emitted from the stove. In this sense, you could think of your skin as being a giant infrared eyeball—it is sensitive to infrared wavelengths. Infrared radiation is also used in television remote controls, and night vision goggles detect infrared radiation from warm objects such as animals. A useful unit for infrared light is the **micron** (abbreviated **μm**, where μ is the Greek letter mu). One micron is 1,000 nm, or one-millionth (10^{-6}) of a meter. Infrared wavelengths are longer than red light and shorter than 500 microns.

Microwave radiation has even longer wavelengths than infrared radiation. The microwave in your kitchen heats the water in food using light of these

Nebraska Simulation: EM Spectrum Module

Figure 5.6 By convention, the electromagnetic spectrum is divided into loosely defined regions ranging from gamma rays to radio waves. Throughout the book, we use the following labels to indicate the form of radiation used to produce astronomical images, with an icon to remind you: G = gamma rays; X = X-rays; U = ultraviolet; V = visible; I = infrared; R = radio. If more than one region is represented, multiple labels are highlighted.

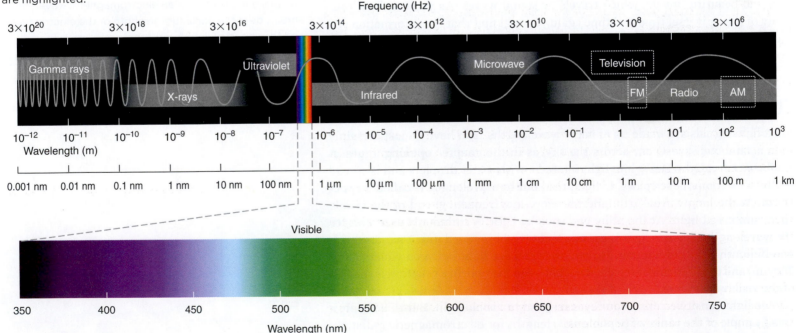

wavelengths. The longest-wavelength light, which has wavelengths longer than a few centimeters, is called **radio waves**. Light of these wavelengths in the form of FM, AM, television, and cell phone signals is used to transmit information from place to place.

CHECK YOUR UNDERSTANDING 5.1

Rank the following in order of decreasing wavelength: (a) gamma rays; (b) visible light; (c) infrared light; (d) ultraviolet light; (e) radio waves.

Light as a Particle

Although the wave theory of light describes many observations, it does not provide a complete picture of the properties of light. Many of the difficulties with the wave model of light have to do with the way in which light interacts with small particles such as atoms and molecules.

The work of Albert Einstein and other scientists modified our understanding of light to show that light sometimes acts like a wave and sometimes acts like a particle. In 1905, Einstein explained the *photoelectric effect*, in which electrons are emitted when surfaces are illuminated by electromagnetic radiation greater than a certain frequency. He showed that the rate at which electrons are emitted depends only on the amount of incoming light, and that the speed of the electrons depends only on the frequency of the incoming radiation. This work earned Einstein a Nobel Prize in 1921. In the particle model, light is made up of massless particles called **photons** (*phot-* means "light," as in *photograph*; and *-on* signifies a particle). Photons always travel at the speed of light (**Process of Science Figure**), and they carry energy. A photon is a *quantum* of light; *quantized* means that something is subdivided into individual units. **Quantum mechanics** is a branch of physics that deals with the quantization of energy and of other properties of matter.

The energy of a photon and the frequency of the electromagnetic wave are directly proportional to each other: the higher the frequency of the light wave, the greater the energy each photon carries. This relationship connects the particle and the wave concepts of light. For example, photons with higher frequencies carry more energy than that carried by photons with lower frequencies. The constant of proportionality between the energy, E, and the frequency, f, is called Planck's constant, h, which is equal to 6.63×10^{-34} joule-seconds (a joule is a unit of energy). Specifically, $E = hf$, where $E =$ the energy of the photon, $h =$ Planck's constant, and $f =$ frequency. Because the wavelength and frequency of electromagnetic waves are inversely proportional, this also means that the photon energy is inversely proportional to the wavelength. In visible light, high-energy light is blue, and low-energy light is red. **Working It Out 5.1** explores the relationships among wavelength, frequency, and energy of light.

In the particle description of light, the electromagnetic spectrum is a spectrum of photon energies. The higher the frequency of the electromagnetic wave, the greater the energy carried by each photon. Photons of shorter wavelength (higher frequency) carry more energy than that carried by photons of longer wavelength (lower frequency). For example, photons of blue light carry more energy than that carried by photons of longer-wavelength red light. Ultraviolet photons carry more energy than that carried by photons of visible light, and X-ray photons carry more energy than that carried by ultraviolet photons. The lowest-energy photons are radio wave photons.

▶❚❚ **AstroTour:** Light as a Wave, Light as a Photon

Process of Science

Scientists working on very different problems in different fields all find the same result: light has a speed and can be measured.

Ole Rømer studies eclipses of Jupiter's moons.

Rømer calculates the speed of light from eclipse delays of Jupiter's moons.

James Bradley studies the apparent motion of stars, which appear to make small circles because of the relative motion of Earth.

A half century after Rømer's measurement, Bradley's motion studies lead to a more accurate measurement of the speed of light.

James Clerk Maxwell studies electricity and magnetism.

Einstein builds on Maxwell's theory in 1905 to claim that the speed of light is the same for all observers.

Astronomers and physicists converge on an understanding that photons travel at the speed of light—a fundamental constant of the universe that is the same for all observers.

5.1 Working It Out Working with Electromagnetic Radiation

Wavelength and Frequency

When you tune to a radio station at, say, 770 AM, you are receiving an electromagnetic signal that travels at the speed of light and is broadcast at a frequency of 770 kilohertz (kHz), or 7.7×10^5 Hz. We can use the relationship between wavelength and frequency, $c = \lambda f$ to calculate the wavelength of the AM signal:

$$\lambda = \frac{c}{f} = \left(\frac{3 \times 10^8 \, \text{m/s}}{7.7 \times 10^5/\text{s}} \right) = 390 \, \text{m}$$

This AM wavelength is about 4 times the length of a football field. FM wavelengths are much shorter than AM wavelengths.

The human eye is most sensitive to light in green and yellow wavelengths, about 500–590 nm. If we examine green light with a wavelength of 530 nm, we can compute its frequency:

$$f = \frac{c}{\lambda} = \left(\frac{3 \times 10^8 \, \text{m/s}}{530 \times 10^{-9} \, \text{m}} \right) = 5.66 \times 10^{14}/\text{s} = 5.66 \times 10^{14} \, \text{Hz}$$

This frequency corresponds to 566 *trillion* wave crests passing by each second.

Photon Energy

Let's compare the energy of an X-ray photon with a wavelength of 1 nm and the energy of a visible light photon with a wavelength of 530 nm as used in the previous calculation. The equation for the energy of a photon is $E = hf$. Because $f = c/\lambda$, substituting c/λ for f yields the inverse relationship, $E = hc/\lambda$, with c = the speed of light = 3×10^8 m/s, and h = Planck's constant. Because we are making a *comparison*, we can take a ratio, and then the constants h and c cancel out:

$$\frac{E_{\text{X-ray photon}}}{E_{\text{visible photon}}} = \frac{hc/\lambda_{1\,\text{nm}}}{hc/\lambda_{500\,\text{nm}}} = \frac{h\cancel{c}}{h\cancel{c}} \times \frac{530 \, \text{nm}}{1 \, \text{nm}} = 530$$

The X-ray photon has 530 times the energy of the visible light photon.

The *total* amount of energy that a beam of the light carries is called its **intensity**. A beam of red light can be just as intense as a beam of blue light—that is, it can carry just as much energy—but because the energy of a red photon is less than the energy of a blue photon, maintaining that same intensity requires more red photons than blue photons. This relationship, illustrated in **Figure 5.7**, is a lot like money: $10 is $10, but it takes a lot more pennies (low-energy photons) than quarters (high-energy photons) to make up $10.

CHECK YOUR UNDERSTANDING 5.2

As wavelength increases, the energy of a photon _____ and its frequency _____. (a) increases; decreases (b) increases; increases (c) decreases; decreases (d) decreases; increases

5.2 The Quantum View of Matter Explains Spectral Lines

Light and matter interact, and this interaction allows us to detect matter even at great distances in space. To understand this interaction, we must understand the building blocks of matter. In this section, we will review atomic structure and the process by which astronomers identify the chemical elements in astronomical objects.

Atomic Structure

Matter is anything that occupies space and has mass. Atoms are composed of a central massive **nucleus**, which contains **protons** with a positive charge and **neutrons**, which have no charge. A cloud of negatively charged **electrons**

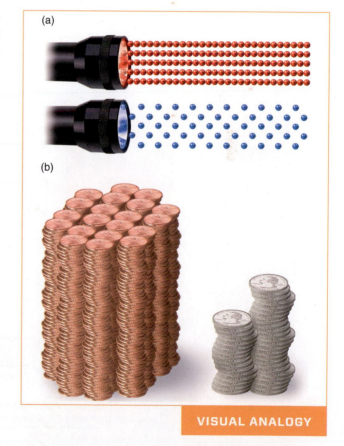

(a)

(b)

VISUAL ANALOGY

Figure 5.7 (a) Red light carries less energy than that carried by blue light, so it takes more red photons than blue photons to make a beam of a particular intensity. (b) Similarly, pennies are worth less than quarters, so it takes more pennies than quarters to add up to $10.

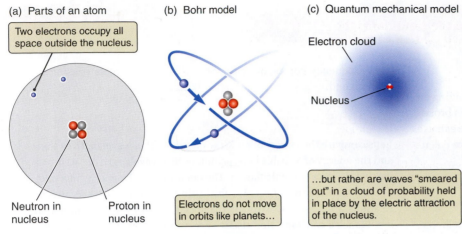

(a) Parts of an atom

Two electrons occupy all space outside the nucleus.

Neutron in nucleus

Proton in nucleus

(b) Bohr model

Electrons do not move in orbits like planets…

(c) Quantum mechanical model

Electron cloud

Nucleus

…but rather are waves "smeared out" in a cloud of probability held in place by the electric attraction of the nucleus.

Figure 5.8 (a) An atom (in this case helium) is made up of a nucleus consisting of positively charged protons and electrically neutral neutrons and surrounded by less massive negatively charged electrons. (b) Atoms are often drawn as miniature "solar systems," but this model is incorrect. (c) Electrons are actually smeared out around the nucleus in quantum mechanical clouds of probability.

▶❚❚ **AstroTour:** Atomic Energy Levels and the Bohr Model

(a)

WAR & PEACE

Energy states of atoms are like shelves in a bookcase.

WAR & PEACE

You can find a book on one shelf or another, but not in between.

(b)

Energy

E_5 ——————
E_4 ———●———
E_3 ——————

E_2 ——————

E_1 ——————
Ground state

Energy

E_5 ——————

E_4 ——————
E_3 ——————

E_2 ———●———

E_1 ——————
Ground state

We use **energy level diagrams** to represent the allowed states of an atom.

Atom in E_4 energy state

Atom in E_2 energy state

Analogously, atoms exist in one allowed energy state or another, but never in between.

VISUAL ANALOGY

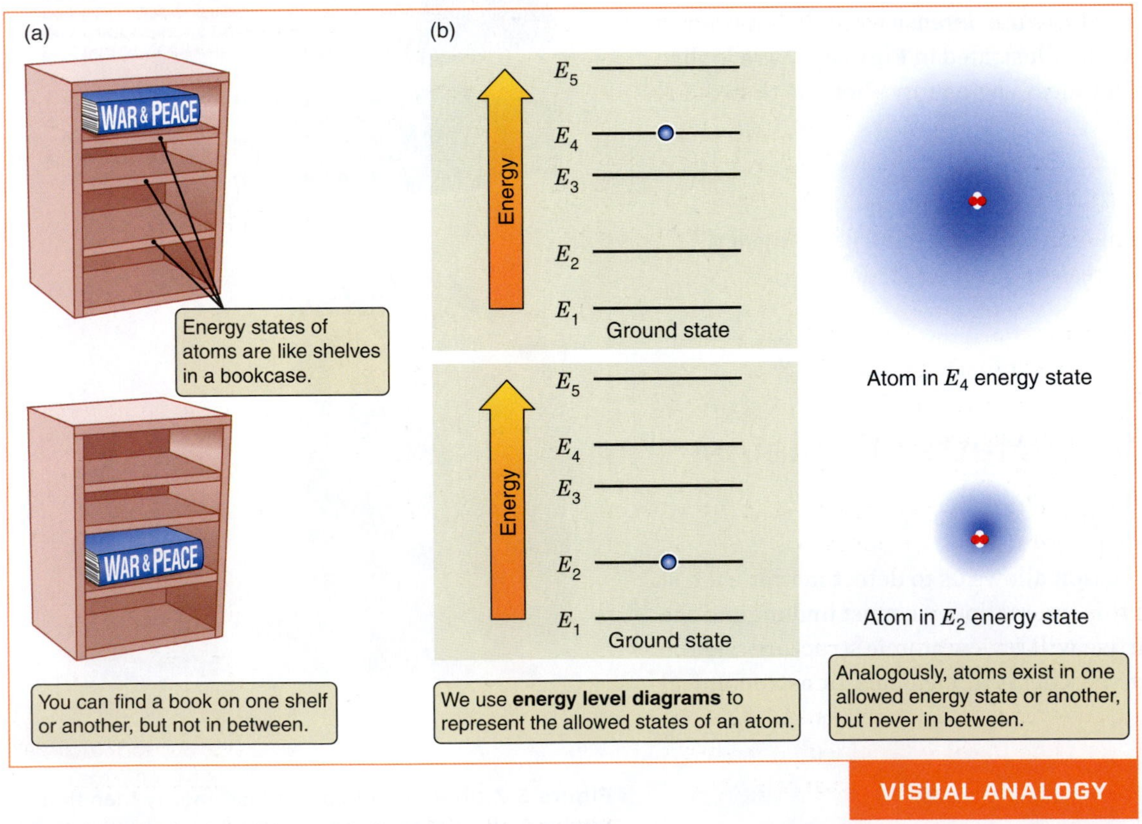

Figure 5.9 (a) Energy states of an atom are analogous to shelves in a bookcase. You can move a book from one shelf to another, but books can never be placed between shelves. (b) Atoms exist in one allowed energy state or another but never in between. There is no level below the ground state.

surrounds the nucleus. Atoms with the same number of protons are all of the same type, known as an **element**. For example, an atom with two protons, shown in **Figure 5.8a**, is the element helium. An atom with six protons is the element carbon; one with eight protons is the element oxygen; and so forth. An element may have many **isotopes**: atoms with the same number of protons but different numbers of neutrons. **Molecules** are groups of atoms bound together by shared electrons. A single teaspoon of water contains about 10^{23} atoms—about as many atoms as there are stars in the observable universe.

For an atom to be electrically neutral, it must have the same number of electrons as protons. Electrons have much less mass than protons or neutrons have, so almost all the mass of an atom is found in its nucleus. This description led to a model of an atom with the massive nucleus sitting in the center and the smaller electrons orbiting around it, much as planets orbit around the Sun (Figure 5.8b). It is called the **Bohr model** after the Danish physicist Niels Bohr (1885–1962), who proposed it in 1913.

However, Bohr's model is not a complete description of an atom. Just as waves of light have particle-like properties, particles of matter also have wavelike properties. With this realization, the Bohr model of the atom was modified so that the positively charged nucleus is surrounded by electron "clouds" or "waves," as shown in Figure 5.8c. In this model, it is not possible to know precisely where the electron is in its orbit. The wave characteristics of particles make it impossible to pin down simultaneously their exact location and their exact velocity; there will always be some uncertainty. This is why a featureless cloud is used to represent electrons in orbit around an atomic nucleus.

Atomic Energy Levels

Because of their wavelike properties, electrons in an atom can take on only certain specific energies that depend on the energy states of the atom. The form that the electron waves take depends on the possible energy states of atoms. We can imagine the energy states of atoms as being like a bookcase with a set of shelves, as depicted in **Figure 5.9a**. The energy of an atom might correspond to the energy of one state or to the energy of the next state, but the energy of the atom is never found between the two states, just as a book can be on only one shelf at a time and cannot be partly on one shelf and partly on another. A given atom may have many different energy states available to it, but these states are *discrete*.

Astronomers keep track of the allowed states of an atom using energy level diagrams, as shown in Figure 5.9b, where each energy level is like a shelf on the bookcase. Both the bookcase and the energy level diagram are simplifications of the possible energies of a three-dimensional system. The lowest possible energy state for a system (or part of a system) such as an atom is called the **ground state**. When the atom is in the ground state, the electron has its minimum energy. It can't give up any more energy to move to a lower state, because there isn't a lower state. An atom will remain in its ground state forever unless it gets energy from outside. In the bookcase analogy, a book sitting on the bottom shelf at the floor is in its ground state. It has nowhere left to fall, and it cannot jump to one of the higher shelves of its own accord.

Energy levels above the ground state are called **excited states**. Just as a book on an upper shelf might fall to a lower shelf, an atom in an excited state might **decay** to a lower state by getting rid of some of its extra energy. An important difference between the atom and the book on the shelf, however, is that whereas a snapshot might catch the book falling between the two shelves, the atom will never be caught between two energy states. When the transition from one higher state to a lower one occurs, the difference in energy between the two states is carried off all at once. A common way for an atom to do this is to emit a photon. The photon emitted by the atom carries away exactly the amount of energy lost by that atom as it goes from the higher energy state to the lower energy state. In a similar fashion, atoms moving from a lower energy state to a higher energy state can absorb only certain specific energies.

To make an analogy with money, suppose you have a penny (1 cent), a nickel (5 cents), and a dime (10 cents), totaling 16 cents. Now imagine that you give away the nickel and are left with 11 cents. You never had exactly 13 cents or 13.6 cents. You had 16 cents, and then 11 cents. Atoms don't accept and give away money to change energy states, but they do accept and give away photons with well-defined energy.

Emission Spectra   Imagine a hypothetical atom that has only two available energy states. The energy of the lower energy state (the ground state) is E_1, and the energy of the higher energy state (the excited state) is E_2. The energy levels of this atom can be represented in an energy level diagram like the one in **Figure 5.10a**. An atom in the excited state moves to the ground state by getting rid of the "extra" energy all at once. It does this when the electron emits a photon. The atom goes from one energy state to another, but it never has an amount of energy in between.

In Figure 5.10b, the downward arrow indicates that the atom went from the higher state with energy E_2 to the lower state with energy E_1. The atom lost an amount of energy equal to the difference between the two states, or $E_2 - E_1$. Because energy is never truly lost or created, the energy lost by the atom has to show up somewhere. In this case, the energy shows up in the form of a photon that is emitted by the atom. The energy of the photon emitted exactly matches the energy lost by the atom; that is, $E_{\text{photon}} = E_2 - E_1$.

An atom can emit photons with energies corresponding *only* to the difference between two of its allowed energy states. Because the energy of a photon is related to the frequency or wavelength of electromagnetic radiation, in the

Astronomy in Action: Emission and Absorption

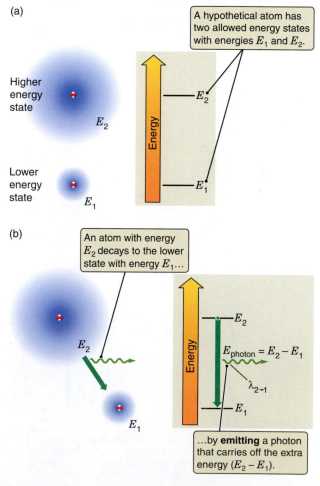

(a)

A hypothetical atom has two allowed energy states with energies E_1 and E_2.

(b)

An atom with energy E_2 decays to the lower state with energy E_1…

$E_{\text{photon}} = E_2 - E_1$

…by **emitting** a photon that carries off the extra energy ($E_2 - E_1$).

Figure 5.10 (a) The energy levels of a hypothetical two-level atom. (b) A photon with energy $E_{\text{photon}} = E_2 - E_1$ is emitted when an atom in the higher energy state decays to the lower energy state.

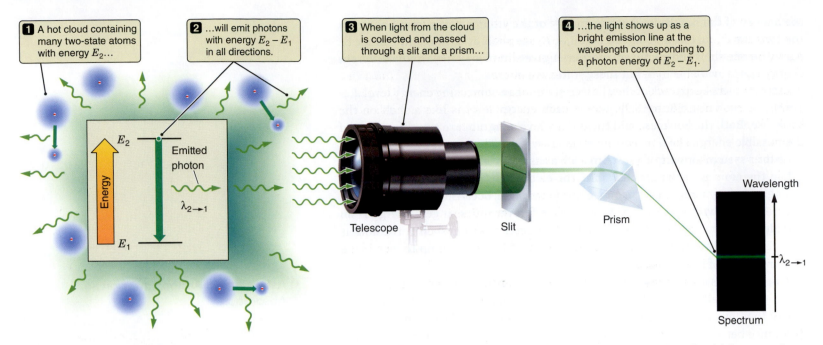

Figure 5.11 A cloud of gas containing atoms with two energy states, E_1 and E_2, emits photons with an energy $E_{photon} = E_2 - E_1$, which appear in the spectrum (right) as a single bright *emission line*.

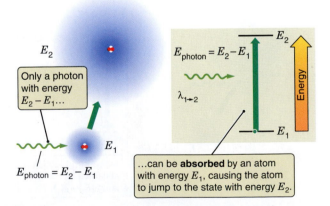

Figure 5.12 An atom in a lower energy state may absorb a photon of energy $E_{photon} = E_2 - E_1$, leaving the atom in a higher energy state.

transition from level 2 to level 1, a photon of energy $E_{photon} = E_2 - E_1$ has a specific wavelength $\lambda_{2 \to 1}$ and a specific frequency $f_{2 \to 1}$. Therefore, these emitted photons have a very specific color, and every photon emitted in any transition from E_2 to E_1 will have this same color defined by a specific wavelength. The energy level structure of an atom determines the wavelengths of the photons it emits—the color of the light that the atom gives off.

Figure 5.11 illustrates the light coming from a cloud of gas consisting of the hypothetical two-state atoms in Figure 5.10. Any atom in the higher energy state (E_2) quickly decays and emits a photon in a random direction, and an enormous number of photons come pouring out of the cloud of gas. Instead of containing photons of all different energies—light of all different colors—this light contains only photons with the specific energy $E_2 - E_1$ and wavelength $\lambda_{2 \to 1}$. In other words, all of the light coming from these atoms in the cloud is the same color. If you spread the light out into its component colors, there would be only one color— a single bright line called an **emission line**.

Why was the atom in the excited state E_2 in the first place? An atom sitting in its ground state will remain there unless it absorbs just the right amount of energy to kick it up to an excited state. In general, the atom either absorbs the energy of a photon or it collides with another atom or an unattached electron and absorbs some of the other particle's energy. In a neon sign, an alternating electric field inside the glass tube pushes electrons in the gas back and forth through the neon gas inside the tube. Some of these electrons crash into atoms of the gas, knocking them into excited states. The atoms then drop back down to their ground states by emitting photons, causing the gas inside the tube to glow.

Absorption Spectra In the opposite process, an atom in a low energy state can absorb the energy of a passing photon and jump up to a higher energy state as shown in **Figure 5.12**. Once again, the energy required to go from E_1 to E_2 is the difference in energy between the two states, $E_2 - E_1$. For a photon to cause an atom to jump from E_1 to E_2, it must provide exactly this much energy. The only photons capable of exciting atoms from E_1 to E_2 are photons with $E_{photon} = E_2 - E_1$. As with emission, these photons have a corresponding frequency and wavelength $f_{1 \to 2} = E_{photon}/h$ and $\lambda_{1 \to 2} = hc/E_{photon}$. These photons have exactly the same energy—the same color of light—emitted by the atoms when they decay from E_2 to E_1. This is not a coincidence. The energy difference between the two levels is the same whether the atom is emitting a photon or absorbing one, so the energy of the photon involved will be the same in either case.

In **Figure 5.13a**, white light (with all wavelengths of photons in it) passes directly though a glass prism, which breaks up the light into a rainbow of colors. However, when the white light passes through a cool cloud composed of our hypothetical gas of two-state atoms, as illustrated in Figure 5.13b, some photons will be absorbed. Almost all of the photons will pass through the cloud of gas unaffected, because they do not have the right energy ($E_2 - E_1$) to be absorbed by atoms of the gas. However, photons with just the right amount of energy can be absorbed, and as a result, these photons will be missing in the light passing

Nebraska Simulation: Three Views Spectrum Demonstrator

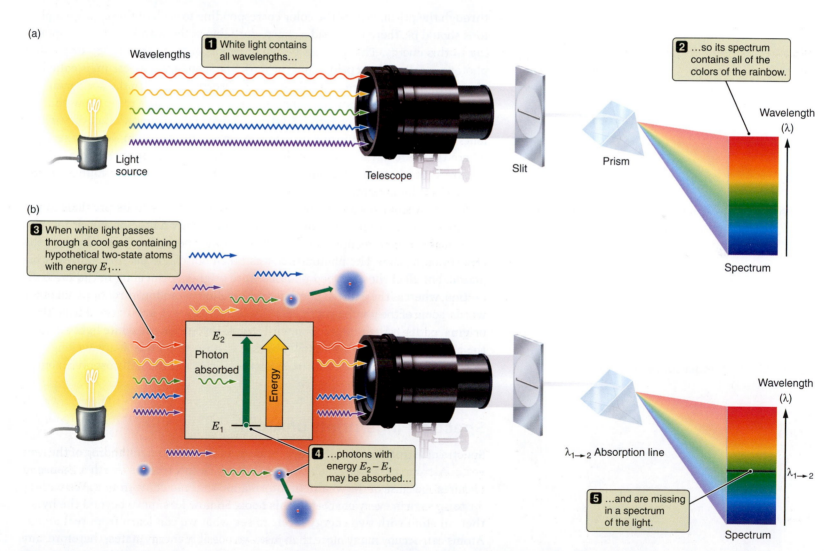

Figure 5.13 (a) When passed through a prism, white light produces a spectrum containing all colors. (b) When light of all colors passes through a cloud of hypothetical two-state atoms, photons with energy $E_{photon} = E_2 - E_1$ may be absorbed, leading to the dark absorption line in the spectrum.

(a)

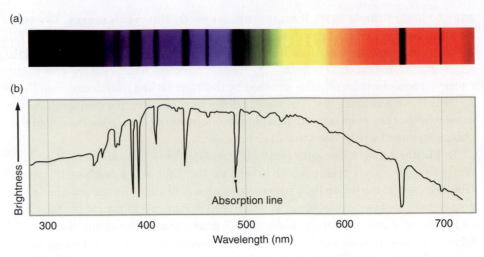

(b)

Figure 5.14 Absorption lines in the spectrum of a star as an image (a) and a graph (b).

▶❚❚ **AstroTour:** Atomic Energy Levels and Light Emission and Absorption

Nebraska Simulation: Hydrogen Atom Simulator

through the prism. Where the color corresponding to each of these missing photons should be, there is instead a sharp, dark line at the wavelength corresponding to this energy. This process by which atoms capture the energy of passing photons is called **absorption**, and the dark line seen in the spectrum is called an **absorption line**. **Figure 5.14a** shows such absorption lines in the spectrum of a star. The spectrum is shown in two different ways here: as a rainbow with light missing, and then again in Figure 5.14b as a graph of the brightness at every wavelength. Comparing the top and bottom versions of the spectrum, you can see that where there are dark lines, the brightness drops abruptly at a particular wavelength. Places between the dark lines are brighter and therefore higher on the graph than the absorption lines.

When an atom absorbs a photon, it may quickly decay to its previous lower energy state, emitting a photon with the same energy as the photon it just absorbed. If the atom reemits a photon just like the one it absorbed, why does the absorption matter? The photon that was taken out of the passing light was replaced, but all of the absorbed photons were originally traveling in the same direction, whereas the emitted photons are traveling in random directions. In other words, some of the photons with energies equal to $E_2 - E_1$ are diverted from their original paths by their interaction with atoms. If you look at a white light through the cloud in Figure 5.13b, you will observe an absorption line at a wavelength of $\lambda_{1\to2}$, but if you look at the cloud from another direction, you will observe an emission line at this same wavelength.

Spectral Fingerprints of Atoms

Spectra of astronomical objects are fundamental to our understanding of the universe. Astronomers who study spectra will say "a spectrum is worth a thousand pictures" because of the wealth of information that can come from it. We will refer to spectra in every chapter in this book. So now let's move beyond the hypothetical atom with two energy levels to see what we can learn from real atoms. Atoms can occupy many more than just two possible energy states; therefore, any given type of atom will be capable of emitting and absorbing photons at many different wavelengths. An atom with three energy states, for example, might jump from state 3 to state 2, or from state 3 to state 1, or from state 2 to state 1. The three

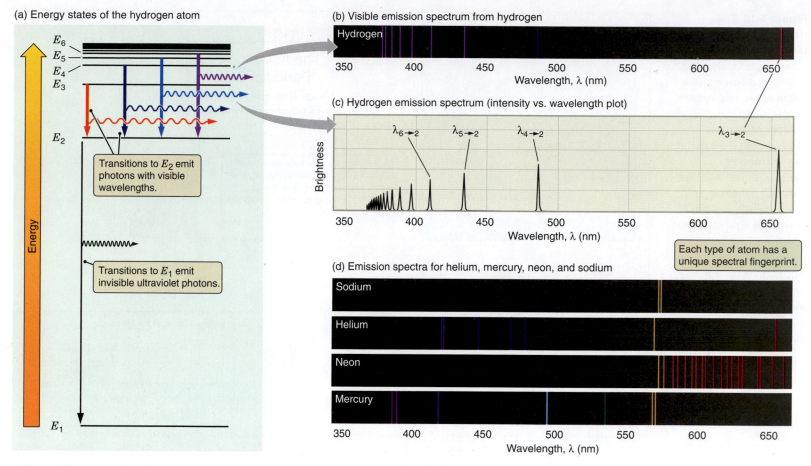

Figure 5.15 (a) The energy states of the hydrogen atom. Decays to level E_2 emit photons in the visible part of the spectrum. (b) This spectrum is what you might see if you looked at the light from a hydrogen lamp projected through a prism onto a screen. (c) This graph of the brightness (intensity) of spectral lines versus their wavelength illustrates how spectra are traditionally plotted. (d) Emission spectra from several other gases: helium, mercury, neon, and sodium.

distinct emission lines in the spectrum from a gas made up of these atoms would have wavelengths of $hc/(E_3 - E_2)$, $hc/(E_3 - E_1)$, and $hc/(E_2 - E_1)$, respectively.

The allowed energy states of an atom are determined by the complex interactions among the electrons and the nucleus. Every neutral hydrogen atom consists of a nucleus containing one proton, plus a single electron in a cloud surrounding the nucleus. Therefore, every hydrogen atom has the same energy states available to it, and all hydrogen atoms have the same emission and absorption lines. **Figure 5.15a** shows the energy level diagram of hydrogen. Figure 5.15b illustrates the visible emission spectrum from hydrogen. Figure 5.15c displays this same information as a graph.

Each different type of atom, that is, each chemical element, has a unique set of available energy states and therefore a unique set of wavelengths at which it can emit or absorb radiation. Figure 5.15d shows the emission spectra of four different kinds of atoms. These unique sets of wavelengths serve as unmistakable spectral "fingerprints" for each chemical element.

Spectral fingerprints are of crucial importance to astronomers. They let astronomers figure out what types of atoms (or molecules) are present in distant objects by simply looking at the spectrum of light from those objects. If the spectral lines of hydrogen, helium, carbon, oxygen, or any other element are visible in

Figure 5.16 The traditional periodic table of the elements (lower right) shows the chemical elements laid out in ascending order according to the number of protons in the nucleus of each. But the "astronomer's periodic table" displays the abundances of the Sun's elements in boxes of relative size, showing hydrogen and helium as the most abundant. See Appendix 3 for a full periodic table of the elements.

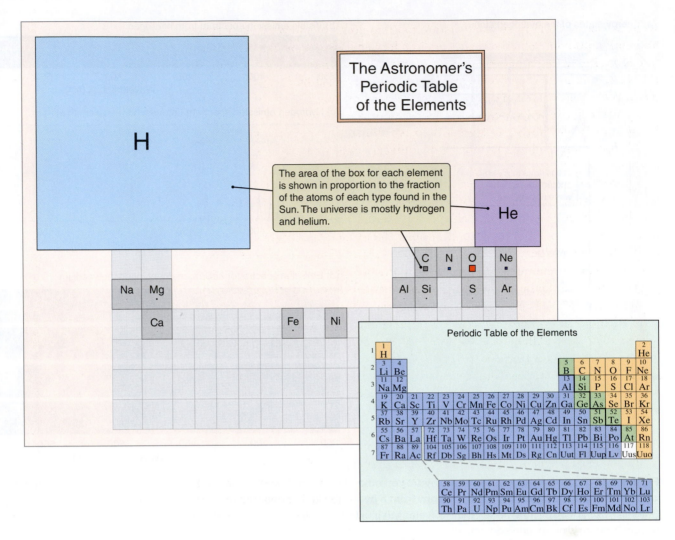

the light from a distant object, then we know that element is present in that object. The strength of a line is determined in part by how many atoms of that type are present in the source. By measuring the strength of the lines from different types of atoms in the spectrum of a distant object, astronomers can often infer the relative amounts of elements that make up the object. Additionally, by looking at the relative strength of different lines from the same element, it is often possible to determine the temperature, density, and pressure of the material as well.

This is how we know what makes up the stars and planets, and that we on Earth are composed of the same elements. Astronomers use the relative abundance of the elements in the Sun as a standard reference, termed **solar abundance**. As illustrated in **Figure 5.16**, hydrogen (H) is the most abundant element in the Sun, followed by helium (He) and 13 others. These 15 elements make up 99.99 percent of the mass of the Sun. The majority of the elements on the regular periodic table (in the lower right) make up less than 0.01 percent of the mass of the Sun.

Excitement and Decay

Return to the analogy between the emission of a photon and a book falling off a shelf. If a book on a level shelf is not disturbed, it will sit there forever; something must *cause* the book to fall off the shelf. Similarly, physicists have wondered what causes an atom in an excited state to jump down to a lower energy state and emit a photon. Sometimes an atom in a higher energy state can be "stimulated" into emitting a photon—but under most circumstances nothing causes the atom to jump to the lower energy state. Instead, the atom decays *spontaneously*. While scientists can determine on average how long the atom is likely to remain in the excited state, exactly when a given atom will decay cannot be known until after the decay has happened. An atom decays at a random moment that is not influenced by anything in the universe and cannot be known ahead of time.

An example of this phenomenon is in toys that glow in the dark. Photons in sunlight or from a lightbulb are absorbed by certain phosphorescent atoms in the toy, knocking those atoms into excited energy states. The excited states of the atoms in the toy live for many seconds, unlike the excited energy states of many atoms that tend to decay in a small fraction of a second. If on average these atoms tend to remain in their excited state for 1 minute before decaying and emitting a photon, then after 1 minute there is a 50-50 chance that any particular atom in the toy will have decayed and a 50-50 chance that the atom will remain in its excited state. Although it is impossible to say exactly which atoms will decay, about half of the trillions and trillions of atoms in the toy will decay within 1 minute, and the brightness of the glow from the toy will have dropped to half of what it was. After each minute, half of the remaining excited atoms decay, and the glow from the toy drops to half of what it was 1 minute earlier. The glow from the toy slowly fades away.

In deep space, where atoms can remain undisturbed for long periods of time, there are certain excited states of atoms that last, on average, for tens of millions of years or even longer. An atom may have been in such an excited energy state for a few seconds, a few hours, or 50 million years when, in an instant, it decays to the lower energy state without anything causing it to do so. Physicists can only calculate the *probabilities* that certain events would take place.

CHECK YOUR UNDERSTANDING 5.3
How can spectra tell us the chemical composition of a distant star?

5.3 The Doppler Shift Indicates Motion Toward or Away from Us

You have already seen that light is a tightly packed bundle of information that can reveal a wealth of information about the physical state of material located tremendous distances away. In this section, we shall see how light can be used to measure one of the most straightforward questions about a distant astronomical object: is it moving away from us or toward us, and at what speed?

Have you ever listened to an ambulance speed by with sirens blaring? As the ambulance comes toward you, its siren has a certain high pitch, but as it passes by, the pitch of the siren drops noticeably. If you close your eyes and listen, you have no trouble knowing when the ambulance passed; the change in the pitch of its siren indicates that it has passed you by. You do not even need an ambulance to hear this effect. The sound of normal traffic behaves in the same way. As a car drives past, the pitch of the sound that it makes suddenly drops.

Astronomy in Action: Doppler Shift

▶❚❚ **AstroTour:** The Doppler Effect

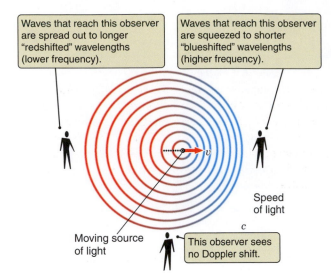

Waves that reach this observer are spread out to longer "redshifted" wavelengths (lower frequency).

Waves that reach this observer are squeezed to shorter "blueshifted" wavelengths (higher frequency).

v

Speed of light

c

Moving source of light

This observer sees no Doppler shift.

Figure 5.17 Motion of a light or sound source relative to an observer may cause waves to be spread out (*redshifted*, or lower in pitch) or squeezed together (*blueshifted*, or higher in pitch). A change in the wavelength of light or the frequency of sound is called a *Doppler shift*.

Nebraska Simulation: Doppler Shift Demonstrator

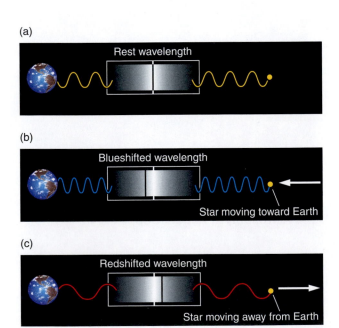

(a)

Rest wavelength

(b)

Blueshifted wavelength

Star moving toward Earth

(c)

Redshifted wavelength

Star moving away from Earth

Figure 5.18 From their rest wavelength (a), spectral lines of astronomical objects are blueshifted if they are moving toward the observer (b) and redshifted if they are moving away from the observer (c).

The pitch of a sound is like the color of light: it is determined by the wavelength or, equivalently, the frequency of the sound wave. What you perceive as higher pitch corresponds to sound waves with higher frequencies and shorter wavelengths. Sounds that you perceive as lower in pitch are waves with lower frequencies and longer wavelengths. When an object is moving toward you, the waves that it emits, whether light or sound or waves in the water, "crowd together" in front of the object. You can see how this works by looking at **Figure 5.17**, which shows the locations of successive wave crests emitted by a moving object. The waves that reach you have a shorter wavelength and therefore a higher frequency than the waves given off by the object when it is not moving. Conversely, if an object is moving away from you, the waves reaching you from the object are spread out. This change in frequency due to motion is known as the **Doppler effect**, named after physicist Christian Doppler (1803–1853).

The Doppler effect causes a shift in the light emitted from a moving object. If the object were at rest, it would emit light with the **rest wavelength** (λ_{rest}), as shown in **Figure 5.18a**. If an object such as a star is moving toward you, the light reaching you from the object has a shorter wavelength than its rest wavelength—so we say the light is "bluer" than the rest wavelength, and the light is described as **blueshifted**, as shown by the blue wave in Figure 5.18b. In contrast, light from a source that is moving away from you is shifted to longer wavelengths. The light that you see is "redder" than if the source were not moving away from you, and is described as **redshifted**, as shown by the red waves in Figure 5.18c. The faster the object is moving with respect to you, the larger the shift. The amount by which the wavelength of light is shifted by the Doppler effect is called the **Doppler shift** of the light, and it depends on the speed of the object emitting the light.

The Doppler shift provides information only about the **radial velocity** (v_r) of the object, which is the part of the motion that is toward you or away from you. The radial velocity is the rate at which the distance between you and the object is changing: if v_r is positive, the object is getting farther away from you; if v_r is negative, the object is getting closer. At the moment the ambulance is passing you, it is getting neither closer nor farther away, so the pitch you hear is the same as the pitch heard by the crew riding on the truck. Similarly, an object moving across the sky does not move toward or away from you, and so its light will not be Doppler shifted from your point of view.

Doppler shifts become especially useful when you are looking at an object that has emission or absorption lines in its spectrum. These spectral lines enable astronomers to determine how rapidly the object is moving toward or away from Earth. To determine this velocity, astronomers first identify the spectral line as being from a certain chemical element, which has a unique rest wavelength (λ_{rest}) measured in a lab on Earth. They then measure the observed wavelength (λ_{obs}) in the spectrum of the distant object. The difference between the rest wavelength and the observed wavelength indicates the object's radial velocity. This is further explored in **Working It Out 5.2**.

CHECK YOUR UNDERSTANDING 5.4

Which of the following Doppler shifts indicates the fastest approaching object (blueshifted)? (a) 0.04 nm; (b) 0.06 nm; (c) −0.04 nm; (d) −0.06 nm

5.2 Working It Out Making Use of the Doppler Effect

The Doppler formula for objects moving at a radial velocity (v_r) that is much less than the speed of light is given by

$$v_r = \frac{\lambda_{obs} - \lambda_{rest}}{\lambda_{rest}} \times c$$

A prominent spectral line of hydrogen atoms has a rest wavelength, λ_{rest}, of 656.3 nm (see Figure 5.15b). Suppose that you measure the wavelength of this line in the spectrum of a distant object and find that instead of seeing the line at 656.3 nm, you see the line at a wavelength, λ_{obs}, of 659.0 nm. How fast is its radial velocity? Using the above equation,

$$v_r = \frac{659.0 \, nm - 656.3 \, nm}{656.3 \, nm} \times (3 \times 10^5 \, km/s)$$

$$v_r = 1,200 \, km/s$$

In this way, you determine that the object is moving away from you with a radial velocity of 1,200 km/s.

For another example, suppose you know the velocity and want to compute the wavelength at which you would observe the spectral line? Earth's nearest stellar neighbor, Proxima Centauri, is moving toward us at a radial velocity of −21.6 km/s. What is the observed wavelength, λ_{obs}, of a magnesium line in Proxima Centauri's spectrum that has a rest wavelength, λ_{rest}, of 517.27 nm? We can rearrange the above equation to solve for λ_{obs}:

$$\lambda_{obs} = \left(1 + \frac{v_r}{c}\right) \times \lambda_{rest}$$

$$\lambda_{obs} = \left(1 + \frac{-21.6 \, km/s}{3 \times 10^5 \, km/s}\right) \times 517.27 \, nm = 517.23 \, nm$$

Although the observed Doppler blueshift ($\lambda_{obs} - \lambda_{rest} = 517.23 - 517.27$) is only −0.04 nm, it is easily measured with modern instrumentation.

5.4 Temperature Affects the Spectrum of Light That an Object Emits

The temperature of any object results from the balance between heating and cooling in an object. If an object's temperature is constant, then these two must be in balance with each other. In this section, we will examine this balance and see how we can use it to predict the temperatures of planets and stars.

Equilibrium and Balance

Your body is heated by the release of chemical energy inside it. Sometimes your body is also heated by energy from your surroundings. If you are standing in sunshine on a hot day, the hot air around you and the sunlight falling on you both heat you. In response to this heating, your body cools itself off by perspiration: water seeps from the pores in your skin and evaporates. The energy to evaporate the water comes from your body. As the perspiration evaporates, it cools your body down. For your body temperature to remain stable, the heating must be balanced by the cooling. If there is more heating than cooling, then your body temperature climbs. If there is more cooling than heating, then your body temperature drops.

Imagine two well-matched teams struggling in a tug-of-war contest. Each team pulls steadfastly on the rope, but the force of one team's pull is only enough to match, not overcome, the force exerted by the other team. A picture taken now and another taken 5 minutes from now would not differ in any significant way. In this static equilibrium, opposing forces balance each other exactly.

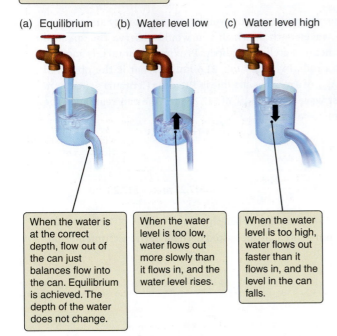

Pressure determines the rate at which water flows out of a hole in a can. The higher the water level, the faster the flow.

(a) Equilibrium   (b) Water level low   (c) Water level high

When the water is at the correct depth, flow out of the can just balances flow into the can. Equilibrium is achieved. The depth of the water does not change.

When the water level is too low, water flows out more slowly than it flows in, and the water level rises.

When the water level is too high, water flows out faster than it flows in, and the level in the can falls.

Figure 5.19 Water flowing into and out of a can determines the water level in the can. This is an example of dynamic equilibrium.

Astronomy in Action: Changing Equilibrium

Static equilibrium can be stable, unstable, or neutral. A nut in a bowl is in a stable equilibrium: if it moves it will return to its original position at the bottom of the bowl. An example of an unstable equilibrium would be a book standing on its edge, unsupported on either side. If you nudged the book, it would fall over rather than settling back into its original position. When an unstable equilibrium is disturbed, it moves further away from equilibrium rather than back toward it.

Equilibrium can also be dynamic, which means the system is constantly changing so that one source of change is exactly balanced by another source of change, and the configuration of the system remains the same. Examine the demonstration illustrated in **Figure 5.19**. Placing a can with a hole cut in the bottom under an open water faucet provides a simple example of dynamic equilibrium. The depth of the water in the can determines how fast water pours out through the hole in the bottom of the can. When the water reaches just the right depth, as shown in Figure 5.19a, water pours out of the hole in the bottom of the can at exactly the same rate it pours into the top of the can from the faucet. The water leaving the can balances the water entering, and equilibrium is established. If you took a picture now and another picture in a few minutes, little of the water in the can would be the same, but the pictures would be indistinguishable.

If a system is not in equilibrium, its configuration will change. If the level of the water in the can is too low, as shown in Figure 5.19b, water will not flow out of the bottom of the can fast enough to balance the water flowing in. The water level will begin to rise until the amount coming in equals the amount going out. A picture taken now and another taken a short time later would not look the same. Conversely, if the water level in the can is too high (Figure 5.19c), water will flow out of the can faster than it flows into the can. The water level will begin to fall until the amount coming in equals the amount going out. Once again, if the system is not in equilibrium, its configuration will change.

When heating is balanced by cooling, we call it **thermal equilibrium**. Planets have a dynamic but stable thermal equilibrium, and electromagnetic radiation plays a crucial role in maintaining this. Energy from sunlight heats the surface of a planet, driving its temperature up, and the planet emits thermal radiation into space, cooling it down. For a planet to remain at the same average temperature over time, the energy it radiates into space must exactly balance the energy it absorbs from the Sun. **Figure 5.20** illustrates that the equilibrium temperature of a planet is analogous to the water level in Figure 5.19. We will return to planetary equilibrium later in the chapter. There are many kinds of equilibrium besides thermal equilibrium, some of which we will encounter later in the book.

Temperature

In everyday life, we define hot and cold subjectively: something is hot when it feels hot or cold. When we measure *temperature*, we use degrees on a thermometer, but the way we define a degree is arbitrary. If you grew up in the United States, you probably think of temperatures in degrees Fahrenheit (°F), whereas if you grew up almost anywhere else in the world, you think of temperatures in degrees Celsius (°C).

Temperature is a measurement of how energetically the atoms that make up an object are moving about. The air around us is composed of vast numbers of atoms and molecules. Those molecules are moving about every which way. Some move slowly; some move more rapidly. All atoms and molecules are constantly in motion. The average kinetic energy (E_K) is given by $E_K = \frac{1}{2}mv^2$, where m is the mass of an atom or molecule and v is its velocity. The more energetically the atoms or molecules are bouncing about, the higher is the object's temperature. In fact, the random motions of atoms and molecules are often called their **thermal motions**, to emphasize the connection between these motions and temperature. **Figure 5.21** illustrates that when the temperature of a gas is increased, the kinetic energy is increased, and therefore the atoms move faster.

The atoms and molecules in a solid body (like you) cannot move about feely but still move back and forth around their normal location, and temperature measures the amount of that movement. If something is hotter than you are, thermal energy flows from that object into you. At the atomic level, that means the object's atoms are bouncing more energetically than are the atoms in your body, so if you touch the object, its atoms collide with your atoms, causing the atoms in your body to move faster. Your body gets hotter as thermal energy flows from the object to you. At the same time, these collisions rob the particles in the object of some of their energy. Their motions slow down, and the hotter object cools. Heating processes increase the average thermal energy of an object's particles, and cooling processes decrease the average thermal energy of those particles.

On the Fahrenheit scale, there are 180 degrees between the freezing point (32°F) and the boiling point (212°F) of water at sea level. On the Celsius scale, water freezes at 0°C and boils at 100°C. Because there is a different number of degrees between freezing and boiling on these two scales, a 1-degree change measured in Fahrenheit is not the same as a 1-degree change measured in Celsius.

There is a lowest possible physical temperature below which no object can fall. As the motions of the particles in an object slow down, the temperature drops lower and lower. The lowest possible temperature, where thermal motions have nearly come to a standstill, is called **absolute zero**. Absolute zero corresponds to −273.15°C, or −459.57°F. Scientists found it useful to define a temperature scale that begins at absolute zero, called the **Kelvin scale**. The size of one unit on the Kelvin scale, called a **kelvin (K)**, is the same as the Celsius degree. Zero kelvin (0 K) is set equal to absolute zero. Other temperatures are equal to Celsius plus 273.15, so water freezes at 273.15 K and water boils at 373.15 K. There are no negative temperatures on the Kelvin scale.

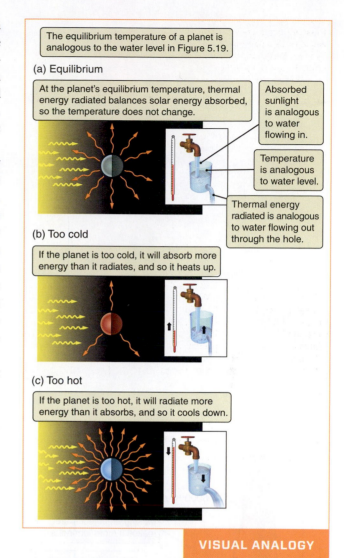

The equilibrium temperature of a planet is analogous to the water level in Figure 5.19.

(a) Equilibrium

At the planet's equilibrium temperature, thermal energy radiated balances solar energy absorbed, so the temperature does not change.

Absorbed sunlight is analogous to water flowing in.

Temperature is analogous to water level.

Thermal energy radiated is analogous to water flowing out through the hole.

(b) Too cold

If the planet is too cold, it will absorb more energy than it radiates, and so it heats up.

(c) Too hot

If the planet is too hot, it will radiate more energy than it absorbs, and so it cools down.

VISUAL ANALOGY

Figure 5.20 Planets are heated by absorbing sunlight (and sometimes by internal heat sources) and cooled by emitting thermal radiation into space. If there are no other sources of heating or means of cooling, then the equilibrium between these two processes determines the temperature of the planet.

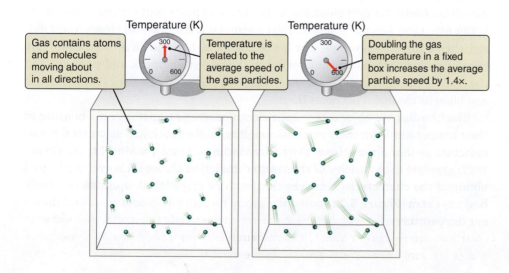

Temperature (K)

Gas contains atoms and molecules moving about in all directions.

Temperature is related to the average speed of the gas particles.

Temperature (K)

Doubling the gas temperature in a fixed box increases the average particle speed by 1.4×.

Figure 5.21 Hotter gas temperatures correspond to faster motions of atoms.

When temperatures are measured in kelvins, the average thermal energy of particles is proportional to the measured temperature. The average thermal energy of the atoms in an object with a temperature of 200 K is twice the average thermal energy of the atoms in an object with a temperature of 100 K.

Temperature, Luminosity, and Color

We have seen the way discrete atoms emit and absorb radiation, which leads to a useful understanding of emission lines and absorption lines that tell us about the physical state and motion of distant objects. But not all objects have spectra that are dominated by discrete spectral lines. As you saw in Figure 5.13a, if you pass the light from a lightbulb through a prism, instead of discrete bright and dark bands you will see light spread out smoothly from the blue end of the spectrum to the red. Similarly, if you look closely at the spectrum of the Sun, you will see absorption lines, but mostly you will see light smoothly spread out across all colors of the spectrum—a type of spectrum called a **continuous spectrum**.

We can think of a dense material as being composed of a collection of charged particles that are being jostled as their thermal motions cause them to run into their neighbors. The hotter the material is, the more violently its particles are being jostled. Recall that any time a charged particle is subjected to an acceleration, it radiates. So the jostling of particles due to their thermal motions causes them to give off a continuous spectrum of electromagnetic radiation. This is why any material that is sufficiently dense for its atoms to be jostled by their neighbors emits light simply because of its temperature. Radiation of this sort is called **thermal radiation**.

The radiation from an object changes as the object heats up or cools down. **Luminosity** is the amount of light *leaving* a source; that is, the total amount of light emitted (energy per second, measured in watts, W). The hotter the object, the more energetically the charged particles within it move, and the more energy they emit in the form of electromagnetic radiation. So as an object gets hotter, the light that it emits becomes more intense. Here is our first point about thermal radiation: *an object is more luminous when it is hotter.*

Now let's move to the question of what color light an object emits. As the object gets hotter, the thermal motions of its particles become more energetic, which produce more energetic photons. The average energy of the photons that it emits increases, the average wavelength of the emitted photons gets shorter, and the light from the object gets bluer. Here is our second point about thermal radiation: *hotter objects are bluer.* If you heat a piece of metal, the metal will glow—first a dull red, then orange, then yellow. The hotter the metal becomes, the more the highly energetic blue photons become mixed with the less energetic red photons, and the color of the light shifts from red toward blue. The light becomes more intense and bluer as the metal becomes hotter.

Blackbodies are objects that emit electromagnetic radiation only because of their temperature, not their composition. Blackbodies emit just as much thermal radiation as they absorb from their surroundings. Physicist Max Planck (1858–1947) graphed the intensity of the emitted radiation across all wavelengths and obtained the characteristic curves that we now call **Planck spectra** or **blackbody spectra**. **Figure 5.22** shows blackbody spectra for objects at several different temperatures.

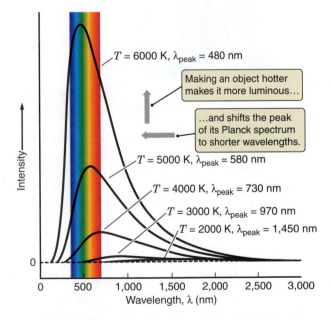

Figure 5.22 This illustration shows blackbody spectra emitted by sources with temperatures of 2000 K, 3000 K, 4000 K, 5000 K, and 6000 K. At higher temperatures, the peak of the spectrum shifts toward shorter wavelengths, and the amount of energy radiated per second from each square meter of the source increases.

Blackbody Laws

In the real world, the light from stars such as the Sun and the thermal radiation from a planet often come close to having blackbody spectra. So these objects follow two blackbody laws that relate luminosity with temperature and temperature with color, respectively.

Stefan-Boltzmann Law As the temperature of an object increases, the object gives off more radiation at every wavelength, so the luminosity of the object should increase. Adding up all of the energy in a blackbody spectrum shows that the increase in luminosity is proportional to the fourth power of the temperature: Luminosity $\propto T^4$, known as the **Stefan-Boltzmann law**. This law was discovered in the laboratory by physicist Josef Stefan (1835–1893) and derived mathematically by his student Ludwig Boltzmann (1844–1906).

It is difficult to measure all of the photons emitted by Earth or the Sun in all possible directions, but it is easier to measure the *flux*. The amount of energy radiated by each square meter of the surface of an object each second is called the **flux**, abbreviated $\mathcal{F}$. The flux is proportional to the luminosity. You can find the luminosity by multiplying the flux by the total surface area. The Stefan-Boltzmann law says that the flux is given by the following equation: $\mathcal{F} = \sigma T^4$. The constant σ (the Greek letter sigma), which is called the **Stefan-Boltzmann constant**, equals 5.67×10^{-8} W/(m² K⁴), where 1 watt = 1 joule per second (J/s).

The Stefan-Boltzmann law says that an object rapidly becomes more luminous as its temperature increases. If the temperature of an object doubles, the amount of energy being radiated each second increases by a factor of 2^4, or 16. If the temperature of an object goes up by a factor of 3, then the energy being radiated by the object each second goes up by a factor of 3^4, or 81. A lightbulb with a filament temperature of 3000 K radiates 16 times as much light as it would if the filament temperature were 1500 K. Even modest changes in temperature can result in large changes in the amount of luminosity radiated by an object.

Wien's Law Look again at Figure 5.22. The wavelength where the blackbody spectrum is at its peak, λ_{peak}, is where the electromagnetic radiation from an object is greatest. As the temperature, T, increases, the peak of the spectrum shifts toward shorter wavelengths. For example, compare the peak wavelengths of a 3000 K object and a 6000 K object. Photon energy and wavelength are inversely related; thus, as the peak wavelength becomes shorter, the average photon energy becomes greater. The object becomes bluer. The physicist Wilhelm Wien (1864–1928) found that the peak wavelength in the spectrum is inversely proportional to the temperature of the object. **Wien's law** states that if you double the temperature, the peak wavelength becomes half of what it was. If you increase the temperature by a factor of 3, the peak wavelength becomes a third of what it was. Stefan-Boltzmann's law and Wien's law are further explored in **Working It Out 5.3**. We will return to these laws later in the chapter when we use them to estimate the temperatures of the planets.

 **Astronomy in Action:** Wien's Law

 Nebraska Simulation: Blackbody Curves

CHECK YOUR UNDERSTANDING 5.5

When you look at the sky on a dark night and see stars of different colors, which are the hottest? (a) orange; (b) red-orange; (c) yellow; (d); red; (e) blue

5.3 | Working It Out   Working with the Stefan-Boltzmann Law and Wien's Law

Stefan-Boltzmann's law can be used to estimate the flux and luminosity of Earth. Earth's average temperature is 288 K, so the flux from its surface is

$$\mathcal{F} = \sigma T^4$$

$$\mathcal{F} = (5.67 \times 10^{-8}\ \text{W/m}^2\ \text{K}^4) \times (288\ \text{K})^4$$

$$\mathcal{F} = 390\ \text{W/m}^2$$

The luminosity is the flux multiplied by the surface area (A) of Earth. Surface area is given by $4\pi R^2$, and the radius of Earth is 6,378 km, or 6.378×10^6 meters. So the luminosity is

$$L = \mathcal{F} \times A = \mathcal{F} \times 4\pi R^2$$

$$L = (390\ \text{W/m}^2) \times [4\pi(6.378 \times 10^6\ \text{m})^2]$$

$$L \approx 2 \times 10^{17}\ \text{W}$$

Earth emits the equivalent of the energy used by 2,000,000,000,000,000 (2 million billion) hundred-watt lightbulbs. This is still not anywhere close to the amount emitted by the Sun.

Wien's law also proves useful to astronomers. If they measure the spectrum of an object emitting thermal radiation and find where the peak in the spectrum is, Wien's law can be used to calculate the temperature of the object. Wien's law can be written as

$$T = \frac{2{,}900{,}000\ \text{nm K}}{\lambda_{\text{peak}}}$$

For example, the spectrum of the light coming from the Sun peaks at a wavelength of $\lambda_{\text{peak}} = 500$ nm, so

$$T = \frac{2{,}900{,}000\ \text{nm K}}{500\ \text{nm}} = 5800\ \text{K}$$

This is how you can know the surface temperature of the Sun.

Suppose you want to calculate the peak wavelength at which Earth radiates. Using Earth's average temperature of 288 K in Wien's law gives

$$\lambda_{\text{peak}} = \frac{2{,}900{,}000\ \text{nm K}}{288\ \text{K}} = 10{,}100\ \text{nm} = 10.1\ \text{microns}$$

Earth's radiation peaks in the infrared region of the spectrum.

5.5 The Brightness of Light Depends on the Luminosity and Distance of the Source

Recall that luminosity refers to the amount of light leaving a source. By contrast, the **brightness** of electromagnetic radiation is the amount of light that is arriving at a particular location. Therefore, brightness depends on the luminosity and the distance of the light source. For example, replacing a 50-W lightbulb with a 100-W bulb makes a room twice as bright because it doubles the light reaching any point in the room. But brightness also depends on the distance from a source of electromagnetic radiation. If you needed more light to read this book, you could replace the bulb in your lamp with a more luminous bulb or you can move the book closer to the light. Conversely, if a light is too bright, you can move away from it. Our everyday experience teaches us that as we move away from a light, its brightness decreases.

The particle description of light provides another way to think about the brightness of radiation and how brightness depends on distance. Suppose you had a piece of cardboard that measured 1 meter by 1 meter. To make the light falling on the cardboard twice as bright, you would need to double the number of photons that hit the cardboard each second. Tripling the brightness of the light would mean increasing the number of photons hitting the cardboard each second by a factor of 3, and so on. Brightness depends on the number of photons falling on each square meter of a surface each second.

Now imagine a lightbulb sitting at the center of a spherical shell, illustrated in **Figure 5.23**. Photons from the bulb travel in all directions and land on the inside of the shell. To find the number of photons landing on each square meter of the shell during each second, that is, to determine the brightness of the light, take the *total* number of photons given off by the lightbulb each second and divide by the number of square meters over which those photons have to be spread. The surface area of a sphere is given by the formula $A = 4\pi r^2$, where r is the distance between the bulb and the surface of the sphere (that is, r = the radius of the sphere). The number of photons striking one square meter each second is equal to the total number of photons emitted each second divided by the surface area $4\pi r^2$.

Now change the size of the spherical shell while keeping the total number of photons given off by the lightbulb each second the same. As the shell becomes larger, the photons from the lightbulb must spread out to cover a larger surface area. Each square meter of the shell receives fewer photons each second, so the brightness of the light decreases. If the shell's surface is moved twice as far from the light, the area over which the light must spread increases by a factor of $2^2 = 2 \times 2 = 4$. The photons from the bulb spread out over 4 times as much area, so the number of photons falling on each square meter each second becomes $\frac{1}{4}$ of what it was. If the surface of the sphere is 3 times as far from the light, the area over which the light must spread increases by a factor of $3^2 = 3 \times 3 = 9$, and the number of photons per second falling on each square meter becomes $\frac{1}{9}$ of what it was originally. This is the same kind of inverse square relationship you saw for gravity in Chapter 4. The brightness of the light from an object is inversely proportional to the square of the distance from the object. Twice as far means one-fourth as bright.

This idea of photons streaming and spreading onto a surface from a light explains why brightness follows an inverse square law. In practice, however, it is usually more convenient to talk about the *energy* coming to a surface each second, rather than the number of photons arriving.

The luminosity of an object is the total number of photons given off by the object multiplied by the energy of each photon. So instead of thinking about how the number of photons must spread out to cover the surface of a sphere, we can think about how the energy carried by the photons must spread out to cover the surface of a sphere. The brightness of the light is the amount of energy falling on a square meter in a second, and it equals the luminosity L divided by the area of the sphere, which depends on the radius squared. This tells us, for example, that the brightness of the Sun will depend on the inverse square of the planet's distance from the Sun. This will factor in as we estimate the equilibrium temperatures of the planets in **Working It Out 5.4**.

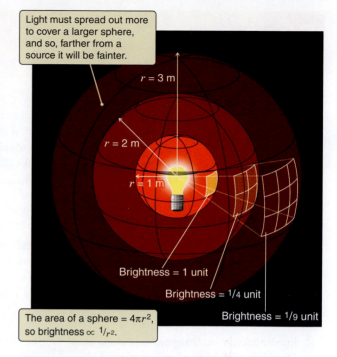

Light must spread out more to cover a larger sphere, and so, farther from a source it will be fainter.

$r = 3$ m

$r = 2$ m

$r = 1$ m

Brightness = 1 unit

Brightness = 1/4 unit

Brightness = 1/9 unit

The area of a sphere = $4\pi r^2$, so brightness $\propto 1/r^2$.

Figure 5.23 Light obeys an inverse square law as it spreads away from a source. Twice as far means one-fourth as bright.

Astronomy in Action: Inverse Square Law

CHECK YOUR UNDERSTANDING 5.6

The average distance of Mars from the Sun is 1.4 AU. How bright is the Sun on Mars compared with its brightness on Earth? (a) 1.4 times brighter; (b) about 2 times brighter; (c) about 2 times fainter; (d) 1.4 times fainter

5.4 Working It Out Using Radiation Laws to Calculate Equilibrium Temperatures of Planets

The temperature of a planet is determined by a balance between the amount of sunlight being absorbed and the amount of energy being radiated back into space. We begin with the amount of sunlight being absorbed. When viewed from the Sun, a planet looks like a circular disk with a radius equal to the radius of the planet, R_{planet}. The area of the planet that is lit by the Sun is

$$\text{Absorbing area of planet} = \pi R_{planet}^2$$

The amount of energy striking a planet also depends on the brightness of sunlight at the distance at which the planet orbits. The brightness of sunlight at a distance d from the Sun is equal to the luminosity of the Sun (L_{Sun}, in watts) divided by $4\pi d^2$ (we use d here to avoid confusion with the planet's radius, R_{planet}):

$$\text{Brightness of sunlight} = \frac{L_{Sun}}{4\pi d^2}$$

A planet does not absorb all the sunlight that falls on it. **Albedo**, a, is the fraction of the sunlight that reflects from a planet. The corresponding fraction of the sunlight that is absorbed by the planet is 1 minus the albedo. A planet covered entirely in snow would have a high albedo (close to 1), while a planet covered entirely by black rocks would have a low albedo, close to 0:

$$\text{Fraction of sunlight absorbed} = 1 - a$$

We can now calculate the energy absorbed by the planet each second. Writing this relationship as an equation, we say that

$$\begin{pmatrix}\text{Energy absorbed}\\ \text{by the planet}\\ \text{each second}\end{pmatrix} = \begin{pmatrix}\text{Absorbing}\\ \text{area of}\\ \text{the planet}\end{pmatrix} \times \begin{pmatrix}\text{Brightness}\\ \text{of sunlight}\end{pmatrix} \times \begin{pmatrix}\text{Fraction}\\ \text{of sunlight}\\ \text{absorbed}\end{pmatrix}$$

$$= \pi R_{planet}^2 \times \frac{L_{Sun}}{4\pi d^2} \times (1 - a)$$

Now let's turn to the other piece of the equilibrium: the amount of energy that the planet radiates away into space each second. We can calculate this amount by multiplying the number of square meters of the planet's total surface area by the energy radiated by each square meter each second. The surface area for the planet is given by $4\pi R_{planet}^2$. The Stefan-Boltzmann law tells us that the energy radiated by each square meter each second is given by σT^4. So we can say that

$$\begin{pmatrix}\text{Energy radiated}\\ \text{by the planet}\\ \text{each second}\end{pmatrix} = \begin{pmatrix}\text{Surface}\\ \text{area of}\\ \text{the planet}\end{pmatrix} \times \begin{pmatrix}\text{Energy radiated}\\ \text{per square meter}\\ \text{per second}\end{pmatrix}$$

$$= 4\pi R_{planet}^2 \times \sigma T^4$$

If the planet's temperature is to remain stable—not heating up or cooling down—then each second the "Energy radiated" must be equal to "Energy absorbed." When we set these two quantities equal to each other, we arrive at the expression

$$\begin{pmatrix}\text{Energy radiated}\\ \text{by the planet}\\ \text{each second}\end{pmatrix} = \begin{pmatrix}\text{Energy absorbed}\\ \text{by the planet}\\ \text{each second}\end{pmatrix}$$

or

$$4\pi R_{planet}^2 \sigma T^4 = \pi R_{planet}^2 \frac{L_{Sun}}{4\pi d^2}(1 - a)$$

Canceling out πR_{planet}^2 on both sides, and rearranging this equation to put T on one side and everything else on the other gives

$$T^4 = \frac{L_{Sun}(1 - a)}{16\sigma\pi d^2}$$

If we take the fourth root of each side, we get

$$T = \left(\frac{L_{Sun}(1 - a)}{16\sigma\pi d^2}\right)^{1/4}$$

Putting in the appropriate numbers for the known luminosity of the Sun, L_{Sun}, and the constants π and σ yields this simpler equation:

$$T = 279\,\text{K} \times \left(\frac{1 - a}{d_{AU}^2}\right)^{1/4}$$

where d_{AU} is the distance of the planet from the Sun in astronomical units.

To use this equation, we would need to know a planet's distance from the Sun and its average albedo. For a blackbody ($a = 0$) at 1 AU from the Sun, the temperature is 279 K. For Earth, with an albedo of 0.3 and a distance from the Sun of 1 AU, the temperature is

$$T = 279\,\text{K} \times \left(\frac{1 - 0.3}{1^2}\right)^{1/4} = 255\,\text{K}$$

(Calculator hint: To take a fourth root, you can take the square root twice, or use the x^y button with $y = 0.25$).

Earth is cooler than a blackbody at 1 AU from the Sun because its average albedo is greater than zero. If Earth's albedo changed or the Sun's luminosity changed, that would affect the result. When we examine planets around other stars, we will need to use the luminosity of the particular star in the equation, instead of the Sun's luminosity, so the temperature at 1 AU will be different than what is it for Earth.

Origins

Temperatures of Planets

In the previous chapters, we discussed how a planet's axial tilt and its orbital shape can affect its temperature, and thus its prospects for life. Now let's get more specific about the temperatures of planets, using what you learned in this chapter about thermal radiation. For a planet at an equilibrium temperature, the energy radiated by a planet exactly balances the energy absorbed by the planet. If the planet is hotter than this equilibrium temperature, it will radiate energy faster than it absorbs sunlight, and its temperature will fall. If the planet is cooler than this temperature, it will radiate energy slower than it absorbs sunlight, and its temperature will rise.

Planets at different distances from the Sun will have different temperatures, and the temperature should be inversely proportional to the square root of the distance, as you saw in

Working It Out 5.4. **Figure 5.24** plots the actual and predicted temperatures of nine solar system objects. Each vertical orange bar shows the range of temperatures found on the surface of the planet or, in the case of the giant planets, at the top of the planet's clouds. The black dots show the predictions made using the equation in Working It Out 5.4. For most planets, the predictions are not too far off; indicating that our basic understanding of *why* planets have the temperatures they have is probably pretty good. The data for Mercury, Mars, and Pluto agree particularly well.

In some cases, however, the predictions are wrong. For Earth, the actual measured temperature is a bit higher than the predicted temperature, and for Venus the actual surface temperature is much higher than the prediction.

The predicted values assume that the temperature of the planet is the

same everywhere. However, planets are likely to be hotter on the day side than on the night side. The predictions also assume that a planet's only source of energy is sunlight, and that the fraction of sunlight reflected is constant over the surface of each planet. There is also the assumption that the planets absorb and radiate energy into space as blackbodies.

The discrepancies between the calculated and the measured temperatures of some of the planets indicate that for these planets, some or all of these assumptions are incorrect. For example, the planet may have its own source of energy besides sunlight, or it may have an atmosphere. Understanding the temperatures of planets makes it possible to hypothesize why life may have evolved here on Earth, instead of on a different planet in the Solar System.

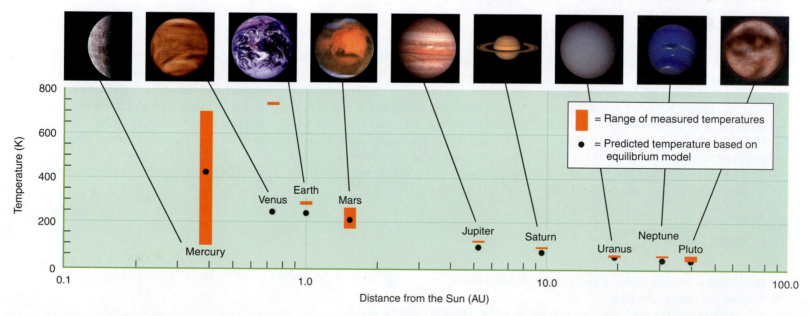

Figure 5.24 Predicted temperatures for the planets and dwarf planet Pluto are based on the equilibrium between absorbed sunlight and thermal radiation into space. These temperatures are compared with ranges of observed surface temperatures.

A Study in Scarlet

ESO

This new image from ESO's La Silla Observatory in Chile reveals a cloud of hydrogen called Gum 41 (**Figure 5.25**). In the middle of this little-known nebula, brilliant hot young stars are giving off energetic radiation that causes the surrounding hydrogen to glow with a characteristic red hue.

This area of the southern sky, in the constellation of Centaurus (The Centaur), is home to many bright nebulae, each associated with hot newborn stars that formed out of the clouds of hydrogen gas. The intense radiation from the stellar newborns excites the remaining hydrogen around them, making the gas glow in the distinctive shade of red typical of star-forming regions. Another famous example of this phenomenon is the Lagoon Nebula, a vast cloud that glows in similar bright shades of scarlet.

The nebula in this picture is located some 7,300 light-years from Earth. Australian astronomer Colin Gum discovered it on photographs taken at the Mount Stromlo Observatory near Canberra, and included it in his catalog of 84 emission nebulae, published in 1955. Gum 41 is actually one small part of a bigger structure called the Lambda Centauri Nebula, also known by the more exotic name of the Running Chicken Nebula. Gum died at a tragically early age in a skiing accident in Switzerland in 1960.

In this picture of Gum 41, the clouds appear to be quite thick and bright, but this is actually

Figure 5.25 The Gum 41 Nebula.

misleading. If a hypothetical human space traveler could pass through this nebula, it is likely that they would not notice it as—even at close quarters—it would be too faint for the human eye to see. This helps to explain why this large object had to wait until the mid-twentieth century to be discovered—its light is spread very thinly and the red glow cannot be well seen visually.

This new portrait of Gum 41—likely one of the best so far of this elusive object—has been created using data from the Wide Field Imager (WFI) on the MPG/ESO 2.2-meter telescope at the La Silla Observatory in Chile. It is a combination of images taken through blue, green, and red filters, along with an image using a special filter designed to pick out the red glow from hydrogen.

1. How long has it taken the light from this nebula to reach us?
2. Why are the young stars blue?
3. What type of spectra would you expect to get from the stars and from the gas?
4. Refer to the spectrum of hydrogen in Figure 5.15. Why is the excited hydrogen gas in the image of Gum 41 glowing red?
5. Would you be able to see Gum 41 from your location? Why or why not?

Summary

Light carries both information and energy throughout the universe. The speed of light in a vacuum is 300,000 km/s; nothing can travel faster. Visible light is only a tiny portion of the entire electromagnetic spectrum. Atoms absorb and emit radiation at unique wavelengths like spectral fingerprints. A planet's temperature depends on its distance from its star, its albedo, and the luminosity of its star.

LG 1 **Describe the wave and particle properties of light, and describe the electromagnetic spectrum.** Light is both a particle and a wave. Light is simultaneously a stream of particles called photons and an electromagnetic wave. Different types of electromagnetic radiation, from gamma rays to visible light to radio waves, are electromagnetic waves that differ in frequency and wavelength.

LG 2 **Describe how to measure the composition of distant objects using the unique spectral lines of different types of atoms.** Nearly all matter is composed of atoms, and light can reveal the identity of the types of atoms that are present in matter. Each type of atom has a different spacing of its electron energy levels, and the emission of photons is related to the electron changing levels. As a result, we can identify different chemical elements and molecules in distant objects.

LG 3 **Describe the Doppler effect and how it can be used to measure the motion of distant objects.** Because of the Doppler effect, light from receding objects is redshifted to longer wavelengths, and light from approaching objects is blueshifted to shorter wavelengths. The wavelength shifts of the spectral lines indicate how fast an astronomical object is moving toward or away from Earth.

LG 4 **Explain how the spectrum of light that an object emits depends on its temperature.** Temperature is a measure of how energetically particles are moving in an object. A light source that emits electromagnetic radiation because of its temperature is called a blackbody. A blackbody emits a continuous spectrum. The total amount of energy emitted is proportional to the temperature to the fourth power, and the peak wavelength, which determines its color, is inversely proportional to the temperature.

LG 5 **Differentiate luminosity from brightness, and illustrate how distance affects each.** The light output, or luminosity, of an object is the amount of light the object emits. The brightness of an object is proportional to its luminosity divided by its distance squared. Thus, the brightness of the Sun is different when measured from each Solar System planet but the luminosity is the same.

❓ UNANSWERED QUESTIONS

- Has the speed of light always been 300,000 km/s? Some theoretical physicists have questioned whether light traveled much faster earlier in the history of our universe. The observational evidence that may test this idea comes from studying the spectra of the most distant objects—whose light has been traveling for billions of years—and determining whether billions of years ago chemical elements absorbed light somewhat differently than they do today. So far, there is no evidence that the speed of light changes.

- Will it ever be possible to travel faster than the speed of light? Our current understanding of the science says no. A staple of science fiction films and stories is spaceships that go into "warp speed" or "hyperdrive"—moving faster than light—to traverse the huge distances of space (and visit a different planetary system every week). If this premise is simply fictional and the speed of light is a true universal limit, then travel between the stars will take many years. Because all electromagnetic radiation travels at the speed of light, even an electromagnetic signal sent to another planetary system would take many years to get there. Interstellar visits (and interstellar conversations) will be quite prolonged.

Questions and Problems

Test Your Understanding

1. If the Sun instantaneously stopped giving off light, what would happen on Earth?
   a. Earth would immediately get dark.
   b. Earth would get dark 8 minutes later.
   c. Earth would get dark 27 minutes later.
   d. Earth would get dark 1 hour later.

2. Why is an iron atom a different element from a sodium atom?
   a. A sodium atom has fewer neutrons in its nucleus than an iron atom has.
   b. An iron atom has more protons in its nucleus than a sodium atom has.
   c. A sodium atom is bigger than an iron atom.
   d. A sodium atom has more electrons.

3. Suppose an atom has three energy levels, specified in arbitrary units as 10, 7, and 5. In these units, which of the following energies might an emitted photon have? (Select all that apply.)
   a. 3
   b. 2
   c. 5
   d. 4

4. When a boat moves through the water, the waves in front of the boat bunch up, while the waves behind the boat spread out. This is an example of
   a. the Bohr model.
   b. the wave nature of light.
   c. emission and absorption.
   d. the Doppler effect.

5. As a blackbody becomes hotter, it also becomes _____ and _____ .
   a. more luminous; redder
   b. more luminous; bluer
   c. less luminous; redder
   d. less luminous; bluer

6. Which of the following factors does *not* directly influence the temperature of a planet?
   a. the luminosity of the Sun
   b. the distance from the planet to the Sun
   c. the albedo of the planet
   d. the size of the planet

7. Two stars are of equal luminosity. Star A is 3 times as far from you as star B. Star A appears _____ star B.
   a. 9 times brighter than
   b. 3 times brighter than
   c. the same brightness as
   d. $\frac{1}{3}$ as bright as
   e. $\frac{1}{9}$ as bright as

8. When less energy radiates from a planet, its _____ increases until a new _____ is achieved.
   a. temperature; equilibrium
   b. size; temperature
   c. equilibrium; size
   d. temperature; size

9. How does the speed of light in a medium compare to the speed in a vacuum?
   a. The speed is the same in both a medium and a vacuum, as the speed of light is a constant.
   b. The speed in the medium is always faster than the speed in a vacuum.
   c. The speed in the medium is always slower than the speed in a vacuum.
   d. The speed in the medium may be faster or slower, depending on the medium.

10. When an electron moves from a higher energy level in an atom to a lower energy level,
    a. a continuous spectrum is emitted.
    b. a photon is emitted.
    c. a photon is absorbed.
    d. a redshifted spectrum is emitted.

11. In Figure 5.15, the red photons come from the transition from E_3 to E_2. These photons will have the _____ wavelengths because they have the _____ energy compared to the other photons.
    a. shortest; least
    b. shortest; most
    c. longest; least
    d. longest; most

12. Star A and star B appear equally bright in the sky. Star A is twice as far away from Earth as star B. How do the luminosities of stars A and B compare?
    a. Star A is 4 times as luminous as star B.
    b. Star A is 2 times as luminous as star B.
    c. Star B is 2 times as luminous as star A.
    d. Star B is 4 times as luminous as star A.

13. What is the surface temperature of a star that has a peak wavelength of 290 nm?
    a. 1000 K
    b. 2000 K
    c. 5000 K
    d. 10,000 K
    e. 100,000 K

14. If a planet is in thermal equilibrium,
    a. no energy is leaving the planet.
    b. no energy is arriving on the planet.
    c. the amount of energy leaving equals the amount of energy arriving.
    d. the temperature is very low.

15. The temperature of an object has a very specific meaning as it relates to the object's atoms. A high temperature means that the atoms
    a. are very large.
    b. are moving very fast.
    c. are all moving together.
    d. have a lot of energy.

Thinking about the Concepts

16. We know that the speed of light in a vacuum is 3×10^5 km/s. Is it possible for light to travel at a lower speed? Explain your answer.

17. Is light a wave or a particle or both? Explain your answer.

18. Referring to the Process of Science Figure, if any of these experiments had *not* agreed with the others, what would that mean for the conclusion that light has a finite, constant speed?

19. If photons of blue light have more energy than photons of red light, how can a beam of red light carry as much energy as a beam of blue light?

20. Patterns of emission or absorption lines in spectra can uniquely identify individual atomic elements. Explain how positive identification of atomic elements can be used as one way of testing the validity of the cosmological principle discussed in Chapter 1.

21. An atom in an excited state can drop to a lower energy state by emitting a photon. Is it possible to predict exactly how long the atom will remain in the higher energy state? Explain your answer.

22. Spectra of astronomical objects show both bright and dark lines. Describe what these lines indicate about the atoms responsible for the spectral lines.

23. Astronomers describe certain celestial objects as being *redshifted* or *blueshifted*. What do these terms indicate about the objects?

24. An object somewhere near you is emitting a pure tone at middle C on the octave scale (262 Hz). You, having perfect pitch, hear the tone as A above middle C (440 Hz). Describe the motion of this object relative to where you are standing.

25. During a popular art exhibition, the museum staff finds it necessary to protect the artwork by limiting the total number of viewers in the museum at any particular time. New viewers are admitted at the same rate that others leave. Is this an example of static equilibrium or of dynamic equilibrium? Explain.

26. A favorite object for amateur astronomers is the double star Albireo, with one of its components a golden yellow and the other a bright blue. What do these colors tell you about the relative temperatures of the two stars?

27. The stars you see in the night sky cover a large range of brightness. What does that range tell you about the distances of the various stars? Explain your answer.

28. Why is it not surprising that sunlight peaks in the "visible"?

29. Study Figure 5.24. For which planet is the range of measured temperatures furthest from the predicted value? What accounts for this difference?

30. Suppose you want to find a planet with the same temperature as Earth. What could you say about the size of the orbit of such a planet if it is orbiting a red star? A yellow star? A blue star?

Applying the Concepts

31. You are tuned to 790 on AM radio. This station is broadcasting at a frequency of 790 kHz (7.90×10^5 Hz). You switch to 98.3 on FM radio. This station is broadcasting at a frequency of 98.3 MHz (9.83×10^7 Hz).
    a. What are the wavelengths of the AM and FM radio signals?
    b. Which broadcasts at higher frequencies: AM or FM?
    c. What are the photon energies of the two broadcasts?

32. Your microwave oven cooks by vibrating water molecules at a frequency of 2.45 gigahertz (GHz), or 2.45×10^9 Hz. What is the wavelength, in centimeters, of the microwave's electromagnetic radiation?

33. You observe a spectral line of hydrogen at a wavelength of 502.3 nm in a distant galaxy. The rest wavelength of this line is 486.1 nm. What is the radial velocity of this galaxy? Is it moving toward you or away from you?

34. Assume that an object emitting a pure tone of 440 Hz is on a vehicle approaching you at a speed of 25 m/s. If the speed of sound at this particular atmospheric temperature and pressure is 340 m/s, what will be the frequency of the sound that you hear? (Hint: Keep in mind that frequency is inversely proportional to wavelength.)

35. If half of the phosphorescent atoms in a glow-in-the-dark toy give up a photon every 30 minutes, how bright (relative to its original brightness) will the toy be after 2 hours?

36. How bright would the Sun appear from Neptune, 30 AU from the Sun, compared to its brightness as seen from Earth? The spacecraft *Voyager 1* is now about 130 AU from the Sun and heading out of the Solar System. Compare the brightness of the Sun seen by *Voyager 1* with that seen from Earth.

37. On a dark night you notice that a distant lightbulb happens to have the same brightness as a firefly that is 5 meters away from you. If the lightbulb is a million times more luminous than the firefly, how far away is the lightbulb?

38. Two stars appear to have the same brightness, but one star is 3 times more distant than the other. How much more luminous is the more distant star?

39. A panel with an area of 1 square meter (m^2) is heated to a temperature of 500 K. How many watts is it radiating into its surroundings?

40. The Sun has a radius of 6.96×10^5 km and a blackbody temperature of 5780 K. Calculate the Sun's luminosity.

41. Some of the hottest stars known have a blackbody temperature of 100,000 K. What is the peak wavelength of their radiation? What type of radiation is this?

42. Your body, at a temperature of about 37°C (98.6°F), emits radiation in the infrared region of the spectrum.
    a. What is the peak wavelength, in microns, of your emitted radiation?
    b. Assuming an exposed body surface area of 0.25 m², how many watts of power do you radiate?

43. A planet with no atmosphere at 1 AU from the Sun would have an average blackbody surface temperature of 279 K if it absorbed all the Sun's electromagnetic energy falling on it (albedo = 0).
    a. What would be the average temperature on this planet if its albedo were 0.1, typical of a rock-covered surface?
    b. What would be the average temperature if its albedo were 0.9, typical of a snow-covered surface?

44. The orbit of Eris, a dwarf planet, carries it out to a maximum distance of 97.7 AU from the Sun. Assuming an albedo of 0.8, what is the average temperature of Eris when it is farthest from the Sun?

45. Suppose our Sun had 10 times its current luminosity. What would be the average blackbody surface temperature of Earth, assuming Earth had the same albedo?

USING THE WEB

46. a. Go to the website for NASA's Astronomy Picture of the Day (http://apod.nasa.gov/apod/ap101027.html) and study the picture of the Andromeda Galaxy in visible light and in ultraviolet light. Which light represents a hotter temperature? What differences do you see in the two images?
    b. Go to the APOD archive (http://apod.nasa.gov/cgi-bin/apod/apod_search) and enter "false color" in the search box. Examine a few images that come up in the search. What does *false color* mean in this context? What wavelength(s) were the pictures exposed in? What is the color coding; that is, what wavelength does each color in the image represent? You can read more about false color here: http://chandra.harvard.edu/photo/false_color.html.

47. Crime scene investigators may use different types of light to examine a crime scene. Search for "forensic lighting" in your browser. What wavelengths of light are used to search for blood and saliva? For fingerprints? Why is it useful for an investigator to have access to different kinds of light? Search on "forensic spectroscopy" and select a recent report. How is spectroscopy being used in crime scene investigations?

48. Using Google Images or an equivalent website, search for "night vision imaging" and "thermal imaging." How do night vision goggles and thermal-imaging devices work differently from regular binoculars or cameras? When are these useful?

49. The Transportation Security Administration (TSA) uses several types of imaging devices to screen passengers in airports. Search for "TSA imaging" in your browser. What wavelengths of light are being used in these devices? What concerns do passengers have about some of these imaging devices?

50. Go to the NASA Earth Observations website (http://neo.sci.gsfc.nasa.gov) and look at the current map of Earth's albedo (click on "albedo" in the menu for "energy" or "land" if it didn't come up). Compare this map with those of 2, 4, 6, 8, and 10 months ago. Which parts of Earth have the lowest and highest albedos? In which parts do the albedos seem to change the most with the time of the year? Would you expect ice, snow, oceans, clouds, forests, and deserts to add or subtract in each case from the total Earth albedo? Which parts of Earth are not showing up on this map?

smartwork

If your instructor assigns homework in SmartWork, access your assignments at digital.wwnorton.com/astro5.

EXPLORATION

Light as a Wave, Light as a Photon

digital.wwnorton.com/astro5

Visit the Student Site at the Digital Landing Page, and open the "Light as a Wave, Light as a Photon" AstroTour in Chapter 5. Watch the first section and then click through, using the "Play" button, until you reach "Section 2 of 3."

Here we will explore the following questions: How many properties does a wave have? Are any of these properties related to each other?

Work your way to the experimental section, where you can adjust the properties of the wave. Watch for a moment to see how fast the frequency counter increases.

1 Increase the wavelength by pressing the arrow key. What happens to the rate of the frequency counter?

_____

_____

_____

2 Reset the simulation and then decrease the wavelength. What happens to the rate of the frequency counter?

_____

_____

3 How are the wavelength and frequency related to each other?

_____

4 Imagine that you increase the frequency instead of the wavelength. How should the wavelength change when you increase the frequency?

_____

_____

5 Reset the simulation, and increase the frequency. Did the wavelength change in the way you expected?

_____

_____

6 Reset the simulation, and increase the amplitude. What happens to the wavelength and the frequency counter?

_____

_____

7 Decrease the amplitude. What happens to the wavelength and the frequency counter?

_____

_____

8 Is the amplitude related to the wavelength or frequency?

_____

_____

9 Why can't you change the speed of this wave?

_____

_____

6 The Tools of the Astronomer

In the previous chapter, you saw that astronomers learn about the physical and chemical properties of distant planets, stars, and galaxies by studying the light from these objects. This electromagnetic radiation must first be collected and processed before it can be analyzed and converted to useful knowledge. In this chapter, you will learn about the tools that astronomers use to capture and scrutinize that information.

LEARNING GOALS

By the conclusion of this chapter, you should be able to:

LG 1 Compare the two main types of optical telescopes and how they gather and focus light.

LG 2 Summarize the main types of detectors that are used on telescopes.

LG 3 Explain why some wavelengths of radiation must be observed from space.

LG 4 Explain the benefits of sending spacecraft to study the planets and moons of our Solar System.

LG 5 Describe other astronomical tools that contribute to the study of the universe.

The twin 10-meter Keck reflectors on Mauna Kea, Hawaii, have a multiple mirror, compact design. ▶▶▶

Why are most
telescopes
on remote
mountaintops?

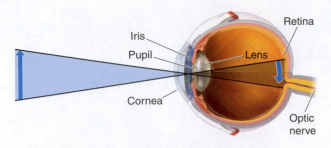

Figure 6.1 A schematic view of the human eye, creating an image of an object (the blue arrow).

Nebraska Simulation: Snell's Law Demonstrator

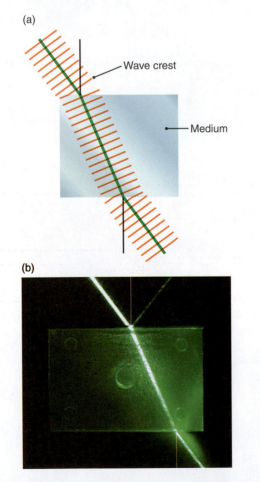

Figure 6.2 (a) When wavefronts enter a new medium, they bend in a new direction relative to a line perpendicular to the surface (black lines). (b) An actual light ray entering and leaving a medium.

6.1 The Optical Telescope Revolutionized Astronomy

Astronomical observations began with the human eye—information about the overall colors of stars and their brightness in the night sky is apparent even to the naked eye, unassisted by binoculars or telescopes or filters. The development of **telescopes**—devices for collecting and focusing light—in the 17th century greatly increased the amount of light that can be collected from astronomical objects. With modern telescopes, astronomers can detect light that has been traveling across space for billions of years—even electromagnetic radiation from soon after the Big Bang, the beginning of the universe itself.

The Eye

Human eyes are sensitive to light with wavelengths ranging from about 350 nanometers (deep violet) to 750 nanometers (far red). A simplified schematic of the human eye is shown in **Figure 6.1**. The part of the human eye that detects light is called the retina, and the individual receptor cells that respond to light falling on the retina are called rods and cones. The center of the human retina consists solely of cones, which detect color and provide the greatest visual acuity. Away from the center, rods and cones intermingle, with rods dominating far from the center, where they are responsible for peripheral vision.

Our vision is limited by the eye's angular **resolution**, which refers to how close two points of light can be to each other before we can no longer distinguish them. Unaided, the best human eyes can resolve objects separated by 1 arcminute (1/60 of a degree), an angular distance of about 1/30 the diameter of the full Moon. (A more in-depth description of angular units—radians, degrees, arcminutes, and arcseconds—can be found in Appendix 1.4.) This may seem small, but when we look at the sky, thousands of stars and galaxies may reside within a patch of sky with this diameter.

Refracting Telescopes

Optical telescopes come in two primary types: **refracting telescopes**, which use lenses; and **reflecting telescopes**, which use mirrors. For all telescopes, the "size" of the telescope refers to the diameter of the largest mirror or lens, which determines the light-collecting area. This diameter is called the **aperture**. The light-gathering power of a telescope is proportional to the area of its aperture; that is, to the square of its diameter. The larger the aperture, the more light the telescope can collect. A "1-meter telescope" has a primary mirror (or lens) that is 1 meter in diameter. The aperture of the human eye is about 6–7 millimeters.

In the late 13th century, craftsmen in Venice were making small lentil-shaped disks of glass that could be mounted in frames and worn over the eyes to improve vision. More than 300 years later, Hans Lippershey (1570–1619), a spectacle maker living in the Netherlands, put two of his lenses together in a tube. With this new instrument, he saw distant objects magnified and could see farther. Galileo Galilei heard news of this invention, and he constructed one of his own. Recall from Chapter 3 that by 1610, Galileo had become the first to see the phases of Venus and the moons of Jupiter, and among the first to see craters on the Moon. He was also the first to realize that the Milky Way is made up of large numbers of individual stars. The refracting telescope—one that uses lenses—quickly revolutionized the science of astronomy.

Refraction is the basis for the refracting telescope. Recall from Chapter 5 that the speed of light is constant in a vacuum, but through a medium such as air or glass, the speed of light is always lower. As light enters a new medium, its speed changes. If the light strikes the surface at an angle, some of the crest of the wave arrives at the surface earlier and some arrives later. You can see this in **Figure 6.2a**, a schematic diagram of wave crests (red lines) striking a medium at an angle. Figure 6.2b shows an actual light ray passing into and out of a medium, in this case glass. The ray bends each time the medium changes. This bending of light when it changes the medium through which it travels is called **refraction**.

The amount of refraction is determined by the properties of the medium. A medium's index of refraction (n) is equal to the ratio of the speed of light in a vacuum (c) to its speed in a medium (v). This can be expressed by the equation $n = c/v$. For example, most glass has an index of refraction of approximately 1.5, so the speed of light in glass is 300,000 kilometers per second (km/s) divided by 1.5, or 200,000 km/s. The light bends by an amount that depends on the index of refraction of the materials involved and the angle at which the light strikes.

The primary lens in a refracting telescope is a simple convex lens, called the **objective lens**, shown in **Figure 6.3**, whose curved surfaces refract the light from a distant object. This refracted light forms an image on the telescope's **focal plane**, which is perpendicular to the *optical axis*—the path that light takes through the center of the lens. Because the telescope's glass lens is curved, light at the outer edges of the lens strikes the surface more obliquely than light near the center. Therefore, light at the outer edges of the lens is refracted more than light near its center. The lens concentrates the light rays entering the telescope, bringing them to a sharp focus at a distance called the **focal length**. Sometimes focal length is specified on a telescope as *focal ratio*, which equals focal length divided by aperture size; this term may be familiar to you from lenses used in photography.

Figure 6.4 illustrates how a telescope uses the light that passes through its lenses. Figure 6.4a shows the light from two stars passing through a lens and converging at the focal plane of the lens. Figure 6.4b shows the same situation for a lens with a longer focal length. Longer focal lengths increase the size and separation of objects in the focal plane. Aperture and focal length are the two most important parameters of a telescope. The image can be viewed with an **eyepiece**—a changeable lens whose focal length determines the magnification (**Working It Out 6.1**). In modern research, however, the images are sent directly to a camera or other detector.

Refracting telescopes have two major shortcomings. First, there are physical limits on the size of refracting telescopes. The larger the area of the objective lens, the more light-gathering power it has and the fainter the stars we can observe. However, as objective lenses get larger, they get heavier, and a massive piece of glass at the end of a very long tube sags too much under the force of gravity. Refracting telescopes grew in size until the 1897 completion of the Yerkes 1-meter refractor (**Figure 6.5**), the world's largest operational refracting telescope. Located in

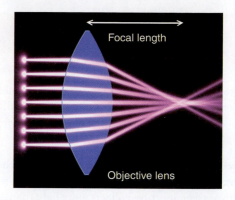

Figure 6.3 For a curved lens like the one shown, refraction causes the light to focus to a point. This point is in a slightly different location for different wavelengths (colors) of light.

Nebraska Simulation: Telescope Simulator

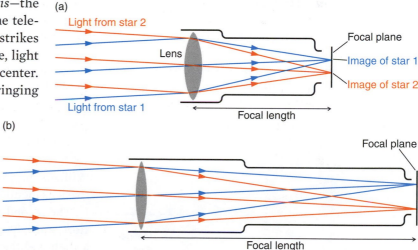

Figure 6.4 (a) A refracting telescope uses a lens to collect and focus light from two stars, forming images of the stars on its focal plane. (b) Telescopes with longer focal length produce larger images.

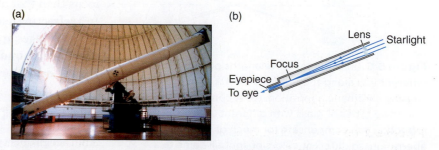

Figure 6.5 (a) The Yerkes 1-meter refractor is the world's largest refracting telescope. (b) This sketch shows the parts of a refractor.

6.1 Working It Out Telescope Aperture and Magnification

If you were shopping for a telescope, you would likely be told to consider aperture and magnification. Here we will look briefly at each.

Aperture

The light-gathering power is proportional to the area of the mirror or lens, and thus to the square of the aperture, $\pi \times (D/2)^2$. A telescope with a larger aperture collects more light than does one with a smaller aperture. We can compare a 200-millimeter (mm), or 8-inch, diameter telescope with the light-gathering power of the pupil of your eye, which is about 6 mm in the dark:

$$\text{Light-gathering power of telescope} = \frac{\pi}{4} \times (200\,\text{mm})^2$$

and

$$\text{Light-gathering power of eye} = \frac{\pi}{4} \times (6\,\text{mm})^2$$

So, to compare:

$$\frac{\text{Light-gathering power of telescope}}{\text{Light-gathering power of eye}} = \frac{\frac{\pi}{4}(200\,\text{mm})^2}{\frac{\pi}{4}(6\,\text{mm})^2} = \left(\frac{200}{6}\right)^2 = 1{,}000$$

A typical 8-inch telescope has more than 1,000 times the light-gathering power of your eye.

Comparing this 8-inch telescope to the Keck 10-meter telescope shows why bigger is better: 200 mm = 0.2 meter, and we cancel out the $\frac{\pi}{4}$ again to obtain

$$\frac{\text{Light-gathering power of Keck}}{\text{Light-gathering power of 8-inch telescope}} = \left(\frac{10\,\text{m}}{0.2\,\text{m}}\right)^2 = 2{,}500$$

Even larger telescopes, 25–40 meters in diameter, are currently under construction.

Magnification

Most telescopes have a set focal length and come with a collection of eyepieces. The magnification of the image in the telescope is given by

$$\text{Magnification} = \frac{\text{Telescope focal length}}{\text{Eyepiece focal length}}$$

Suppose the focal length of the 200-mm telescope in the preceding example is 2,000 mm. Combined with the focal length of a standard eyepiece, 25 mm, this telescope will give the following magnification:

$$\text{Magnification} = \frac{2{,}000\,\text{mm}}{25\,\text{mm}} = 80$$

This telescope and eyepiece combination has a magnifying power of 80, meaning that a crater on the Moon will appear 80 times (80×) larger in the telescope's eyepiece than it does when viewed by the naked eye. An eyepiece that has a focal length of 8 mm will have about 3 times more magnifying power, or 250.

A higher magnification will not necessarily let you see the object better. A faint and fuzzy image will not look clearer when magnified.

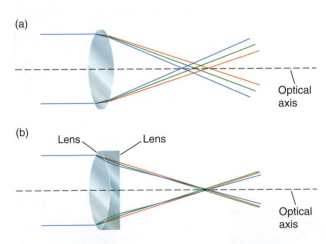

(a)

Optical axis

(b)

Lens   Lens

Optical axis

Figure 6.6 (a) Different wavelengths of light come to different foci along the optical axis of a simple lens, causing chromatic aberration. (b) A compound lens using two types of glass with different indices of refraction can compensate for much of the chromatic aberration, so different colors of light all come to a focus at the same point.

Williams Bay, Wisconsin, the Yerkes telescope carries a 450-kilogram (kg) objective lens mounted at the end of a 19.2-meter tube.

The second major shortcoming of refracting telescopes is **chromatic aberration**. Starlight is made up of all the colors of the rainbow, and each color refracts at a slightly different angle because the index of refraction depends on the wavelength of the light. As seen in **Figure 6.6a**, shorter (bluer) wavelengths are refracted more strongly than longer (redder) wavelengths. This wavelength-dependent difference in refraction, which spreads the white light out into its spectral colors, is called **dispersion**. Dispersion causes bluer light to come to a shorter focus than that of the longer visible wavelengths, creating chromatic aberration. In a refracting telescope with a simple convex lens, chromatic aberration produces haloed images around the star. Manufacturers of quality cameras and telescopes use a **compound lens** composed of two types of glass to correct for chromatic aberration (Figure 6.6b).

Reflecting Telescopes

Another property of light is **reflection**, the basis for reflecting telescopes. When light encounters a different medium—in this case going from air to glass—there

(a)

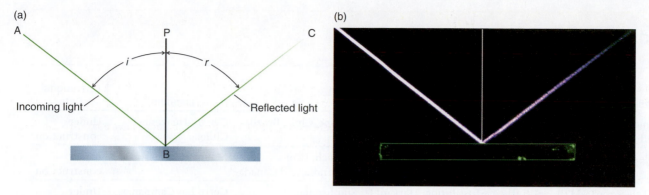

(b)

Figure 6.7 (a) When a ray of incoming light (AB) shines on a flat surface, it reflects from the surface, becoming the reflected ray BC. The angle between AB and PB, the perpendicular to the surface, is the angle of incidence (*i*). The angle between BC and PB is the angle of reflection (*r*). The angles of incidence and reflection are always equal. (b) Light from a laser beam is reflected from a flat glass surface.

will be an amount of light reflected from the surface of the new medium. In other words, some of the light will reverse its direction of travel. The most common example occurs when light encounters an ordinary flat mirror. As shown in **Figure 6.7**, the angle of the incoming light and the angle of outgoing light are always equal. A reflected image from a mirror is a good representation of what falls on it, although left and right are interchanged.

In 1668, Isaac Newton designed a reflecting telescope, which uses mirrors instead of lenses (**Figure 6.8a**). The direction of reflected light does not depend on the wavelength of light; therefore, chromatic aberration is not a problem in reflecting telescopes. A sketch of parts of Newton's reflecting telescope is presented in Figure 6.8b. To make this reflecting telescope, Newton cast a 2-inch primary mirror made of copper and tin and polished it to a special curvature. He then placed this primary mirror at the bottom of a tube with a secondary flat mirror mounted above it at a 45° angle. The second mirror directed the focused light to an eyepiece on the outside of the tube.

Astronomers use mirrors with a surface that curves inward toward the incoming light, called concave mirrors. The same rules of incidence and reflection hold here for each ray of light, but in this case the reflected rays do not maintain the same angle with respect to each other as they do with a flat mirror. Concave mirrors will reflect the rays so that they converge to form an image, as shown in **Figure 6.9**. If the incoming light rays are parallel, as from a distant source like a star, the reflected light rays cross at the focal length of the mirror. The light path from the primary mirror to the focal plane can be "folded" by using a secondary mirror, which enables a significant reduction in the length and weight of the telescope. In many modern telescopes, the primary mirror has a hole so that light can pass back through it; the eyepiece is on the back of the tube of the telescope, and the tube can be shortened.

Large reflecting telescopes did not become common until the latter half of the 18th century. But then the size of the primary mirrors in reflecting telescopes continued to grow; and they became larger every decade. Primary mirrors can be supported from the back, and they can be made thinner and therefore less massive than the objective lenses found in refracting telescopes. The limitation on the size of reflecting telescopes is the cost of their fabrication and support structure.

(a)

(b)

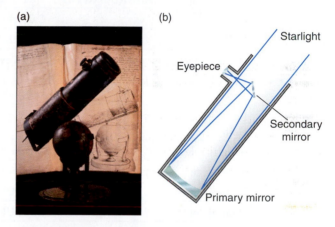

Figure 6.8 Newton's reflecting telescope (a) has parts shown in the sketch (b).

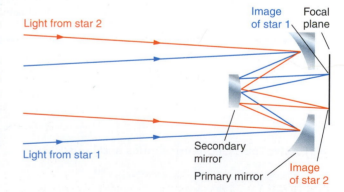

Figure 6.9 Large reflecting telescopes often use a secondary mirror that directs the light back through a hole in the primary mirror to an accessible focal plane behind the primary mirror. Parallel rays of light that strike a concave parabolic mirror are brought to a focus in the mirror's focal plane.

| TABLE 6.1 | The World's Largest Optical Telescopes |

Mirror Diameter (meters)	Telescope	Sponsor(s)	Location	Operational Date
39.3	European Extremely Large Telescope (E-ELT)	European Southern Observatory (Europe, Chile, Brazil)	Cerro Armazones, Chile	Under construction
30.0	Thirty Meter Telescope (TMT)	International collaboration led by Caltech, U. of California, U. of Hawaii, China, Japan, India, and Canada	Mauna Kea, Hawaii	Under construction
24.5	Giant Magellan Telescope (GMT)	Carnegie Institution, Harvard U., Smithsonian Institution, U. of Arizona, U. of Texas, Texas A&M, U. of Chicago, Australian National U., Astronomy Australia Ltd., Korea Astronomy and Space Science Institute	Cerro Las Campanas, Chile	Under construction
11.0	South African Large Telescope (SALT)	South Africa, USA, UK, Germany, Poland, New Zealand, India	Sutherland, South Africa	2005
10.4	Gran Telescopio CANARIAS (GTC)	Spain, Mexico, U. of Florida	Canary Islands	2007
10	Keck I	Caltech, U. of California, NASA	Mauna Kea, Hawaii	1993
10	Keck II	Caltech, U. of California, NASA	Mauna Kea, Hawaii	1996
9.2	Hobby-Eberly Telescope (HET)	U. of Texas, Penn State U., Stanford U., Germany	Mount Fowlkes, Texas	1999
8.4 × 2	Large Binocular Telescope (LBT)	U. of Arizona, Ohio State U., Italy, Germany, Arizona State, and others	Mount Graham, Arizona	2008
8.4	Large Synoptic Survey Telescope (LSST)	Many partners	Cerro Pachón	Under construction
8.3	Subaru Telescope	Japan	Mauna Kea, Hawaii	1999
8.2 × 4	Very Large Telescope (VLT)	European Southern Observatory	Cerro Paranal, Chile	2000
8.1	Gemini North	USA, UK, Canada, Chile, Brazil, Argentina, Australia	Mauna Kea, Hawaii	1999
8.1	Gemini South	USA, UK, Canada, Chile, Brazil, Argentina, Australia	Cerro Pachón, Chile	2000
6.5	MMT	Smithsonian Institution, U. of Arizona	Tucson, Arizona	2000
6.5	Magellan I	Carnegie Institution, U. of Arizona, Harvard U., U. of Michigan, MIT	Cerro Las Campanas, Chile	2000
6.5	Magellan II	Carnegie Institution, U. of Arizona, Harvard U., U. of Michigan, MIT	Cerro Las Campanas, Chile	2002

▶‖ **AstroTour:** Geometric Optics and Lenses

Table 6.1 lists the world's largest optical telescopes. All are reflecting telescopes. The largest single mirrors constructed today are 8 meters in diameter, but reflecting telescopes even bigger than this are designed to make use of an array of smaller segments. The primary mirror of each of the 10-meter, twin Keck telescopes is made up of 36 hexagon-shaped segments that are 1.8 meters in diameter (**Figure 6.10**). Located on 4,100-meter-high Mauna Kea in Hawaii, the Keck telescopes are among the world's largest reflecting telescopes. Each one has 4 million times the light-gathering power of the human eye.

CHECK YOUR UNDERSTANDING 6.1

Which of the following is a reason that all large astronomical telescopes are reflectors (choose all that apply): (a) chromatic aberration is minimized; (b) they are not as heavy; (c) they can be shorter; (d) the glass doesn't need to be curved.

Optical and Atmospheric Limitations

Another important characteristic of a telescope is its *resolution*—how close two points of light can be to each other before they are indistinguishable. The concept of resolution is illustrated in **Figure 6.11**. Review Figure 6.4a to see the path followed by rays of light from two distant stars as they pass through the lens of a refracting telescope. Figure 6.4b illustrated that increasing the focal length increases the size of and separation between the images that a telescope produces. This is one reason why telescopes provide a much clearer view of the stars than that obtained with the naked eye. The focal length of a human eye is typically about 20 mm, whereas telescopes used by professional astronomers often have focal lengths of tens or even hundreds of meters. Such telescopes make images that are far larger than those formed by the human eye, and consequently they contain far more detail.

Focal length explains only one difference between the resolution of telescopes and that of the unaided eye. The other difference results from the wave nature of light. **Figure 6.12** shows what happens when light waves pass through the aperture of a telescope: they spread out from the edges of the lens or mirror. The distortion that occurs as light passes the edge of an opaque object is called **diffraction**. Diffraction "diverts" some of the light from its path, slightly blurring the image made by the telescope. The degree of blurring depends on the wavelength of the light and the telescope's aperture. The larger the aperture, the smaller the problem posed by diffraction. The best resolution that a given telescope can achieve is known as the **diffraction limit** (**Working It Out 6.2**).

Larger telescopes have better resolution and can distinguish objects that appear closer together. Theoretically, the 10-meter Keck telescopes have a diffraction-limited resolution of 0.0113 arcseconds (arcsec) in visible light, which would be good enough for you to read newspaper headlines 60 km away. But for telescopes with apertures larger than about a meter, Earth's atmosphere stands in the way of better resolution. If you have ever looked out across a large asphalt parking lot on a summer day, you have seen the distant horizon shimmer as light is bent this way and that by turbulent bubbles of warm air rising off the hot

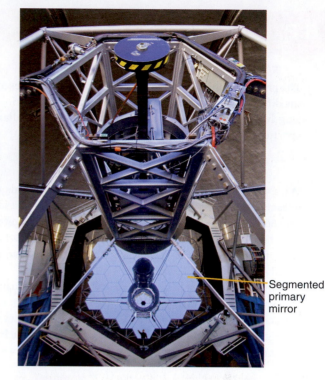

Figure 6.10 Each of the Keck 10-meter reflectors uses an aligned group of 36 hexagonal mirrors to collect light.

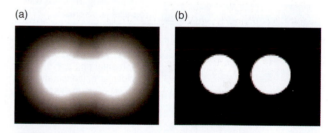

(a)   (b)

Figure 6.11 Resolution is the ability to separate two images that appear close together. When resolution is lower (a), the two images blend together. When resolution is higher (b), individual images can be seen.

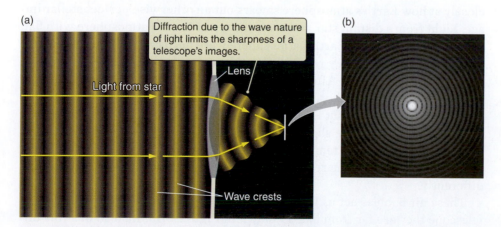

(a)

Diffraction due to the wave nature of light limits the sharpness of a telescope's images.

Lens

Light from star

Wave crests

(b)

Figure 6.12 (a) Light waves from a star are diffracted by the edges of a telescope's lens or mirror. (b) This diffraction causes the stellar image to be blurred, limiting a telescope's ability to resolve objects.

6.2 Working It Out Diffraction Limit

The practical limit on the angular resolution, θ, of a telescope is called the diffraction limit. This limit is determined by the ratio of the wavelength of light, λ, passing through the telescope to the diameter of the aperture, D:

$$\theta = 2.06 \times 10^5 \left(\frac{\lambda}{D}\right) \text{arcsec}$$

With the constant, 2.06×10^5, the units are arcseconds (arcsec). An **arcsecond** is a tiny angular measure found by first dividing the sky into 360 degrees, and then dividing a degree by 60 to get arcminutes, and then by 60 again to get arcseconds. An arcsecond is 1/1,800 of the size of the Moon in the sky, or about the size of a tennis ball if you could see it from 8 miles away.

Both λ and D must be expressed in the same units, usually meters. The smaller the ratio of λ/D, the better the resolution. For example, the size of the human pupil (see Figure 6.1) ranges from about 2 mm in bright light to 8 mm in the dark. A typical pupil size in the dark is about 6 mm, or 0.006 meter. Visible (green) light has a wavelength (λ)

of 550 nanometers (nm); that is, 550×10^{-9} meter, or 5.5×10^{-7} meter. Using these values for the aperture and the wavelength gives

$$\theta = 2.06 \times 10^5 \left(\frac{5.5 \times 10^{-7}\,\text{m}}{0.006\,\text{m}}\right) \text{arcsec} = 19\,\text{arcsec}$$

or about 0.5 arcmin. The typical resolution of the human eye is 2 arcmin. We do not achieve the theoretical resolution with our eyes because the physical properties of our eyes are not perfect.

How does the resolution of the human eye compare to that of a telescope? Consider the Hubble Space Telescope, when operating in the visible part of the spectrum. Its primary mirror has a diameter of 2.4 meters. Substituting this value for D and again using visible (green) light gives

$$\theta = 2.06 \times 10^5 \left(\frac{5.5 \times 10^{-7}\,\text{m}}{2.4\,\text{m}}\right) \text{arcsec} = 0.047\,\text{arcsec}$$

or about 600 times better than the theoretical resolving power of the human eye.

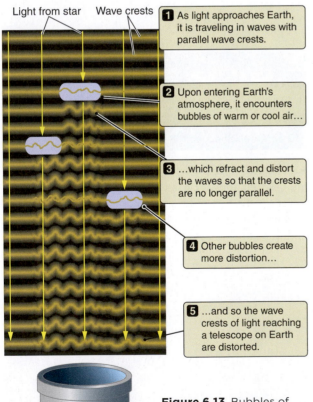

Light from star   Wave crests

1 As light approaches Earth, it is traveling in waves with parallel wave crests.

2 Upon entering Earth's atmosphere, it encounters bubbles of warm or cool air…

3 …which refract and distort the waves so that the crests are no longer parallel.

4 Other bubbles create more distortion…

5 …and so the wave crests of light reaching a telescope on Earth are distorted.

Telescope

Figure 6.13 Bubbles of warmer or cooler air in Earth's atmosphere distort the wavefront of light from a distant object.

pavement. The problem of the shimmering atmosphere is less pronounced when we look overhead, but the twinkling of stars in the night sky is caused by the same phenomenon. As telescopes magnify the angular diameter of an object, they also magnify the shimmering effects of the atmosphere. The limit on the resolution of a telescope on the surface of Earth caused by this atmospheric distortion is called **astronomical seeing**. One advantage of launching telescopes such as the Hubble Space Telescope into orbit around Earth above the atmosphere is that they are not hindered by astronomical seeing.

Modern technology has improved ground-based telescopes with computer-controlled **adaptive optics** that compensate for much of the atmosphere's distortion. To understand how adaptive optics work, we need to look more closely at how Earth's atmosphere smears out an otherwise perfect stellar image. Light from a distant star arrives at the top of Earth's atmosphere with flat, parallel wave crests. If Earth's atmosphere were perfectly uniform, the crests would remain flat as they reached the objective lens or primary mirror of a ground-based telescope. After making its way through the telescope's optical system, the crests would produce a tiny diffraction disk in the focal plane, as shown in Figure 6.12b. But Earth's atmosphere is not uniform. It is filled with bubbles of air that have slightly different temperatures than those of their surroundings. Different temperatures mean different densities, and different densities mean different refractive properties, so each bubble bends light differently.

These air bubbles act as weak lenses, and by the time the waves reach the telescope they are far from flat, as shown in **Figure 6.13**. Instead of a tiny diffraction disk, the image in the telescope's focal plane is distorted and swollen, degrading the resolution. Adaptive optics flatten out this distortion. First, an optical device within the telescope constantly measures the wave crests. Then, before reaching the telescope's focal plane, the light is reflected off yet another mirror, which has a flexible surface. A computer analyzes the light and bends the flexible mirror so that it accurately corrects for the distortion caused by the air bubbles. **Figure 6.14** shows an example of an image corrected by adaptive optics. The widespread use of adaptive optics has made the image quality of ground-based telescopes competitive with the quality of Hubble images from space at some wavelengths.

Observatory Locations

What makes a good location for a telescope on Earth? Look back at Table 6.1—what do these locations have in common? Astronomers look for sites that are high, dry, and dark. The best sites are far away from the lights of cities, in locations with little moisture, humidity, or rain, and where the atmosphere is relatively still. Telescopes are located as high as possible so that they get above a significant part of Earth's atmosphere, which distorts images and blocks infrared light. Many telescopes are situated on remote, high mountaintops surrounded by desert or ocean. Recall from Chapter 2 that the stars that can be seen throughout the year depend on latitude, and only at the equator would a telescope have access to all of the stars in the sky. But equatorial latitudes have tropical weather—wet, humid, and stormy—and thus are poor locations for a telescope. So, to cover the entire sky, astronomers have built telescopes in both northern and southern locations. In the United States, large telescopes are located in California, Arizona, New Mexico, Texas, and Hawaii. The largest southern-sky observatories are found in Chile, South Africa, and Australia. The twin Gemini telescopes, designed to be a matched pair, are located in Hawaii in the Northern Hemisphere and in Chile in the Southern Hemisphere.

Newer and larger telescopes are planned for many of the same locations that are listed in Table 6.1. The 8-meter Large Synoptic Survey Telescope (LSST) is headed for Cerro Pachón in Chile, current site of the Gemini South telescope. The Giant Magellan Telescope (GMT), consisting of seven 8-meter mirrors in a pattern equivalent to a 24.5-meter mirror, will be constructed at Cerro Las Campanas in Chile. The Thirty Meter Telescope (TMT) (**Figure 6.15**) is planned for Mauna Kea in Hawaii, current site of the twin Keck telescopes; and the European Southern Observatory (ESO) is building the 39-meter European Extremely Large Telescope (E-ELT) at Cerro Armazones in Chile. As telescopes get larger—and more expensive—international collaboration becomes even more important.

Today's professional astronomers rarely look through the eyepiece of a telescope because they learn much more and make better use of observing time by permanently recording an object's image at a variety of wavelengths or seeing its light spread out into a revealing spectrum. Some astronomers no longer travel to telescopes at all, instead observing remotely from the base of the mountain or far away at their own institutions.

Professional and amateur astronomers alike are concerned about loss of the dark sky. As cities and suburbs grow and expand around the world, the use of outdoor artificial light becomes more widespread. Pictures from space show how bright many areas of Earth are at night. In the United States, two-thirds of the population resides in an area that is too bright to see the Milky Way in the sky at night (**Figure 6.16**), and it has been estimated that by 2025 there will be almost no dark skies in the continental United States. Increased air pollution also dims the

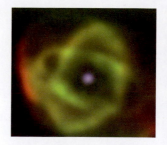
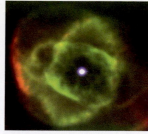

Figure 6.14 These images of the Cat's Eye Nebula from the Palomar Observatory telescope without (left) and with (right) adaptive optics show the benefit of the technique.

Figure 6.15 This is an artist's rendering of the Thirty Meter Telescope, a planned reflecting telescope.

Figure 6.16 This satellite image of the United States at night shows that few populated areas are free from light pollution.

view of the night sky in many locations. The U.S. National Park Service now advertises evening astronomy programs in natural, unpolluted dark skies as one of the reasons to visit some parks. Several international astronomy associations are working with UNESCO (the United Nations Educational, Scientific and Cultural Organization) to promote the "right to starlight," arguing that for historical, cultural, and scientific reasons, it would be a huge loss if humanity could no longer view the stars. These organizations are encouraging countries to create starlight reserves and starlight parks where people can experience increasingly rare dark skies and a natural nocturnal environment.

CHECK YOUR UNDERSTANDING 6.2

In practice, the smallest angular size that one can resolve with a 10-inch telescope is governed by the: (a) blurring caused by Earth's atmosphere; (b) diffraction limit of the telescope; (c) size of the primary mirror; (d) magnification of the telescope.

6.2 Optical Detectors and Instruments Used with Telescopes

Beginning in the 1800s, the development of film photography, and later digital photography, revolutionized astronomy, allowing astronomers to detect fainter and more distant objects than possible to detect with the eye alone. In this section, we will examine some of the more common types of detectors.

Integration Time and Quantum Efficiency

Originally, the retina of the human eye was the only astronomical detector. The limit of the faintest stars we can see with our unaided eyes is determined in part by two factors that are characteristic of all detectors: *integration time* and *quantum efficiency*.

Integration time is the limited time interval during which the eye can add up photons—this is analogous to leaving the shutter open on a camera. The brain "reads out" the information gathered by the eye about every 100 milliseconds (ms). Anything that happens faster than that appears to happen all at once. If two images on a computer screen appear 30 ms apart, you will see them as a single image because your eyes will add up (or integrate) whatever they see over an interval of 100 ms or less. However, if the images occur 200 ms apart, you will see them as separate images. This relatively brief integration time is the most important factor limiting our nighttime vision. Stars too faint to be seen with the unaided eye are those from which you receive too few photons for your eyes to process in 100 ms.

Quantum efficiency determines how many responses occur for each photon received. For the human eye, 10 photons must strike a cone within 100 ms to activate a single response. So the quantum efficiency of our eyes is about 10 percent: for every 10 events, the eye sends one signal to the brain. Together, integration time and quantum efficiency determine the rate at which photons must arrive at the retina before the brain says, "Aha, I see something." Astronomers seek to use detectors with longer integration times and higher quantum efficiency than those of our eyes.

From Photographic Plates to Charge-Coupled Devices

For more than two centuries after the invention of the telescope, astronomers struggled with the problem of surface brightness. Only *point sources* such as stars appear brighter in a telescope; extended astronomical objects like the Moon appear bigger in the eyepiece, but their surfaces are no brighter than they appear to the unaided eye. Even when astronomers built larger telescopes, nebulae and galaxies appeared larger, but the details of these faint objects remained elusive. The problem was not with the telescopes but with the limitations of optics and the human eye. Only with the longer exposure times made possible by the invention of photography and the later development of electronic cameras were astronomers finally able to discern intricate details in faint objects.

In 1840, John W. Draper (1811–1882), a New York chemistry professor, created the earliest known astronomical photograph (**Figure 6.17**). By the late 1800s, astronomers had created thousands of photographic plates with permanent images of planets, nebulae, and galaxies. The quantum efficiency of most photographic systems used in astronomy was poorer than that of the human eye—typically 1–3 percent. But unlike the eye, photography can overcome poor quantum efficiency by leaving the shutter open on the camera, increasing the integration time to many hours of exposure. Photography made it possible for astronomers to record and study objects that were invisible to the human eye. However, one problem is that the response of photography to light is not linear, especially at long exposures, so if you doubled the exposure time, you did not get twice as much light on your image. By the middle of the 20th century, the search was on for electronic detectors that would overcome the sensitivity, spectral range, and nonlinearity problems of photography.

In 1969, scientists at Bell Laboratories invented a detector called a **charge-coupled device**, or **CCD**. By the late 1970s, the CCD had become the detector of choice in almost all astronomical-imaging applications. CCDs are linear, so doubling the exposure means you record twice as much light. Therefore, they are good for measuring objects that vary in brightness, as well as for faint objects that require long exposures. CCDs have a quantum efficiency far superior to that of photography or the eye, up to 80 percent at some wavelengths. This improvement dramatically increases the ability to view faint objects with short exposure times.

A CCD is an ultrathin wafer of silicon—less than the thickness of a human hair—that is divided into a two-dimensional array of picture elements, or **pixels**, as seen in **Figure 6.18a**. When a photon strikes a pixel, it creates a small electric charge within the silicon. As each CCD pixel is read out, the digital signal that flows to the computer is nearly proportional to the accumulated charge. This is what we mean when we say that the CCD is a linear device. However, if a CCD is exposed to too much light, it can lose its linearity.

Liquid nitrogen or helium is used to cool the CCDs down to very low temperatures to reduce noise caused by the movement of the charge-carrying atoms within the silicon wafer. The first astronomical CCDs were small arrays containing a few hundred thousand pixels. The larger CCDs used in astronomy today may contain more than 100 million pixels (Figure 6.18b). Still larger arrays are under development as ever-faster computing power keeps up with image-processing demands.

Figure 6.17 A photograph of the Moon taken by John W. Draper in 1840.

(a)

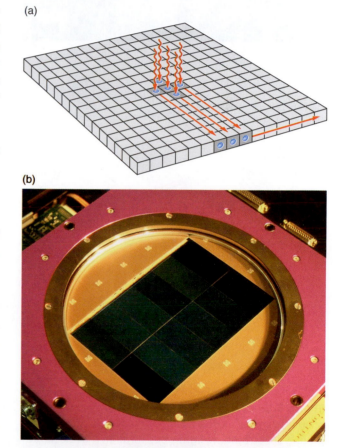

(b)

Figure 6.18 (a) In this simplified diagram of a charge-coupled device (CCD), photons from a star land on pixels (represented by gray squares) and produce free electrons within the silicon. The electron charges are electronically moved sequentially to the collecting register at the bottom. Each row is then moved out to the right to an electronic amplifier, which converts the electric charge of each pixel into a digital signal. (b) This large CCD (about 6 inches across) contains 12,288 × 8,192 pixels.

Nebraska Simulation: CCD Simulator

(a)

(b)

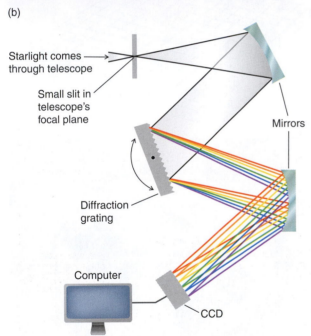

Starlight comes through telescope

Small slit in telescope's focal plane

Mirrors

Diffraction grating

Computer

CCD

Figure 6.19 (a) A spectrum is created by the reflection of light from the closely spaced tracks of a CD. (b) In a grating spectrograph, light goes through the telescope and then a slit, where it is reflected to the diffraction grating and split into components. The spectrum is recorded on the CCD.

The output from a CCD is a digital signal that can be sent directly from the telescope to image-processing software or stored electronically for later analysis. Nearly every spectacular astronomical image in ultraviolet, visible, or infrared wavelength that you find online was recorded by a CCD in a telescope either on the ground or in space. CCDs are found in many common devices such as digital cameras, digital video cameras, and camera phones.

Your cell phone takes color pictures by using a grid of CCD pixels arranged in groups of three. Each pixel in a group is constructed to respond to only a particular range of colors—only to red light, for example. This is also true for digital image displays. You can see this for yourself if you place a small drop of water on the screen of your smartphone or tablet and turn it on. The water magnifies the grid of pixels so that you can see them individually. This grid degrades the angular resolution of the camera because each spot in the final image requires three pixels of information. Astronomers choose instead to use all the pixels on the camera to measure the number of photons that fall on each pixel, without regard to color. They put filters in front of the camera to allow light of only particular wavelengths to pass through, such as the light of a specific spectral line. Color pictures like those from the Hubble Space Telescope are constructed by taking multiple pictures, coloring each one, and then carefully aligning and overlapping them to produce beautiful and informative images. Sometimes the colors are "true"; that is, they are close to the colors you would see if you were actually looking at the object with your eyes. Other times "false" colors represent different portions of the electromagnetic spectrum and tell you the temperature or composition of different parts of the object. Using changeable filters instead of designated color pixels gives astronomers greater flexibility and greater resolution.

Spectrographs

Spectroscopy is the study of an object's *spectrum* (plural: *spectra*)—its electromagnetic radiation split into component wavelengths. **Spectrographs** (sometimes called **spectrometers**) are instruments that take the spectrum of an object and then record it. The first spectrographs used prisms to disperse the light. Modern spectrographs use a **diffraction grating**, which is made by engraving closely spaced lines on glass to disperse incoming light into its constituent wavelengths. **Figure 6.19a** shows the light reflected from a CD or DVD; the closely spaced tracks act as a grating and create a spectrum. Figure 6.19b shows a grating spectrograph: light from an astronomical object enters a telescope and passes through a slit. The light is reflected onto the diffraction grating, which creates a spectrum like the one shown in Figure 5.14. The spectra are recorded on a CCD and then analyzed. Some modern spectrographs use bundles of optical fibers, or masks with multiple slits, to obtain spectra simultaneously from multiple objects in the field of view of the telescope.

CHECK YOUR UNDERSTANDING 6.3

CCD cameras have much higher quantum efficiency than other detectors. This means that CCD cameras: (a) can collect photons for longer times; (b) can collect photons of different energies; (c) can generate a signal from fewer photons; (d) can split light into different colors.

6.3 Astronomers Observe in Wavelengths Beyond the Visible

Recall from Chapter 5 that an object's temperature can be found from the peak wavelength of its continuous spectrum. Extending beyond visible light, radio or infrared telescopes are used to study cool objects, like clouds of dust, whereas X-ray or gamma-ray telescopes are used to study violently hot gas. Therefore, astronomers must utilize telescopes that observe at all the wavelengths of the electromagnetic spectrum. However, not all of these wavelengths reach Earth, so some telescopes must be put into space. **Figure 6.20** shows that Earth has a few **atmospheric windows** that let in parts of the spectrum. The largest window is in radio wavelengths, including microwaves at the short-wavelength end of the radio window. These telescopes can be built on the ground. However, gamma-ray, X-ray, ultraviolet, and most of the infrared light arriving at Earth fails to reach the ground because it is partially or completely absorbed by ozone, water vapor, carbon dioxide, and other molecules in Earth's atmosphere. Light at these wavelengths has to be observed from space.

 Nebraska Simulation: EM Spectrum Module

Radio Telescopes

Karl Jansky (1905–1950), a young physicist working for Bell Laboratories in the early 1930s, identified a radio source in the Milky Way in the direction of the galactic center, in the constellation Sagittarius. Jansky's discovery marked the birth

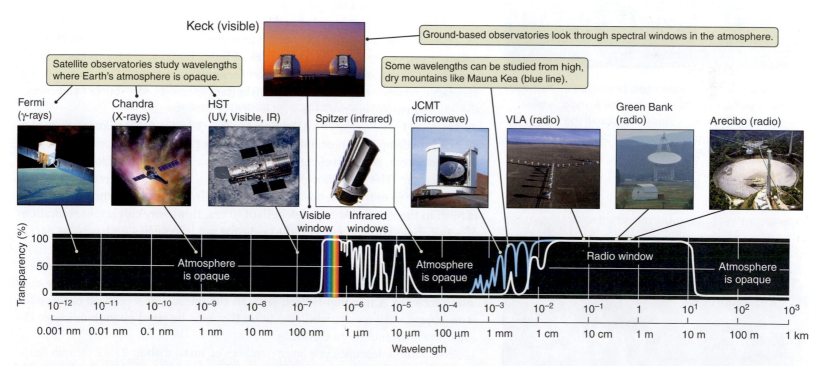

Figure 6.20 Earth's atmosphere blocks most electromagnetic radiation. Fermi = Fermi Gamma-ray Space Telescope (orbiting); Chandra = Chandra X-ray Observatory (orbiting); HST = Hubble Space Telescope (orbiting); Keck = Keck Observatory (Hawaii); Spitzer = Spitzer Space Telescope (orbiting); JCMT = James Clerk Maxwell Telescope (Hawaii); VLA = Very Large Array (New Mexico); Green Bank = Robert C. Byrd Green Bank Telescope (West Virginia); Arecibo = Arecibo Observatory (Puerto Rico).

(a)

(b)

Figure 6.21 (a) The Parkes radio telescope in Australia. (b) The Arecibo radio telescope is the world's largest. The steerable receiver suspended above the dish permits limited pointing toward celestial targets as they pass close to the zenith.

Figure 6.22 The VLA in New Mexico combines signals from 27 different telescopes so that they act as one "very large" telescope.

of radio astronomy, and in his honor, the basic unit for the strength of a radio source is called the **jansky (Jy)**. A few years later, Grote Reber (1911–2002), a radio engineer and ham radio operator, built his own radio telescope and conducted the first survey of the sky at radio frequencies. Reber was largely responsible for the rapid advancement in radio astronomy that blossomed in the post–World War II era.

Most radio telescopes are large, steerable dishes, typically tens of meters in diameter, like the one shown in **Figure 6.21a**. The world's largest single-dish radio telescope is the 305-meter Arecibo dish built into a natural bowl-shaped depression in Puerto Rico (Figure 6.21b). (China is constructing a 500-meter single-dish radio telescope with a similar design.) The Arecibo telescope is not steerable, so it can only observe sources that pass within 20° of the zenith as Earth's rotation carries them overhead.

As large as radio telescopes are, they have relatively poor angular resolution. Recall that a telescope's angular resolution is determined by the ratio λ/D, so a larger ratio means poorer resolution. Radio telescopes have diameters much larger than the apertures of most optical telescopes. However, the wavelengths of radio waves range from about 1 centimeter (cm) to 10 meters, or up to several hundred thousand times greater than the wavelengths of visible light, which makes the ratio larger. Radio telescopes are thus limited by the very long wavelengths they are designed to receive. For example, the resolution of the huge Arecibo dish in Figure 6.21b is typically about 1 arcmin, little better than the unaided human eye.

Radio astronomers have had to develop ways to improve resolution. Mathematically combining the signals from two radio telescopes turns them into a telescope with a diameter equal to the separation between them. For example, if two 10-meter telescopes are located 1,000 meters apart, the D in λ/D is 1,000, not 10. This combination of two (or more) telescopes is called an **interferometer**, and it makes use of the wavelike properties of light. Usually, several telescopes are used in an arrangement called an **interferometric array**. Through the use of very large arrays, radio astronomers can better observe bright sources and exceed the angular resolution possible with optical telescopes.

The Very Large Array (VLA) in New Mexico (**Figure 6.22**) is an interferometric array made up of 27 movable dishes spread out in a Y-shaped configuration up to 36 km across. At a wavelength of 10 cm, this array reaches resolutions of less than 1 arcsec. The Very Long Baseline Array (VLBA) uses 10 radio telescopes spread out over more than 8,000 km from the Virgin Islands in the Caribbean to Hawaii in the Pacific. At a wavelength of 10 cm, this array can attain resolutions of better than 0.003 arcsec. A radio telescope put into near-Earth orbit as part of a Space Very Long Baseline Interferometer (SVLBI) overcomes even this limit. The new Event Horizon Telescope will combine many of the most advanced existing radio telescopes, from Greenland to the South Pole, to make an *Earth-sized* interferometer. A few nights each year, all the telescopes would all observe the same object, with a combined resolution that may be good enough to image the objects near the center of the Milky Way.

Some radio telescopes use large numbers of small dishes. The Atacama Large Millimeter/submillimeter Array (ALMA; **Figure 6.23**), located at an elevation of 5,000 meters in the Atacama Desert in Chile, was completed in 2013. This project, an international collaboration of astronomers from Europe, North America, East Asia, and Chile, consists of sixty-six 12- and 7-meter dishes for observations in the 0.3- to 9.6-mm wavelength range. The Square Kilometre Array (SKA) is designed to have *thousands* of small radio dishes, which together will act as one dish with a collecting area of 1 square kilometer (km²). Twenty countries are supporting this telescope, which will be located in Australia and South Africa and is scheduled to be built by 2024.

Optical telescopes can also be combined in an array to yield resolutions greater than those of single telescopes, although for technical reasons the individual units cannot be spread as far apart as radio telescopes. The Very Large Telescope Interferometer (VLTI) in Chile, operated by ESO, combines the four Very Large Telescope (VLT) 8-meter telescopes with four movable 1.8-meter auxiliary telescopes. It has a baseline of up to 200 meters, yielding angular resolution of about 0.001 arcsec. The six-telescope Center for High Angular Resolution Astronomy (CHARA) array in California works in visible and near-infrared and has a baseline of 330 meters with similar resolution.

Infrared Telescopes

Molecules such as water vapor in Earth's atmosphere block infrared (IR) photons from reaching astronomical telescopes on the ground, so telescopes that observe in the infrared (0.75–30 microns; μm) are at the highest locations. Mauna Kea, a dormant volcano and home of the Mauna Kea Observatories (MKO), rises 4,200 meters above the Pacific Ocean. At this altitude, the MKO telescopes sit above 40 percent of Earth's atmosphere; but more important, 90 percent of Earth's atmospheric water vapor lies below. Still, for the infrared astronomer the remaining 10 percent is troublesome.

Airborne observatories overcome atmospheric absorption of infrared light by placing telescopes above most of the water vapor in the atmosphere. NASA's Stratospheric Observatory for Infrared Astronomy (SOFIA) (**Figure 6.24**), a joint project with the German Aerospace Center (DLR), is a modified 747 airplane that carries a 2.5-meter telescope and works in the far-infrared region of the spectrum, from 1 to 650 μm. It flies in the stratosphere at an altitude of about 12 km, above 99 percent of the water vapor in Earth's lower atmosphere. Because airplanes are highly mobile, SOFIA can observe in both the Northern and Southern hemispheres. Other infrared wavelengths must be observed from space.

Orbiting Observatories

Gaining full access to the complete electromagnetic spectrum requires getting completely above Earth's atmosphere. The first astronomical satellite was the British Ariel 1, launched in 1962 to study solar UV and X-ray radiation. Today, a multitude of orbiting astronomical telescopes cover the electromagnetic spectrum from gamma rays to microwaves, with more in the planning stage (**Table 6.2**). Optical telescopes, such as the 2.4-meter Hubble Space Telescope (HST), operate successfully at low Earth orbit, 600 km above Earth's surface. Launched in 1990, HST has been the workhorse for UV, visible, and IR space astronomy for more than 25 years. Low Earth orbit is also the region where the International Space Station (ISS) and many scientific satellites orbit. For certain other satellites and space telescopes, 600 km is not high enough.

The Chandra X-ray Observatory, NASA's X-ray telescope, cannot see through even the tiniest traces of atmosphere and therefore orbits more than 16,000 km above Earth's surface. NASA's Spitzer Space Telescope, an infrared telescope, is so sensitive that it needs to be completely free from Earth's own infrared

Figure 6.23 The new Atacama Large Millimeter/submillimeter Array (ALMA) telescope in the Atacama Desert in northern Chile has many international partners.

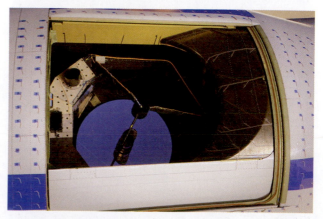

Figure 6.24 SOFIA is a 2.5-meter infrared telescope that is mounted in a Boeing 747 aircraft.

TABLE 6.2	Selected Current and Future Space Observatories

Telescope	Sponsor(s)	Description	Launch Year
Hubble Space Telescope (HST)	NASA, ESA	Optical, infrared, ultraviolet observations	1990
Chandra X-ray Observatory	NASA	X-ray imaging and spectroscopy	1999
X-ray Multi-Mirror Mission (XMM-Newton)	ESA	X-ray spectroscopy	1999
Galaxy Evolution Explorer (GALEX)	NASA	Ultraviolet observations	2003
Spitzer Space Telescope	NASA	Infrared observations	2004
Swift Gamma-Ray Burst Mission	NASA	Gamma-ray bursts	2004
Convection Rotation and Planetary Transits (COROT) space telescope	CNES (France)	Planet finder	2006
Fermi Gamma-ray Space Telescope	NASA, European partners	Gamma-ray imaging and gamma-ray bursts	2008
Planck telescope	ESA	Cosmic microwave background radiation	2009
Herschel Space Observatory	ESA	Far-infrared and submillimeter observations	2009
Kepler telescope	NASA	Planet finder	2009
Solar Dynamics Observatory (SDO)	NASA	Sun, solar weather	2010
RadioAstron	Russia	Very-long-baseline interferometry in space	2011
Nuclear Spectroscopic Telescopic Array (NuSTAR)	NASA	High-energy X-ray	2012
Gaia	ESA	Optical, digital 3D space camera	2013
James Webb Space Telescope (JWST)	NASA, ESA, Canadian Space Agency	Optical and infrared; replacement for HST	2018

radiation. The solution was to put it into a *solar* orbit, trailing tens of millions of kilometers behind Earth. The James Webb Space Telescope, scheduled to replace the HST, will observe primarily in infrared wavelengths. It will be located 1.5 million miles away from Earth, orbiting the Sun at a fixed distance from the Sun and Earth.

Orbiting telescopes located above the atmosphere are not affected by atmospheric image distortions, weather, or brightening night skies. But space observatories are much more expensive than ground-based observatories and can be difficult or impossible to repair. The HST required several servicing missions, but such missions are not possible for the observatories in more distant Earth orbits. Ground-based telescopes at even the most remote mountaintop locations can receive shipments of replacement parts in a few days; space telescopes cannot. Of course, some wavelengths can be observed from space only. But issues of cost and repair are the reason why ground-based telescopes are much more prevalent.

CHECK YOUR UNDERSTANDING 6.4

Which of the following is the biggest disadvantage of putting a telescope in space? (a) Astronomers don't have as much control in choosing what to observe. (b) Astronomers have to wait until the telescopes come back to Earth to get their images. (c) Space telescopes can only observe in certain parts of the electromagnetic spectrum. (d) Space telescopes are much more expensive than similar ground-based telescopes.

6.4 Planetary Spacecraft Explore the Solar System

Recall from Chapter 2 that everyone always sees the same face of the Moon from Earth because the Moon's orbital and rotational periods are equal. The first view of the "far" side was in 1959, when the Soviet flyby mission *Luna 3* sent back pictures showing that the far side of the Moon was very different from its Earth-facing half. No matter how powerful our ground-based or Earth-orbiting telescopes, sometimes we need to send a spacecraft for a different view.

Only in the past half century has the technology existed to explore the Solar System. Spacecraft have now visited all of the planets and some of their moons, as well as a few comets and asteroids, providing the first close-up views of these distant worlds. The study of the Solar System from space is an international collaboration involving NASA, the European Space Agency (ESA), the Russian Federal Space Agency (Roscosmos), the Japan Aerospace Exploration Agency (JAXA), the China National Space Administration (CNSA), and the Indian Space Research Organisation (ISRO). Other countries may soon join the endeavor. In this section, we will look at the different types of spacecraft used to explore our Solar System.

Flybys and Orbiters

Exploration of the Solar System began with a reconnaissance phase, using spacecraft to fly by or orbit a planet or other body. A **flyby** is a spacecraft that first approaches and then continues flying past the target. As these spacecraft speed by, instruments aboard them briefly probe the physical and chemical properties of the target and its environment.

Flyby missions are the most common first phase of exploration. They cost less than orbiters or landers and are easier to design and execute. Flyby spacecraft such as *Voyager* are sometimes able to visit several different worlds during their travels (**Figure 6.25**). The downside of flyby missions is that because of the physics of orbits, these spacecraft must move by very swiftly. They are limited to just a few hours or at most a few days in which to conduct close-up studies of their targets. Flyby spacecraft provide astronomers with their first close-up views of Solar System objects, and sometimes the data obtained are then used for planning follow-up studies.

More detailed reconnaissance work is done by spacecraft known as **orbiters** because they orbit around their target. These missions are intrinsically more difficult than flyby missions because they have to make risky maneuvers and use up fuel to change their speed to enter an orbit. But orbiters can linger, looking in detail at more of the surfaces of the objects they are orbiting and studying things that change with time, like planetary weather.

Orbiters use remote-sensing instrumentation like that used by Earth-orbiting satellites to study our own planet. These instruments include tools such as cameras that take images at different wavelength ranges, radar that can map surfaces hidden beneath obscuring layers of clouds, and spectrographs that analyze the electromagnetic spectrum. These instruments enable planetary scientists to map other worlds, measure the heights of mountains, identify geological features and rock types, watch weather patterns develop, measure the composition of atmospheres, and get a general sense of the place. Additional instruments make

(a)

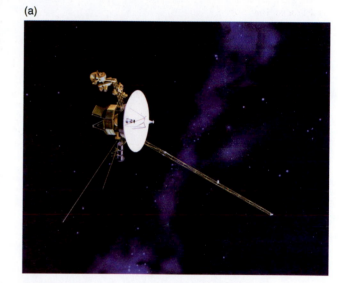

(b)

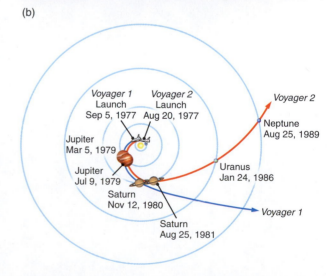

Figure 6.25 The *Voyager* spacecraft (a) flew past the outer planets and (b) are now near the boundary of our Solar System.

Figure 6.26 The robotic rover *Curiosity* took this photograph of itself on the surface of Mars.

measurements of the extended atmospheres and space environment through which they travel.

Landers, Rovers, and Atmospheric Probes

Reconnaissance spacecraft provide a wealth of information about a planet, but there is no better way of exploring a planet than within a planet's atmosphere or on solid ground. Spacecraft have landed on the Moon, Mars, Venus, Saturn's large moon Titan, several asteroids, and a comet. These spacecraft have taken pictures of planetary surfaces, measured surface chemistry, and conducted experiments to determine the physical properties of the surface rocks and soils.

There are several disadvantages of using **landers**—spacecraft that touch down and remain on the surface. Because of the expense, only a few landings in limited areas are practical. Given this limitation, the results may apply only to the small area around the landing site. Imagine, for example, what a different picture of Earth you might get from a spacecraft that landed in Antarctica, as opposed to a spacecraft that landed in a volcano or on the floor of a dry riverbed. Sites to be explored with landed spacecraft must be very carefully chosen on the basis of reconnaissance data. Some landers have wheels and can explore the vicinity of the landing site. Such remote-controlled vehicles, called **rovers**, were used first by the Soviet Union on the Moon four decades ago and more recently by the United States on Mars. **Figure 6.26** shows a self-portrait of the *Curiosity* rover on Mars.

Atmospheric probes descend into the atmospheres of planets and continually measure and send back data on temperature, pressure, and wind speed, along with other properties, such as chemical composition. Atmospheric probes have survived all the way to the solid surfaces of Venus and Saturn's moon Titan, sending back streams of data during their descent. An atmospheric probe sent into Jupiter's atmosphere never reached that planet's surface because, as we will discuss later in the book, Jupiter does not have a solid surface in the same sense that terrestrial planets and moons do. After sending back its data, the Jupiter probe eventually melted and vaporized as it descended into the hotter layers of the planet's atmosphere.

Sample Returns

If you pick up a rock from the side of a mountain road, you might learn a lot from the rock using tools that you could carry in your pocket or your car. But it would be much better to pick up a few samples and carry them back to a laboratory equipped with a full range of state-of-the-art instruments capable of measuring chemical composition, mineral type, age, and other information needed to reconstruct the story of your rock sample's origin and evolution. The same is true of Solar System exploration. One of the most powerful methods for investigating remote objects is to collect samples of the objects and bring them back to Earth for detailed study. So far, only samples of the Moon, a comet, and streams of charged particles from the Sun have been collected and returned to Earth. Scientists have found meteorites on Earth that are likely pieces of Mars that were blasted loose by objects that crashed into Mars. Someday, there may be unmanned "sample and return" missions to Mars.

The missions discussed so far in this section have all been conducted with robotic spacecraft. The only spacecraft that took people to another world were the *Apollo* missions to the Moon. This program ran from 1961 to 1972 and included several missions before the actual Moon landings. The *Apollo 8* astronauts brought back the famous picture of Earth viewed over the surface of the Moon (see the opening figure of Chapter 1). Each mission from *Apollo 11* through *Apollo 17* had three astronauts—two to land on the Moon and one to remain in orbit. One mission (*Apollo 13*) did not reach the Moon but returned to Earth safely. Twelve American astronauts walked on the Moon between 1969 and 1972 and brought back a total of 382 kg of rocks and other material.

The return of extraterrestrial samples to Earth is governed by international treaties and standards to ensure that these samples do not contaminate Earth. For example, before the lunar samples brought back by the *Apollo* missions could be studied, they (and the astronauts) had to be placed in quarantine and tested for alien life-forms. The same international standards apply to spacecraft landing on other planets. The goal of these standards is to avoid transporting life-forms from Earth to another planet. If there is life on other planets, there is concern about introducing potential harm, and we do not want to "discover" life that we, in fact, introduced.

With numerous missions under way and others on the horizon, unmanned exploration of the Solar System is an ongoing, dynamic activity. Appendix 5 summarizes some recent and current missions. Information on the latest discoveries can be found on mission websites and in science news sources.

CHECK YOUR UNDERSTANDING 6.5

Spacecraft are the most effective way to study planets in our Solar System because: (a) planets move too fast across the sky for us to image them well from Earth; (b) planets cannot be imaged from Earth; (c) they can collect more information than is available just from images from Earth; (d) space missions are easier than long observing campaigns.

6.5 Other Tools Contribute to the Study of the Universe

High-profile space missions have sent back stunning images and data from across the electromagnetic spectrum, but astronomers use other tools as well, including particle accelerators and colliders, neutrino and gravitational-wave detectors, and high-speed computers.

Particle Accelerators

Ever since the early years of the 20th century, physicists have been peering into the structure of the atom by observing what happens when small particles collide. By the 1930s, physicists had developed the technology to accelerate charged subatomic particles such as protons to very high speeds and then observe what happens when they slam into a target. From such experiments, physicists have discovered many kinds of subatomic particles and learned about their physical properties. High-energy particle colliders have proved to be an essential tool for physicists studying the basic building blocks of matter.

Astronomers have realized that to understand the very largest structures seen in the universe, it is important to understand the physics that took place during

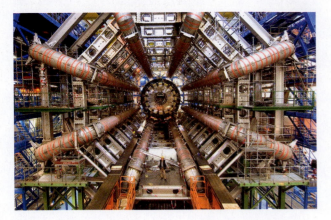

Figure 6.27 The ATLAS particle detector at CERN's Large Hadron Collider near Geneva, Switzerland. The enormous size of this instrument is evident from the person standing near the bottom center of the picture.

the earliest moments in the universe, when everything was extremely hot and dense. High-energy particle colliders that physicists use today are designed to approach the energies of the early universe. The effectiveness of particle accelerators is determined by the energy they can achieve and the number of particles they can accelerate. Modern particle colliders such as the Large Hadron Collider near Geneva, Switzerland (**Figure 6.27**), reach very high energies. Particles can also be studied from space. The Alpha Magnetic Spectrometer, installed on the International Space Station in 2011, searches for some of the most exotic forms of matter, such as dark matter, antimatter, and high-energy particles called cosmic rays.

Neutrinos and Gravitational Waves

The **neutrino** is an elusive elementary particle that plays a major role in the physics of the interiors of stars. Neutrinos are extremely difficult to detect. In less time than it takes you to read this sentence, a thousand trillion (10^{15}) solar neutrinos from the Sun are passing through your body, even during the night. Neutrinos are so nonreactive with matter that they can pass right through Earth (and you) as though it (or you) weren't there at all. A neutrino has to interact with a detector to be observed. Neutrino detectors typically record only one out of every 10^{22} (10 billion trillion) neutrinos passing though them, but that's enough to reveal processes deep within the Sun or the violent death of a star 160,000 light-years away.

Experiments designed to look for neutrinos originating outside of Earth are buried deep underground in mines or caverns or under the ocean or ice to ensure that only neutrinos are detected. For example, the ANTARES experiment uses the Mediterranean Sea as a neutrino telescope. Detectors located 2.5 km under the sea, off the coast of France, observe neutrinos that originated in objects visible in southern skies and passed through Earth. In the IceCube neutrino observatory located at the South Pole in Antarctica, the neutrino detectors are 1.5–2.5 km under the ice, and they observe neutrinos that originated in objects visible in northern skies (**Figure 6.28**).

Another elusive phenomenon is the **gravitational wave**. Gravitational waves are disturbances in a gravitational field, similar to the waves that spread out from the disturbance you create when you toss a pebble onto the quiet surface of a pond. There is strong, although indirect, observational evidence for the existence of gravitational waves, but they are so elusive that they have not yet actually been detected (**Process of Science Figure**). Several facilities, including the Laser Interferometer Gravitational-Wave Observatory (LIGO), have been constructed to detect gravitational waves. Scientists are eager to detect gravitational waves—to confirm their existence and to study the physical phenomena they are likely to reveal, such as the birth and evolution of the universe, stellar evolution, or the very force of gravity itself.

Computers

Astronomers use powerful computers for data gathering, analysis, and interpretation. A single CCD image may contain as many as 100 million pixels, with each pixel displaying roughly 30,000 levels of brightness. That adds up to several trillion pieces of information in each image. To analyze their data, astronomers typically do calculations on *every single pixel* of an image in order to remove unwanted contributions from Earth's atmosphere or to correct for instrumental effects. Astronomers conduct many different types of sky surveys—in which one or more

Figure 6.28 The IceCube neutrino telescope at the South Pole, Antarctica.

Process of Science

TECHNOLOGY AND SCIENCE ARE SYMBIOTIC

Scientists have been searching for waves that carry
gravitational information for nearly 100 years,
but the accuracy of their measurements is limited by the available technology.

Take 1

Weber Bar

Precision-machined bars of metal that should
"ring" as a gravitational wave passes by.

Sensitive only to extremely powerful
gravitational waves.

No detection.

Take 2

LIGO

New Technology: Lasers

Lasers should interfere as
gravitational waves pass by.

Roughly 100 times more sensitive
than Weber bar measurements.

No detection (yet).

Take 3

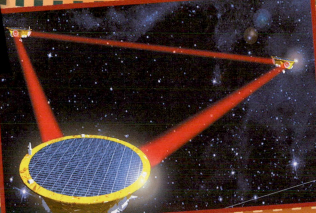

Future Science Mission

Lasers will interfere as
gravitational waves pass by.

Sensitive to more types of objects
than LIGO is.

Technology and science develop together. New technologies enable humans to ask new scientific questions.
Asking scientific questions pushes the development of better instrumentation. Deeper scientific understanding
leads to new technologies.

Figure 6.29 These images show supercomputer simulations of the collision of two galaxies. Astronomers compare simulations like these with telescopic observations.

telescopes survey a specific part of the sky—yielding thousands of images that need to be analyzed.

High-speed computers also play an essential role in generating and testing theoretical models of astronomical objects. Even when we completely understand the underlying physical laws that govern the behavior of a particular object, often the object is so complex that it would be impossible to calculate its properties and behavior without the assistance of high-speed computers. For example, as you learned in Chapter 4, you can use Newton's laws to compute the orbits of two stars that are gravitationally bound to one another, because their orbits take the form of simple ellipses. However, it is not so easy to understand the orbits of the several hundred billion stars that make up the Milky Way Galaxy, even though the underlying physical laws are the same.

Computer modeling is used to determine the interior properties of stars and planets, including Earth. Although astronomers cannot see beneath the surfaces of these bodies, they have a surprisingly good understanding of the interiors of these bodies, as we will describe in later chapters. Astronomers start a model by assigning well-understood physical properties to tiny volumes within a planet or star. The computer assembles an enormous number of these individual elements into an overall representation. When it is all put together, the result is a rather good picture of what the interior of the star or planet is like.

Astronomers also use supercomputers to study the evolution of astronomical objects, systems of objects, or the universe as a whole over time. For example, astronomers create models of galaxies, and then run computer simulations to study how those galaxies might change over billions of years. **Figure 6.29** shows a simulation of the collision of two galaxies. The results of the computer simulations are then compared with telescopic observations. If the simulations do not match the observations, then adjustments are made to the model and the simulations are run again until there is general agreement between them.

CHECK YOUR UNDERSTANDING 6.6

High-speed computers have become one of an astronomer's most important tools. Which of the following require the use of a high-speed computer? (Choose all that apply.) (a) analyzing images taken with very large CCDs; (b) generating and testing theoretical models; (c) pointing a telescope from object to object; (d) studying the evolution of astronomical objects or systems over time.

Origins

Microwave Telescopes Detect Radiation from the Big Bang

In this chapter, we explored the tools of the astronomer, from basic optical telescopes to instruments that observe in different wavelengths. Now let's examine in more detail one type of telescope that has aided the study of the origin of the universe. Recall from Chapter 1 that astronomers think the universe originated with a hot Big Bang. The multiple strands of evidence for this conclusion will be discussed in Chapter 21. Here, we look at one piece: the observation of faint microwave radiation left over from the early hot universe. Two Bell Laboratories physicists, Arno Penzias (1933–) and Robert Wilson (1936–), were working on satellite communications when they first accidentally detected this radiation in 1964 with a microwave dish antenna in New Jersey. Today, we routinely use cell phones and handheld GPS systems that communicate directly with satellites, but at the time, this capability was at the limit of technology.

Penzias and Wilson needed a very sensitive microwave telescope for the work they were doing for Bell Labs, because any spurious signals coming from the telescope itself might wash out the faint signals bounced off a satellite. To that end, they were working very hard to eliminate all possible sources of interference originating from within their instrument, including keeping the telescope free of bird droppings. They found that no matter how carefully they tried to eliminate sources of extraneous noise, they always still detected a faint signal at microwave wavelengths. This faint signal was the same in every direction and

turned out to be from the Big Bang. Penzias and Wilson shared the 1978 Nobel Prize in Physics for the discovery of this **cosmic microwave background radiation (CMB)** left over from the Big Bang itself.

Since 1964, astronomers from around the world have designed increasingly precise instruments to measure this radiation from the ground, from high-altitude balloons, from rockets, and from satellites. The Russian experiment RELIKT-1, launched in 1983, found some limits on the variation of the CMB. The COBE (Cosmic Background Explorer) satellite, launched in 1989, showed that the spectrum of this radiation precisely matched that of a blackbody with a temperature of 2.73 K—exactly what was predicted for the radiation left over from the Big Bang. (Compare **Figure 6.30** with the curves in Figure 5.22.) The data also showed some slight differences in temperature—small fractions of a degree—over the map of the sky. These slight variations tell us about how the universe evolved from one that was dominated by radiation to one that contains structures such as galaxies, stars, planets, and us. John Mather and George Smoot shared the 2006 Nobel Prize in Physics for this work.

In 1998 and 2003, a high-altitude balloon experiment called BOOMERANG (short for "balloon observations of millimetric extragalactic radiation and geophysics") flew over Antarctica at an altitude of 42 km to study CMB variations and estimate the overall geometry of the universe. The *WMAP* (Wilkinson Microwave Anisotropy

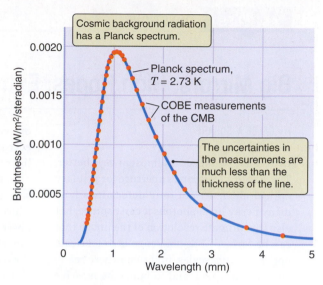

Figure 6.30 This graph shows the spectrum of the cosmic microwave background radiation (CMB) as measured by the COBE satellite (red dots). A steradian is a unit of solid angle. The uncertainty in the measurement at each wavelength is much less than the size of a dot. The line running through the data is a Planck blackbody spectrum with a temperature of 2.73 K.

Probe) satellite, launched in 2001, created an even more detailed map of the temperature variations in this radiation, yielding more precise values for the age and shape of the universe and the presence of dark matter and dark energy. The Atacama Cosmology Telescope (Chile) and the South Pole Telescope (Antarctica) study this radiation to look for evidence of when galaxy clusters formed. The newest microwave observatory in space, the Planck telescope, was launched in 2009 by the European Space Agency. Planck has much greater sensitivity than *WMAP* and studied these CMB variations in even more precise detail. These experiments and observations have opened up the current era of precision cosmology, in which astronomers can make detailed models of how the universe was born, eventually leading to stars, planets, and us.

ARTICLES   QUESTIONS

Planning to build a new large telescope can involve many complications, including negotiations among different countries and among people who live near the site of the telescope.

Big Mirrors, High Hopes: Extremely Large Telescope Is a Go

By **CALEB A. SCHARF**

In astronomy, bigger is almost always better. The size of a telescope's aperture (or primary optical element) not only determines how many pesky little photons it can capture, but also the ultimate resolution of the image that can be formed. The challenge is to fabricate optics on large scales, find somewhere really good to put them, and to build the massive structure to house them and the sensitive instruments to analyze the photons that come out of the pipeline.

Now one of the world's next generation of huge telescopes has been given the green light to push ahead with construction and operation. At an astonishing 39.3 meters in diameter, the European Extremely Large Telescope, or E-ELT (one can only hope a more poetic name is eventually chosen), will hold the crown for sheer girth among optical and infrared sensitive telescopes. With an authorized spending of about a billion euros (about $1.24B) the project can move ahead, with an anticipated "first light" sometime in 2024.

Although E-ELT is not alone in the race for so-called "30-meter class" telescopes (with the Thirty Meter Telescope and Giant Magellan observatories also on track, and also extraordinarily powerful), it's definitely the Hulk of the bunch.

Observatories on this scale will have profound impact on how we study the universe around us, from cosmology to planets.

A telescope like E-ELT will peer at the faintest, most distant objects in the young cosmos, and it'll be capable of sensing the atmospheric signatures of life in a few nearby terrestrial-type exoplanets (including the presence of oxygen). These tasks require an enormous "light bucket" to catch enough photons, but perhaps one of the most vivid examples of the gain of a big telescope comes from considering the resolving power.

Equipped with adaptive optics, E-ELT should, for example, be able to routinely study Jupiter down to scales of about 20 kilometers—by comparison the Great Red Spot is at present about 20,000 kilometers across. Mars can be imaged to roughly 5-kilometer resolution (depending of course on the relative separation of Earth). In other words, on a nightly basis we will be able to monitor the worlds in our Solar System with a fidelity comparable to flyby missions of yore.

Preparations have already begun. In June 2014, explosives were used to help level the mountaintop where E-ELT will sit—Cerro Armazones in northern Chile, a dry peak at about 3,000 meters altitude where the night skies are cloudless 89 percent of the time.

What rises there over the next 10 years will help take our understanding of the universe to a whole new level.

ARTICLES   QUESTIONS

1. Why do astronomers want to build at this particular location?
2. What are the advantages of larger telescopes?
3. How can adaptive optics yield images as good as those from old space missions?
4. Why will this telescope observe only in optical and infrared wavelengths?
5. Do a Web search to see the status of this project. What countries are partners? Is there any local opposition to the project in the host country?

Summary

Earth's atmosphere blocks many spectral regions and distorts telescopic images. Telescopes are sited to be above as much of the atmosphere as possible. Telescopes are matched to the wavelengths of observation, with different technologies required for each region of the spectrum. The aperture of a telescope both determines its light-gathering power and limits its resolution; larger telescopes are better in both measures. Modern CCD cameras have improved quantum efficiency and longer integration times, which allow astronomers to study fainter and more distant objects than were observable with prior detectors. Telescopes observing at microwave wavelengths have detected radiation left over from the Big Bang.

LG 1 **Compare the two main types of optical telescopes and how they gather and focus light.** The telescope is the astronomer's most important tool. Ground-based telescopes that observe in visible wavelengths come in two basic types: refractors (lenses) and reflectors (mirrors). All large astronomical telescopes are reflectors. Large telescopes collect more light and have greater resolution. The diffraction limit is the limiting resolution of a telescope.

LG 2 **Summarize the main types of detectors that are used on telescopes.** Photography improved the ability of astronomers to record details of faint objects seen in telescopes. CCDs are today's astronomical detector of choice because they are much more linear, have a broader spectral response, and can send electronic images directly to a computer. Spectrographs are specialized instruments that take the spectrum of an object to reveal what the object is made of and many other physical properties.

LG 3 **Explain why some wavelengths of radiation must be observed from space.** Radio telescopes are able to see through our atmosphere. Radio, near-infrared, and optical telescopes can be arrayed to greatly increase angular resolution. Putting telescopes in space solves problems created by Earth's atmosphere.

LG 4 **Explain the benefits of sending spacecraft to study the planets and moons of our Solar System.** Most of what is known about the planets and moons comes from observations by spacecraft. Flyby and orbiting missions obtain data from space, and landers and rovers collect data from the ground.

LG 5 **Describe other astronomical tools that contribute to the study of the universe.** Astronomers also use particle accelerators, neutrino detectors, and gravitational-wave detectors to study the universe. High-speed computers are essential to the acquisition, analysis, and interpretation of astronomical data.

? UNANSWERED QUESTIONS

- Will telescopes be placed on the Moon? The Moon has no atmosphere to make stars twinkle, cause weather, or block certain wavelengths of light from reaching its surface. The far side of the Moon faces away from the light and radio radiation of Earth, and all parts of the Moon have nights that last for two Earth weeks. One proposal calls for a Lunar Array for Radio Cosmology (LARC), an array of hundreds of radio telescopes that would be deployed on the Moon—after the year 2025—to study the earliest formation of stars and galaxies. Another proposal is for the Lunar Liquid Mirror Telescope (LLMT), with a diameter of 20–100 meters, to be located at one of the Moon's poles. Gravity would settle the rotating liquid into the necessary parabolic shape, and these liquid mirror telescopes are much simpler than are arrays of telescopes with large glass mirrors. The LLMT would observe extremely distant protostars and protogalaxies in infrared wavelengths. Astronomers debate whether telescopes on the Moon would be easier to service and repair than those in space and whether problems caused by lunar dust would outweigh any advantages.

- Will there be human exploration of the Solar System within your lifetime? Since the *Apollo* program, humans have not returned to the Moon or traveled to other planets or moons in the Solar System. Sending humans to the worlds of the Solar System is much more complicated, risky, and more expensive than sending robotic spacecraft. Humans need life support such as air, water, and food. Radiation in space can be dangerous. Furthermore, human explorers would expect to return to Earth, whereas most spacecraft do not come back. Astronomers and space scientists have heated debates about human spaceflight versus robotic exploration. Some argue that true exploration requires that human eyes and brains actually go there; others argue that the costs and risks are too high for the potential additional scientific knowledge. Beyond basic exploration, we also do not know whether humans will ever permanently colonize space.

Questions and Problems

Test Your Understanding

1. You are shopping for telescopes online. You find two in your price range. One of these has an aperture of 20 cm, and the other has an aperture of 30 cm. If aperture size is the only difference, which should you choose, and why?
   a. The 20 cm, because the light-gathering power will be better.
   b. The 20 cm, because the image size will be larger.
   c. The 30 cm, because the light-gathering power will be better.
   d. The 30 cm, because the image size will be larger.

2. Which of the following can be observed from Earth's surface? (Choose all that apply.)
   a. radio waves
   b. gamma radiation
   c. far UV light
   d. X-ray light
   e. visible light

3. Match the following properties of telescopes (lettered) with their corresponding definitions (numbered).
   a. aperture
   b. resolution
   c. focal length
   d. chromatic aberration
   e. diffraction
   f. interferometer
   g. adaptive optics

   (1) two or more telescopes connected to act as one
   (2) distance from lens to focal plane
   (3) diameter
   (4) ability to distinguish close objects
   (5) computer-controlled atmospheric distortion correction
   (6) color-separating effect
   (7) smearing effect due to sharp edge

4. The two Keck 10-meter telescopes, separated by a distance of 85 meters, can operate as an optical interferometer. What is its resolution when it observes in the infrared at a wavelength of 2 microns?
   a. 0.01 arcsec
   b. 0.005 arcsec
   c. 0.2 arsec
   d. 0.05 arcsec

5. Arrays of radio telescopes can produce much better resolution than single-dish telescopes because they work based on the principle of
   a. reflection.
   b. refraction.
   c. diffraction.
   d. interference.

6. Refraction is caused by
   a. light bouncing off a surface.
   b. light changing colors as it enters a new medium.
   c. light changing speed as it enters a new medium.
   d. two light beams interfering.

7. The light-gathering power of a 4-meter telescope is _____ than that of a 2-meter telescope.
   a. 8 times larger
   b. 4 times larger
   c. 16 times smaller
   d. 2 times smaller

8. Improved resolution is helpful to astronomers because
   a. they often want to look in detail at small features of an object.
   b. they often want to look at very distant objects.
   c. they often want to look at many objects close together.
   d. all of the above

9. The part of the human eye that acts as the detector is the
   a. retina.
   b. pupil.
   c. lens.
   d. iris.

10. Cameras that use adaptive optics provide higher-spatial-resolution images primarily because
    a. they operate above Earth's atmosphere.
    b. deformable mirrors are used to correct the blurring due to Earth's atmosphere.
    c. composite lenses correct for chromatic aberration.
    d. they simulate a much larger telescope.

11. The advantage of an interferometer is that
    a. the resolution is dramatically improved.
    b. the focal length is dramatically increased.
    c. the light-gathering power is dramatically increased.
    d. diffraction effects are dramatically decreased.
    e. chromatic aberration is dramatically decreased.

12. The angular resolution of a ground-based telescope is usually determined by
    a. diffraction.
    b. the focal length.
    c. refraction.
    d. atmospheric seeing.

13. A grating is able to spread white light out into a spectrum of colors because of the property of
    a. reflection.
    b. diffraction.
    c. dispersion.
    d. interference.

14. Why would astronomers put telescopes in airplanes?
    a. to get the telescopes closer to the stars
    b. to get the telescopes above the majority of the water vapor in Earth's atmosphere
    c. to be able to observe one object for more than 24 hours without stopping
    d. to allow the telescopes to observe the full spectrum of light

15. If we could increase the quantum efficiency of the human eye, it would
    a. allow humans to see a larger range of wavelengths.
    b. allow humans to see better at night or in other low-light conditions.
    c. increase the resolution of the human eye.
    d. decrease the resolution of the human eye.

Thinking about the Concepts

16. Galileo's telescope used simple lenses. What is the primary disadvantage of using a simple lens in a refracting telescope?

17. The largest astronomical refractor has an aperture of 1 meter. List several reasons why it would be impractical to build a larger refractor with twice this aperture.

18. Your camera may have a zoom lens, ranging between wide angle (short focal length) and telephoto (long focal length). How does the size of an object in the camera's focal plane differ between wide angle and telephoto?

19. Optical telescopes reveal much about the nature of astronomical objects. Why do astronomers also need information provided by gamma-ray, X-ray, infrared, and radio telescopes?

20. For light reflecting from a flat surface, the angles of incidence and reflection are the same. This is also true for light reflecting from the curved surface of a reflecting telescope's primary mirror. Sketch a curved mirror and several of these reflecting rays.

21. Consider two optically perfect telescopes having different diameters but the same focal length. Is the image of a star larger or smaller in the focal plane of the larger telescope? Explain your answer.

22. Study the Process of Science Figure. Make a flowchart for the symbiosis between technology and science that led to the development of the CCD camera as discussed in Section 6.2.

23. Explain adaptive optics and how they improve a telescope's image quality.

24. Explain integration time and quantum efficiency and how each contributes to the detection of faint astronomical objects.

25. Some people believe that we put astronomical telescopes on high mountaintops or in orbit because doing so gets them closer to the objects they are observing. Explain what is wrong with this popular misconception, and give the actual reason telescopes are located in these places.

26. Humans have sent various kinds of spacecraft—including flybys, orbiters, and landers—to all of the planets in our Solar System. Explain the advantages and disadvantages of each of these types of spacecraft.

27. If there are meteorites that are pieces of Mars on Earth, why is it so important to go to Mars and bring back samples of the martian surface?

28. Humans had a first look at the far side of the Moon as recently as 1959. Why had we not seen it earlier—when Galileo first observed the Moon with his telescope in 1610?

29. Where are neutrino detectors located? Why are neutrinos so difficult to detect?

30. Why do telescopes in space give a better picture of the leftover radiation from the Big Bang?

Applying the Concepts

31. Compare the light-gathering power of the Thirty Meter Telescope with that of the dark-adapted human eye (aperture 8 mm) and with that of one of the 10-meter Keck telescopes.

32. Study the photograph of light entering and leaving a block of refractive material in Figure 6.2b. Use a protractor to measure the angles of the green light as it enters the block and as it leaves the block. How are these angles related?

33. Many amateur astronomers start out with a 4-inch (aperture) telescope and then graduate to a 16-inch telescope. By what factor does the light-gathering power of the telescope increase with this upgrade? How much fainter are the faintest stars that can be seen in the larger telescope?

34. The resolution of the human eye is about 1.5 arcmin. What would the aperture of a radio telescope (observing at 21 cm) have to be to have this resolution? Even though the atmosphere is transparent at radio wavelengths, humans do not see light in the radio range. Using your calculations and logic, explain why.

35. Assume that you have a telescope with an aperture of 1 meter. Compare the telescope's theoretical resolution when you are observing in the near-infrared region of the spectrum ($\lambda = 1,000$ nm) with that when you are observing in the violet region of the spectrum ($\lambda = 400$ nm).

36. Assume that the maximum aperture of the human eye, D, is approximately 8 mm and the average wavelength of visible light, λ, is 5.5×10^{-4} mm.
    a. Calculate the diffraction limit of the human eye in visible light.
    b. How does the diffraction limit compare with the actual resolution of 1–2 arcmin (60–120 arcsec)?
    c. To what do you attribute the difference?

37. The diameter of the full Moon in the focal plane of an average amateur's telescope (focal length 1.5 meters) is 13.8 mm. How big would the Moon be in the focal plane of a very large astronomical telescope (focal length 250 meters)?

38. One of the earliest astronomical CCDs had 160,000 pixels, each recording 8 bits (256 levels of brightness). A new generation of astronomical CCDs may contain a billion pixels, each recording 15 bits (32,768 levels of brightness). Compare the number of bits of data that each of these two CCD types produces in a single image.

39. Consider a CCD with a quantum efficiency of 80 percent and a photographic plate with a quantum efficiency of 1 percent. If an exposure time of 1 hour is required to photograph a celestial object with a given telescope, how much observing time would be saved by substituting a CCD for the photographic plate?

40. The VLBA uses an array of radio telescopes ranging across 8,000 km of Earth's surface from the Virgin Islands to Hawaii.
    a. Calculate the angular resolution of the array when radio astronomers are observing interstellar water molecules at a microwave wavelength of 1.35 cm.
    b. How does this resolution compare with the angular resolution of two large optical telescopes separated by 100 meters and operating as an interferometer at a visible wavelength of 550 nm?

41. When operational, the SVLBI may have a baseline of 100,000 km. What will be the angular resolution when studying interstellar molecules emitting at a wavelength of 17 mm from a distant galaxy?

42. The *Mars Reconnaissance Orbiter* (*MRO*) flies at an average altitude of 280 km above the martian surface. If its cameras have an angular resolution of 0.2 arcsec, what is the size of the smallest objects that the *MRO* can detect on the martian surface?

43. *Voyager 1* is now about 125 astronomical units (AU) from Earth, continuing to record its environment as it approaches the limits of our Solar System.
    a. How far away is *Voyager 1*, in kilometers?
    b. How long does it take observational data to come back to us from *Voyager 1*?
    c. How does *Voyager 1*'s distance from Earth compare with that of the nearest star (other than the Sun)?

44. Gravitational waves travel at the speed of light. Their speed, wavelength, and frequency are related as $c = \lambda \times f$. If we were to observe a gravitational wave from a distant cosmic event with a frequency of 10 hertz (Hz), what would be the wavelength of the gravitational wave?

45. Compute the peak of the blackbody spectrum with a temperature of 2.73 K. What region of the spectrum is this?

USING THE WEB

46. A webcast for the International Year of Astronomy 2009 called "Around the World in 80 Telescopes" can be accessed at http://eso.org/public/events/special-evt/100ha.html. The 80 telescopes are situated all over, including Antarctica and space. Pick two of the telescopes and watch the videos. Do you think these videos are effective for public outreach for the observatory in question or for astronomy in general? For each telescope you choose, answer the following questions: Does the telescope observe in the Northern Hemisphere or the Southern Hemisphere? What wavelengths does the telescope observe? What are some of the key science projects at the telescope?

47. Most major observatories have their own websites. Use the link in question 46 to find a master list of telescopes, and click on a telescope name to link to an observatory website (or run a search on names from Tables 6.1 and 6.2). For the telescope you choose, answer the following questions: (a) What is this telescope's "claim to fame"—is it the largest? at the highest altitude? at the driest location? with the darkest skies? the newest? (b) Does the observatory website have news releases? What is a recent discovery from this telescope?

48. Go to the website for the International Dark Sky Association (http://www.darksky.org/). Click on "Night Sky Conservation" and then "Do you live under light pollution?" Is your location dark? From the menu on the left, is there a "dark sky park" near you? What are the ecological arguments against too much light at night?

49. What is the current status of the James Webb Space Telescope (http://jwst.nasa.gov)? How will this telescope be different from the Hubble Space Telescope? What are some of the instruments for the JWST and its planned projects? What is the current estimated cost of the JWST?

50. Pick a mission from Appendix 5, go to its website, and see what's new. For the mission you choose, answer the following questions: Is the spacecraft still active? Is it sending images? What new science is coming from this mission?

smartwork

If your instructor assigns homework in SmartWork, access your assignments at digital.wwnorton.com/astro5.

Visit the Student Site at the Digital Landing Page, and open the "Geometric Optics and Lenses" animation in Chapter 6. Read through the animation until you reach the optics simulation, pictured in **Figure 6.31**. The simulator shows a converging lens and a pencil. Rays come from the pencil on the left of the converging lens, pass through the lens, and make an image to the right of the lens. The view that would be seen by an observer at the position of the eye is shown in the circle at upper right. Initially, when the pencil is at position 2.3 and the eye is at position 2.0, the pencil is out of focus and blurry.

1 Is the eraser at the top or the bottom of the actual pencil? (This becomes important later.)

_____

_____

Using the red slider in the upper left of the window, try moving the pencil to the right. Pause when the observer's eye sees a recognizable pencil (even if it's still blurry).

2 Does the eye see the pencil right side up or upside down?

_____

_____

3 This is somewhat analogous to the view through a telescope. The objects are very far from the lenses, and the observer sees things upside down in the telescope. If an object in your field of view is at the top of the field and you want it in the center, which way should you move the telescope: up or down?

_____

_____

Now return the pencil to position 2.3. Use the red slider in the lower right of the window to move the eye closer to the lens (to the left).

4 At what distance does the image of the pencil first become crisp and clear?

_____

_____

5 Is the pencil right side up or upside down?

_____

_____

6 In practice at the telescope, we do not move the observer back (away from the eyepiece) to bring the image into focus. Why not?

_____

_____

7 Instead of moving the observer, we use a focusing knob to move the lens in the eyepiece, which brings the image into focus. Imagine that you are looking through the eyepiece of a telescope and the image is blurry. You turn the focusing knob and things get blurrier! What should you try next?

_____

_____

_____

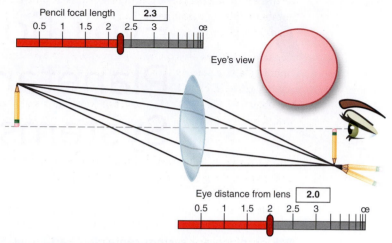

Figure 6.31 Use this simulation to change the position of the object of the eye to explore what will be seen for various configurations.

8 Now imagine that you get the image focused just right, so it is crisp and sharp. The next person to use the telescope wears glasses and insists that the image is blurry. But when you look through the telescope again, the image is still crisp. Explain why your experiences differ.

_____

_____

_____

Step through the animation to the next picture. Carefully study the two telescopes shown and the path the light takes through them.

9 Which telescope has a longer focal length: the top one or the bottom one?

_____

10 Which telescope produces an image with the red and the blue stars more separated: the top one or the bottom one?

_____

11 A longer focal length is an advantage in one sense, but it's not the entire story. What are some disadvantages of a telescope with a very long focal length?

_____

_____

_____

7

The Birth and Evolution of Planetary Systems

The planetary system containing Earth—our Solar System—is a by-product of the birth of the Sun. But the physical processes that shaped the formation of the Solar System are not unique to it. The same processes have formed numerous multiplanet systems. In this chapter, we will examine how planetary systems are born and evolve.

LEARNING GOALS

By the conclusion of this chapter, you should be able to:

LG 1 Describe how our understanding of planetary system formation developed from the work of both planetary and stellar scientists.

LG 2 Discuss the role of gravity and angular momentum in explaining why planets orbit the Sun in a plane and why they revolve in the same direction that the Sun rotates.

LG 3 Explain how temperature at different locations in the protoplanetary disk affects the composition of planets, moons, and other bodies.

LG 4 Discuss the processes that resulted in the formation of planets and other objects in our Solar System.

LG 5 List how astronomers find planets around other stars, and explain how we know that planetary systems around other stars are common.

From clouds of gas and dust, planetary systems are born. ▶ ▶ ▶

How was our
Solar System
born?

7.1 Planetary Systems Form around a Star

Earth is part of a collection of **planets**—large, round bodies that orbit a star in individual orbits. Astronomers call a system of planets surrounding a star a **planetary system**. The Solar System, shown in **Figure 7.1**, is the planetary system that includes Earth, other planets, and the Sun. It also includes moons that orbit planets and small bodies that occupy particular regions of the Solar System; for example, in the asteroid belt or in the Kuiper Belt. Our Solar System is a tiny part of our galaxy, which is a tiny part of the universe. Review Figure 1.3 in Chapter 1 to remind yourself of the size scales involved. Light takes about 4 hours to travel to Earth from Neptune, the outermost planet in the Solar System, but light from the most distant galaxies takes nearly *14 billion years* to reach Earth.

Until the latter part of the 20th century, the origin of the Solar System remained speculative. Over the past century, with the aid of spectroscopy, astronomers have determined that the Sun is a typical star, one of hundreds of billions in its galaxy, the Milky Way, and that the Milky Way is a typical galaxy, one of hundreds of billions in the universe. In the past few decades, stellar astronomers studying the formation of stars and planetary scientists analyzing clues about the history of the Solar System have found themselves arriving at the same picture of the early Solar System—but from two very different directions. This unified understanding provides the foundation for the way astronomers now think about the Sun and the myriad objects that orbit it. In this section, we will look at how the work of stellar and planetary scientists converged to inform our understanding of planetary system formation.

The Nebular Hypothesis

The first plausible theory for the formation of the Solar System, the **nebular hypothesis**, was proposed in 1734 by the German philosopher Immanuel Kant (1724–1804) and conceived independently a few years later by the French astronomer Pierre-Simon Laplace (1749–1827). Kant and Laplace argued that a rotating cloud of interstellar gas, or **nebula** (Latin for "cloud"), gradually collapsed and flattened to form a disk with the Sun at its center. Surrounding the Sun were rings of material from which the planets formed. This configuration would explain why the planets orbit the Sun in the same direction in the same plane. The nebular hypothesis remained popular throughout the 19th century, and these basic principles of the hypothesis are still retained today.

Our modern theory of planetary system formation calculates under what conditions clouds of interstellar gas collapse under the force of their own self-gravity to form stars. Recall from Chapter 4 that self-gravity is the gravitational attraction between the parts of an object such as a planet or star that pulls all the parts toward the object's center. This inward force is opposed by either structural strength (in the case of

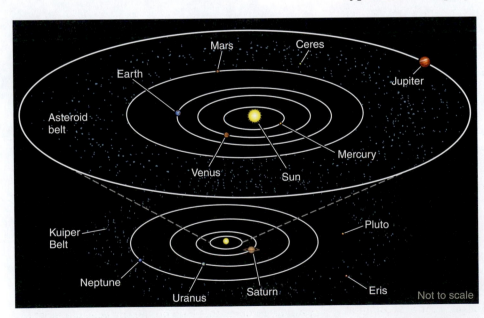

Figure 7.1 Our Solar System includes planets, moons, and other small bodies. Sizes and distances are not to scale in this sketch.

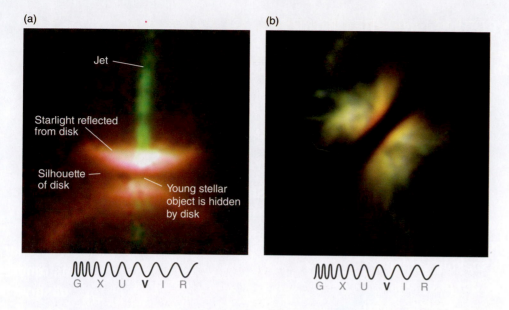

(a)

Jet

Starlight reflected from disk

Silhouette of disk

Young stellar object is hidden by disk

G X U V I R

(b)

G X U V I R

Figure 7.2 Hubble Space Telescope images show disks around newly formed stars. (a) The dark band is the silhouette of the disk seen more or less edge-on. Bright regions are dust illuminated by starlight. Some disk material may be expelled in a direction perpendicular to the plane of the disk in the form of violent jets. (b) In this image, the disk is seen in silhouette. Planets may be forming or have already formed in this disk.

rocks that make up terrestrial planets) or the outward force resulting from gas pressure within a star. If the outward force is less than self-gravity, the object contracts; if it is greater, the object expands. In a stable object, the inward and outward forces are balanced.

In support of the nebular hypothesis, disks of gas and dust have been observed surrounding young stellar objects (**Figure 7.2**). From this observational evidence, stellar astronomers have shown that, much like a spinning ball of pizza dough spreads out to form a flat crust, the cloud that produces a star—the Sun, for example—collapses first into a rotating disk. Material in the disk eventually suffers one of three fates: it travels inward onto the forming star at its center, it remains in the disk itself to form planets and other objects, or it is ejected back into interstellar space.

▶‖ **AstroTour:** Solar System Formation

Planetary Scientists and the Convergence of Evidence

While astronomers were working to understand star formation, other groups of scientists with very different backgrounds were piecing together the history of the Solar System. Planetary scientists, geochemists, and geologists looking at the current structure of the Solar System inferred what some of its early characteristics must have been. The orbits of all the planets lie very close to a single plane, so the early Solar System must have been flat. Additionally, all the planets orbit the Sun in the same direction, so the material from which the planets formed must have been orbiting the Sun in the same direction as well.

To find out more, scientists study samples of the very early Solar System. Rocks that fall to Earth from space, known as **meteorites**, include pieces of material that are left over from the Solar System's youth. Many meteorites, such as the one in **Figure 7.3**, resemble a piece of concrete in which pebbles and sand are mixed with a much finer filler, suggesting that the larger bodies in the Solar System must have grown from the aggregation of smaller bodies. This chain of thought suggests an early Solar System in which the young Sun was surrounded by a flattened

Figure 7.3 Meteorites are the surviving pieces of Solar System fragments that land on the surfaces of planets. This meteorite formed from many smaller components that stuck together.

Process of Science

CONVERGING LINES OF INQUIRY

Astronomers asked: Why is the Solar System a disk, with all planets orbiting in the same direction?

Stellar astronomers find dust and gas around young stars.

Stellar astronomers observe this gas and dust to be in the shape of disks.

Stellar astronomers test the nebular hypothesis, seeking evidence for or against.

Mathematicians suggest the nebular hypothesis: a collapsing rotating cloud formed the Solar System.

Planetary scientists test the nebular hypothesis, seeking evidence for or against.

Planetary scientists study meteorites that show the Solar System bodies formed from many smaller bodies.

Beginning from the same fundamental observations about the shape of the Solar System, theorists, planetary scientists, and stellar astronomers converge in the nebular theory that stars and planets form together from a collapsing cloud of gas and dust.

disk of both gaseous and solid material. Our Solar System formed from this swirling disk of gas and dust.

As astronomers and planetary scientists compared notes, they realized they had arrived at the same picture of the early Solar System from two completely different directions. The rotating disk from which the planets formed was the remains of the disk that had accompanied the formation of the Sun. Earth, along with all the other orbiting bodies that make up the Solar System, formed from the remnants of an *interstellar cloud* that collapsed to form the local star, the Sun. The connection between the formation of stars and the origin and subsequent evolution of the Solar System is one of the cornerstones of both astronomy and planetary science—a central theme of our understanding of our Solar System (**Process of Science Figure**).

CHECK YOUR UNDERSTANDING 7.1

Which of the following pieces of evidence supports the nebular hypothesis? (Choose all that apply.) (a) Planets orbit the Sun in the same direction. (b) The Solar System is relatively flat. (c) Earth has a large Moon. (d) We observe disks of gas and dust around other stars.

7.2 The Solar System Began with a Disk

Now that we've noted the general idea that planets formed in a disk around young stars, let's look at some of the specifics. **Figure 7.4** illustrates the young Solar System as it appeared roughly 5 billion years ago. At that time, the Sun was still a **protostar**—a large ball of gas but not yet hot enough in its center to be a star. As the cloud of interstellar gas collapsed to form the protostar, its gravitational energy was converted into heat energy and radiation. Surrounding the protostellar Sun was a flat, orbiting disk of gas and dust. Each bit of the material in this thin disk orbited the Sun according to the same laws of motion and gravitation that govern the orbits of the planets. The disk around the Sun, like the disks that astronomers see today surrounding protostars elsewhere in our galaxy, is called a **protoplanetary disk**. The disk probably contained less than 1 percent of the mass of the star forming at its center, but this amount was more than enough to account for the bodies that make up the Solar System today.

The Collapsing Cloud and Angular Momentum

The Solar System formed from a protoplanetary disk, and similar disks are seen around newly formed stars. *Angular momentum* causes these disks to form. **Angular momentum** is a conserved property of a revolving or rotating system with a value that depends on both the velocity and distribution of the system's mass. The angular momentum of an isolated object is always conserved; that is, it remains unchanged unless acted on by an external force. You have likely seen

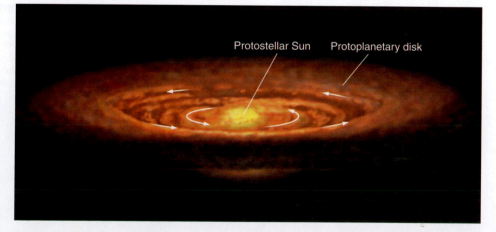

Figure 7.4 Think of the young Sun as being surrounded by a flat, rotating disk of gas and dust that was flared at its outer edge.

a figure-skater spinning on the ice like the one shown in **Figure 7.5**. Like any other rotating object, the spinning ice-skater has some amount of angular momentum. Unless an external force acts on her, such as the ice pushing on her skates, she will always have the same amount of angular momentum.

The amount of angular momentum depends on three factors:

1. How fast the object is rotating. The faster an object is rotating, the more angular momentum it has.

2. The mass of the object. If a bowling ball and a basketball are spinning at the same speed, the bowling ball has more angular momentum because it has more mass.

3. How the mass of the object is distributed relative to the spin axis; that is, how far the object is from the spin axis, or how spread out the object is. For an object of a given mass and rate of rotation, the more spread out it is, the more angular momentum it has. A spread-out object that is rotating slowly might have the same angular momentum as a compact object rotating rapidly.

Both an ice-skater and a collapsing interstellar cloud are affected by **conservation of angular momentum**: the angular momentum must remain the same in the absence of an external force. In order for angular momentum to be conserved, a change in one of the above quantities (the rate of spin, mass, or distribution of mass) must be accompanied by a change of another quantity. For example, an ice-skater can control how rapidly she spins by pulling in or extending her arms or legs. As she pulls in her arms to become more compact, she decreases her distribution of mass and must spin faster to maintain the same angular momentum. When her arms are held tightly in front of her and one leg is wrapped around the other, the skater's spin becomes a blur. She finishes with a flourish by throwing her arms and leg out—an action that abruptly slows her spin by spreading out her mass. Despite the changes in her spin, the skater's angular momentum remains constant throughout the entire maneuver. Similarly—as shown in **Figure 7.6**—the cloud that formed our Sun rotated faster and faster as it collapsed, just as the ice-skater speeds up when she pulls in her arms.

VISUAL ANALOGY

Figure 7.5 A figure-skater relies on the principle of conservation of angular momentum to change the speed with which she spins.

Astronomy in Action: Angular Momentum

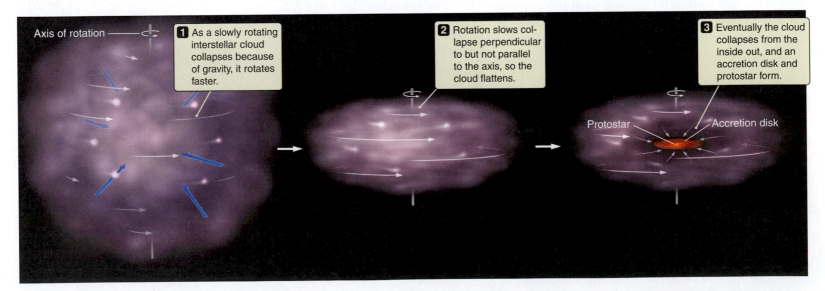

Axis of rotation

1 As a slowly rotating interstellar cloud collapses because of gravity, it rotates faster.

2 Rotation slows collapse perpendicular to but not parallel to the axis, so the cloud flattens.

3 Eventually the cloud collapses from the inside out, and an accretion disk and protostar form.

Protostar   Accretion disk

Figure 7.6 A rotating interstellar cloud collapses in a direction parallel to its axis of rotation, thus forming an accretion disk.

However, this description presents a puzzle. Suppose the Sun formed from a typical cloud—one that was about a light-year across and was rotating so slowly that it took a million years to complete one rotation. By the time such a cloud collapsed to the size of the Sun today, it would have been spinning so fast that one rotation would occur every 0.6 second. This is more than 3 million times faster than our Sun actually spins. At this rate of rotation, the Sun would tear itself apart. It appears that angular momentum was not conserved in the actual formation of the Sun—but that can't be right, because angular momentum must be conserved. We must be missing something. Where did the angular momentum go?

The Formation of an Accretion Disk

To understand how angular momentum is conserved in disk formation, we must think in three dimensions. Imagine that the ice-skater bends her knees, compressing herself downward instead of bringing her arms toward her body. As she does this, she again makes herself less spread out, but her rate of spin does not change because no part of her body has become any closer to the axis of spin. Similarly, as shown in Figure 7.6, a clump of a molecular cloud can flatten out without speeding up by collapsing parallel to its axis of rotation. Instead of collapsing into a ball, the interstellar cloud flattens into a disk. As the cloud collapses, its self-gravity increases, and the inner parts begin to fall freely inward, raining down on the growing object at the center. The outer portions of the cloud lose the support of the collapsed inner portion, and they start falling inward, too. As this material makes its final inward plunge, it lands on a thin, rotating disk that forms from the accretion of material around a massive object, called an **accretion disk**. The formation of accretion disks is common in the universe.

A visual analogy might be helpful for understanding how interstellar material collects on the accretion disk. Imagine a huge traffic circle, or roundabout, with multiple entrances but with all exits blocked by incoming traffic, as shown in **Figure 7.7a**. As traffic flows into the traffic circle, it has nowhere else to go, resulting in a continual, growing line of traffic driving around and around in an increasingly crowded circle. Eventually, as more and more cars try to pack in, the traffic piles up. This situation is roughly analogous to an accretion disk, shown in Figure 7.7b. Of course, traffic in a roundabout moves on a flat surface, whereas the accretion disk around a protostar forms from material coming in from all directions in three-dimensional space.

As material falls onto the disk, its motion perpendicular to the disk stops abruptly, but its mass motion *parallel* to the surface of the disk adds to the disk's total angular momentum. In this way, the angular momentum of the infalling material is transferred to the accretion disk. The rotating accretion disk has a radius of hundreds of astronomical units and is thousands of times greater than the radius of the star that will eventually form at its center. Therefore, most of the angular momentum in the original interstellar cloud ends up in the accretion disk rather than in the central protostar (see **Working It Out 7.1** for an example of the relevant calculation).

Now we can explain why the Sun does not have the same angular momentum that was present in the original clump of cloud. The radius of a rotating accretion disk is thousands of times greater than the radius of the star that will form at its center. Much of the angular momentum in the original interstellar clump is conserved in its accretion disk rather than in its central protostar.

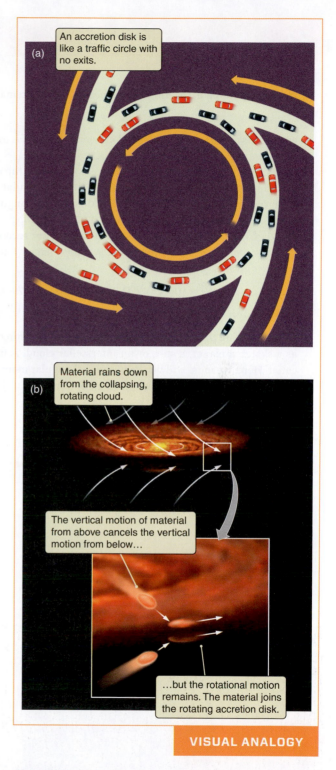

(a) An accretion disk is like a traffic circle with no exits.

(b) Material rains down from the collapsing, rotating cloud.

The vertical motion of material from above cancels the vertical motion from below…

…but the rotational motion remains. The material joins the rotating accretion disk.

VISUAL ANALOGY

Figure 7.7 (a) Traffic piles up in a traffic circle with entrances but no exits. (b) Similarly, gas from a rotating cloud falls inward from opposite sides, piling up onto a rotating disk.

▶❙❙ AstroTour: Traffic Circle Analogy

7.1 Working It Out Angular Momentum

In its simplest form, the angular momentum of a system is given by

$$L = m \times v \times r$$

where m is the mass, v is the speed at which the mass is moving, and r represents how spread out the mass is.

As an example, we can apply this relationship to the angular momentum of Jupiter in its orbit about the Sun. The angular momentum from one body orbiting another is called *orbital* angular momentum, $L_{orbital}$. The mass (m) of Jupiter is 1.90×10^{27} kilograms (kg), the speed of Jupiter in orbit (v) is 1.31×10^4 meters per second (m/s), and the radius of Jupiter's orbit (r) is 7.79×10^{11} meters. Putting all this together gives

$$L_{orbital} = (1.90 \times 10^{27} \text{ kg}) \times (1.31 \times 10^4 \text{ m/s}) \times (7.79 \times 10^{11} \text{ m})$$

$$L_{orbital} = 1.94 \times 10^{43} \text{ kg m}^2/\text{s}$$

Calculating the *spin* angular momentum of a spinning object, such as a skater, a planet, a star, or an interstellar cloud, is more complicated. Here, we must add up the individual angular momenta of *every tiny mass element* within the object. In the case of a uniform sphere, the spin angular momentum is

$$L_{spin} = \frac{4\pi m R^2}{5P}$$

where R is the radius of the sphere, and P is the rotation period of its spin.

Let's compare Jupiter's orbital angular momentum with the Sun's spin angular momentum to investigate the distribution of angular momentum in the Solar System. Appendix 2 provides the Sun's radius (6.96×10^8 meters), mass (1.99×10^{30} kg), and rotation period (24.5 days = 2.12×10^6 seconds). Assuming that the Sun is a uniform sphere, the spin angular momentum of the Sun is

$$L_{spin} = \frac{4 \times \pi \times (1.99 \times 10^{30} \text{ kg}) \times (6.96 \times 10^8 \text{ m})^2}{5 \times (2.12 \times 10^6 \text{ s})}$$

$$L_{spin} = 1.14 \times 10^{42} \text{ kg m}^2/\text{s}$$

Jupiter's orbital angular momentum is about 17 times greater than the spin angular momentum of the Sun. Most of the angular momentum of the Solar System now resides in the orbits of its major planets.

For a collapsing sphere to conserve spin angular momentum, its rotation period P must be proportional to R^2. Like with the skater, when a sphere decreases in radius, its rotation period decreases; that is, it spins faster.

Most of the matter that lands on the accretion disk either becomes part of the star or is ejected back into interstellar space, sometimes in the form of jets or other outflows, as seen in Figure 7.2a. Material swirling in the bipolar jets carries angular momentum away from the accretion disk in the general direction of the poles of the rotation axis. However, a small amount of material is left behind in the disk. It is the objects in this leftover disk—the dregs of the process of star formation—that form planets and other objects that orbit the star. Look again at Figure 7.2 showing images of edge-on accretion disks around young stars. The dark bands are the shadows of the edge-on disks, the top and the bottom of which are illuminated by light from the forming star. Our Sun and Solar System formed from a protostar and disk much like those in these pictures.

Formation of Large Objects

The chain of events that connects the accretion disk around a young star to a planetary system such as the Solar System begins with random motions of the gas within the protoplanetary disk. As shown in **Figure 7.8**, these motions push the smaller grains of solid material back and forth past larger grains, and as this happens, the smaller grains stick to the larger grains. The "sticking" process among smaller grains is due to the same static electricity that causes dust and hair to cling to plastic surfaces. Starting out at only a few microns (μm) across—about the size of particles in smoke—the slightly larger bits of dust grow to the size of pebbles and then to clumps the size of boulders, which are not as easily pushed around by gas. When clumps grow to about 100 meters across, the objects are so few and far between that they collide less frequently, and their growth rate slows down but does not stop. Within a protoplanetary disk, the larger dust grains become larger at the expense of the smaller grains.

For two large clumps to stick together rather than explode into many small pieces, they must bump into each other very gently: collision speeds must be about 0.1 m/s or less for colliding boulders to stick together. Your stride is probably about a meter, so to walk as slowly as the collision speed of 0.1 m/s, you would take one step every 10 seconds. The process is not a uniform movement toward larger and larger bodies. Violent collisions do occur in an accretion disk, and larger clumps break back into smaller pieces. But over a long period, large bodies do form.

Objects continue to grow by "sweeping up" smaller objects that get in their way. These objects can eventually measure up to a couple hundred meters across. As the clumps reach the size of about a kilometer, they are massive enough that their gravity begins to pull on nearby bodies, as shown in **Figure 7.9**. These bodies of rock and ice, 100 meters or more in diameter, are known as **planetesimals** ("tiny planets") and eventually combine with each other to form planets. The growth of planetesimals is not fed only by chance collisions with other objects: a planetesimal's gravity can now pull in and capture small objects outside its direct path. The growth of planetesimals speeds up, and larger planetesimals quickly consume most of the remaining bodies in the vicinity of their orbits to become small planets.

CHECK YOUR UNDERSTANDING 7.2

Where does the majority of the angular momentum of the original cloud go? (a) into the orbital angular momentum of planets; (b) into the star; (c) into the spin of the planets; (d) lost along the jets from the star

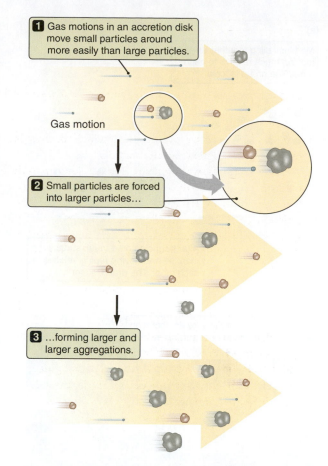

1 Gas motions in an accretion disk move small particles around more easily than large particles.

Gas motion

2 Small particles are forced into larger particles…

3 …forming larger and larger aggregations.

Figure 7.8 Motions of gas in a protoplanetary disk blow smaller particles of dust into larger particles, making the larger particles larger still. This process continues, eventually creating objects many meters in size.

7.3 The Inner Disk and Outer Disk Formed at Different Temperatures

In the Solar System, the inner planets are small and mostly rocky, while the outer planets are very large and mostly gaseous. This distinct difference between the inner and the outer Solar System can be explained by how the local disk environment affects the formation process. In this section, we will examine these differences.

Energy in the Disk

The accretion disks surrounding young stars form from interstellar material that may have a temperature of only a few kelvins, but the disks themselves reach temperatures of hundreds of kelvins or more. Astronomers want to understand what heats up the disk around a forming star so that we can calculate how hot these disks get.

Imagine dumping a box of marbles from the top of a tall ladder onto a rough, hard floor below. The marbles fall, picking up speed as they go. Even though the falling marbles are speeding up, they are all speeding up *together*. If you were riding on one of the marbles, the other marbles would not appear to you to be moving very much; it would be the rest of the room that was whizzing

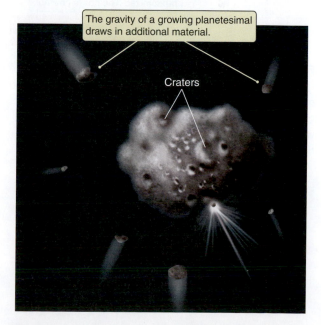

The gravity of a growing planetesimal draws in additional material.

Craters

Figure 7.9 The gravity of a planetesimal is strong enough to attract surrounding material, which causes the planetesimal to grow more rapidly.

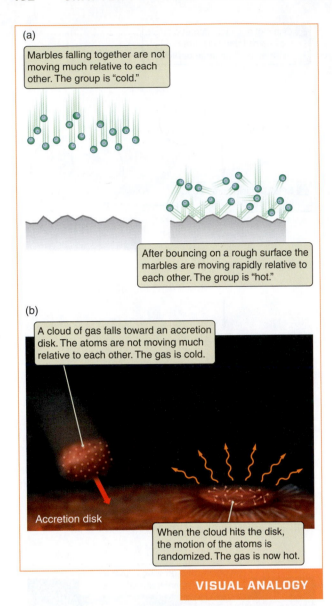

(a)

Marbles falling together are not moving much relative to each other. The group is "cold."

After bouncing on a rough surface the marbles are moving rapidly relative to each other. The group is "hot."

(b)

A cloud of gas falls toward an accretion disk. The atoms are not moving much relative to each other. The gas is cold.

Accretion disk

When the cloud hits the disk, the motion of the atoms is randomized. The gas is now hot.

VISUAL ANALOGY

Figure 7.10 (a) Marbles dropped as a group fall together until they hit a rough floor, at which point their motions become randomized. (b) Similarly, atoms in a gas fall together until they hit the accretion disk, at which point their motions become randomized, raising the temperature of the gas.

by (**Figure 7.10a**). The atoms and molecules in the gas falling toward the proto-star are like these marbles. They are picking up speed as they fall as a group toward the protostar, but the gas is still cold because the random thermal velocities of atoms and molecules with respect to each other are still low. Now imagine what happens when the marbles hit the rough floor. They bounce every which way. They are still moving rapidly, but they are no longer moving together. A change has taken place from the ordered motion of marbles falling together to the random motions of marbles traveling in all directions.

Like the falling marble, the atoms and molecules in the gas falling toward the central star scatter in the same fashion when they encounter the uneven gravitational field of the dusty accretion disk. As shown in Figure 7.10b, they are no longer moving as a group, but their random thermal velocities are now very large. The gas is now hot. Similarly, material from the collapsing interstellar cloud falls inward toward the protostar, but because of its angular momentum it misses the protostar and instead falls onto the disk. When this material reaches the disk, its infalling motion comes to an abrupt halt, and the velocity that the atoms and molecules in the gas had before hitting the disk is suddenly converted into random *thermal* velocities instead. The cold gas that was falling toward the disk heats up when it lands on the disk.

Another way to think about why the gas falling on the disk makes the disk hot is to apply another conservation law. The law of **conservation of energy** states that unless energy is added to or taken away from a system from the outside, the total amount of energy in the system must remain constant. But the form the energy takes can change. Imagine you are working against gravity by lifting a heavy object; for example, a brick. It takes energy to lift the brick, and the law of conservation of energy states that energy is never lost. Where does that energy go? The energy is stored and changed into a form called **gravitational potential energy**. If you drop the brick it falls, and as it falls it speeds up. The gravitational potential energy that was stored is converted to energy of motion, which is called **kinetic energy**. When the brick hits the floor, it stops suddenly. The brick loses its energy of motion, so what form does this energy take now?

If the brick cracks, part of the energy goes into breaking the chemical bonds that hold it together. Some of the energy is converted into the sound the brick makes when it hits the floor. Some goes into heating and distorting the floor. But much of the energy is converted into thermal energy. The atoms and molecules that make up the brick are moving about within the brick a bit faster than they were before the brick hit, so the brick and its surroundings, including the floor, grow a tiny bit warmer. Similarly, as gas falls toward the disk surrounding a protostar, gravitational potential energy is converted first to kinetic energy, causing the gas to pick up speed. When the gas hits the disk and stops suddenly, that kinetic energy is turned into thermal energy.

Similarly, material falling onto the accretion disk around a forming star causes the disk to heat up. The amount of heating depends on *where* the material hits the disk. Material hitting the inner part of the disk (the *inner disk*) has fallen farther and picked up greater speed within the gravitational field of the forming star than has material hitting the disk farther out. Like a brick dropped from a tall building, material striking the inner disk is moving quite rapidly when it hits, so it heats the inner disk to high temperatures. In contrast, material falling onto the outer part of the disk (the *outer disk*) is moving much more slowly, like a brick dropped from just a foot or so above the ground. So the temperature at the outermost parts of the disk is not much higher than that of the original interstellar cloud. Stated another way, material falling onto the inner disk converts more gravitational potential energy into thermal energy than does material falling onto the outer disk.

The energy released as material falls onto the disk is not the only source of thermal energy in the disk. Even before the nuclear reactions that will one day power the new star have ignited, conversion of gravitational energy into thermal energy drives the temperature at the surface of the protostar to several thousand kelvins, and it also drives the luminosity of the huge ball of glowing gas to many times the luminosity of the present-day Sun. For the same reasons that Mercury is hot while Pluto is not (see Chapter 5), the radiation streaming outward from the protostar at the center of the disk drives the temperature in the inner parts of the disk even higher, increasing the difference in temperature between the inner and outer parts of the disk.

The Compositions of Planets

Temperature affects whether or not a material exists in a solid form. On a hot summer day, ice melts and water quickly evaporates; on a cold winter night, water in your breath freezes into tiny ice crystals. Some materials remain solid even at higher temperatures. These include metals and rocky materials, such as iron, **silicates**—which are minerals containing silicon and oxygen—and carbon. Substances that are capable of withstanding high temperatures without melting or being vaporized are called **refractory materials**. Other materials, such as water, ammonia, and methane, remain in a solid form only if their temperature is very low. These materials, which become gases at moderate temperatures, are called **volatile materials** (or *volatiles* for short). Astronomers generally call the solid form of any volatile material an **ice**.

Differences in temperature from place to place within the protoplanetary disk have a significant effect on the makeup of the dust grains in the disk. As **Figure 7.11** illustrates, in the hottest parts of the disk—closest to the protostar—only the most refractory substances can exist in solid form. In the inner disk, dust grains are composed almost entirely of refractory materials. Some substances can survive in solid form somewhat farther out, including some hardier volatiles, such as water ice and certain chemical compounds that are **organic**, meaning that they contain carbon. These solids add to the materials that make up dust grains. In the coldest, outermost parts of the accretion disk, far from the central protostar, highly volatile components such as methane, ammonia, and carbon monoxide ices and other organic molecules survive only in solid form. The different composition of dust grains within the disk determines the composition of the planetesimals formed from the dust. Planets that form closer to the central star tend to be made up mostly of refractory materials such as rock and metals. Planets that form farther from the central star contain refractory materials, but they also contain large quantities of ices and organic materials.

In the Solar System, the inner planets are composed of rocky material surrounding metallic cores of iron and nickel. Objects in the outer Solar System, including moons, giant planets, and comets, are composed largely of ices of various types. But not all planetary systems are so neatly organized as our Solar System. When planets around other stars were first discovered, they appeared to be very different, with large planets close in to their respective stars. Astronomers now think that **chaotic** encounters, in which a small change in the initial state of a system can lead to a large change in the final state of the system, may change the organization of planetary compositions. In a process called **planet migration**, the

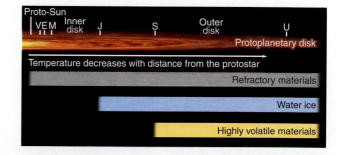

Figure 7.11 Differences in temperature within a protoplanetary disk determine the composition of dust grains that then evolve into planetesimals and planets. The colored bars show that refractory materials are found throughout the disk, while water ice is found only outside Jupiter's orbit, and highly volatile materials are found only outside Saturn's orbit. Shown here are the proto-Sun (PS) and the orbits of Venus (V), Earth (E), Mars (M), Jupiter (J), Saturn (S), and Uranus (U).

force of gravity from all of the nearby objects can move some planets so that they end up far from the place of their birth. For example, in our Solar System, Uranus and Neptune originally may have formed nearer to the orbits of Jupiter and Saturn, but were then driven outward to their current locations by gravitational encounters with Jupiter and Saturn. A planet can also migrate when it gives up some of its orbital angular momentum to the disk material that surrounds it. Such a loss of angular momentum causes the planet slowly to spiral inward toward the central star. Thus, the order of planets in a system can change over time.

Formation of an Atmosphere

Once a solid planet has formed, it may continue growing by capturing gas from the protoplanetary disk. To do so, it must act quickly. Young stars and protostars emit fast-moving particles and intense radiation that can quickly disperse the gaseous remains of the accretion disk. Gaseous planets such as Jupiter probably have only about 10 million years or so to form and to grab whatever gas they can. Because of their strong gravitational fields, more massive young planets can capture more of the hydrogen and helium gas that makes up the bulk of the disk. What follows is much like the formation of a star and protoplanetary disk, but on a smaller scale. Just as happens in the accretion disk around the star, gas from a mini accretion disk moves inward and falls onto the planet.

The gas that is captured by a planet at the time of its formation is called the planet's **primary atmosphere**. The primary atmosphere of a large planet can be more massive than the solid body, as in the case of Jupiter. Some of the solid material in the mini accretion disk might stay behind to coalesce into larger bodies in much the same way that particles of dust in the protoplanetary disk came together to form planets. The result is a mini "solar system"—a group of moons that orbit about the planet.

A less massive planet may also capture some gas from the protoplanetary disk, only to lose it later. The gravity of small planets may be too weak to hold low-mass gases such as hydrogen or helium. Even if a small planet is able to gather some hydrogen and helium from its surroundings, this primary atmosphere will not last long. In the inner solar system, the temperatures are higher, so the hydrogen and helium atoms are moving faster than in the outer solar system and will escape from a small planet. The atmosphere that remains around a small planet like Earth is a **secondary atmosphere**, which forms later in the life of a planet. Volcanism is one important source of a secondary atmosphere because it releases heavier and thus slower-moving gases such as carbon dioxide, water vapor, and other gases from the planet's interior. In addition, volatile-rich comets that formed in the outer parts of the disk continue to fall inward toward the new star long after its planets have formed, and they sometimes collide with planets. **Comets** are icy planetesimals that survive planetary accretion. They may provide a significant source of water, organic compounds, and other volatile materials on planets close to the central star.

CHECK YOUR UNDERSTANDING 7.3

In our Solar System, the inner planets are rocky because: (a) the original cloud had more rocky material near the center; (b) warm temperatures in the inner disk caused the inner planetesimals to be formed of only rocky material; (c) the inner disk filled a smaller volume so it was denser; (d) the hydrogen and helium atoms were too low mass to remain in the inner disk.

7.4 The Formation of Our Solar System

We have seen that nearly 5 billion years ago, the Sun was still a protostar surrounded by a protoplanetary disk of gas and dust. During the next few hundred thousand years, much of the dust in the disk had collected into planetesimals—clumps of rock and metal near the emerging Sun and aggregates of rock, metal, ice, and organic materials farther from the Sun. In this section, we will look at the formation of the different types of planets in our own Solar System.

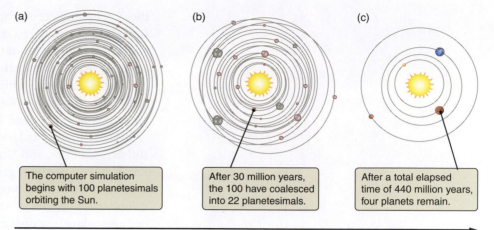

(a) The computer simulation begins with 100 planetesimals orbiting the Sun.

(b) After 30 million years, the 100 have coalesced into 22 planetesimals.

(c) After a total elapsed time of 440 million years, four planets remain.

Time

Figure 7.12 Computer models simulate how material in the protoplanetary disk became clumped into the planets over time. Only a few planets remain at the end.

The Terrestrial Planets

Within the inner 5 astronomical units (AU) of the disk, several rock and metal planetesimals quickly grew in size to become the dominant masses in their orbits. With their ever-strengthening gravitational fields, they either captured most of the remaining planetesimals or ejected them from the inner part of the disk. **Figure 7.12** shows some results from a computer simulation of how this might have happened. The dominant planetesimals became planet-sized bodies with masses ranging between 5 percent and 100 percent of Earth's mass. These dominant planetesimals evolved into the **terrestrial planets**, which are planets that are Earth-like, or rocky. Today, the surviving terrestrial planets are Mercury, Venus, Earth, and Mars. Earth's Moon is often grouped with these terrestrial planets because of its similar physical and geological properties, even though it is not a planet itself and formed in a different way. One or two other planets or large moons may have formed in the young Solar System but were later destroyed.

For several hundred million years after the formation of the four surviving terrestrial planets, leftover pieces of debris still in orbit around the Sun continued to rain down on the surfaces of these planets. Today, we can still see the scars of these early impacts on the cratered surfaces of all the terrestrial planets (**Figure 7.13**). This rain of debris continues even today, but at a much lower rate.

Before the proto-Sun became a true star, gas in the inner part of the protoplanetary disk was still plentiful. During this early period the two larger terrestrial planets, Earth and Venus, may have held on to weak primary atmospheres of hydrogen and helium, but these thin atmospheres were soon lost to space. The terrestrial planets did not develop thick atmospheres until the formation of the secondary atmospheres that now surround Venus, Earth, and Mars. Mercury's proximity to the Sun and the Moon's small mass prevented these bodies from retaining significant secondary atmospheres.

The Giant Planets

Beyond 5 AU from the Sun, in a much colder part of the accretion disk, planetesimals combined to form a number of bodies with masses about 5–20 times that of Earth. These planet-sized objects formed from planetesimals containing volatile ices and organic compounds in addition to rock and metal. In a process astronomers call *core accretion–gas capture*, mini accretion disks formed around

Figure 7.13 Large impact craters on Mercury (and on other solid bodies throughout the Solar System) record the final days of the Solar System's youth, when planets and planetesimals grew as smaller planetesimals rained down on their surfaces.

these planetary cores, capturing massive amounts of hydrogen and helium and funneling this material onto the planets. Four such massive bodies became the cores of the **giant planets**—Jupiter, Saturn, Uranus, and Neptune. These giant planets are many times the mass of any terrestrial planet.

Jupiter's massive solid core captured and retained the most gas—roughly 300 times the mass of Earth (300 M_{Earth}). The other outer planetary cores captured less hydrogen and helium, perhaps because their cores were less massive or because there was less gas available to them. Saturn ended up with less than 100 M_{Earth} of gas, and Uranus and Neptune were able to grab less than twenty Earth masses worth of gas.

The core accretion model indicates that it could take up to 10 million years for a Jupiter-like planet to accumulate. Some planetary scientists do not think that our protoplanetary disk could have survived long enough to form gas giants such as Jupiter through the general process of core accretion. All the gas may have dispersed in roughly half that time, cutting off Jupiter's supply of hydrogen and helium. An alternative explanation is a process called *disk instability*, in which the protoplanetary disk suddenly and quickly fragments into massive clumps equivalent to those of a large planet. It is possible that both core accretion and disk instability played a role in the formation of our own and other planetary systems.

During the formation of the planets, gravitational energy was converted into thermal energy as the individual atoms and molecules moved faster. This conversion warmed the gas surrounding the cores of the giant planets. Proto-Jupiter and proto-Saturn probably became so hot that they actually glowed a deep red color, similar to the heating element on an electric stove. Their internal temperatures may have been even higher.

Some of the material remaining in the mini accretion disks surrounding the giant planets combined into small bodies, which became moons. A **moon** is any natural satellite in orbit about a planet or asteroid. The composition of the moons that formed around the giant planets followed the same trend as that of the planets that formed around the Sun: the innermost moons formed under the hottest conditions and therefore contained the smallest amounts of volatile material. For example, the closest of Jupiter's many moons may have experienced high temperatures from nearby Jupiter glowing so intensely that it would have evaporated most of the volatile substances in the inner part of its mini accretion disk.

Remaining Planetesimals

Not all planetesimals in the disk went on to become planets. For example, dwarf planets orbit the Sun but have not cleared other, smaller bodies from their orbits. Ceres and Pluto, which are shown in Figure 7.1, are dwarf planets. More dwarf planets, along with a large number of smaller bodies, are found in the Kuiper Belt, beyond Pluto's orbit. Asteroids are small bodies found interior to Jupiter's orbit; most are located in the main asteroid belt between Mars and Jupiter. Jupiter's gravity kept the region between Jupiter and Mars so stirred up that most planetesimals there never formed a large planet.

Planetesimals persist to this day in the outermost part of the Solar System as well. Formed in a deep freeze, these objects have retained most of the highly volatile materials found in the grains present at the formation of the accretion disk. Unlike the crowded inner part of the disk, the outermost parts of the disk had planetesimals that were too sparsely distributed for large planets to grow. Icy planetesimals in the outer Solar System that survived planetary accretion remain today as **comet nuclei**. The frozen, distant dwarf planets Pluto and Eris are especially large examples of these residents of the outer Solar System.

Many Solar System objects show evidence of cataclysmic impacts that reshaped worlds, suggesting that the early Solar System must have been a remarkably violent and chaotic place. The dramatic difference in the terrain of the northern and southern hemispheres on Mars, for example, has been interpreted as the result of one or more colossal collisions. The leading theory for the origin of our Moon is that it resulted from the collision of an object with Earth. Mercury has a crater on its surface from an impact so devastating that it caused the crust to buckle on the opposite side of the planet. In the outer Solar System, one of Saturn's moons, Mimas, has a crater roughly one-third the diameter of the moon itself. Uranus suffered one or more collisions that were violent enough literally to knock the planet on its side. Today, as a result, its equatorial plane is tilted at almost a right angle to its orbital plane. We will see other examples in subsequent chapters.

CHECK YOUR UNDERSTANDING 7.4

Suppose that astronomers found a rocky, terrestrial planet beyond the orbit of Neptune. What is the most likely explanation for its origin? (a) It formed close to the Sun and migrated outward. (b) It formed in that location and was not disturbed by migration. (c) It formed later in the Sun's history than other planets. (d) It is a captured planet that formed around another star.

7.5 Planetary Systems Are Common

When astronomers turn their telescopes to young nearby stars, they see disks of the same type from which the Solar System formed. As illustrated in **Figure 7.14**, when the light from the central star is blocked, evidence of the planetary disk is observed. The physical processes that led to the formation of the Solar System should be commonplace wherever new stars are being born. Compared to stars, however, planets are small and dim objects. They shine primarily by reflection and therefore are millions to billions of times fainter than their host stars. Thus, they were difficult to identify until advances in telescope detector technology enabled astronomers to discover them in the 1990s through indirect methods. In 1995, astronomers announced the first confirmed **extrasolar planet**, also called an **exoplanet**—a planet orbiting around a star other than the Sun. Today, the number of known extrasolar planets has grown to the thousands, and new discoveries are occurring almost daily.

The discovery of extrasolar planets raised the question of what we mean by the term *planet*. The full International Astronomical Union (IAU) definitions for planets and dwarf planets within the Solar System are provided in Appendix 9. The IAU defines an extrasolar planet as an object that orbits a star and has a mass less than 13 Jupiter masses (13 M_{Jup}). Objects more massive than 13 M_{Jup} but less massive than 0.08 solar masses (0.8 M_{Sun}; about 80 M_{Jup}) are **brown dwarfs**. Objects more massive than 0.08 M_{Sun} are defined as stars. **Figure 7.15** compares the diameters of these different objects.

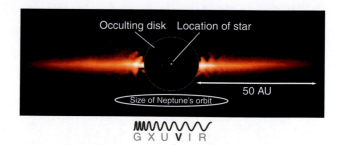

Occulting disk   Location of star

50 AU

Size of Neptune's orbit

G X U V I R

Figure 7.14 An edge-on circumstellar dust disk is seen extending outward to 60 AU from the young (12-million-year-old) star AU Microscopii. The star itself, whose brilliance would otherwise overpower the circumstellar disk, is hidden behind an occulting disk (opaque mask) placed in the telescope's focal plane. Its position is represented by the dot.

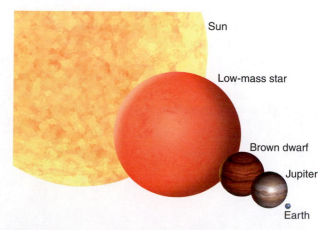

Sun

Low-mass star

Brown dwarf

Jupiter

Earth

Figure 7.15 A comparison of the diameters of the Sun, a low-mass star, a brown dwarf, Jupiter, and Earth.

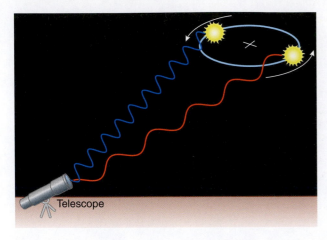

Figure 7.16 Doppler shifts observed in the spectrum of a star are due to the wobble of the star caused by its planet.

 Nebraska Simulation: Influence of Planets on the Sun

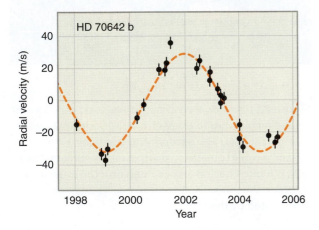

Figure 7.17 Radial velocity data for a star with a planet. A positive number is motion away from the observer; a negative number is motion toward the observer. The plot repeats as the planet completes another orbit around the star.

 **Nebraska Simulation:** Radial Velocity Graph

 Astronomy in Action: Doppler Shift

The Search for Extrasolar Planets

Currently, more than 100 projects are focused on searching for extrasolar planets from the ground and from space. The first planets were discovered indirectly, by observing their gravitational tug on the central star. As technology has improved, other methods have become more productive. Astronomers now have direct images of planets orbiting stars and have also been able to take spectra of planets to observe the composition of their atmospheres. Almost certainly, between the time we write this and the time you read it, there will be new discoveries. The field is advancing extremely quickly. We will now look at each discovery method.

The Radial Velocity Method As a planet orbits a star, the planet's gravity tugs the star around ever so slightly. This motion toward or away from us, its radial velocity, creates an observable Doppler shift in the spectrum of the star. **Figure 7.16** illustrates this method. When the star is moving toward us (negative radial velocity), the light is blueshifted; when the star is moving away from us (positive radial velocity), the light is redshifted. This pattern of radial velocity repeats over time. After detecting this wobble (**Figure 7.17**), astronomers can infer the planet's mass and its distance from the star.

We can see how this works by using the Solar System as an example. Jupiter's mass is greater than the mass of all the other planets, asteroids, and comets combined, so this is the planet most likely to be detected. Both the Sun and Jupiter orbit a common center of gravity (sometimes called center of mass; this is the location where the effect of one mass balances the other) that lies just outside the surface of the Sun, as shown in **Figure 7.18**. Alien astronomers would find that the Sun's radial velocity varies by ±12 m/s, with a period equal to Jupiter's orbital period of 11.86 years. From this information, the astronomers would rightly conclude that the Sun has at least one planet with a mass comparable to Jupiter's but, without greater precision, would be unaware of the other less massive major planets. If the alien astronomers could improve the sensitivity of their instruments to measure radial velocities as small as 2.7 m/s, Saturn would be detectable, and if the precision of their spectrograph extended to radial velocities as small as 0.09 m/s, Earth would be detectable.

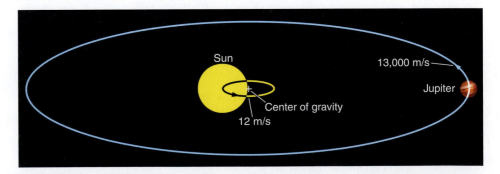

Figure 7.18 The Sun and Jupiter orbit around a common center of gravity, which lies just outside the Sun's surface. Spectroscopic measurements made by an extrasolar astronomer would reveal the Sun's radial velocity varying by ±12 m/s over 11.86 years, which is Jupiter's orbital period. Jupiter travels around its orbit at a speed of 13,000 m/s. (The orbit is shown in perspective and is not actually very elliptical.)

7.2 Working It Out Estimating the Size of the Orbit of a Planet

In the spectroscopic radial velocity method, the star is moving about its center of mass, and its spectral lines are Doppler-shifted accordingly. Recall from Figure 7.18 that the alien astronomer looking toward the Solar System would observe a shift in the wavelengths of the Sun's spectral lines—caused by the presence of Jupiter—of about 12 m/s.

Figure 7.17 showed the radial velocity data for a star with a planet discovered by this method. How do astronomers use this method to estimate the distance (A) of the planet from the star? Recall from Chapter 4 that Newton generalized Kepler's law relating the period of an object's orbit to the orbital semimajor axis:

$$P^2 = \frac{4\pi^2}{G} \times \frac{A^3}{M}$$

where A is the semimajor axis of the orbit, P is its period, and M is the combined mass of the two objects. To find A, we rearrange the equation as follows:

$$A^3 = \frac{G}{4\pi^2} \times M \times P^2$$

From the graph of radial velocity observations in Figure 7.17, we can determine that the period of the orbit is 5.7 years. There are 3.16×10^7 seconds in a year, so

$$P = 5.7 \text{ yr} \times (3.16 \times 10^7 \text{ s/yr})$$

$$P = 1.8 \times 10^8 \text{ s}$$

The mass of the star is much greater than the mass of the planet, so the combined masses of the star and the planet can be approximated as the mass of the star, which in this case is about equal to the mass of the Sun, 2×10^{30} kg. (Stellar masses can be estimated from their spectra.) The gravitational constant $G = 6.67 \times 10^{-20}$ km³/(kg s²). Putting in the numbers gives

$$A^3 = \frac{6.67 \times 10^{-20} \dfrac{\text{km}^3}{\text{kg s}^2}}{4\pi^2} \times (2 \times 10^{30} \text{ kg}) \times (1.8 \times 10^8 \text{ s})^2$$

$$A^3 = 1.1 \times 10^{26} \text{ km}^3$$

Taking the cube root of 1.1×10^{26} km³ solves for A, which is equal to 4.8×10^8 km. To get a better feel for this number, we might put it into astronomical units (where 1 AU = 1.5×10^8 km). The semimajor axis of the orbit of this planet is given by

$$A = \frac{4.8 \times 10^8 \text{ km}}{1.5 \times 10^8 \text{ km/AU}} = 3.2 \text{ AU}$$

This planet is more than 3 times farther from its star than Earth is from the Sun.

 Nebraska Simulation: Exoplanet Radial Velocity Simulator

Current technology limits the precision of radial velocity instruments to about 0.3 m/s, but to date it has been the most successful ground-based approach to finding extrasolar planets. This technique enables astronomers to detect giant planets around solar-type stars, but not yet to find planets with masses similar to Earth's. Finding the signal of the Doppler shift in the noise of the observation requires the star to be quite bright in our sky. So this method is limited to nearby stars, within about 160 light-years from Earth. **Working It Out 7.2** provides additional explanation of the spectroscopic radial velocity method.

The Transit Method

Another technique for finding extrasolar planets is the **transit method**, in which we observe the effect of a planet passing in front of its parent star. From Earth it is sometimes possible to see the inner planets Mercury and Venus transit, or pass in front of, the Sun. An alien located somewhere in the plane of Earth's orbit would see Earth pass in front of the Sun and could infer the existence of Earth by detecting the 0.009 percent drop in the Sun's brightness during the transit. Similarly, for astronomers on Earth to observe a planet passing in front of a star, Earth must lie nearly in the orbital plane of that planet. When an extrasolar planet passes in front of its parent star, the light from the star diminishes by a tiny amount, as seen in **Figure 7.19**. Whereas the radial velocity method gives us the mass of the planet and its orbital distance from a star, the transit method provides the size of a planet. **Working It Out 7.3** demonstrates how the radii are estimated.

 Nebraska Simulation: Exoplanet Transit Simulator

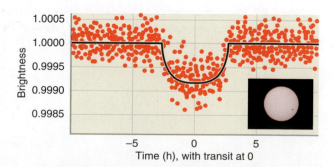

Figure 7.19 The data show the light curve for Kepler-11c. The inset photograph shows Venus passing in front of the Sun in June 2012, similar to this transit of Kepler-11c.

The masses of extrasolar planets can often be estimated using Kepler's laws and the conservation of angular momentum. When planets are detected by the transit method, astronomers can estimate the radius of an extrasolar planet. In this method, astronomers look for planets that eclipse their stars and observe how much the star's light decreases during this eclipse (see Figure 7.19). In the Solar System when Venus or Mercury transits the Sun, a black circular disk is visible on the face of the circular Sun. During the transit, the amount of light from the transited star is reduced by the area of the circular disk of the planet divided by the area of the circular disk of the star:

$$\text{Percentage reduction in light} = \frac{\text{Area of disk of planet}}{\text{Area of disk of star}}$$

$$= \frac{\pi R_{\text{planet}}^2}{\pi R_{\text{star}}^2} = \frac{R_{\text{planet}}^2}{R_{\text{star}}^2}$$

Then, to solve for the radius of the planet, astronomers need an estimate of the radius of the star and a measurement of the percentage reduction in light during the transit. The radius of a star is estimated from the surface temperature and the luminosity of the star.

Let's consider an example. Kepler-11 is a system of at least six planets that transit a star. The radius of the star, R_{star}, is estimated to be 1.1 times the radius of the Sun, or $1.1 \times (7.0 \times 10^5 \text{ km}) = 7.7 \times 10^5 \text{ km}$. The light from planet Kepler-11c is observed to decrease by 0.077 percent, or 0.00077 (see Figure 7.19). What is Kepler-11c's size?

$$0.00077 = \frac{R_{\text{Kepler-11c}}^2}{R_{\text{star}}^2} = \frac{R_{\text{Kepler-11c}}^2}{(7.7 \times 10^5 \text{ km})^2}$$

$$R_{\text{Kepler-11c}}^2 = 4.5 \times 10^8 \text{ km}^2$$

$$R_{\text{Kepler-11c}} = 2.1 \times 10^4 \text{ km}$$

Dividing Kepler-11c's radius by the radius of Earth (6,400 km) shows that the planet Kepler-11c has a radius of 3.3 R_{Earth}.

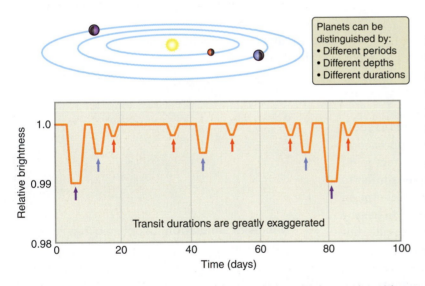

Figure 7.20 Multiple planets can be detected by multiple transits with different brightness changes. The arrows point to the changes in the total light as the three planets transit the star.

More than a thousand extrasolar planets have been detected from ground-based and space telescopes using the transit method. Current ground-based technology limits the sensitivity of the transit method to about 0.1 percent of a star's brightness. Amateur astronomers have confirmed the existence of several extrasolar planets by observing transits using charge-coupled device (CCD) cameras mounted on telescopes with apertures as small as 20 centimeters (cm). Telescopes in space improve the sensitivity because smaller dips in brightness can be measured. The small French COROT telescope (27 cm) discovered 32 planets during its 6 years of operation (2007–2013). NASA's 0.95-meter Kepler telescope has discovered many planets and has found thousands more candidates that are being investigated further. **Figure 7.20** illustrates how multiplanet systems are identified with this method: if one planet is found, then observations of the variations in timing of the transit can indicate that there are other planets orbiting the same star.

Microlensing The gravitational field of an unseen planet can act like a lens, bending the light from a distant star in such a way that it causes the star to brighten temporarily while the planet is passing in front of it. Because the effect is small, it is usually called microlensing. Like the radial velocity method, microlensing provides an estimate of the mass of the planet. To date, several dozen extrasolar planets have been found with this technique.

The Astrometric Method Planets may also be detected by astrometry—precisely measuring the position of a star in the sky. If the system is viewed from "above," the star moves in a mini-orbit as the planet pulls it around. This motion is generally tiny and therefore very difficult to measure. However, for systems viewed from above the plane of the planet's orbit, none of the prior methods will work because the planet neither passes in front of the star nor causes a shift in its speed along the line of sight. Space missions such as the Gaia observatory, launched in 2013 by the European Space Agency, is conducting observations of this kind.

Direct Imaging Direct imaging means taking a picture of the planet directly. This technique is conceptually straightforward but is technically difficult because it involves searching for a relatively faint planet in the overpowering glare of a bright star—a challenge far more difficult than looking for a star in a clear, bright daytime sky. Even when an object is detected by direct imaging, an astronomer must still determine whether the observed object is actually a planet. Suppose we detect a faint object near a bright star. Could it be a more distant star that just happens to be in the line of sight? Future observations could tell if the object shares the bright star's motion through space. But it could also be a brown dwarf rather than a true planet. An astronomer would need to make further observations to determine the object's mass.

Some planets have been discovered by this method with large ground-based telescopes operating in the infrared region of the spectrum using adaptive optics. **Figure 7.21** is an infrared image of Beta Pictoris b. The first visible-light discovery was made from space while the Hubble Space Telescope was observing Fomalhaut, a bright naked-eye star only 25 light-years away. The planet Fomalhaut b is shown in **Figure 7.22**. It has a mass no more than 3 times that of Jupiter and orbits within a dusty debris ring about 17 billion km from the central star. A related form of direct observation involves separating the spectrum of a planet from the spectrum of its star to obtain information about the planet directly. Large ground-based telescopes have been able to obtain spectra of the atmospheres of some extrasolar planets and have found, for example, carbon monoxide and water in these atmospheres.

The Discovery of Extrasolar Planets

Searches for extrasolar planets have been remarkably successful. Between the discovery of the first (in 1995) and this writing, nearly 2,000 more have been confirmed, and thousands more candidates are under investigation. As the number of observed systems with single and multiple planets increases, astronomers can compare them with those of the Solar System, and they have found more variation than they expected. The field is changing so fast that the most up-to-date information can only be found online or through mobile applications such as the Kepler App.

The first discoveries included many **hot Jupiters**, which are Jupiter-sized planets orbiting solar-type stars in circular or highly eccentric orbits that bring them closer to their parent stars than Mercury is to our own Sun. These planets were among the first to be detected because they are relatively easy targets for the spectroscopic radial velocity method. The large mass of a nearby hot Jupiter tugs the star very hard, creating large radial velocity variations in the star. In addition, these large planets orbiting close to their parent stars are more likely to pass in front of the star periodically and reveal themselves via the transit method. Therefore, these hot Jupiter systems are easier to find than smaller, more distant planets. Astronomers realized that these hot Jupiter systems are not representative of most planetary systems; they were just easier to find. Scientists call this bias a *selection effect*.

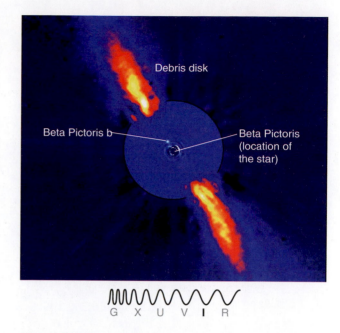

Figure 7.21 Beta Pictoris b is seen orbiting within a dusty debris disk that surrounds the bright naked-eye star Beta Pictoris. The planet's estimated mass is 8 times that of Jupiter. The star is hidden behind an opaque mask, and the planet appears through a semitransparent mask used to subdue the brightness of the dusty disk.

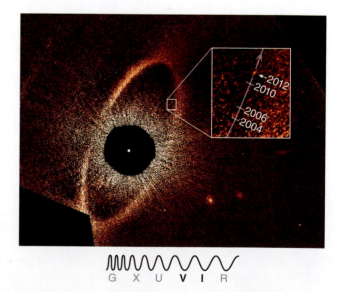

Figure 7.22 A Hubble Space Telescope image of Fomalhaut b, seen here moving in its orbit around Fomalhaut, a nearby star easily visible to the naked eye. The parent star, hidden by an obscuring mask, is about a billion times brighter than the planet, which is located within a dusty debris ring that surrounds the star.

Astronomers were surprised by the hot Jupiters because, according to the planetary system formation theory available at the time (based only on the Solar System), these giant, volatile-rich planets should not have been able to form so close to their parent stars. The expectation was that Jupiter-type planets should form in the more distant, cooler regions of the protoplanetary disk, where the volatiles that make up much of their composition are able to survive. So astronomers suggested that perhaps hot Jupiters formed much farther away from their parent stars and subsequently migrated inward to a closer orbit. The mechanism by which a planet could migrate over such a distance must involve an interaction with gas or planetesimals in which orbital angular momentum is somehow transferred from the planet to its surroundings, allowing it to spiral inward.

Many of the new planets being discovered by Kepler are mini-Neptunes (gaseous planets with masses of 2–10 M_{Earth}) or super-Earths (rocky planets more massive than Earth), but each month brings an announcement of the discovery of smaller planets. Planets with longer orbital periods, and therefore larger orbits, can be discovered only when the observations have gone on long enough to observe more than one complete orbit. Some of the extrasolar planets have highly elliptical orbits compared with those in the Solar System. Planets have been found with orbits that are highly tilted compared with the plane of the rotation of their star, and some planets move in orbits whose direction is opposite that of their star's rotation. Multiple-planet systems have been observed in which the larger mini-Neptunes alternate with smaller super-Earths. The multiple-planet systems that have been found by the transit method reside in flat systems like our own, offering further evidence that the planets formed in a flat protoplanetary disk around a young star. But the current hypothesis to explain the Solar System's inner, small rocky planets and outer, large gaseous planets may not be applicable in these other planetary systems.

In addition, some planets detected by microlensing seem to be wandering freely through the Milky Way. These planets may have been ejected from their solar systems after their formation and are no longer in gravitationally bound orbits around their stars. The frequent new discoveries requiring revisions of existing theories make extrasolar planets one of the most exciting topics in astronomy today.

CHECK YOUR UNDERSTANDING 7.5

Suppose you hear of the discovery of an Earth-mass planet around a star. This planet was most likely discovered through the _____ method. (a) Doppler spectroscopy; (b) direct imaging; (c) transit; (d) astrometric

Origins

Kepler's Search for Earth-Sized Planets

The discovery of planetary systems, many different from the Solar System, shows us that the formation of planets frequently, and perhaps always, accompanies the formation of stars. The implications of this conclusion are profound. Planets are a common by-product of star formation. In a galaxy of 200 billion stars and a universe of hundreds of billions of galaxies, how many planets (and also moons) might exist? And with all of these planets in the universe, how many might have suitable conditions for the particular category of chemical reactions that we refer to as "life"? (We will return to this point in Chapter 24.)

The Kepler Mission was developed by NASA to find Earth-sized and larger planets in orbit about a variety of stars. Kepler is a 1-meter telescope with 42 CCD detectors and is designed to observe approximately 150,000 stars in 100 square degrees of sky and look for planetary transits. To confirm a planetary detection, the transits need to be observed three times with repeatable changes in brightness, duration of transit times, and computed orbital period.

Kepler can detect a dip in the brightness of a star of 0.01 percent—which is sensitive enough to detect an Earth-sized planet. Kepler identified the first Earth-sized planets in 2011. Stars with transiting planets detected by Kepler are also observed spectroscopically to obtain radial velocity measurements that can lead to an estimate of a planet's mass. If a planet's radius and mass are known, the planet's density (mass per volume) can be estimated, too. From the density, astronomers can get a sense of whether the planet is composed primarily of gas, rock, ice, water, or a mixture of some of these.

On Earth, liquid water was essential for the formation and evolution of life. Because life on Earth is the only example of life for which we have evidence, we do not know whether liquid water is a cosmic requirement, but it is a place to start. The primary scientific goal of the Kepler Mission is to look for rocky planets at the right distance from their stars to permit the existence of liquid water, a distance known as the **habitable zone**. If a planet is too close to its star, water will exist only as a vapor; if it is too far, water will be frozen as ice. In the Solar System, Earth is the only planet in the habitable zone. Although announcements of new planets often state whether the planet is in the habitable zone, just being in the zone doesn't guarantee that the planet actually has liquid water—or that the planet is inhabited! An example of an Earth-sized planet in a habitable zone is shown in **Figure 7.23**.

In 2013, Kepler suffered a mechanical failure that stopped observations, but a work-around was approved, and observations resumed in 2014. Kepler has identified thousands of planet candidates, some in the habitable zones of their respective stars. The candidates must be confirmed by follow-up observations of more transits or of radial velocities before they are officially announced as planet detections. Amateur astronomers can access the candidate lists online (at the "Exoplanet Transit Database") and conduct their own observations. Anyone with Internet access can go to planethunters.org, examine some Kepler data, and contribute to the search.

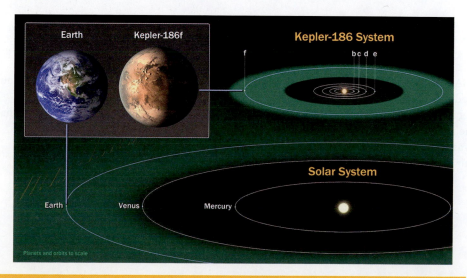

Figure 7.23 An artist's conception of the Kepler-186 system. Located about 500 light-years from Earth, this system has a planet in its habitable zone. The Solar System is shown for comparison.

ARTICLES   QUESTIONS

A system with five planets was observed by NASA's Kepler space telescope. Ⓧ

Earth-Size Planet Found in the "Habitable Zone" of Another Star

By **Science@NASA**

Using NASA's Kepler space telescope, astronomers have discovered the first Earth-size planet orbiting in the "habitable zone" of another star (see Figure 7.23). The planet, named "Kepler-186f," orbits an M dwarf, or red dwarf, a class of stars that makes up 70 percent of the stars in the Milky Way Galaxy. The discovery of Kepler-186f confirms that planets the size of Earth exist in the habitable zone of stars other than our Sun.

The "habitable zone" is defined as the range of distances from a star where liquid water might pool on the surface of an orbiting planet. While planets have previously been found in the habitable zone, the previous finds are all at least 40 percent larger in size than Earth, and understanding their makeup is challenging. Kepler-186f is more reminiscent of Earth.

Kepler-186f orbits its parent M dwarf star once every 130 days and receives one-third the energy that Earth gets from the Sun, placing it nearer the outer edge of the habitable zone. On the surface of Kepler-186f, the brightness of its star at high noon is only as bright as our Sun appears to us about an hour before sunset.

"M dwarfs are the most numerous stars," said Elisa Quintana, research scientist at the SETI Institute at NASA's Ames Research Center in Moffett Field, California, and lead author of the paper published today in the journal *Science*. "The first signs of other life in the galaxy may well come from planets orbiting an M dwarf."

However, "being in the habitable zone does not mean we know this planet is habitable," cautions Thomas Barclay, a research scientist at the Bay Area Environmental Research Institute at Ames, and coauthor of the paper. "The temperature on the planet is strongly dependent on what kind of atmosphere the planet has. Kepler-186f can be thought of as an Earth-cousin rather than an Earth-twin. It has many properties that resemble Earth."

Kepler-186f resides in the Kepler-186 system, about 500 light-years from Earth in the constellation Cygnus. The system is also home to four companion planets: Kepler-186b, Kepler-186c, Kepler-186d, and Kepler-186e, whiz around their sun every four, seven, 13, and 22 days, respectively, making them too hot for life as we know it. These four inner planets all measure less than 1.5 times the size of Earth.

Although the size of Kepler-186f is known, its mass and composition are not. Previous research, however, suggests that a planet the size of Kepler-186f is likely to be rocky.

"The discovery of Kepler-186f is a significant step toward finding worlds like our planet Earth," said Paul Hertz, NASA's Astrophysics Division director at the agency's headquarters in Washington.

The next steps in the search for distant life include looking for true Earth-twins—Earth-size planets orbiting within the habitable zone of a Sun-like star—and measuring their chemical compositions. The Kepler space telescope, which simultaneously and continuously measured the brightness of more than 150,000 stars, is NASA's first mission capable of detecting Earth-size planets around stars like our Sun.

Looking ahead, Hertz said, "future NASA missions, like the Transiting Exoplanet Survey Satellite and the James Webb Space Telescope, will discover the nearest rocky exoplanets and determine their composition and atmospheric conditions, continuing humankind's quest to find truly Earth-like worlds."

ARTICLES   QUESTIONS

1. This NASA press release was picked up by business and international news feeds. Why do you think coverage of this discovery was so widespread?
2. The planet is closer to its star than Earth is to the Sun yet receives much less energy. What does that imply about the temperature of the star?
3. Why is the mass of this planet not yet known? What method will be used to find its mass?
4. How will astronomers estimate the planet's composition?
5. Why is this planet called a "cousin" of Earth?

Summary

Stars and their planetary systems form from collapsing interstellar clouds of gas and dust, following the laws of gravity and conservation of angular momentum. Conservation of angular momentum produces an accretion disk around a protostar that often fragments to form multiple planets, as well as smaller objects such as asteroids and dwarf planets, through the gradual accumulation of material into larger and larger objects. There are multiple methods for finding planets around other stars, and these planets are now thought to be very common. This field of study is evolving very quickly as technology advances.

LG 1 **Describe how our understanding of planetary system formation developed from the work of both planetary and stellar scientists.** Planets are a common by-product of star formation, and many stars are surrounded by planetary systems. Gravity pulls clumps of gas and dust together, causing them to shrink and heat up. Angular momentum must be conserved, leading to both a spinning central star and an accretion disk that rotates and revolves in the same direction as the central star. Solar System meteorites show that larger objects build up from smaller objects.

LG 2 **Discuss the role of gravity and angular momentum in explaining why planets orbit the Sun in a plane and why they revolve in the same direction that the Sun rotates.** As particles orbit the forming star, the cloud of dust and gas flattens into a plane. Conservation of angular momentum determines both the speed and the direction of the revolution of the objects in the forming system. Dust grains in the protoplanetary disk first stick together because of collisions and static electricity. As these objects grow, they eventually have enough mass to attract other objects gravitationally. Once this occurs, they begin emptying the space around them. Collisions of planetesimals lead to the formation of planets.

LG 3 **Explain how temperature at different locations in the protoplanetary disk affects the composition of planets, moons, and other bodies.** Near the central protostar, the temperature is higher. This forces volatile elements, such as water, to evaporate and leave the inner part of the disk. Planets in the inner part of the disk will have fewer volatiles than those in the outer part of the disk. The gas that is captured by a planet at the time of its formation is the planet's primary atmosphere. Less massive planets lose their primary atmospheres and then form secondary atmospheres.

LG 4 **Discuss the processes that resulted in the formation of planets and other objects in our Solar System.** In the current model of the formation of the Solar System, solid terrestrial planets formed in the inner disk, where temperatures were high, and giant gaseous planets formed in the outer disk, where temperatures were low. Dwarf planets such as Pluto formed in the asteroid belt and in the region beyond the orbit of Neptune. Asteroids and comet nuclei remain today as leftover debris.

LG 5 **List how astronomers find planets around other stars, and explain how we know that planetary systems around other stars are common.** Astronomers find planets around other stars using a variety of methods: the radial velocity method, the transit method, microlensing, astrometry, and direct imaging. As technology has improved, the number and variety of known extrasolar planets has increased dramatically, with thousands of planets and planet candidates discovered orbiting other stars near the Sun within the Milky Way Galaxy in just the past few years.

? UNANSWERED QUESTIONS

- How typical is the Solar System? Only within the past few years have astronomers found other systems containing four or more planets, and so far the observed distributions of large and small planets in these multiplanet systems have looked different from those of the Solar System. Computer simulations of planetary system formation suggest that a system with an orbital stability and a planetary distribution like those of the Solar System may develop only rarely. Improved supercomputers can run more complex simulations, which can be compared with the observations to understand better how solar systems are configured.

- How Earth-like must a planet be before scientists declare it to be "another Earth"? An editorial in the science journal *Nature* cautioned that scientists should define "Earth-like" in advance—before multiple discoveries of planets "similar" to Earth are announced and a media frenzy ensues. Must a planet be of similar size and mass, be located in the habitable zone, and have spectroscopic evidence of liquid water before we call it "Earth 2.0"?

Questions and Problems

Test Your Understanding

1. Place the following events in the order that corresponds to the formation of a planetary system.
   a. Gravity collapses a cloud of interstellar gas.
   b. A rotating disk forms.
   c. Small bodies collide to form larger bodies.
   d. A stellar wind "turns on" and sweeps away gas and dust.
   e. Primary atmospheres form.
   f. Primary atmospheres are lost.
   g. Secondary atmospheres form.
   h. Dust grains stick together by static electricity.

2. If the radius of an object's orbit is halved, and angular momentum is conserved, what must happen to the object's speed?
   a. It must be halved.   c. It must be doubled.
   b. It must stay the same. d. It must be squared.

3. Unlike the giant planets, the terrestrial planets formed when
   a. the inner Solar System was richer in heavy elements than the outer Solar System.
   b. the inner Solar System was hotter than the outer Solar System.
   c. the outer Solar System took up more volume than the inner Solar System, so there was more material to form planets.
   d. the inner Solar System was moving faster than the outer Solar System.

4. The terrestrial planets and the giant planets have different compositions because
   a. the giant planets are much larger.
   b. the terrestrial planets formed closer to the Sun.
   c. the giant planets are made mostly of solids.
   d. the terrestrial planets have few moons.

5. The spectroscopic radial velocity method preferentially detects
   a. large planets close to the central star.
   b. small planets close to the central star.
   c. large planets far from the central star.
   d. small planets far from the central star.
   e. the method detects all of these equally well

6. The concept of disk instability was developed to solve the problem that
   a. Jupiter-like planets migrate after formation.
   b. there was not enough gas in the Solar System to form Jupiter.
   c. the early solar nebula likely dispersed too soon to form Jupiter.
   d. Jupiter consists mostly of volatiles.

7. Because angular momentum is conserved, an ice-skater who throws her arms out will
   a. rotate more slowly.   c. rotate at the same rate.
   b. rotate more quickly.   d. stop rotating entirely.

8. Clumps grow into planetesimals by
   a. gravitationally pulling in other clumps.
   b. colliding with other clumps.
   c. attracting other clumps with opposite charge.
   d. conserving angular momentum.

9. The transit method preferentially detects
   a. large planets close to the central star.
   b. small planets close to the central star.
   c. large planets far from the central star.
   d. small planets far from the central star.
   e. the method detects all of these equally well

10. If the radius of a spherical object is halved, what must happen to the period so that the spin angular momentum is conserved?
    a. It must be divided by 4.   d. It must double.
    b. It must be halved.   e. It must be multiplied by 4.
    c. It must stay the same.

11. The amount of angular momentum in an object does *not* depend on
    a. its radius.   c. its rotation speed.
    b. its mass.   d. its temperature.

12. The planets in the inner part of the Solar System are made primarily of refractory materials; the planets in the outer Solar System are made primarily of volatiles. The difference occurred because
    a. refractory materials are heavier than volatiles, so they sank farther into the nebula.
    b. there were no volatiles in the inner part of the accretion disk.
    c. the volatiles on the inner planets were lost soon after the planet formed.
    d. the outer Solar System has gained more volatiles from space since formation.

13. If scientists want to find out about the composition of the early Solar System, the best objects to study are
    a. the terrestrial planets.   c. the Sun.
    b. the giant planets.   d. asteroids and comets.

14. The direction of revolution in the plane of the Solar System was determined by
    a. the plane of the galaxy in which the Solar System sits.
    b. the direction of the gravitational force within the original cloud.
    c. the direction of rotation of the original cloud.
    d. the amount of material in the original cloud.

15. A planet in the "habitable zone"
    a. is close to the central star.
    b. is far from the central star.
    c. is the same distance from its star as Earth is from the Sun.
    d. is at a distance where liquid water can exist on the surface.

Thinking about the Concepts

16. What is the source of the material that now makes up the Sun and the rest of the Solar System?

17. Describe the different ways by which stellar astronomers and planetary scientists each came to the same conclusion about how planetary systems form.

18. What is a protoplanetary disk? What are two reasons that the inner part of the disk is hotter than the outer part?

19. Physicists describe certain properties, such as angular momentum and energy, as being *conserved*. What does this mean? Do conservation laws imply that an individual object can never lose or gain angular momentum or energy? Explain your reasoning.

20. The Process of Science Figure in this chapter makes the point that different areas of science must agree with one another. Suppose that a handful of new exoplanets are discovered that appear not to have formed from the collapse of a stellar nebula (for example, the planetary orbits might be in random orientations). What will scientists do with this new information?

21. How does the law of conservation of angular momentum control a figure-skater's rate of spin?

22. What is an accretion disk?

23. Describe the process by which tiny grains of dust grow to become massive planets.

24. Look under your bed, the refrigerator, or any similar place for dust bunnies. Once you find them, blow one toward another. Watch carefully and describe what happens as they meet. What happens if you repeat this action with additional dust bunnies? Will these dust bunnies ever have enough gravity to begin pulling themselves together? If they were in space instead of on the floor, might that happen? What force prevents their mutual gravity from drawing them together into a "bunny-tesimal" under your bed?

25. Why do we find rocky material everywhere in the Solar System but large amounts of volatile material only in the outer regions?

26. Why were the four giant planets able to collect massive gaseous atmospheres, whereas the terrestrial planets could not? Explain the source of the secondary atmospheres surrounding the terrestrial planets.

27. Describe four methods that astronomers use to search for extrasolar planets. What are the limitations of each method; that is, what circumstances are necessary to detect a planet by each method?

28. Why is it so difficult for astronomers to obtain an image of an extrasolar planet?

29. Many of the first exoplanets that astronomers found orbiting other stars are giant planets with Jupiter-like masses and with orbits located very close to their parent stars. Explain why these characteristics are a selection effect of the discovery method.

30. How does Kepler find Earth-like planets, and what do astronomers mean by "Earth-like"?

Applying the Concepts

31. Study Figure 7.17. What is the maximum radial velocity of HD 70642b in meters per second? Convert this number to miles per hour (mph). How does this compare to the speed at which Earth orbits the Sun (67,000 mph)?

32. Use Appendix 4 to answer the following:
    a. What is the total mass of all the planets in the Solar System, expressed in Earth masses (M_{Earth})?
    b. What fraction of this total planetary mass is Jupiter?
    c. What fraction does Earth represent?

33. Compare Earth's orbital angular momentum with its spin angular momentum using the following values: $m = 5.97 \times 10^{24}$ kg, $v = 29.8$ kilometers per second (km/s), $r = 1$ AU, $R = 6,378$ km, and $P = 1$ day. Assume Earth to be a uniform body. What fraction does each component (orbital and spin) contribute to Earth's total angular momentum? Refer to Working It Out 7.1.

34. Venus has a radius 0.949 times that of Earth and a mass 0.815 times that of Earth. Its rotation period is 243 days. What is the ratio of Venus's spin angular momentum to that of Earth? Assume that Venus and Earth are uniform spheres.

35. Jupiter has a mass equal to 318 times Earth's mass, an orbital radius of 5.2 AU, and an orbital velocity of 13.1 km/s. Earth's orbital velocity is 29.8 km/s. What is the ratio of Jupiter's orbital angular momentum to that of Earth?

36. In the text, we give an example of an interstellar cloud having a diameter of 10^{13} km and a rotation period of 10^6 years collapsing to a sphere the size of the Sun (1.4×10^6 km in diameter). We point out that if all the cloud's angular momentum went into that sphere, the sphere would have a rotation period of only 0.6 second. Do the calculation to confirm this result.

37. The asteroid Vesta has a diameter of 530 km and a mass of 2.7×10^{20} kg.
    a. Calculate the density (mass/volume) of Vesta.
    b. The density of water is 1,000 kg/m³, and that of rock is about 2,500 kg/m³. What does this difference tell you about the composition of this primitive body?

38. Study Figure 7.20.
    a. Recalling Kepler's Laws, put the three planets in order, from fastest to slowest.
    b. Compare the duration of the transits. Why does the outermost planet have the longest duration?

39. The best current technology can measure radial velocities of about 0.3 m/s. Suppose you are observing a spectral line with a wavelength of 575 nanometers (nm). How large a shift in wavelength would a radial velocity of 0.3 m/s produce?

40. Earth tugs the Sun around as it orbits, but it has a much smaller effect (only 0.09 m/s) than that of any known extrasolar planet. How large a shift in wavelength does this effect cause in the Sun's spectrum at 500 nm?

41. If an alien astronomer observed a plot of the light curve as Jupiter passed in front of the Sun, by how much would the Sun's brightness drop during the transit?

42. A planet has been found to orbit a 1-M_{Sun} in 200 days.
    a. What is the orbital radius of this extrasolar planet?
    b. Compare its orbit with that of the planets around our own Sun. What environmental conditions must this planet experience?

43. One of the planets orbiting the star Kepler-11 with an orbital radius of radius 1.1 solar radii, or R_{Sun} has a radius of 4.5 Earth radii (R_{Earth}). By how much does the brightness of Kepler-11 decrease when this planet transits the star?

44. Kepler detected a planet with a diameter of 1.7 Earth (D_{Earth}).
    a. How much larger is the volume of this planet than Earth's?
    b. Assume that the density of the planet is the same as Earth's. How much more massive is this planet than Earth?

45. The planet COROT-11b was discovered using the transit method, and astronomers have followed up with radial velocity measurements, so both its size (radius 1.43 R_{Jup}) and its mass (2.33 M_{Jup}) are known. The density provides a clue about whether the object is gaseous or rocky.
    a. What is the mass of this planet in kilograms?
    b. What is the planet's radius in meters?
    c. What is the planet's volume?
    d. What is the planet's density? How does this density compare to the density of water (1,000 kg/m³)? Is the planet likely to be rocky or gaseous?

USING THE WEB

46. Go to the "Extrasolar Planets Global Searches" Web page (http://exoplanet.eu/searches.php) of the Extrasolar Planets Encyclopedia. Click on one ongoing project under "Ground" and one ongoing project under "Space." What method is used to detect planets in each case? Has the selected project found any planets, and if so, what type are they? Now click on one of the future projects. When will the one you chose be ready to begin? What will be the method of detection?

47. Using the exoplanet catalogs:
    a. Go to the "Catalog" Web page (http://exoplanet.eu/catalog) of the Extrasolar Planets Encyclopedia and set to "All Planets detected." Look for a star that has multiple planets. Make a graph showing the distances of the planets from that star, and note the masses and sizes of the planets. Put the Solar System planets on the same axis. How does this extrasolar planet system compare with the Solar System?
    b. Go to the "Exoplanets Data Explorer" website (http://exoplanets.org) and click on "Table." This website lists planets that have detailed orbital data published in scientific journals, and it may have a smaller total count than the website in part (a). Pick a planet that was discovered this year or last, as specified in the "First Reference" column. What is the planet's minimum mass? What is its semimajor axis and the period of its orbit? What is the eccentricity of its orbit? Click on the star name in the first column to get more information. Is there a radial velocity curve for this planet? Was it observed in transit, and if so, what is the planet's radius and density? Is it more like Jupiter or more like Earth?

48. Space missions:
    a. Go to the website for the Kepler Mission (http://kepler.nasa.gov). How many confirmed planets has Kepler discovered? Mouse over "confirmed planets": How many planet candidates are there? What kinds of follow-up observations are being done to verify whether the candidates are planets? What is new?
    b. Search for the latest version of the "Kepler Orrery," an animation that shows multiplanet systems discovered by Kepler. Do most of these systems look like our own?
    c. Go to the website for the European Space Agency (ESA) mission Gaia (http://sci.esa.int/gaia). This mission was launched in 2013. Click on the "Exoplanets" link on the left-hand side. What method(s) will *GAIA* use to look for planets? What are the science goals? Have some planets been found?

49. Citizen science projects:
    a. Go to the "PlanetHunters" website at http://planethunters.org. PlanetHunters is part of the Zooniverse, a citizen science project that invites individuals to participate in a major science project using their own computers. To participate in this or any of the other Zooniverse projects mentioned in later chapters, you will need to sign up for an account. Read through the sections under "About," including the FAQ. What are some of the advantages to crowdsourcing Kepler data analysis? Back on the PlanetHunters home page, click on "Tutorial" and watch the "Introduction" and "Tutorial Video." When you're ready to try looking for planets, click on "Classify" and begin. Save a copy of your stars for your homework.
    b. Go to the "Disk Detective" website at http://www.diskdetective.org/, another Zooniverse project for which you will need to make an account as in part (a). In this project, you will look at observations of young stars to see if there is evidence for a planetary disk. Under "Menu," read "Science" and "About," and then "Classify." Work through an example, and then classify a few images.

50. Go to the "Super Planet Crash" Web page (http://www.stefanom.org/spc/ or http://apod.nasa.gov/apod/ap150112.html). Read "Help" to see the rules. First build a system like ours with four Earth-sized planets in the inner 2 AU—is this stable? What happens if you add in super-Earths or "ice giants"? Build up a few completely different planetary systems and see what happens. What types of situations cause instability in the inner 2 AU of these systems?

smartwork

If your instructor assigns homework in SmartWork, access your assignments at digital.wwnorton.com/astro5.

digital.wwnorton.com/astro5

Visit the Student Site at the Digital Landing Page, and open the Exoplanet Radial Velocity Simulator in Chapter 7. This applet has a number of different panels that allow you to experiment with the variables that are important for measurement of radial velocities. First, in the window labeled "Visualization Controls," check the box to show multiple views. Compare the views shown in panels 1–3 with the colored arrows in the last panel to see where an observer would stand to see the view shown. Start the animation (in the "Animation Controls" panel), and allow it to run while you watch the planet orbit its star from each of the views shown. Stop the animation, and in the "Presets" panel, select "Option A" and then click "set."

1 Is Earth's view of this system most nearly like the "side view" or most nearly like the "orbit view"?

_____

2 Is the orbit of this planet circular or elongated?

_____

3 Study the radial velocity graph in the upper right panel. The blue curve shows the radial velocity of the star over a full period. What is the maximum radial velocity of the star?

_____

4 The horizontal axis of the graph shows the "phase," or fraction of the period. A phase of 0.5 is halfway through a period. The vertical red line indicates the phase shown in views in the upper left panel. Start the animation to see how the red line sweeps across the graph as the planet orbits the star. The period of this planet is 365 days. How many days pass between the minimum radial velocity and the maximum radial velocity?

_____

5 When the planet moves away from Earth, the star moves toward Earth. The sign of the radial velocity tells the direction of the motion (toward or away). Is the radial velocity of the star positive or negative at this time in the orbit? If you could graph the radial velocity of the planet at this point in the orbit, would it be positive or negative?

_____

In the "Presets" window, select "Option B" and then click "set."

6 What has changed about the orbit of the planet as shown in the views in the upper left panel?

_____

7 When is the planet moving fastest: when it is close to the star or when it is far from the star?

_____

8 When is the star moving fastest: when the planet is close to it or when the planet is far away?

_____

9 Explain how an astronomer would determine, from a radial velocity graph of the star's motion, whether the orbit of the planet was in a circular or elongated orbit.

_____

_____

10 Study the "Earth view" panel at the top of the window. Would this planet be a good candidate for a transit observation? Why or why not?

_____

_____

In the "System Orientation" panel, change the inclination to 0.0.

11 Now is Earth's view of this system most nearly like the "side view" or most nearly like the "orbit view"?

_____

12 How does the radial velocity of the star change as the planet orbits?

_____

_____

13 Click the box that says "show simulated measurements," and change the "noise" to 1.0 m/s. The gray dots are simulated data, and the blue line is the theoretical curve. Use the slider bar to change the inclination. What happens to the radial velocity as the inclination increases? (Hint: Pay attention to the vertical axis as you move the slider, not just the blue line.)

_____

_____

_____

14 What is the smallest inclination for which you would find the data convincing? That is, what is the smallest inclination for which the theoretical curve is in good agreement with the data?

_____

_____

13

Taking the Measure of Stars

To all but the largest telescopes, even nearby stars are just points of light in the night sky. Astronomers study the stars by observing their light, by using the laws of physics discussed in earlier chapters, and by finding patterns in subgroups of stars that are extrapolated to other stars. Astronomers use knowledge of geometry, radiation, and orbits to begin to answer basic questions about stars, such as how they are similar to or different from the Sun, and whether they might have planets orbiting around them as the Sun does.

LEARNING GOALS

By the conclusion of this chapter, you should be able to:

LG 1 Explain how the brightness of nearby stars and their distances from Earth are used to determine how luminous they are.

LG 2 Explain how astronomers obtain the temperatures, sizes, and composition of stars.

LG 3 Describe how astronomers estimate the masses of stars.

LG 4 Classify stars, and organize this information on a Hertzsprung-Russell (H-R) diagram.

LG 5 Explain how the mass and composition of a main-sequence star determine its luminosity, temperature, and size.

The constellation Orion. The reddish object in the middle of the vertical line of stars is a nebula. ▶ ▶ ▶

Why do stars
have different
colors?

13.1 Astronomers Measure the Distance, Brightness, and Luminosity of Stars

When looking up at the stars in the sky, it is immediately noticeable that they differ in brightness and color. However, we don't know if one star appears brighter than another because it has a higher luminosity and is emitting more light or because it is closer to us. In this section, you will learn how to find the distances to nearby stars and how to use distance and apparent brightness to find the luminosity of a star.

Stereoscopic Vision

Your two eyes have different views that depend on the distance to the object you are viewing. Hold up your finger in front of you, quite close to your nose. View it with your right eye only and then with your left eye only. Each eye sends a slightly different image to your brain, so your finger *appears* to move back and forth relative to the background behind it. Now hold up your finger at arm's length, and blink your right eye, then your left. Your finger appears to move much less. The way your brain combines the different information from your eyes to perceive the distances to objects around you is called **stereoscopic vision**. **Figure 13.1a** shows an overhead view of the experiment you just performed with your finger. The left eye sees the blue pencil almost directly between the green balls on the bookshelf. But the right eye sees the blue pencil to the left of both balls. Similarly, the position of the pink pencil appears to vary. Because the pink pencil is closer to the observer, its position appears to change more than the position of the blue pencil—it seems to move from the right of the blue pencil to the left of the blue pencil.

Stereoscopic vision enables you to judge the distances of objects as far away as a few hundred meters, but beyond that it is of little use. Your right eye's view of a mountain several kilometers away is indistinguishable from the view seen by your left eye—all you can determine is that the mountain is too far away for you to judge its distance stereoscopically. The distance over which your stereoscopic vision works is limited by the separation between your eyes, about 6 centimeters (cm). If you could separate your eyes by several meters, the view from each eye would be different enough for you to judge the distances to objects that are kilometers away.

Of course, you cannot separate your eyes, but you can compare pictures taken with a camera from two widely separated locations. The greatest separation we can obtain without leaving Earth is to let Earth's orbital motion carry us from one side of the Sun to the other. If you take a picture of the sky tonight and then wait 6 months and take another picture, the distance between the two locations is the diameter of Earth's orbit (2 astronomical units [AU]), which gives us more powerful stereoscopic vision.

Distances to Nearby Stars

Figure 13.1b shows how astronomers apply this concept of stereoscopic vision to measure the distances to stars. This illustration shows Earth's orbit as viewed from far above the Solar System. The change in position of Earth over 6 months is like the distance between the right eye and the left eye in Figure 13.1a. The nearby (pink and blue) stars are like the pink and blue pencils, while the distant yellow stars are like the green balls on the bookcase. Because of the shift in perspective as Earth orbits the Sun, nearby stars appear to shift their positions. The pink star, which is closer, appears to move farther than the more distant blue star. Over the

Astronomy in Action: Parallax

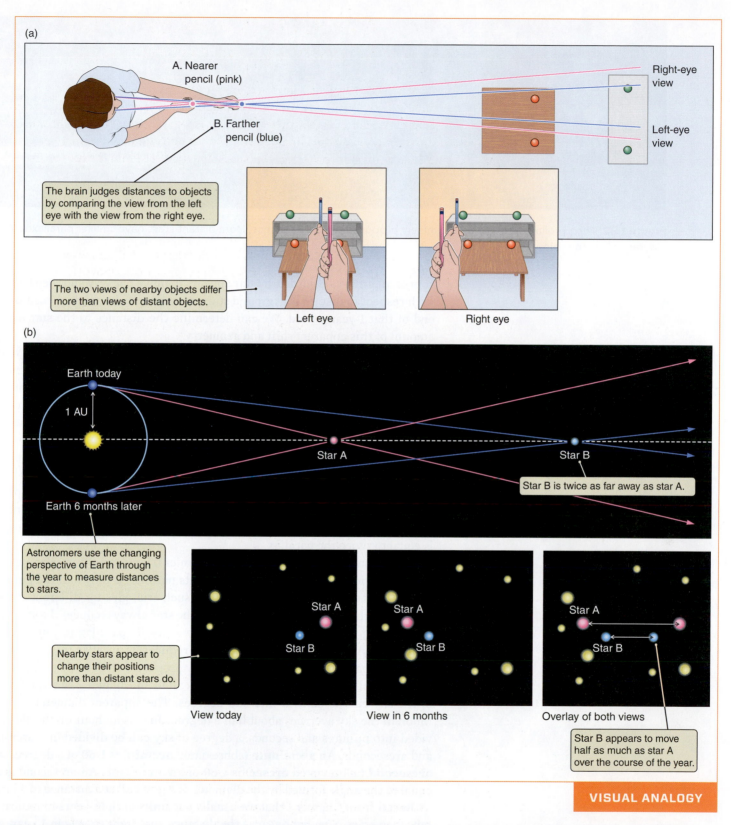

VISUAL ANALOGY

Figure 13.1 (a) Stereoscopic vision enables you to determine the distance to an object by comparing the view from each eye. (b) Similarly, comparing views from different places in Earth's orbit enables astronomers to determine the distances to stars. As Earth moves around the Sun, the apparent positions of nearby stars change more than the apparent positions of more distant stars. (The diagram is not to scale.)

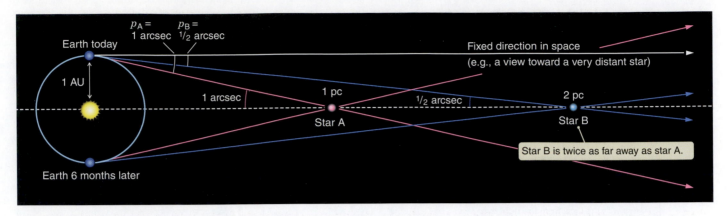

Figure 13.2 The parallax (*p*) of a star is inversely proportional to its distance. More distant stars have smaller parallaxes. (The diagram is not to scale.)

Nebraska Simulation: Parallax Calculator

course of 1 full year, the nearby star appears to move one way and then back again with respect to distant background stars, returning to its original position at the end of that 1-year period. We can determine the distance to the star using the amount of this apparent shift and geometry.

The eye cannot detect the changes in position of a nearby star throughout the year, but telescopes can reveal these small shifts relative to the background stars. **Figure 13.2** shows Earth, the Sun, and two stars in a similar configuration as that in Figure 13.1b. Look first at star A, the closest star. When Earth is at the top of the figure, it forms a right triangle with the Sun and star A at the other corners. (Remember that a right triangle is one with a 90° angle in it.) The short leg of the triangle is the distance from Earth to the Sun, which is 1 AU. The long leg of the triangle is the distance from the Sun to star A. The small angle near star A is called the *parallactic angle*, or simply **parallax**, of the star. As Earth completes an orbit around the Sun, the star's position in the sky appears to shift back and forth, returning to its original position at the end of that year. The amount of this shift is equal to twice the parallax.

More distant stars make longer and skinnier triangles with smaller parallaxes. Star B is twice as far away as star A, and its parallax is only half the parallax of star A. If you were to draw a number of such triangles for different stars, you would find that increasing the distance to the star always reduces the star's parallax. Moving a star 3 times farther away reduces its parallax to $\frac{1}{3}$ of its original value. Moving a star 10 times farther away reduces its parallax to $\frac{1}{10}$ of its original value. The parallax of a star (*p*) is inversely proportional to its distance (*d*).

The parallaxes of real stars are tiny. Recall from Chapter 2 that the full circle of the sky can be divided into 360 degrees. The apparent diameter of the full Moon in the sky averages about half a degree. Just as an hour on the clock is divided into minutes and seconds, a degree of sky can be divided into arcminutes and arcseconds. An **arcminute** (abbreviated **arcmin**) is 1/60 of a degree, and an **arcsecond** (abbreviated **arcsec**) is 1/60 of an arcminute. An arcsecond is about equal to the angle formed by the diameter of a golf ball at a distance of 9 km.

Recall from Chapter 1 that we usually use units of light-years to indicate distances to stars. One light-year is the distance that light travels in 1 year—about 9.5 trillion kilometers (km). We use this unit because it is the unit you are most likely to see online or in a popular book about astronomy. When astronomers discuss distances to stars and galaxies, however, the unit they often use is the **parsec (pc)**, which is equal to 3.26 light-years (or 206,265 AU). The term is short for *parallax second*—a star at a distance of 1 parsec has a parallax of 1 arcsecond.

13.1 Working It Out Parallax and Distance

As seen in Figure 13.2, if star B is twice as far away from us as star A, then star B will have half the parallax of star A. The parallax of a star (p) is inversely proportional to its distance (d):

$$p \propto \frac{1}{d} \quad \text{or} \quad d \propto \frac{1}{p}$$

If the angle at the apex of a triangle is 1 arcsec, and the base of the triangle is 1 AU, then the length of the triangle is 1 parsec. One parsec equals 206,265 AU (see Appendix 1), or 3.09×10^{16} meters, or 3.26 light-years. Astronomers use the unit of parsecs because it makes the relationship between distance and parallax easier than that with the use of light-years.

As illustrated in Figure 13.2, a star with a parallax of 1 arcsec is at a distance of 1 pc. The inverse proportionality between distance and parallax becomes

$$\left(\begin{array}{c} \text{Distance measured} \\ \text{in parsecs} \end{array} \right) = \frac{1}{\left(\begin{array}{c} \text{Parallax measured} \\ \text{in arcseconds} \end{array} \right)}$$

or

$$d(\text{pc}) = \frac{1}{p(\text{arcsec})}$$

Suppose that the parallax of a star is measured to be 0.5 arcsec. The distance can be found by

$$d(\text{pc}) = \frac{1}{0.5} = 2 \text{ pc}$$

Similarly, a star with a measured parallax of 0.01 arcsec is located at a distance of 1/0.01 = 100 pc.

After the Sun, the next closest star to Earth is Proxima Centauri. Located at a distance of 4.24 light-years, Proxima Centauri is a faint member of a system of three stars called Alpha Centauri. What is this star's parallax? First, we convert the distance to parsecs:

$$d = 4.24 \text{ light-years} \times \frac{1 \text{ parsec}}{3.26 \text{ light-years}} = 1.30 \text{ parsecs}$$

Then,

$$p(\text{arcsec}) = \frac{1}{1.30 \text{ pc}} = 0.77 \text{ arcsec}$$

Even the closest star to the Sun has a parallax of only about $\frac{3}{4}$ arcsec.

When astronomers began to measure the parallax angles of stars, they discovered that stars are very distant objects (**Working It Out 13.1**). The first successful measurement of the parallax of a star was made by F. W. Bessel (1784–1846), who in 1838 reported a parallax of 0.314 arcsec for the star 61 Cygni. This finding implied that 61 Cygni was 3.2 pc away, or 660,000 times as far away as the Sun. With this one measurement, Bessel increased the known volume of the universe by a factor of 10,000. Today, astronomers know of about 60 stars in 54 single-, double-, or triple-star systems within 5 pc (16.3 light-years) of the Sun. In the neighborhood of the Sun, each star or star system has on average a volume of about 260 cubic light-years of space to itself.

Most stars are so far away that the parallax angle is too small to measure using ground-based telescopes, which are limited by Earth's atmosphere. In the 1990s, the European Space Agency's Hipparcos satellite measured the positions and parallaxes of 120,000 stars, thus greatly improving our picture of the Sun's stellar neighborhood. The accuracy of any given Hipparcos parallax measurement is about ±0.001 arcsec. Because of this observational uncertainty, measurements of the distances to stars are not perfect. For example, a star with a Hipparcos-measured parallax of 0.004±0.001 arcsec really has a parallax between 0.003 and 0.005 arcsec. This gives a corresponding distance range of 200–330 pc from Earth. As an analogy, consider your speed while driving down the road. If your digital speedometer says 10 kilometers per hour (km/h), you might actually be traveling 10.4 km/h or 9.6 km/h. The precision of your speedometer is limited to the nearest 1 km/h.

A successor to Hipparcos is Gaia, a space mission to study stellar parallaxes that was launched at the end of 2013. Gaia is expected to observe 1 billion stars and measure the parallaxes of 20 million of them with a high precision. Other methods of measuring distance to more remote stars will be discussed later.

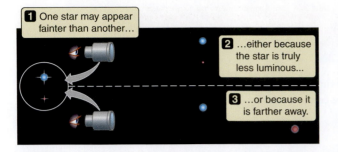

1 One star may appear fainter than another…

2 …either because the star is truly less luminous…

3 …or because it is farther away.

Figure 13.3 The brightness of a star visible in our sky depends on both its luminosity—how much light it emits—and its distance.

Luminosity, Brightness, and Distance

The stars in Earth's sky are of different brightnesses. In Chapter 5, you saw that brightness corresponds to the amount of energy falling on a square meter of area each second in the form of electromagnetic radiation. Although the brightness of a star can be measured directly, it does not immediately give much information about the star itself. As illustrated in **Figure 13.3**, a bright star in the night sky may in fact only appear bright because it is nearby. Conversely, a faint star may be a powerful beacon, still visible despite its tremendous distance.

Astronomers measure the brightness of stars by comparing them to one another. The system they use dates back 2,100 years, when the Greek astronomer Hipparchus classified stars according to their brightness. The details of his system, which is still in use today, are discussed in **Working It Out 13.2** and Appendix 7.

To learn about the actual properties of a star, astronomers need to know the total energy radiated by a star each second—the star's luminosity. Recall from

---

13.2 Working It Out The Magnitude System

The **magnitude** system of brightness for celestial objects can be traced back 2,100 years to the ancient Greek astronomer Hipparchus, who classified the brightest stars he could see as being "of the first magnitude" and the faintest as being "of the sixth magnitude." Later, astronomers defined Hipparchus's 1st magnitude stars as being exactly 100 times brighter than his 6th magnitude stars. Hipparchus must have had typical eyesight, because an average person under dark skies can see stars only as faint as 6th magnitude. Today, telescopes extend our vision far into space. The Hubble Space Telescope can integrate for long exposures and detect stars as faint as 30th magnitude. The limits have also been extended to stars brighter than 1st magnitude using zero and negative numbers. A negative magnitude signifies that an object is *brighter* than an object at zero magnitude. For example, Sirius, the brightest star in the sky, has a magnitude of −1.46. Venus can be as bright as magnitude −4.4, or about 15 times brighter than Sirius and bright enough to cast a shadow. The magnitude of the full Moon is −12.6, and that of the Sun is −26.7 (**Figure 13.4**).

Looking at this mathematically, with five steps between the 1st and 6th magnitudes, each step is equal to the fifth root of 100, or $100^{1/5}$, which is approximately 2.512. This system is logarithmic, but instead of the usual base 10, it is base 2.512. Thus, a 5-magnitude difference in brightness equals $(2.512)^5$, or 100, times difference in brightness. Fifth magnitude stars are 2.512 times brighter than 6th magnitude stars, and 4th magnitude stars are $2.512 \times 2.512 = 6.310$ times brighter than 6th magnitude stars. A 2.5-magnitude difference equals $(2.512)^{2.5}$, or 10, times difference in brightness. The brightness ratio between any two stars is equal to $(2.512)^N$, where N is the magnitude difference between them.

Since the limit of the Hubble Space Telescope (HST) is 30 and $30 − 6 = 24$, HST can detect stars that are $(2.512)^{24} = 4 \times 10^9$, or 4 billion, times fainter than the magnitude 6 that the naked eye can see. Or if we compare the Sun and the Moon, the Sun is 14 magnitudes, or $(2.512)^{14} = 4 \times 10^5$ (400,000) times brighter than the full Moon. (More detailed calculations and a table of magnitudes and brightness differences are located in Appendix 7.)

The magnitude of a star, as we have discussed it, is called the star's **apparent magnitude** because it is the brightness of the star as it *appears* in Earth's sky. Now imagine that all stars were located at exactly 10 pc (32.6 light-years) away from Earth. The brightness of each star would then reflect its luminosity. If the distance from Earth to a star is known, astronomers compute how bright the star would appear if it were located at 10 pc. The **absolute magnitude** of a star—its apparent magnitude at a distance of 10 pc—measures the star's luminosity.

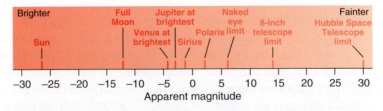

Figure 13.4 Apparent magnitude indicates the apparent brightness of an object in our sky. The brightest objects have a negative apparent magnitude, while telescopes have extended the observable range to fainter objects with higher magnitudes.

Chapter 5 that the brightness of an object that has a known luminosity and is located at a distance d is given by the following equation:

$$\text{Brightness} = \frac{\text{Total light emitted per second}}{\text{Area of a sphere of radius } d} = \frac{\text{Luminosity}}{4\pi d^2}$$

You can rearrange this equation, moving the quantities you know how to measure (distance and brightness) to the right-hand side and the quantity you would like to know (luminosity) to the left, to get

$$\text{Luminosity} = 4\pi d^2 \times \text{Brightness}$$

This equation is used to find how much total light a star must be giving off in order to appear as bright as it does when seen from Earth.

Different stars have different luminosities. The Sun provides a convenient comparison when measuring the properties of stars, including their luminosity. The luminosity of the Sun is measured at $L_{\text{Sun}} = 3.9 \times 10^{26}$ watts (W). The most luminous stars exceed a million times the luminosity of the Sun ($10^6\,L_{\text{Sun}}$). The least luminous stars have luminosities less than 1/10,000 that of the Sun ($10^{-4}\,L_{\text{Sun}}$). The most luminous stars are therefore more than 10 billion (10^{10}) times more luminous than the least luminous stars. Only a very small fraction of stars are near the upper end of this range of luminosities. The vast majority of stars are at the faint end of this distribution, less luminous than the Sun. **Figure 13.5** shows the relative number of stars compared to their luminosities in solar units. (Distances for the nearest stars are obtained from their parallaxes; other methods—to be discussed later—are used for the more distant stars.)

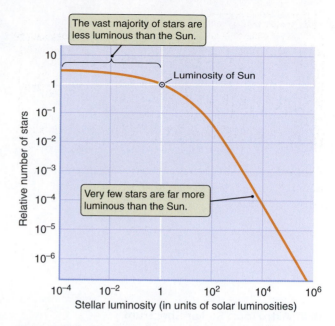

Figure 13.5 The distribution of the luminosities of stars is plotted logarithmically in this graph, so that increments are in powers of 10.

CHECK YOUR UNDERSTANDING 13.1

Stars A and B appear equally bright, but star A is twice as far away from us as star B. Which of the following is true? (a) Star A is twice as luminous as star B. (b) Star A is 4 times as luminous as star B. (c) Star B is twice as luminous as star A. (d) Star B is 4 times as luminous as star B. (e) Star A and star B have the same luminosity because they have the same brightness.

Nebraska Simulation: Stellar Luminosity Calculator

13.2 Astronomers Can Determine the Temperature, Size, and Composition of Stars

Two everyday concepts—stereoscopic vision and the fact that objects appear brighter when closer—have provided the tools needed to measure the distance and luminosity of the closest stars. Stars that appear to be faint points of light in the night sky are in fact luminous beacons located at great distances. The laws of radiation that we described in Chapter 5 reveal still more about stars.

Stars are gaseous, but they are dense enough that the radiation from a star comes close to obeying the same laws as the radiation from solid objects like the heating element on an electric stove. We can therefore use our knowledge of Planck blackbody radiation to understand the radiation from stars. Recall both the Stefan-Boltzmann law from Chapter 5, which states that among same-sized objects, the hotter objects are more luminous, and Wien's law, which states that hotter objects are bluer. In this section, we will use these two laws to measure the temperatures and sizes of stars. We will also develop a more detailed understanding of the line emission mentioned in Chapter 5 to obtain information about the composition of stars.

Wien's Law Revisited: The Color and Surface Temperature of Stars

Wien's law (see Working It Out 5.3) shows that the temperature of an object determines the peak wavelength of its spectrum. The hotter the surface of an object, the bluer the light that it emits. Stars with especially hot surfaces are blue, stars with especially cool surfaces are red, and yellow-white stars such as the Sun are in-between. If you obtain a spectrum of a star and measure the wavelength at which the spectrum peaks, then Wien's law will tell you the temperature of the star's surface. The color of a star tells you about the temperature only at the surface, because this layer is giving off most of the radiation that we see. (Stellar interiors are far hotter than this, as we will discuss in the next chapter.)

In practice, it is not necessary to obtain a complete spectrum of a star to determine its temperature. Astronomers often measure the colors of stars by comparing the brightness at two different, specific wavelengths. The brightness of a star is usually measured through an optical **filter**—sometimes just a piece of colored glass—that lets through only a small range of wavelengths. Two of the most common filters are a blue filter that allows light with wavelengths of about 440 nanometers (nm) to pass through and a "visual" (yellow-green) filter that allows light with wavelengths of about 550 nm to pass through. The ratio of brightness between the blue and visual filters is called the *color index* of the star (more details are discussed in Appendix 7). From a pair of pictures of a group of stars, each taken through a different filter, we can find an approximate value of the surface temperature of every star in the picture—perhaps hundreds or even thousands— all at once. This type of analysis shows that there are many more cool stars than hot stars, and most stars have surface temperatures lower than that of the Sun.

Classification of Stars by Surface Temperature

Although the hot "surface" of a star emits radiation with a spectrum very close to a smooth Planck blackbody curve, this light must then escape through the outer layers of the star's atmosphere. The atoms and molecules in the cooler layers of the star's atmosphere leave their absorption line fingerprints in the escaping light, as shown in **Figure 13.6**. Under some circumstances, the atoms and molecules in the star's atmosphere, along with any gas that might be found near the star, can produce emission lines in stellar spectra. Absorption and emission lines complicate how astronomers use the laws of Planck blackbody radiation to interpret light from stars, but spectral lines provide a wealth of information about the state of the gas in a star's atmosphere.

The spectra of stars were first classified during the late 1800s, long before stars, atoms, or radiation were well understood. Stars were classified by the appearance of the dark bands (now known as absorption lines) seen in their spectra. The original ordering of this classification was arbitrarily based on the prominence of particular absorption lines known to be associated with the element hydrogen. Stars with the strongest hydrogen lines were denoted *A stars*, stars with somewhat weaker hydrogen lines were denoted *B stars*, and so on.

▶❙❙ **AstroTour:** Stellar Spectrum

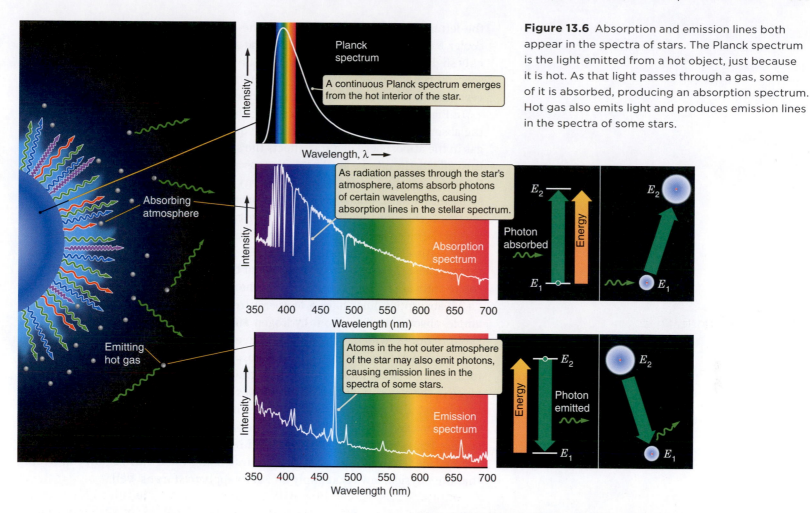

Figure 13.6 Absorption and emission lines both appear in the spectra of stars. The Planck spectrum is the light emitted from a hot object, just because it is hot. As that light passes through a gas, some of it is absorbed, producing an absorption spectrum. Hot gas also emits light and produces emission lines in the spectra of some stars.

Planck spectrum

A continuous Planck spectrum emerges from the hot interior of the star.

As radiation passes through the star's atmosphere, atoms absorb photons of certain wavelengths, causing absorption lines in the stellar spectrum.

Absorption spectrum

Photon absorbed

Atoms in the hot outer atmosphere of the star may also emit photons, causing emission lines in the spectra of some stars.

Emission spectrum

Photon emitted

Annie Jump Cannon (1863–1941) led an effort at the Harvard College Observatory to examine and classify the spectra of hundreds of thousands of stars systematically. She dropped many of the earlier spectral types, keeping only seven that were subsequently reordered based on surface temperatures. Spectra of stars of different spectral types are shown between the horizontal bars in **Figure 13.7**. The hottest stars, with surface temperatures above 30,000 K, are denoted *O stars*. O stars have only weak absorption lines from hydrogen and helium. The coolest stars—*M stars*—have temperatures as low as about 2800 K. M stars show absorption lines from many different types of atoms and molecules. The sequence of **spectral types** of stars, from hottest to coolest, is O, B, A, F, G, K, M. This sequence has undergone several modifications over time, most recently to add cooler objects known as brown dwarfs with spectral types L, T, and Y.

Astronomers divide the main spectral types into a finer sequence of subclasses by adding numbers to

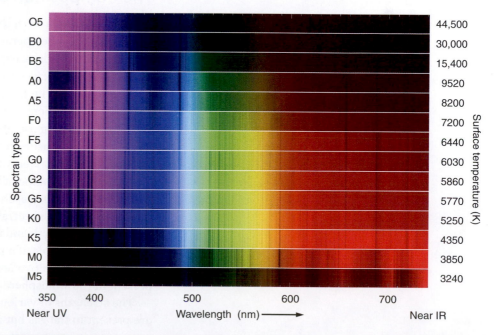

Figure 13.7 Spectra of stars with different spectral types, ranging from hot, blue O stars to cool, red M stars. Hotter stars are more luminous at shorter wavelengths. The dark lines are absorption lines.

the letter designations. For example, the hottest B stars are B0 stars, slightly cooler B stars are B1 stars, and so on. The coolest B stars are B9 stars, which are only slightly hotter than A0 stars. The Sun is a G2 star. The boundaries between spectral types are not always easy to determine. A hotter-than-average G star is very similar to a cooler-than-average F star.

In Figure 13.7, notice that not only are hot stars bluer than cool stars but also the absorption lines in their spectra are quite different. The temperature of the gas in the atmosphere of a star affects the state of the atoms in that gas, which in turn affects the energy level transitions available to absorb radiation (see Section 5.2 to review the concept of atomic energy levels). In O stars, the temperature is so high that most atoms have had one or more electrons stripped from them by energetic collisions within the gas. Few transitions are available in the visible part of the electromagnetic spectrum, so the visible spectrum of an O star is relatively featureless. At lower temperatures, there are more atoms in energy states that can absorb light in the visible part of the spectrum, so the visible spectra of cooler stars are far more complex than the spectra of O stars.

All absorption lines have a temperature at which they are strongest. For example, absorption lines from hydrogen are most prominent at temperatures of about 10,000 K, which is the surface temperature of an A star. At the very lowest stellar temperatures, atoms in the atmosphere of a star react with each other, forming molecules. Molecules such as titanium oxide (TiO) are responsible for much of the absorption in the atmospheres of cool M stars.

Because different spectral lines are formed at different temperatures, astronomers can use these absorption lines to measure a star's temperature. The surface temperatures of stars measured in this way agree extremely well with the surface temperatures of stars measured using Wien's law, again confirming that the physical laws that apply on Earth apply to stars as well.

The Composition of Stars

Most of the variations in the lines of a particular chemical element seen in stellar spectra are due to temperature, but the details of the absorption and emission lines found in starlight also carry a wealth of other information. By applying the physics of atoms and molecules to the study of stellar absorption lines, astronomers can accurately determine not only surface temperatures of stars but also pressures, chemical compositions, magnetic-field strengths, and other physical properties. In addition, by making use of the Doppler shift of emission and absorption lines, astronomers can measure rotation rates, atmospheric motions, expansion and contraction, "winds" driven away from stars, and other dynamic properties of stars.

The chemical composition of a star is also found from its spectra. In Chapter 5, you saw that because each type of atom has different energy levels, each type of atom has different spectral lines. The patterns of spectral lines are measured in laboratories on Earth and then used to identify the types of atoms (or molecules) in stars. For example, if a star has absorption lines that correspond to the energy difference between two levels in the calcium atom, then we know that calcium is present in the atmosphere of the star.

The strengths of various absorption lines tell us not only what kinds of atoms are present in the gas but also the amount of each. However, we must take great care in interpreting spectra to account properly for the temperature and density of the gas in the atmosphere of a star. Recall the "astronomer's periodic table of the elements" from Figure 5.16. Typically, hydrogen composes more than 90 percent of the atoms in the atmosphere of a star, while helium accounts for most of what remains. All of the other chemical elements, which collectively are called **heavy elements** or **massive elements**, are present in only very small amounts.

Table 13.1 shows the chemical composition of the atmosphere of the Sun. The Sun's composition is fairly typical for stars in its vicinity, but the percentages of various heavy elements can vary tremendously from star to star. Some stars have lower amounts of heavier elements than the Sun. The existence of such stars, all but devoid of more massive elements, provides important clues about the origin of chemical elements and the chemical evolution of the universe. Note that many of the atoms that make up Earth and its atmosphere (for example, iron, silicon, nitrogen, oxygen, and carbon) exist as only a small percentage of the Sun.

The Stefan-Boltzmann Law and Finding the Sizes of Stars

Stars are so far away that only two can be imaged as more than point sources of light. To determine the size of a star, astronomers must use other measurements: the temperature and the luminosity.

The temperature of a star can be found directly, either from its color through Wien's law (**Figure 13.8a**) or from the strength of its spectral lines. The temperature of the surface of a star is one factor that influences its luminosity. If a large star and a small star are the same temperature, they will emit the same energy from every patch of surface, but the large star has more patches, so it is more luminous altogether. Conversely, if two stars are the same size, the hotter one will be more luminous than the cooler one. This is an application of the Stefan-Boltzmann law, shown in Figure 13.8b. A small, hot star might even be more luminous than a larger cool star.

| TABLE 13.1 | The Relative Amounts of Different Chemical Elements in the Atmosphere of the Sun |

Element	Percentage of Atoms in the Sun This Element Represents	Percentage of Sun's Mass This Element Represents
Hydrogen	92.5	74.5
Helium	7.4	23.7
Oxygen	0.064	0.82
Carbon	0.039	0.37
Neon	0.012	0.19
Nitrogen	0.008	0.09
Silicon	0.004	0.09
Magnesium	0.003	0.06
Iron	0.003	0.16
Sulfur	0.001	0.04
Total of others	0.001	0.03

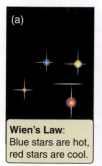

Wien's Law: Blue stars are hot, red stars are cool.

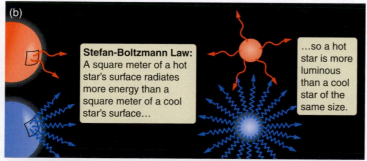

Stefan-Boltzmann Law: A square meter of a hot star's surface radiates more energy than a square meter of a cool star's surface…

…so a hot star is more luminous than a cool star of the same size.

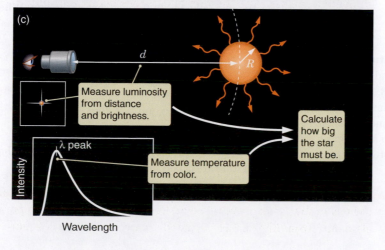

Measure luminosity from distance and brightness.

λ peak

Measure temperature from color.

Calculate how big the star must be.

Intensity

Wavelength

Figure 13.8 (a) The temperature of a star can be found from its color through Wien's law. (b) The luminosity depends on both the temperature and the size of the star. (c) Once the temperature and the luminosity are known, the size of the star can be calculated.

The luminosity of a star can be found from its brightness and its distance. Because the luminosity depends on both the temperature and the size of the star, we can use the luminosity and the temperature to determine the radius of the star, as shown in Figure 13.8c (**Working It Out 13.3**). Astronomers have used the luminosity-temperature-radius relationship to estimate the radii of many thousands of stars. The radius of the Sun, written R_{Sun}, is 696,000 km. One of the

13.3 Working It Out Estimating the Sizes of Stars

In Chapter 5, you learned that according to the Stefan-Boltzmann law, the amount of energy radiated each second by each square meter of the surface of a star is equal to the constant σ multiplied by the surface temperature of the star raised to the fourth power. Written as an equation, this relationship says:

$$\begin{pmatrix} \text{Energy radiated each} \\ \text{second by 1 m}^2 \text{ of surface} \end{pmatrix} = \sigma T^4$$

To find the total amount of light radiated each second by the star, we need to multiply the radiation per second from each square meter by the number of square meters of the star's surface:

$$\begin{pmatrix} \text{Energy radiated} \\ \text{each second} \end{pmatrix} = \begin{pmatrix} \text{Energy radiated each} \\ \text{second by 1 m}^2 \text{ of surface} \end{pmatrix} \times \begin{pmatrix} \text{Surface} \\ \text{area} \end{pmatrix}$$

The left-hand term in this equation—the total energy emitted by the star per second (in units of joules per second [J/s] = watts [W])—is the star's luminosity, L. The middle term—the energy radiated by each square meter of the star per second (in units of joules per square meter per second; J/m²/s)—can be replaced with the σT^4 factor from the Stefan-Boltzmann law. The remaining term—the number of square meters covering the surface of the star—is the surface area of a sphere, $A_{sphere} = 4\pi R^2$ (in units of square meters; m²), where R is the radius of the star.

If we replace the words in the equation with the appropriate mathematical expressions for the Stefan-Boltzmann law and the area of a sphere, our equation for the luminosity of a star looks like this:

$$\text{Luminosity} = \sigma T^4 \times 4\pi R^2$$

Combining gives

$$L = 4\pi R^2 \sigma T^4 \text{ J/s (W)}$$

This last equation is called the **luminosity-temperature-radius relationship** for stars. Because the constants (4, π, and σ) do not change, the luminosity of a star is proportional only to $R^2 T^4$. Make a star 3 times as large, and its surface area becomes $3^2 = 9$ times as large. There is 9 times as much area to radiate, so there is 9 times as much radiation. Make a star twice as hot, and each square meter of the star's surface radiates $2^4 = 16$ times as much energy. Larger, hotter stars are more luminous than smaller, cooler stars.

Now turn this question around and ask, How large must a star of a given temperature be to have a total luminosity of L? The star's luminosity (L) and temperature (T) are quantities that we can measure, and the star's radius (R) is what we want to know. We can rearrange the previous equation, moving the properties that we know how to measure (temperature and luminosity) to the right-hand side of the equation and the property that we would like to know (the radius of the star) to the left-hand side. After a couple of steps of algebra, we find:

$$R = \sqrt{\frac{L}{4\pi\sigma T^4}} = \frac{1}{T^2}\sqrt{\frac{L}{4\pi\sigma}}$$

Again, the right-hand side of the equation contains only things that we know or can measure. The constants 4, π, and σ are always the same. We can find L, the luminosity of the star, by use of the measurements of the star's brightness and parallax (although only for nearby stars with known parallax). T is the surface temperature of the star, which can be measured from its color. From the relationship of these measurements, we now know something new: the size of the star.

Often, we compare two stars and the constants all cancel out, leaving L, T, and R:

$$\frac{L_{star1}}{L_{star2}} = \frac{R_{star1}^2}{R_{star2}^2} \times \frac{T_{star1}^4}{T_{star2}^4} \quad \text{or} \quad \frac{R_{star1}}{R_{star2}} = \sqrt{\frac{L_{star1}}{L_{star2}} \times \frac{T_{star2}^2}{T_{star1}^2}}$$

Suppose we compare the Sun to the second brightest star in the constellation Orion, a red star called Betelgeuse. From its spectrum, we know that Betelgeuse's surface temperature T is about 3500 K. Its distance is about 200 pc, and from that and its brightness, its luminosity is estimated to be 140,000 times that of the Sun. What can we say about the size of Betelgeuse? Using the preceding equation, we can determine the following:

$$\frac{R_{Betelgeuse}}{R_{Sun}} = \sqrt{\frac{L_{Betelgeuse}}{L_{Sun}} \times \frac{T_{Sun}^2}{T_{Betelgeuse}^2}}$$

$$\frac{R_{Betelgeuse}}{R_{Sun}} = \sqrt{\frac{140,000}{1} \times \frac{5,800^2}{3,500^2}} = 374 \times 2.7 = 1,010$$

Betelgeuse has a radius more than 1,000 times larger than that of the Sun; such stars are called *supergiants*.

smallest types of stars, called white dwarfs, have radii that are only about 1 percent of the Sun's radius—about the size of Earth. The largest stars, called red supergiants, can have radii more than 1,000 times that of the Sun. There are many more stars toward the small end of this range—smaller than the Sun—than there are giant stars.

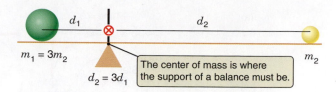

Figure 13.9 The center of mass of two objects is the "balance" point on a line joining the centers of two masses.

CHECK YOUR UNDERSTANDING 13.2

If star A has twice the surface temperature of the Sun but has the same luminosity as the Sun, the diameter of star A must be _____ the diameter of the Sun.
(a) 16 times; (b) 4 times; (c) $\frac{1}{2}$; (d) $\frac{1}{4}$

13.3 Measuring the Masses of Stars in Binary Systems

Determining the mass of a star is difficult. Astronomers cannot use the amount of light from a star or the star's size as a measure of its mass. Stars can be large or small, faint or luminous. However, more massive stars *always* have stronger gravity. When astronomers are trying to determine the masses of astronomical objects, they almost always wind up looking for the effects of gravity. In Chapter 4, you learned that Kepler's laws of planetary motion are the result of gravity, and that the properties of the orbit of a planet can be used to measure the mass of the Sun. Similarly, astronomers can study two stars that orbit each other to determine their masses.

About half of the higher-mass stars in the sky are actually systems consisting of several stars moving about under the influence of their mutual gravity. Most of these are **binary stars** in which two stars orbit each other in elliptical orbits as predicted by Newton's version of Kepler's laws. This version of Kepler's laws can be used to find the mass of a star, as we will demonstrate. However, most low-mass stars are single, and low-mass stars far outnumber higher-mass stars, so most stars are single and their mass cannot be found this way. In this section, we will look at how astronomers measure the masses of stars in binary systems.

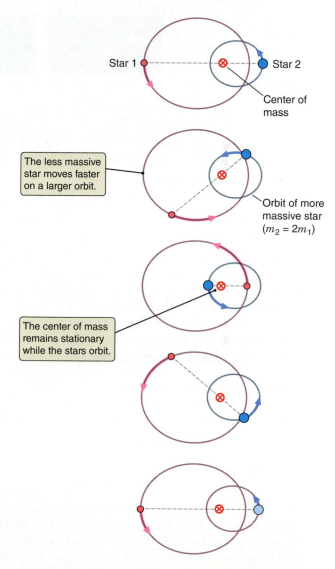

Figure 13.10 In a binary star system, the two stars orbit on elliptical paths about their common center of mass. In this case, the blue star has twice the mass of the red one. The eccentricity of the orbits is 0.5. There are equal time steps between the frames.

Binary Star Orbits

The **center of mass** is the balance point of a system. If the two objects were sitting on a seesaw in a gravitational field, the support of the seesaw would have to be directly under the center of mass for the objects to balance, as shown in **Figure 13.9**. When Newton applied his laws of motion to the problem of orbits, he found that two objects must move in elliptical orbits around each other, and that their common center of mass lies at one focus shared by both of the ellipses, as shown in **Figure 13.10**. The center of mass, which lies along the line between the two objects, remains stationary. The two objects will always be found on exactly opposite sides of the center of mass.

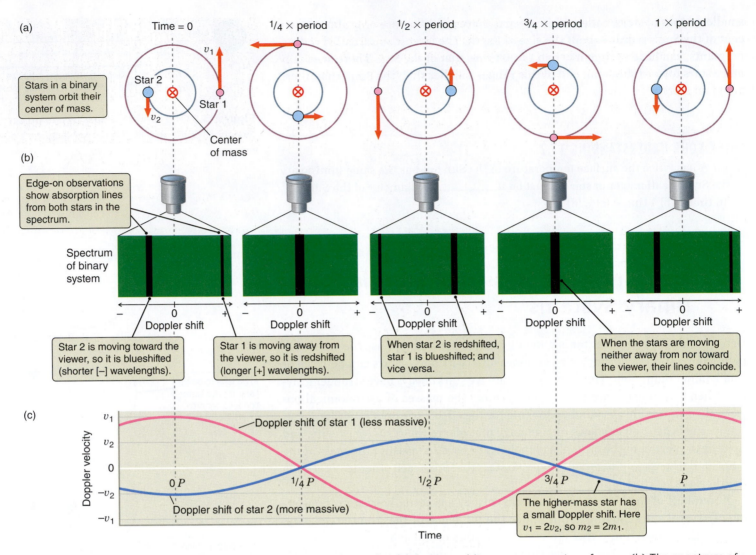

(a)

Time = 0   ¼ × period   ½ × period   ¾ × period   1 × period

Stars in a binary system orbit their center of mass.

v_1

Star 2

Star 1

Center of mass

v_2

(b)

Edge-on observations show absorption lines from both stars in the spectrum.

Spectrum of binary system

− 0 + Doppler shift

Star 2 is moving toward the viewer, so it is blueshifted (shorter [−] wavelengths).

Star 1 is moving away from the viewer, so it is redshifted (longer [+] wavelengths).

When star 2 is redshifted, star 1 is blueshifted; and vice versa.

When the stars are moving neither away from nor toward the viewer, their lines coincide.

(c)

Doppler velocity

v_1
v_2
0
$-v_2$
$-v_1$

Doppler shift of star 1 (less massive)

$0\,P$ $1/4\,P$ $1/2\,P$ $3/4\,P$ P

Doppler shift of star 2 (more massive)

The higher-mass star has a small Doppler shift. Here $v_1 = 2v_2$, so $m_2 = 2m_1$.

Time

Figure 13.11 (a) The view from "above" the binary system shows that both stars orbit a common center of mass. (b) The spectrum of the combined system (seen edge-on) shows the spectral lines of each star shift back and forth. (c) Graphing the Doppler shift of star 1 with star 2 versus time reveals that star 2 has half the maximum Doppler shift, so star 2 is twice as massive as star 1. *P* is the period of the orbit.

Nebraska Simulation: Center of Mass Simulator

Imagine that you are watching a binary star as shown in **Figure 13.11a**. As seen from above, two stars orbit the common center of mass. Star 1, which is less massive, must complete its orbit in the same time as star 2, which is more massive. Because the less massive star has farther to go around the center of mass, it must be moving *faster* than the more massive star. In this view, no determination of the Doppler shift (Chapter 5) can be made because all the motion is in the plane of the sky, and none is toward or away from the observer.

When a system is edge-on to the observer, however, the observer can take advantage of the Doppler shift to find out about the motion. Figure 13.11b shows observations of the spectrum of the combined system associated with each position in Figure 13.11a. The spectral lines of the stars shift back and forth as they move toward and away from the observer. Because the two stars are always exactly on opposite sides of their center of mass, they are always moving in opposite directions. When star 2 approaches, star 1 recedes. The light coming from star 2 will be shifted to shorter wavelengths by the Doppler effect as it approaches, so the light will be blueshifted, and the light coming from star 1 will be shifted to longer wavelengths as it recedes, so the light will be redshifted. Half an orbital period later, the situation is reversed: lines from star 1 are blueshifted, and lines from star 2 are redshifted.

The less massive star has a larger orbit—and consequently moves more quickly—than the more massive star. Figure 13.11c compares the velocity obtained from the maximum Doppler shift for star 1 with the velocity obtained from maximum Doppler shift for star 2. This comparison gives the ratio of the masses of the two stars:

$$\frac{v_1}{v_2} = \frac{m_2}{m_1}$$

By observing the spectrum of the system, we can find from this equation the relative masses of the two stars; star 2 is two times as massive as star 1. But we can't find the actual mass of either star from these observations alone.

The Masses of Binary Stars

In Chapter 4, we ignored the complexity of the motion of two objects around their common center of mass. Now, however, this very complexity enables us to measure the masses of the two stars in a binary system. If we can measure the period of the binary system and the average separation between the two stars, then Kepler's third law gives us the total mass in the system: the sum of the two masses. Because the analysis in the previous subsection gives us the ratio of the two masses, we now have two different relationships between two different unknowns. We have all we need to determine the mass of each star separately. In other words, if we know that star 2 is 2 times as massive as star 1, and we know that star 1 and star 2 together are 3 times as massive as the Sun, then we can calculate separate values for the masses of star 1 and star 2.

Depending on the type of system, there are several ways to measure the average separation and the orbital period. In a **visual binary** system, the system is close enough to Earth, and the stars are far enough from each other, that we can take pictures that show the two stars separately (**Figure 13.12**). Then, astronomers can directly measure the shapes and periods of the orbits of the two stars just by watching them as they orbit each other. These can be used with Doppler measurements of the radial (line-of-sight) velocities of the stars to solve for the ratio of the two masses.

In most binary systems, however, the two stars are so close together and far away from us that we cannot actually see the stars separately. The identification of these stars as binary systems is more indirect and comes from observing periodic variations in the *light* from the star or from observing periodic changes in the *spectrum* of the star. If we view a binary system nearly edge-on, so that one star passes in front of the other, it is called an **eclipsing binary**. An observer will see a repeating dip in brightness as one star passes in front of (eclipses) the other. If the stars are of different temperatures, there will be a repeating pattern of a smaller dip in brightness when the hotter star eclipses the cooler one, followed by a larger dip in brightness when the cooler star eclipses the hotter one, as shown in **Figure 13.13**. The pattern of these dips also gives an estimate of the relative sizes (radii) of the two stars. This procedure for identifying binary systems is similar to the transit method for finding extrasolar planets discussed in Chapter 7, and it works only when the system is viewed nearly edge-on. The Kepler space telescope

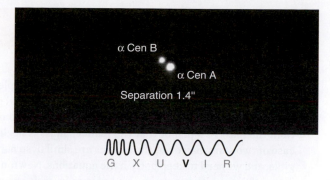

Figure 13.12 The two stars of a visual binary are resolved. These stars are two components of Alpha Centauri, the nearest star system to the Solar System.

Nebraska Simulation: Eclipsing Binary Simulator

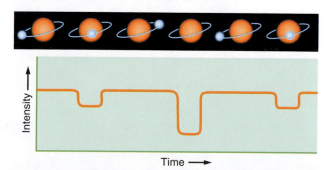

Figure 13.13 In an eclipsing binary system, the system is viewed nearly edge-on, so that the stars repeatedly pass behind one another, blocking some of the light. When the blue star passes in front of the larger, cooler star, less light is blocked than when it passes behind the red star. The shape of the dips in the light curve of an eclipsing binary can reveal information about the relative size and surface brightness of the two stars.

13.4 Working It Out Measuring the Mass of an Eclipsing Binary Pair

In Working It Out 4.3, we used Newton's version of Kepler's third law to calculate the mass of the Sun by observing the orbital period of one of its planets. In that special case, the Sun's mass is so much greater than the mass of a planet that the planet's mass is negligible. In the case of two stars, however, the masses are similar, so neither is negligible, and we need to keep both in the equations. Newton showed that if two objects with masses m_1 and m_2 are in orbit about each other, then the period of the orbit, P, is related to the average distance between the two masses, the semimajor axis A, by the equation

$$P^2 = \frac{4\pi^2 A^3}{G(m_1 + m_2)}$$

Rearranging this equation a bit turns it into an expression for the sum of the masses of the two objects:

$$m_1 + m_2 = \frac{4\pi^2}{G} \times \frac{A^3}{P^2}$$

We could use the equation this way, with the masses of the two stars in kilograms (kg), the distance between them in kilometers, the period of their orbit in seconds, and the gravitational constant G in kilometers, kilograms, and seconds. However, astronomers often think about stellar masses in units of the Sun's mass. If we divide this equation by the "mass of the Sun" equation in Working It Out 4.3,

$$M_{Sun} = \frac{4\pi^2}{G} \times \frac{A^3}{P^2}$$

(where M_{Sun} = mass of the Sun, $A = 1$ AU, and $P = 1$ year), then the constants cancel out and this equation simplifies to

$$\frac{m_1}{M_{Sun}} + \frac{m_2}{M_{Sun}} = \frac{A_{AU}^3}{P_{years}^2}$$

Therefore, if we know both m_1/m_2 from measuring velocities by Doppler shifts and $m_1 + m_2$ from the observed orbital properties, we can solve for the separate values of m_1 and m_2.

Suppose you are an astronomer studying a binary star system. After observing the star for several years, you accumulate the following information about the system:

1. The star is an eclipsing binary.

2. The period of the orbit is 2.63 years.

3. Star 1 has a Doppler velocity that varies between +20.4 and −20.4 km/s.

4. Star 2 has a Doppler velocity that varies between +6.8 and −6.8 km/s.

5. The stars are in circular orbits. You know this because the Doppler velocities about the star are symmetric; the approach and recession speeds of the star are equal.

These data are summarized in **Figure 13.14.** You begin your analysis by noting that the star is an eclipsing binary, which tells you that the orbit of the star is edge-on to your line of sight. The Doppler velocities tell you the total orbital velocity of each star, and you determine the size of the orbits using the relationship

$$\text{Distance} = \text{Speed} \times \text{Time}$$

In one orbital period, star 1 travels around a circle—a distance of

$$d = (20.4 \text{ km/s}) \times (2.63 \text{ yr}) = 53.7 \text{ km} \times \text{yr/s}$$

Multiply by the number of seconds in a year:

$$d = 53.7 \frac{\text{km} \times \text{yr}}{\text{s}} \times \frac{3.16 \times 10^7 \text{ s}}{\text{yr}} = 1.70 \times 10^9 \text{ km}$$

This distance is the circumference of the star's orbit, or 2π times the radius of the star's orbit, A_1. Thus, star 1 is following an orbit with a radius of

$$A_1 = \frac{d}{2\pi} = \frac{1.70 \times 10^9 \text{ km}}{2\pi} = 2.7 \times 10^8 \text{ km}$$

has been observing and discovering thousands of eclipsing binaries in addition to finding new extrasolar planets.

If a binary system is a **spectroscopic binary**, the spectral lines of the two stars show periodic changes as they are Doppler-shifted away from each other, first in one direction and then in the other, as shown in Figure 13.11. The period of the orbit is determined from the time it takes for a set of spectral lines to go from approaching to receding and back again. The orbital velocities of the stars and the period of the orbit give the size of the orbit because distance equals velocity multiplied by time. Consequently, astronomers can estimate the combined masses of the two stars. To calculate the individual masses, an estimate of the tilt of the orbit is needed. Thus, spectroscopic binary masses are more approximate than those in eclipsing binary systems.

To convert this to astronomical units, use the relation 1 AU = 1.50×10^8 km:

$$A_1 = 2.7 \times 10^8 \, \text{km} \times \frac{1 \, \text{AU}}{1.50 \times 10^8 \, \text{km}} = 1.8 \, \text{AU}$$

A similar analysis of star 2 shows that its orbit has a radius of $A_2 = 0.6$ AU.

Next, apply Newton's version of Kepler's third law. Because the stars are always on opposite sides of the center of mass, $A_{AU} = 1.8$ AU + 0.6 AU = 2.4 AU. Because you know A and the period P (measured as 2.63 years), you can calculate the total mass of the two stars:

$$\frac{m_1}{M_{Sun}} + \frac{m_2}{M_{Sun}} = \frac{(A_{AU})^3}{(P_{years})^2} = \frac{(2.4)^3}{(2.63)^2} = 2.0$$

So you have learned that the combined mass of the two stars is twice the mass of the Sun. To sort out the individual masses of the stars, use the measured velocities and the fact that the mass and velocity are inversely proportional:

$$\frac{m_2}{m_1} = \frac{v_1}{v_2} = \frac{20.4 \, \text{km/s}}{6.8 \, \text{km/s}} = 3.0$$

Star 2 is 3 times as massive as star 1. In mathematical terms, $m_2 = 3 \times m_1$. Substituting into the equation

$$m_1 + m_2 = 2.0 \, M_{Sun}$$

gives

$$m_1 + 3m_1 = 2.0 \, M_{Sun}$$

or $4m_1 = 2.0 \, M_{Sun}$, so $m_1 = 0.5 \, M_{Sun}$. Because $m_2 = 3 \times m_1$, then $m_2 = 1.5 \, M_{Sun}$.

Star 1 has a mass of $0.5 \, M_{Sun}$, and star 2 has a mass of $1.5 \, M_{Sun}$. You have just found the masses of two distant stars.

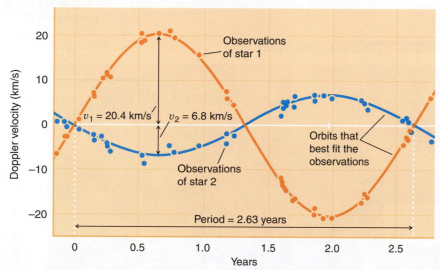

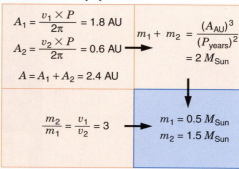

Masses are calculated from observations of stars in a binary system.

$$A_1 = \frac{v_1 \times P}{2\pi} = 1.8 \, \text{AU}$$

$$A_2 = \frac{v_2 \times P}{2\pi} = 0.6 \, \text{AU}$$

$$A = A_1 + A_2 = 2.4 \, \text{AU}$$

$$m_1 + m_2 = \frac{(A_{AU})^3}{(P_{years})^2} = 2 \, M_{Sun}$$

$$\frac{m_2}{m_1} = \frac{v_1}{v_2} = 3$$

$$m_1 = 0.5 \, M_{Sun}$$
$$m_2 = 1.5 \, M_{Sun}$$

Figure 13.14 Doppler velocities of the stars in an eclipsing binary are used to measure the masses of the stars.

A binary system can fall into more than one of these three categories, regardless of how it was originally discovered. If a spectroscopic binary system is also a visual or eclipsing binary, then the orbit and masses of the stars can be completely solved (**Working It Out 13.4**). Historically, most stellar masses were measured for stars in eclipsing binary systems, rather than for those in visual or spectroscopic binaries. But new observational capabilities have increased the number of known visual binaries by greatly improving the ability to see the stars in a binary directly. Accurate measurements of masses have been obtained for several hundred binary stars, about half of which are eclipsing binaries. The range of stellar masses found in this way is not nearly as great as the range of stellar luminosities. The least massive stars have masses of about $0.08 \, M_{Sun}$; the most massive stars appear to have masses up to about $200 \, M_{Sun}$.

CHECK YOUR UNDERSTANDING 13.3

Which of the following properties must be measured to determine the masses of stars in a typical binary system? (Choose all that apply.) (a) the period of the orbits of the two stars; (b) the average separation between the two stars; (c) the radii of the two stars; (d) the velocities of the two stars.

13.4 The Hertzsprung-Russell Diagram Is the Key to Understanding Stars

Nebraska Simulation: Hertzsprung-Russell Diagram Explorer

Measuring some basic properties of stars is only the first step in understanding; the next step is to look for patterns in their properties. In the early part of the 20th century, Ejnar Hertzsprung (1873–1967) and Henry Norris Russell (1877–1957) independently studied the properties of stars. Both astronomers plotted the

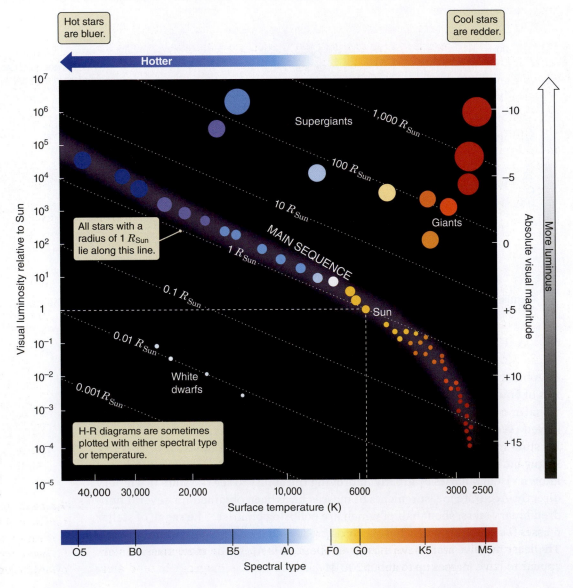

Figure 13.15 The Hertzsprung-Russell, or H-R, diagram is used to plot the properties of stars. More luminous stars are at the top of the diagram. Hotter stars are on the left. Stars of the same radius (R) lie along the dotted lines moving from upper left to lower right. Absolute magnitudes are discussed in Working It Out 13.2 and Appendix 7.

luminosities of stars versus their surface temperatures—a diagram that came to be known as the Hertzsprung-Russell diagram, or simply **H-R diagram**. We will use this diagram often for the study of stars. In this section, we take a first look at this important diagram and the way stars are organized within it.

The H-R Diagram

We begin with the layout of the H-R diagram, shown in **Figure 13.15**. The spectral type is plotted on the horizontal axis (the *x*-axis), along with the surface temperature plotted backward: temperature is higher on the left and lower on the right. Hot blue stars are on the left side of the H-R diagram; cool red stars are on the right. Temperature is plotted logarithmically. This means that the size of an interval along the axis from a point representing a star with a surface temperature of 40,000 K to one with a surface temperature of 20,000 K—a temperature change by a factor of 2—is the same as the size of an interval between points representing a star with a temperature of 10,000 K and a star with a temperature of 5000 K, which is also a temperature change by a factor of 2. The horizontal axis is sometimes labeled with another characteristic that corresponds to temperature, such as the color.

The luminosity of stars is plotted along the vertical axis (the *y*-axis). More luminous stars are toward the top of the diagram, and less luminous stars are toward the bottom. Sometimes, the luminosity axis is labeled with the absolute visual magnitude instead of luminosity, as shown on the right-hand *y*-axis. As with the temperature axis, luminosities are plotted logarithmically. In this case, each step along the left-hand *y*-axis corresponds to a multiplicative factor of 10 in the luminosity. To understand why the plotting is done this way, recall that the most luminous stars are 10 billion times more luminous than the least luminous stars, yet all of these stars must fit on the same plot.

Each point on the H-R diagram is specified by a surface temperature and luminosity. Therefore, we can use the luminosity-temperature-radius relationship described earlier in the chapter to find the radius of a star at that point as well. A star in the upper right corner of the H-R diagram is very cool, so each square meter of its surface radiates only a small amount of energy. But this star is also extremely luminous. It must be huge to account for its high luminosity, despite the feeble radiation coming from each square meter of its surface. Conversely, a star in the lower left corner of the H-R diagram is very hot, which means that a large amount of energy is coming from each square meter of its surface. However, this star has a very low overall luminosity, so it must be very small. Moving up and to the right takes you to larger and larger stars. Moving down and to the left takes you to smaller and smaller stars. All stars of the same radius lie along slanted lines across the H-R diagram. Astronomers can note the properties of a star—its temperature, color, size, and luminosity—from a glance at its position on the H-R diagram. The discovery and study of these patterns led to an understanding of the astrophysics of stars (**Process of Science Figure**).

The Main Sequence

Figure 13.16 shows 16,600 nearby stars plotted on an H-R diagram. The data are based on observations of stars near enough for parallax measurements obtained by the Hipparcos satellite. A quick look at this diagram immediately

▶‖ **AstroTour:** H-R Diagram

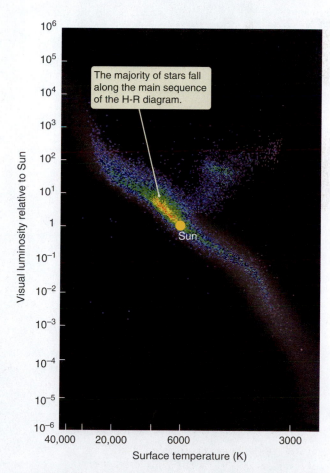

The majority of stars fall along the main sequence of the H-R diagram.

Figure 13.16 An H-R diagram for 16,600 stars plotted from data obtained by the Hipparcos satellite clearly shows the main sequence. Most of the stars lie along this band running from the lower right of the diagram toward the upper left.

Process of Science

An understanding of the meaning behind stellar data took decades, and the contributions of dozens of people, all working toward a common goal.

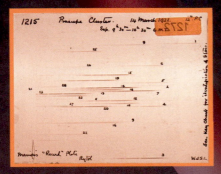

The Observations (~1880s):
About 500,000 photographs of stellar spectra are obtained, by many astronomers at many telescopes.

The Classification (~1900s):
Annie Jump Cannon leads a team that classifies all the spectra according to the strengths of particular absorption lines at particular wavelengths.

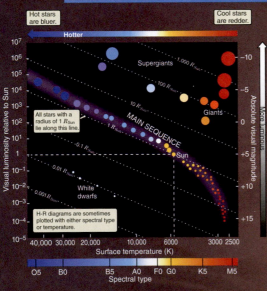

The Graph (~1910s):
Hertzsprung and Russell independently develop what will later be called the H-R diagram. They do not understand why the x-axis, when ordered O-B-A-F-G-K-M, gives such a nice band across the middle. Russell hypothesizes this must come from a single stellar characteristic.

The Understanding (~1920s):
Meghnad Saha shows that the stellar characteristic in question is temperature. Cecilia Payne-Gaposchkin shows that stars are mostly composed of hydrogen and helium. Modern astrophysics is born; others go on to develop the understanding of stellar atmospheres.

Scientific discoveries sometimes seem to occur suddenly. However, new scientific knowledge is usually the effort of many minds working for years to solve a problem.

shows what was first discovered in the original diagrams of Hertzsprung and Russell. About 90 percent of the stars in the sky lie in a well-defined region running across the H-R diagram from lower right to upper left, known as the **main sequence**. On the left end of the main sequence are the O stars: hotter, larger, and more luminous than the Sun. On the right end of the main sequence are the M stars: cooler, smaller, and fainter than the Sun. If you know where a star lies on the main sequence, then you know its approximate luminosity, surface temperature, and size.

The H-R diagram supplies a useful method for finding the distance to main-sequence stars. Astronomers can determine whether a star is on the main sequence by looking at the absorption lines in its spectrum. The spectral type is also determined from the spectral lines, and this spectral type indicates the star's temperature. Once this value on the x-axis is known, we can then read up to the main sequence and then across to the y-axis to find the star's luminosity. Recall that the luminosity, brightness, and distance are all connected. We can think about how far away a star of a particular luminosity must be to have the observed brightness. So we can find the star's distance by comparing a star's luminosity, obtained from the H-R diagram, with its apparent brightness. This method of determining distances to main-sequence stars from the spectra, luminosity, and brightness of stars is called **spectroscopic parallax**. Details of this method are discussed in Appendix 7. Despite the similarity between the names, this method is very different from the parallax method using trigonometry. Spectroscopic parallax is useful to much larger distances than trigonometric parallax, although it is less precise.

From a combination of observations of binary star masses, parallax, luminosity measurements, and mathematical models, astronomers have determined that stars of different masses lie on different parts of the main sequence. Stellar mass increases smoothly from the lower right to the upper left along the main sequence. If a main-sequence star is less massive than the Sun, it is also smaller, cooler, redder, and less luminous than the Sun, and it is located to the lower right of the Sun on the main sequence. Conversely, if a main-sequence star is more massive than the Sun, it is also larger, hotter, bluer, and more luminous than the Sun, and it is located to the upper left of the Sun on the main sequence, as illustrated in **Figure 13.17**. The mass of a star determines where on the main sequence the star will lie.

Table 13.2 summarizes the properties of the different spectral classes of main-sequence stars. *All* main-sequence stars with a mass of 1 M_{Sun} are G2 stars like the Sun and have the *same* surface temperature, size, and luminosity as the Sun. Similarly, if a main-sequence star is classified as B0, it has these properties: a surface temperature of about 30,000 K, a luminosity about 32,500 times that of the Sun, a mass of about 17.5 M_{Sun}, and a radius of about 6.7 R_{Sun}. If a different main-sequence star is classified as M5, it has a surface temperature of 3,170 K, a luminosity of about 0.008 L_{Sun}, a mass of about 0.21 M_{Sun}, and a radius of about 0.29 R_{Sun}.

The relationship between the mass and the luminosity of stars is very sensitive. Relatively small differences in the masses of stars result in large differences in their main-sequence luminosities. From determining the luminosities of binary stars with measured mass, a relationship between the mass and luminosity was found. This

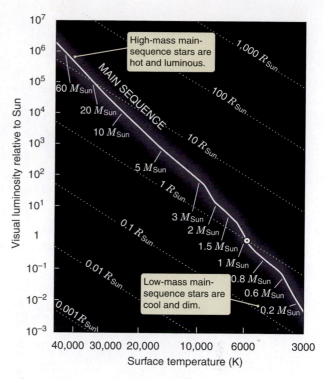

Figure 13.17 Mass determines the location of a star along the main sequence..

 Nebraska Simulation: Spectroscopic Parallax Simulator

TABLE 13.2	The Properties of Main Sequence Stars			
Spectral Type	Temperature (K)	Mass (M_{Sun})	Radius (R_{Sun})	Luminosity (L_{Sun})
O5	42,000	60	13	500,000
B0	30,000	17.5	6.7	32,500
B5	15,200	5.9	3.2	480
A0	9800	2.9	2.0	39
A5	8200	2.0	1.8	12.3
F0	7300	1.6	1.4	5.2
F5	6650	1.4	1.2	2.6
G0	5940	1.05	1.06	1.25
G2 (Sun)	5780	1.00	1.00	1.0
G5	5560	0.92	0.93	0.8
K0	5150	0.79	0.93	0.55
K5	4410	0.67	0.80	0.32
M0	3840	0.51	0.63	0.08
M5	3170	0.21	0.29	0.008

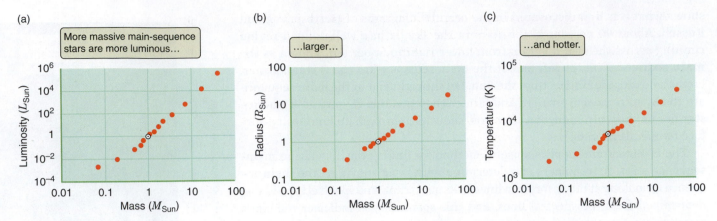

Figure 13.18 These graphs plot luminosity (a), radius (b), and temperature (c) versus mass for stars along the main sequence. The mass of a main-sequence star determines its other properties.

mass-luminosity relationship, usually expressed as $L \propto M^{3.5}$, is shown in **Figure 13.18a**. The exact exponent varies somewhat for different ranges of stellar masses, but this method is useful for estimating masses of single stars. The mass also correlates to the size of a star, shown in Figure 13.18b, and to the temperature of a star, shown in Figure 13.18c.

The mass and chemical composition of a main-sequence star determine its other characteristics: how large it is, what its surface temperature is, how luminous it is, what its internal structure is, how long it will live, how it will evolve, and what its final fate will be. A star must have a balance between gravity trying to hold the star together and the energy released by nuclear reactions in the interior of the star trying to blow it apart. The mass of the star determines the strength of its gravity, which in turn determines how much energy must be generated in its interior to prevent it from collapsing under its own weight. The mass of a star determines where the balance is struck.

Stars Not on the Main Sequence

Although 90 percent of stars are main-sequence stars, some stars are found in the upper right portion of the H-R diagram, well above the main sequence (see Figure 13.15). So they must be luminous, cool, and large, with radii hundreds or thousands of times the radius of the Sun. These stars are called giants. At the other extreme are stars found in the far lower left corner of the H-R diagram. These stars are the tiny white dwarfs, comparable to the size of Earth. Their small surface areas explain why they have such low luminosities, despite having high temperatures.

Stars that lie off the main sequence on the H-R diagram can be identified by their luminosities (determined by their distance) or by slight differences in their spectral lines. The width of a star's spectral lines is an indicator of the density and surface pressure of gas in the star's atmosphere. In general, denser stars have broader lines. Puffed-up stars above the main sequence have lower densities and lower surface pressure and narrower absorption lines compared to main-sequence stars.

When using the H-R diagram to estimate the distance to a star by the spectroscopic parallax method, astronomers must know whether the star is on, above, or below the main sequence in order to find the star's luminosity. The spectral line widths of stars both on and off the main sequence indicate **luminosity class**, which tells us the relative *size* of the star within each spectral class. Supergiant stars, which are the largest stars that we see, are luminosity class I, bright giants are class II, giants are class III, subgiants are class IV, main-sequence stars are class V, and white dwarfs are class WD. Luminosity classes I–IV lie above the main sequence, while class WD falls below and to the left of the main sequence, as shown in **Figure 13.19**. Thus, the complete spectral classification of a star includes both its spectral type (which indicates temperature and color) and its luminosity class (which indicates relative size).

The existence of the main sequence, together with the fact that the mass of a main-sequence star determines where on the sequence it will lie, is a grand pattern that points to the possibility of a deep understanding of stars. The existence of stars that do *not* follow this pattern raises yet more questions. In the coming chapters, you will learn that the main sequence tells us what stars are and how they work, and that stars off the main sequence reveal how stars form, how they evolve, and how they die. **Table 13.3** summarizes the techniques that astronomers use to determine some of the basic properties of stars. Of the properties listed in the table, only temperature, distance, and composition can be *measured*. Luminosity must be *inferred* from the H-R diagram or calculated from distance and brightness, and size and mass must be *calculated*. Other properties that can be measured include brightness, color, spectral type, and parallax shift.

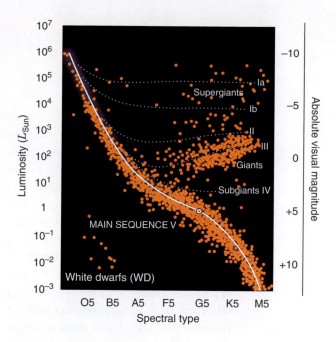

Figure 13.19 Stellar luminosity classes indicate the size (radius) of a star at each spectral type.

CHECK YOUR UNDERSTANDING 13.4

Choose the two qualities that describe a star located in the lower right of the H-R diagram: (a) hot; (b) cold; (c) high luminosity; (d) low luminosity.

TABLE 13.3	Taking the Measure of Stars

Property	Methods
Luminosity	• For a star with a known distance, measure the brightness, then apply the inverse square law of radiation: $$\text{Luminosity} = 4\pi \times \text{Distance}^2 \times \text{Brightness}$$ • For a star without a known distance, take a spectrum of the star to determine its spectral and luminosity classes, plot them on an H-R diagram, and read the luminosity from the diagram.
Temperature	• Measure the color index of the star using blue and visual filters. Use Wien's law to relate the color to a temperature. • Take a spectrum of the star, and estimate the temperature from its spectral class by noting which spectral lines are present.
Distance	• For a relatively nearby star (within a few hundred parsecs), measure the parallax shift of the star over the course of the year. • For a more distant star, find the luminosity using the H-R diagram as noted earlier, and then use the spectroscopic parallax method to relate luminosity, distance, and brightness.
Size	• For a few of the largest and closest stars, measure the size directly or by the length of eclipse in eclipsing binary stars. • From the width of the star's spectral lines, estimate the luminosity class (supergiant, giant, or main sequence). • For a star with known luminosity and temperature, use the Stefan-Boltzmann law to calculate the star's radius (the luminosity-temperature-radius relationship).
Mass	• Measure the motions of the stars in a binary system, and use these to determine the orbits of the stars, then apply Newton's form of Kepler's third law. • For a non-binary star, use the mass-luminosity relationship to estimate the mass from the luminosity.
Composition	• Analyze the lines in the star's spectrum to measure chemical composition.

Origins

Habitable Zones

How might a basic property of a star, such as its luminosity, color, mass, or surface temperature, affect the chance of there being a planet with life in orbit around that star? The only known life is that on planet Earth, where liquid water was essential for its formation and evolution. Whether liquid water is an absolute requirement for life elsewhere is not known, but the presence of water is a good starting point for determining where to look. So astronomers look for planets that are at the right distance from their stars to have a planetary temperature that permits water to exist in a liquid state on their surfaces—a range of distances known as the **habitable zone**. On planets that lie inside the habitable zone, water would exist only as a vapor—if at all. On planets that lie outside the habitable zone, water would be permanently frozen as ice.

Recall from the Chapter 5 "Origins" and Working It Out 5.4 that an important factor for estimating the temperature of a planet is the brightness of the sunlight that falls on that planet. This factor depends on the luminosity of the star and the planet's distance from the star. In the Solar System, the habitable zone ranges from about ~0.9 to ~1.4 AU, which includes Earth but just misses Venus and Mars. Main-sequence stars that are less luminous than the Sun are cooler and have narrower habitable zones, minimizing the chance that a planet will form within that slender zone. Main-sequence stars that are more massive than the Sun are hotter and have larger habitable zones. **Figure 13.20** illustrates these zones around Sun-like, hotter, and cooler stars.

In the past few years, astronomers have started to find planets in the habitable zones of their respective stars. Methods of planet detection, as discussed in Chapter 7, work best when the planet is close to its star. Using the transit method, the Kepler Mission has identified and confirmed 1–2 dozen planets in habitable zones as of this writing and has found more candidate planets that need to be confirmed.

The distance from a star at a certain temperature is not the only consideration for whether a planet has water. The presence of a planetary atmosphere is also a factor. More massive planets can retain their atmospheres, which can trap heat and raise the planet's temperature, as we saw in Chapter 5 for Venus and Earth. Smaller planets have a lower gravitational pull and may not be able to keep an atmosphere. Additionally, some habitable zones may be near planets, not stars. Some of the giant planets in the cold outer part of our own Solar System have moons with liquid water. The heat keeping the water liquid comes from the nearby planet, not from the Sun.

Finally, we note that *habitable* does not mean *inhabited*; it only means the planet is at the right distance from its star that it could have liquid water. Identifying planets in their habitable zone is a first step to selecting which planets are most interesting for further study.

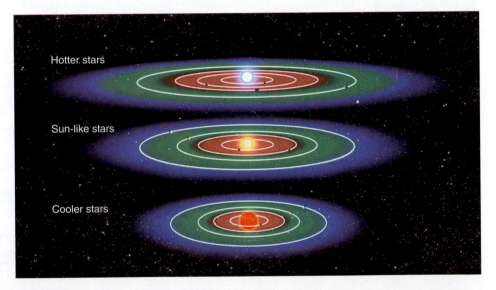

Figure 13.20 The distance and extent of a habitable zone (green) surrounding a star depends on the star's temperature. Regions too close to the star are too hot (red) and those too far away are too cold (blue) to permit the existence of liquid water. The orbits of Mercury, Venus, Earth, and Mars have been drawn around these stars for scale.

NASA reports on a new technique for using the Hubble Space Telescope to measure parallax angles.

NASA's Hubble Extends Stellar Tape Measure 10 Times Farther into Space

NASA Press Release

Using NASA's Hubble Space Telescope, astronomers now can precisely measure the distance of stars up to 10,000 light-years away—10 times farther than previously possible.

Astronomers have developed yet another novel way to use the 24-year-old space telescope by employing a technique called spatial scanning, which dramatically improves Hubble's accuracy for making angular measurements. The technique, when applied to the age-old method for gauging distances called astronomical parallax, extends Hubble's tape measure 10 times farther into space (**Figure 13.21**).

"This new capability is expected to yield new insight into the nature of dark energy, a mysterious component of space that is pushing the universe apart at an ever-faster rate," said Nobel laureate Adam Riess of the Space Telescope Science Institute (STScI) in Baltimore, Maryland.

Parallax, a trigonometric technique, is the most reliable method for making astronomical distance measurements, and a practice long employed by land surveyors here on Earth. The diameter of Earth's orbit is the base of a triangle, and the star is the apex where the triangle's sides meet. The lengths of the sides are calculated by accurately measuring the three angles of the resulting triangle.

Astronomical parallax works reliably well for stars within a few hundred light-years of Earth. For example, measurements of the distance to Alpha Centauri, the star system closest to our Sun, vary only by 1 arcsecond. This variance in distance is equal to the apparent width of a dime seen from 2 miles away.

Stars farther out have much smaller angles of apparent back-and-forth motion that are extremely difficult to measure. Astronomers have pushed to extend the parallax yardstick ever deeper into our galaxy by measuring smaller angles more accurately.

This new long-range precision was proven when scientists successfully used Hubble to measure the distance of a special class of bright stars called Cepheid variables, approximately 7,500 light-years away in the northern constellation Auriga. The technique worked so well, they are now using Hubble to measure the distances of other far-flung Cepheids.

Such measurements will be used to provide firmer footing for the so-called cosmic "distance ladder." This ladder's "bottom rung" is built on measurements to Cepheid variable stars that, because of their known brightness, have been used for more than a century to gauge the size of the observable universe. They are the first step in calibrating far more distant extra-galactic milepost markers such as Type Ia supernovae.

Riess and the Johns Hopkins University in Baltimore, Maryland, in collaboration with Stefano Casertano of STScI, developed a technique to use Hubble to make measurements as small as five-billionths of a degree.

To make a distance measurement, two exposures of the target Cepheid star were taken 6 months apart, when Earth was on opposite sides of the Sun. A very subtle shift in the star's position was measured to an accuracy of 1/1,000 the width of a single image pixel in Hubble's Wide Field Camera 3, which has 16.8 megapixels total. A third exposure was taken after another 6 months to allow for the team to subtract the effects of the subtle space motion of stars, with additional exposures used to remove other sources of error.

Riess shares the 2011 Nobel Prize in Physics with another team for his leadership in the 1998 discovery that the expansion rate of the universe is accelerating—a phenomenon widely attributed to a mysterious, unexplained dark energy filling the universe. This new high-precision distance measurement technique is enabling Riess to gauge just how much the universe is stretching. His goal is to refine estimates of the universe's expansion rate to the point where dark energy can be better characterized.

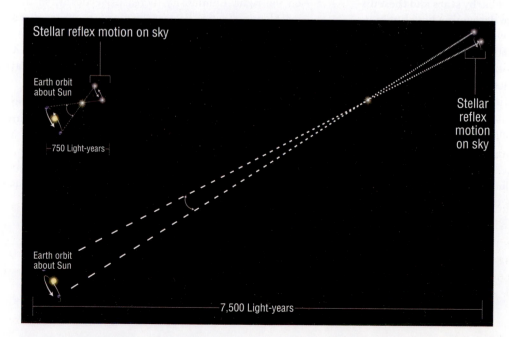

Figure 13.21 By applying a technique called spatial scanning to an age-old method for gauging distances called astronomical parallax, scientists now can use NASA's Hubble Space Telescope to make precision distance measurements 10 times farther into our galaxy than previously possible.

1. How many parsecs are in 7,500 light-years? in 10,000 light-years?
2. What is the parallax angle of a star that is 7,500 light-years away? 10,000 light-years away?
3. Why were the exposures taken 6 months apart?
4. Why is it better to make parallax measurements from space than from the ground on Earth?
5. How does improving the accuracy of the distances to nearby stars from trigonometric parallax affect astronomers' estimates of the distances to farther stars using spectroscopic parallax?

Summary

Finding the distances to stars is a difficult but important task for astronomers. Parallax and spectroscopic parallax are two of the methods that astronomers use to determine distances to stars. Brightness and distance can be used to obtain the luminosity. Careful study of the light from a star, including its spectral lines, gives the temperature, size, and composition of the star. Study of binary systems gives the masses of stars of various spectral types, which we can extend to all stars of the same spectral type. The H-R diagram shows the relationship among the various physical properties of stars. The mass of a star will be the major determining factor in its changes over time. The habitable zone is the distance from a star in which a planet could have the right temperature for liquid water to exist on its surface. Stars of different luminosities and temperatures have habitable zones of different widths at different distances from the star.

LG 1 Explain how the brightness of nearby stars and their distances from Earth are used to determine how luminous they are. The distance to a nearby star is measured by finding its parallax—by measuring how the star's apparent position changes in the sky over the course of a year. The nearest star (other than the Sun) is about 4 light-years (1.3 parsecs) away. The brightness of a star in the sky can be measured directly, and brightness and distance can be used to obtain the star's luminosity—how much light the star emits.

LG 2 Explain how astronomers obtain the temperatures, sizes, and composition of stars. The temperature of a star is determined by its color, with blue stars being hotter and red stars being cooler. The radius can be computed from the temperature and luminosity of the star. Small, cool stars greatly outnumber large, hot stars. Spectral lines carry a great deal of information about a star, including what chemical elements and molecules are present in the star.

LG 3 Describe how astronomers estimate the masses of stars. Masses of stars are measured in binary star systems by observing the effects of the gravitational pull between the stars. Newton's universal law of gravitation and Kepler's laws connect the motion of the star to the forces they experience, and thus to their masses.

LG 4 Classify stars, and organize this information on a Hertzsprung-Russell (H-R) diagram. The H-R diagram shows the relationship among the various physical properties of stars. Temperature increases to the left, so that hotter stars lie on the left side of the diagram, while cooler stars lie on the right. Luminosity increases vertically, so that the most luminous stars lie near the top of the diagram. A star's luminosity class and temperature indicate its size. The mass and composition of a main-sequence star determine its luminosity, temperature, and size. Ninety percent of stars lie along the main sequence.

LG 5 Explain how the mass and composition of a main-sequence star determine its luminosity, temperature, and size. The mass and composition of a main-sequence star determine its original position on the H-R diagram. The main sequence on the H-R diagram is actually a sequence of masses. This position connects its other properties such as its luminosity, temperature, and size.

? UNANSWERED QUESTIONS

- What is the upper limit for stellar mass? Both theory and observation have shown that there is a lower limit for stellar mass, approximately 0.08 M_{Sun} with temperature about 2000 K. However, neither theory nor observation has provided a definitive value for the upper limit. Many astronomers believe that the upper limit lies somewhere around 150–200 M_{Sun}. The very first stars that formed in the universe might have been even larger, but they are no longer around to observe.

- Are there likely to be planets with life orbiting around main-sequence stars of type M? These low-luminosity stars are the most common type of stars in the Milky Way, and the Kepler telescope has detected many planets orbiting these stars. However, the habitable zone of an M star is very close to the star, so that the planet may be strongly affected by streams of charged particles blowing off of the star. This radiation may decrease or completely strip a planet of its atmosphere, unless it has a strong protective magnetic field. Another complication is that a close-in planet is likely to be tidally locked to the star, so that one hemisphere of the planet receives light and the other hemisphere is permanently dark. This imbalance of light and heat might make the planet uninhabitable.

Questions and Problems

Test Your Understanding

1. Star A and star B are nearly the same distance from Earth. Star A is half as bright as star B. Which of the following statements must be true?
   a. Star B is farther away than star A.
   b. Star B is twice as luminous as star A.
   c. Star B is hotter than star A.
   d. Star B is larger than star A.

2. Star A and star B are two nearby stars. Star A is blue, and star B is red. Which of the following statements must be true?
   a. Star A is hotter than star B.
   b. Star A is cooler than star B.
   c. Star A is farther away than star B.
   d. Star A is more luminous than star B.

3. Star A and star B are two stars nearly the same distance from Earth. Star A is blue, and star B is red, but they have equal brightness. Which of the following statements is true?
   a. Star A is more luminous than star B.
   b. Star A is larger than star B.
   c. Star A is smaller than star B.
   d. Star A is less luminous than star B.

4. If a star has very weak hydrogen lines and is blue, what does that most likely mean?
   a. The star is too hot for hydrogen lines to form.
   b. The star has no hydrogen.
   c. The star is too cold for hydrogen lines to form.
   d. The star is moving too fast to measure the lines.

5. Star A and star B are a binary system. The Doppler shift of star A's absorption lines is 3 times the Doppler shift of star B's absorption lines. Which of the following statements is true?
   a. Star A is 3 times as massive as star B.
   b. Star A is one-third as massive as star B.
   c. Star A is closer than star B.
   d. The binary pair is moving toward Earth, but star A is farther away.

6. Star A and star B are two red stars at nearly the same distance from Earth. Star A is many times brighter than star B. Which of the following statements is true?
   a. Star A is a main-sequence star, and star B is a red giant.
   b. Star A is a red giant, and star B is a main-sequence star.
   c. Star A is hotter than star B.
   d. Star A is a white dwarf, and star B is a red giant.

7. Star A and star B are two blue stars at nearly the same distance from Earth. Star A is many times brighter than star B. Which of the following statements is true?
   a. Star A is a main-sequence star, and star B is a red giant.
   b. Star A is a main-sequence star, and star B is a blue giant.
   c. Star A is a white dwarf, and star B is a blue giant.
   d. Star A is a blue giant, and star B is a white dwarf.

8. In which region of an H-R diagram would you find the main-sequence stars with the widest habitable zones?
   a. upper left
   b. upper right
   c. center
   d. lower left
   e. lower right

9. Star A is more massive than star B. Both are main-sequence stars. Therefore, star A is _____ than star B. (Choose all that apply.)
   a. more luminous
   b. less luminous
   c. hotter
   d. colder
   e. larger
   f. smaller

10. A telescope on Mars would be able to measure the distances to more stars than can be measured from Earth because
    a. the resolution of the telescope would be better.
    b. Mars has a thin atmosphere.
    c. it would be closer to the stars.
    d. the parallax "baseline" would be longer.

11. Star A and star B are two nearby stars. Star A has a parallactic angle 4 times as large as star B's. Which of the following statements is true?
    a. Star A is one-quarter as far away as star B.
    b. Star A is 4 times as far away as star B.
    c. Star A has moved through space one-quarter as far as star B.
    d. Star A has moved through space 4 times as far as star B.

12. Star A appears twice as bright as star B, but is also twice as far away. Star A is _____ as luminous as star B.
    a. 8 times
    b. 4 times
    c. twice
    d. half

13. Table 13.1 shows two ways of reporting the amount of an element in the Sun. The percentage of hydrogen drops when changing from percentage by number of atoms to percentage by mass. But the percentage of helium grows. Why?
    a. Hydrogen is more massive than helium.
    b. Helium is more massive than hydrogen.
    c. Hydrogen is located in a different part of the Sun.
    d. It is difficult to measure the mass of hydrogen.

14. Capella (in the constellation Auriga) is the sixth brightest star in the sky. When viewed with a high-power telescope, it is clear that Capella is actually two pairs of binary stars: the first pair are G-type giants; the second pair are M-type main-sequence stars. What color does Capella appear to be?
    a. red
    b. yellow
    c. blue
    d. color cannot be determined from this information

15. An eclipsing binary system has a primary eclipse (star A is eclipsed by star B) that is deeper (more light is removed from the light curve) than the secondary eclipse (star B is eclipsed by star A). What does this tell you about stars A and B?
    a. Star A is hotter than star B.
    b. Star B is hotter than star A.
    c. Star B is larger than star A.
    d. Star B is moving faster than star A.

Thinking about the Concepts

16. The distances of nearby stars are determined by their parallaxes. Why is there greater uncertainty in the distances of stars that are farther from Earth?

17. To know certain properties of a star, you must first determine the star's distance. For other properties, knowledge of distance is not necessary. Explain why an astronomer does or does not need to know a star's distance to determine each of the following properties: size, mass, temperature, color, spectral type, and chemical composition? In each case, state your reason(s).

18. Albireo, in the constellation Cygnus, is a visual binary system whose two components can easily be seen with even a small, amateur telescope. Viewers describe the brighter star as "golden" and the fainter one as "sapphire blue."
    a. What does this description tell you about the relative temperatures of the two stars?
    b. What does this description tell you about their respective sizes?

19. Very cool stars have temperatures around 2500 K and emit Planck spectra with peak wavelengths in the red part of the spectrum. Do these stars emit any blue light? Explain your answer.

20. The stars Betelgeuse and Rigel are both in the constellation Orion. Betelgeuse appears red, and Rigel is bluish white. To the eye, the two stars seem equally bright. If you can compare the temperature, luminosity, or size from just this information, do so. If not, explain why.

21. Explain why the stellar spectral types (O, B, A, F, G, K, M) are not in alphabetical order. What sequence of temperatures is defined by these spectral types?

22. Other than the Sun, the only stars whose mass astronomers can measure *directly* are those in eclipsing or visual binary systems. Why? How do astronomers estimate the masses of stars that are not in eclipsing or visual binary systems?

23. Once the mass of a certain spectral type of star located in a binary system has been determined, it can be assumed that all other stars of the same spectral type and luminosity class have the same mass. Why is this a reasonable assumption?

24. Explain why the Kepler Mission is finding eclipsing binary stars while it is searching for extrasolar planets using the transit method.

25. Scientific advances often require the participation of scientists from all over the world, working on the same problem over many decades, even centuries. Compare and contrast this mode of "collaboration" with collaborations in your courses (perhaps on final projects or papers). What mechanisms must be in place to allow scientists to collaborate across space and time in this way?

26. What would happen to our ability to measure stellar parallax if we were on the planet Mars? What if we were on Venus or Jupiter?

27. In Figure 13.7, there is an absorption line at about 410 nm that is weak for O stars and weak for G stars but very strong in A stars. This particular line comes from the transition from the second excited state of hydrogen up to the sixth excited state. Why is this line weak in O stars? Why is it weak in G stars? Why is it strongest in the middle of the range of spectral types?

28. Which kinds of binary systems are best observed edge-on? Which kind are best observed face-on?

29. In Figure 13.10, two stars orbit a common center of mass.
    a. Explain why star 2 has a smaller orbit than star 1.
    b. Re-sketch this picture for the case where star 1 has a very low mass, perhaps close to that of a planet.
    c. Re-sketch this picture for the case where star 1 and star 2 have the same mass.

30. If our Sun were a blue main-sequence star, and Earth was still 1 AU from the Sun, would you expect Earth to be in the habitable zone? What about if our Sun were a red main-sequence star?

Applying the Concepts

31. Look at Figure 13.1b. Suppose the figure included a third star, located 4 times as far away as star A. How much less than star A would it appear to move each year? How much less than star B?

32. Suppose you see an object jump from side to side by half a degree as you blink back and forth between your eyes. How much farther away is an object that moves only one-third of a degree?

33. Logarithmic (log) plots show major steps along an axis scaled to represent equal factors, most often factors of 10. Why do astronomers sometimes use a log plot instead of the more conventional linear plot? Is the horizontal axis of the H-R diagram in Figure 13.15 logarithmic or linear?

34. Examine Figure 13.5. This figure is plotted logarithmically on both axes. The luminosities are in units of solar luminosities.
    a. How much more luminous than the Sun is a star on the far right side of the plot?
    b. How much less luminous than the Sun is a star on the far left side of the plot?

35. Study Figure 13.17. Compared to the Sun, how luminous, large, and hot is a star that has 10 times the mass of the Sun?

36. Sirius, the brightest star in the sky, has a parallax of 0.379 arcsec. What is its distance in parsecs? in light-years? How long does it take light to reach Earth?

37. Sirius is actually a binary pair of two A-type stars. The brighter of the two stars is called the "Dog Star" and the fainter is called the "Pup Star" because Sirius is in the constellation Canis Major (meaning "Big Dog"). The Dog Star appears about 6,800 times brighter than the Pup Star, even though both stars are at the same distance from Earth. Compare the temperatures, luminosities, and sizes of these two stars.

38. Sirius and its companion orbit around a common center of mass with a period of 50 years. The mass of Sirius is 2 times the mass of the Sun.
    a. If the orbital velocity of the companion is 2.35 times greater than that of Sirius, what is the mass of the companion?
    b. What is the semimajor axis of the orbit?

39. Sirius is 25 times more luminous than the Sun, and Polaris (the "North Pole Star") is 2,500 times more luminous than the Sun. Sirius appears 24 times brighter than Polaris. How much farther away is Polaris than Sirius? Use your answer from problem 36 to find the distance of Polaris in light-years.

40. Betelgeuse (in Orion) has a parallax of 0.00763±0.00164 arcsec, as measured by the Hipparcos satellite. What is the distance to Betelgeuse, and what is the uncertainty in that measurement?

41. Rigel (also in Orion) has a Hipparcos parallax of 0.00412 arcsec. Given that Betelgeuse and Rigel appear equally bright in the sky, which star is actually more luminous. Knowing that Betelgeuse appears reddish while Rigel appears bluish white, which star would you say is larger and why?

42. The Sun is about 16 trillion (1.6×10^{13}) times brighter than the faintest stars visible to the naked eye.
    a. How far away (in astronomical units) would an identical solar-type star be if it were just barely visible to the naked eye?
    b. What would be its distance in light-years?

43. Study Figure 13.9. If $m_1 = m_2$, where would the center of mass be located? If $m_1 = 2m_2$, where would the center of mass be located?

44. Find the peak wavelength of blackbody emission for a star with a temperature of about 10,000 K. In what region of the spectrum does this wavelength fall? What color is this star?

45. About 1,470 watts (W) of solar energy hits each square meter of Earth's surface. Use this value and the distance to the Sun to calculate the Sun's luminosity.

USING THE WEB

46. Go to the European Space Agency's Gaia mission website (http://esa.int/science/gaia). How will it help astronomers determine the distances to more stars? Why is it better to make parallax measurements from space than from the ground on Earth? Have any data been released?

47. Go to the "Eclipsing Binary Stars Lab" website (http://astro.unl.edu/naap/ebs/ebs.html). Click on "Eclipsing Binary Simulator." Select preset Example 1, in which the two stars are identical. The animation will run with inclination 90° and show a 50 percent eclipse. What happens when you slowly change your viewing angle to the system—the inclination. How does this change the eclipse? At what value of inclination do you no longer see eclipses? What does the system look like at 0°? Reset the inclination to 90° and adjust the separation of the two stars. How does the light curve change when the separation is larger or smaller? Now make the two stars different. Change star 2 so that its radius is 3.0 R_{Sun} and its temperature is 4000 K. At what value of inclination do you no longer see eclipses? What types of eclipsing binary systems do you think are the easiest to detect?

48. Go to the Kepler home page (http://kepler.nasa.gov) and mouse over "Confirmed Planets" on the upper right. How many eclipsing binary stars has Kepler found? Go to the Kepler Eclipsing Binary Catalog (http://keplerebs.villanova.edu) to see what new observations look like. Pick a few stars to study. What is the inclination ("sin i")? Look at the last 2 columns ("Figures"). The "raw" and "dtr" figures are rough, but the "pf" figure shows a familiar light curve. How deep is the eclipse; that is, how much lower is the "normalized flux" during maximum eclipse?

49. Do a search for a photograph of your favorite constellation (or go outside and take a picture yourself). Can you see different colors in the stars? What do the colors tell you about the surface temperatures of the stars? From your photograph, can you tell which are the three brightest stars in the constellation? These stars will be named "alpha" (α), "beta" (β), and "gamma" (γ) for that constellation. Look up the constellation online and see if you chose the right stars. What are their temperatures and luminosities? What are their distances?

50. Citizen science: Go to the website for the Stellar Classification Online Public Exploration (SCOPE) project (http://scope.pari.edu/takepart.php). This project uses crowdsourcing to classify stars seen on old photographic plates of photographs taken in the Southern Hemisphere. Create an account, review the science and the FAQ, and then click on "To Take Part" to see some practice examples. Then go to "Classify," choose a photographic plate, and classify a few stars.

smartwork

If your instructor assigns homework in SmartWork, access your assignments at digital.wwnorton.com/astro5.

digital.wwnorton.com/astro5

Open the "HR Explorer" interactive simulation for this chapter at the Student Site at the Digital Landing Page. This simulation enables you to compare stars on the H-R diagram in two ways. You can compare an individual star (marked by a red *X*) to the Sun by varying its properties in the box in the left half of the window. Or you can compare groups of the nearest and brightest stars. Play around with the controls for a few minutes to familiarize yourself with the simulation.

Begin by exploring how changes to the properties of the individual star change its location on the H-R diagram. First, press the "Reset" button at the top right of the window.

Decrease the temperature of the star by dragging the temperature slider to the left. Notice that the luminosity remains the same. Because the temperature has decreased, each square meter of star surface must be emitting less light. What other property of the star changes in order to keep the total luminosity of the star constant?

Predict what will happen when you slide the temperature slider all the way to the right. Now do it. Did the star behave as you expected?

1 As you move to the left across the H-R diagram, what happens to the radius?

_____

_____

2 What happens as you move to the right?

_____

_____

Press "Reset" and experiment with the luminosity slider.

3 As you move up on the H-R diagram, what happens to the radius?

_____

_____

4 What happens as you move down?

_____

_____

Press "Reset" again and predict how you would have to move the slider bars to move your star into the red giant portion of the H-R diagram (upper right). Adjust the slider bars until the star is in that area. Were you correct?

5 How would you adjust the slider bars to move the star into the white dwarf area of the H-R diagram?

_____

_____

Press the "Reset" button and explore the right-hand side of the window. Add the nearest stars to the graph by clicking their radio button under "Plotted Stars." Using what you learned above, compare the temperatures and luminosities of these stars to the Sun (marked by the *X*).

6 Are the nearest stars generally hotter or cooler than the Sun?

_____

7 Are the nearest stars generally more or less luminous than the Sun?

_____

Press the radio button for the brightest stars. This action will remove the nearest stars and add the brightest stars in the sky to the plot. Compare these stars to the Sun.

8 Are the brightest stars generally hotter or cooler than the Sun?

_____

9 Are the brightest stars generally more or less luminous than the Sun?

_____

10 How do the temperatures and luminosities of the brightest stars in the sky compare to the temperatures and luminosities of the nearest stars? Does this information support the claim in the chapter that there are more low-luminosity stars than high-luminosity stars? Explain.

_____

_____

_____

_____

14 Our Star— The Sun

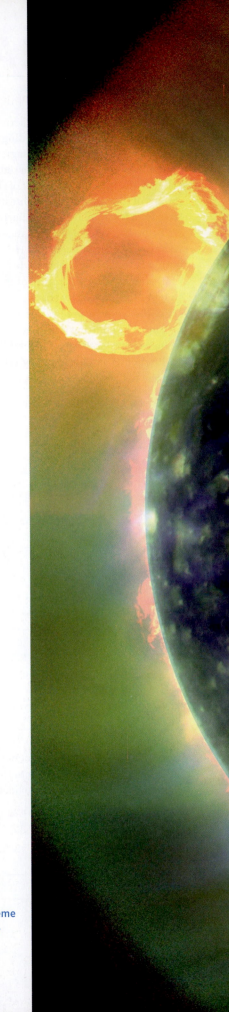

Because the Sun is the only star close to Earth, much of the detailed knowledge about stars has come from studying the Sun. In Chapter 13, we looked at the physical properties of distant stars, including their mass, luminosity, size, temperature, and chemical composition. In this chapter, we ask fundamental questions about Earth's local star. How does the Sun work? Where does it get its energy? How has its luminosity been able to remain so constant over the billions of years since the Solar System formed?

LEARNING GOALS

By the conclusion of this chapter, you should be able to:

LG 1 Describe the balance between the forces that determine the structure of the Sun.

LG 2 Explain how mass is converted into energy in the Sun's core and how long it will take the Sun to use up its fuel.

LG 3 Sketch a physical model of the Sun's interior, and list the different ways that energy moves outward from the Sun's core toward its surface.

LG 4 Describe how observations of solar neutrinos and seismic vibrations on the surface of the Sun test astronomers' models of the Sun.

LG 5 Describe the solar activity cycles of 11 and 22 years, and explain how these cycles are related to the Sun's changing magnetic field.

This image is a combination of several extreme ultraviolet images of the Sun from the Solar Dynamics Observatory. ▶ ▶ ▶

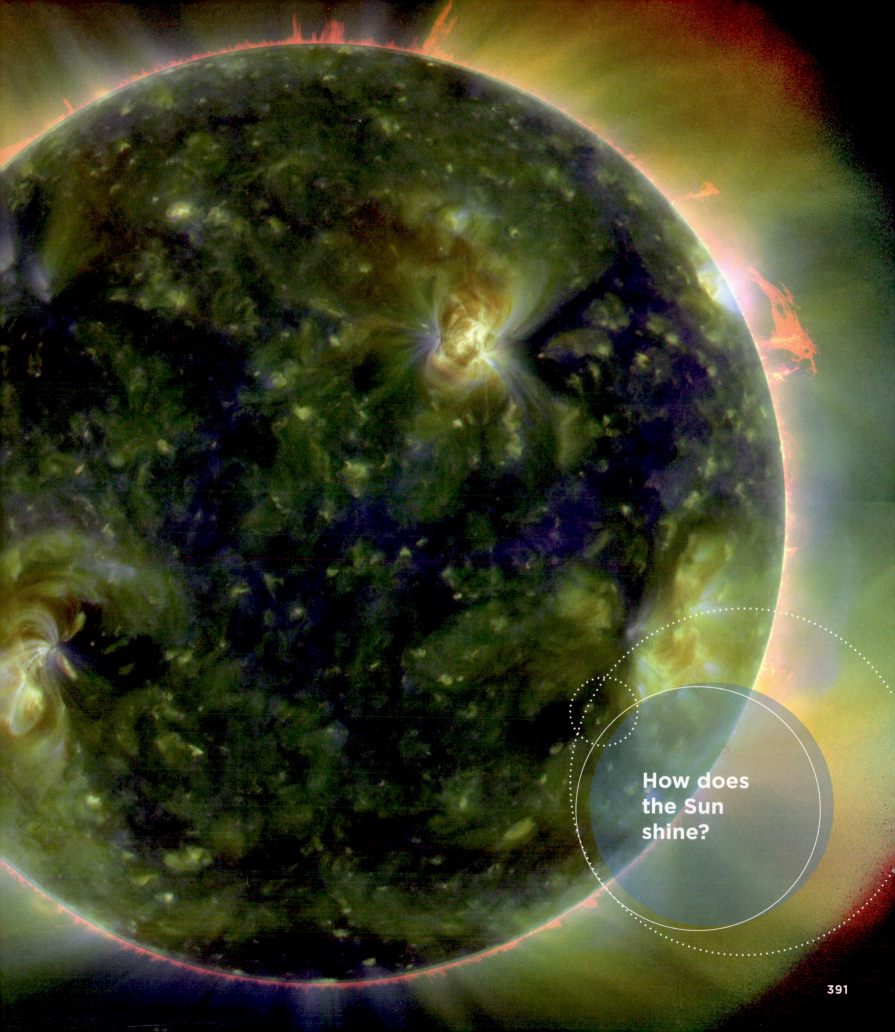

How does
the Sun
shine?

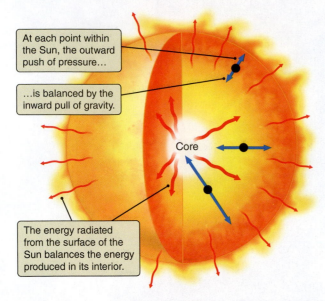

At each point within the Sun, the outward push of pressure…

…is balanced by the inward pull of gravity.

Core

The energy radiated from the surface of the Sun balances the energy produced in its interior.

Figure 14.1 The structure of the Sun is determined by the balance between the forces of pressure and gravity and the balance between the energy generated in its core and the energy radiated from its surface.

(a)

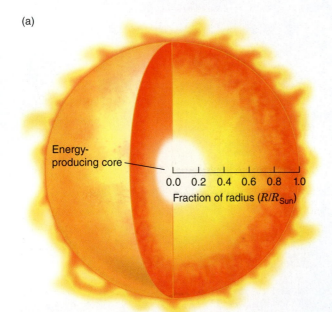

Energy-producing core

0.0 0.2 0.4 0.6 0.8 1.0
Fraction of radius (R/R_{Sun})

Figure 14.2 (a) This cutaway figure shows how the fraction of radius given in the x-axis of the graphs in (b) is measured. The energy produced by the Sun is generated in the Sun's core. (b) Pressure, density, and temperature increase toward the center of the Sun.

14.1 The Sun Is Powered by Nuclear Fusion

Energy from the Sun is responsible for daylight, for Earth's weather and seasons, and for terrestrial life itself. At a luminosity of 3.85×10^{26} watts (W), the Sun produces more energy in a second than all of the electric power plants on Earth could generate in a half-million years. In this section, we will look at energy production in the Sun.

Hydrostatic Equilibrium

Geologists learn about the interior of Earth by using a combination of physics, detailed computer models, and experiments that test the predictions of those models. The task of exploring the interior of the Sun is much the same. Like Earth's structure, the structure of the Sun is governed by a number of physical processes and relationships. Using physics, chemistry, and the properties of matter and radiation, astronomers can express these processes and relationships as mathematical equations. They then solve these equations and create a model of the Sun. One of the great successes of 20th century astronomy was the construction of a physical model of the Sun that agrees with observations of the mass, composition, size, temperature, and luminosity of the real thing.

The structure of the Sun is a matter of balance between the pressure outward and the force of gravity inward: this balance is known as **hydrostatic equilibrium**. The pressure results from energy finding its way to the surface of the Sun from deep in its interior. To understand how hydrostatic equilibrium affects the Sun, we need to know how these forces are produced and how they continually change to balance each other.

The balance between the forces due to pressure and gravity is illustrated in **Figure 14.1**. The Sun is a huge ball of hot gas. Deep in the Sun's interior, the outer layers press downward because of gravity, producing a large inward force. To maintain balance, the outward force due to pressure must be equally large. If gravity were not balanced by pressure, the Sun would collapse. If pressure were not balanced by gravity, the Sun would blow itself apart. At every point within the Sun's interior, the pressure must be just enough to hold up the weight of all the layers above that point. If the Sun were not in a stable hydrostatic equilibrium, forces within it would not be in balance, and the size of the Sun would change.

Hydrostatic equilibrium becomes an even more powerful concept when combined with the way gases behave. Deeper in the interior of the Sun, the weight of the material above becomes greater, and hence the pressure must increase. In a gas, higher pressure means higher density and/or higher temperature. **Figure 14.2** shows how conditions vary inside the Sun. As illustrated in the graphs in Figure 14.2b, calculations show that toward the center of the Sun, the pressure, the density, and the temperature of the gas increase.

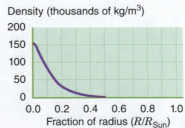

(b) Pressure (billions of atmospheres)

250
200
150
100
50
0
0.0 0.2 0.4 0.6 0.8 1.0
Fraction of radius (R/R_{Sun})

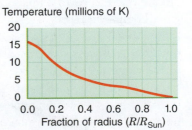

Density (thousands of kg/m³)

200
150
100
50
0
0.0 0.2 0.4 0.6 0.8 1.0
Fraction of radius (R/R_{Sun})

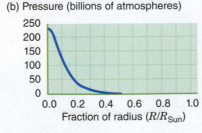

Temperature (millions of K)

20
15
10
5
0
0.0 0.2 0.4 0.6 0.8 1.0
Fraction of radius (R/R_{Sun})

CHECK YOUR UNDERSTANDING 14.1

Hydrostatic equilibrium in the Sun means that: (a) the Sun does not change over time; (b) the Sun absorbs and emits equal amounts of energy; (c) pressure balances the weight of overlying layers; (d) energy produced in the core per unit time equals energy emitted at the surface per unit time.

Nuclear Fusion

A second fundamental balance within the Sun is the balance of energy (see Figure 14.1). Stars like the Sun are remarkably stable objects. Geological records show that the luminosity of the Sun has remained nearly constant for billions of years. To remain in balance, the Sun must produce just enough energy in its interior to replace the energy radiated away from its surface. This energy balance tells us how much energy must be produced in the interior of the Sun and how that energy finds its way from the interior to the Sun's surface, where it is radiated away. Models of stellar evolution indicate that the luminosity of the Sun is increasing with time, but very, very slowly. The Sun's luminosity 4.5 billion years ago was about 70 percent of its current luminosity.

The amount of energy produced by the Sun each second is truly astronomical: 3.85×10^{26} W. One of the most basic questions facing the pioneers of stellar astrophysics was how the Sun and other stars get their energy. In the 19th century, physicists proposed that the Sun was slowly shrinking, and that the core was heating up as a result of this gravitational contraction. However, calculations soon showed that this would power the Sun for only millions of years. Geological and biological evidence available at the time suggested that Earth was tens of millions or hundreds of millions of years old. By the early 20th century, radiometric dating suggested that Earth was more than a billion years old, and therefore gravitational contraction could not be the source of the Sun's energy. In the 1930s, using theoretical and laboratory physics, nuclear physicists concluded that the Sun's energy comes from nuclear reactions at its core, capable of powering the star for billions of years.

Recall from Chapter 5 that the nucleus of most hydrogen atoms consists of a single proton. Nuclei of all other atoms are built from a mixture of protons and neutrons. Most helium nuclei, for example, consist of two protons and two neutrons. Protons have a positive electric charge, and neutrons have no electric charge. Because like charges repel, and the closer they are the stronger the force, all of the protons in an atomic nucleus are continually repelling each other with a tremendous force. The nuclei of atoms should fly apart due to electric repulsion—yet atomic nuclei are held together by the **strong nuclear force**, which overcomes this repulsion. However, the strong nuclear force acts only over very short distances, of the order 10^{-15} meter, about the size of the atomic nucleus, or about a hundred-thousandth the size of an atom.

Compared to the energy required to free an electron from an atom, the amount of energy required to tear a nucleus apart is enormous. Conversely, when an atomic nucleus (with a mass up to and including the nucleus of iron) is formed from component parts, energy is released. **Nuclear fusion**—the process of combining two less massive atomic nuclei into a single more massive atomic nucleus—occurs when atomic nuclei are brought close enough together for the strong nuclear force to overcome the force of electric repulsion, as illustrated in **Figure 14.3**. Many kinds of nuclear fusion can occur in stars. In main-sequence stars like the Sun, the primary energy generation process is the fusion of hydrogen into helium—a process often called **hydrogen burning** (even though it has nothing to

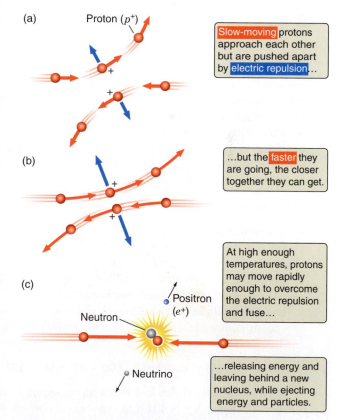

(a) Proton (p^+)

Slow-moving protons approach each other but are pushed apart by electric repulsion…

(b)

…but the faster they are going, the closer together they can get.

(c)

Positron (e^+)

Neutron

Neutrino

At high enough temperatures, protons may move rapidly enough to overcome the electric repulsion and fuse…

…releasing energy and leaving behind a new nucleus, while ejecting energy and particles.

Figure 14.3 (a) Atomic nuclei are positively charged and electrically repel each other. (b) The faster that two nuclei are moving toward each other, the closer they will get before veering away. (c) At the temperatures and densities found in the centers of stars, nuclei can overcome this electric repulsion, so fusion takes place.

do with fire or other chemical combustion). The fusion of hydrogen into helium always takes several steps, but the net result is that four hydrogen nuclei become the one helium nucleus plus energy.

The energy produced in nuclear reactions comes from the conversion of mass into energy. The exchange rate between mass and energy is given by Einstein's famous equation, $E = mc^2$, in which E is energy, m is mass, and c^2 is the speed of light squared. For any nuclear reaction, we can determine the mass that is turned into energy by calculating the mass that is lost. To find this lost mass, we subtract the mass of the outputs from the mass of the inputs. In hydrogen burning, the inputs are four hydrogen nuclei, and the output is a helium nucleus plus energy. The mass of four separate hydrogen nuclei is 1.007 times greater than the mass of a single helium nucleus; so when hydrogen fuses to make helium, 0.7 percent of the mass of the hydrogen is converted to energy.

Although each fusion reaction produces a small amount of energy, the total mass of the Sun is very large, so there is much hydrogen to "burn." When the amount of energy produced by nuclear burning is compared with the luminosity of the Sun, we see that these reactions can power the Sun for 10 billion years, a time frame that is longer than the 4.6-billion-year age of the Solar System measured from radioactive dating. Details of this calculation are provided in **Working It Out 14.1**.

14.1 Working It Out The Source of the Sun's Energy

Like all stars, the Sun's lifetime is limited by the amount of fuel available to it. We can calculate how long the Sun will live by comparing the mass involved in nuclear fusion with the amount of mass available. Converting four hydrogen nuclei (protons) into a single helium nucleus results in a loss of mass. The mass of a single hydrogen nucleus is 1.6726×10^{-27} kilogram (kg). So, four hydrogen nuclei have a mass of 4 times that, or 6.6904×10^{-27} kg. The mass of a helium nucleus is 6.6447×10^{-27} kg, which is less than the mass of the four hydrogen nuclei. The amount of mass lost, m, is

$$m = 6.6904 \times 10^{-27}\,\text{kg} - 6.6447 \times 10^{-27}\,\text{kg} = 0.0457 \times 10^{-27}\,\text{kg}$$

We can write this as 4.57×10^{-29} kg—a mass loss of about 0.7 percent. Conversion of 0.7 percent of the mass of the hydrogen into energy might not seem very efficient—until we compare it with other sources of energy and discover that it is millions of times more efficient than even the most efficient chemical reactions.

Using Einstein's equation $E = mc^2$, where c is the speed of light (3×10^8 m/s), along with the definition of a joule ($1\,\text{J} = 1\,\text{kg m}^2/\text{s}^2$), we can calculate the energy released by this mass-to-energy conversion:

$$E = mc^2 = (4.57 \times 10^{-29}\,\text{kg}) \times (3.00 \times 10^8\,\text{m/s})^2 = 4.11 \times 10^{-12}\,\text{J}$$

Each reaction that takes four hydrogen nuclei and turns them into a helium nucleus releases 4.11×10^{-12} J of energy, which doesn't seem like very much. But atoms are very small. Fusing a single kilogram of hydrogen into helium releases about 6.3×10^{14} J of energy—about the equivalent of the chemical energy released in burning 100,000 barrels of oil. To see how much the Sun must be fusing per second to produce its current luminosity, we divide the luminosity of the Sun by this amount of energy per kilogram:

$$\frac{\text{Luminosity of Sun}}{\text{Energy per kilogram}} = \frac{4 \times 10^{26}\,\text{J/s}}{6.3 \times 10^{14}\,\text{J/kg}} = 6.2 \times 10^{11}\,\text{kg/s}$$

For the Sun to produce as much energy as it does, it must convert roughly 620 billion kg of hydrogen into helium every second (and about 4 billion kg of matter is converted to energy in the process). The Sun has been burning hydrogen at this rate for at least the age of Earth and the Solar System—4.6 billion years. How much longer will the Sun last?

Astronomers estimate that only 10 percent of the Sun's total mass will ever be involved in fusion, because the other 90 percent will never get hot enough or dense enough for the strong nuclear force to make fusion happen. Ten percent of the mass of the Sun is $(0.1) \times (2 \times 10^{30})$ kg, or 2×10^{29} kg. That is the amount of "fuel" the Sun has available. The Sun consumes hydrogen at a rate of 620 billion kg/s, so each year the Sun consumes:

$$M_{\text{year}} = (6.2 \times 10^{11}\,\text{kg/s}) \times (3.16 \times 10^7\,\text{s/yr}) = 2 \times 10^{19}\,\text{kg/yr}$$

If we know how much fuel the Sun has (2×10^{29} kg), and we know how much the Sun burns each year (2×10^{19} kg/yr), then we can divide the amount by the rate to find the lifetime of the Sun:

$$\text{Lifetime} = \frac{M_{\text{fuel}}}{M_{\text{year}}} = \frac{2 \times 10^{29}\,\text{kg}}{2 \times 10^{19}\,\text{kg/yr}} = 10^{10}\,\text{yr}$$

When the Sun was formed, it had enough fuel to power it for about 10 billion years. The Sun is nearly halfway through its lifetime of hydrogen burning.

Energy is produced in the Sun's innermost region, the **core**, where the conditions are the most extreme. The density of matter in the core is about 150 times the density of water, and the temperature is about 15 million kelvins (K). Under these conditions, the atomic nuclei have tens of thousands of times more kinetic energy than that of atoms at room temperature and can slam into each other hard enough to overcome the electric repulsion, allowing the strong nuclear force to act (Figure 14.3c). In hotter and denser gases, such collisions happen more frequently. For this reason, the rate of nuclear fusion reactions is extremely sensitive to the temperature and the density of the gas, which is why these energy-producing collisions are concentrated in the Sun's core. Half of the energy produced by the Sun is generated within the inner 9 percent of the Sun's radius, or less than 0.1 percent of the volume of the Sun.

▶️II **AstroTour:** The Solar Core

The conversion of four hydrogen nuclei to one helium nucleus is the most significant source of energy in main-sequence stars. Hydrogen is the most abundant element in the universe, so it is the most abundant source of nuclear fuel at the beginning of a star's lifetime. Hydrogen is also the easiest type of atom to fuse. Hydrogen nuclei—protons—have an electric charge of +1. The electric barrier that must be overcome to fuse protons is the repulsion of one proton against another. To fuse 2 carbon nuclei together, for example, the repulsion of the six protons in one carbon nucleus pushing against the six protons in another carbon nucleus must be overcome. The repulsion between two carbon nuclei is 36 times stronger than that between two hydrogen nuclei. Therefore, hydrogen fusion occurs at a much lower temperature than any other type of nuclear fusion.

The Proton-Proton Chain

To test the theory that the Sun shines because of nuclear fusion, astronomers can analyze the predicted by-products of the nuclear reactions. In the core of the Sun and in other low-mass stars, hydrogen burning takes place in a series of nuclear reactions called the **proton-proton chain**, which has three different branches. The most important branch, responsible for about 85 percent of the energy generated in the Sun, consists of three steps, illustrated in **Figure 14.4**. Each step produces particles and/or energy in the form of light. We will begin by following the creation of the helium nucleus, and then go back to find out what happens to the other products of the reaction.

Follow along in Figure 14.4 as we step through the proton-proton chain. The nucleus of hydrogen consists of one proton. In the first step, two protons fuse. During this process, one of the protons is transformed into a neutron. To conserve spin and charge, two particles are emitted: a positively charged particle called a **positron** and a neutral particle called a **neutrino**. Energy is also emitted in the form of photons carrying electromagnetic radiation. The new atomic nucleus formed by the first

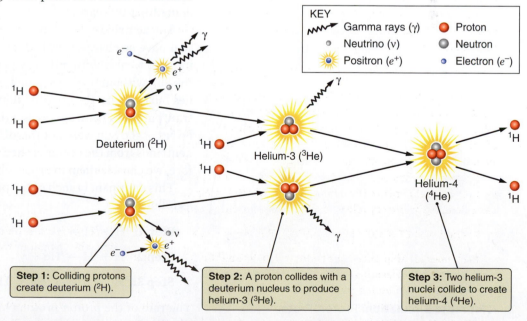

Figure 14.4 The Sun and all other main-sequence stars get their energy by fusing the nuclei of four hydrogen atoms together to make a single helium atom. In the Sun, about 85 percent of the energy produced comes from the branch of the proton-proton chain illustrated here.

step in the chain consists of a proton and a neutron. Recall from Chapter 5 that an *isotope* of an element has the same number of protons and a different number of neutrons. Thus, the new atomic nucleus is still hydrogen because it has only one proton. This particular isotope of hydrogen is common enough that it has its own name—deuterium, written as ^{2}H (H is the element symbol for hydrogen, which always has one proton, and 2 is the *atomic mass number*—the total number of protons and neutrons in the nucleus).

In the second step of the proton-proton chain, another proton slams into the deuterium nucleus, forming the nucleus of an isotope of helium, ^{3}He, consisting of two protons and a neutron. The energy released in this step is carried away as a highly energetic gamma-ray photon. Notice that the first two steps are shown twice, along the top of the figure and the bottom, because these steps must occur twice to produce a single ^{4}He nucleus.

In the third and final step of the proton-proton chain, two ^{3}He nuclei collide and fuse, producing an ordinary ^{4}He nucleus and ejecting two protons in the process. The energy released in this step is the kinetic energy of the helium nucleus and two ejected protons. Overall, four hydrogen nuclei have combined to form one helium nucleus.

Now let's go back and look at what happens to the other products of the reaction. In step 1, a positron—a particle of antimatter—is produced. **Antimatter** particles have the same mass as a corresponding matter particle but have opposite values of other properties, such as charge. The positron (e^+) is the antimatter counterpart of an electron (e^-). When matter (electrons) and antimatter (positrons) meet, they annihilate each other, and their total mass is converted to energy in the form of gamma-ray photons (γ). This happens to the emitted positrons inside the Sun, and the emitted photons from the annihilation carry away part of the energy released when the two protons fused. These photons heat the surrounding gas. The gamma rays emitted in step 2 similarly heat the gas. The thermal energy produced in the core of the Sun takes 100,000 years to find its way to the Sun's surface, and so the light we see from the Sun indicates what the Sun was doing 100,000 years ago.

The neutrino emitted in step 1 has a very different fate. Neutrinos are particles that have no charge, very little mass, and travel at nearly the speed of light. They interact weakly with ordinary matter, so weakly that the neutrino escapes from the Sun without further interactions with any other particles. The core of the Sun lies buried beneath 700,000 kilometers (km) of dense, hot matter, yet the Sun is transparent to neutrinos—essentially all of them travel into space as if the outer layers of the Sun were not there. Because they travel at nearly the speed of light, neutrinos from the center of the Sun arrive at Earth after only $8\frac{1}{3}$ minutes. Therefore, we can use them to probe what the Sun is doing today.

This dominant branch of the proton-proton chain can be written symbolically as follows:

Step 1: ^{1}H + ^{1}H $\rightarrow$ ^{2}H + e^+ + ν and then e^+ + e^- $\rightarrow$ γ + γ

Step 2: ^{2}H + ^{1}H $\rightarrow$ ^{3}He + γ

Step 3: ^{3}He + ^{3}He $\rightarrow$ ^{4}He + ^{1}H + ^{1}H

Nebraska Simulation: Proton-Proton Animation

The rate of the proton-proton chain reaction depends on both temperature and density. At the temperature and pressure that exist within the Sun's core, the reaction rate is relatively slow—in fact, extremely slow compared to a nuclear bomb explosion. The Sun's slow nuclear fusion is fortunate for life on Earth: if its hydrogen burned quickly, the Sun would have exhausted its supply long ago, and life might not have had time to evolve.

The other 15 percent of the Sun's energy is generated by variations of the proton-proton chain. The most common variation happens in step 3, where ^{3}He fuses with an existing ^{4}He to create beryllium (^{7}Be), which decays to lithium (^{7}Li) and energy, and then the ^{7}Li plus one ^{1}H become two ^{4}He. In a less common variation, the beryllium combines with hydrogen to become boron (^{8}B), which then decays to beryllium and then to two ^{4}He. In both of these variations, ultimately four hydrogen nuclei become one helium nucleus.

CHECK YOUR UNDERSTANDING 14.2

When hydrogen is fused into helium, energy is released from: (a) gravitational collapse; (b) conversion of mass to energy; (c) the increase in pressure; (d) the decrease in the gravitational field.

14.2 Energy Is Transferred from the Interior of the Sun

Although geologists cannot travel deep inside Earth to find out how it is structured, they are able to build a model of its interior using data on how seismic waves travel during earthquakes. Similarly, astronomers can create a model of the Sun's interior using their knowledge of the balance of forces and energy within the Sun and an understanding of how energy moves from one place to another. These models can be tested by observations of waves traveling through the Sun and by studying neutrinos from the Sun.

Energy Transport

Some of the energy released by hydrogen burning in the core of the Sun escapes directly into space in the form of neutrinos. However, most of the energy heats the solar interior and then moves outward through the Sun to the surface, a process known as **energy transport**. Energy transport, a key determinant of the Sun's structure, can occur by *conduction*, *convection*, or *radiation*.

Conduction is important primarily in solids. For example, when you pick up a hot object, your fingers are heated by conduction. This happens because energetic thermal vibrations of atoms and molecules cause neighboring atoms and molecules to vibrate more rapidly as well. Conduction is typically ineffective in a gas because the atoms and molecules are too far apart to transmit vibrations to one another efficiently. Conduction does not play a key role in the transport of energy from the core of the Sun to its surface, but it will be relevant later when we discuss dying stars.

In the Sun, energy is transported by convection and radiation through different zones, as shown in **Figure 14.5**. The mechanism of energy transport from the center of the Sun outward depends on the decreasing temperature and density as the radius increases. First, energy moves outward through the inner layers of the

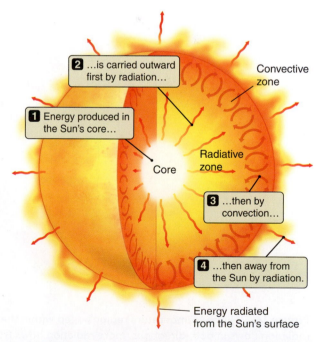

1 Energy produced in the Sun's core…

2 …is carried outward first by radiation…

Convective zone

Core

Radiative zone

3 …then by convection…

4 …then away from the Sun by radiation.

Energy radiated from the Sun's surface

Figure 14.5 The interior structure of the Sun is divided into zones on the basis of where energy is produced and how it is transported outward.

Sun as radiation in the form of photons. Next, energy moves by convection in parcels of gas. Finally, energy radiates from the Sun's surface as light. We will look at each process in turn.

Near the core, **radiation** transfers energy from hotter to cooler regions via photons, which carry the energy with them. Consider a hotter region of the Sun located next to a cooler region, as shown in **Figure 14.6**. Recall from your study of radiation in Chapter 5 that the hotter region contains more (and more energetic) photons than the cooler region. More photons move from the hotter, very crowded region to the cooler, less crowded region than in the reverse direction. A net transfer of photons and photon energy occurs from the hotter region to the cooler region, and radiation carries energy outward from the Sun's core.

The transfer of energy from one point to another by radiation also depends on how freely radiation can move from one point to another within a star. The degree to which matter blocks the flow of photons through it is called **opacity**. The opacity of a material depends on many things, including the density of the material, its composition, its temperature, and the wavelength of the photons moving through it.

Energy transfer by radiation is most efficient in regions with low opacity. The radiative zone (see Figure 14.5) is the region in the inner part of the Sun where the opacity is relatively low, and radiation carries the energy produced in the core outward through the star. This radiative zone extends about 70 percent of the way out toward the surface of the Sun. Even though this region's opacity is low enough for radiation to dominate convection as an energy transport mechanism, photons still travel only a short distance within the region before being absorbed, emitted, or deflected by matter, much like a beach ball being batted about by a

Astronomy in Action: Random Walk

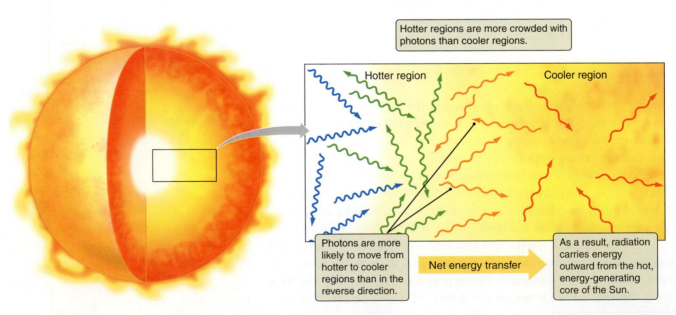

Figure 14.6 Higher-temperature regions deep within the Sun produce more radiation than do lower-temperature regions farther out. Although radiation flows in both directions, more radiation flows from the hotter regions to the cooler regions than from the cooler regions to the hotter regions. Therefore, radiation carries energy outward from the inner parts of the Sun. For simplicity, we have only included in this illustration the most common photons (those at the peak of the blackbody curve). Photons of all colors are present in all regions, and there are more of all kinds in the hotter regions and fewer of all kinds in the cooler regions.

crowd of people as illustrated in **Figure 14.7**. Each interaction sends the photon in an unpredictable direction—not necessarily toward the surface of the star. The distances between interactions are so short that, on average, it takes the energy of a gamma-ray photon produced in the interior of the Sun about 100,000 years to find its way to the outer layers of the Sun. Opacity holds energy within the interior of the Sun and lets it seep away only slowly. As it travels, the gamma-ray photon gradually becomes converted to lower-energy photons, emerging mostly as optical and infrared radiation from the surface.

From a peak of 15 million K at the center of the Sun, the temperature falls to about 100,000 K at the outer margin of the radiative zone. At this cooler temperature, the opacity is higher, so radiation is less efficient at carrying energy from one place to another. The energy that is flowing outward through the Sun "piles up" against this edge of the radiative zone.

Nearer the surface of the Sun, transfer by radiation becomes inefficient and the temperature changes quickly. Instead, **convection** takes over. Convection carries energy from the interior of a planet to its surface or from the Sun-heated surface of Earth upward through Earth's atmosphere. Convection also plays an important role in the transport of energy outward from the interior of the Sun. It transports energy by moving packets of hot gas, like hot-air balloons, which become buoyant and rise up through the lower-temperature gas above them, carrying energy with them. The solar convective zone (see Figure 14.5) extends from the outer boundary of the radiative zone outward to just below the visible surface of the Sun where evidence of convection can be seen in the bubbling surface (**Figure 14.8**).

In the outermost layers of the Sun, radiation again takes over as the primary mode of energy transport, and it is radiation that transports energy from the Sun's outermost layers off into space.

Observing Neutrinos from the Core of the Sun

The model of energy production and energy transport in the Sun discussed above correctly matches observed global properties of the Sun such as its size, temperature, and luminosity. The nuclear fusion model of the Sun predicts exactly which nuclear reactions should be occurring in the core of the Sun and at what rate. The nuclear reactions that make up the proton-proton chain produce a vast number of neutrinos. Because neutrinos barely interact with other ordinary matter, almost all of the neutrinos produced in the heart of the Sun travel freely through the outer parts of the Sun and on into space as if the outer layers of the Sun were not there. Solar neutrinos produced in the core of the Sun, traveling at nearly the speed of light, take only $8\frac{1}{3}$ minutes to reach Earth, much quicker than the 100,000-year journey of photons.

Neutrinos interact so weakly with matter that they are extremely difficult to observe. However, the extremely large number of nuclear reactions in the Sun means that the Sun produces an enormous number of neutrinos. As you read this sentence, about 400 trillion solar neutrinos pass through your body. This happens even at night, as neutrinos easily pass through Earth. With this many neutrinos about, a neutrino detector does not have to detect a very large percentage of them to be effective.

A neutrino telescope looks very different from other telescopes. The first apparatus designed to detect solar neutrinos was built 1,500 meters underground,

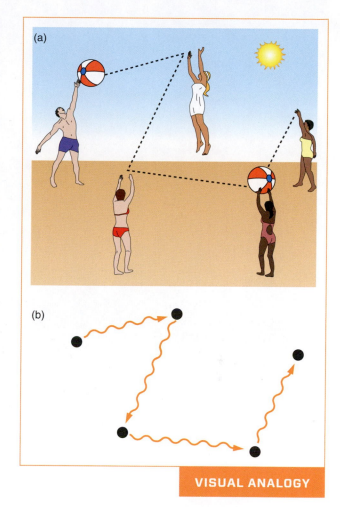

(a)

(b)

VISUAL ANALOGY

Figure 14.7 (a) When a crowd of people plays with a beach ball, the ball never travels very far before someone hits it, turning it in another direction. It often takes a ball a long time to make its way from one edge of the crowd to the other. (b) Similarly, when a photon travels through the Sun, it takes a long time for a photon to make its way out of the Sun.

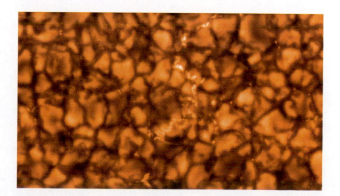

Figure 14.8 The top of the Sun's convective zone shows the bubbling of the surface caused by rising and falling packets of gas.

within the Homestake Mine in Lead, South Dakota. Astronomers filled a tank with 100,000 gallons of dry-cleaning fluid—C_2Cl_4, or perchloroethylene. Over the course of 2 days, astronomers predicted that roughly 10^{22} solar neutrinos passed through the Homestake detector. Of these, on average only one neutrino interacted with a chlorine atom within the fluid to form a radioactive isotope of argon. Over time, a measurable amount of argon was produced.

The Homestake experiment operated from the late 1960s to the early 1990s and detected this argon isotope—evidence of neutrinos from the Sun, confirming that nuclear fusion powers the Sun. However, astronomers noticed that they were measuring only one-third to one-half as many solar neutrinos as predicted by solar models. The difference between the predicted and measured number of solar neutrinos was called the **solar neutrino problem**.

One possible explanation of the solar neutrino problem was that the working model of the structure of the Sun was somehow wrong. This possibility seemed unlikely, however, because of the many other successful predictions of the solar model. A second possibility was that an understanding of the neutrino itself was incomplete. The neutrino was long thought to have zero mass, like photons, and to travel at the speed of light. But if neutrinos actually do have a tiny amount of mass, then particle physics suggests that solar neutrinos should oscillate—alternate back and forth among three different kinds of neutrinos: the electron, muon, and tau neutrinos. According to this explanation, early neutrino experiments could detect only the electron neutrino and consequently observed only about a third of the expected number of neutrinos. Since then, many other neutrino detectors have been built, each using different reactions to detect neutrinos of different energies or different types. Experiments at high-energy physics labs, nuclear reactors, and neutrino telescopes around the world have shown that neutrinos do have a nonzero mass and do oscillate among neutrino types.

Solving the solar neutrino problem is a good example of how science works—how a better model of the neutrino showed that the solar neutrino problem was real and not merely an experimental mistake, and how a single set of anomalous observations was later confirmed by other, more sophisticated experiments. All of this effort has led to a better understanding of basic physics (**Process of Science Figure**).

Probing the Sun's Interior

Models of Earth's interior predict how density and temperature change from place to place. These differences affect the seismic waves traveling through Earth, bending the paths that they travel. Geologists test models of Earth's interior by comparing measurements of seismic waves from earthquakes with model predictions of how seismic waves should travel through the planet.

Just as geologists use seismic waves from earthquakes to probe the interior of Earth, solar physicists use the surface oscillations of the Sun to probe the solar interior. The science that uses solar oscillations to study the Sun is called **helioseismology**. Detailed observations of motions of material from place to place across the surface of the Sun show that the Sun vibrates or rings, something like a bell that has been struck. Unlike a well-tuned bell—which vibrates primarily at one frequency—the vibrations of the Sun are very complex. In the Sun, many different frequencies of vibrations occur simultaneously, which cause some parts of

LEARNING FROM FAILURE
The first detections of solar neutrinos raised more questions than they answered.

The Hypothesis:
The Sun's energy comes from nuclear fusion, which produces neutrinos.

The Test:
A specific number of neutrinos must be produced each day to account for the brightness of the Sun.

The Experiment:
Homestake detects one-third as many neutrinos as predicted.

The Conclusion:
One of these things is true...

Scientists don't understand nuclear fusion.

But thousands of experiments on Earth support our understanding!

Scientists don't understand neutrinos.

New Hypothesis:
What if neutrinos come in three types and Homestake can detect only one type?

Newer laboratory and solar measurements confirmed the new hypothesis. Part of the "scientific attitude" is to find failure exciting. When experiments do not turn out as expected, good scientists get excited—there is something new to understand!

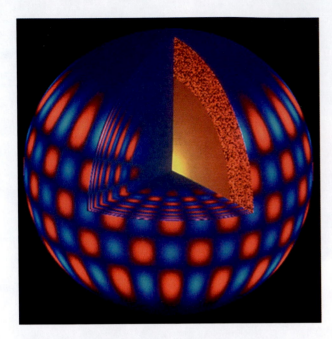

Figure 14.9 The interior of the Sun rings like a bell as helioseismic waves move through it. This figure shows one particular mode of the Sun's vibration. Red indicates regions where gas is traveling inward; blue indicates regions where gas is traveling outward. Astronomers observe these motions via Doppler shifts.

the Sun to bulge outward and some to draw inward. These motions help us to probe what lies below. **Figure 14.9** illustrates the motions of the different parts of the Sun, with red and blue areas moving in opposite directions. Some waves are amplified and some are suppressed, depending on how they overlap as they travel through the Sun. Astronomers study these waves using the Doppler effect (see Chapter 5), which distinguishes between parts of the Sun that move toward the observer and parts that move away.

To detect the disturbances of helioseismic waves on the surface of the Sun, astronomers must measure Doppler shifts of less than 0.1 m/s while detecting changes in brightness of only a few parts per million at any given location on the Sun. Tens of millions of different wave motions are possible within the Sun. Some waves travel around the circumference of the Sun, providing information about the density of the upper convection zone. Other waves travel through the interior of the Sun, revealing the density structure of the Sun close to its core. Still others travel inward toward the center of the Sun, until they are bent by the changing solar density and return to the surface.

All of these wave motions are going on at the same time, so sorting them out requires computer analysis of long, unbroken strings of solar observations from several sources. The Global Oscillation Network Group (GONG) is a network of six solar observation stations spread around the world that enables astronomers to observe the surface of the Sun approximately 90 percent of the time.

To interpret helioseismology data, scientists compare the measurements of the strength, frequency, and wavelengths of the waves against predicted vibrations calculated from models of the solar interior. This technique provides a powerful test of models of the solar interior, and it has led both to some surprises and to improvements in the models. For example, some scientists proposed that the solar neutrino problem might be solved if the models had overestimated the amount of helium in the Sun. This explanation was ruled out by analysis of the waves that penetrate to the core of the Sun. Helioseismology showed that the value for opacity used in early solar models was too low. This realization led astronomers to recalculate the location of the bottom of the convective zone. Both theory and observation now put the base of the convective zone at 70 percent of the way out from the center of the Sun, with an uncertainty in this number of less than half a percent.

Working back and forth between observation and theory has enabled astronomers to probe the otherwise inaccessible interior of the Sun. We now know that the energy is produced by nuclear fusion deep in the core and that it moves outward by radiation to a point about 70 percent of the radius of the Sun. Then it travels outward by convection to the surface. We also know how the temperature, density, and pressure change with radius and how these factors change the opacity at different distances from the center. Even though it is usually not possible to sample directly or to set up controlled experiments, this kind of collaboration between theory and observation is essential to observational sciences like astronomy.

CHECK YOUR UNDERSTANDING 14.3

How do neutrinos help us understand what is going on in the core of the Sun?
(a) Neutrinos from distant objects pass through the Sun, probing the interior.
(b) Neutrinos from the Sun pass easily through Earth. (c) Neutrinos from the interior of the Sun easily escape. (d) Neutrinos change form on their way to Earth.

14.3 The Atmosphere of the Sun

Beyond the convective zone lie the outer layers of the Sun, which are collectively known as the Sun's atmosphere. These layers, shown in **Figure 14.10a**, include the *photosphere*, the *chromosphere*, and the *corona*. We can observe these layers of the Sun directly using telescopes and satellites. Observations of the Sun's atmosphere are important because activity in the Sun's atmosphere has consequences for human infrastructure such as power grids and satellites in orbit around Earth.

The Sun is a large ball of gas, and so, unlike Earth, it has no solid surface. Its apparent surface is like a fog bank on Earth. Imagine watching some people walking into a fog bank. After they disappear from view, you would say they were inside the fog bank, even though they never passed through a definite boundary. The apparent surface of the Sun is similar. Light from the Sun's surface can escape directly into space, so we can see it. Light from below the Sun's surface cannot escape directly into space, so we cannot see it.

At the base of the atmosphere is the **photosphere**: the Sun's apparent surface. This is where features such as sunspots can be seen. Above this photosphere is the **chromosphere**, a region of strong emission lines. The top layer is the **corona**, which can be viewed during a solar eclipse as a halo around the Sun. In the Sun's atmosphere, the density of the gas drops very rapidly with increasing altitude. Figure 14.10b shows how density and temperature change across the atmosphere of the Sun. In this section, we will explore each of these layers, beginning at the bottom with the photosphere.

The Photosphere

The **effective temperature** of the photosphere is calculated from the Sun's luminosity and radius using the Stefan-Boltzmann law (see Chapter 5). The photosphere has an effective temperature of 5780 K, ranging from 6600 K to 4500 K over a 500-km-thick zone. As you can see in the graphs in Figure 14.10b, the temperature

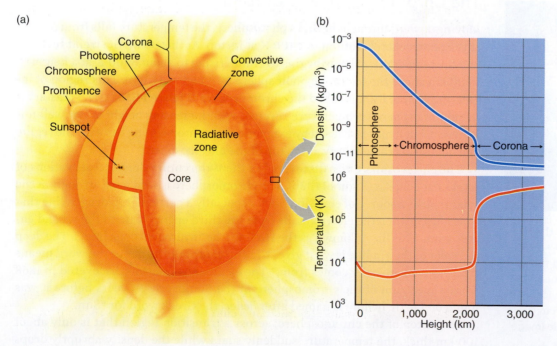

Figure 14.10 (a) The components of the Sun's atmosphere are located above the convective zone. (b) The density and temperature of the Sun's atmosphere change abruptly at the boundary between the chromosphere and corona. Note that the *y*-axes are logarithmic.

(a)   (b)

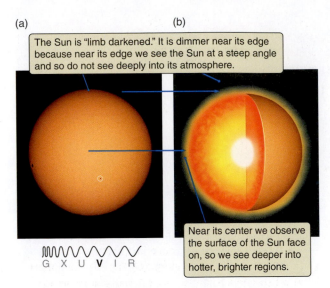

The Sun is "limb darkened." It is dimmer near its edge because near its edge we see the Sun at a steep angle and so do not see deeply into its atmosphere.

Near its center we observe the surface of the Sun face on, so we see deeper into hotter, brighter regions.

G X U V I R

Figure 14.11 (a) When viewed in visible light, the Sun appears to have a sharp outline, even though it has no true surface. The center of the Sun appears brighter, while the limb of the Sun is darker—an effect known as limb darkening. (b) Looking at the center of the Sun allows us to see deeper into the Sun's interior than we do when looking at the edge of the Sun. Because higher temperature means more luminous radiation, the center of the Sun appears brighter than its limb.

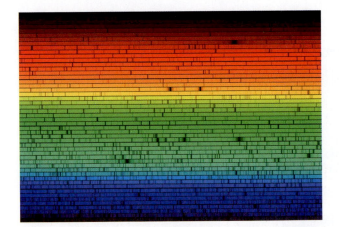

Figure 14.12 This high-resolution spectrum of the Sun stretches from 400 nanometers (nm) in the lower left corner to 700 nm in the upper right corner and shows dark absorption lines. This spectrum was produced by passing the Sun's light through a prism-like device, and then cutting and folding the single long spectrum (from blue to red) into rows so that it will fit in a single image taken by a camera.

increases sharply across the boundary between the chromosphere and the corona, while the density falls sharply across the same boundary. The Sun appears to have a well-defined surface and a sharp outline when viewed from Earth because 500 km does not look very thick when viewed from a distance of 150 million km.

In **Figure 14.11a**, the Sun appears fainter near its edges than near its center, an effect called **limb darkening**. This effect is an artifact of the structure of the Sun's photosphere. When looking near the edge of the Sun, you are looking through the photosphere at a steep angle. As a result, you do not see as deeply into the interior of the Sun as when you are looking directly down through the photosphere near the center of the Sun's disk. The light from the limb of the Sun comes from a shallower layer that is cooler and fainter, as shown in Figure 14.11b.

In the Sun's atmosphere, the density of the gas drops very rapidly with increasing altitude. All visible solar phenomena take place in the Sun's atmosphere. Most of the radiation from below the Sun's photosphere is absorbed by matter and reemitted at the photosphere as a blackbody spectrum.

As we examine the structure of the Sun in more detail, however, we see that this simple description of the spectra of stars is incomplete. Light from the solar photosphere must escape through the upper layers of the Sun's atmosphere, which affects the spectrum we observe. In Chapter 13, we discussed the presence of absorption lines in the spectra of stars. Now we can take a closer look at how these absorption lines form. As photospheric light travels upward, atoms in the solar atmosphere absorb the light at discrete wavelengths, forming absorption lines. Because the Sun appears so much brighter than any other star, its spectrum can be studied in far more detail, so specially designed telescopes and high-resolution spectrometers have been built specifically to study the Sun's light. The solar spectrum is shown in **Figure 14.12**. Absorption lines from more than 70 elements have been identified. Analysis of these lines forms the basis for much of astronomers' knowledge of the solar atmosphere, including the composition of the Sun. This is also the starting point for an understanding of the atmospheres and spectra of other stars.

The Chromosphere and Corona

Moving upward through the Sun's photosphere, the temperature falls from 6600 K at the photosphere's bottom to 4400 K at its top. At this point, the trend reverses and the temperature slowly begins to climb, rising to about 6000 K at a height of 1,500 km above the top of the photosphere (see Figure 14.10b). This region of increasing temperature is called the chromosphere (**Figure 14.13a**). The reason for the chromosphere's temperature reversal with increasing height is not well understood, but it may be caused by magnetic waves propagating through the region and depositing their energy at the top of the chromosphere.

The chromosphere was discovered in the 19th century during observations of total solar eclipses (Figure 14.13b). The chromosphere is seen most strongly at the solar limb as a source of emission lines, especially a particular hydrogen line that is produced when the electron falls from the third energy state to the second energy state. This line is known as the Hα line (the "hydrogen alpha line"). The deep red color of the Hα line is what gives the chromosphere its name; the word means "the place where color comes from." The element helium was discovered in 1868 from a spectrum of the chromosphere of the Sun nearly 30 years before it was found on Earth: helium is named after *helios*, the Greek word for "Sun."

At the top of the chromosphere, across a transition region that is only about 100 km thick, the temperature suddenly soars while the density abruptly drops

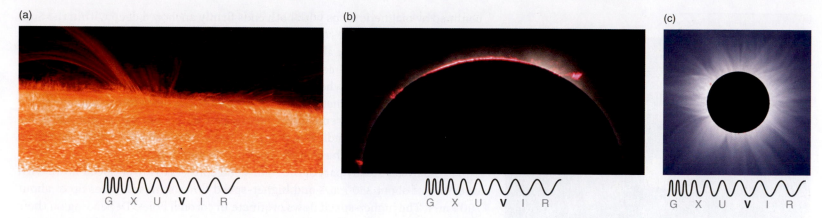

(a)

(b)

(c)

G X U V I R   G X U V I R   G X U V I R

Figure 14.13 (a) This spacecraft image of the Sun shows fine structure in the chromosphere extending outward from the photosphere. (b) The chromosphere is visible during a total eclipse. (c) This eclipse image shows the Sun's corona, consisting of million-kelvin gas that extends for millions of kilometers beyond the surface of the Sun.

(see Figure 14.10b). Above this transition lies the outermost region of the Sun's atmosphere, the corona, where temperatures reach 1 million to 2 million K. The corona is thought to be heated by magnetic fields and micro solar flares.

The Sun's corona has been known since ancient times: it is visible during total solar eclipses as an eerie outer glow stretching a distance of several solar radii beyond the Sun's surface (Figure 14.13c). Because it is so hot, the solar corona is also a strong source of X-rays, and there is so much energy in these X-ray photons that many electrons are stripped away from nuclei, leaving atoms in the corona highly ionized.

CHECK YOUR UNDERSTANDING 14.4

The surface of the Sun appears sharp in visible light because: (a) the photosphere is cooler than the layers below it; (b) the photosphere is thin compared to the other layers in the Sun; (c) the photosphere is less dense than the convection zone; (d) the Sun has a distinct surface.

14.4 The Atmosphere of the Sun Is Very Active

The atmosphere of the Sun is a very turbulent place. The best-known features on the surface of the Sun are relatively dark blemishes in the solar photosphere, called **sunspots**. Sunspots come and go over time, though they remain long enough for us to determine the rotation rate of the Sun. These spots are associated with active regions: loops of material and explosions that fling particles far out into the Solar System. Long-term patterns have been observed in the variations of sunspots and active regions, revealing that the magnetic field of the Sun is constantly changing.

Solar Activity Is Caused by Magnetic Effects

The magnetic field (see Chapter 5) of the Sun causes virtually all of the structure seen in the Sun's atmosphere. High-resolution images of the Sun show *coronal loops* that make up much of the Sun's lower corona (**Figure 14.14**). This texture is the result of magnetic structures called flux tubes. Magnetic fields are responsible for much of the structure of the corona as well. The corona is far too hot to be held in by the Sun's gravity, but over most of the surface of the Sun, coronal gas is

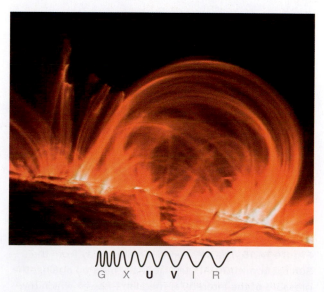

G X U V I R

Figure 14.14 This close-up image of the Sun shows the tangled structure of coronal loops.

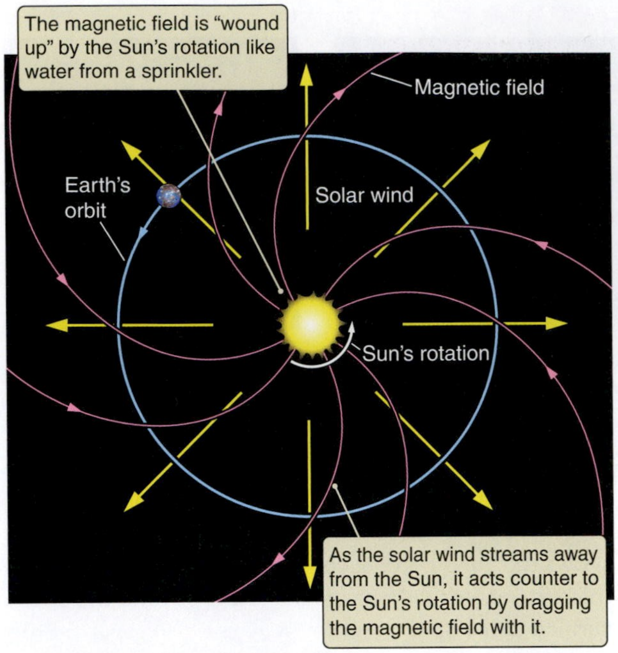

The magnetic field is "wound up" by the Sun's rotation like water from a sprinkler.

Magnetic field

Earth's orbit

Solar wind

Sun's rotation

As the solar wind streams away from the Sun, it acts counter to the Sun's rotation by dragging the magnetic field with it.

Figure 14.15 The solar wind streams away from active areas and coronal holes on the Sun. As the Sun rotates, the solar wind takes on a spiral structure, much like the spiral of water that streams away from a rotating lawn sprinkler.

confined by magnetic loops with both ends firmly anchored deep within the Sun. The magnetic field in the corona acts almost like a network of rubber bands that coronal gas is free to slide along but cannot cross. In contrast, about 20 percent of the surface of the Sun is covered by an ever-shifting pattern of **coronal holes**, which are large regions where the magnetic field points outward, away from the Sun, and where coronal material is free to stream away into interplanetary space as the solar wind. In extreme ultraviolet images of the Sun we see coronal holes as dark regions, which indicates that they are cooler and lower in density than their surroundings (see the chapter-opening photograph).

The relatively steady part of the solar wind consists of lower-speed flows with velocities of about 350 km/s and higher-speed flows with velocities up to about 700 km/s. The higher-speed flows originate in coronal holes. Depending on their speed, particles in the solar wind take about 2–5 days to reach Earth. Frequently, 2–5 days after a coronal hole passes across the center of the face of the Sun, the speed and density of the solar wind reaching Earth increases. As you can see in **Figure 14.15**, the solar wind drags the Sun's magnetic field along with it. The magnetic field in the solar wind gets "wound up" by the Sun's rotation. Consequently, the solar wind has a spiral structure resembling the stream of water from a rotating lawn sprinkler.

The effects of the solar wind are felt throughout the Solar System. The solar wind blows the tails of comets away from the Sun, shapes the magnetospheres of the planets, and provides the energetic particles that power Earth's spectacular auroral displays. Using space probes, astronomers have been able to observe the solar wind extending out to 100 astronomical units (AU) from the Sun. But the solar wind does not go on forever. The farther it gets from the Sun, the more it has to spread out. Just like radiation, the density of the solar wind follows an inverse square law. At a distance of about 100 AU from the Sun, the solar wind stops abruptly. Here it piles up against the pressure of the **interstellar medium**, which is the gas and dust that lie between stars in a galaxy. **Figure 14.16** shows the

Figure 14.16 The solar wind streams away from the Sun for about 100 AU, until it finally piles up against the pressure of the interstellar medium through which the Sun is traveling. The *Voyager 1* spacecraft has recently crossed this boundary.

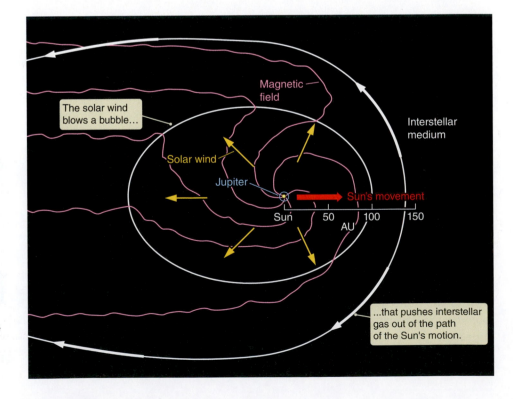

The solar wind blows a bubble…

Magnetic field

Solar wind

Interstellar medium

Jupiter

Sun's movement

Sun

50   100   150

AU

…that pushes interstellar gas out of the path of the Sun's motion.

region of space over which the solar wind is measured. The *Voyager 1* spacecraft has crossed the outer edge of this boundary and sent back the first direct measurements of true interstellar space. The *Interstellar Boundary Explorer* spacecraft, launched in 2008, is also exploring this region.

Sunspots and Changes in the Sun

Sunspots have been noted since antiquity. Telescopic observations of sunspots date back almost 400 years, and there are records of naked-eye observations by Chinese, Greek, and medieval astronomers centuries before that. *But remember that you should never look directly at the Sun!* Direct viewing through a commercial solar filter is safe, as is projecting the image through a telescope or binoculars onto a surface such as paper and looking only at the projection. Many websites have live images of the Sun viewed through ground and space telescopes (see the "Using the Web" problems at the end of the chapter). Sunspots are places where material is trapped at the surface of the Sun by magnetic-field lines. When this material cools, convection cannot carry it downward, so it makes a cooler (therefore darker) spot on the surface of the Sun. **Figure 14.17** shows a large sunspot group. Sunspots appear dark, but only in contrast to the brighter surface of the Sun (**Working It Out 14.2**).

Early telescopic observations of sunspots made during the 17th century led to the discovery of the Sun's rotation, which has an average period of about 27 days as seen from Earth and 25 days relative to the stars. Because Earth orbits the Sun in the same direction that the Sun rotates, observers on Earth see a slightly longer rotation period. Observations of sunspots also show that the Sun's rotation period

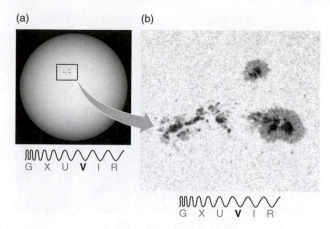

Figure 14.17 (a) This image from the Solar Dynamics Observatory (SDO), taken in 2010, shows a large sunspot group. Sunspots are magnetically active regions that are cooler than the surrounding surface of the Sun. (b) This high-resolution view shows the sunspots in this group.

14.2 | Working It Out Sunspots and Temperature

Sunspots are about 1500 K cooler than their surroundings. What does this lower temperature tell us about their luminosity? Think back to the Stefan-Boltzmann law in Chapter 5. The flux, $\mathcal{F}$, from a blackbody is proportional to the fourth power of the temperature, T. The constant of proportionality is the Stefan-Boltzmann constant, σ, which has a value of 5.67×10^{-8} W/(m^2 K^4). We write this relationship as

$$\mathcal{F} = \sigma T^4$$

Remember that the flux is the amount of energy coming from a square meter of surface every second. How much less energy comes out of a sunspot than out of the rest of the Sun? Let's take round numbers for the temperature of a typical sunspot and the surrounding photosphere: 4500 K and 6000 K, respectively. We can set up two equations:

$$\mathcal{F}_{spot} = \sigma T^4_{spot} \quad \text{and} \quad \mathcal{F}_{surface} = \sigma T^4_{surface}$$

We could solve each of these separately, and then divide the value of $\mathcal{F}_{spot}$ by $\mathcal{F}_{surface}$ to find out how much fainter the sunspot is, but it's much easier to solve for the ratio of the fluxes:

$$\frac{\mathcal{F}_{spot}}{\mathcal{F}_{surface}} = \frac{\sigma T^4_{spot}}{\sigma T^4_{surface}} = \frac{T^4_{spot}}{T^4_{surface}} = \left(\frac{T_{spot}}{T_{surface}}\right)^4$$

Plugging in our values for T_{spot} and $T_{surface}$ gives

$$\frac{\mathcal{F}_{spot}}{\mathcal{F}_{surface}} = \left(\frac{4500\,\text{K}}{6000\,\text{K}}\right)^4 = 0.32$$

and multiplying both sides by $\mathcal{F}_{surface}$ gives

$$\mathcal{F}_{spot} = 0.32\mathcal{F}_{surface}$$

So the amount of energy coming from a square meter of sunspot every second is about one-third as much as the amount of energy coming from a square meter of surrounding surface every second. In other words, the sunspot is about one-third as bright as the surrounding photosphere. If you could cut out the sunspot and place it elsewhere in the sky, it would be brighter than the full Moon.

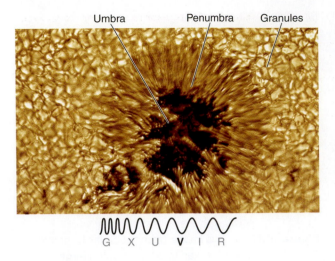

Umbra   Penumbra   Granules

G X U V I R

Figure 14.18 This very high-resolution view of a sunspot shows the dark umbra surrounded by the lighter penumbra. The solar surface around the sunspot bubbles with separate cells of hot gas called *granules*. The smallest features are about 100 km across.

is shorter at its equator than at higher latitudes, an effect called **differential rotation**. Differential rotation is possible only because the Sun is a large ball of gas rather than a solid object.

Figure 14.18 shows the structure of a sunspot on the surface of the Sun. A sunspot consists of an inner dark core called the **umbra**, which is surrounded by a less dark region called the **penumbra**, which shows an intricate radial pattern, reminiscent of the petals of a flower. Sunspots are caused by magnetic fields thousands of times greater than the magnetic field at Earth's surface. They occur in pairs that are connected by loops in the magnetic field. Sunspots range in size from a few tens of kilometers across up to complex groups that may contain several dozen individual spots and span as much as 150,000 km. The largest sunspot groups are so large that they can be seen without a telescope.

Although sunspots occasionally last 100 days or longer, half of all sunspots come and go in about 2 days, and 90 percent are gone within 11 days. The number and distribution of sunspots change over time in a pattern averaging 11 years called the **sunspot cycle**. **Figure 14.19a** shows data for several recent cycles. At the beginning of a cycle, sunspots appear at solar latitudes of about 30° north and south of the solar equator. Over the following years, sunspots are found closer to the equator as their number increases to a maximum and then declines.

Figure 14.19 (a) The number of sunspots varies with time, as shown in this graph of the past few solar cycles. (b) The "solar butterfly" diagram shows the fraction of the Sun covered by sunspots at each latitude. The data are color coded to show the percentage of the strip at that latitude that is covered in sunspots at that time: black, 0 to 0.1 percent; red, 0.1–1.0 percent; yellow, greater than 1.0 percent. (c) The Sun's magnetic poles flip every 11 years. Yellow indicates magnetic south; blue indicates magnetic north.

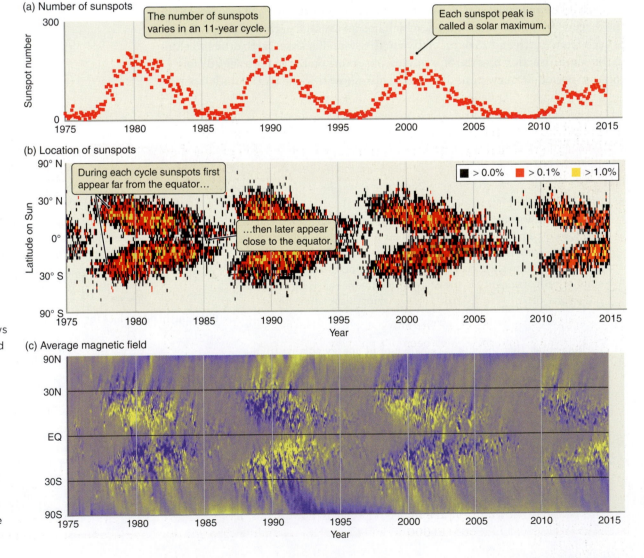

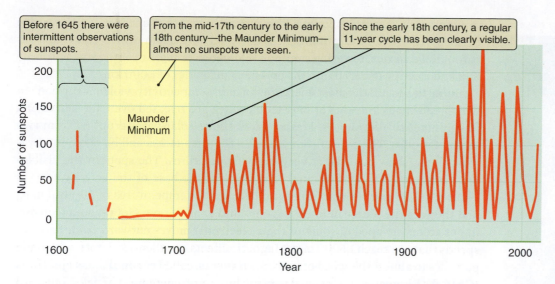

Before 1645 there were intermittent observations of sunspots.

From the mid-17th century to the early 18th century—the Maunder Minimum—almost no sunspots were seen.

Since the early 18th century, a regular 11-year cycle has been clearly visible.

Figure 14.20 Sunspots have been observed for hundreds of years. In this plot, the 11-year cycle in the number of sunspots (half of the 22-year solar magnetic cycle) is clearly visible. Sunspot activity varies greatly over time. The period from the middle of the 17th century to the early 18th century, when almost no sunspots were seen, is called the *Maunder Minimum*.

As the last few sunspots approach the equator, new sunspots again begin appearing at middle latitudes, and the next cycle begins. Figure 14.19b shows the number of sunspots at a given latitude plotted against time: this diagram of opposing diagonal bands is called the sunspot "butterfly diagram."

In the early 20th century, solar astronomer George Ellery Hale (1868–1938) was the first to show that the 11-year sunspot cycle is actually half of a 22-year magnetic cycle during which the direction of the Sun's magnetic field reverses after each 11-year sunspot cycle. Figure 14.19c shows how the average strength of the magnetic field at every latitude has changed over more than 35 years. The direction of the Sun's magnetic field flips at the maximum of each sunspot cycle. Sunspots tend to come in pairs, with one spot (the leading sunspot) in front of the other with respect to the Sun's rotation. In one 11-year sunspot cycle, the leading sunspot in each pair tends to be a north magnetic pole, whereas the trailing sunspot tends to be a south magnetic pole. In the next 11-year sunspot cycle, this polarity is reversed: the leading sunspot in each pair is a south magnetic pole, whereas the trailing sunspot tends to be a north magnetic pole. The transition between these two magnetic polarities occurs near the peak of each sunspot cycle. Magnetic activity on the Sun affects the photosphere, chromosphere, and corona.

Telescopic observations of sunspots date back almost 400 years. As you can see in **Figure 14.20**, the 11-year cycle is neither perfectly periodic nor especially reliable. The time between peaks in the number of sunspots actually varies between about 9.7 and 11.8 years. The number of spots seen during a given cycle fluctuates as well, and there have been periods when sunspot activity has disappeared almost entirely. An extended lull in solar activity, called the **Maunder Minimum**, lasted from 1645 to 1715. Typically, there are about six peaks of solar activity in 70 years, but virtually no sunspots were seen during the Maunder Minimum, and auroral displays were less frequent than usual.

Sunspots are only one of several phenomena that follow the Sun's 22-year cycle of magnetic activity. The peaks of the cycle, called **solar maxima**, are times of intense activity. Sunspots are often accompanied by a brightening of the solar chromosphere that is seen most clearly in emission lines such as Hα. These bright regions are known as solar active regions. The magnificent loops arching through the solar corona, shown in **Figure 14.21**, are solar **prominences**, magnetic flux tubes of relatively cool (5000–10,000 K) but dense gas extending through the

Approx. size of Earth

Figure 14.21 Solar prominences are magnetically supported arches of hot gas that rise high above active regions on the Sun. Here, you can see a close-up view at the base of a large prominence. An image of Earth is included for scale (it is not actually that close to the Sun).

(a)   (b)   (c)

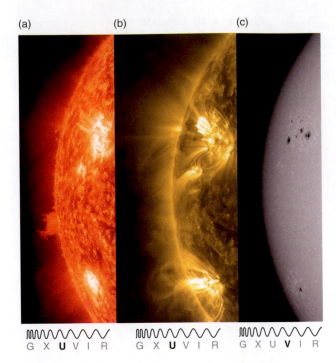

G X U V I R   G X U V I R   G X U V I R

Figure 14.22 The Solar Dynamics Observatory (SDO) observed these active regions of the Sun that produced solar flares in August 2011. (a) Activity near the surface at 60,000 K is visible in extreme ultraviolet light (along with a prominence rising up from the Sun's edge). (b) Viewed at other ultraviolet wavelengths, many looping arcs and plasma heated to about 1 million K become visible. (c) The dark spots in this image are the magnetically intense sunspots that are the sources of all the activity.

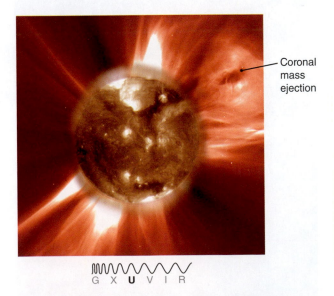

Coronal mass ejection

G X U V I R

Figure 14.23 This Solar and Heliospheric Observatory (*SOHO*) image shows a coronal mass ejection (upper right): a simultaneously recorded ultraviolet image of the solar disk is superimposed.

million-kelvin gas of the corona. These prominences are anchored in the active regions. Although most prominences are relatively quiet, others can erupt out through the corona, towering a million kilometers or more over the surface of the Sun and ejecting material into the corona at velocities of 1,000 km/s.

Solar flares are the most energetic form of solar activity, violent eruptions in which enormous amounts of magnetic energy are released over the course of a few minutes to a few hours. **Figure 14.22** shows solar flares erupting from two sunspot groups. The two left-hand images (Figures 14.22a and b), taken in ultraviolet light, show material at very high temperatures. The spots in the visible-light image on the right (Figure 14.22c) are at the base of the activity seen in Figures 14.22a and b. Solar flares can heat gas to temperatures of 20 million K, and they are the source of intense X-rays and gamma rays. Hot **plasma** (consisting of atoms stripped of some of their electrons) moves outward from flares at speeds that can reach 1,500 km/s. Magnetic effects can then accelerate subatomic particles to almost the speed of light. Such events, called **coronal mass ejections (CMEs)** (**Figure 14.23**), send powerful bursts of energetic particles outward through the Solar System. Coronal mass ejections occur about once per week during the minimum of the sunspot cycle and as often as several times per day near the maximum of the cycle.

Solar Activity Affects Earth

The amount of solar radiation received at the distance of Earth from the Sun has been measured to be, on average, 1,361 watts per square meter (W/m²). As you can see in **Figure 14.24**, satellite measurements of the amount of radiation coming from the Sun show that this value varies by as much as 0.2 percent over periods of a few weeks, as dark sunspots in the photosphere and bright spots in the chromosphere move across the disk. Overall, however, the increased radiation from active regions on the Sun more than makes up for the reduction in radiation from sunspots. On average, the Sun seems to be about 0.1 percent brighter during the peak of a solar cycle than it is at its minimum.

Solar activity affects Earth in many ways. Solar active regions are the source of most of the Sun's extreme ultraviolet and X-ray emissions, energetic radiation that heats Earth's upper atmosphere and, during periods of increased solar activity, causes Earth's upper atmosphere to expand. When this happens, the swollen upper atmosphere can significantly increase the atmospheric drag on spacecraft orbiting at relatively low altitudes, such as that of the Hubble Space Telescope, causing their orbits to decay. Periodic boosts have been necessary to keep the Hubble Space Telescope in its orbit.

Earth's magnetosphere is the result of the interaction between Earth's magnetic field and the solar wind. Increases in the solar wind accompanying solar activity, especially coronal mass ejections directed at Earth, can disrupt Earth's magnetosphere. Spectacular auroras can accompany such events, as can magnetic storms that have been known to disrupt electric power grids and cause blackouts across large regions. Coronal mass ejections that are emitted in the direction of Earth also hinder radio communication and navigation, and they can damage sensitive satellite electronics, including communication satellites. In addition, energetic particles accelerated in solar flares pose one of the greatest dangers to human exploration of space.

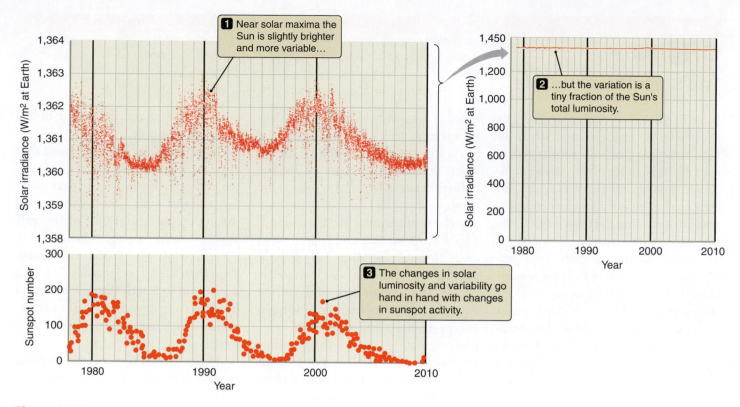

Figure 14.24 Measurements taken by satellites above Earth's atmosphere show that the amount of light from the Sun changes slightly over time.

Detailed observations from the ground and from space help astronomers understand the complex nature of the solar atmosphere. The Solar and Heliospheric Observatory (*SOHO*) spacecraft is a joint mission between NASA and the European Space Agency (ESA). *SOHO* moves in lockstep with Earth at a location approximately 1,500,000 km from Earth that is almost directly in line between Earth and the Sun. *SOHO* carries 12 scientific instruments that monitor the Sun and measure the solar wind upstream of Earth. Additionally, NASA's Solar Dynamics Observatory (SDO) studies the solar magnetic field in order to predict when major solar events will occur, rather than simply responding after they happen.

CHECK YOUR UNDERSTANDING 14.5

Sunspots appear dark because: (a) they have very low density; (b) magnetic fields absorb most of the light that falls on them; (c) they are regions of very high pressure; (d) they are cooler than their surroundings.

Origins

The Solar Wind and Life

Solar flares and coronal mass ejections can affect the space around Earth. In fact, energetic particles accelerated in solar flares pose one of the greatest dangers to human exploration of space, and they need to be considered when astronauts are orbiting Earth in a space station or, someday, traveling to the Moon or farther. Earth's magnetic field protects life on the surface from these energetic particles: the particles travel along the magnetic-field lines to Earth: poles, creating the auroras, without bombarding the surface, which could be harmful to life on Earth. But the Moon does not have this protection because its magnetic field is very weak. Astronauts on the lunar surface would be exposed to as much radiation as astronauts traveling in space. The strength of the solar wind varies with the solar cycle, as noted in Section 14.4, so exposure danger varies as well.

As illustrated in **Figure 14.25**, the Solar System is surrounded by the **heliosphere**, in which the solar wind blows against the interstellar medium and clears out an area like the inside of a bubble. As the Sun and Solar System move through the Milky Way Galaxy, passing in and out of interstellar clouds, this heliosphere protects the entire Solar System from galactic high-energy particles known as cosmic rays that originate primarily in high-energy explosions of massive dying stars. When the Sun is in its lower-activity state, the heliosphere is weaker, so more galactic cosmic rays enter the Solar System. In addition, the intensity of these cosmic rays depends on where the Sun and Solar System are located in their orbit about the center of the Milky Way Galaxy.

Some scientists have theorized that at times when the Sun was quiet and the heliosphere was weaker than average, and the Solar System was passing through a particular part of the galaxy, the cosmic-ray flux in the Solar System—and on Earth—increased. This increased flux possibly led to a disruption in Earth's ozone layer and possibly contributed to a mass extinction in which many species died out on Earth.

Thus, in addition to the obvious contribution of the Sun to heat and light on Earth, the extension of the Sun through the solar wind may have affected the evolution of life on Earth—and it may also affect the ability of humans to live and work in space.

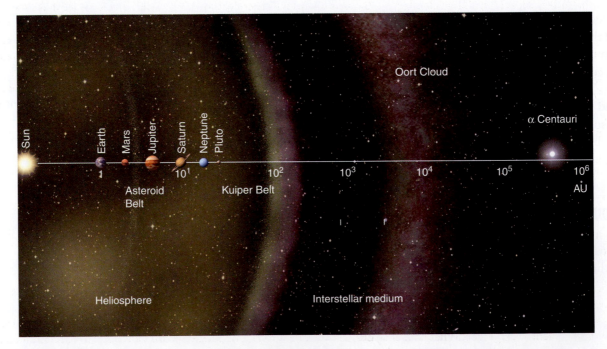

Figure 14.25 The heliosphere of the Sun, a bubble of charged particles partially covering the Solar System that is formed by the solar wind blowing against the interstellar medium. The *Voyager* spacecraft are just past 100 AU. Notice that the scale is logarithmic.

Carrington-Class CME Narrowly Misses Earth

By **DR. TONY PHILLIPS, Science@NASA**

Last month (April 8–11, 2014), scientists, government officials, emergency planners, and others converged on Boulder, Colorado, for NOAA's Space Weather Workshop—an annual gathering to discuss the perils and probabilities of solar storms.

The current solar cycle is weaker than usual, so you might expect a correspondingly low-key meeting. On the contrary, the halls and meeting rooms were abuzz with excitement about an intense solar storm that narrowly missed Earth.

"If it had hit, we would still be picking up the pieces," says Daniel Baker of the University of Colorado, who presented a talk entitled The Major Solar Eruptive Event in July 2012: Defining Extreme Space Weather Scenarios.

The close shave happened almost two years ago. On July 23, 2012, a plasma cloud or "CME" rocketed away from the Sun as fast as 3,000 km/s, more than four times faster than a typical eruption. The storm tore through the Earth's orbit, but fortunately Earth wasn't there. Instead it hit the *STEREO-A* spacecraft. Researchers have been analyzing the data ever since, and they have concluded that the storm was one of the strongest in recorded history.

"It might have been stronger than the Carrington Event itself," says Baker.

The Carrington Event of September 1859 was a series of powerful CMEs that hit Earth head-on, sparking Northern Lights as far south as Tahiti. Intense geomagnetic storms caused global telegraph lines to spark, setting fire to some telegraph offices and disabling the "Victorian Internet." A similar storm today could have a catastrophic effect on modern power grids and telecommunication networks. According to a study by the National Academy of Sciences, the total economic impact could exceed $2 trillion or 20 times greater than the costs of a Hurricane Katrina. Multi-ton transformers fried by such a storm could take years to repair and impact national security.

A recent paper in *Nature Communications* authored by UC Berkeley space physicist Janet G. Luhmann and former postdoc Ying D. Liu describes what gave the July 2012 storm Carrington-like potency. For one thing, the CME was actually *two* CMEs separated by only 10 to 15 minutes. This double storm cloud traveled through a region of space that had been cleared out by another CME only four days earlier. As a result, the CMEs were not decelerated as much as usual by their transit through the interplanetary medium.

Had the eruption occurred just one week earlier, the blast site would have been facing Earth, rather than off to the side, so it was a relatively narrow escape.

When the Carrington Event enveloped Earth in the 19th century, technologies of the day were hardly sensitive to electromagnetic disturbances. Modern society, on the other hand, is deeply dependent on Sun-sensitive technologies such as GPS, satellite communications, and the Internet.

"The effect of such a storm on our modern technologies would be tremendous," says Luhmann.

During informal discussions at the workshop, Nat Gopalswamy of the Goddard Space Flight Center noted that "without NASA's *STEREO* probes, we might never have known the severity of the 2012 superstorm. This shows the value of having 'space weather buoys' located all around the Sun."

It also highlights the potency of the Sun even during so-called "quiet times." Many observers have noted that the current solar cycle is weak, perhaps the weakest in 100 years. Clearly, even a weak solar cycle can produce a very strong storm. Says Baker, "We need to be prepared."

1. What is a CME?
2. Why would a CME cause disruptions on Earth?
3. Explain how this storm missed Earth by 1 week.
4. More sensationalistic headlines for this story claimed that Earth almost "was sent back to the Dark Ages." What did they mean by this exaggeration?
5. Go to the NASA press release for this story (http://science.nasa.gov/science-news/science-at-nasa/2014/23jul_superstorm/), and watch the 4-minute "ScienceCast" video. What happened during the Carrington CME in 1859? Is this video effective at communicating the science information to the nonspecialist?

Summary

The forces due to pressure and gravity balance each other in hydrostatic equilibrium, maintaining the Sun's structure. Nuclear reactions converting hydrogen to helium are the source of the Sun's energy. Energy created in the Sun's core moves outward to the surface, first by radiation and then by convection. The solar wind may adversely affect astronauts located in space or on planets that lack a protective magnetic field, but it also has protected the Solar System from galactic, high-energy, cosmic-ray particles.

LG 1 **Describe the balance between the forces that determine the structure of the Sun.** The outward pressure of the hot gas inside the Sun balances the inward pull of gravity at every point. This balance is dynamically maintained. An energy balance is also maintained, with the energy produced in the core of the Sun balancing the energy lost from the surface.

LG 2 **Explain how mass is converted into energy in the Sun's core and how long it will take the Sun to use up its fuel.** In the core of the Sun, mass is converted to energy via the proton-proton cycle. When four hydrogen atoms fix to one helium atom, some mass is lost. This mass is released as energy, nearly all of which leaves the Sun either as photons or as neutrinos. Neutrinos are elusive, almost massless particles that interact only very weakly with other matter. Observations of neutrinos confirm that nuclear fusion is the Sun's primary energy source.

LG 3 **Sketch a physical model of the Sun's interior, and list the different ways that energy moves outward from the Sun's core toward its surface.** The interior of the Sun is divided into zones that are defined by how energy is transported in that region. Energy moves outward through the Sun by radiation and by convection.

LG 4 **Describe how observations of solar neutrinos and seismic vibrations on the surface of the Sun test astronomers' models of the Sun.** The Sun has multiple layers, each with a characteristic pressure, density, and temperature. Neutrinos directly probe the interior of the Sun. This model of the interior of the Sun has been tested by helioseismology, in much the same way that the model of Earth's interior has been tested by seismology.

LG 5 **Describe the solar activity cycles of 11 and 22 years, and explain how these cycles are related to the Sun's changing magnetic field.** Activity on the Sun follows a cycle that peaks every 11 years but takes 22 full years for the magnetic field to reverse. Sunspots are photospheric regions that are cooler than their surroundings, and they reveal the cycles in solar activity. Material streaming away from the Sun's corona creates the solar wind, which moves outward through the Solar System until it meets the interstellar medium. Solar storms, including ejections of mass from the corona, produce auroras and can disrupt power grids and damage satellites.

? UNANSWERED QUESTIONS

- Will nuclear fusion become a major source of energy production on Earth? Scientists have been working on controlled nuclear fusion for more than 60 years, since the first hydrogen bombs were developed. But so far there are too many difficulties in replicating the conditions inside the Sun. Nuclear fusion requires that we have hydrogen isotopes at very high temperature, density, and pressure, just as is the case when a hydrogen bomb explodes. However, controlled nuclear fusion requires that we confine this material sufficiently long to get more energy out than we put in. Several major experiments have attempted to fuse isotopes of hydrogen. An alternative approach is to fuse an isotope of helium, ^{3}He, which has only three particles in the nucleus (two protons and one neutron). On Earth, ^{3}He is found in very limited supply. But ^{3}He is in much greater abundance on the Moon, so some people propose setting up mining colonies on the Moon to extract ^{3}He for use in fusion reactions on Earth or possibly even on the Moon (see question 49b at the end of the chapter).

- Are variations in Earth's climate related to solar activity? Solar activity affects Earth's upper atmosphere, and it may affect weather patterns as well. It has also been suggested that variations in the amount of radiation from the Sun might be responsible for past variations in Earth's climate. Current models indicate that observed variations in the Sun's luminosity could account for only about 0.1 K differences in Earth's average temperature—much less than the effects due to the ongoing buildup of carbon dioxide in Earth's atmosphere. Triggering the onset of an ice age may require a sustained drop in global temperatures of only about 0.2–0.5 K, so astronomers are continuing to investigate a possible link between solar variability and changes in Earth's climate.

Questions and Problems

Test Your Understanding

1. The physical model of the Sun's interior has been confirmed by observations of
   a. neutrinos and seismic vibrations.
   b. sunspots and solar flares.
   c. neutrinos and positrons.
   d. sample returns from spacecraft.
   e. sunspots and seismic vibrations.

2. Place in order the following steps in the fusion of hydrogen into helium. If two or more steps happen simultaneously, use an equals sign (=).
   a. A positron is emitted.
   b. One gamma ray is emitted.
   c. Two hydrogen nuclei are emitted.
   d. Two ^{3}He collide and become ^{4}He.
   e. Two hydrogen nuclei collide and become ^{2}H.
   f. Two gamma rays are emitted.
   g. A neutrino is emitted.
   h. One deuterium nucleus and one hydrogen nucleus collide and become ^{3}He.

3. Sunspots, flares, prominences, and coronal mass ejections are all caused by
   a. magnetic activity on the Sun.
   b. electrical activity on the Sun.
   c. the interaction of the Sun's magnetic field and the interstellar medium.
   d. the interaction of the solar wind and Earth's magnetic field.
   e. the interaction of the solar wind and the Sun's magnetic field.

4. The structure of the Sun is determined by both the balance between the forces due to _____ and gravity and the balance between energy generation and energy _____.
   a. pressure; production
   b. pressure; loss
   c. ions; loss
   d. solar wind; production

5. In the proton-proton chain, four hydrogen nuclei are converted to a helium nucleus. This does not happen spontaneously on Earth because the process requires
   a. vast amounts of hydrogen.
   b. very high temperatures and densities.
   c. hydrostatic equilibrium.
   d. very strong magnetic fields.

6. The solar neutrino problem pointed to a fundamental gap in our knowledge of
   a. nuclear fusion.
   b. neutrinos.
   c. hydrostatic equilibrium.
   d. magnetic fields.

7. Sunspots change in number and location during the solar cycle. This phenomenon is connected to
   a. the rotation rate of the Sun.
   b. the temperature of the Sun.
   c. the magnetic field of the Sun.
   d. the tilt of the axis of the Sun.

8. Suppose an abnormally large amount of hydrogen suddenly burned in the core of the Sun. Which of the following would be observed first?
   a. The Sun would become brighter.
   b. The Sun would swell and become larger.
   c. The Sun would become bluer.
   d. The Sun would emit more neutrinos.

9. The solar corona has a temperature of 1 million to 2 million K; the photosphere has a temperature of only about 6000 K. Why isn't the corona much, much brighter than the photosphere?
   a. The magnetic field traps the light.
   b. The corona emits only X-rays.
   c. The photosphere is closer to us.
   d. The corona has a much lower density.

10. The Sun rotates once every 25 days relative to the stars. The Sun rotates once every 27 days relative to Earth. Why are these two numbers different?
    a. The stars are farther away.
    b. Earth is smaller.
    c. Earth moves in its orbit during this time.
    d. The Sun moves relative to the stars.

11. Place the following regions of the Sun in order of increasing radius.
    a. corona
    b. core
    c. radiative zone
    d. convective zone
    e. chromosphere
    f. photosphere
    g. a sunspot

12. Coronal mass ejections
    a. carry away 1 percent of the mass of the Sun each year.
    b. are caused by breaking magnetic fields.
    c. are always emitted in the direction of Earth.
    d. are unimportant to life on Earth.

13. As energy moves out from the Sun's core toward its surface, it first travels by _____, then by _____, and then by _____.
    a. radiation; conduction; radiation
    b. conduction; radiation; convection
    c. radiation; convection; radiation
    d. radiation; convection; conduction

14. Energy is produced primarily in the center of the Sun because
    a. the strong nuclear force is too weak elsewhere.
    b. that's where neutrinos are created.
    c. that's where most of the helium is.
    d. the outer parts have lower temperatures and densities.

15. The solar wind pushes on the magnetosphere of Earth, changing its shape, because
    a. the solar wind is so dense.
    b. the magnetosphere is so weak.
    c. the solar wind contains charged particles.
    d. the solar wind is so fast.

Thinking about the Concepts

16. Explain how hydrostatic equilibrium acts as a safety valve to keep the Sun at its constant size, temperature, and luminosity.

17. Two of the three atoms in a molecule of water (H_2O) are hydrogen. Why are Earth's oceans not fusing hydrogen into helium and setting Earth ablaze?

18. Why are neutrinos so difficult to detect?

19. Explain the proton-proton chain through which the Sun generates energy by converting hydrogen to helium.

20. On Earth, nuclear power plants use *fission* to generate electricity. In fission, a heavy element like uranium is broken into many atoms, where the total mass of the fragments is less than that of the original atom. Explain why fission could not be powering the Sun today.

21. If an abnormally large amount of hydrogen suddenly burned in the core of the Sun, what would happen to the rest of the Sun? Would the Sun change as seen from Earth?

22. Study the Process of Science Figure. If the follow-up experiments did not detect the other types of neutrinos, what would have been the next step for scientists at that point?

23. What is the solar neutrino problem, and how was it solved?

24. The Sun's visible "surface" is not a true surface, but a feature called the photosphere. Explain why the photosphere is not a true surface.

25. How are orbiting satellites and telescopes affected by the Sun?

26. Describe the solar corona. Under what circumstances can it be seen without special instruments?

27. In the proton-proton chain, the mass of four protons is slightly greater than the mass of a helium nucleus. Explain what happens to this "lost" mass.

28. What have sunspots revealed about the Sun's rotation?

29. Why are different parts of the Sun best studied at different wavelengths? Which parts are best studied from space?

30. Why is it important to study the interaction of the solar wind with the interstellar medium?

Applying the Concepts

31. In Figure 14.10, density and temperature are both graphed versus height.
    a. Is the height axis linear or logarithmic? How do you know?
    b. Is the density axis linear or logarithmic? How do you know?
    c. Is the temperature axis linear or logarithmic? How do you know?

32. Using the data in Figures 14.19b and c, present an argument that sunspots occur in regions of strong magnetic field.

33. Study Figure 14.17a and the graph in Figure 14.20.
    a. Estimate the fraction of the Sun's surface that is covered by the large sunspot group in the image. (Remember that you are seeing only one hemisphere of the Sun.)
    b. From the graph, estimate the average number of sunspots that occurs at solar maximum.
    c. On average, what fraction of the Sun could be covered by sunspots at solar maximum? Is this a large fraction?
    d. Compare your conclusion to the graph of irradiance in Figure 14.24. Does this graph make sense to you?

34. The Sun has a radius equal to about 2.3 light-seconds. Explain why a gamma ray produced in the Sun's core does not emerge from the Sun's surface 2.3 seconds later.

35. Assume that the Sun's mass is about 300,000 Earth masses and that its radius is about 100 times that of Earth. The density of Earth is about 5,500 kg/m^3.
    a. What is the average density of the Sun?
    b. How does this compare with the density of Earth? With the density of water?

36. The Sun shines by converting mass into energy according to $E = mc^2$. Show that if the Sun produces 3.85×10^{26} J of energy per second, it must convert 4.3 million metric tons (4.3×10^9 kg) of mass per second into energy.

37. Assume that the Sun has been producing energy at a constant rate over its lifetime of 4.6 billion years (1.4×10^{17} seconds).
    a. How much mass has it lost creating energy over its lifetime?
    b. The current mass of the Sun is 2×10^{30} kg. What fraction of its current mass has been converted into energy over the lifetime of the Sun?

38. Suppose our Sun was an A5 main-sequence star, with twice the mass and 12 times the luminosity of the Sun, a G2 star. How long would this A5 star burn hydrogen to helium? What would this mean for Earth?

39. Imagine that the source of energy in the interior of the Sun changed abruptly.
    a. How long would it take before a neutrino telescope detected the event?
    b. When would a visible-light telescope see evidence of the change?

40. On average, how long does it take particles in the solar wind to reach Earth from the Sun if they are traveling at an average speed of 400 km/s?

41. A sunspot appears only 70 percent as bright as the surrounding photosphere. The photosphere has a temperature of approximately 5780 K. What is the temperature of the sunspot?

42. The hydrogen bomb represents an effort to create a similar process to what takes place in the core of the Sun. The energy released by a 5-megaton hydrogen bomb is 2×10^{16} J.
    a. This textbook, *21st Century Astronomy*, has a mass of about 1.6 kg. If all of its mass were converted into energy, how many 5-megaton bombs would it take to equal that energy?
    b. How much mass did Earth lose each time a 5-megaton hydrogen bomb was exploded?

43. Verify the claim made at the start of this chapter that the Sun produces more energy per second than all the electric power plants on Earth could generate in a half-million years. Estimate or look up how many power plants there are on the planet and how much energy an average power plant produces. Be sure to account for different kinds of power; for example, coal, nuclear, wind.

44. Let's examine the reason that the Sun cannot power itself by chemical reactions. Using Working It Out 14.1 and the fact that an average chemical reaction between two atoms releases 1.6×10^{-19} J, estimate how long the Sun could emit energy at its current luminosity. Compare that estimate to the known age of Earth.

45. The Sun could get energy from gravitational contraction for a time period of ($GM_{Sun}/R_{Sun}L_{Sun}$). How long would the Sun last at its current luminosity? (Be careful with units!)

46. a. Go to *QUEST*'s "Journey into the Sun" Web page (http://science.kqed.org/quest/video/journey-into-the-sun) to watch a short video on the Solar Dynamics Observatory (SDO), launched in 2010. Why is studying the magnetic field of the Sun so important? What is new and different about this observatory? What is the "Music of the Sun"?
    b. Go to the SDO website (http://sdo.gsfc.nasa.gov). Under "Data," select "The Sun Now" and view the Sun at many wavelengths. What activity do you observe in the images at the location of any sunspots seen in the "HMI Intensitygram" images? (You can download a free SDO app by Astra to get real-time images on your mobile device.) Look at a recent news story from the SDO website. What was observed, and why is it newsworthy?
    c. Go to the *STEREO* mission's website (http://stereo.gsfc.nasa.gov). What is *STEREO*? Where are the spacecraft located? How does this configuration enable observations of the entire Sun at once? (You can download the app "3-D Sun" to get the latest images on your mobile device.)

47. a. What are the science goals of NASA's *Interface Region Imaging Spectrograph* (*IRIS*) mission (www.nasa.gov/mission_pages/iris/)? What has it discovered?
    b. An older space mission, *SOHO* (Solar and Heliospheric Observatory; http://sohowww.nascom.nasa.gov), was launched in 1995 by NASA and ESA. Click on "The Sun Now" to see today's images. The Extreme Ultraviolet imaging Telescope (EIT) images are in the far ultraviolet and show violent activity. How do these images differ from the ones of SDO in question 46b?
    c. Go to the Daniel K. Inouye Solar Telescope (DKIST) website (http://atst.nso.edu). This adaptive-optics telescope under construction on Haleakala, Maui, will be the largest solar telescope. Why is it important to study the magnetic field of the Sun? What are some of the advantages of studying the Sun from a ground-based telescope instead of a space-based telescope? What wavelengths does the DKIST observe? Why is Maui a good location? When is the telescope scheduled to be completed?

48. a. Go to the Space Weather website (http://spaceweather.com). Are there any solar flares today? What is the sunspot number? Is it about what you would expect for this year? (Click on "What is the sunspot number?" to see a current graph.) Are there any coronal holes today?
    b. Citizen science: Go to the website for Sunspotter (http://www.sunspotter.org/), a Zooniverse project that evaluates the complexity of sunspots and how they change over time. Zooniverse projects offer an opportunity for people to

contribute to science by analyzing pieces of data. Create an account for Zooniverse if you don't already have one (you will use it again in this course). Log in and click on "Science" and skim through the sections. What are the goals of this project? Why is it useful to have multiple people looking at these data? Click on "Classify" and analyze some sunspots. Save a screen shot for your homework.

c. Citizen science: Go to the Solar Stormwatch website (http://solarstormwatch.com), a Zooniverse project from the Royal Observatory in Greenwich, England. Create an account for Zooniverse if you don't already have one (you will use it again in this course). Log in and click on "Spot and Track Storms" and go through the Spot and Track training exercises. You are now ready to look at some real data. Click on an image to do the classification. Save a screen shot for your homework.

49. a. Go to the National Ignition Facility (NIF) website (https://lasers.llnl.gov/about/). Under "Science," click on "How to Make a Star." How are lasers used in experiments to develop controlled nuclear fusion on Earth? How does the fusion reaction here differ from that in the Sun?

b. An alternative approach is to fuse ^{3}He + ^{3}He instead of the hydrogen isotopes. But on Earth, ^{3}He is in limited supply. Helium-3 is in much greater abundance on the Moon, so some people propose setting up mining colonies on the Moon to extract ^{3}He for fusion reactions on Earth. Do a search on "helium 3 Moon." Which countries are talking about going to the Moon for this purpose? What is the timeline for when this might happen? What are the difficulties?

50. a. Go to http://voyager.jpl.nasa.gov/where/. Where are the *Voyager* spacecraft now? Has *Voyager 2* crossed into interstellar space?

b. Go to the website for the Interstellar Boundary Explorer (*IBEX*) (http://www.nasa.gov/mission_pages/ibex/.) What has *IBEX* learned about the solar wind and the interstellar medium?

smartwork

If your instructor assigns homework in SmartWork, access your assignments at digital.wwnorton.com/astro5.

The proton-proton chain powers the Sun by fusing hydrogen into helium. This fusion process produces several different particles as byproducts, as well as energy. In this Exploration, we will explore the steps of the proton-proton chain in detail, with the intent of helping you keep them straight.

Visit the Student Site at the Digital Landing Page, and open the "Proton-Proton Animation" for this chapter.

Watch the animation all the way through once.

Play the animation again, pausing after the first collision. Two hydrogen nuclei (both positively charged) have collided to produce a new nucleus with only one positive charge.

1 Which particle carried away the other positive charge?

_____

_____

2 What is a neutrino? Did the neutrino enter the reaction or was the neutrino produced in the reaction?

_____

_____

Compare the interaction on the top with the interaction on the bottom.

3 Did the same reaction occur in each instance?

_____

_____

Resume playing the animation, pausing it after the second collision.

4 What two types of nuclei entered the collision? What type of nucleus resulted?

_____

_____

5 Was charge conserved in this reaction or was it necessary for a particle to carry charge away?

_____

_____

6 What is a gamma ray? Did the gamma ray enter the reaction or was it produced by the reaction?

_____

_____

Resume the animation again, and allow it to run to the end.

7 What nuclei enter the final collision? What nuclei are produced?

_____

_____

8 In chemistry, a catalyst facilitates the reaction but is not used up in the process. Do any nuclei act like catalysts in the proton-proton chain?

_____

_____

Make a table of inputs and outputs. Which of the particles in the final frame of the animation were inputs to the reaction? Which were outputs? Fill in your table with these inputs and outputs.

9 Which outputs are converted into energy that leaves the Sun as light?

_____

_____

10 Which outputs could become involved in another reaction immediately?

_____

_____

11 Which output is likely to stay in that form for a very long time?

_____

_____

15

The Interstellar Medium and Star Formation

The birth of a star—from a cloud of gas and dust to nuclear burning—is a process that can happen within tens of thousands of years (for the most massive stars) or require hundreds of millions of years (for the least massive). Astronomers have come to understand the process by observing many different stars at various stages of development. In this chapter, we will look at the interstellar environment from which stars both large and small form. Then we will focus our attention on the forming star—the *protostar*—and discuss how it becomes a star.

LEARNING GOALS

By the conclusion of the chapter, you should be able to:

LG 1 Describe the types and states of material that exist in the space between the stars and how this material is detected.

LG 2 Explain the conditions under which a cloud of gas can contract into a stellar system and the role that gravity and angular momentum play in the formation of stars and planets.

LG 3 List the steps in the evolution of a protostar.

LG 4 Describe the track of a protostar as it evolves to a main-sequence star on the Hertzsprung-Russell (H-R) diagram.

Stars are forming in this giant molecular cloud. ▶ ▶ ▶

How do
stars form?

15.1 The Interstellar Medium Fills the Space between the Stars

The space between the stars is far from empty. It contains giant clouds of cool gas and dust, which can be observed in the visible part of the spectrum as they absorb the light from objects behind them, emit light from excited atoms, or reflect starlight from nearby stars. In turn, the space between these giant clouds is filled with tenuous gas. Together, all the material between the stars is known as the **interstellar medium**. Stars supply the energy that heats and stirs the interstellar medium. Warm gas heated by ultraviolet radiation from massive, hot stars pushes outward into its surroundings. Blast waves from dying stars sweep out vast, hot "cavities," piling up material in their path like snow in front of a snowplow. Interstellar clouds are both destroyed and formed by these violent events. Swept-up gas becomes the next generation of clouds. Hot bubbles of high-pressure gas crush molecular clouds, driving up their densities and sometimes triggering the formation of new generations of stars. Stars come from, and return much of their material to, the interstellar medium. In this section, we survey these varied states of the interstellar medium.

The Composition and Density of the Interstellar Medium

The Sun formed from the interstellar medium, so it is not too surprising that the chemical composition of the interstellar medium in the region of the Sun is similar to the chemical composition of the Sun (see Table 13.1). In the interstellar medium, hydrogen accounts for about 90 percent of the atomic nuclei, and the remaining 10 percent is almost all helium. The more massive elements account for only 0.1 percent of the atomic nuclei, or about 2 percent of the mass in the interstellar medium. Roughly 99 percent of that interstellar matter is gaseous, consisting of individual atoms or molecules moving about freely, as the molecules in the air do.

However, interstellar gas is far less dense than the air that you breathe. Each cubic centimeter (cm^3) of the air around you contains about 2.7×10^{19} molecules. A good vacuum pump on Earth can reduce this density down to about 10^{10} molecules/cm^3—approximately a billionth as dense. By comparison, the interstellar medium has an average density of about 0.1 atom/cm^3—one 100-billionth as dense as the vacuum pump can attain. Stated another way, there is about as much material in a column of air between your eye and the floor as there is interstellar gas in a column of the same diameter that stretches from the Solar System to the center of our galaxy 26,000 light-years away.

Interstellar Dust and Its Effects on Light

About 1 percent of the material in the interstellar medium is in the form of solid grains, called **interstellar dust**. Ranging in size from little more than large molecules up to particles about 300 nanometers (nm) across, these solid grains more closely resemble the particles of soot from a candle flame than the dust that collects on a windowsill. It would take several hundred "large" interstellar grains to span the thickness of a single human hair. Interstellar dust begins to form when materials such as iron, silicon, and carbon stick together to form grains in dense, relatively cool environments such as the outer atmospheres and "stellar winds" of cool, red giant stars or in dense material thrown into space by stellar explosions. Once these grains are in the interstellar medium, other atoms and molecules stick

(a)

G X U V I R

(b)

G X U V I R

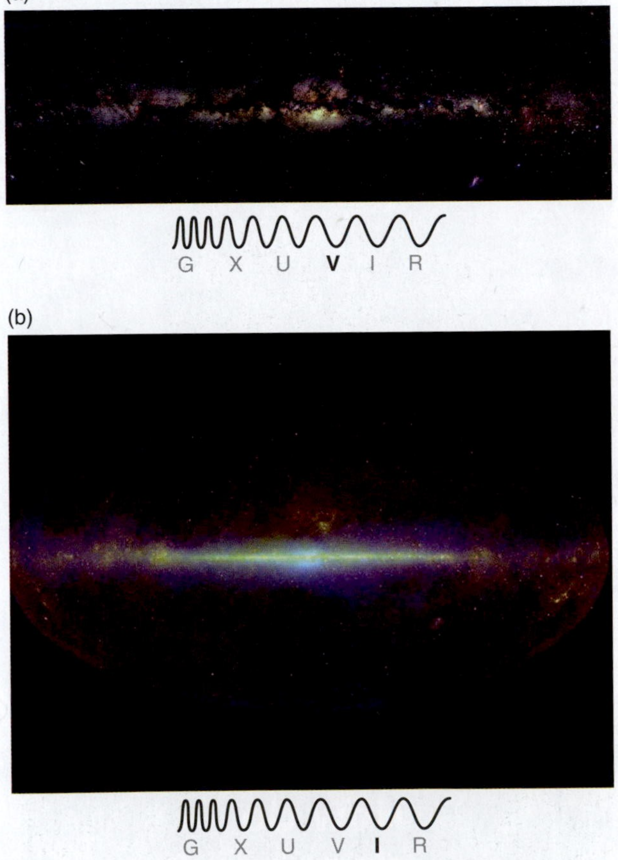

Figure 15.1 (a) This all-sky picture of the Milky Way was taken in visible light. The dark splotches blocking the view are dusty interstellar clouds. The center of this and all the other all-sky images in this chapter is the center of the Milky Way. (b) This all-sky picture was taken in the near infrared. Infrared radiation penetrates the interstellar dust, providing a clearer view of the stars in the disk of the Milky Way.

to them. This process is remarkably efficient: about half of the interstellar matter that is more massive than helium (1 percent of the total mass of the interstellar medium) is found in interstellar grains.

Because interstellar dust is extremely effective at blocking and diverting light, the view of distant objects is affected even by the low-density interstellar material. Recall the comparison of the air column between you and the floor and the column of interstellar space stretching to the center of the galaxy. If the interstellar column were compressed to the same density as air, it would be so dirty that it would be difficult to see your hand 10 centimeters (cm) in front of your face. Go out on a dark summer night in the Northern Hemisphere (or on a dark winter night in the Southern Hemisphere) and look closely at the Milky Way, visible as a faint band of diffuse light running through the constellation Sagittarius. You will see a dark "lane" running roughly down the middle of this bright band, splitting it in two. This dark band is a vast expanse of interstellar dust and blocks astronomers' views of distant stars.

When interstellar dust gets in the way of radiation from distant objects, the effect is called **interstellar extinction**. Not all electromagnetic radiation is affected equally by interstellar extinction. **Figure 15.1** shows two images of the Milky Way Galaxy: one taken in visible light, the other taken in the infrared (IR). The dark clouds that block the shorter-wavelength visible light (Figure 15.1a) seem to have vanished in the longer-wavelength IR image (Figure 15.1b), enabling observations through the clouds to the center of the galaxy and beyond. These two images are *all-sky images*: they portray the entire sky surrounding Earth. They have been oriented so that the disk of the Milky Way, in which the Solar System is embedded, runs horizontally across the center of the image.

To understand why short-wavelength radiation is obscured by dust while long-wavelength radiation is not, think about a different kind of wave—waves on the surface of the ocean, as shown in **Figure 15.2a**. Imagine you are on the ocean in a boat in a strong swell. If the waves are much bigger than your small boat, the swell causes you to bob gently up and down. But that is about all: there is no other interaction between the waves and the small boat, and no energy is lost from the wave to the boat. The situation is quite different if the waves are closer in size to your boat. To picture this, imagine waves roughly half the size of your boat. Now, the front of the boat may be on a wave crest while the back of the boat is in a trough or vice versa. The boat tips wildly back and forth as the waves go by. If the size of the boat and the wavelength of the waves are the right match, even fairly modest waves will rock the boat. You might have noticed this if you were ever in a canoe or rowboat when the wake from a speedboat came by. Now imagine viewing these two situations from the perspective of the wave. The wave is hardly affected when it is much bigger than the boat, but it is strongly affected when it is comparable in size or small compared to the boat. The energy to drive the wild motions of the boat comes from the wave, so the wave loses energy.

The interaction of electromagnetic waves with matter is more involved than that of a boat rocking on the ocean, but the same basic idea often applies, as shown in Figure 15.2b. Tiny interstellar dust grains effectively block the transmission of ultraviolet light and blue light, which have wavelengths comparable to or smaller than the typical size of dust grains. In contrast, longer-wavelength infrared and radio radiation does not interact strongly with the tiny interstellar dust grains. Therefore, at visible and ultraviolet wavelengths, most of the Milky Way is hidden from view by dust, but in the infrared and radio portions of the spectrum, we get a far more complete view.

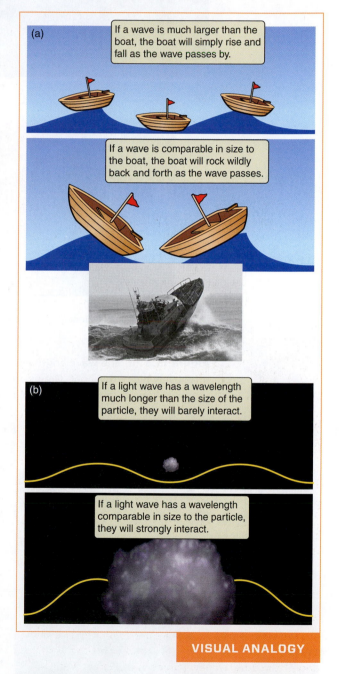

(a) If a wave is much larger than the boat, the boat will simply rise and fall as the wave passes by.

If a wave is comparable in size to the boat, the boat will rock wildly back and forth as the wave passes.

(b) If a light wave has a wavelength much longer than the size of the particle, they will barely interact.

If a light wave has a wavelength comparable in size to the particle, they will strongly interact.

VISUAL ANALOGY

Figure 15.2 (a) Just as boats interact most strongly with ocean waves that are similar in size, (b) particles interact most strongly with wavelengths of light of similar size.

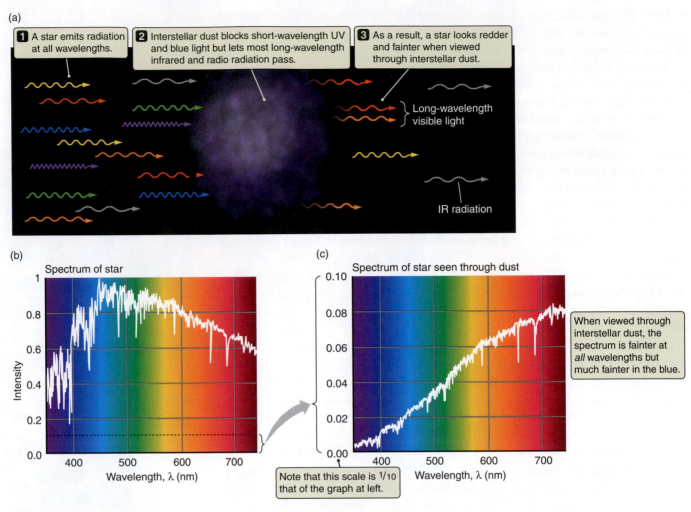

Figure 15.3 (a) The wavelengths of ultraviolet and blue light are close to the size of interstellar grains, so the grains effectively block this light. Grains are less effective at blocking longer-wavelength light. As a result, the spectrum of a star (b) when seen through an interstellar cloud (c) appears fainter and redder.

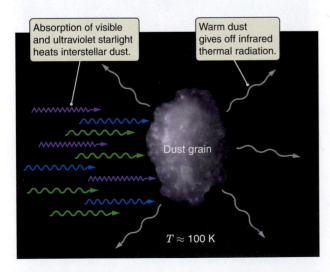

Figure 15.4 The temperature of interstellar grains is determined by the equilibrium between absorbed and emitted radiation.

As shown in **Figure 15.3**, extinction occurs at all wavelengths, causing an object viewed through dust to be fainter than it would be otherwise. However, extinction affects short-wavelength blue light more than it affects long-wavelength red light. This makes the object appear less blue than it really is. This effect of dust, of removing the blue light to cause an object to be more red, is called **reddening**. The presence of dust significantly affects the spectrum of an object, as shown in Figures 15.3b and c. Correcting for the presence of dust can be one of the most difficult parts of interpreting astronomical observations, often adding to uncertainty in the measurement of an object's properties.

Interstellar extinction is less of a concern at infrared wavelengths, but dust still plays an important role in infrared observations. Like any other solid object, a large grain of dust glows at a wavelength determined by its temperature. In Chapter 5, we discussed how the equilibrium between absorbed sunlight and emitted thermal radiation determines the temperatures of the terrestrial planets (see Figure 5.20). As **Figure 15.4** illustrates, a similar equilibrium is at work in interstellar space, where dust is heated both by starlight and by the gas in which it is immersed to temperatures of tens to hundreds of kelvins. At a temperature of 100 kelvins (K), Wien's law says that dust will glow most strongly at a wavelength

15.1   Working It Out   Dust Glows in the Infrared

The temperature of interstellar dust can be found from its spectrum. Wien's law, discussed in Chapter 5, relates the temperature of an object to the peak wavelength (λ_{peak}) of its emitted radiation. For warm dust at a temperature of 100 K (recall that 1 μm = 10^{-6} meter = 1,000 nm):

$$\lambda_{\text{peak}} = \frac{2{,}900\,\mu\text{m K}}{T} = \frac{2{,}900\,\mu\text{m K}}{100\,\text{K}} = 29\,\mu\text{m}$$

For cooler dust, at a temperature of 10 K:

$$\lambda_{\text{peak}} = \frac{2{,}900\,\mu\text{m K}}{T} = \frac{2{,}900\,\mu\text{m K}}{10\,\text{K}} = 290\,\mu\text{m}$$

The temperature and the peak wavelength are inversely proportional, so if the temperature drops, the peak wavelength gets longer. For the temperatures that are common for dust in the interstellar medium, the peak wavelength is in the far-infrared part of the electromagnetic spectrum.

of 29 microns (μm), whereas cooler dust—at a temperature of 10 K—glows most strongly at a longer wavelength (**Working It Out 15.1**). Much of the light in infrared observations is thermal radiation from this dust. **Figure 15.5a**, from NASA's Wide-field Infrared Survey Explorer (WISE), shows the sky at combined wavelengths of 3.4, 12, and 22 μm. In Figure 15.5b, a far-infrared image of the sky at 100 μm from the Infrared Astronomical Satellite (IRAS) telescope is combined with a microwave image from the Cosmic Background Explorer (COBE) telescope to show the Milky Way's dark clouds glowing brilliantly in infrared radiation from dust.

Temperatures and Densities of Interstellar Gas

About half of the gas and dust that fills interstellar space is concentrated in dense regions called **interstellar clouds**, in which the interstellar gas is more concentrated than in surrounding regions. These clouds fill only about 2 percent of the volume of interstellar space. The other half of the interstellar gas and dust is spread out through 98 percent of the volume of interstellar space and is called **intercloud gas**.

The properties of intercloud gas vary from place to place (**Table 15.1**). About half of the volume of interstellar space is filled with an intercloud gas that is extremely hot, heated primarily by the energy of tremendous stellar explosions

(a)

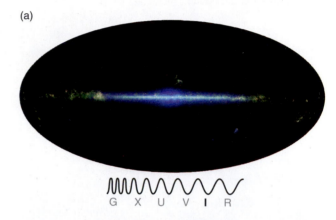

(b)

Figure 15.5 (a) This WISE infrared image shows the plane of the Milky Way at combined wavelengths of 3.4, 12, and 22 μm. (b) This all-sky image, in the far-infrared wavelength of 100 μm (from IRAS) combined with a microwave image from COBE, shows dust throughout the galaxy.

TABLE 15.1   Typical Properties of Components of the Interstellar Medium

Component	Temperature (K)	Number Density (atoms/cm³)	Size of Cube per Gram* (km)	State of Hydrogen
Hot intercloud gas	~1 million	~0.005	~8,000	Ionized
Warm intercloud gas	~8000	0.01–1	~800	Ionized or neutral
Cold intercloud gas	~100	1–100	~80	Neutral
Interstellar clouds	~10	100–1,000	~8	Molecular or neutral

*This is the length of one side of the cube of space you would need to search to find 1 gram of the material. It is another way of thinking about density.

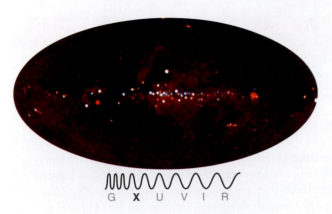

Figure 15.6 Bright spots in this image are distant X-ray sources, including objects such as bubbles of very hot, high-pressure gas surrounding the sites of recent explosions of supernovae.

called *supernovae* (**Figure 15.6**). This hot intercloud gas has temperatures like those found in the cores of stars—in the millions of kelvins. Because the temperature is so high, the atoms in the gas are moving very rapidly. But the gas also has an extremely low density. Typically, you would have to search a liter (1,000 cm³) or more of hot intercloud gas to find a single atom. If you were adrift in an expanse of this million-kelvin intercloud gas, it would do little to keep you warm. Even though the atoms are moving quickly and each one would impart a lot of energy if it collided with you, there are so few atoms that collisions would be rare. You would radiate energy away and cool off much faster than the gas around you could replace the lost energy.

The Solar System is passing through a bubble of this hot intercloud gas that has a density of about 0.005 hydrogen atom/cm³ and is at least 650 light-years across. This may be the remnant of the hot bubble produced by a supernova explosion 300,000 years ago. Like the million-kelvin gas in the corona of the Sun, hot intercloud gas glows faintly in the energetic X-ray portion of the electromagnetic spectrum. Orbiting X-ray telescopes observe the entire sky aglow with faint X-rays coming from our local bubble of million-kelvin gas.

Not all intercloud gas is as hot as that of the local bubble. Most other intercloud gas is "warm" and has a temperature of about 8000 K and a density ranging from about 0.01 to 1 atom/cm³. About half of the volume of warm intercloud gas is kept ionized by starlight. Ultraviolet starlight with wavelengths shorter than 91.2 nm has enough energy to **ionize** hydrogen, which means the electron has been stripped away from the nucleus. But if there is a large enough expanse of warm intercloud gas, then the outermost atoms shield the neutral gas nearer the center by "using up" all the ultraviolet light, much as Earth's ozone layer shields the surface of Earth from harmful ultraviolet radiation from the Sun. This shielding leaves much gas in an unionized state.

One way to look for interstellar gas, including both warm and hot interstellar gas, is to study the spectra of distant stars. Most commonly, this gas is observed in absorption lines in the spectra of distant stars. Atoms in the gas absorb starlight at particular wavelengths, indicating the temperature, density, and chemical composition of the gas. Intercloud gas can also sometimes produce emission lines in regions of warm, ionized gas, for example in supernova remnants, where protons and electrons constantly recombine into hydrogen atoms. When a proton and an electron combine to form a neutral hydrogen atom, energy must be given up in the form of electromagnetic radiation. Typically, the resulting hydrogen atom is left in an excited state (see Chapter 5). The atom then drops down to lower and lower energy states, emitting a photon at each step, so warm, ionized interstellar gas glows in emission lines characteristic of hydrogen. Usually, the strongest emission line given off by warm interstellar gas in the visible part of the spectrum is the Hα (hydrogen alpha) line, which is seen in the red part of the spectrum at a wavelength of 656.3 nm. Other elements undergo a similar process.

The faint, diffuse emission in **Figure 15.7** comes mostly from warm (about 8000 K), ionized intercloud gas, glowing in Hα. The bright spots are called **H II regions** ("H two") because the hydrogen atoms are ionized; that is, they are in the second state after neutral. In H II regions, intense ultraviolet radiation from massive, hot, luminous O and B stars is able to ionize even relatively dense interstellar clouds, causing Hα emission in a rough sphere around each star. O stars live only a few million years, so they usually do not move very far from where they formed. The glowing clouds seen as H II regions are the very clouds from which these stars were born and indicate active star formation.

Figure 15.7 Warm interstellar gas (about 8000 K) glows in the Hα line of hydrogen. This image of the Hα emission from much of the northern sky reveals the complex structure of the interstellar medium.

Figure 15.8 The Orion Nebula is only a small part of the larger Orion star-forming region. The dark Horsehead Nebula is seen at the lower left of image (b) and in (a). The circular halos around the bright stars in (a) are a photographic artifact. (c) The Orion Nebula is seen as a glowing region of interstellar gas surrounding a cluster of young, hot stars. New stars are still forming in the dense clouds surrounding the nebula.

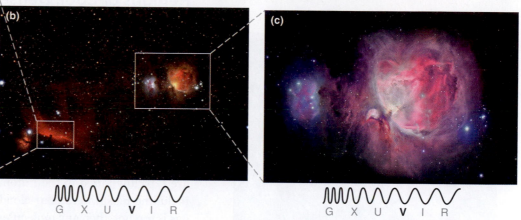

One of the closest H II regions to the Sun is the Orion Nebula, located 1,340 light-years from the Sun in the constellation Orion (**Figure 15.8**). Almost all of the ultraviolet light that powers this nebula comes from a single hot star, and only a few hundred stars are forming in its immediate vicinity. In contrast, a dense star cluster containing thousands of hot, luminous stars powers a giant H II region called 30 Doradus, located in the Large Magellanic Cloud, a small companion galaxy to the Milky Way 160,000 light-years away. If 30 Doradus were as close as the Orion Nebula, it would be bright enough in the nighttime sky to cast shadows.

Warm, *neutral* hydrogen gas gives off radiation in a different way from warm, *ionized* hydrogen. Many subatomic particles, including protons and electrons, have a property called *spin* that causes them to behave as though each particle has a bar magnet, with a north and a south pole, built into it. As demonstrated in **Figure 15.9**, a hydrogen atom can exist in only two configurations: either the magnetic "poles" of the proton and electron point in opposite directions or they are aligned. These configurations have different energies. When the two "magnets" point in the same direction, the atom has slightly less energy then when they point in the opposite direction. If left undisturbed long enough, a hydrogen atom in the higher energy state will spontaneously jump to the lower energy state, emitting a photon in the process. The energy difference between the two magnetic spin states of a hydrogen atom is extremely small, so the emitted photon has a wavelength of 21 cm, in the radio region of the spectrum. Later, interactions between atoms in the gas will bump the hydrogen atoms back to the higher energy state, refreshing the supply of atoms that can produce the 21-cm line.

The tendency for hydrogen atoms to emit 21-cm radiation is extremely weak. On average, you would have to wait about 11 million years for a hydrogen atom in the higher energy state to jump spontaneously to the lower energy state and give off a photon. But there is a lot of hydrogen in the universe, so at any given time, some atoms are making this transition. In **Figure 15.10**, the sky is aglow with 21-cm radiation from neutral hydrogen. Because of its long wavelength, 21-cm radiation freely penetrates dust in the interstellar medium, enabling astronomers to see neutral hydrogen throughout the galaxy, while measurements of the Doppler shift of

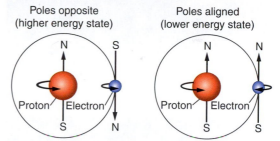

A 21-cm photon is emitted when poles go from being opposite to aligned (a spin flip).

Figure 15.9 There is a slight difference in energy when the poles of the proton and electron are aligned compared to when they are opposite. This energy difference corresponds to a photon of 21 cm.

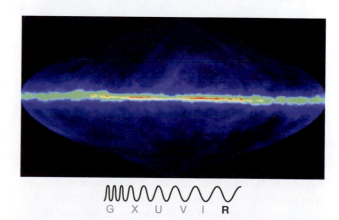

Figure 15.10 This radio image of the sky shows the distribution of neutral hydrogen gas throughout the galaxy. Red indicates directions of the highest hydrogen density, and blue and black show areas with little hydrogen. Radio waves penetrate interstellar dust, and probe of the structure of the galaxy.

the line indicate how fast the emitting gas is moving toward us or away from us. These two attributes make the 21-cm line of neutral hydrogen important for understanding the structure of our galaxy (**Process of Science Figure**).

Regions of Cool, Dense Gas

Most interstellar clouds are composed primarily of isolated neutral hydrogen atoms and are much cooler and denser than the warm intercloud gas. They have temperatures of about 100 K and densities in the range of about 1–100 atoms/cm³. On average, atoms in interstellar clouds are moving around with random velocities of approximately 20 kilometers per second (km/s). The fastest motions of interstellar material are those of very hot gas and are measured in thousands of kilometers per second.

On Earth, it is uncommon to find atoms in isolation; most atoms are in molecules. For example, in Earth's atmosphere, only nonreactive gases such as argon are typically found in their atomic form. In most of interstellar space, however, including most interstellar clouds, molecules do not survive long. If interstellar gas is too hot, then any molecules that do exist soon collide with other molecules or atoms that have enough energy to break the molecules apart. The temperature in a neutral hydrogen cloud may be low enough for some molecules to survive, but photons of starlight with enough energy to break molecules apart can penetrate neutral hydrogen clouds. The hearts of the densest of interstellar clouds are known as **molecular clouds** because in these regions, dust effectively blocks even relatively low-energy photons, and so molecules persist.

Molecular cloud masses range from a few solar masses to 10 million solar masses. The smallest molecular clouds may be less than half a light-year across; the largest may be more than a thousand light-years in size. **Giant molecular clouds** are typically about 100–200 light-years across and have masses a few hundred thousand times that of the Sun. The Milky Way Galaxy contains several thousand giant molecular clouds and a much larger number of smaller ones. Molecular clouds fill only about 0.1 percent of interstellar space. These clouds may be rare, but they are extremely important because they are the cradles of star formation.

In images such as those of **Figure 15.11**, the dust in molecular clouds causes a silhouette in visible light against a background of stars. Infrared radiation passes through this dust, however, so we can see through the dust to sources inside and behind the cloud. Inside such clouds it is dark and usually very cold, with a typical temperature of only about 10 K. Most of these clouds have densities of about 100–1,000 molecules/cm³, but densities as high as 10^{10} molecules/cm³ have been observed. Even at 10^{10} molecules/cm³, this gas is still less than a billionth as dense as the air around you, making it an extremely good vacuum by terrestrial standards. In this cold, relatively dense environment, atoms combine to form a wide variety of molecules.

By far the most common component of molecular clouds is molecular hydrogen. Molecular hydrogen (H_2) consists of two hydrogen atoms and is the smallest possible molecule. Molecules radiate as they make transitions between energy states: these states are determined by the way the molecules rotate or vibrate, for example. Molecular emission lines are useful in the same way that atomic emission lines are useful. Each type of molecule is unique in its properties, and thus unique in its energy states. The wavelengths of emission lines from molecules are an unmistakable fingerprint of the kinds of molecules responsible for them.

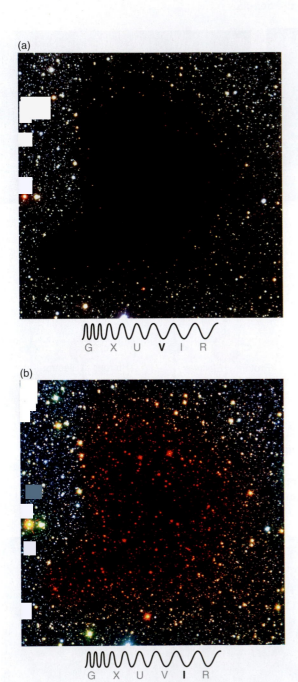

(a)

G X U V I R

(b)

G X U V I R

Figure 15.11 In visible light, interstellar molecular clouds are seen in silhouette against a background of stars and glowing gas. (a) Light from background stars is blocked by dust and gas in nearby Barnard 68, a dense, dark molecular cloud. (b) Infrared wavelengths can penetrate much of this gas and dust, as seen in this false-color image of Barnard 68.

ALL BRANCHES OF SCIENCE ARE INTERCONNECTED

Studies of the natural world on the smallest and largest scales inform one another.

Atomic physics: Scientists studying atoms and quantum mechanics—the underlying principles that govern the behavior of atoms—find that the electron in the hydrogen atom makes a rare transition, releasing radiation with a wavelength of 21 cm. This is so rarely observed on Earth that it is known as a *forbidden transition*.

Radio astronomy: Because space is so large, there are vast numbers of hydrogen atoms along the line of sight in every direction. There are so many that even exceptionally rare events happen often enough to be detectable. The forbidden transition is commonly observed by radio astronomers and maps out the neutral hydrogen in the Milky Way.

Radio astronomers use atomic physics to understand the behavior of the entire galaxy, which is examined at a scale 10^{30} times larger.

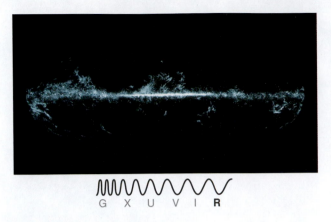

Figure 15.12 This all-sky image from the Planck observatory shows the distribution of carbon monoxide (CO), which traces molecular clouds where stars are born.

Because some of these transitions are in the radio or infrared portions of the spectrum, these molecules can be detected even deep inside a molecular cloud.

In addition to molecular hydrogen, approximately 150 other molecules have been observed in interstellar space. These molecules range from very simple structures such as carbon monoxide (CO), to complex organic compounds such as methanol (CH_3OH) and acetone ($(CH_3)_2CO$), to some with large carbon chains. Very large carbon molecules, made of hundreds of individual atoms, bridge the gap between large interstellar molecules and small interstellar grains. Visible light cannot escape from the molecular clouds where interstellar molecules are concentrated. The radio waves emitted by molecules, however, are unaffected by interstellar dust, so they escape easily from dark molecular clouds. Observations of molecular lines reveal the innermost workings of the densest and most opaque interstellar clouds. Among the more important molecules is CO (**Figure 15.12**). The ratio of CO to H_2 is relatively constant, as far as has been tested, and interstellar carbon monoxide observed from Earth is often used to estimate the amounts and distribution of interstellar neutral molecular hydrogen, which is more difficult to observe directly.

CHECK YOUR UNDERSTANDING 15.1

When radiation from an object passes through the interstellar medium: (a) the object appears dimmer; (b) the object appears bluer; (c) the object appears bluer *and* dimmer; (d) the object appears redder *and* dimmer.

15.2 Molecular Clouds Are the Cradles of Star Formation

Stars—and planets—form from large clouds of dust and gas in the interstellar medium. This star formation is triggered by nearby energetic events such as the explosion of a dying massive star. In this section, we explore the first steps of this process as a cloud begins to contract and fragment to form stars.

Self-Gravity in the Molecular Cloud

Recall that in our discussion of the Sun in Chapter 14, we referred to the balance between gravity and pressure in a stable object as *hydrostatic equilibrium*. Interstellar clouds are not always in hydrostatic equilibrium—in most interstellar clouds, internal pressure is much stronger than the gravity that holds a cloud together (the "self-gravity"). Because gravity follows an inverse square law, the more spread out an object's mass, the weaker its self-gravity. In most interstellar clouds, the internal gas pressure pushing out is much stronger than self-gravity, so a cloud should expand. But the much hotter gas surrounding the clouds also exerts a pressure inward on a cloud. This external hot gas helps hold a cloud together and often provides a trigger for collapse.

If a cloud is massive enough and dense enough (or becomes so after a triggering event), self-gravity becomes important. As shown in **Figure 15.13**, each part of the cloud feels a gravitational attraction from every other part of the cloud. The sum of all these forces acting on a particular parcel of gas will always point toward the center of the cloud's mass. This is the *net force* and indicates the direction in which the parcel will begin to move. In massive, dense clouds, self-gravity

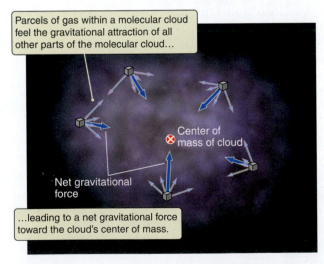

Parcels of gas within a molecular cloud feel the gravitational attraction of all other parts of the molecular cloud…

Center of mass of cloud

Net gravitational force

…leading to a net gravitational force toward the cloud's center of mass.

Figure 15.13 Self-gravity causes a molecular cloud to collapse, drawing parcels of gas toward a single point inside the cloud. The lighter blue arrows are examples of forces on the parcel due to other parcels of gas. The darker blue arrows show the sum of all these forces. This net force always points toward the center of mass of the cloud.

is much greater than pressure, so the clouds collapse under their own weight, beginning a chain of events that will form a new generation of stars.

If self-gravity in a molecular cloud is much greater than internal pressure, gravity should win outright, and the cloud should rapidly collapse toward its center. In practice, the process goes very slowly because several other effects stand in the way of the collapse. One effect that slows the collapse of a cloud is conservation of angular momentum, described in Chapter 7. Other effects that slow the collapse are turbulence, the effects of magnetic fields, and thermal pressure. Even though these effects may slow the collapse of a molecular cloud, in the end gravity will predominate. One part of the cloud can lose angular momentum to another part of the cloud, allowing the part of the cloud with less angular momentum to collapse further. Neutral matter crosses magnetic field lines, gradually increasing the gravitational pull toward the center, until the force on the charged particles is large enough to drag the magnetic field toward the center as well. Turbulence ultimately fades away. The details of these processes are complex and are the subject of much current research. For our purpose, the important point is that effects that prevent the collapse of a molecular cloud are temporary, and gravity is persistent. As the forces that oppose the cloud's self-gravity gradually fade, the cloud slowly shrinks.

Astronomy in Action: Angular Momentum

▶❙❙ **AstroTour:** Star Formation

Molecular Clouds Fragment as They Collapse

Molecular clouds are clumpy. Some regions within the cloud are denser and collapse more rapidly than surrounding regions. As these regions collapse, their self-gravity becomes stronger because they are more compact, so they collapse even faster. **Figure 15.14** shows the process of collapse in a molecular cloud. Slight variations in the density of the cloud grow to become very dense concentrations of gas. Instead of collapsing into a single object, the molecular cloud fragments into very dense **molecular-cloud cores**. A single molecular cloud may form hundreds or thousands of molecular-cloud cores, each of which is typically a few light-months in size. Some of these dense cores will eventually form stars.

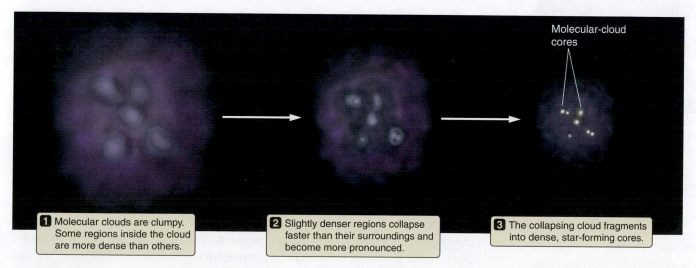

Molecular-cloud cores

1 Molecular clouds are clumpy. Some regions inside the cloud are more dense than others.

2 Slightly denser regions collapse faster than their surroundings and become more pronounced.

3 The collapsing cloud fragments into dense, star-forming cores.

Figure 15.14 When a molecular cloud collapses, denser regions within the cloud collapse more rapidly than less dense regions. As this process continues, the cloud fragments into a number of very dense molecular-cloud cores that are embedded within the large cloud. These cloud cores may go on to form stars.

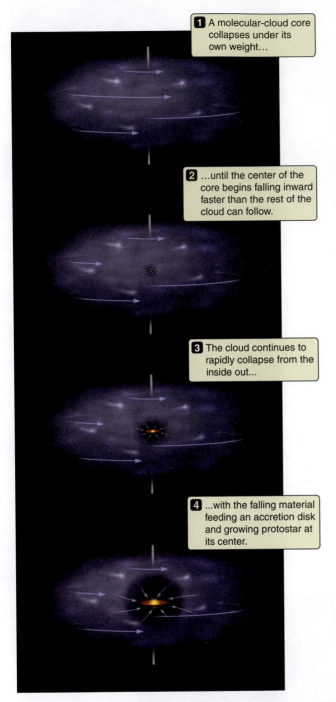

1 A molecular-cloud core collapses under its own weight…

2 …until the center of the core begins falling inward faster than the rest of the cloud can follow.

3 The cloud continues to rapidly collapse from the inside out…

4 …with the falling material feeding an accretion disk and growing protostar at its center.

Figure 15.15 When a molecular-cloud core gets very dense, it collapses from the inside out. Conservation of angular momentum causes the infalling material to form an accretion disk that feeds the growing protostar.

As a molecular-cloud core collapses, the gravitational forces grow stronger still, because the force of gravity is inversely proportional to the square of the radius. Suppose a cloud is 4 light-years across. When the cloud has collapsed to 2 light-years across, the different parts of the cloud are, on average, only half as far apart as when the collapse started. As a result, the gravitational attraction the parts of the cloud feel toward each other is 4 times stronger. When the cloud is one-fourth as large as it was at the beginning of the collapse, the force of gravity is 16 times stronger. As a core collapses, the inward force of gravity increases; as gravity increases, the collapse speeds up; as the collapse speeds up, the gravitational force increases even faster.

Eventually, gravity is able to overwhelm the opposing forces due to pressure, magnetic fields, and turbulence. This happens first near the center of the cloud core where the cloud material is most strongly concentrated. The pressure from the central part of the cloud core supports the weight of the layers above it. The inner parts of the cloud core start to fall rapidly inward, removing support from the more distant parts of the cloud. Without the support of that inner material, the more distant material begins to fall freely toward the center. The cloud core collapses from the inside out, as shown in **Figure 15.15**.

CHECK YOUR UNDERSTANDING 15.2

Molecular clouds fragment as they collapse because: (a) the rotation of the cloud throws some mass to the outer regions; (b) the density increases fastest in the center of the cloud; (c) density variations from place to place grow larger as the cloud collapses; (d) the interstellar wind is stronger in some places than others.

15.3 Formation and Evolution of Protostars

Material from the collapsing molecular-cloud core falls inward. Because of conservation of angular momentum, this material accumulates in a flat, rotating accretion disk. Most of this material eventually finds its way inward to the center of the disk. The object that is forming there, on its way to becoming a star, is called a **protostar**. In this section, we follow the evolution of this object on the way to becoming a star.

A Protostar Forms

As the cloud collapses, gravitational energy is converted to thermal energy. Particles are pulled toward the center by gravity. As they fall, they move faster and faster. As they become more densely packed, they begin to crash into each other, causing random motions and raising the temperature of the core. These random motions of particles are collectively known as the *thermal energy*. When the particles are hotter, they are moving faster, and the thermal energy is higher. In the collapsing protostar, the thermal energy comes from the gravitational energy that was stored in the cloud when the particles were far apart. The collapse of the cloud converts gravitational energy to thermal energy, and the gas in the outer layers of the protostar is heated to a temperature of thousands of kelvins, causing the protostar to shine.

Because of this accumulation of thermal energy, the core gradually reaches more than 1 million K. At this temperature, deuterium can fuse with hydrogen to create helium-3. (Notice that this reaction occurs at a lower temperature than that at which hydrogen burning occurs.) Once deuterium burning begins, it drives convection in the core. You may recall from Chapter 14 that convection is the transport of energy by moving packets of gas. This turbulent motion temporarily keeps more gas from falling in and creates a "surface"—more properly called a photosphere. This photosphere radiates away energy. The hotter it gets, the more energy it radiates, and the bluer that radiation becomes (see Working It Out 5.3). The photosphere of a protostar is tens of thousands of times larger than the surface of the Sun today, and each square meter radiates away energy. As a result, the protostar is thousands of times more luminous than the Sun.

Although the protostar is extremely luminous, astronomers often cannot see it in visible light for two reasons: First, the photosphere of the protostar is relatively cool, so most of its radiation is in the infrared part of the spectrum. Second, and even more important, the protostar is buried deep in the heart of a dense and dusty molecular cloud. However, astronomers are able to view protostars in the infrared part of the spectrum because much of the longer-wavelength infrared light from a protostar is able to escape through the cloud. In addition, as the dust absorbs the visible light, it warms up, and this heated dust also glows in the infrared.

Sensitive infrared instruments developed since the 1980s have revolutionized the study of protostars and other young stellar objects. Dark clouds have revealed themselves to be clusters of dense cloud cores, young stellar objects, and glowing dust when viewed in the infrared. Stars are forming in nodules attached to the tops of the columns of dust and gas in the Eagle Nebula, shown in **Figure 15.16**.

The Evolving Protostar

At any given moment, the protostar is in balance: the forces from hot gas pushing outward and the force of gravity pulling inward exactly oppose each other. However, this balance is constantly changing. Once the core switches from convection to radiation, the deuterium in the core becomes depleted. This allows material to resume falling onto the protostar, adding to its mass and gravitational pull inward and therefore increasing the weight that underlying layers of the protostar must

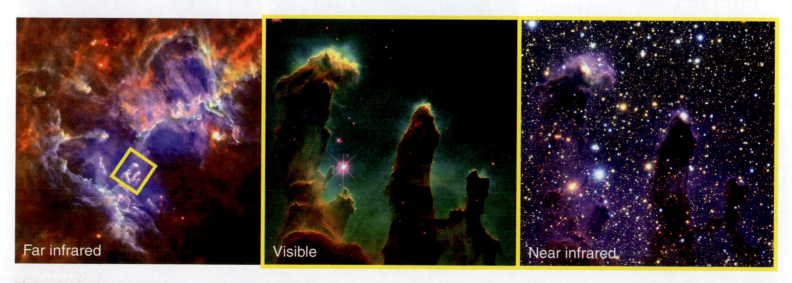

Far infrared   Visible   Near infrared

Figure 15.16 The Eagle Nebula contains dense columns of molecular gas and dust at the edge of an H II region. The yellow box in the left image identifies the region that is magnified in the second and third images.

support. The protostar also slowly loses its internal thermal energy by radiating it away.

How can an object be in perfect balance and yet be changing at the same time? Consider an everyday example. **Figure 15.17a** shows a simple spring balance, which works on the principle that the more a spring is compressed, the harder it pushes back. You can measure the weight of an object by determining the point at which the pull of gravity and the push of the spring are equal.

When sand is poured slowly onto the spring balance, at any instant the downward weight of the sand is balanced by the upward force of the spring. As the weight of the sand increases, the spring is slowly compressed. The spring and the weight of the sand are always in balance, but this balance is changing with time as more sand is added. Turning to Figure 15.17b, you can see that the situation is analogous to that of the protostar, in which the outward pressure of the gas behaves like the spring. The inward force of gravity is always matched by the internal pressure pushing out.

Material falls onto the protostar, adding to its mass and gravitational pull. Even though the protostar slowly loses internal thermal energy by radiating it away, the material that has fallen onto the protostar also compresses the protostar and heats it up. The interior becomes denser and hotter, and the pressure rises—just enough

Figure 15.17 (a) A spring balance comes to rest at the point where the downward force of gravity is matched by the upward force of the compressed spring. As sand is added, the location of this balance point shifts. (b) Similarly, the structure of a protostar is determined by a balance between pressure and gravity. Like the spring balance, the structure of the protostar constantly shifts as additional material falls onto its surface and as the protostar radiates energy away.

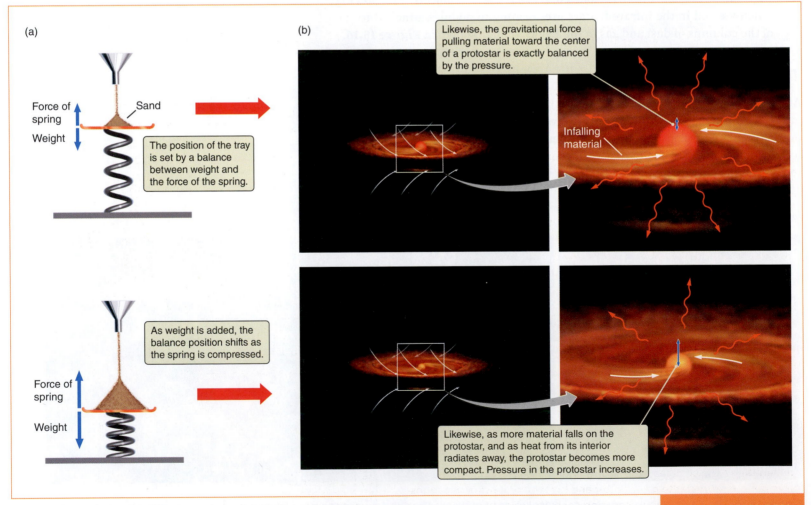

(a)

Force of spring

Weight

Sand

The position of the tray is set by a balance between weight and the force of the spring.

As weight is added, the balance position shifts as the spring is compressed.

Force of spring

Weight

(b)

Likewise, the gravitational force pulling material toward the center of a protostar is exactly balanced by the pressure.

Infalling material

Likewise, as more material falls on the protostar, and as heat from its interior radiates away, the protostar becomes more compact. Pressure in the protostar increases.

VISUAL ANALOGY

to balance the increased weight of the material above it. Dynamic balance is always maintained as the protostar slowly contracts.

Figure 15.18 illustrates this chain of events as the protostar shrinks. Gravitational energy is converted to thermal energy, which heats the core, raising the pressure to oppose gravity. This process continues, with the protostar becoming smaller and smaller and its interior growing hotter and hotter. If the protostar is massive enough, its interior will eventually become so hot that nuclear fusion of hydrogen to helium can begin. This is the point at which the transition from protostar to star takes place. The distinction between the two is that a protostar draws its energy from gravitational collapse, whereas a star draws its energy from thermonuclear reactions in its interior.

The protostar's mass determines whether it will actually become a star. As the protostar slowly collapses, the temperature at its center rises. If the protostar's mass is greater than about 0.08 times the mass of the Sun (0.08 M_{Sun}), the temperature in its core will eventually reach 10 million K, and fusion of hydrogen into helium will begin. The newly born star will once again adjust its structure until it is radiating energy away from its surface at just the rate that energy is being liberated in its interior. As it does so, it achieves hydrostatic and thermal equilibrium and "settles" onto the main sequence of the Hertzsprung-Russell (H-R) diagram, where it will spend the majority of its life.

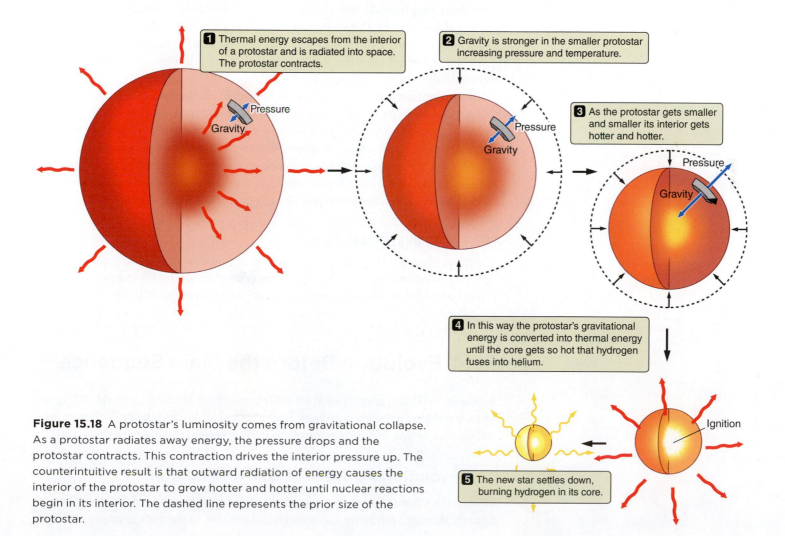

1 Thermal energy escapes from the interior of a protostar and is radiated into space. The protostar contracts.

2 Gravity is stronger in the smaller protostar increasing pressure and temperature.

3 As the protostar gets smaller and smaller its interior gets hotter and hotter.

4 In this way the protostar's gravitational energy is converted into thermal energy until the core gets so hot that hydrogen fuses into helium.

5 The new star settles down, burning hydrogen in its core.

Pressure
Gravity
Ignition

Figure 15.18 A protostar's luminosity comes from gravitational collapse. As a protostar radiates away energy, the pressure drops and the protostar contracts. This contraction drives the interior pressure up. The counterintuitive result is that outward radiation of energy causes the interior of the protostar to grow hotter and hotter until nuclear reactions begin in its interior. The dashed line represents the prior size of the protostar.

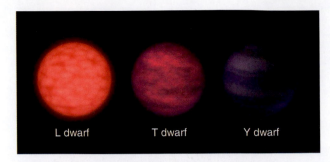

Figure 15.19 This artist's conception shows the three types of brown dwarf stars: L dwarfs ($T \sim 1700$ K); T dwarfs ($T \sim 1200$ K); and Y dwarfs ($T \sim 500$ K).

Brown Dwarfs

If the mass of the protostar is less than 0.08 M_{Sun}, it will never reach the point at which sustained nuclear burning takes place. An object of roughly this mass is called a **brown dwarf**, or sometimes it is called a *substellar object*. A brown dwarf forms in the same way a star forms, yet in many respects it is like a giant planet Jupiter. The International Astronomical Union (IAU) has somewhat arbitrarily set the boundary between a brown dwarf and a supermassive giant planet at 13 Jupiter masses (13 M_{Jup}), although some astronomers think that 10 M_{Jup} would be more appropriate. That is, if the mass of the object is greater than 10 M_{Jup}, it must be a brown dwarf, not a supermassive planet. The upper limit of brown dwarf masses is about 75–80 M_{Jup}. Despite this range of brown dwarf masses, they all have radii about the same as Jupiter's radius—more massive brown dwarfs are denser.

Brown dwarf spectral types L, T, and Y have been added to the sequence of spectral classes to represent stars that are even cooler than M stars (**Figure 15.19**). A brown dwarf never grows hot enough to burn the most common hydrogen nuclei consisting of a single proton, but instead glows primarily by continually cannibalizing its own gravitational energy. The cores of brown dwarfs larger than 13 M_{Jup} can get hot enough to burn deuterium (^{2}H), and those with a mass greater than 65 M_{Jup} can burn lithium. But both of these energy sources are in very short supply, and after a brief period of deuterium or lithium fusion, brown dwarfs shine only by the energy of their own gravitational contraction. As the years pass, a brown dwarf becomes progressively smaller and fainter. The coldest Y dwarfs observed with the WISE infrared space telescope are colder than the human body, which radiates at 310 K.

Since the first brown dwarfs were identified in the mid-1990s, more than a thousand have been found. The cooler among them have methane and ammonia in their atmospheres, similar to what is found in the atmospheres of the giant planets of the Solar System. Winds on brown dwarfs can be very high, producing weather far more violent than storms observed in the atmospheres of the giant planets. Planets have been found orbiting a few brown dwarfs. The nearest one to us is a binary brown dwarf system about 2 parsecs (6.5 light-years) away.

CHECK YOUR UNDERSTANDING 15.3

The energy required to begin nuclear fusion in a protostar originally came from: (a) the gravitational potential energy of the protostar; (b) the kinetic energy of the protostar; (c) the wind from nearby stars; (d) the pressure from the interstellar medium.

....

15.4 Evolution Before the Main Sequence

Protostars and young evolving stars will change their location on the H-R diagram as they settle into the main sequence, where they will spend the bulk of their lives. In this section, we examine some of the early stages in the life of the new stars.

The Evolutionary Track of an Evolving Star

Within the young protostar, convection carries energy outward, keeping the protostar's interior well stirred. Although the interior of the protostar grows hotter and hotter as it contracts, its photosphere stays at about the same, much cooler, temperature through most of this phase of its evolution. This temperature difference is similar to that between the photosphere of the Sun, which is about 5780 K, and its interior, which is millions of kelvins.

In the 1960s, the theoretical physicist Chushiro Hayashi (1920–2010) explained the difference between the surface temperature of a star or protostar and the temperature deep in its interior. Hayashi showed that the atmospheres of stars and protostars contain a natural thermostat: the H^- ion. (An H^-, or "H minus," ion is a hydrogen atom that has acquired an extra electron and therefore has a negative charge.) The amount of H^- in the atmosphere of a protostar is highly sensitive to the temperature at the protostar's surface. The cooler the atmosphere of a star, the more slowly atoms and electrons are moving, and the easier it is for a hydrogen atom to hold on to an extra electron. As a result, the cooler the atmosphere of the star, the more H^- there is.

The H^- ion, in turn, helps control how much energy a star or protostar radiates away. The more H^- there is in the atmosphere of the star or protostar, the more opaque the atmosphere is, and the more effectively the thermal energy of the protostar is trapped in its interior. Imagine that the surface of the protostar is "too cool," meaning that extra H^- forms in the atmosphere and makes the atmosphere of the protostar more opaque. The atmosphere thus traps more of the radiation that is trying to escape, and the trapped energy heats up the star. As the temperature climbs, the H^- ions are changed to neutral H atoms. Now imagine the other possibility—that the protostar is too hot. In this case, H^- in the protostar's atmosphere is destroyed, so the atmosphere becomes more transparent, allowing radiation to escape more freely from the interior. Because the protostar cannot hold on to enough of its energy to stay warm, the surface cools. In either case—too cold or too hot—H^- is formed or destroyed until the star's atmosphere once again traps just the right amount of escaping radiation. The H^- ion is basically doing the same thing that you do with your bedcovers at night. If you get too cold, you pile on extra covers to trap your body's thermal energy and keep warm (more H^- ions). If you get too hot, you kick off some covers to cool off (fewer H^- ions).

The amount of H^- in the atmosphere keeps the surface temperature of the protostar somewhere between about 3000 and 5000 K, depending on the protostar's mass and age. Because the surface temperature of the protostar is not changing much, the amount of energy per unit time (power) radiated away by each square meter of the surface of the protostar does not change much either. Recall the Stefan-Boltzmann law from Chapter 5, which says that the amount radiated by each square meter of an object's surface is determined by its temperature. As the protostar shrinks, the area of its surface shrinks as well. There are fewer square meters of surface from which to radiate, so the luminosity of the protostar drops. As viewed from the outside, the protostar stays at nearly the same temperature and color but gradually gets fainter as it evolves toward its eventual life as a main-sequence star.

In Chapter 13, we introduced the H-R diagram and used it to help explain how the properties of stars differ. For the next several chapters, we will use the H-R diagram to keep track of how stars change as they evolve through their lifetimes. The path on the H-R diagram that a star follows as it goes through the different stages of its life is called the star's **evolutionary track**. The protostar is brighter than it will be as a true star on the main sequence, so a protostar's track is located above the main sequence on the H-R diagram. **Figure 15.20** shows the pre-main-sequence evolutionary tracks of stars of several different

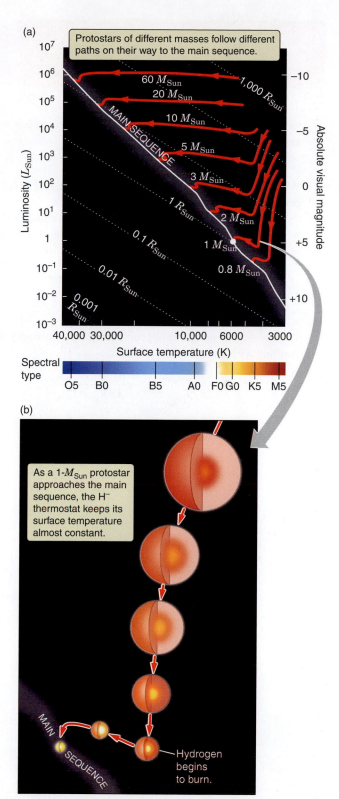

Figure 15.20 (a) The evolution of pre-main-sequence stars can be followed on the H-R diagram. More massive protostars in the upper right portion of the diagram follow horizontal tracks. (b) The roughly vertical, constant-temperature part of the evolutionary track of a low-mass protostar is called the Hayashi track.

15.2 Working It Out Luminosity, Surface Temperature, and Radius of Protostars

In Chapter 13, you learned how the luminosity, surface temperature, and radius of a star are related:

$$L = 4\pi R^2 \sigma T^4$$

What can this equation reveal about the changing properties of the protostar as it shrinks its radius?

Suppose that when the Sun was a protostar, it had a radius 10 times what it is now and a surface temperature of 3300 K. What would its luminosity have been? The equations for each are

$$L_{\text{protostar}} = 4\pi R^2_{\text{protostar}} \sigma T^4_{\text{protostar}}$$

and

$$L_{\text{Sun}} = 4\pi R^2_{\text{Sun}} \sigma T^4_{\text{Sun}}$$

We can set this up as a ratio, comparing the luminosity of the protostar Sun with its luminosity now, L_{Sun}:

$$\frac{L_{\text{protostar}}}{L_{\text{Sun}}} = \frac{4\pi R^2_{\text{protostar}} \sigma T^4_{\text{protostar}}}{4\pi R^2_{\text{Sun}} \sigma T^4_{\text{Sun}}}$$

We rewrite this as follows, grouping like terms together:

$$\frac{L_{\text{protostar}}}{L_{\text{Sun}}} = \frac{4\pi\sigma}{4\pi\sigma} \times \left(\frac{R_{\text{protostar}}}{R_{\text{Sun}}}\right)^2 \times \left(\frac{T_{\text{protostar}}}{T_{\text{Sun}}}\right)^4$$

Then we cancel out the constants, $4\pi\sigma$, and use the value for $T_{\text{Sun}} = 5780$ K from Chapter 14. We know that the protostar's radius is 10 times that of the Sun, so $R_{\text{protostar}}/R_{\text{Sun}} = 10$. Then the equation becomes

$$\frac{L_{\text{protostar}}}{L_{\text{Sun}}} = \left(\frac{10}{1}\right)^2 \times \left(\frac{3300}{5780}\right)^4 = 10^2 \times (0.57)^4 = 10.6$$

So the Sun was about 10.6 times more luminous as a protostar than it is now. We see this on the H-R diagram of protostars (see Figure 15.20). As a 1-M_{Sun} star approaches the main sequence on the diagram, it moves down (toward lower luminosity) and to the left (toward higher surface temperature).

---

masses. In a higher-mass protostar, the luminosity stays about the same, but the surface temperature increases; on the H-R diagram, its track moves left horizontally to the main sequence. In a lower-mass protostar, the temperature stays about the same, but the luminosity decreases. On the H-R diagram, its track moves vertically down to the main sequence; this is called its **Hayashi track** (Figure 15.20b). The relationship between surface temperature, luminosity, and radius of protostars is further explored in **Working It Out 15.2**.

Bipolar Outflow

As shown in **Figure 15.21a**, material falls onto the accretion disk around a young stellar object and moves inward toward the equator of the star, while at the same time other material is blown away from the protostar and disk in two opposite directions from the plane of the disk. The resulting stream of material away from the protostar is called a **bipolar outflow**. Powerful outflows can disrupt the cloud core and accretion disk from which the protostar formed, shutting down the flow of material onto the protostar.

Some bipolar outflows from young stellar objects are slow and fairly disordered, but others produce remarkable **jets** of material that move away from the central protostar and disk at velocities of hundreds of kilometers per second (Figure 15.21b). The material in these jets flows out into the interstellar medium, where it heats, compresses, and pushes away surrounding interstellar gas. Knots of glowing gas accelerated by jets are referred to as **Herbig-Haro objects** (or **HH**

Material moves in toward the protostar in the accretion disk.

Bipolar jets and outflows flow away from the young star and disk.

Wind

Jets

This structure is seen in images of forming stars.

Figure 15.21 (a) Material falls onto an accretion disk around a protostar and then moves inward, eventually falling onto the protostar. In the process, some of this material is driven away in powerful jets that stream perpendicular to the disk. (b) This infrared Spitzer Space Telescope image shows jets streaming outward from a young, developing star. Note the nearly edge-on, dark accretion disk surrounding the young star.

objects for short), named after the two astronomers who first identified them and associated them with star formation. An example is shown in **Figure 15.22**.

The origin of outflows from protostars is not well understood, but current models suggest that they are the result of magnetic interactions between the protostar and the disk. The interior of a protostar on its Hayashi track is convective. Great cells of hot gas are rising from the interior, while other cells of cooler gas are falling toward the center. This convection, coupled with the protostar's rapid rotation, can lead to the formation of a dynamo, similar to the dynamo that drives the Sun's magnetic field. The dynamo in the center of a protostar would be much more powerful than the Sun's dynamo, however. The protostar's resulting strong magnetic field might cause the protostar to begin blowing a powerful wind. It might also act something like the blade in a blender, tearing at the inner edge of the accretion disk and flinging material off into interstellar space.

Until the protostellar wind begins, the protostar is enshrouded in the dusty molecular-cloud core from which it was born. As the wind from the protostar disperses this obscuring envelope, the first direct, visible-light view of the protostar emerges: the protostar is "revealed." Some of these protostars of lower mass are called **T Tauri stars**. This name comes from the first recognized member of this class of objects, the star labeled T in the constellation Taurus. Higher-mass

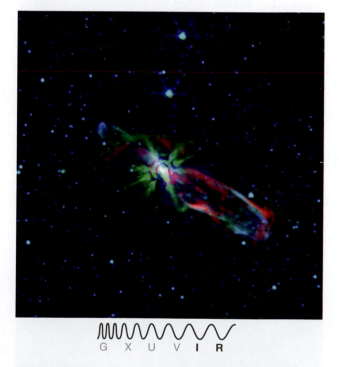

G X U V I R

Figure 15.22 In this combined Spitzer infrared and Atacama Large Millimeter/submillimeter Array (ALMA) observation of Herbig-Haro object HH 46/47, twin supersonic jets originate from the newborn central star, blasting away surrounding gas and forming two lobes.

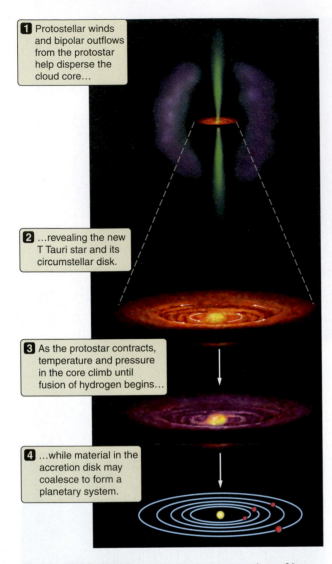

1 Protostellar winds and bipolar outflows from the protostar help disperse the cloud core…

2 …revealing the new T Tauri star and its circumstellar disk.

3 As the protostar contracts, temperature and pressure in the core climb until fusion of hydrogen begins…

4 …while material in the accretion disk may coalesce to form a planetary system.

Figure 15.23 This figure presents an overview of how stars like the Sun form, beginning with the onset of stellar winds and ending with the ignition of a star sitting at the center of a revolving system of planets.

G X U V I R

Figure 15.24 The Pleiades or "Seven Sisters" is a cluster of young stars. The diffuse blue light around the stars is starlight scattered by interstellar dust.

protostars of spectral type B or A are called Herbig Be/Ae stars. **Figure 15.23** summarizes the evolution of a protostar from the onset of winds to the formation of planets.

The Influence of Mass

Astronomers are very interested in how and why molecular clouds subdivide themselves into stars with a range of masses. The details of this division—specifically, what fraction of newly formed stars will have which masses—are crucial if observations of the stars in the Sun's vicinity today are to help untangle the history of star formation in our galaxy. Astronomers do not understand why some cloud cores become 1-M_{Sun} stars while others become 5- or 10-M_{Sun} stars.

A look around our galaxy at large reveals a variety of stars—some very old and others very young. If these were the only stars available to study, it would be extremely difficult to learn much about how stars evolve. But astronomers have long known that stars are often found together in close collections called **star clusters**. **Figure 15.24** shows one such star cluster: a group called the Pleiades or "Seven Sisters." Star clusters are collections of stars that all formed in the same place, from the same material, and at about the same time. They provide extremely useful samples for studying star formation. Even though the few brightest and most massive stars in a cluster dominate any observation of a cluster, most of the stars in a cluster are less massive than the Sun. In fact, some star-forming regions do not seem to form any especially massive stars at all.

After a cloud-core collapse, the evolution of a protostar is determined largely by its mass. Calculations suggest that a star like the Sun probably takes about 10 million years or so to descend its Hayashi track and become a star on the main sequence. Taking into account the entire history of protostellar formation, including the collapse and fragmentation of the molecular cloud itself, the total time for initial fragmentation into cloud core up to the ignition of hydrogen burning might be more like 30 million years. Because the self-gravity of a more massive core is stronger, more massive cores collapse to form stars more quickly. A 10-M_{Sun} star might go from the stage of being a molecular-cloud core to that of burning hydrogen in its interior in only 100,000 years. A 100-M_{Sun} star might take less than 10,000 years. By comparison, a 0.1-M_{Sun} star might take 100 million years finally to reach the main sequence.

The 30 million years or so that it took for the Sun to form is a long time, but it is a tiny fraction of the 10 billion years during which the Sun will steadily fuse hydrogen into helium as a main-sequence star. It is no wonder that so few among the many stars visible in the sky are young. But every star was young at one time, including the Sun.

CHECK YOUR UNDERSTANDING 15.4

Why are so few of the many stars that astronomers see in the sky protostars?
(a) Protostars are hidden in giant molecular clouds. (b) Protostars are small.
(c) Protostars are dim. (d) Protostars are short-lived.

Origins

Star Formation, Planets, and Life

When astronomers consider the possibility of other life in the universe, one of the first things they think about is the formation of stars and planets. Life probably needs planets, and planets form along with stars. The conditions under which a star is born, and the mass and chemical composition that it has when it begins its nuclear fusion, set the stage for the rest of its life. In Chapter 7, we noted that planet formation can be a common consequence of star formation; thus, if a star is going to have planets, they form at about the same time as the star. If the star is going to have rocky planets with hard surfaces (such as the planets of the inner Solar System) or gaseous planets with rocky cores and rocky moons (such as the planets of the outer Solar System), then the material from which the star and planets form must be "enriched" with the heavy elements that make up these rocky surfaces.

These enriched clouds would also provide elements that are essential to life on Earth. In addition to the presence of organic molecules mentioned earlier in the chapter, astronomers have detected water in star-forming regions such as W3 IRS5 (**Figure 15.25**). Water exists as ice mixed with dust grains in the cool molecular clouds or as vapor when it is closer to a protostar and the dust grains and ice evaporate. In 2011, the Herschel Space Observatory detected oxygen molecules (O_2, the type we breathe) in a star-forming complex in Orion. Oxygen is the third most common element in the universe, yet it had not been decisively observed before in molecular form. This oxygen also may have come from the melting and evaporation of water ice on the tiny dust grains.

As noted in Chapter 13, astronomers had doubted that planets could exist in stable orbits in binary star systems, but now a few such circumbinary systems have been found. Planets that form within associations of O and B stars may be too unstable to last very long. Isolated planets unattached to any star likely move through the Milky Way. Perhaps these isolated planets were gravitationally ejected soon after their formation in a multiple system. But these planets do not have a source of energy like Earth's Sun. Astronomers theorize that only planets that orbit stars will be able to support life. So when they try to estimate the possibility of life in the galaxy, astronomers include estimations of the rate of

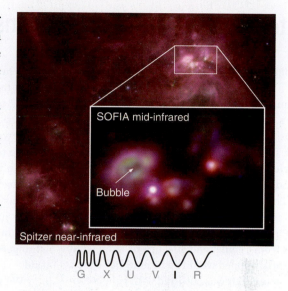

Figure 15.25 Water was detected in the W3 IRS5 star-forming complex. Here, W3 IRS5 is observed with the Spitzer Space Telescope in near infrared and with the Stratospheric Observatory for Infrared Astronomy (SOFIA) telescope in mid-infrared. A massive star has cleared the dust and gas from a small bubble, sweeping it into a dense shell (green).

formation of stars in the galaxy, along with the fraction of stars that have planets. Advances in the study of star formation and planet detection help astronomers to understand better the conditions under which life might develop elsewhere.

In this article, astronomers report on their search for dust from outside of the Solar System.

Interstellar Dust Discovered Inside NASA Spacecraft

By **IRENE KLOTZ, Discovery News**

Thanks to a massive effort by 30,716 volunteers, scientists have pinpointed what appear to be seven precious specks of dust from outside the Solar System, each bearing unique stories of exploded stars, cold interstellar clouds, and other past cosmic lives.

The Herculean effort began eight years ago after NASA's *Stardust* robotic probe flew by Earth to deposit a capsule containing samples from a comet and dust grains from what scientists hoped would be interstellar space. The spacecraft was outfitted with panels containing a smoke-like substance called aerogel that could trap and preserve fast-moving particles.

Stardust twice put itself into position to fish for interstellar grains, which are so small that a trillion of them would fit in a teaspoon. The only way scientists back on Earth would be able to find them was by the microscopic trails the grains made as they plowed into the aerogel.

"When we did the math we realized it would take us decades to do the search ourselves," physicist Andrew Westphal, with the University of California, Berkeley, told Discovery News.

The team used an automated microscope to scan the collector and put out a call for volunteers.

"This whole approach was treated with pretty justifiable criticism by people in my community. They said, 'How can you trust total strangers to take on this project?'" Westphal said.

"We really didn't know how else to do it. We still don't," he added.

Recruits were trained and had to pass a test before they were given digital scans to peruse. Scientists sometimes inserted images with known trails just to see if the volunteers, known as "dusters," would spot them.

"We were very pleased to see that people are really good at finding these tracks, even really, really difficult things to find," Westphal said.

More than 50 candidate dust motes turned out to be bits of the spacecraft itself, but scientists found seven specks that bear chemical signs of interstellar origin and travel.

The grains are surprisingly diverse in shape, size, and chemical composition. The larger ones, for example, have a fluffy, snowflake-like structure.

Additional tests are needed to verify the grains' interstellar origins and ferret out their histories. But the grains are so tiny that with currently available technology, additional analysis would mean their demise.

"It'll probably be years before we can do a lot more with these samples," said space scientist Mike Zolensky, who oversees NASA's collection of cosmic dust, moon rocks, and other extraterrestrial samples at the Johnson Space Center in Houston.

"But we've got them safely tucked away and we can hang on to them until those techniques come along," Zolensky said.

1. Why do scientists want to identify interstellar dust grains?
2. How were these dust grains distinguished from Solar System grains?
3. Why can't the scientists do a complete analysis on these particles?
4. How did volunteer "citizen scientists" assist with this project?
5. Visit the website noted in question 49 in the "Questions and Problems" section later in this chapter. Is this project continuing? What else has been discovered?

"Interstellar dust discovered inside NASA spacecraft," by Irene Klotz. Discovery News, August 14, 2014. Reprinted by permission of Discovery Access.

Summary

Astronomers have never watched the full process of a star forming from beginning to end. Instead, they have observed many different stars at different stages in their formation and evolution at different wavelengths, and they have used their knowledge of physical laws to tie these observations together into a coherent, consistent description of how, why, and where stars form. Stars form from material in the interstellar medium, under the influence of gravity, often triggered by external factors such as the winds of hot stars nearby. The conditions under which a star is born determine whether it will have planets along with the chemical elements required by life as it exists on Earth.

LG 1 **Describe the types and states of material that exist in the space between the stars and how this material is detected.** The interstellar medium is complex, ranging from cold, relatively dense molecular clouds to hot, tenuous intercloud gas heated and ionized by energy from stars and stellar explosions. Dust and gas in the interstellar medium blocks much visible light but becomes more transparent at longer, infrared wavelengths. Different phases of the interstellar medium emit various types of radiation and can be observed at different wavelengths, ranging from radio waves to X-rays. Neutral hydrogen cannot be detected at visible and infrared wavelengths, but it is revealed by its 21-cm emission.

LG 2 **Explain the conditions under which a cloud of gas can contract into a stellar system and the role that gravity and angular momentum play in the formation of stars and planets.** Star formation begins when the self-gravity of dense clouds exceeds outward pressure. The clouds collapse, heat up, and fragment to form stars. The conservation of angular momentum is important to the formation of disks during the collapse. Forming stars are detected from their infrared emission and from the effects they have on their surroundings.

LG 3 **List the steps in the evolution of a protostar.** Protostars collapse, radiating away their gravitational energy until fusion starts in their cores. When they reach hydrostatic and thermal equilibrium, they settle onto the main sequence. Stars form in clusters from dense cores buried within giant molecular clouds. A protostar must have a mass of at least $0.08\,M_{Sun}$ to become a true star. Brown dwarfs are neither stars nor planets, but something in between.

LG 4 **Describe the track of a protostar as it evolves to a main-sequence star on the Hertzsprung-Russell (HR) diagram.** Because star formation takes tens of thousands to millions of years, what astronomers know about the evolution of the birth of stars comes from observations of many protostars at various stages of their development. Protostars of different masses follow different paths to the main sequence, but in general there is a stage at which the protostar moves horizontally, at roughly constant luminosity, followed by a constant temperature drop nearly straight down toward the main sequence. Once hydrogen burning begins, the star again moves nearly horizontally until it reaches the main sequence.

? UNANSWERED QUESTIONS

- Many questions about star formation remain. For example, how must theories be modified to explain the formation of binary stars or other multiple-star systems? At what point during star formation is it determined that a collapsing cloud core will form several stars instead of just one? Some models suggest that this split may happen early in the process, during the fragmentation and collapse of the molecular cloud. The advantage of these ideas is that they provide a natural way of dealing with much of the angular momentum of the cloud core: it goes into the orbital angular momentum of the stars around each other. Other models suggest that additional stars may form from the accretion disk around an initially single protostar.

- Do high-mass and low-mass stars form in very different ways? The smallest stars, spectral type M, are most likely to form as single stars, but a high fraction of medium-mass stars are formed in binary pairs. One theory is that these binaries actually start out as triple systems, from which the smallest star is gravitationally ejected, leading to a remaining pair and a single star. The highest-mass stars are less likely to form alone; many form in OB associations—larger groups of massive stars in which the formation of one large star may stimulate the formation of another nearby in the molecular cloud.

- How common are brown dwarfs? As with extrasolar planets, they have been observed only recently, so their space density is not well known. The Kepler Mission, which finds extrasolar planets (see Chapter 7) and eclipsing binaries (see Chapter 13), also detects brown dwarfs, so there may be an answer to this question in a few years.

Questions and Problems

Test Your Understanding

1. Phases of the interstellar medium include (choose all that apply)
   a. hot, low-density gas.
   b. cold, high-density gas.
   c. hot, high-density gas.
   d. cold, low-density gas.

2. Dust in the interstellar medium can be observed in
   a. visible light.
   b. infrared radiation.
   c. radio waves.
   d. X-rays.

3. The interstellar medium in the Sun's region of the galaxy is closest in composition to
   a. the Sun.
   b. Jupiter.
   c. Earth.
   d. comets in the Oort Cloud.

4. Interstellar dust is effective at blocking visible light because
   a. the dust is so dense.
   b. dust grains are so few.
   c. dust grains are so small.
   d. dust grains are so large.

5. Hot intercloud gas is heated primarily by
   a. starlight.
   b. protostars.
   c. supernova explosions.
   d. neutrinos.

6. Astronomers determined the composition of the interstellar medium from
   a. observing its emission and absorption lines.
   b. measuring the composition of the planets.
   c. return samples from spacecraft.
   d. composition of meteorites.

7. In astronomy, the term *bipolar* refers to outflows that
   a. point in opposite directions.
   b. alternate between expanding and collapsing.
   c. rotate about a polar axis.
   d. show spiral structure.

8. Which of the following has contributed most to our understanding of the process of star formation?
   a. Astronomers have observed star formation as it happens for a small number of stars.
   b. Astronomers have observed star formation as it happens for a large number of stars.
   c. Astronomers have observed many different stars at each step of the formation process.
   d. Theoretical models predict the way stars form.

9. Cold neutral hydrogen can be detected because
   a. it emits light when electrons drop through energy levels.
   b. it blocks the light from more distant stars.
   c. it is always hot enough to glow in the radio and infrared wavelengths.
   d. the atoms in the gas change spin states.

10. The Hayashi track is a nearly vertical evolutionary track on the H-R diagram for low-mass protostars. Which of the following would you expect from a protostar moving along a vertical track?
    a. The star remains the same brightness.
    b. The star remains the same luminosity.
    c. The star remains the same color.
    d. The star remains the same size.

11. Which two forces establish hydrostatic equilibrium in an evolving protostar?
    a. the force from pressure and gravity
    b. the force from pressure and the strong nuclear force
    c. gravity and the strong nuclear force
    d. energy emitted and energy produced

12. Suppose you are studying a visible-light image of a distant galaxy, and you see a dark lane cutting across the bright disk. This dark line is most likely caused by
    a. gravitational instabilities that clear the area of stars.
    b. dust in the Milky Way blocking the view of the distant galaxy.
    c. dust in the distant galaxy blocking the view of stars in the disk.
    d. a flaw in the instrumentation.

13. What causes a hydrogen atom to radiate a photon of 21-cm radio emission?
    a. The electron drops down one energy level.
    b. The formerly free electron is captured by the proton.
    c. The electron flips to an aligned spin state.
    d. The electron flips to an unaligned spin state.

14. Astronomers know that there are dusty accretion disks around protostars because
    a. there is often a dark band across the protostar.
    b. there is often a bright band across the protostar.
    c. theory says accretion disks should be there.
    d. there are planets in the Solar System.

15. What is the single most important property of a star that will determine its evolution?
    a. temperature
    b. composition
    c. mass
    d. radius

Thinking about the Concepts

16. The interstellar medium is approximately 99 percent gas and 1 percent dust. Why does dust and not gas block a visible-light view of the galactic center?

17. Explain why observations in the infrared are necessary for astronomers to study the detailed processes of star formation.

18. How does the material in interstellar clouds and intercloud gas differ in density and distribution?

19. When a star forms inside a molecular cloud, what happens to the cloud? Is it possible for a molecular cloud to remain cold and dark with one or more stars inside it? Explain your answer.

20. If you placed your hand in boiling water (100°C) for even 1 second, you would get a very serious burn. If you placed your hand in a hot oven (200°C) for a second or two, you would hardly feel the heat. Explain this difference and how it relates to million-kelvin regions of the interstellar medium.

21. How do astronomers know that the Sun is located in a "local bubble" formed by a supernova?

22. Interstellar gas atoms typically cool by colliding with other gas atoms or grains of dust; during the collision, each gas atom loses energy and hence its temperature is lowered. How does this explain why very low-density gases are generally so hot, while dense gases tend to be so cold?

23. Explain how the 21-cm line discussed in the Process of Science Figure supports the cosmological principle (which states that the laws of physics must be the same everywhere).

24. Molecular hydrogen is very difficult to detect from the ground, but astronomers can easily detect carbon monoxide (CO) by observing its 2.6-cm microwave emission. Describe how observations of CO might help astronomers infer the amounts and distribution of molecular hydrogen within giant molecular clouds.

25. The Milky Way contains several thousand giant molecular clouds. Describe a giant molecular cloud and its role in star formation.

26. As a cloud collapses to form a protostar, the forces of gravity felt by all parts of the cloud (which follow an inverse square law) become stronger and stronger. One might argue that under these conditions, the cloud should keep collapsing until it becomes a single massive object. Why doesn't this happen?

27. The internal structure of a protostar maintains hydrostatic equilibrium even as more material is falling onto it. Explain how this can be.

28. What are the similarities and differences between a brown dwarf and a giant planet such as Jupiter? Would you classify a brown dwarf as a supergiant planet? Explain your answer.

29. The H^- ion acts as a thermostat in controlling the surface temperature of a protostar. Explain the process.

30. How does the composition of the molecular cloud affect the type of planets and stars that form within it?

Applying the Concepts

31. In Chapter 13, you learned that astronomers can measure the temperature of a star by comparing its brightness in blue and yellow light. Does reddening by interstellar dust affect a star's temperature measurement? If so, how?

32. When a hydrogen atom is ionized, it splits into two components.
    a. Identify the two components.
    b. If both components have the same kinetic energy, which moves faster?

33. Estimate the typical density of dust grains (grains per cubic centimeter) in the interstellar medium. A typical grain has a mass of about 10^{-17} kilogram (kg). (Hint: You know the typical density of gas, and the fraction of the interstellar medium's mass that is made of dust.)

34. Referring to Figure 15.3, estimate the blackbody temperature of the star as shown in part (b) (without dust) and part (c) (with dust). How significant are the effects of interstellar dust when observed data are used to determine the properties of a star?

35. A typical temperature of intercloud gas is 8000 K. Using Wien's law (see Working It Out 15.1 and Chapter 5), calculate the wavelength at which this gas would radiate.

36. Some parts of the Orion Nebula have a blackbody peak wavelength of 0.29 μm. What is the temperature of these parts of the nebula?

37. Stellar radiation can convert atomic hydrogen (H I) to ionized hydrogen (H II).
    a. Why does a B8 main-sequence star ionize far more interstellar hydrogen in its vicinity than does a K0 giant of the same luminosity?
    b. What properties of a star are important in determining whether it can ionize large amounts of nearby interstellar hydrogen?

38. The mass of a proton is 1,850 times the mass of an electron. If a proton and an electron have the same kinetic energy $(E_K = \frac{1}{2}mv^2)$, how many times greater is the velocity of the electron than that of the proton?

39. If a typical hydrogen atom in a collapsing molecular-cloud core starts at a distance of 1.5×10^{12} km (10,000 AU) from the core's center and falls inward at an average velocity of 1.5 km/s, how many years does it take to reach the newly forming protostar? Assume that a year is 3×10^7 seconds.

40. The ratio of hydrogen atoms (H) to carbon atoms (C) in the Sun's atmosphere is approximately 2,400:1 (see Table 13.1). It would be reasonable to assume that this ratio also applies to molecular clouds. If 2.6-cm radio observations indicate 100 M_{Sun} of carbon monoxide (CO) in a giant molecular cloud, what is the implied mass of molecular hydrogen (H_2) in the cloud? (Carbon represents $\frac{3}{7}$ of the mass of a CO molecule.)

41. Neutral hydrogen emits radiation at a radio wavelength of 21 cm when an atom drops from a higher-energy spin state to a lower-energy spin state. On average, each atom remains in the higher energy state for 11 million years (3.5×10^{14} seconds).
    a. What is the probability that any given atom will make the transition in 1 second?
    b. If there are 6×10^{59} atoms of neutral hydrogen in a 500-M_{Sun} cloud, how many photons of 21-cm radiation will the cloud emit each second?
    c. How does this number compare with the 1.8×10^{45} photons emitted each second by a solar-type star?

42. The Sun took 30 million years to evolve from a collapsing cloud core to a star, with 10 million of those years spent on its Hayashi track. It will spend a total of 10 billion years on the main sequence. Suppose the Sun's main-sequence lifetime were compressed into a single day.
    a. How long would the total collapse phase last?
    b. How long would the Sun spend on its Hayashi track?

43. A protostar with the mass of the Sun starts out with a temperature of about 3500 K and a luminosity about 200 times larger than the Sun's current value. Estimate this protostar's size and compare it to the size of the Sun today.

44. The star-forming region 30 Doradus is 160,000 light-years away in the nearby galaxy called the Large Magellanic Cloud and appears about one-sixth as bright as the faintest stars visible to the naked eye. If it were located at the distance of the Orion Nebula (1,300 light-years away), how much brighter than the faintest visible stars would it appear?

45. Assume a brown dwarf has a surface temperature of 1000 K and approximately the same radius as Jupiter. What is its luminosity compared to that of the Sun? How many brown dwarfs like this one would be needed to produce the luminosity of a star like the Sun?

USING THE WEB

46. Go to the Astronomy Picture of the Day (APOD) website (http://apod.nasa.gov/apod), do a search on "molecular clouds," and pick out a few images. Were these pictures obtained from space or on the ground, and at what wavelengths? With which telescopes? What wavelengths do the colors in the images represent? Are they "real" or "false-color" images?

47. Go to NASA's Spitzer Space Telescope website (http://www.spitzer.caltech.edu). Click on "News" and find a recent story about star formation. What did Spitzer observe? What wavelengths do the colors in the picture represent? How does this "false color" help astronomers to analyze these images? Why do astronomers study star formation in the infrared rather than in the visual part of the spectrum?

48. The Stratospheric Observatory for Infrared Astronomy (SOFIA) is a 2.5-meter telescope on a modified Boeing 747 aircraft. Go to the SOFIA website (http://sofia.usra.edu). Why have astronomers put an infrared telescope on an airplane? What has been detected with this telescope?

49. Citizen science: Go to the website for *Stardust* (http://stardustathome.ssl.berkeley.edu), a project in which volunteers use a virtual microscope to analyze digital scans of particles collected by the *Stardust* mission in 2006. The goal is to identify tiny interstellar dust grains. Follow the steps under "Get Started" (you need to create a log-in account) and help search for stardust. Click on "News." What has been learned from this project? Remember to save the images for your homework, if required.

50. Do a news search for a story about brown dwarfs. Is this story from an observatory? A NASA mission? A press release? What is new, and why is it interesting?

smartwork

If your instructor assigns homework in SmartWork, access your assignments at digital.wwnorton.com/astro5.

EXPLORATION

digital.wwnorton.com/astro5

In this Exploration, you will see how the H⁻ thermostat works in the formation of stars. You will need about 20 coins (they do not have to be all the same type).

Place your coins on a sheet of paper and draw a circle around them—the smallest possible circle that will fit all the coins. Then divide the circle into three parts as shown in **Figure 15.26**. This circle represents a star with a changing temperature. The coins represent H⁻ ions. Removing a coin from the circle means that the H⁻ ion has turned into a neutral hydrogen atom. Placing a coin in the circle means that the neutral hydrogen atom has become an H⁻ ion.

Place all the coins back on the circle.

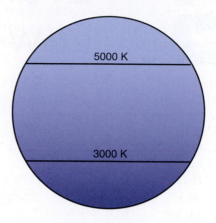

Figure 15.26

1 How many "H⁻ ions" are now in the star?

_____

_____

The "blanket" of H⁻ ions holds heat in the star, so the star begins to heat up until it reaches about 5000 K. At that surface temperature, the H⁻ ions begin to be destroyed. Now that the star is hot, begin removing coins one at a time, starting from the top of the circle and working downward. When you see the line marking 3000 K, stop removing coins.

2 How many " H⁻ ions" are now in the star?

_____

_____

3 What will happen to the surface temperature of the star, now that there are fewer ions?

_____

_____

When the star cools off to about 3000 K, H⁻ ions begin to form. Place the coins back on the circle, starting from the bottom and working your way up to the line at 5000 K.

4 How many " H⁻ ions" are now in the star?

_____

_____

5 What will happen to the surface temperature of the star, now that there are more ions?

_____

_____

Now that the star is hot, begin removing coins one at a time, starting from the top of the circle and working downward. When you see the line marking 3000 K, stop removing coins.

6 What should happen next?

_____

_____

7 Make a circular flowchart that includes the following steps, in the proper order: the star heats up; the star cools down; H⁻ is formed; H⁻ is destroyed.

_____

_____

16

Evolution of Low-Mass Stars

Within its core, the Sun fuses more than 4 billion kilograms of hydrogen to helium each second, and although the Sun may seem immortal by human standards, eventually it will run out of fuel. When it does, some 5 billion years from now, the Sun's time on the main sequence will come to an end. In this chapter, we look at how the mass of a star affects the length of its life. Then we examine what happens when a low-mass star like the Sun nears the end of its life.

LEARNING GOALS

By the conclusion of this chapter, you should be able to:

LG 1 Estimate the main-sequence lifetime of a star from its mass.

LG 2 Explain why low-mass stars grow larger and more luminous as they run out of fuel.

LG 3 Sketch post-main-sequence evolutionary tracks on a Hertzsprung-Russell (H-R) diagram, and list the stages of evolution for low-mass stars.

LG 4 Describe how planetary nebulae and white dwarfs form.

LG 5 Explain how some close binary systems evolve differently than single stars.

When low-mass stars die, they leave behind an expanding nebula of gas and dust as seen in the Hubble Space Telescope image of the Ring Nebula. ▶ ▶ ▶

Is this the
future of
our Sun?

449

16.1 The Life of a Main-Sequence Star Depends on Its Mass

A star cannot remain on the main sequence forever. It will eventually exhaust the hydrogen fuel in its core, and when it does, its structure will begin to change dramatically. Just as the balance between pressure and gravity within a protostar constantly changes as it evolves toward the main sequence, the balance between pressure and gravity changes as a star evolves beyond the main sequence. The mass and the composition of a star determine the star's life on the main sequence, and these two qualities remain at center stage as the star begins to die. In this section, we explore the changes in the star that trigger its departure from the main sequence.

Observing Stellar Evolution

Suppose you had 1 minute to observe all of the people in a crowded stadium and, from that minute of observation, draw a conclusion about the life cycle of humans. It is possible, but highly unlikely, that you might observe a significant life change such as a birth or a death. More probably, you would observe people of different ages and note some properties of people of various ages that would indicate that some people are young, some are old, and most are in between. But you wouldn't see individual people change over the course of the minute, because a minute is a very small fraction of a typical human lifetime.

Similarly, astronomers observe only a very brief fraction of each star's lifetime. They would have to observe a star like the Sun for several hundred years to be watching it for the equivalent of 1 minute in a human life span. Thus, astronomers do not see individual stars age. But they can observe many, many stars at each stage to piece together an evolutionary picture. Sometimes, just by chance, astronomers observe a star undergoing a dramatic change.

To understand what is happening in the core of the star, theorists use the most powerful computers available to model the nuclear reactions that take place there. These models make predictions about how a star of a given mass and chemical composition will change over its lifetime. These predictions are compared to what astronomers observe. The study of stellar evolution involves back-and-forth between observation and theory, which has led to a general understanding of how and when stars die.

On the basis of this type of analysis, astronomers have found that relatively minor differences in the masses and chemical compositions of two stars can sometimes result in significant, and possibly even dramatic, differences in their fates. Nevertheless, stars can be divided roughly into four broad categories whose members evolve in qualitatively different ways. Stars of spectral types O and B are hot and luminous. These stars have masses above 8 M_{Sun} and are called **high-mass stars**. Other B stars with masses between 3 and 8 M_{Sun} are **medium-mass** stars. Stars of spectral types A, F, G and K, which have masses between 0.5 and 3 M_{Sun}, are considered **low-mass stars** and are typified by the Sun. M stars with masses less than 0.5 M_{Sun} are considered very low-mass stars. In this chapter, we are primarily concerned with the more common stars of mass less than 8 M_{Sun}. In the next chapter, we will consider the fate of more massive stars.

Main-Sequence Lifetime

Think about how long you can drive your car before it runs out of gas: it depends on how much gas your tank holds and on the size and efficiency of your engine. The amount of time your car will run is determined by a competition between these two effects. An SUV uses gas faster than a subcompact, and so it might run out of gas faster, even though it has a much larger gas tank.

The competition between these two effects—tank size and engine size—is most readily expressed as a ratio. How long your car runs is given by the amount of gas in the tank, divided by how quickly the car uses it:

$$\text{Lifetime of tank of gas (hours)} = \frac{\text{Amount of fuel (gallons)}}{\text{Rate at which fuel is used (gallons/hour)}}$$

For example, if you have a 15-gallon tank and your engine is burning fuel at a rate of 3 gallons each hour, then your car will use up all of the gas in 5 hours.

The same principle works for main-sequence stars. The mass of the star determines how much fuel is available. The more massive the star, the more hydrogen available to power nuclear burning. The luminosity of the star indicates the rate at which fuel is used: energy is radiated into space from the surface of a main-sequence star at the same rate at which energy is being generated in its core. If one main-sequence star has twice the luminosity of another, then it must be burning hydrogen at twice the rate of the other.

The **main-sequence lifetime** of a star is the amount of time that it spends on the main sequence, burning hydrogen as its primary source of energy. An expression for the main-sequence lifetime of the star looks very similar to the expression for the time it takes your car to run out of fuel:

$$\text{Lifetime of star} = \frac{\text{Amount of fuel } (\propto \text{ mass of star})}{\text{Rate fuel is used } (\propto \text{ luminosity of star})}$$

The graph in **Figure 16.1** shows that as the mass of a main-sequence star increases, so does the luminosity: this occurs because the mass of the star governs the rate at which nuclear reactions occur in the core. More mass means stronger gravity; stronger gravity means higher pressure in the interior. This higher pressure compresses the core, so more atomic nuclei are packed together into a smaller volume, and it is more likely that they will run into each other and fuse. The increased pressure increases the rate of fuel use. Stronger gravity also increases the temperature in the core and speeds up the atomic nuclei. The nuclei collide more violently, increasing the chances that they will overcome the electric repulsion that pushes the positively charged nuclei apart. The increased temperature increases the rate of fuel use. As a result of the combined effects of temperature and pressure, modest increases in mass can sometimes lead to dramatic increases in the amount of energy released by nuclear burning. Stars with higher masses live shorter lives, not longer ones, because they burn their fuel faster. **Table 16.1** shows the main-sequence lifetimes for stars of different spectral types and masses. This concept is developed further in **Working It Out 16.1**.

Changes in Structure

As discussed in Chapter 14, at the end of the proton-proton chain in main-sequence stars, two ^{3}He nuclei fuse together to form ^{4}He (and two ^{1}H). However,

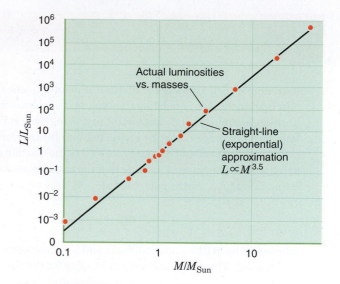

Figure 16.1 This graph plots the mass-luminosity relationship for main-sequence stars: $L \propto M^{3.5}$. The exponent can vary from 2.5 to 5.0, depending on the mass of the star. The average value, over the wide range of main-sequence star masses, is 3.5. Observational data show that the deviation of stars from the average relationship depends on their composition.

TABLE 16.1	Main-Sequence Lifetimes		
Spectral Type	Mass (M_{Sun})	Luminosity (L_{Sun})	Main-Sequence Lifetime (years)
O5	60	500,000	3.6×10^5
B0	17.5	32,500	7.8×10^6
B5	5.9	480	1.2×10^8
A0	2.9	39	7×10^8
A5	2.0	12.3	1.8×10^9
F0	1.6	5.2	3.1×10^9
F5	1.4	2.6	4.3×10^9
G0	1.05	1.25	8.9×10^9
G2 (Sun)	1.0	1.0	1.0×10^{10}
G5	0.92	0.8	1.2×10^{10}
K0	0.79	0.55	1.8×10^{10}
K5	0.67	0.32	2.7×10^{10}
M0	0.51	0.08	5.4×10^{10}
M5	0.14	0.008	4.9×10^{11}
M8	~0.08	0.0003	1.1×10^{12}

16.1 Working It Out Estimating Main-Sequence Lifetimes

Astronomers can estimate the lifetime of main-sequence stars either observationally or by modeling the evolution of stars of a given composition, assuming a constant luminosity. Using what is known about how much hydrogen must be converted into helium each second to produce a given amount of energy, as well as the fraction of its hydrogen that a star burns, we can state that the main-sequence lifetime of a star, Lifetime$_{MS}$, can be expressed as:

$$\text{Lifetime}_{MS} \propto \frac{M_{MS}}{L_{MS}}$$

where M is mass (the amount of fuel), and L is luminosity (the rate the fuel is used). The same equation would apply for the Sun:

$$\text{Lifetime}_{Sun} \propto \frac{M_{Sun}}{L_{Sun}}$$

We can express the lifetime as a ratio, adding in that the computed lifetime of a 1-M_{Sun} star like the Sun is 10 billion (1.0×10^{10}) years:

$$\frac{\text{Lifetime}_{MS}}{\text{Lifetime}_{Sun}} = \frac{\text{Lifetime}_{MS}}{10^{10}\,\text{yr}} = \frac{M_{MS}/L_{MS}}{M_{Sun}/L_{Sun}}$$

Multiplying through by 10^{10} years and rearranging the fractions yields

$$\text{Lifetime}_{MS} = 10^{10}\,\text{yr} \times \frac{M_{MS}/L_{MS}}{M_{Sun}/L_{Sun}} = 10^{10} \times \frac{M_{MS}/M_{Sun}}{L_{MS}/L_{Sun}}\,\text{yr}$$

Now let's compare the lifetime of a star with that of the Sun. The relationship between the mass and the luminosity of stars is such that relatively small differences in the masses of stars result in large differences in their main-sequence luminosities. Figure 16.1 shows the **mass-luminosity relationship**, $L \propto M^{3.5}$, for main-sequence stars. As above, we can express this relationship relative to the Sun's mass and luminosity:

$$\frac{L_{MS}}{L_{Sun}} = \left(\frac{M_{MS}}{M_{Sun}}\right)^{3.5}$$

Substituting the mass-luminosity relationship into the lifetime equation gives us

$$\text{Lifetime}_{MS} = 10^{10} \times \frac{M_{MS}/M_{Sun}}{(M_{MS}/M_{Sun})^{3.5}}\,\text{yr} = 10^{10} \times \left(\frac{M_{MS}}{M_{Sun}}\right)^{-2.5}\,\text{yr}$$

For example, let's look at a K5 main-sequence star. According to Table 16.1, a K5 star has a mass that is equal to about 0.67 M_{Sun}:

$$\text{Lifetime}_{KS} = 10^{10} \times (0.67)^{-2.5}\,\text{yr} = 2.7 \times 10^{10}\,\text{yr}$$

Instead of the 10-billion-year life span of the Sun, a K5 star has a main-sequence lifetime 2.7 times larger than the Sun's. Even though the K5 star starts out with less fuel than the Sun, it burns that fuel more slowly, so it lives longer.

at the temperatures found at the centers of main-sequence stars, collisions are not energetic or frequent enough for ^{4}He nuclei to overcome the electric repulsion and fuse into more massive elements.

Hydrogen burns most rapidly at the center of a main-sequence star because the temperature and pressure are highest there. Thus, helium, the product of hydrogen burning, accumulates most rapidly at the center of the star. If we could cut open a star and watch it evolve, we would see its chemical composition changing most rapidly at its center and less rapidly as we move outward. **Figure 16.2** shows how the chemical composition inside a star like the Sun changes throughout its main-sequence lifetime. When the Sun formed, it had a uniform composition of about 70 percent hydrogen and 30 percent helium by mass (Figure 16.2a). Since then, the Sun has produced its energy by converting hydrogen into helium via the proton-proton chain. As hydrogen fused into helium, the helium fraction in the core of the Sun climbed. Today, roughly 5 billion years later, only about 35 percent of the mass in the core of the Sun is hydrogen (Figure 16.2b). Five billion years from now, as the Sun begins to leave the main sequence, the very center of the core will have no hydrogen left to burn (Figure 16.2c).

A main-sequence star slowly changes in response to these changes in structure. As a main-sequence star converts the fuel in its core from hydrogen to helium, its structure must continually shift in response to the changing core composition to maintain energy balance. Between the time the Sun was born and the time it will leave the main sequence, its luminosity will roughly double, with most of this change occurring during the last billion years of its life on the main sequence. Main-sequence evolution is slow and modest in comparison with the events that follow the departure of the star from the main sequence.

CHECK YOUR UNDERSTANDING 16.1

Why does mass determine the main-sequence lifetime of a star? (a) Because more massive stars burn fuel faster and therefore have shorter lives. (b) Because more massive stars have more fuel and therefore have longer lives. (c) Because more massive stars burn different fuels and therefore have longer lives. (d) Because more massive stars have different initial compositions and therefore have shorter lives.

16.2 The Star Leaves the Main Sequence

Once the star has used up the hydrogen in the innermost core, thermal energy leaks out of the helium core into the surrounding layers of the star, but no more energy is generated within the core to replace it. The balance that has maintained the structure of the star throughout its life is now broken. The star's life on the main sequence has come to an end, and its further evolution depends on core temperature changes, which govern fusion reactions.

Electron-Degenerate Matter in the Helium Core

At the enormous internal temperatures of a star, almost all of the electrons have been stripped away from their nuclei by energetic collisions. In other words, the gas is completely ionized—a mixture of electrons and atomic nuclei all flying about freely. Because the size of atomic nuclei is very much less than the distance between nuclei in this gas, most of the space inside the star is empty, with electrons and atomic nuclei filling only a tiny fraction of the star's volume. However, once a low-mass star like the Sun exhausts the hydrogen at its center, the situation changes. Because the star is no longer generating energy to keep it from collapsing, gravity begins to "win," and the helium core begins to collapse and become denser. But because only one electron can occupy a single state at a single time, there is a limit to how dense the core can get. As the matter in the core of the star is compressed further and further, it finally reaches this limit. The electrons are smashed tightly together. This matter at the center of the star is now so dense that a single cubic centimeter has a mass of 1,000 kilograms (kg) or more. Matter in which electrons are packed as closely as possible is called **electron-degenerate** matter.

Electron-degenerate matter has a number of fascinating properties. For example, as more and more helium piles up on the electron-degenerate core, the core *shrinks* in size. This is one of the ways that degenerate matter differs from more normal matter: the more massive it is, the smaller it is. This is noticeably different from normal matter, such as cows: more massive cows are bigger, not smaller. The reason the core shrinks is that the added mass increases the strength of gravity and therefore the weight bearing down on the core, so the electrons are smashed together into a smaller volume. The presence of the electron-degenerate core triggers a chain of events that will dominate the evolution of a 1-M_{Sun} star for the next 50 million years after the hydrogen runs out.

Because the electrons resist being packed more closely together, they produce a type of pressure, known as degeneracy pressure, which pushes outward against gravity. Degeneracy pressure keeps the core from further collapse.

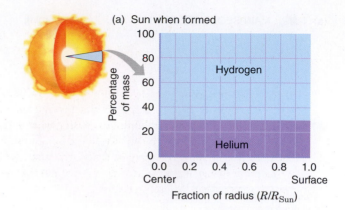

(a) Sun when formed

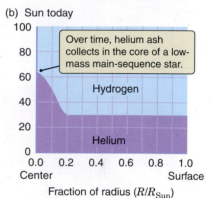

(b) Sun today

> Over time, helium ash collects in the core of a low-mass main-sequence star.

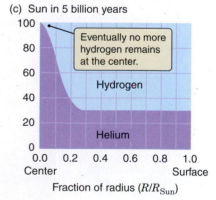

(c) Sun in 5 billion years

> Eventually no more hydrogen remains at the center.

Figure 16.2 Chemical composition of the Sun is plotted here as a percentage of mass against distance from the center of the Sun. (a) When the Sun formed 5 billion years ago, it was evenly mixed: about 30 percent of its mass was helium and 70 percent was hydrogen. (b) Today, the material at the center of the Sun is about 65 percent helium and 35 percent hydrogen. (c) The Sun's main-sequence life will end in about 5 billion years, when all of the hydrogen at its center will be exhausted.

(a) 1-M_{Sun} MAIN-SEQUENCE STAR

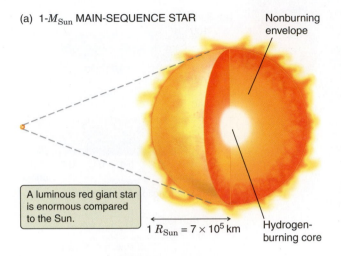

Nonburning envelope

A luminous red giant star is enormous compared to the Sun.

1 R_{Sun} = 7 × 10⁵ km

Hydrogen-burning core

(b) 1-M_{Sun} RED GIANT STAR

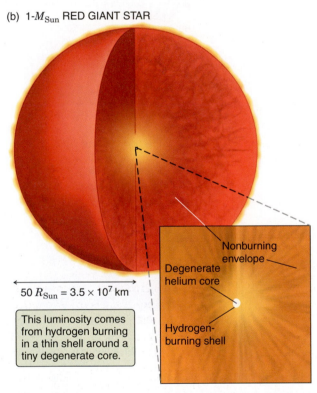

50 R_{Sun} = 3.5 × 10⁷ km

This luminosity comes from hydrogen burning in a thin shell around a tiny degenerate core.

Nonburning envelope

Degenerate helium core

Hydrogen-burning shell

Figure 16.3 The size of the Sun (a) is compared with the size of a star near the top of the red giant branch of the Hertzsprung-Russell (H-R) diagram (b). The structure of the Sun and the core of the red giant are compared in the 50-times larger views identified by the dashed lines.

Hydrogen Shell Burning

After a low-mass star has exhausted the hydrogen at its center, nuclear burning in the core pauses. The layers surrounding the degenerate core still contain hydrogen, and this hydrogen continues to burn. Astronomers call this **hydrogen shell burning** because the hydrogen is burning in a shell surrounding a core of inert helium. This layered structure is like that of a plum, with an internal seed (inert helium), a thin seed coat (hydrogen-burning shell), and a large sphere of flesh (inert hydrogen).

The changes that occur in the heart of a star with a hydrogen-burning shell around a degenerate helium core are reflected in changes in the overall structure of the star. A degenerate core is very compact, so it has very strong gravity, which leads to higher pressure. Faster nuclear reactions in the shell release more energy, so the luminosity of the star increases. With time, the mass of the degenerate helium core grows as more and more hydrogen is converted into helium in the surrounding shell. As the mass of the degenerate helium core grows, so too does its gravitational pull, further increasing the rate of energy generation in the surrounding hydrogen-burning shell. The liberated energy heats the overlying layers of the star, causing them to expand to form a large, luminous giant.

The star becomes larger and more luminous, but also cooler and redder. The enormous surface area of the star allows it to cool very efficiently, so the outer layers become cooler. The relation among radius, temperature, and luminosity that we discussed in Working It Out 13.2 and Chapter 15 still applies ($L = 4\pi\sigma R^2 T^4$). In this case, the radius and the luminosity increase, but the temperature surface falls. A red giant star fuses hydrogen in a shell around a degenerate helium core and is larger, more luminous, and redder than it was on the main sequence.

As illustrated in **Figure 16.3**, the internal structure of the main-sequence star (Figure 16.3a) changes as the star evolves to a red giant (Figure 16.3b). The giant can grow to have a luminosity hundreds of times the luminosity of the Sun and a radius of more than 50 solar radii (50 R_{Sun}). Yet at the same time, the core of the giant star is compact: much of the star's mass becomes concentrated into a volume that is only a few times the size of Earth.

The Red Giant Branch

The Hertzsprung-Russell (H-R) diagram shows the changes in a protostar on its way to the main sequence and is also a handy device for keeping track of the star as it evolves away from the main sequence. As soon as the star exhausts the hydrogen in its core, it becomes a **subgiant**: somewhat more luminous, cooler, and larger than it was on the main sequence. Because it grows more luminous but cooler, its position on the H-R diagram travels upward and to the right. As the subgiant continues to evolve, it grows even larger and cooler. When the surface temperature of the subgiant star has dropped by about 1000 kelvins (K) relative to its temperature on the main sequence, H⁻ ions start to form in great abundance in its atmosphere. Recall from Chapter 15 that H⁻ ions act as a thermostat, regulating temperature in a protostar by efficiently absorbing and scattering outgoing radiation, trapping this radiation and thus preventing the star from cooling down. The H⁻ ions serve exactly the same role here as a thermostat regulating how much radiation can escape from the subgiant star and preventing the star from becoming any cooler.

Once the subgiant can cool no further, it becomes a **red giant**—a star that moves almost vertically upward on the H-R diagram as it grows larger and more luminous but remains about the same temperature. A red giant is both redder and larger than the star was on the main sequence. You can think of the path that a star follows on the H-R diagram as it leaves the main sequence as being a tree "branch" growing out of the "trunk" of the main sequence, as shown in **Figure 16.4**. Astronomers call the lower part of this track (which moves somewhat horizontally) the **subgiant branch**. The vertical part is the **red giant branch**. During this time, the luminosity increases, but the mass does not, so the main-sequence mass-luminosity relation no longer applies. The path that a red giant follows on the H-R diagram closely parallels the path that it followed earlier as a collapsing protostar on its way toward the main sequence, except in reverse: this time, the star is moving up that path rather than coming down it. This similarity is not a coincidence. The same physical processes (such as the H⁻ thermostat) that give rise to the vertical Hayashi track followed by a collapsing protostar also control the relationship of luminosity, size, and surface temperature in an expanding red giant.

As the star leaves the main sequence, the changes in its structure occur slowly at first, but then the star moves up the red giant branch faster and faster. It takes several hundred million years for a star like the Sun to go from the main sequence to the top of the red giant branch. Roughly the first half of this time is spent on the subgiant branch as the star's luminosity increases to about 10 times the luminosity of the Sun (L_{Sun}). During the second half of this time, the helium core of the star grows in mass—but not in radius—as hydrogen is converted to helium in the hydrogen-burning shell and the helium adds to the degenerate core. The increasing mass of the ever more compact helium core increases the force of gravity in the heart of the star. This increased gravity once again increases the pressure and the temperature and thus the rate of nuclear burning, this time in the hydrogen-burning shell. Faster nuclear reactions in the shell convert hydrogen into helium more quickly, so the core grows more rapidly. The star has entered a cycle that feeds on itself. Increasing core mass leads to faster shell burning, and faster shell burning leads to faster core growth. As the core gains mass and the shell becomes more luminous, the outer layers swell. As a result, the star's luminosity climbs at

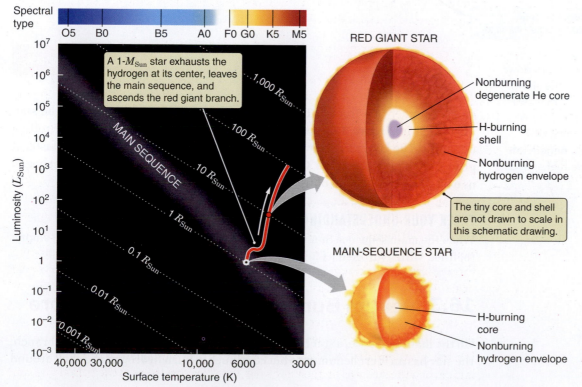

Figure 16.4 A red giant star consists of a degenerate core of helium surrounded by a hydrogen-burning shell. As the star moves up the red giant branch, it comes close to retracing the Hayashi track that it followed when it was a protostar collapsing toward the main sequence.

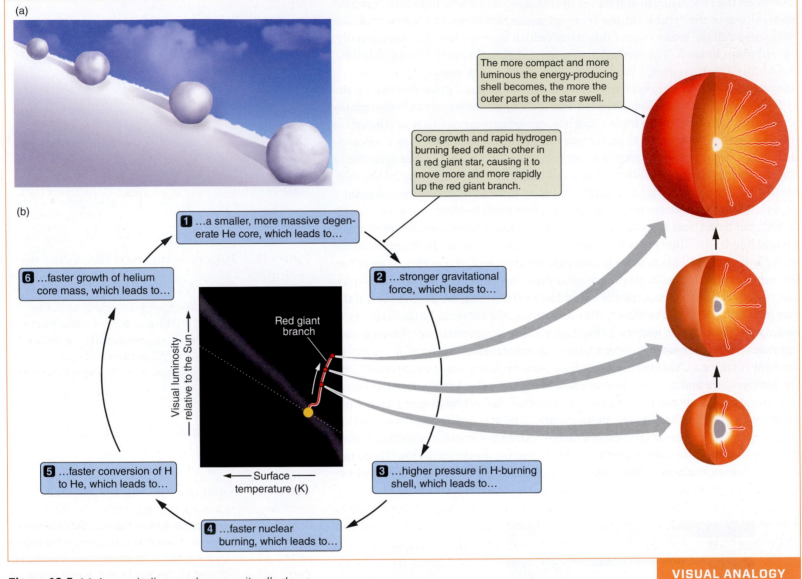

Figure 16.5 (a) A snowball grows larger as it rolls down a hill. The rate at which it grows increases as the size of the snowball increases. (b) Similarly, as a star moves up the red giant branch in the H-R diagram, the luminosity of the star grows faster and faster. The burning of hydrogen to helium in a shell surrounding a degenerate helium core feeds on itself, creating a cycle that speeds up as time goes on.

an ever-higher rate from 10 L_{Sun} to almost 1,000 L_{Sun}. The evolution of the star, illustrated in **Figure 16.5**, is reminiscent of the growth of a snowball rolling downhill. The larger the snowball becomes, the faster it grows; and the faster it grows, the larger it becomes. Growth and size feed off each other, and what began as a bit of snow at the top of the mountain soon becomes a huge ball.

CHECK YOUR UNDERSTANDING 16.2

When the main sequence Sun runs out of fuel in its core, the core will be: (a) empty; (b) filled with hydrogen; (c) filled with helium; (d) filled with carbon.

16.3 Helium Burns in the Degenerate Core

On the main sequence, a star burns hydrogen in the core. On the red giant branch, the star has an inert helium core but continues to burn hydrogen in a shell around this core. Once the hydrogen in the shell is used up, the core of the star contracts and heats. Eventually, the temperature and pressure rise enough to start the next stage of stellar evolution: helium burning.

Helium Burning and the Triple-Alpha Process

As the star evolves up the red giant branch, its helium core grows not only smaller and more massive but also hotter. This increase in temperature is due partly to the gravitational energy released as the core shrinks and partly to the energy released by hydrogen burning in the surrounding shell. The thermal motions of the atomic nuclei in the core become more and more energetic. Eventually, at a temperature of about 100 million (10^8) K, the collisions among helium nuclei in the core become energetic enough to overcome the electric repulsion. Helium nuclei are slammed together hard enough for the strong nuclear force to act, and helium burning begins.

Helium burns in a two-stage process called the **triple-alpha process**, which takes its name from the fact that it involves the fusion of three ^{4}He nuclei, which are called **alpha particles**. The process, illustrated in **Figure 16.6**, begins when two helium-4 (^{4}He) nuclei fuse to form a beryllium-8 (^{8}Be) nucleus consisting of four protons and four neutrons. The ^{8}Be nucleus is extremely unstable, with a very short lifetime before it decays. But if, in that short time, it collides with another ^{4}He nucleus, the two nuclei will fuse into a stable nucleus of carbon-12 (^{12}C) consisting of six protons and six neutrons. The reaction rate is very temperature-dependent: higher temperatures enable more reactions and increase the number of ^{8}Be nuclei that collide with a ^{4}He nucleus.

Recall that the core of the red giant star is electron-degenerate, which means that as many electrons are packed into that space as possible. These degenerate electrons prevent the core from collapsing. The atomic nuclei, however, behave like a normal gas, moving through the sea of degenerate electrons almost as if the electrons were not there. The atomic nuclei in the core move freely about as shown in **Figure 16.7**, just as they do throughout the rest of the star. We can understand the fusion of helium by treating the nuclei as matter in a normal state. Once the pressure and temperature are high enough, these nuclei begin

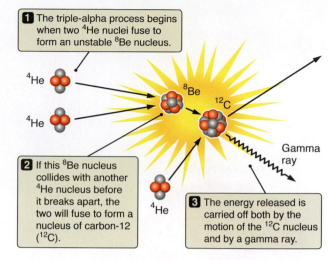

1 The triple-alpha process begins when two ^{4}He nuclei fuse to form an unstable ^{8}Be nucleus.

2 If this ^{8}Be nucleus collides with another ^{4}He nucleus before it breaks apart, the two will fuse to form a nucleus of carbon-12 (^{12}C).

3 The energy released is carried off both by the motion of the ^{12}C nucleus and by a gamma ray.

Figure 16.6 The triple-alpha process produces a stable nucleus of carbon-12. Two helium-4 (^{4}He) nuclei fuse to form an unstable beryllium-8 (^{8}Be) nucleus. If this nucleus collides with another ^{4}He nucleus before it breaks apart, the two will fuse to form a stable nucleus of carbon-12 (^{12}C). The energy produced is carried off both by the motion of the ^{12}C nucleus and by a high-energy gamma ray emitted in the second step of the process.

Figure 16.7 In a red giant star, the weight of the overlying layers is supported by electron degeneracy pressure in the core arising from the fact that electrons are packed together as tightly as quantum mechanics allows. Atomic nuclei in the core are able to move freely about within the sea of degenerate electrons, so they behave as a normal gas.

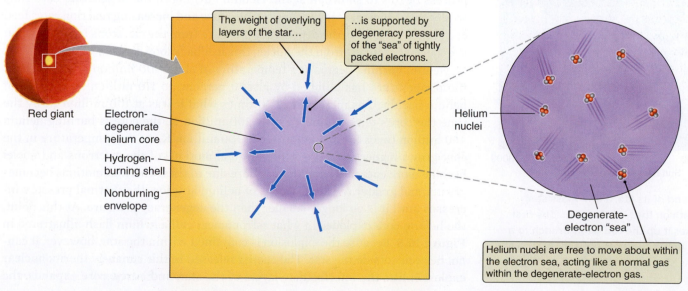

The weight of overlying layers of the star…

…is supported by degeneracy pressure of the "sea" of tightly packed electrons.

Red giant

Electron-degenerate helium core

Hydrogen-burning shell

Nonburning envelope

Helium nuclei

Degenerate-electron "sea"

Helium nuclei are free to move about within the electron sea, acting like a normal gas within the degenerate-electron gas.

to fuse, as the hydrogen nuclei do in main-sequence stars. But the energy released will not affect the degenerate-electron core in the same way as in a main-sequence star.

The Helium Flash

Degenerate material is a very good conductor of thermal energy, so any differences in temperature within the core rapidly disperse. As a result, when helium burning begins at the center of the core, the energy released quickly heats the entire core. Within a few minutes, the entire core is burning helium into carbon by the triple-alpha process. Some of the carbon will fuse with an additional ^{4}He nucleus to form stable oxygen-16 (^{16}O).

In a normal gas such as the air around you or the core of a main-sequence star, the pressure of the gas comes from the random thermal motions of the atoms. Increasing the temperature of such a gas means that the motions of the atoms become more energetic, so the pressure of the gas increases. If the helium core of a red giant star were a normal gas, the increase in temperature that accompanies the onset of helium burning would lead to an increase in pressure. The core of the star would expand; the temperature, density, and pressure would decrease; nuclear reactions would slow; and the star would settle into a new balance between gravity and pressure. These are exactly the sorts of changes that are steadily occurring within the core of a main-sequence star like the Sun as the structure of the star shifts in response to the changing composition in the star's core.

However, the degenerate core of a red giant is not a normal gas. The pressure in a red giant's degenerate core comes from how tightly the electrons in the core are packed together. Heating the core does not change the number of electrons that can be packed into its volume, so the core's pressure does not respond to changes in temperature. Because the pressure does not increase, the core does not expand when heated, as a normal gas would.

Yet even though the higher temperature does not change the pressure, it does cause the helium nuclei to collide with more frequency and greater force, so the nuclear reactions become more vigorous. More vigorous reactions mean higher temperature, and higher temperature means even more vigorous reactions. The process begins to snowball again. Helium burning in the degenerate core runs wildly out of control as increasing temperature and increasing reaction rates feed each other. As long as the degeneracy pressure from the electrons is greater than the thermal pressure from the nuclei, this feedback loop continues.

Helium burning begins at a temperature of about 100 million K. By the time the temperature has climbed by just 10 percent, to 110 million K, the rate of helium burning has increased to 40 times what it was at 100 million K. By the time the core's temperature reaches 200 million K, the core is burning helium 460 million times faster than it was at 100 million K. As the temperature in the core grows higher and higher, the thermal motions of the electrons and nuclei become more energetic, and the pressure due to these thermal motions becomes greater and greater. Within seconds of helium ignition, the thermal pressure increases until it is no longer smaller than the degeneracy pressure. At this point, the helium core explodes in what astronomers call a helium flash, illustrated in **Figure 16.8**. Because the explosion is contained within the star, however, it cannot be seen outside the star. The energy released in this runaway thermonuclear explosion lifts the intermediate layers of the star, and as the core expands, the

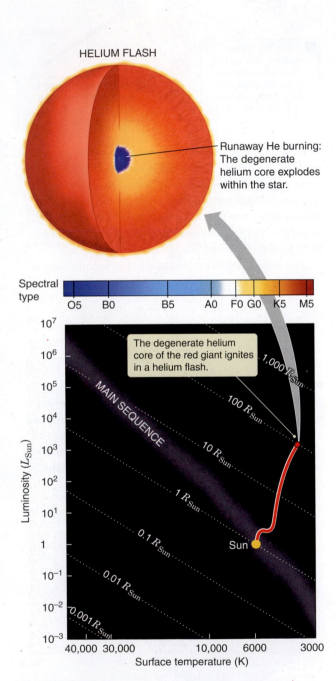

HELIUM FLASH

Runaway He burning: The degenerate helium core explodes within the star.

The degenerate helium core of the red giant ignites in a helium flash.

Figure 16.8 At the end of its life, a low-mass star travels a complex path on the H-R diagram. The first part of that path takes it up the red giant branch to a point where helium ignites in a helium flash. After a few hours, the core of the star begins to inflate, ending the helium flash.

electrons are able to spread out. The drama is over within a few hours because the expanded helium-burning core is no longer degenerate, and the star is on its way toward a new equilibrium.

Helium burning in the core does not cause the star to grow more luminous. The tremendous energy released during the helium flash goes into fighting gravity and puffing up the core. After the helium flash, the core (which is no longer degenerate) is much larger, so the force of gravity within it and the surrounding shell are much smaller. Weaker gravity means less weight pushing down on the core and the shell, which means lower pressure. Lower pressure, in turn, slows the nuclear reactions. The net result is that after the helium flash, core helium burning keeps the core of the star puffed up, and the star becomes less luminous than it was as a red giant.

The Horizontal Branch

The star takes about 100,000 years or so to settle into stable helium burning. It then spends about 100 million years burning helium into carbon in a normal, non-degenerate core while hydrogen burns to helium in a surrounding shell. The star is now about a hundred times less luminous than it was when the helium flash occurred. The lower luminosity means that the outer layers of the star are not as puffed up as they were when the star was a red giant. Gravity becomes stronger than the outward pressure of the escaping radiation and pulls the outer layers back in. The star shrinks, and its surface temperature climbs as gravitational energy is converted to thermal energy. The star moves horizontally across the H-R diagram, remaining at the same luminosity, but increasing in surface temperature. This portion of a star's path on the H-R diagram is called the horizontal branch, shown in **Figure 16.9**. At this point in their evolution, low-mass stars with chemical compositions similar to that of the Sun lie on the H-R diagram just to the left of the red giant branch. Stars that contain much less iron than the Sun tend to distribute themselves away from the red giant branch along the horizontal branch.

The structure and behavior of a star on the horizontal branch are similar to those of a main-sequence star. However, instead of burning hydrogen into helium in the core, the horizontal branch star burns helium into carbon. In addition, in the horizontal branch star, hydrogen continues to burn in a shell surrounding the core.

The star's time on the horizontal branch, however, is much shorter than its time on the main sequence. There is now less fuel to burn in its core. In addition, the star is more luminous than it was on the main sequence, so it is consuming fuel more rapidly. Helium is a much less efficient nuclear fuel than hydrogen, so the star has to burn fuel even faster to maintain equilibrium. Even so, for about a hundred million years the horizontal branch star remains stable, burning helium to carbon in its core and hydrogen to helium in a shell around the core.

The temperature at the center of a horizontal branch star is not high enough for carbon to burn, so carbon builds up in the heart of the star. When the horizontal branch star has burned all of the helium at its core, gravity once again begins to overwhelm the pressure of the escaping radiation. The nonburning carbon core is crushed by the weight of the layers of the star above it until once again the electrons in the core are packed together as tightly as possible, given its pressure. The carbon core is now electron-degenerate, with physical properties much like those of the degenerate helium core at the center of a red giant.

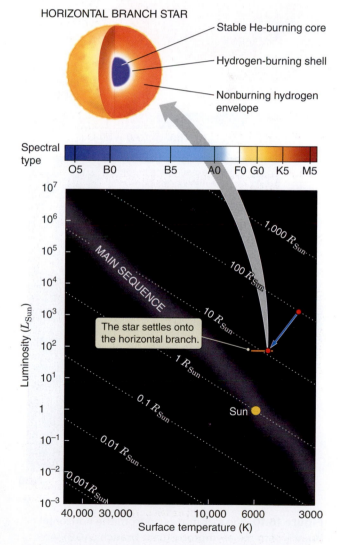

Figure 16.9 The star moves down from the red giant branch onto the horizontal branch. For about 100 million years or less, the star will remain on the horizontal branch and burn helium in its core and hydrogen in a surrounding shell.

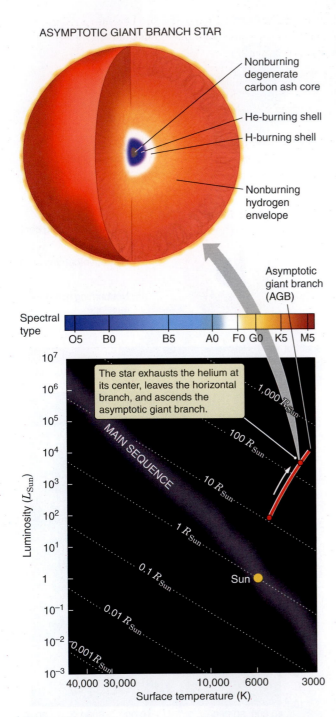

ASYMPTOTIC GIANT BRANCH STAR

Nonburning degenerate carbon ash core

He-burning shell

H-burning shell

Nonburning hydrogen envelope

Asymptotic giant branch (AGB)

Spectral type

O5 B0 B5 A0 F0 G0 K5 M5

The star exhausts the helium at its center, leaves the horizontal branch, and ascends the asymptotic giant branch.

MAIN SEQUENCE

1,000 R_{Sun}

100 R_{Sun}

10 R_{Sun}

1 R_{Sun}

0.1 R_{Sun}

Sun

0.01 R_{Sun}

0.001 R_{Sun}

Luminosity (L_{Sun})

10^7
10^6
10^5
10^4
10^3
10^2
10^1
1
10^{-1}
10^{-2}
10^{-3}

40,000 30,000   10,000 6000   3000

Surface temperature (K)

Figure 16.10 The star moves up from the horizontal branch onto the asymptotic giant branch (AGB). An AGB star consists of a degenerate carbon core surrounded by helium-burning and hydrogen-burning shells. As the carbon core grows, the star brightens, accelerating up the asymptotic giant branch just as it earlier accelerated up the red giant branch while its degenerate helium core grew.

The Lowest-Mass Stars

Brown dwarfs have masses less than 0.08 M_{Sun}. They do not become main-sequence stars, so they do not follow post-main-sequence steps. Small red dwarf stars with masses lower than about 0.4–0.5 M_{Sun} may be the majority of the stars in our galaxy, but they are small and faint and hard to detect. These stars have main-sequence lifetimes longer than the 13.8-billion-year age of the universe, so we have not actually observed any of these stars in post-main-sequence stages. There has not been enough time for these stars to evolve off of the main sequence.

CHECK YOUR UNDERSTANDING 16.3

Stars begin burning helium to carbon when the temperature rises in the core. This temperature increase is caused by (choose all that apply): (a) gravitational collapse; (b) fusion of hydrogen into helium in the core; (c) fusion of hydrogen into helium in a shell around the core; (d) electron degeneracy pressure.

16.4 Dying Stars Shed Their Outer Layers

After the horizontal branch, small changes in the properties of a star—mass, chemical composition, strength of the star's magnetic field, or even the rate at which the star is rotating—can lead to noticeable differences in how the star (and especially its outer envelope) evolves. In this section, we follow a 1-M_{Sun} star with solar composition as it concludes its evolutionary stages, losing its outer layers and leaving behind a cooling carbon core.

Stellar-Mass Loss and the Asymptotic Giant Branch

After a time on the horizontal branch, a small, dense, electron-degenerate carbon core remains. This core is very compact, causing the gravity in the inner parts of the star to be very high, which in turn drives up the pressure, which speeds up the nuclear reactions, which causes the degenerate core to grow more rapidly. The internal changes occurring within the star are similar to the changes that took place at the end of the star's main-sequence lifetime, and the path the star follows as it leaves the horizontal branch echoes that earlier phase of evolution. Just as the star accelerated up the red giant branch as its degenerate helium core grew, the star now leaves the horizontal branch and once again begins to grow larger, redder, and more luminous as its degenerate carbon core grows. As shown in **Figure 16.10**, the path that the star follows, called the **asymptotic giant branch (AGB)** of the H-R diagram, parallels the path it followed as a red giant, approaching the red giant branch as the star grows more luminous. An AGB star burns helium and hydrogen in nested concentric shells surrounding a degenerate carbon core as the star moves once again up the H-R diagram. Before the temperature in the carbon core becomes high enough for carbon to burn, the pressure of the escaping radiation exceeds the gravitational pull on the outer layers of the star. The star begins to lose these outer layers to space.

AGB stars are huge objects. When the Sun becomes an AGB star, its outer layers will swell to the point that they engulf the orbits of the inner planets, possibly including Earth and maybe even Mars. When a star expands to such a size, the gravitational force at its surface is only 1/10,000 as strong as the gravity at the surface of the present-day Sun. It takes little extra energy to push surface

16.2 Working It Out Escaping the Surface of an Evolved Star

Why are giant stars likely to lose mass? The escape velocity from the surface of a planet or star was given in Working It Out 4.2:

$$v_{esc} = \sqrt{2GM/R}$$

How does v_{esc} change when a star becomes a red giant? Let's look at the Sun as an example. When the Sun is on the main sequence, its escape velocity can be calculated using $M_{Sun} = 1.99 \times 10^{30}$ kg, $R_{Sun} = 6.96 \times 10^5$ km, and $G = 6.67 \times 10^{-20}$ km³/(kg s²):

$$v_{esc} = \sqrt{\frac{2 \times [6.67 \times 10^{-20}\,\text{km}^3/(\text{kg s}^2)] \times (1.99 \times 10^{30}\,\text{kg})}{6.96 \times 10^5\,\text{km}}}$$

$$v_{esc} = \sqrt{3.81 \times 10^5\,\text{km}^2/\text{s}^2} = 618\,\text{km/s}$$

What will the escape velocity be when the Sun becomes a red giant, with a radius 50 times greater than the radius it has today and a mass 0.9 times its current mass:

$$v_{esc} = \sqrt{\frac{2 \times [6.67 \times 10^{-20}\,\text{km}^3/(\text{kg s}^2)] \times 0.9 \times (1.99 \times 10^{30}\,\text{kg})}{50 \times (6.96 \times 10^5\,\text{km})}}$$

$$v_{esc} = \sqrt{6.86 \times 10^3\,\text{km}^2/\text{s}^2} = 83\,\text{km/s}$$

The escape velocity from the surface of a red giant star is only 13 percent that of a main-sequence star. This is part of the reason that red giant and AGB stars lose mass. The Sun may eventually lose half of its mass.

material away from the star. **Stellar-mass loss**—the loss of mass from the outer layers of the star as it evolves—actually begins when the star is still on the red giant branch: by the time a 1-M_{Sun} main-sequence star reaches the horizontal branch, it may have lost 10–20 percent of its total mass. As the star ascends the asymptotic giant branch, it loses another 20 percent or even more of its total mass. By the time it is well up on this branch, a star that began as a 1-M_{Sun} star may have lost more than half of its original mass. Stellar-mass loss is further explored in **Working It Out 16.2**.

Mass loss on the asymptotic giant branch can be spurred on by the star's unstable interior. The extreme sensitivity of the triple-alpha process to temperature in the core can lead to episodes of rapid energy release, which can provide the extra kick needed to expel material from the star's outer layers. Even stars that are initially quite similar can behave very differently when they reach this stage in their evolution.

Planetary Nebula

Toward the end of an AGB star's life, mass loss itself becomes a runaway process. When a star loses a bit of mass from its outermost layers, the weight pushing down on the underlying layers of the star is reduced. Without this weight holding them down, the remaining outer layers of the star puff up further. The post-AGB star, which is now both less massive and larger, is even less tightly bound by gravity, so less energy is needed to push its outer layers away. Mass loss leads to weaker gravity, which leads to faster mass loss, which leads to weaker gravity, and so on. When the end comes, much of the remaining mass of the star is ejected into space, typically at speeds of 20–30 kilometers per second (km/s).

After ejection of its outer layers, all that is left of the low-mass star is a tiny, very hot, electron-degenerate carbon core surrounded by a thin envelope in which hydrogen and helium are still burning. This star is now somewhat less luminous than when it was at the top of the asymptotic giant branch, but it is still much more luminous than a horizontal branch star. The remaining hydrogen and helium in the star rapidly burn to carbon, and as more and more of the mass of the star ends up in the carbon core, the star itself shrinks and becomes hotter and

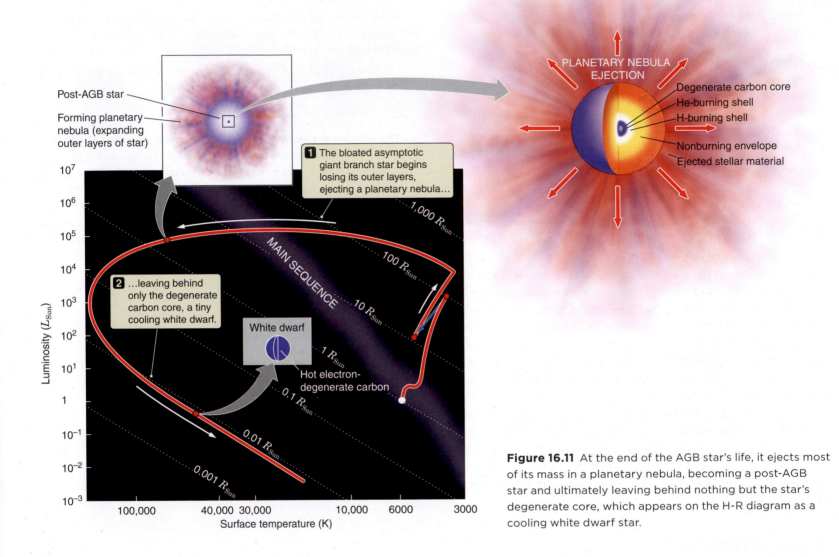

Post-AGB star

Forming planetary nebula (expanding outer layers of star)

1 The bloated asymptotic giant branch star begins losing its outer layers, ejecting a planetary nebula…

2 …leaving behind only the degenerate carbon core, a tiny cooling white dwarf.

White dwarf

Hot electron-degenerate carbon

MAIN SEQUENCE

1,000 R_{Sun}
100 R_{Sun}
10 R_{Sun}
1 R_{Sun}
0.1 R_{Sun}
0.01 R_{Sun}
0.001 R_{Sun}

Luminosity (L_{Sun})

Surface temperature (K)

PLANETARY NEBULA EJECTION

Degenerate carbon core
He-burning shell
H-burning shell
Nonburning envelope
Ejected stellar material

Figure 16.11 At the end of the AGB star's life, it ejects most of its mass in a planetary nebula, becoming a post-AGB star and ultimately leaving behind nothing but the star's degenerate core, which appears on the H-R diagram as a cooling white dwarf star.

hotter. Over the course of only about 30,000 years after the beginning of runaway mass loss, the star moves from right to left across the top of the H-R diagram as shown in **Figure 16.11**.

The surface temperature of the star may eventually rise above 100,000 K. At such temperatures, the peak wavelength of the radiation is in the high-energy ultraviolet (UV) part of the spectrum, as determined by Wien's law. This intense UV light heats and ionizes the ejected, expanding shell of gas, causing it to glow in the same way that UV light from an O star causes an H II region to glow. When these glowing shells were first observed in small telescopes, they were named "planetary nebulae" because they appeared fuzzy like nebular clouds of dust and gas, but they were approximately round, like planets. Later imagery, such as the images in **Figure 16.12**, showed that these objects are not like planets at all. Rather, a **planetary nebula** is the remaining outer layers of a star, ejected into space at the end of the star's ascent of the asymptotic giant branch. A planetary nebula may be visible for 50,000 years or so before the gas ejected by the star disperses so far that the nebula is too faint to be seen. Not all stars form planetary

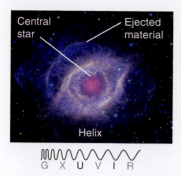
Central star — Ejected material
Helix
G X U V I R

Butterfly
G X U V I R

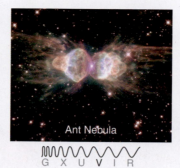
Ant Nebula
G X U V I R

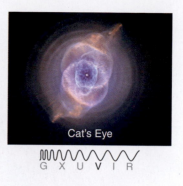
Cat's Eye
G X U V I R

nebulae. Stars more massive than about 8 M_{Sun} pass through the post-AGB stage too quickly. Stars with insufficient mass take too long in the post-AGB stage, so their envelope evaporates before they can illuminate it. Astronomers do not know if our own Sun will retain enough mass during its post-AGB phase to form a planetary nebula.

The detailed structure of a planetary nebula may contain concentric rings of varied density, indicating that the stellar-mass loss was slower and faster at various times. In the image of the Ring Nebula in the chapter-opening photograph, the colored rings are emission lines from different ions in different places. During the formation of this nebula, a lot of mass was lost nearly all at once. Then the mass loss ceased, resulting in a hollow shell around the central star. In other planetary nebula, the material may be concentrated parallel to the equator or poles of the star, indicating that the stellar-mass loss was blocked in some directions. You can see this in the images of the Butterfly, Ant, and Cat's Eye nebulae in Figure 16.12.

The gas in a planetary nebula carries the chemical elements enriching the star's outer layers off into interstellar space. Planetary nebulae often show an overabundance of elements such as carbon, nitrogen, and oxygen compared to the abundance of these elements in the outer layers of the Sun. These elements are by-products of nuclear burning either from the star that produced the planetary nebula or from the stars of earlier generations. Once this chemically enriched material leaves the star, it mixes with interstellar gas, increasing the chemical diversity of the universe.

Figure 16.12 At the end of its life, a low-mass star ejects its outer layers and may form a planetary nebula consisting of an expanding shell of gas surrounding the white-hot remnant of the star. Planetary nebulae are not all simple spherical shells around their parent stars. These images of planetary nebulae from the Hubble Space Telescope and the Spitzer Space Telescope show the wealth of structures that result from the complex processes by which low-mass stars eject their outer layers.

White Dwarfs

Within about 50,000 years, a post-AGB star burns all of the fuel remaining on its surface, leaving behind a nonburning ball of carbon with a mass less than 70 percent of the mass of the original star. In the process, the post-AGB star becomes smaller and fainter, and its position on the H-R diagram falls down the left side. Within a few thousand years, the burned-out core shrinks to about the size of Earth, at which point it has become fully electron-degenerate and can shrink no further. This remnant of stellar evolution is called a **white dwarf**.

The white dwarf, composed of nonburning electron-degenerate carbon and maybe some oxygen, continues to radiate energy away into space. As it does so it cools, just like the heating coil on an electric stove once it is turned off. Because the white dwarf is electron-degenerate, its size does not change much as it cools, so it moves down and to the right on the H-R diagram, following a line of constant radius. The white dwarf will remain very hot for 10 million years or so, but its tiny size means it is a thousand times less luminous than a main-sequence star like the Sun. Many white dwarfs are known, but none can be seen without a telescope. Sirius, the brightest star in Earth's sky, has a faint white dwarf as a binary companion.

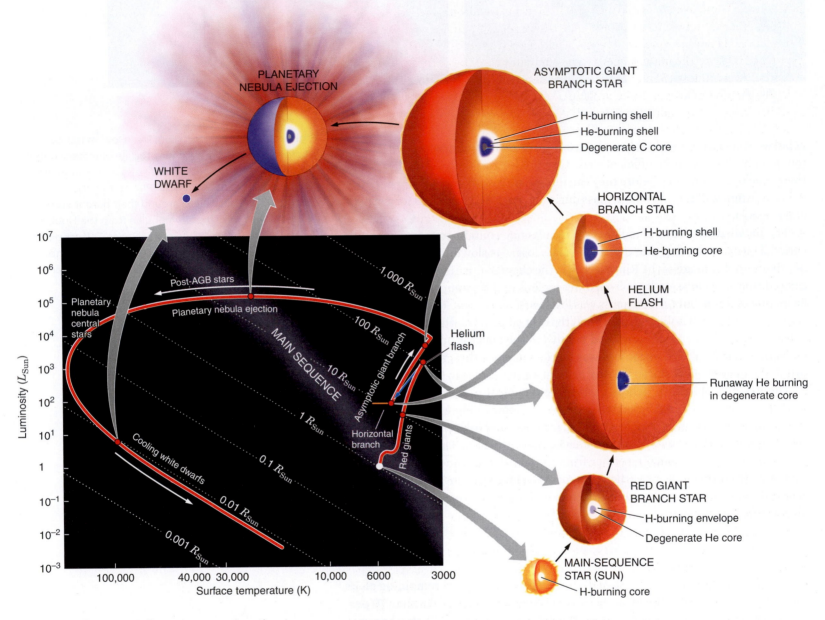

Figure 16.13 This H-R diagram summarizes the stages in the post-main-sequence evolution of a 1-M_{Sun} star.

Figure 16.13 summarizes the evolution of a solar-type, 1-M_{Sun} main-sequence star through to its final existence as a 0.6-M_{Sun} white dwarf. The star leaves the main sequence, climbs the red giant branch, falls to the horizontal branch, climbs back up the asymptotic giant branch, takes a left across the top of the diagram while ejecting a planetary nebula, and finally falls to its final resting place in the bottom left of the diagram. This process is representative of the fate of low-mass stars. Although every low-mass star forms a white dwarf at the end point of its evolution, the precise path a low-mass star follows from core hydrogen burning on the main sequence to white dwarf depends on many details particular to the star.

Some stars less massive than the Sun may become white dwarfs composed largely of helium rather than carbon. Conversely, temperatures in the cores of evolved 2- to 3-M_{Sun} stars are high enough to allow additional nuclear reactions to occur, leading to the formation of somewhat more massive white dwarfs composed of materials such as oxygen, neon, and magnesium. Differences in chemical composition of a star can also lead to differences in its post-main-sequence evolution.

In our discussion of stellar evolution, we have focused on what happens after a star leaves the main sequence. The spectacle of a red giant or AGB star is ephemeral. Once it leaves the main sequence, the Sun will travel the path from red giant to white dwarf in less than one-tenth of the time it spent on the main sequence steadily burning hydrogen to helium in its core. Stars spend most of their luminous lifetimes on the main sequence, which is why most of the stars in the sky are main-sequence stars. The fainter white dwarfs constitute the final resting place for the vast majority of stars that have been or ever will be formed.

The Fate of the Planets

What happens to the planets orbiting a low-mass star as it goes through these post-main-sequence stages? Because the first decade of extrasolar planet discoveries primarily yielded planets with orbits that are very close to their respective stars, astronomers assumed that any planets closer than 1–2 astronomical units (AU) would not survive the post-main-sequence expansion of the star. Therefore, they did not expect to find planets in orbit around evolved stars. However, many planets have been discovered in orbit around red giants, AGB stars, and horizontal branch stars.

Astronomers cannot be sure whether the extrasolar planets they observe have remained in the same orbital locations as when their stars were on the main sequence or if the planets migrated to different orbits. The surface gravity of a red giant is low, and some of the mass in its outermost layers will blow away. This decrease in mass reduces the gravitational force between the star and its planets, which could lead to planetary orbits evolving outward away from the star. There are some models of planetary migration in which the tidal forces between planets or between a star and a planet are significant factors too, suggesting that some planetary orbits could evolve inward.

As a star loses mass during the evolutionary process, stellar or planetary companions may change their orbits. Rocky asteroids or smaller planets could migrate inward past the Roche limit (see Chapter 4) of the white dwarf and break up. Some of the material could remain in orbit around the white dwarf as a debris disk, similar to those seen in main-sequence stars with a planetary system. These dusty debris disks have been observed around white dwarf stars with the Hubble Space Telescope and the Spitzer Space Telescope. Some of this dusty or rocky material may also fall onto the white dwarf, "polluting" its spectrum with heavy elements that were not produced in the stellar core.

What will happen to Earth? The age of the Sun from radioactive dating of meteorites indicates that the Solar System formed 4.6 billion years ago. In Chapter 14, we estimated how long the Sun might last by calculating its rate of hydrogen burning. The Sun's luminosity is about 30 percent higher now than it was early in the history of the Solar System, and it will continue to increase at a steady rate over the rest of the Sun's main-sequence lifetime of another 5 billion years or so. Models estimate that the Sun's luminosity may increase enough—even while the Sun is a main-sequence star—that Earth will heat up to the point where the oceans evaporate, perhaps as soon as 1 billion to 2 billion years from now. By the time the Sun leaves the main sequence, the habitable zone may have moved out to Mars and no longer include Earth.

It is not certain whether the radius of the red giant Sun will extend past 1 AU so that Earth becomes completely engulfed. The red giant Sun will have low surface gravity and will lose mass, which could cause Earth's orbit to expand, thereby enabling Earth to escape the encroaching solar surface. Alternatively, as the Sun expands in radius and its rotation rate slows (see Working It Out 7.1), tidal forces might pull Earth inward. The habitable zone of a red giant Sun might be in the vicinity of Jupiter or Saturn. Eventually, depending on how much mass the Sun loses in the red giant and possibly AGB stages, the outer layers may or may not form a planetary nebula. The solar core will become a white dwarf, perhaps with a dusty disk and a "polluted" atmosphere as the only remaining evidence of our rocky planet.

CHECK YOUR UNDERSTANDING 16.4

A planetary nebula forms from: (a) the ejection of mass from a low-mass star; (b) the collision of planets around a dying star; (c) the collapse of the magnetosphere of a high-mass star; (d) the remainders of the original star-forming nebula.

16.5 Binary Star Evolution

So far in this chapter, we have discussed the evolution of single stars in isolation. But many stars are members of binary systems. How is evolution different for stars in these systems? In some cases, if the stars are close together and one star is more massive than the other, their evolution after the main sequence may be linked. In this section, we will trace the steps from binary star system through to the end point of their evolution: nova or supernova.

Mass Flows from an Evolving Star onto Its Companion

Think for a moment about what would happen if you were to travel in a spacecraft from Earth toward the Moon. When you are still near Earth, the force of Earth's gravity is far stronger than that of the Moon. As you move away from Earth and closer to the Moon, the gravitational attraction of Earth weakens, and the gravitational attraction of the Moon becomes stronger. You eventually reach an intermediate zone where neither body has the stronger pull. If you continue beyond this point, the lunar gravity begins to dominate until you find yourself firmly in the gravitational grip of the Moon. The regions surrounding the two objects—their gravitational domains—are called the **Roche lobes** of the system.

Exactly the same situation exists between two stars, as shown in **Figure 16.14**. Gas near each star clearly belongs to that star. When one star leaves the main sequence and swells up, its outer layers may cross that gravitational dividing line separating the star from its companion. Once a star expands past the boundary of its Roche lobe, its material begins to fall onto the other star. This exchange of material from one star to the other is called **mass transfer**.

Evolution of a Close Binary System

The best way to understand how mass transfer affects the evolution of stars in a binary system is to apply what is known from studying the evolution of single low-mass stars. Figure 16.14a shows a binary system consisting of two low-mass stars of somewhat different mass. Star 1 is the more massive of the two stars, and star 2

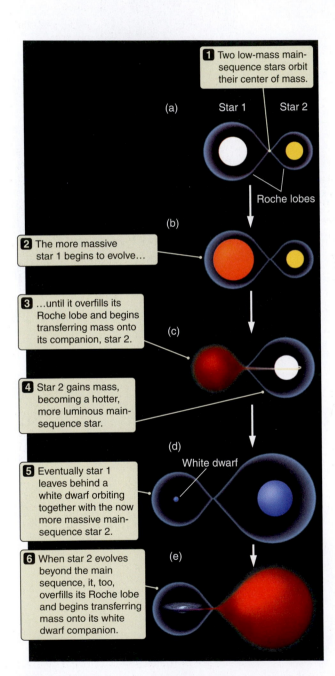

1 Two low-mass main-sequence stars orbit their center of mass.

(a)   Star 1   Star 2

Roche lobes

(b)

2 The more massive star 1 begins to evolve…

3 …until it overfills its Roche lobe and begins transferring mass onto its companion, star 2.

(c)

4 Star 2 gains mass, becoming a hotter, more luminous main-sequence star.

(d)   White dwarf

5 Eventually star 1 leaves behind a white dwarf orbiting together with the now more massive main-sequence star 2.

6 When star 2 evolves beyond the main sequence, it, too, overfills its Roche lobe and begins transferring mass onto its white dwarf companion.

(e)

Figure 16.14 A compact binary system consisting of two low-mass stars passes through a sequence of stages as the stars evolve and mass is transferred back and forth.

is the less massive of the two stars. This is an ordinary binary system, and each of these stars is an ordinary main-sequence star for most of the system's lifetime.

More massive main-sequence stars evolve more rapidly than less massive main-sequence stars. Therefore, star 1 will be the first to use up the hydrogen at its center and begin to evolve off the main sequence, as shown in Figure 16.14b. If the two stars are close enough to each other, star 1 will eventually grow to overfill its Roche lobe, and material will transfer onto star 2, as shown in Figure 16.14c. The transfer of mass between the two stars can result in a sort of "drag" that causes the orbits of the two stars to shrink, bringing the stars closer together and further enhancing mass loss. In addition, as star 1 loses mass, its Roche lobe shrinks, further enhancing the mass transfer. The two stars can even reach the point where they are effectively two cores sharing the same extended envelope of material.

Despite these complexities, star 2 probably remains a basically normal main-sequence star throughout this process, burning hydrogen in its core. However, over time, the mass of star 2 increases because of the accumulation from its companion. As it does so, the structure of star 2 must change to accommodate its new status as a higher-mass star. If we plotted star 2's position on the H-R diagram during this period, we would see it move up and to the left along the main sequence, becoming larger, hotter, and more luminous.

Star 1, because it is losing mass to star 2, never grows larger than its Roche lobe, so it does not become an isolated red giant or AGB star at the top of the H-R diagram. Yet star 1 continues to evolve, burning helium in its core on the horizontal branch, proceeding through a stage of helium shell burning, and finally losing its outer layers and leaving behind a white dwarf. Figure 16.14d shows the binary system after star 1 has completed its evolution. All that remains of star 1 is a white dwarf, orbiting about star 2, its bloated main-sequence companion.

Novae

As star 2 begins to evolve off the main sequence, it expands to fill its Roche lobe, as shown in Figure 16.14e. Like star 1 before it, star 2 grows to fill its Roche lobe: material from star 2 begins to pour through the "neck" connecting the Roche lobes of the two stars. However, this time the mass is not being added to a normal star but is drawn toward the tiny white dwarf left behind by star 1. Because the system is revolving and the white dwarf is so small, the infalling material generally misses the star, instead landing on an accretion disk around the white dwarf. This disk is similar to the accretion disk that forms around a protostar. As in the process of star formation, the accretion disk accumulates material that has too much angular momentum to hit the white dwarf directly.

A white dwarf has a large mass and a small radius; therefore, it has strong gravity. The material streaming toward the white dwarf in the binary system falls into an incredibly deep gravitational "well." The depth of this well affects the amount of energy with which matter impacts the white dwarf. A kilogram of material falling from space onto the surface of a white dwarf releases 100 times more energy than a kilogram of material falling from the outer Solar System onto the surface of the Sun. All of this energy is turned into thermal energy. The spot where the stream of material from star 2 hits the accretion disk can be heated to millions of kelvins, where it glows in the far-ultraviolet and X-ray parts of the electromagnetic spectrum.

The infalling material accumulates on the surface of the white dwarf (**Figure 16.15a**), where it is compressed by the enormous gravitational pull of the white dwarf to a density close to that of the white dwarf itself. As more and more material builds up on the surface of the white dwarf, the white dwarf shrinks (just as

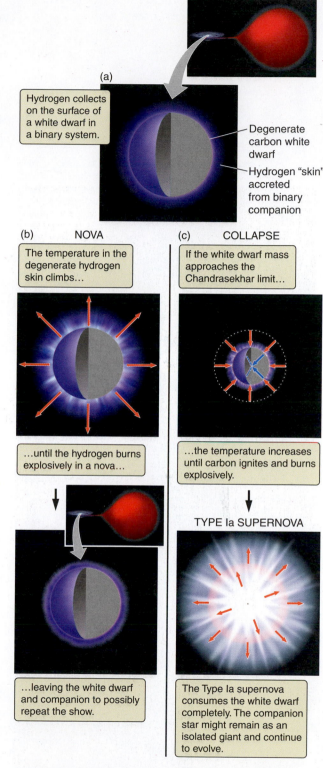

Figure 16.15 (a) In a binary system in which mass is transferred onto a white dwarf, a layer of hydrogen builds up on the surface of the degenerate white dwarf. (b) If hydrogen burning ignites on the surface of the white dwarf, the result is a nova. (c) If enough hydrogen accumulates to raise the core temperature high enough, carbon ignites and the result is a Type Ia supernova.

the core of a red giant shrinks as it grows more massive). The density increases more and more, and at the same time the release of gravitational energy drives the temperature of the white dwarf higher and higher. The infalling material comes from the outer, unburned layers of star 2, so it is composed mostly of hydrogen. Hydrogen, the best nuclear fuel around, is being compressed to higher and higher densities and heated to higher and higher temperatures on the surface of the white dwarf.

Once the temperature at the base of the white dwarf's surface layer of hydrogen reaches about 10 million K, this hydrogen begins to fuse to helium. But this is not the contained hydrogen burning that takes place in the center of the Sun; this is explosive hydrogen burning in a degenerate gas. Energy released by hydrogen burning drives up the temperature. Because the surface is degenerate, this rising temperature does not cause an expansion, as in normal matter, but instead drives up the rate of hydrogen burning. This runaway thermonuclear reaction is much like the runaway helium fusion that takes place during the helium flash, except now there are no overlying layers of a star to absorb the energy liberated by fusion. The result is a tremendous explosion that blows part of the layer covering the white dwarf out into space at speeds of thousands of kilometers per second, as shown in Figure 16.15b. An exploding white dwarf of this kind is called a **nova**.

The explosion of a nova does not destroy the underlying white dwarf star. In fact, much of the material that had built up on the white dwarf may remain behind after the explosion. Afterward, the binary system is in much the same configuration as before: material from star 2 is still pouring onto the white dwarf, as seen in Figure 16.15a. The nova can repeat many times, as material builds up and ignites again and again on the surface of the white dwarf. If the underlying white dwarf is old and cooler, or the mass accumulates slowly on its surface, outbursts are separated by thousands of years, so most novae have been seen only once in historical times. But if the white dwarf is hotter, or mass is transferring quickly, such explosions can happen every few years or even every few months. These recurring novae are called dwarf novae because they are generally less luminous than those that occur less frequently.

About 50 novae occur in our galaxy each year, but we observe only some of them because dust and gas obscure the view from Earth. Novae get bright quickly—typically reaching their peak brightness in only a few hours—and for a brief time they can be several hundred thousand times more luminous than the Sun. Although the brightness of a nova sharply declines in the weeks after the outburst, it can sometimes still be seen for years. During this time, the glow from the expanding cloud of ejected material is powered by the decay of radioactive isotopes created in the explosion. Large bursts of gamma rays have been observed to be associated with several novae.

Supernovae

Novae are spectacular events, but they pale in comparison with another observed phenomenon: the supernova. A **supernova** is an extraordinary nova—a newly visible star that is much more luminous than a nova. Tycho Brahe observed one in 1572, and Kepler observed one in 1604. Supernovae occur in several types, which are distinguished by both their spectra and the way in which they brighten and fade over time. The peak luminosity of a **Type Ia supernova** can be inferred from the rate at which the brightness changes after the explosion. This peak luminosity can then be combined with the peak apparent brightness to find the distance to the supernova. Type Ia supernovae occur in a galaxy the size of the Milky Way about once a century on average. For a brief time they can shine with a luminosity billions of times that of the Sun, in some cases outshining the galaxy itself. Type Ia supernovae are so luminous that they can be seen at great distances, so they can be useful for estimating distances to very distant galaxies. The origins of Type Ia supernovae lie in the evolution of binary systems, and so we discuss them here. Type II supernovae will be discussed in the next chapter.

There is a limit to how massive a white dwarf can be, called the **Chandrasekhar limit**. Above this mass, gravity is stronger than the pressure supplied by degenerate electrons, and the white dwarf collapses. As a white dwarf in a binary system gains mass and approaches the Chandrasekhar limit, the increase in pressure and density causes the temperature of the core to rise. When the core temperature reaches about 6×10^8 K, carbon fusion begins. The star may then have a "simmering" phase with a growing central convective region that prevents thermonuclear runaway for a while. After about 1,000 years, when the temperature reaches about 8×10^8 K, runaway carbon burning begins throughout the entire white dwarf, and the star explodes (Figure 16.15c). The white dwarf is consumed within about a second. This likely happens at about 1.38 M_{Sun}, just before the white dwarf actually reaches the ultimate mass limit. The luminosity of the explosion is about 5 billion times the luminosity of the Sun. Runaway fusion reactions convert a large fraction of the mass of the star into elements such as iron and nickel, and the explosion blasts the remains of the white dwarf into space at speeds in excess of 20,000 km/s, enriching the interstellar medium with these heavier elements. Explosive carbon burning in a white dwarf is the leading theory used to explain Type Ia supernovae.

In this scenario, the explosion completely destroys star 1 of the binary system, but star 2 can be left behind to continue its evolution. Surveys at different wavelengths have found companions to very few Type Ia supernovae, so some scientists think that another process may be responsible for as many as 80 percent of these supernovae. For example, eventually star 2 will go on to form a white dwarf too, leaving behind a binary system consisting of two degenerate white dwarfs. These two white dwarfs might orbit around each other quickly, getting closer and closer. Eventually, tidal forces will disrupt the smaller one, and material will fall onto the larger one until they merge, creating one object with a larger mass, as shown in **Figure 16.16**. In this case, the massive white dwarf came from the merger of two smaller white dwarfs, and so a supernova explosion destroys both stars. This could be why astronomers see so few companions remaining. However, it is also possible that the companions are "missing" because the explosion of one star will eject the other from the system, making it difficult to find. This topic is still an active area of exploration in astronomy (**Process of Science Figure**).

These explosions leave behind expanding shells of dust and gas called **supernova remnants**, as shown in **Figure 16.17**. The material is heated by the expanding blast wave from the supernova and glows in X-rays.

CHECK YOUR UNDERSTANDING 16.5

A white dwarf will become a supernova if: (a) the original star was more than 1.38 M_{Sun}; (b) it accretes an additional 1.38 M_{Sun} from a companion; (c) some mass falls on it from a companion; (d) enough mass accretes from a companion to give the white dwarf a total mass of 1.38 M_{Sun}.

TODAY

40 million years
from now

60 million years
from now

61 million years
from now

Figure 16.16 Two white dwarfs in close orbit about each other get closer and move faster over time, eventually merging into a single, more massive white dwarf. If the mass of the combined white dwarfs is above the Chandrasekhar limit, it will explode as a Type Ia supernova.

SCIENCE IS NOT FINISHED

Understanding Type Ia supernova is critical for measuring distances to the farthest galaxies and for our conclusions about the universe as a whole. Yet the observational evidence remains incomplete.

Supernovae have been observed since antiquity. As technology developed, it became clear there are several types of supernovae.

Type Ia supernova are thought to come from white dwarfs which accumulate mass.

?

But data suggest…

If the white dwarf accretes mass from a red giant or other large star, and explodes at exactly 1.38 M_{Sun}, then the peak luminosity of the explosion should be the same for all Type Ia supernovae.

But searches for potential companions that survived the explosion have found only a few candidates

If the explosion arises from 2 white dwarfs merging together, then the mass of the explosion is variable and the peak luminosity may also vary.

But searches for binary white dwarfs suggest they are not very common

When observational evidence is inconclusive, each scientist adds a piece to the puzzle. Some will conduct larger observational studies, and others will create new theoretical models.

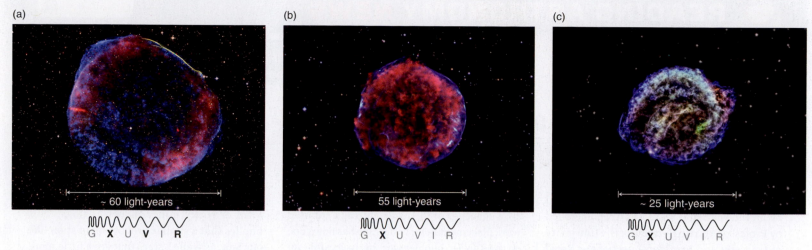

(a)   ~ 60 light-years   G X U V I R

(b)   55 light-years   G X U V I R

(c)   ~ 25 light-years   G X U V I R

Figure 16.17 These images show Type Ia supernovae. The material heated by the expanding blast wave from a supernova glows in the X-ray portion of the spectrum. (a) SN 1006 is the brightest recorded supernova and was observed in China, Japan, Europe, and the Middle East in 1006. X-ray data are shown in blue; radio data are shown in red. (b) Tycho's supernova was observed in 1572. Low-energy X-rays are red; high-energy X-rays are blue. (c) Kepler's supernova, observed in 1604, is shown in five different X-ray wavelengths.

Origins

Stellar Lifetimes and Biological Evolution

From fossil records and DNA analysis, scientists estimate that life appears to have taken hold on Earth within 1 billion years after the Solar System and Earth formed 4.6 billion years ago. It took another 1.5 billion years for more complex cells to develop and another billion years to develop multicellular life. The first animals didn't appear on Earth until 600 million years ago, 4 billion years after the formation of the Sun and the Solar System.

The only known example of biology is the life on Earth. It is always risky to extrapolate from one data point, and it is not known whether biology is widespread in the universe or whether Earth's biological timeline is "typical." Still, reasoning from this one example is the only way to begin thinking about life in the universe. How does the preceding timeline of the evolution of life on Earth compare with the lifetimes of main-sequence stars? Table 16.1 indicates that the lifetime of an O5 star is less than half a million years; of a B5 star, about 120 million years; and of an A0 star, about 700 million years. These stars would have run out of hydrogen in the core and started post-main-sequence evolution in less time than it took for Earth to settle down after its periods of heavy bombardment by debris early in the Solar System.

The 1-billion-year main-sequence lifetime of an A5 star corresponds to the amount of time it took for the simplest life-forms to develop on Earth. The 3-billion-year lifetime for F0 stars corresponds to the amount of time it took for photosynthetic bacteria to develop on Earth. Only stars cooler and less massive than F5 stars have main-sequence lifetimes longer than the 4 billion years it took for life to evolve into animals on Earth. Thus, searches for extrasolar planets survey stars that are F5 or cooler—because hotter and more massive stars probably don't live long enough on the main sequence for complex life to develop.

After a star leaves the main sequence, the helium-burning red giant stage is estimated to last for about 1/10 of the main-sequence lifetime, so this doesn't help stars with short lifetimes to last long enough for complex biology. Could life survive the transition of its star to a red giant? As noted earlier in the chapter, even if a planet is not destroyed, its orbit, temperature, and atmospheric conditions will drastically change, and any life might have to relocate if it is to survive.

ARTICLES   QUESTIONS

Scientists report that the spectra of some dead stars show evidence of rocky material left over from planetary systems.

Scientists Solve Riddle of Celestial Archaeology

By **University of Leicester Press Office**

A decades old space mystery has been solved by an international team of astronomers led by Professor Martin Barstow of the University of Leicester and President-elect of the Royal Astronomical Society.

Scientists from the University of Leicester and University of Arizona investigated hot, young, white dwarfs—the super-dense remains of Sun-like stars that ran out of fuel and collapsed to about the size of the Earth. Their research is featured in MNRAS—the *Monthly Notices of the Royal Astronomical Society*, published by Oxford University Press.

It has been known that many hot white dwarf atmospheres, essentially of pure hydrogen or pure helium, are contaminated by other elements—like carbon, silicon, and iron. What was not known, however, was the origins of these elements, known in astronomical terms as metals.

"The precise origin of the metals has remained a mystery and extreme differences in their abundance between stars could not be explained," said Professor Barstow, a Pro-Vice-Chancellor at the University of Leicester whose research was assisted by his daughter Jo, a coauthor of the paper, during a summer work placement in Leicester. She has now gone on to be an astronomer working in Oxford—on extrasolar planets.

"It was believed that this material was 'levitated' by the intense radiation from deeper layers in the star," said Professor Barstow.

Now the researchers have discovered that many of the stars show signs of contamination by rocky material, the leftovers from a planetary system.

The researchers surveyed 89 white dwarfs, using the Far Ultraviolet Spectroscopic Explorer to obtain their spectra (dispersing the light by color) in which the "fingerprints" of carbon, silicon, phosphorus, and sulfur can be seen when these elements are present in the atmosphere.

"We found that in stars with polluted atmospheres the ratio of silicon to carbon matched that seen in rocky material, much higher than found in stars or interstellar gas."

"The new work indicates that around one-third of all hot white dwarfs are contaminated in this way, with the debris most likely in the form of rocky minor planet analogs. This implies that a similar proportion of stars like our Sun, as well as stars that are a little more massive like Vega and Fomalhaut, build systems containing terrestrial planets. This work is a form of celestial archaeology where we are studying the 'ruins' of rocky planets and/or their building blocks, following the demise of the main star.

"The mystery of the composition of these stars is a problem we have been trying to solve for more than 20 years. It is exciting to realize that they are swallowing up the leftovers from planetary systems, perhaps like our own, with the prospect that more detailed follow-up work will be able to tell us about the composition of rocky planets orbiting other stars," said Professor Barstow.

The study also points to the ultimate fate of the Earth billions of years from now—ending up as a contamination within the white dwarf Sun.

ARTICLES   QUESTIONS

1. This communication refers to the white dwarfs in the study as "hot, young, white dwarfs." What does "young" mean in this context?
2. The spectra described are compared to fingerprints. In what ways are white dwarf spectra like fingerprints?
3. Why were scientists surprised to find elements other than hydrogen and helium in the atmospheres of white dwarf stars? Where do they think these other elements originate?
4. Why does a telescope need to be in space to observe far-ultraviolet wavelengths?
5. How common are contaminated white dwarfs in the sample in this study? Compare this to the percentage of Sun-like stars with planets (22 percent). Does the finding in this communication seem like a sensible number?

Summary

Stars with masses similar to the Sun form planetary nebulae like the one shown in the chapter-opening photograph. This is a likely fate for our own Sun at the end of its evolution. After leaving the main sequence, low-mass stars follow a convoluted path along the H-R diagram that includes the red giant branch, the horizontal branch, the asymptotic giant branch, and a path across the top and then down to the lower left of the diagram. These stages of evolution are dominated by the balance between gravity and energy production from various fusion processes that "turn on and off" in the core of the star. This entire process takes much less time than the main-sequence lifetime of the star. Stars that follow this path have main-sequence lifetimes comparable to the evolutionary timescales of life on Earth. If life on Earth is typical, then more massive stars with shorter lifetimes will not be stable long enough for complex life to evolve.

LG 1 **Estimate the main-sequence lifetime of a star from its mass.** All stars eventually exhaust their nuclear fuel as hydrogen fuses to helium in the cores of main-sequence stars. Less massive stars exhaust their fuel more slowly and have longer lifetimes than more massive stars.

LG 2 **Explain why low-mass stars grow larger and more luminous as they run out of fuel.** When a low-mass star uses up the hydrogen in its core, it begins to burn hydrogen in a shell around the core, heating the gaseous interior. The star expands in size and becomes a red giant.

LG 3 **Sketch post-main-sequence evolutionary tracks on a Hertzsprung-Russell (H-R) diagram, and list the stages of evolution for low-mass stars.** After exhausting its hydrogen, a low-mass star leaves the main sequence and swells to become a red giant, with a helium core made of electron-degenerate matter. The red giant fuses helium to carbon via the triple-alpha process, and quickly the core ignites in a helium flash. The star then moves onto the horizontal branch. A horizontal branch star accumulates carbon and sometimes oxygen in its core and then moves up the asymptotic giant branch. The star may become an AGB star and lose some of its mass.

LG 4 **Describe how planetary nebulae and white dwarfs form.** In their dying stages, some stars eject their outer layers to form planetary nebulae. All low-mass stars eventually become white dwarfs, which are very hot but very small.

LG 5 **Explain how some close binary systems evolve differently than single stars.** Transfer of mass within some binary systems can lead to a nuclear explosion. A nova occurs when hydrogen collects and ignites on the surface of a white dwarf in a binary system. If the mass of the white dwarf approaches $1.38\ M_{Sun}$, the entire star explodes in a Type Ia supernova.

? UNANSWERED QUESTIONS

- Why do planetary nebulae have different shapes? Some are not simply chaotic but are well organized, with varying types of symmetry. Some are spherically symmetric (like a ball), some have bipolar symmetry (like a long hollow tube, pinched in the middle), and some are even point-symmetric (like the letter *S*). How can an essentially spherically symmetric object such as a star produce such beautifully organized outflows? Because these are three-dimensional objects, and astronomers can view each object from only one direction, it is difficult to determine how much of this variation is due to orientation and how much is due to actual differences in the shape of the object. For example, a bipolar nebula, viewed from one end, would appear spherically symmetric. This orientation effect, among other problems, complicates efforts to understand how these shapes are formed. No single explanation has yet satisfactorily covered all the object types.

- Could Earth be moved farther from the Sun to accommodate the Sun's inevitable changes in luminosity, temperature, and radius? One proposal suggests that Earth could capture energy from a passing asteroid and migrate outward, thus staying in the habitable zone while moving farther from the Sun as the Sun ages. Or a huge, thin "solar sail" could be constructed so that radiation pressure from the Sun would slowly push Earth into a larger orbit. These feats of "astronomical engineering" are not feasible anytime in the near future, but perhaps in millions of years (when they will be needed) this could be accomplished.

Questions and Problems

Test Your Understanding

1. Place the main-sequence lifetimes of the following stars in order from shortest to longest.
   a. the Sun: mass 1 M_{Sun}, luminosity 1 L_{Sun}
   b. Capella Aa: mass 3 M_{Sun}, luminosity 76 L_{Sun}
   c. Rigel: mass 24 M_{Sun}, luminosity 85,000 L_{Sun}
   d. Sirius A: mass 2 M_{Sun}, luminosity 25 L_{Sun}
   e. Canopus: mass 8.5 M_{Sun}, luminosity 13,600 L_{Sun}
   f. Achernar: mass 7 M_{Sun}, luminosity 3,150 L_{Sun}

2. Place the following steps in the evolution of a low-mass star in order.
   a. main-sequence star
   b. planetary nebula ejection
   c. horizontal branch
   d. helium flash
   e. red giant branch
   f. asymptotic giant branch
   g. white dwarf

3. If a star follows a horizontal path across the H-R diagram, the star
   a. maintains the same temperature.
   b. stays the same color.
   c. maintains the same luminosity.
   d. keeps the same spectral type.

4. Degenerate matter is different from normal matter because as the mass goes up,
   a. the radius goes down.
   b. the temperature goes down.
   c. the density goes down.
   d. the luminosity goes down.

5. The most massive stars have the shortest lifetimes because
   a. the temperature is higher in the core, so they burn their fuel faster.
   b. they have less fuel in the core when the star forms.
   c. their fuel is located farther from the core.
   d. the temperatures are lower in the core, so they burn their fuel slower.

6. If a main-sequence star suddenly started burning hydrogen at a faster rate in its core, it would become
   a. larger, hotter, and more luminous.
   b. larger, cooler, and more luminous.
   c. smaller, hotter, and more luminous.
   d. smaller, cooler, and more luminous.

7. It is rare to see a helium flash because
   a. few stars go through this stage.
   b. stars that go through this stage are all far away.
   c. the flash only glows at infrared wavelengths.
   d. the flash does not take very long.

8. Post-main-sequence stars lose up to 50 percent of their mass because
   a. jets from the poles release material at an increasing rate.
   b. the mass of the star drops because of mass loss from fusion.
   c. the magnetic field causes increasing numbers of coronal mass ejections.
   d. the star swells until the surface gravity is too weak to hold material.

9. A planetary nebula glows because
   a. it is hot enough to emit UV radiation.
   b. fusion is happening in the nebula.
   c. it is heating up the interstellar medium around it.
   d. light from the central star causes emission lines.

10. As an AGB star evolves into a white dwarf, it runs out of nuclear fuel, and one might guess that the star should cool off and move to the right on the H-R diagram. Why does the star move instead to the left?
    a. It becomes larger.
    b. More of the star is involved in fusion.
    c. As outer layers are lost, deeper layers are exposed.
    d. The temperature of the core rises.

11. When compressed, ordinary gas heats up but degenerate gas does not. Why, then, does a degenerate core heat up as the star continues shell burning around it?
    a. It is heated by the radiation from fusion.
    b. It is heated by the gravitational collapse of the shell.
    c. It is heated by the weight of helium falling on it.
    d. It is insulated by the shell.

12. All Type Ia supernovae
    a. are at the same distance from Earth.
    b. always involve two stars of identical mass.
    c. have identical peak luminosities.
    d. always release the same amount of energy in fusion.

13. In Latin, *nova* means "new." This word is used for novae and supernovae because they are
    a. newly formed stars.
    b. newly dead stars.
    c. newly visible stars.
    d. new main-sequence stars.

14. When the Sun runs out of hydrogen in its core, it will become larger and more luminous because
    a. it will start fusing hydrogen in a shell around a helium core.
    b. it will start fusing helium in a shell and hydrogen in the core.
    c. infalling material will rebound off the core and puffs up the star.
    d. energy balance will no longer hold, and the star will drift apart.

15. A white dwarf is located in the lower left of the H-R diagram. From this information alone, you can determine that
    a. it is very massive.
    b. it is very dense.
    c. it is very hot.
    d. it is very bright.

Thinking about the Concepts

16. Is it possible for a star to skip the main sequence and immediately begin burning helium in its core? Explain your answer.

17. Suppose a main-sequence star suddenly started burning hydrogen at a faster rate in its core. How would the star react? Discuss changes in size, temperature, and luminosity.

18. Describe some possible ways in which the temperature in the core of a star might increase while the density decreases.

19. Astronomers typically say that the mass of a newly formed star determines its destiny from birth to death. However, there is a common environmental circumstance for which this statement is not true. Identify this circumstance and explain why the birth mass of a star might not fully account for the star's destiny.

20. Study the Process of Science Figure. Suppose that a new mechanism is found to explain Type Ia supernovae. In this mechanism, all Type Ia supernovae are more luminous than previously thought. Would the derived distances to galaxies be larger or smaller than we currently understand them to be?

21. Do stars change structure while on the main sequence? Why or why not?

22. Suppose Jupiter is not a planet but a G5 main-sequence star with a mass of 0.8 M_{Sun}.
    a. How will life on Earth be affected, if at all?
    b. How will the Sun be affected as it comes to the end of its life?

23. Explain the similarity in the paths that a star follows along the H-R diagram as it forms from a protostar and as it leaves the main sequence to climb the red giant branch?

24. Why is a horizontal branch star (which burns helium at a high temperature) less luminous than a red giant branch star (which burns hydrogen at a lower temperature)?

25. Suppose the core temperature of a star is high enough for the star to begin fusing oxygen. Predict how the star will continue to evolve, including its path on the H-R diagram.

26. Why does a white dwarf move down and to the right along the H-R diagram?

27. Why does fusion in degenerate material always lead to a runaway reaction?

28. Suppose the more massive red giant star in a binary system engulfs its less massive main-sequence companion, and their nuclear cores combine. What structure will the new star have? Where will the star lie on the H-R diagram?

29. T Coronae Borealis is a well-known recurrent nova.
    a. Is it a single star or a binary system? Explain.
    b. What mechanism causes a nova to flare up?
    c. How can a nova flare-up happen more than once?

30. Why do astronomers prefer to search for planets around low-mass stars?

Applying the Concepts

31. Figure 16.1 contains the text "Straight-line (exponential) approximation." What does this text tell you about the axes on the graph: are they linear or logarithmic? Explain why these data are plotted this way.

32. Use Figure 16.2 to estimate the percentage of the Sun's mass that is turned from hydrogen into helium over its lifetime.

33. Study Figure 16.13. How many times brighter is a star at the top of the giant branch than the same star (a) when it was on the main sequence? and (b) when it was on the horizontal branch?

34. Study Figure 16.13. Make a graph of surface temperature versus time for the evolutionary track shown—from the time the star leaves the main sequence until it arrives at the dot showing that it is a white dwarf. Your time axis may be approximate, but it should show that the star spends different amounts of time in the different phases.

35. For most stars on the main sequence, luminosity scales with mass as $M^{3.5}$ (see Working It Out 16.1). What luminosity does this relationship predict for (a) 0.5-M_{Sun} stars, (b) 6-M_{Sun} stars, and (c) 60-M_{Sun} stars? Compare these numbers to values given in Table 16.1.

36. Calculate the main-sequence lifetimes for (a) 0.5-M_{Sun} stars, (b) 6-M_{Sun} stars, and (c) 60-M_{Sun} stars. (See Working It Out 16.1.) Compare them to the values given in Table 16.1.

37. What will the escape velocity be when the Sun becomes an AGB star with a radius 200 times greater and a mass only 0.7 times that of today? How will these changes in escape velocity affect mass loss from the surface of the Sun as an AGB star?

38. Each form of energy generation in stars depends on temperature.
    a. The rate of hydrogen fusion (proton-proton chain) near 10^7 K increases with temperature as T^4. If the temperature of the hydrogen-burning core is raised by 10 percent, how much does the hydrogen fusion energy increase?
    b. Helium fusion (the triple-alpha process) at 10^8 K increases with an increase in temperature at a rate of T^{40}. If the temperature of the helium-burning core is raised by 10 percent, how much does the helium fusion energy increase?

39. A planetary nebula has an expansion rate of 20 km/s and a lifetime of 50,000 years. Roughly how large will this planetary nebula grow before it disperses?

40. Suppose a companion star transferred mass onto a white dwarf at a rate of about $10^{-9}\ M_{Sun}$ per year. Roughly how long after mass transfer begins will the white dwarf explode as a Type Ia supernova? How does this length of time compare to the typical lifetime of a low-mass star? Assume that the white dwarf started with a mass of 0.6 M_{Sun}.

41. Use Kepler's third law to estimate how fast material in an accretion disk orbits around a white dwarf.

42. A white dwarf has a density of approximately 10^9 kilograms per cubic meter (kg/m^3). Earth has an average density of 5,500 kg/m^3 and a diameter of 12,700 km. If Earth were compressed to the same density as a white dwarf, what would its radius be?

43. What is the density of degenerate material? Calculate how large the Sun would be if all of its mass were degenerate.

44. Recall from Chapter 5 that the luminosity of a spherical object at temperature T is given by $L = 4\pi R^2 \sigma T^4$, where R is the object's radius. If the Sun became a white dwarf with a radius of 10^7 meters, what would its luminosity be at the following temperatures: (a) 10^8 K; (b) 10^6 K; (c) 10^4 K; (d) 10^2 K?

45. According to current astronomical evidence, our universe is approximately 13.8 billion years old. If this is correct, what is the least mass that a star could possess in order to have already evolved into a red giant? What spectral type of star is this?

USING THE WEB

46. Go to the website for the Katzman Automatic Imaging Telescope at Lick Observatory (http://astro.berkeley.edu/bait/public_html/kait.html). What is this project? Why can a search for supernovae be automated? Pick a recent year. How many supernovae were discovered? Look at some of the images. How bright do the supernovae look compared to their galaxies?

47. Go to the American Association of Variable Star Observers (AAVSO) website (http://www.aavso.org). What does this 100-year-old organization do? Read about the types of intrinsic variable stars. Click on "Getting Started." If you have access to dark skies, you can contribute to the study of variable stars. Go to the page for observers (http://www.aavso.org/observers) and click on each item in the "For New Observers" list, including the list of stars that are easy to observe. Assemble a group and observe a variable star from this list.

48. Go to the University of Washington's "Properties of Planetary Nebulae" Web page (https://sites.google.com/a/uw.edu/introductory-astronomy-clearinghouse/assignments/labs-exercises/properties-of-planetary-nebulae) and download and complete the lab exercise.

49. Go to the Hubble Space Telescope's planetary nebula gallery (http://hubblesite.org/gallery/album/nebula/planetary). For each of the three types of symmetry, find an example of a nebula that shows clearly the type of symmetry: spherical (being symmetric in every direction, like a circle), bipolar (having an axis about which they are symmetric, like a person's face), and point-symmetric (being symmetric about a point, like the letter S). Print each of the three images you chose, and label the type of symmetry each one represents. For all three nebulae, identify the location of the central star. For bipolar symmetry, draw a line that shows the axis about which the nebula is symmetric. For point symmetry, identify several features that are symmetric across the location of the central star.

50. In the Hubble telescope news archive, look up press releases on planetary nebulae (http://hubblesite.org/newscenter/archive/releases/nebula/planetary) and white dwarf stars (http://hubblesite.org/newscenter/archive/releases/star/white-dwarf). Pick a story for each. What observations were reported, and why were they important?

smartwork

If your instructor assigns homework in SmartWork, access your assignments at digital.wwnorton.com/astro5.

digital.wwnorton.com/astro5

The evolution of a low-mass star, as discussed in this chapter, corresponds to many twists and turns on the H-R diagram. In this Exploration, we return to the "H-R Diagram Explorer" interactive simulation to investigate how these twists and turns affect the appearance of the star.

Open the "H-R Explorer" simulation, linked from the Chapter 16 area on the Student Site of the Digital Landing Page. The box labeled "Size Comparison" shows an image of both the Sun and the test star. Initially, these two stars have identical properties: the same temperature, the same luminosity, and the same size.

Examine the box labeled "Cursor Properties." This box shows the temperature, luminosity, and radius of a test star located at the "X" in the H-R diagram. Before you change anything, answer the first three questions.

1 What is the temperature of the test star?

_____

2 What is the luminosity of the test star?

_____

3 What is the radius of the test star?

_____

As a star leaves the main sequence, it moves up and to the right on the H-R diagram. Move the cursor (the X on the diagram) up and to the right.

4 What changes about the image of the test star next to the Sun?

_____

_____

5 What is the test star's temperature? What property of the image of the test star indicates that its temperature has changed?

_____

_____

6 What is the test star's luminosity?

_____

7 What is the test star's radius?

_____

8 Ordinarily, the hotter an object is, the more luminous it is. In this case, the temperature has gone down, but the luminosity has gone up. How can this be?

_____

_____

_____

The star then moves around quite a lot in that part of the H-R diagram. Look at Figure 16.13, and then use the cursor to approximate the motion of the star as it moves up the red giant branch, back down and onto the horizontal branch, and then back to the right and up the asymptotic giant branch.

9 Are the changes you observe in the image of the star as dramatic as the ones you observed for question 4?

_____

_____

10 What is the most noticeable change in the star as it moves through this portion of its evolution?

_____

_____

Next, the star begins moving across the H-R diagram to the left, maintaining almost the same luminosity. Drag the cursor across the top of the H-R diagram to the left, and study what happens to the image of the star in the "Size Comparison" box.

11 What changed about the star as you dragged it across the H-R diagram?

_____

_____

12 How does the star's size now compare to that of the Sun?

_____

_____

Finally, the star drops to the bottom of the H-R diagram and then begins moving down and to the right. Move the cursor toward the bottom of the H-R diagram, where the star becomes a white dwarf.

13 What changed about the star as you dragged it down the H-R diagram?

_____

_____

14 How does its size now compare to that of the Sun?

_____

_____

To thoroughly cement your understanding of stellar evolution, press the "Reset" button and then move the star from main sequence to white dwarf several times. This exercise will help you remember how this part of a star's life appears on the H-R diagram.

17

Evolution of High-Mass Stars

The vast majority of stars are smaller and less massive than the Sun and live a relatively long time. The most massive stars, the O and B stars, are rarer and live a much shorter time. When these massive stars die, the result is far more spectacular than when lower-mass stars die. In this chapter, we will look at the lives of these high-mass stars.

LEARNING GOALS

By the conclusion of this chapter, you should be able to:

LG 1 Describe how the death of high-mass stars differs from that of low-mass stars.

LG 2 List the sequence of stages for evolving high-mass stars.

LG 3 Explain the origin of chemical elements up to and heavier than iron.

LG 4 Identify how Hertzsprung-Russell (H-R) diagrams of clusters enable astronomers to measure the ages of stars and test theories of stellar evolution.

The Crab Nebula is the remains of a massive star whose explosion was observed in 1054. The center is an X-ray image of the pulsar. ▶▶▶

What caused
this star to
explode?

17.1 High-Mass Stars Follow Their Own Path

We have seen that low-mass stars live for billions of years. High-mass stars have masses greater than about 8 M_{Sun}. These stars have more mass, but they also have luminosities thousands or even millions of times greater than the Sun's luminosity. Even though they have more fuel to begin with, high-mass stars use it up faster than low-mass stars and therefore have shorter lives. High-mass stars live only hundreds of thousands to millions of years. In this section, we will discuss the stages in the evolution of high-mass (greater than 8 M_{Sun}) and medium-mass (3–8 M_{Sun}) stars.

The CNO Cycle

A high-mass star has greater gravitational force pressing down on the interior than that of a low-mass star. This greater force leads to higher temperature and pressure, increasing the rate of nuclear reactions, and therefore generating greater luminosities. In addition, at the much higher temperatures at the center of a high-mass star, additional nuclear reactions become possible. Recall from Chapter 14 that the hydrogen nucleus has only one proton—a single positive charge—so hydrogen fuses at lower temperatures (a few million kelvins) than any other atomic nucleus. However, the probability that any two hydrogen atoms will fuse is low. The low probability of this first step in the proton-proton chain limits how rapidly the entire process can move forward.

If the star contains elements from a previous generation of stars that lived and died, it will have carbon and other elements mixed in with its hydrogen. At the high temperatures in the core of a massive star, the hydrogen and carbon nuclei can interact in a series of reactions called the carbon-nitrogen-oxygen (CNO) cycle. The **CNO cycle** is a nuclear fusion process that converts hydrogen to helium in the presence of carbon. This process is illustrated in **Figure 17.1a**, which shows each step. Through most of the process, only one of two reactions happens at each step: either a hydrogen nucleus fuses with another nucleus to create a new element with higher atomic number or a proton spontaneously decays to a neutron to create a new element with lower atomic number. First, a hydrogen nucleus fuses with a carbon-12 (^{12}C) nucleus to form nitrogen-13 (^{13}N). Second, a proton in this ^{13}N nucleus decays, so that the atom is once again carbon. This carbon nucleus has an "extra" neutron, so it is now carbon-13 (^{13}C), not carbon-12. Third and fourth, two more hydrogen nuclei then fuse with this ^{13}C nucleus, creating nitrogen-14 (^{14}N) and then oxygen-15 (^{15}O). Fifth, a proton in the oxygen nucleus decays, leaving behind nitrogen-15 (^{15}N). One more proton enters the nucleus, causing the ejection of a helium-4 (^{4}He) nucleus, and leaving behind a ^{12}C nucleus, which can participate in the cycle again. Along the way, several by-products are produced: a positron and a neutrino are ejected each time a proton decays to form a neutron. Each positron subsequently annihilates with an electron to produce a gamma ray. Additional gamma rays are released each time fusion occurs. Figure 17.1b shows the net reaction: a carbon-12 nucleus and four hydrogen nuclei

Nebraska Simulation: CNO Cycle Animation

Figure 17.1 (a) In high-mass stars, carbon serves as a catalyst for the fusion of hydrogen to helium. This process is the carbon-nitrogen-oxygen (CNO) cycle. (b) The CNO cycle takes carbon-12, hydrogen, and electrons as inputs and produces carbon-12, helium, neutrinos, and gamma rays.

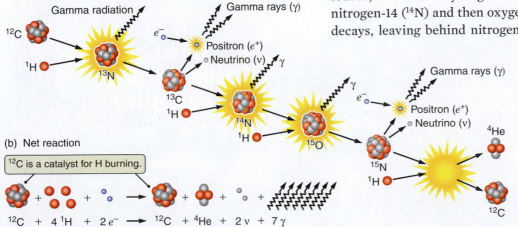

(a) CNO cycle

Gamma radiation   Gamma rays (γ)

^{12}C e^- Positron (e^+)
   Neutrino (ν) γ
^{1}H ^{13}N

^{13}C
^{1}H
^{14}N
^{1}H ^{15}O

γ e^- Gamma rays (γ)
   Positron (e^+)
   Neutrino (ν) ^{4}He
^{15}N
^{1}H
     ^{12}C

(b) Net reaction

^{12}C is a catalyst for H burning.

$$^{12}C + 4\,^1H + 2\,e^- \longrightarrow \,^{12}C + \,^4He + 2\,\nu + 7\,\gamma$$

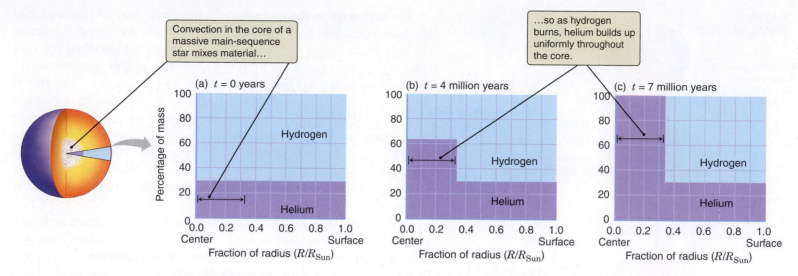

Convection in the core of a massive main-sequence star mixes material…

…so as hydrogen burns, helium builds up uniformly throughout the core.

(a) $t = 0$ years

(b) $t = 4$ million years

(c) $t = 7$ million years

Figure 17.2 Convection keeps the core of a high-mass main-sequence star well mixed, so the composition remains uniform throughout the core as it evolves from zero age in the top graph to age 7 million years in the bottom graph. (Evolution times are for a 25-M_{Sun} star.)

combine with two electrons to produce a carbon-12 nucleus, a helium-4 nucleus, two neutrinos, and seven gamma rays. It takes a lot of energy to get a hydrogen nucleus past the electric barrier set up by a carbon nucleus, with its six protons, but when this barrier can be overcome, fusion is much more probable than it is in the interaction between two hydrogen nuclei. The CNO cycle is far more efficient than the proton-proton chain in stars more massive than about 1.3–1.5 M_{Sun}.

The different ways that hydrogen burning takes place in high-mass stars and low-mass stars are reflected in the different core structures of the two types of stars. The temperature gradient in the core of a high-mass star is so steep that convection sets in within the core itself, "stirring" the core like the water in a boiling pot. As shown in **Figure 17.2**, helium ash spreads uniformly throughout the core of a high-mass star as the star consumes its hydrogen. This is unlike the case for low-mass stars, in which case the helium ash builds up from the center outward (see Figure 16.2).

The High-Mass Star Leaves the Main Sequence

As the high-mass star runs out of hydrogen in its core, the weight of the overlying star compresses the core, just as it does in a low-mass star. Yet long before the core of the high-mass star becomes electron-degenerate, the pressure and temperature in the core become high enough (10^8 K) for helium burning to begin. This rapid increase in pressure and temperature prevents the growth of a degenerate core in the high-mass star. The star makes a fairly smooth transition from hydrogen burning to helium burning as it leaves the main sequence.

Recall that when a low-mass star leaves the main sequence, the path it follows on the Hertzsprung-Russell (H-R) diagram is nearly vertical (see Figure 16.4), going to higher and higher luminosities at roughly constant temperature. But as a high-mass star leaves the main sequence, it grows in size while its surface temperature falls, so it moves nearly horizontally on the H-R diagram, as shown in **Figure 17.3**. The massive star has the same structure as a low-mass horizontal branch star, burning helium in its core and hydrogen in a surrounding shell. Stars more massive than 10 M_{Sun} become red supergiants during their helium-burning phase. They have very cool surface temperatures (about 4000 K) and radii as much as 1,000 times that of the Sun.

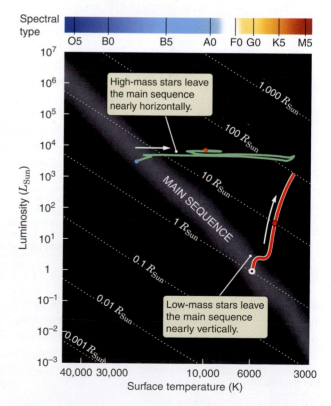

High-mass stars leave the main sequence nearly horizontally.

Low-mass stars leave the main sequence nearly vertically.

Figure 17.3 When high-mass stars leave the main sequence, they move horizontally across the H-R diagram, unlike low-mass stars, which move nearly vertically.

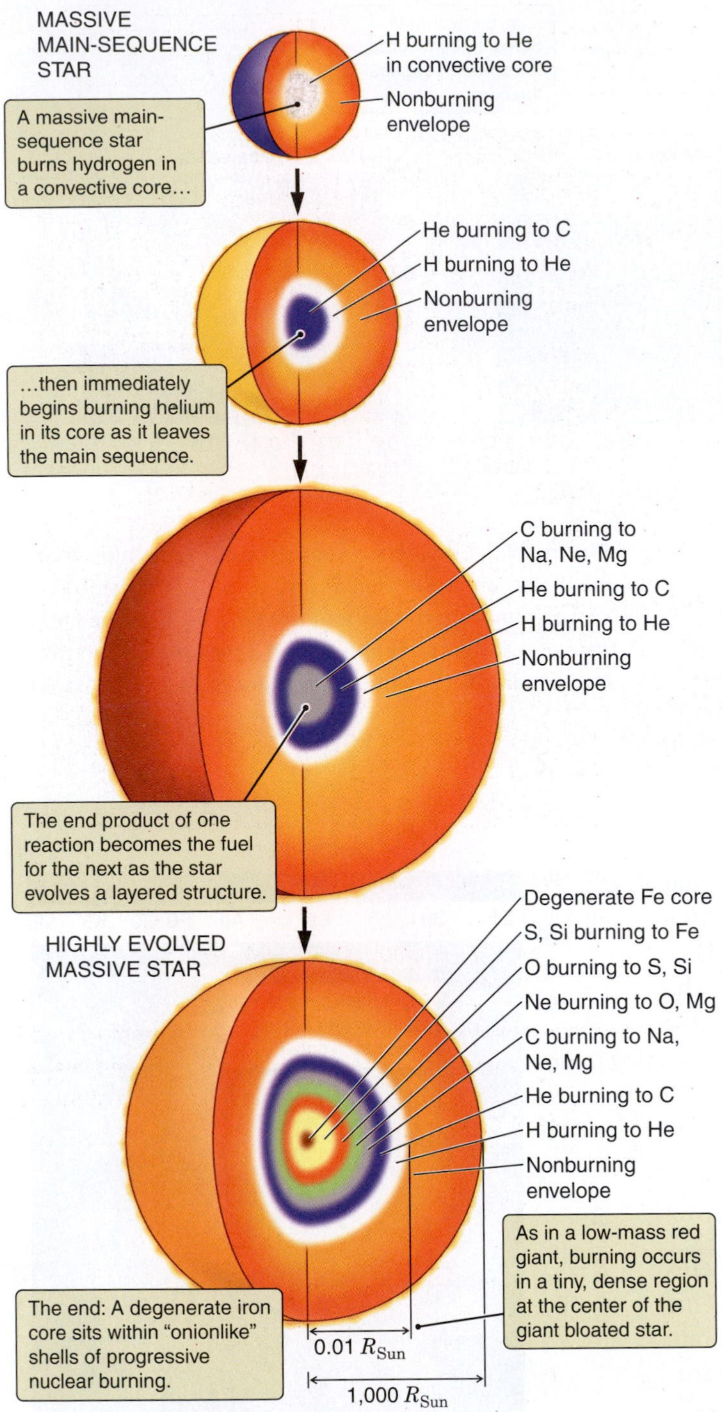

MASSIVE MAIN-SEQUENCE STAR

H burning to He in convective core

Nonburning envelope

A massive main-sequence star burns hydrogen in a convective core...

He burning to C
H burning to He
Nonburning envelope

...then immediately begins burning helium in its core as it leaves the main sequence.

C burning to Na, Ne, Mg
He burning to C
H burning to He
Nonburning envelope

The end product of one reaction becomes the fuel for the next as the star evolves a layered structure.

HIGHLY EVOLVED MASSIVE STAR

Degenerate Fe core
S, Si burning to Fe
O burning to S, Si
Ne burning to O, Mg
C burning to Na, Ne, Mg
He burning to C
H burning to He
Nonburning envelope

The end: A degenerate iron core sits within "onionlike" shells of progressive nuclear burning.

As in a low-mass red giant, burning occurs in a tiny, dense region at the center of the giant bloated star.

$0.01\ R_{Sun}$

$1,000\ R_{Sun}$

Figure 17.4 As a high-mass star evolves, it builds up a layered structure like that of an onion, with progressively more advanced stages of nuclear burning found deeper and deeper within the star. Note that the bottom image has been reduced in size in order to fit on the page.

Nebraska Simulation: *H-R Explorer*

The next stage in the evolution of a high-mass star has no analog in low-mass stars. When the high-mass star exhausts the helium in its core, the core again begins to collapse. Carbon fusion begins when the core reaches temperatures of 8×10^8 K or higher. This produces more massive elements, including oxygen, neon, sodium, and magnesium. The star at this time consists of a carbon-burning core surrounded by a helium-burning shell surrounded by the outward-moving, hydrogen-burning shell. When carbon is exhausted as a nuclear fuel at the center of the star, neon breaks down to oxygen and helium or fuses to magnesium; and when neon is exhausted, oxygen begins to fuse. The structure of the evolving high-mass star, shown in **Figure 17.4**, is like an onion, with many concentric layers.

Medium-mass stars with masses between 3 and 8 M_{Sun} burn hydrogen via the CNO cycle like massive stars. They also leave the main sequence as massive stars do, burning helium in their cores immediately after their hydrogen is exhausted and skipping the helium flash phase of low-mass star evolution. When helium burning in the core is complete, however, the temperature at the center of a medium-mass star is too low for carbon to burn. From this point on, the star evolves more like a low-mass star, ascending the asymptotic giant branch, burning helium and hydrogen in shells around a degenerate core, and then ejecting its outer layers and leaving behind a white dwarf.

Stars on the Instability Strip

As a star undergoes post-main-sequence evolution, it may make one or more passes through a region of the H-R diagram known as the **instability strip**, as shown by the dashed-line region in **Figure 17.5**. **Variable stars** are stars that vary in brightness over time. In this region of the H-R diagram, pulsating variable stars can be found. **Pulsating variable stars** do not achieve a steady balance between pressure and gravity; rather, they alternately grow larger and smaller. The pulsating variable star repeatedly overshoots the equilibrium point, shrinking too far before being pushed back out by pressure or expanding too far before being pulled back in by gravity.

The most luminous pulsating variable stars are **Cepheid variables**, named after the prototype star Delta Cephei. Type I, or Classical, Cepheids are massive and luminous yellow supergiants. A Cepheid variable completes one cycle of its pulsation in anywhere from about 1 to 100 days, depending on its luminosity, as shown in **Figure 17.6**. This type of relationship is called a **period-luminosity relationship**; in this case, the more luminous the star, the longer its period of variation. This period-luminosity relationship for Cepheid variables, first discovered experimentally by Henrietta Leavitt (1868–1921) in 1912, allows astronomers to use Cepheid variables to find the distances to galaxies beyond our own. By observing the period, astronomers can determine the luminosity. Combining this luminosity with the brightness in the sky gives the distance to the star.

Thermal energy powers the pulsations of stars like Cepheid variables. Ionization of helium atoms in the star alternately traps and releases thermal energy, causing the star to expand and contract. An ionized gas is nearly opaque, because ions can interact with light of all wavelengths, scattering many of them back toward the interior. A neutral gas can absorb only light with energies that match the electron transitions. As the helium atoms alternate between ionized and neutral states, the atmosphere of the star alternates between opaque and transparent. This alternately traps and releases light from the star. It is much like a lid on a pot of boiling water. The pot builds up pressure enough to pop open the lid and let steam escape. Then gravity pulls the lid back down, and pressure builds again to repeat the process. These pulsations do not affect the nuclear burning in the star's interior. However, they do affect the light escaping from the star. From Chapter 13, recall the luminosity-temperature-radius relationship for stars: both the luminosity and the temperature (and hence color) of the star change periodically as the star expands and shrinks. The star is at its brightest and bluest while it expands through its equilibrium size and at its faintest and reddest while it falls back inward.

Type I Cepheid variables are not the only type of variable star. The instability strip on the H-R diagram also intersects the low-mass horizontal branch. Low-mass ($\sim$0.8-M_{Sun}) stars can be Type II Cepheid variables or **RR Lyrae variables**, which have periods of less than a day. RR Lyrae stars pulsate by the same mechanism as Cepheid variables but are typically hundreds of times less luminous. They, too, follow a period-luminosity relationship, illustrated in Figure 17.6b.

High-mass stars also change their composition by expelling a significant percentage of their mass back into space throughout their lifetimes. Even while on

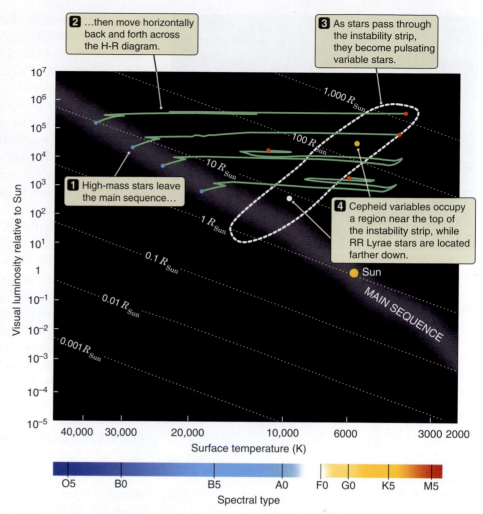

Figure 17.5 The paths of high-mass stars along the H-R diagram takes them through a region known as the instability strip, shown by the region surrounded by the white dashed line. Pulsating variable stars such as Cepheid variables and RR Lyrae stars are found in this region of the H-R diagram.

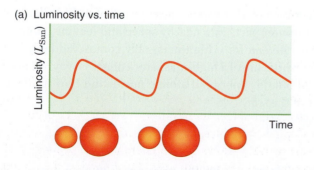

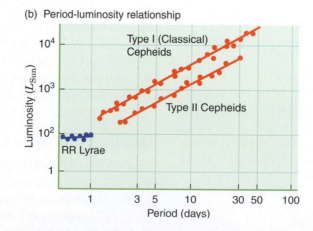

Figure 17.6 (a) Cepheid variable stars pulsate in size and therefore their luminosity changes with time in a periodic way. The shape of the resulting curve is distinctive. (b) The length of the period of pulsation is related to the star's luminosity at maximum.

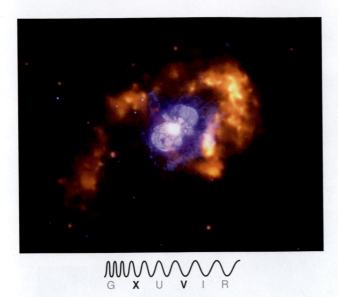

G   **X**   U   **V**   I   R

Figure 17.7 In this image of the luminous blue variable (LBV) star Eta Carinae, an expanding cloud of ejected dusty material is seen in optical (blue) and X-ray (yellow) light. The star itself, which is largely hidden by the surrounding dust, has a luminosity of 5 million L_{Sun} and a mass probably in excess of 120 M_{Sun}. Dust is created when volatile material ejected from the star condenses.

the main sequence, massive O and B stars have low-density winds with velocities as high as 3,000 kilometers per second (km/s). These winds are pushed outward by the pressure of the radiation from the star. The pressure of the intense radiation at the surface of a massive star overcomes the star's gravity and drives away material in its outermost layers. Main-sequence O and B stars lose mass at variable rates ranging from 10^{-7} up to 10^{-5} M_{Sun} of material per year. The fastest mass loss occurs in the most massive stars. These numbers may sound tiny, but over millions of years mass loss plays a prominent role in the evolution of high-mass stars. O stars with masses of 20 M_{Sun} or more may lose about 20 percent of their mass while on the main sequence and possibly more than 50 percent of their mass over their entire lifetimes. Even an 8-M_{Sun} star may lose 5–10 percent of its mass.

In general, stars with masses below 15 M_{Sun} go through a red supergiant stage, perhaps with a Cepheid variable phase. Stars under 30 M_{Sun} move back and forth on the H-R diagram—becoming red supergiants, then blue supergiants, then red supergiants again (sometimes with a brief period as a yellow supergiant), depending on what is burning in their cores. Luminous blue variable (LBV) stars are hot, luminous, extremely rare stars that may be as massive as 150 M_{Sun}. An example is Eta Carinae (**Figure 17.7**), a binary system with a 120-M_{Sun} star and a luminosity (summed over all wavelengths) of 5 million L_{Sun}. Currently, Eta Carinae is losing mass at a rate of 1 M_{Sun} every 1,000 years. However, during a 19th century eruption, when Eta Carinae became the second brightest star in the sky, it shed 10 M_{Sun} of material in only 20 years. Eta Carinae is expected to explode in the astronomically near future.

CHECK YOUR UNDERSTANDING 17.1

How does energy production in a high-mass, main-sequence star differ from energy production in the Sun? (Choose all that apply.) (a) High-mass stars get a lot of energy through non-nuclear processes. (b) High-mass stars produce energy at a faster rate. (c) High-mass stars burn carbon on the main sequence. (d) High-mass stars use carbon in a process that fuses hydrogen to helium.

17.2 High-Mass Stars Go Out with a Bang

You have already seen that a low-mass star approaches the end of its life relatively slowly and gently, ejecting its outer parts into nearby space and leaving behind a degenerate core. In contrast, the end for a high-mass star comes suddenly and explosively. To understand why the life of a high-mass star ends this way, we need to understand the concept of binding energy in atomic nuclei and how it affects nuclear fusion in the center of the star.

Binding Energy

An evolving high-mass star builds up its onionlike structure as nuclear burning in its interior proceeds to more and more advanced stages (see Figure 17.4). Hydrogen burns to helium, helium burns to carbon and oxygen, carbon burns to magnesium, oxygen burns to sulfur and silicon, and then silicon and sulfur burn to iron. Many different types of nuclear reactions occur up to this point, forming almost all of the different stable isotopes of elements less massive than iron. The chain of nuclear fusion stops with iron.

Recall that when four hydrogen atoms combine to form a helium atom, the resulting helium atom has less mass than the sum of the four individual hydrogen atoms. This difference in mass is converted to energy, which maintains the temperature of the gas at the high levels needed to sustain the reaction. The same happens in the triple-alpha process when three helium nuclei combine to form a ^{12}C nucleus: The net energy produced by the reaction is the difference between the mass of the three alpha particles and the mass of the carbon nucleus. This energy helps sustain the nuclear reactions in the core.

The **binding energy** of an atomic nucleus is the energy required to break the nucleus into its constituent parts. A nuclear reaction that increases a nucleus's binding energy releases energy. Conversely, decreasing the binding energy absorbs energy. **Figure 17.8** shows the binding energy per nucleon (that is, per each proton or neutron in the nucleus) for different atomic nuclei. Moving up the plot from helium to carbon increases the binding energy, so fusing helium to carbon releases energy. Iron is at the peak of the binding-energy curve, so moving up the plot from lighter elements to iron (Fe) also releases energy; conversely, moving down the plot from iron to heavier elements absorbs energy. Iron fusion absorbs energy in the reaction. This reaction, therefore, does not sustain itself (**Working It Out 17.1**).

The Final Days in the Life of a Massive Star

Fusion of helium to carbon produces about one-quarter as much energy per reaction as the fusion of hydrogen to helium. This lower energy production rate results in lower pressure inside the star. As gravity compresses the star, this less efficient nuclear fuel is consumed more rapidly. Although fusion of hydrogen into helium can provide the energy needed to support the high-mass star against the force of gravity for millions of years, helium fusion can support the star for only a

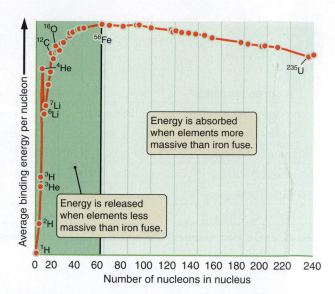

Figure 17.8 The binding energy per nucleon is plotted against the number of nucleons for each element. This is the energy it would take to break the atomic nucleus apart into protons and neutrons. Energy is released by nuclear fusion only if the resulting element is higher on the curve.

17.1 Working It Out Binding Energy of Atomic Nuclei

The net energy released by a nuclear reaction is the difference between the binding energy of the products and the binding energy of the reactants:

$$\text{Net energy} = \begin{pmatrix} \text{Binding energy} \\ \text{of products} \end{pmatrix} - \begin{pmatrix} \text{Binding energy} \\ \text{of reactants} \end{pmatrix}$$

In the example of the triple-alpha process, the binding energy of a helium nucleus is 6.824×10^{14} joules (J) per kilogram (kg) of helium, and the binding energy of the produced ^{12}C nuclei is 7.402×10^{14} J. The amount of energy available from fusing 1 kg of helium nuclei into carbon is given by

$$\begin{pmatrix} \text{Net energy from} \\ \text{burning 1 kg He} \end{pmatrix} = \begin{pmatrix} \text{Binding energy} \\ \text{of C formed} \end{pmatrix} - \begin{pmatrix} \text{Binding energy} \\ \text{of He burned} \end{pmatrix}$$

$$= (7.402 \times 10^{14}\,\text{J}) - (6.824 \times 10^{14}\,\text{J})$$

$$= 5.780 \times 10^{13}\,\text{J}$$

This release of net energy indicates that helium is a good nuclear fuel, as Figure 17.8 shows.

What about fusing iron into more massive elements? Because iron is at the peak of the binding-energy curve, the products of iron burning will have less binding energy than the initial reactants. Going from iron to more massive elements means moving down on the binding-energy curve in Figure 17.8, so the net energy in the reaction will be negative. Rather than producing energy, fusion of iron uses energy.

TABLE 17.1	Burning Stages in High-Mass Stars		
Core Burning Stage	15-M_{Sun} Star	25-M_{Sun} Star	Typical Core Temperatures
Hydrogen (H) burning	11 million years	7 million years	$(3–10) \times 10^7$ K
Helium (He) burning	2 million years	800,000 years	$(1–7.5) \times 10^8$ K
Carbon (C) burning	2,000 years	500 years	$(0.8–1.4) \times 10^9$ K
Neon (Ne) burning	8 months	11 months	$(1.4–1.7) \times 10^9$ K
Oxygen (O) burning	2.6 years	5 months	$(1.8–2.8) \times 10^9$ K
Silicon (Si) burning	18 days	0.7 day	$(2.8–4) \times 10^9$ K

few hundred thousand years. **Table 17.1** shows that the star proceeds from helium burning to the end of its life at a much faster pace.

Maintaining the balance in a star is like trying to keep a leaky balloon inflated. The larger the leak, the more rapidly air must be pumped into the balloon to keep it inflated. A star that is fusing hydrogen or helium is like a balloon with a slow leak. At the temperatures generated by hydrogen or helium fusion, energy leaks out of the interior of a star primarily by radiation and convection. Neither of these processes is very efficient, because the outer layers of the star act like a thick, warm blanket. Much of the energy is kept in the star, so nuclear fuels do not need to burn very fast to support the weight of the outer layers of the star while keeping up with the energy escaping from the surface.

Once carbon burning begins, this balance shifts. Energy is carried away primarily by neutrinos—referred to as neutrino cooling—rather than by radiation and convection. Recall from your study of the Sun in Chapter 14 that neutrinos escape easily, carrying energy away from the core. Like air pouring out through a huge hole in the side of a balloon, neutrinos produced in the interior of the star stream through the overlying layers of the star as if they were not even there, carrying the energy from the stellar interior out into space. As thermal energy pours out of the interior of the star, the outer layers of the star fall inward, driving up the density and temperature and increasing the rate of nuclear reactions.

Once this process of neutrino cooling becomes significant, the star begins evolving much more rapidly. Carbon fusion supports the star for less than about a thousand years. Oxygen fusion holds the star up for only about a year. Silicon fusion lasts only a few days. A silicon-burning star is not much more luminous than it was while burning helium. But because of neutrino cooling, the silicon-burning star actually releases about 200 million times more energy per second than it did while it was fusing helium.

The Collapse of the Core and Subsequent Explosion

As shown in Figure 17.4, an evolving high-mass star builds up its onionlike structure as hydrogen fuses to helium, helium fuses to carbon, carbon fuses to sodium, neon, and magnesium, oxygen fuses to sulfur and silicon, and silicon and sulfur fuse to iron. Many different types of nuclear reactions occur up to this point, forming almost all of the different stable isotopes of elements less massive than iron. However, because iron does not release energy when it fuses but rather absorbs energy, the chain of nuclear fusion stops with iron. After silicon fuses to form an iron core in the star, the end comes suddenly and dramatically. For iron, once the reaction starts, energy is absorbed. No source of nuclear energy remains to replenish the energy that is being taken away by escaping neutrinos. The high-mass star's life balancing gravity and controlled nuclear energy production is over. No longer supported by thermonuclear fusion, the iron core of the massive star begins to collapse.

Figure 17.9 shows the stages a high-mass star passes through at the end of its life. The early stages of collapse of the iron core of an evolved massive star are much the same as in the collapse of a nonburning core in a low-mass star. As the core collapses, the force of gravity increases and the density and temperature skyrocket. The gas in the core becomes electron-degenerate when it is about the size of Earth. Unlike the electron-degenerate core of a low-mass red giant, however, the weight bearing down on the interior of the iron core is too great to be held up by electron degeneracy pressure (step 1 in Figure 17.9). As the core collapses, the core temperature climbs to 10 billion K (10^{10} K) and higher, while the density exceeds 10^{10} kilograms per cubic meter (kg/m³)—10 times the density of an electron-degenerate white dwarf.

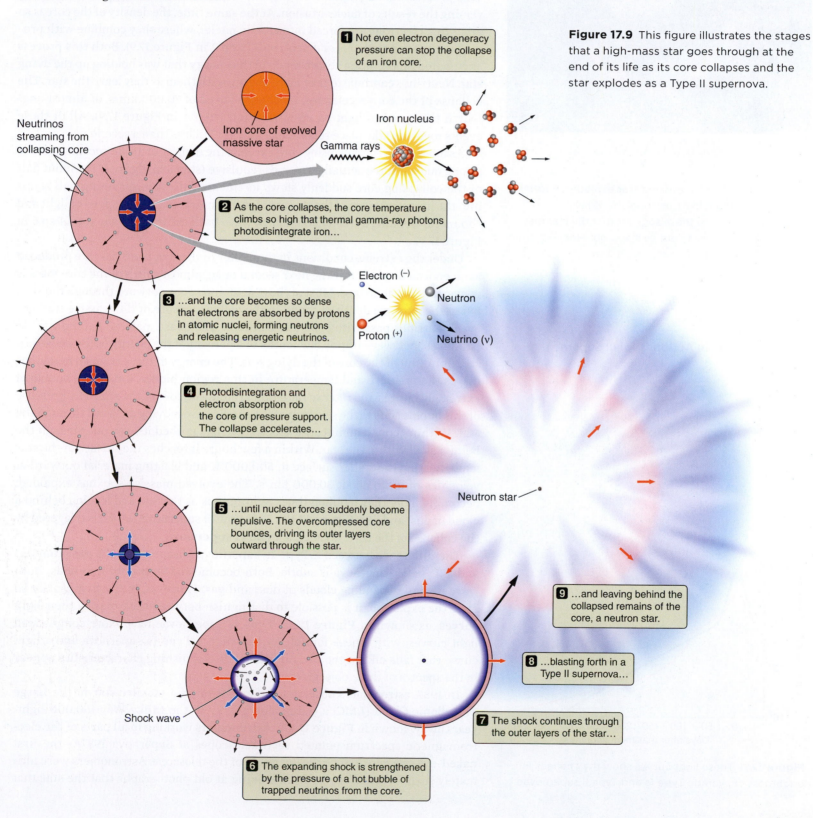

Figure 17.9 This figure illustrates the stages that a high-mass star goes through at the end of its life as its core collapses and the star explodes as a Type II supernova.

1 Not even electron degeneracy pressure can stop the collapse of an iron core.

Iron core of evolved massive star

Iron nucleus

Gamma rays

Neutrinos streaming from collapsing core

2 As the core collapses, the core temperature climbs so high that thermal gamma-ray photons photodisintegrate iron…

Electron (−)

Proton (+)

Neutron

Neutrino (ν)

3 …and the core becomes so dense that electrons are absorbed by protons in atomic nuclei, forming neutrons and releasing energetic neutrinos.

4 Photodisintegration and electron absorption rob the core of pressure support. The collapse accelerates…

5 …until nuclear forces suddenly become repulsive. The overcompressed core bounces, driving its outer layers outward through the star.

Neutron star

9 …and leaving behind the collapsed remains of the core, a neutron star.

8 …blasting forth in a Type II supernova…

Shock wave

7 The shock continues through the outer layers of the star…

6 The expanding shock is strengthened by the pressure of a hot bubble of trapped neutrinos from the core.

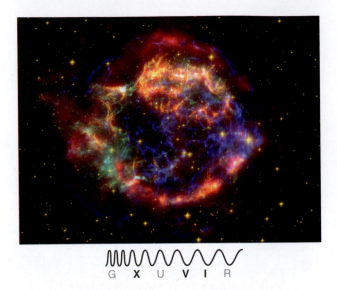

Figure 17.10 When an evolved star explodes, it forms a supernova remnant like Cassiopeia A. X-ray observations such as these suggest that the star may have turned itself inside out as it spit out elements from its core.

Astronomy in Action: Type II Supernova

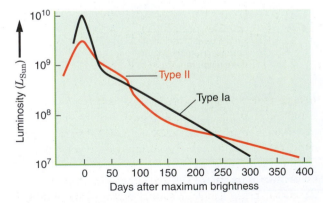

Figure 17.11 These light curves show the changes in brightness of average Type Ia and Type II supernovae.

These phenomenal temperatures and pressures trigger fundamental changes in the core. The laws describing thermal radiation say that at these temperatures, the nucleus of the star is awash in extremely energetic thermal radiation. This radiation is so energetic that thermal gamma-ray photons are produced with enough energy to break iron nuclei apart into helium nuclei (step 2 in Figure 17.9). This process, called photodisintegration, absorbs thermal energy and begins reversing the results of nuclear fusion. At the same time, the density of the core is so great that electrons are forced into atomic nuclei, where they combine with protons to produce neutrons and neutrinos (step 3 in Figure 17.9). Both this process and photodisintegration absorb much of the energy that was holding up the dying star. Neutrinos continue to take more energy with them as they leave the star. The collapse of the core accelerates, reaching a speed of 70,000 km/s, or almost one-fourth the speed of light on its inward fall (step 4 in Figure 17.9). All of these events together take place remarkably quickly—in less than a second.

As material in the collapsing core exceeds the density of an atomic nucleus, the strong nuclear force actually becomes repulsive (step 5 in Figure 17.9). About half of the collapsing core suddenly slows its inward fall. The remaining half slams into the innermost part of the star at a significant fraction of the speed of light and "bounces," sending a tremendous shock wave back out through the star (step 6 in Figure 17.9).

Under the extreme conditions in the center of the star, neutrinos are produced at an enormous rate. Over the next second or so, almost a fifth of the core mass is converted into neutrinos. Most of these neutrinos pour outward through the star; but at the extreme densities found in the collapsing core of the massive star, not even neutrinos pass with complete freedom. The dense material behind the expanding shock wave traps a few tenths of a percent of the energy of the neutrinos streaming out of the core of the dying star. The energy of these trapped neutrinos drives the pressure and temperature in this region higher, inflating a bubble of extremely hot gas and intense radiation around the core of the star. The pressure of this bubble adds to the strength of the shock wave moving outward through the star. Within about a minute, the shock wave has pushed its way out through the helium shell within the star. Within a few hours it reaches the surface of the star itself, heating the stellar surface to 500,000 K and blasting material outward at velocities of up to about 30,000 km/s. The evolved massive star has exploded, becoming more than a billion times as luminous as the Sun and leaving behind a cloud of dust and gas (**Figure 17.10**). This type of supernova, which is triggered by the collapse of the core, is called a **Type II supernova**.

The difference between a Type Ia supernova (discussed in the past chapter) and a Type II supernova is subtle. Both become suddenly very luminous. Both leave behind expanding clouds of dust and gas. However, in the first year or so after the explosion, it is possible to distinguish between the types by their light curves, as shown in **Figure 17.11**. Type II supernovae have more complicated light curves, with a peak luminosity less than a Type Ia supernova and a light curve that falls off less rapidly. Additional distinguishing characteristics appear in the spectra of these objects.

In 1987, astronomers observed the explosion of a massive star in the Large Magellanic Cloud (LMC; a companion galaxy to the Milky Way 160,000 light-years away, shown in **Figure 17.12**). Astronomers working in all parts of the electromagnetic spectrum pointed their telescopes at Supernova 1987A—the first naked-eye supernova since the invention of the telescope. Astronomers were ultimately surprised to discover from looking at old photographs that the star that blew up was not a red supergiant, but a 20-M_{Sun} B3 I blue supergiant now classified as a luminous blue variable star. Neutrino telescopes recorded a burst of neutrinos passing through Earth from this tremendous stellar explosion that had occurred in the LMC. The detection of neutrinos from SN 1987A provided astronomers with a rare and crucial glimpse of the very heart of a massive star at the moment of its death, confirming a fundamental prediction of theories about the collapse of the core and its effects.

CHECK YOUR UNDERSTANDING 17.2

What causes a high-mass star to explode as a Type II supernova? (a) The high-mass star merges with another star. (b) Iron absorbs energy when it fuses. (c) The high-mass star runs out of mass in the core. (d) The CNO cycle uses up all the carbon.

17.3 The Spectacle and Legacy of Supernovae

Supernova explosions leave a rich and varied legacy in the universe. Explosions that occurred thousands of years ago have created huge expanding bubbles of million-kelvin gas that glow in X-rays and ultraviolet radiation and drive visible shock waves into the surrounding interstellar medium. The energy and matter flowing out from supernova explosions also compresses nearby clouds, triggering the initial collapse that begins star formation. In this section, we discuss what happens after a massive star explodes as a supernova.

The Energetic and Chemical Legacy of Supernovae

The energy carried away by light from a supernova represents only about 1 percent of the kinetic energy being carried away by the outer parts of the star. This ejected material contains approximately 10^{47} J of kinetic energy—enough energy to accelerate the entire Sun to a speed of 10,000 km/s. The kinetic energy of the material ejected from both Type Ia and Type II supernovae heats the hottest phases of the interstellar medium and pushes around the clouds in the interstellar medium. Yet even this amount of energy is small by comparison with the energy carried away from the supernova explosion by neutrinos—an amount of energy at least 100 times larger.

Perhaps even more important is the chemical legacy left behind by supernova explosions. Only the least massive chemical elements were present at the beginning of the universe: hydrogen, helium, and trace amounts of lithium and beryllium. All of the rest of the chemical elements formed in stars, either during nuclear burning in the core or in the rapid nuclear reactions that occur during a supernova explosion. These elements were released to the interstellar medium when stars died. The process of forming more massive atomic nuclei from less massive nuclei is called **nucleosynthesis**. Nucleosynthesis is responsible for the progressive chemical enrichment of the universe.

Elements up to carbon, oxygen, and small amounts of neon and magnesium form from nuclear fusion in the cores of low-mass stars and travel to the interstellar medium in asymptotic giant branch (AGB) winds and planetary nebulae. A look at the periodic table of the elements (see Appendix 3) shows that many naturally occurring elements are more massive than iron. Recall that fusion up to iron

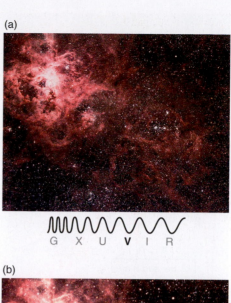
(a)
G X U V I R

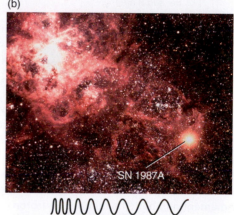
(b)
SN 1987A
G X U V I R

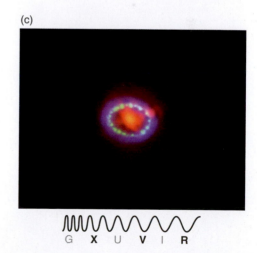
(c)
G X U V I R

Figure 17.12 Supernova 1987A was a supernova that exploded in a small companion galaxy of the Milky Way called the Large Magellanic Cloud (LMC). These images show the LMC before the explosion (a) and while the supernova was near its peak (b). Notice the "new" bright star at lower right. A 2013 image (c) from the Atacama Large Millimeter/submillimeter Array (ALMA) telescope shows freshly formed dust inside the glowing rings of gas.

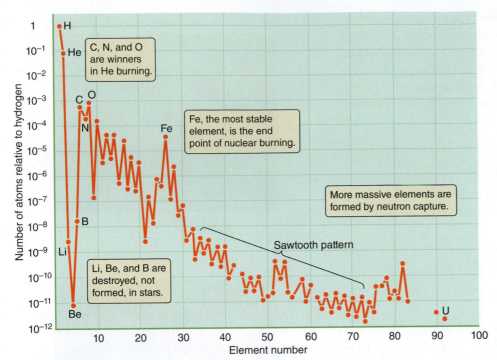

Figure 17.13 This graph plots the observed relative abundances of different elements in the Solar System against the element number of each element's nucleus. This pattern can be understood as a result of nucleosynthesis in stars. The periodic table of the elements in Appendix 3 identifies individual elements by their element number (the number of protons).

creates energy, but fusion beyond iron absorbs energy (see Figure 17.8). So elements heavier than iron fuse only under conditions in which there is abundant energy to be absorbed—such as in the enormous energies of supernova explosions. The naturally occurring elements heavier than iron were all produced in the deaths of high-mass stars.

Normally, electric repulsion keeps positively charged atomic nuclei far apart. Extreme temperatures are needed to slam nuclei together hard enough to overcome this electric repulsion. Free neutrons, however, have no net electric charge, so there is no electric repulsion to prevent them from simply colliding with an atomic nucleus, regardless of how many protons that nucleus contains. Under normal conditions in nature, free neutrons are rare. In the interiors of evolved stars, however, a number of nuclear reactions produce free neutrons, and under some circumstances—including those shortly before and during a supernova—free neutrons are produced in very large numbers. Free neutrons are easily captured by atomic nuclei and later decay to become protons. This process of neutron capture and decay forms the elements with atomic numbers and masses higher than those of iron.

Nuclear physics predicts the abundances of the elements. These predictions agree with abundances that have been measured on Earth, in the Solar System (**Figure 17.13**), and in the atmospheres of stars and their remains. Less massive elements are far more abundant than more massive elements because more massive elements are progressively built up from less massive elements. An exception to this pattern is the dip in the abundances of the light elements lithium (Li), beryllium (Be), and boron (B). Nuclear burning easily destroys these elements, and they are not produced by the common reactions involved in burning hydrogen (H) and helium (He). Conversely, carbon (C), nitrogen (N), and oxygen (O) are produced in quantity in the triple-alpha process, so they are very abundant. The spike in the abundances of elements near iron is evidence of processes that favor these tightly bound nuclei. Even the sawtooth pattern observed in the plot of the abundances of even-numbered and odd-numbered elements is a consequence of the way atomic nuclei form in stars. By comparing the predictions of nuclear physics with observations of elemental abundances, astronomers repeatedly test the theory of stellar evolution.

Neutron Stars and Pulsars

In the explosion of a Type II supernova, the outer parts of the star are blasted back into interstellar space—but what remains of the core that was left behind? The matter at the center of the massive star has collapsed to the point where it has about the same density as the nucleus of an atom. For cores less than about 3 M_{Sun}, the collapse is halted when neutrons are packed as tightly together as the rules of quantum mechanics permit. The *neutron-degenerate* core left behind by the explosion of a Type II supernova is a **neutron star**. It has a radius of 10–20 km, making it roughly the size of a small city, but into that volume is packed a mass between 1.4 and 3 M_{Sun}. (Cores greater than 3 M_{Sun} cannot be supported by

17.2 | Working It Out Gravity on a Neutron Star

Neutron stars are incredibly dense objects. As a result, the surface gravity and escape velocity of neutron stars are very high. For example, let's look at a typical neutron star with a radius of 15 km and a mass of $2\,M_{\text{Sun}}$.

Recall from Working It Out 4.1 that the acceleration due to gravity on the surface—in this case the surface of a neutron star (NS)—is given by

$$g = \frac{GM_{\text{NS}}}{R_{\text{NS}}^2}$$

$$g = 6.67 \times 10^{-20}\,\frac{\text{km}^3}{\text{kg\,s}^2} \times \frac{2.0 \times (1.99 \times 10^{30}\,\text{kg})}{(15\,\text{km})^2}$$

$$g = 1.2 \times 10^9\,\frac{\text{km}}{\text{s}^2}$$

Dividing this number by the gravitational acceleration on Earth, $9.8\,\text{m/s}^2 = 0.0098\,\text{km/s}^2$, shows that the gravitational acceleration on a neutron star is more than 100 billion times as large as that on Earth.

What about the escape velocity from a neutron star? From Working It Out 16.2, we know that the escape velocity is given by

$$v_{\text{esc}} = \sqrt{2GM/R}$$

Putting in the above numbers for a typical neutron star yields

$$v_{\text{esc}} = \sqrt{\frac{2 \times \left[6.67 \times 10^{-20}\,\text{km}^3/(\text{kg\,s}^2)\right] \times 2.0 \times (1.99 \times 10^{30}\,\text{kg})}{(15\,\text{km})}}$$

$$v_{\text{esc}} = 190{,}000\,\text{km/s}$$

Dividing this result by the speed of light gives

$$\frac{v_{\text{esc}}}{c} = \frac{190{,}000\,\text{km/s}}{300{,}000\,\text{km/s}} = 0.63$$

The escape velocity from this neutron star is more than 60 percent of the speed of light and almost 17,000 times greater than the escape velocity from Earth (11.2 km/s). The physicist Albert Einstein showed that strange things happen at velocities near the speed of light (including modifications to Newton's equations), which we will discuss in detail in Chapter 18.

neutron degeneracy and become black holes. These will be discussed in the next chapter.) At a density of about $10^{18}\,\text{kg/m}^3$, the neutron star is a billion times denser than a white dwarf and a thousand trillion (10^{15}) times denser than water. Imagine the entire Earth were crushed down to the size of a football stadium. Earth would then have about the same density as a neutron star. Neutron stars are so compact that the acceleration due to gravity on the surface is more than 100 billion times the acceleration on Earth, as shown in **Working It Out 17.2**. This extremely high surface gravity implies a very large escape velocity. A spacecraft would need to be traveling at $0.63c$ to escape the gravity of a typical neutron star.

The massive star from which the neutron star formed may have been part of a binary system. Unless the massive star loses so much mass that gravity no longer holds the two stars together, then the neutron star is left with a binary companion. Processes like those in the white dwarf binary systems responsible for novae and Type Ia supernovae are possible. **Figure 17.14** illustrates an **X-ray binary**, a binary system in which mass from an evolving star spills over onto a collapsed companion such as a white dwarf, neutron star, or black hole. As the lower-mass star in such a binary system evolves and overfills its Roche lobe, matter falls toward the accretion disk around the neutron star, heating it to millions of kelvins and causing it to glow brightly in X-rays. X-ray binaries sometimes develop powerful jets of material that are perpendicular to the accretion disk and carry material away at speeds close to the speed of light.

Figure 17.14 X-ray binaries are systems consisting of a normal evolving star with a white dwarf, a neutron star, or a black hole. As the evolving star overflows its Roche lobe, mass falls toward the collapsed object. The gravitational well of the collapsed object is so deep that when the material hits the accretion disk, it is heated to such high temperatures that it radiates away most of its energy as X-rays.

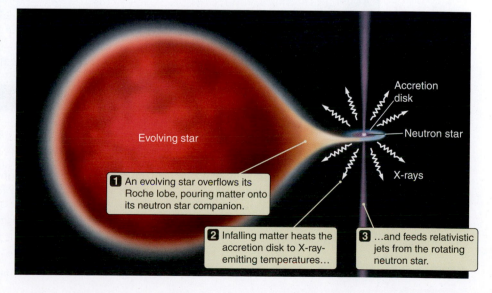

Accretion disk

Evolving star

Neutron star

X-rays

1 An evolving star overflows its Roche lobe, pouring matter onto its neutron star companion.

2 Infalling matter heats the accretion disk to X-ray-emitting temperatures…

3 …and feeds relativistic jets from the rotating neutron star.

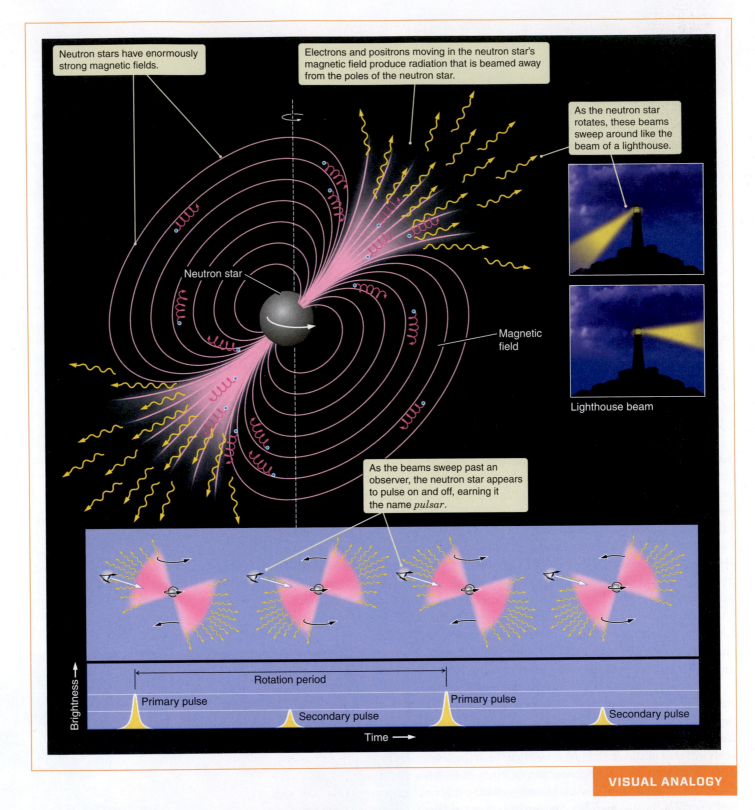

Neutron stars have enormously strong magnetic fields.

Electrons and positrons moving in the neutron star's magnetic field produce radiation that is beamed away from the poles of the neutron star.

As the neutron star rotates, these beams sweep around like the beam of a lighthouse.

Neutron star

Magnetic field

Lighthouse beam

As the beams sweep past an observer, the neutron star appears to pulse on and off, earning it the name *pulsar*.

Brightness

Rotation period

Primary pulse

Secondary pulse

Primary pulse

Secondary pulse

Time

VISUAL ANALOGY

Figure 17.15 When a highly magnetized neutron star rotates rapidly, light is given off, much like the beams from a rotating lighthouse lamp. As these beams sweep past Earth, the star will appear to pulse on and off, earning it the name *pulsar*.

Recall from Chapter 15 that the conservation of angular momentum requires a collapsing molecular cloud to spin faster as it shrinks into a protostar. Similarly, as the core of a massive star collapses, it spins faster because angular momentum is conserved. A massive main-sequence O star rotates perhaps once every few days. As a neutron star, it might rotate tens or even hundreds of times each second. The collapsing star also concentrates the magnetic field to strengths trillions of times greater than the magnetic field at Earth's surface. A neutron star has a magnetosphere just like Earth and several other planets do, except that the neutron star's magnetosphere is much stronger and is whipped around many times a second by the spinning star. As in planets, the magnetic axis in stars is often not aligned with the rotation axis.

Electrons and positrons circle around the magnetic-field lines of the neutron star and are "funneled" along the field toward the magnetic poles of the system. Any accelerating charged particle produces radiation, so these particles produce beams of radiation along the magnetic poles of the neutron star as shown in **Figure 17.15**. As the neutron star rotates, these beams sweep through space like the rotating beams of a lighthouse. When Earth is located in the paths of these beams, the neutron star appears to flash on and off with a regular period equal to the period of rotation of the star (or half the rotation period, if both beams are seen).

Many of the unusual objects discussed in this and the previous chapter—such as pulsating stars, supernovae, and planetary nebulae—puzzled astronomers when they were first observed, but were later understood to be associated with the end points of stellar evolution. In contrast, neutron stars were predicted in 1934, not long after neutrons themselves were discovered. Astronomers Walter Baade and Fritz Zwicky proposed that supernova explosions could lead to the formation of a neutron star. But neutron stars were not actually observed for another 30 years. In 1967, rapidly pulsing objects were first discovered by people observing in radio wavelengths (**Process of Science Figure**). These objects, which blinked like very fast, regularly ticking clocks, puzzled astronomers. Today, these rotating neutron stars are called **pulsars**. More than 2,000 pulsars are known, and more are being discovered all the time.

Astronomy in Action: Pulsar Rotation

The Crab Nebula

In 1054, Chinese astronomers noticed a "guest star" in the direction of the constellation Taurus. The new star was so bright that it could be seen during the daytime for 3 weeks, and it did not fade away altogether for many months. From the Chinese description of the changing brightness and color of the object, modern-day astronomers concluded that the guest star of 1054 was a fairly typical Type II supernova. Today, an expanding cloud of debris from the explosion occupies this place in the sky—forming an object called the Crab Nebula (see the chapter-opening photograph).

The Crab Nebula has filaments of glowing gas expanding away from the central star at 1,500 km/s—50 times faster than the expansion rate of a planetary nebula. These filaments contain anomalously high abundances of helium and other more massive chemical elements—the products of the nucleosynthesis that took place in the supernova and its progenitor star.

The Crab pulsar at the center of the nebula flashes 60 times a second: first with a main pulse associated with one of the "lighthouse" beams, then with a

Process of Science

OCCAM'S RAZOR

Occam's razor is a guiding principle in science: when scientists consider two hypotheses that explain a phenomenon equally well, they should give preference to the simpler theory and prioritize testing it. Simpler does not mean that the math is easier, or even that the concept is easy to understand. It means that the fewest number of other, new assumptions need to be made.

Jocelyn Bell, a student at Cambridge, has a "mystery signal" in her data.

Her adviser, Anthony Hewish, half-jokingly suggests "little green men" as the cause of the signal.

Bell and Hewish find four more such signals. It is unlikely that the same "little green men" would be sending the same signal from four separate locations in the sky.

They suggest the signals are from pulsating white dwarfs or neutron stars.

Franco Pacini and Thomas Gold each develop a detailed explanation involving rotating neutron stars.

This explanation relies entirely on previously understood physical phenomena: rotation, magnetic fields, and neutron stars.

It does not require assumptions about the existence of extra-terrestrials.

The neutron star explanation is "simpler."

Because simpler theories are easier to rule out, if they survive testing then scientists are more likely to consider them more seriously.

fainter secondary pulse associated with the other beam. As the Crab pulsar spins 30 times a second, it whips its powerful magnetosphere around with it. A few thousand kilometers from the pulsar, material in its magnetosphere must move at almost the speed of light to keep up with this rotation. Like a tremendous slingshot, the rotating pulsar magnetosphere flings particles away from the neutron star in a powerful wind moving at nearly the speed of light. This wind fills the space between the pulsar and the expanding shell. The Crab Nebula is almost like a big balloon, but instead of being filled with hot air, it is filled with a mix of very fast particles and strong magnetic fields. The energy that accelerates these particles is exactly equal to the energy lost as the pulsar's rotation slows down. Images of the Crab Nebula show this bubble as a glow from synchrotron radiation—a type of beamed radiation that is emitted as very fast moving particles spiral around the magnetic field.

CHECK YOUR UNDERSTANDING 17.3

One reason astronomers think neutron stars were formed in supernova explosions is that: (a) all supernova remnants contain pulsars; (b) pulsars are made of heavy elements, such as those produced in supernova explosions; (c) pulsars spin very rapidly, as did the massive star just before it exploded; (d) pulsars sometimes have material around them that looks like the ejecta from supernovae.

17.4 Star Clusters Are Snapshots of Stellar Evolution

Recall from Chapter 15 that when an interstellar cloud collapses, it breaks into pieces, forming not one star but many stars of different masses. These large groups of gravitationally bound stars are called star clusters. **Globular clusters** are densely packed collections of hundreds of thousands to millions of stars (**Figure 17.16a**). **Open clusters** are much less tightly bound collections of a few dozen to a few thousand stars (Figure 17.16b). Because stars in a cluster are formed together at nearly the same time, observations of star clusters at different ages provide evidence for the evolution of stars of different masses.

Cluster Ages

In the 1920s, astronomers plotted the observed brightness versus the spectral type for as many stars as possible in each cluster. The resulting cluster H-R diagrams showed stars of all the categories in the "textbook" H-R diagram (see Figure 13.15). Because all of the stars in a cluster are at approximately the same distance from Earth, the effect of distance on the brightness of each star is the same. By matching the main sequence on the observed cluster H-R diagram to the main sequence on the "textbook" H-R diagram, astronomers could estimate the distance to a cluster.

Astronomers also realized that the cluster H-R diagrams offered clues to the newly developing theories of stellar evolution. All of the stars in a cluster formed together at nearly the same time, so a look at a cluster that is 10 million years old shows what the stars of different masses evolve into during the first 10 million years after they form. A look at a cluster 10 billion years after it formed shows

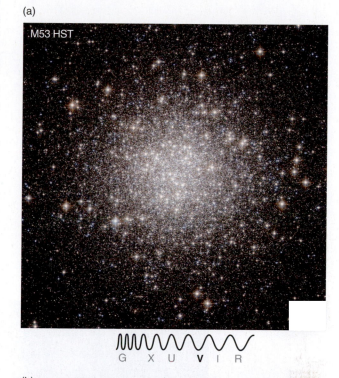

(a)

M53 HST

G   X   U   V   I   R

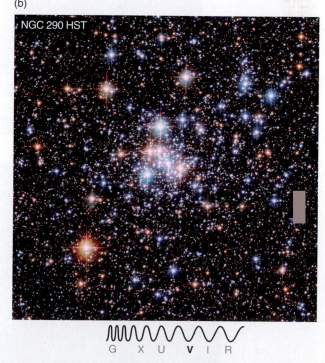

(b)

NGC 290 HST

G   X   U   V   I   R

Figure 17.16 (a) Globular clusters can have hundreds of thousands of stars. (b) Open clusters have up to hundreds or thousands of stars.

what becomes of stars of different masses after 10 billion years. **Figure 17.17a** shows an H-R diagram of a very young cluster, NGC 6530. O, B, and A stars are on the main sequence; F through M stars with lower masses are still evolving *to* the main sequence. There are no red giants or white dwarfs. In contrast, Figure 17.17b shows the H-R diagram of a very old cluster, M55. There are no high-mass stars on the main sequence because they have evolved off of it, but there are stars on the horizontal, red giant, and asymptotic giant branches and in the lower part of the main sequence. This globular cluster is about 12 billion years old.

Astronomers cannot watch an individual cluster age over millions of years, but they can observe different clusters of different ages. Astronomers explore cluster evolution by examining H-R diagrams of a *simulated* cluster of 40,000 stars as it would appear at several different ages and then comparing it to observed H-R diagrams of actual clusters. In **Figure 17.18a**, most stars have not yet reached the main sequence. Star formation in a molecular cloud is spread out over several million years, and it takes considerable time for lower-mass stars to contract to reach the main sequence. The H-R diagram of a very young cluster typically shows many lower-mass stars located well above the main sequence; eventually, they move onto the main sequence.

The more massive a star is, the shorter its life on the main sequence will be. Figure 17.18b shows that after only 4 million years, all stars with masses greater

Figure 17.17 This figure shows H-R diagrams of (a) a very young cluster (2 million years old), NGC 6530, and (b) a very old cluster (12 billion years old), M55. In (a), some of the stars haven't yet arrived on the main sequence. In (b), more than the top half of the main sequence has already evolved. Note that the vertical scales are logarithmic, (b) is zoomed in compared with (a).

(a)

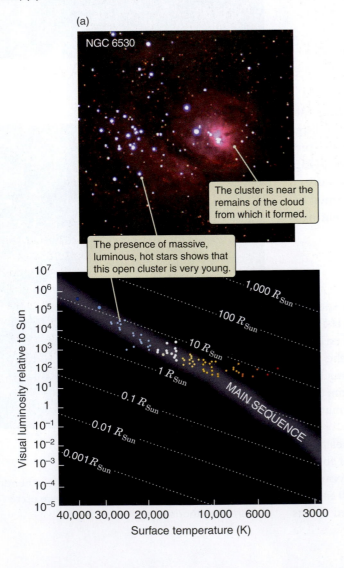

(b)

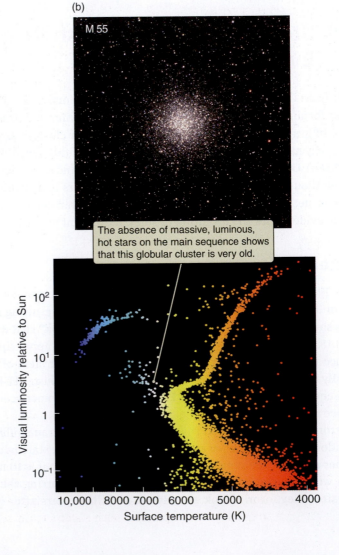

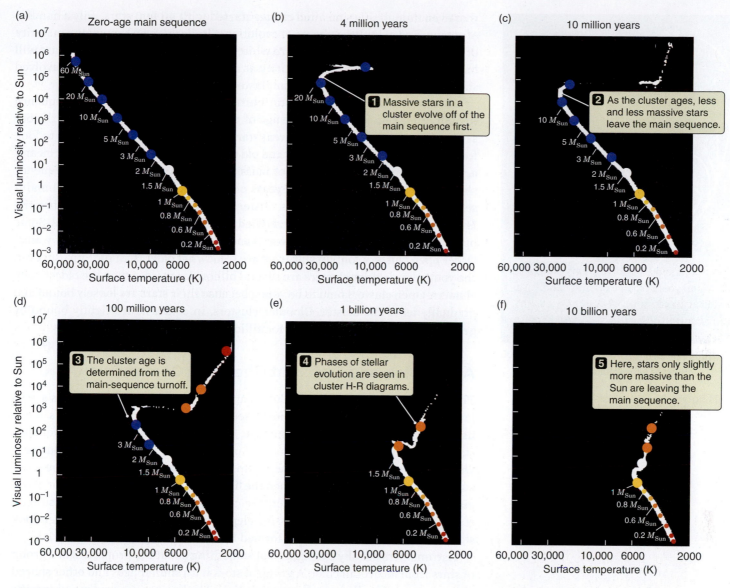

Figure 17.18 H-R diagrams of star clusters are snapshots of stellar evolution. These H-R diagrams of a simulated cluster of 40,000 stars of solar composition are shown at different times after the birth of the cluster. Note the progression of the main-sequence turnoff to lower and lower masses. For the purposes of the simulation, the stars are all placed on the main sequence at zero age. In reality, however, the lowest-mass stars have not yet reached the main sequence by the time the most massive stars have left it.

Nebraska Simulation: H-R Diagram Star Cluster Fitting Explorer

than about 20 M_{Sun} have evolved off the main sequence and are now spread out across the top of the H-R diagram. The most massive stars have already disappeared from the H-R diagram entirely, having vanished in supernovae. As time goes on, stars of lower and lower mass evolve off the main sequence, and the turnoff point moves toward the bottom right in the H-R diagram. By the time the cluster is 10 million years old, illustrated in Figure 17.18c, only stars with masses less than about 15 M_{Sun} remain on the main sequence. The location of the most massive star that remains on the main sequence is called the **main-sequence turnoff**. As the cluster ages, the main-sequence turnoff moves farther and farther down the main sequence to stars of lower and lower mass.

As a cluster ages further, shown in Figures 17.18d and e, we see the details of all stages of stellar evolution. By the time the star cluster is 10 *billion* years old (Figure 17.18f), stars with masses of only 1 M_{Sun} are beginning to die. Stars slightly more massive than this are seen as giant stars of various types. Note how few supergiant and giant stars are present in any of the cluster H-R diagrams. The supergiant, giant, horizontal, and asymptotic giant branch phases in the evolution of stars pass so quickly in comparison with a star's main-sequence lifetime

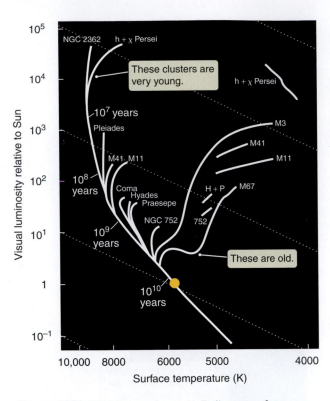

Figure 17.19 This figure shows H-R diagrams for clusters having a range of different ages. The ages associated with the different main-sequence turnoffs are indicated.

that even though this simulated cluster started with 40,000 stars, only a handful of stars are seen in these phases of evolution. Similarly, even though the majority of evolved stars in an old cluster are white dwarfs, all but a few of these stars will have cooled and faded into obscurity at any given time. More of these evolved stars are seen in the larger globular cluster M55 in Figure 17.17b.

To astronomers observing a star cluster, the location of the main-sequence turnoff immediately indicates the age of the cluster. **Figure 17.19** traces the observed H-R diagrams for several real star clusters. Once you know what to look for, the difference between young and old clusters is obvious. NGC 2362 is clearly a young cluster. Its complement of massive, young stars on the main sequence shows it to be only a few million years old. In contrast, cluster M3 has a main-sequence turnoff that indicates its cluster age is about 11 billion years. When the H-R diagrams of open clusters are studied, a wide range of ages is observed. Some open clusters contain the short-lived O and B stars and are therefore very young. Other open clusters contain stars that are somewhat older than the Sun. But even the youngest globular clusters are several billion years older than the oldest open clusters. Open clusters tend to be young because their stars are loosely bound and gradually leave the cluster. Globular clusters, in contrast, are tightly bound by gravity, allowing them to survive for billions of years.

Age, Color, and Different Chemical Composition

This understanding of stellar evolution applies even when groups of stars are so far away that individual stars cannot be seen. The light from a star cluster is dominated by its most luminous stars: massive, short-lived blue O and B main-sequence stars and evolved post-main-sequence supergiant and giant stars. If the cluster is young, then most of the light comes from luminous hot, blue stars and some red supergiants. If the cluster is old, then the light from the cluster has the color of red giants and red dwarf stars. This can be complicated by chemical composition—stars with lower amounts of massive elements in their atmospheres often look significantly bluer than stars that formed from chemically enriched material.

Astronomers usually can figure out something about the properties of a group of stars from its overall color. A group of stars with similar ages and other shared characteristics is called a **stellar population**. The link between color and specific characteristics will be useful when we begin to discuss the much larger collections of stars called galaxies. An especially bluish color to a galaxy or a part of a galaxy often signifies that the galaxy contains a young stellar population that still includes hot, luminous, blue stars that formed recently. In contrast, a galaxy or part of a galaxy that has a reddish color is usually composed primarily of an old stellar population.

All elements more massive than boron are formed in stars and are expelled into the interstellar medium when a star dies. New stars form from this material, thus the abundance of massive elements in the atmosphere of a star provides a snapshot of the chemical composition of the interstellar medium *at the time the star formed*. In main-sequence stars, material from the core does not mix with material in the atmosphere, so the abundances of chemical elements inferred from the spectrum of a star are the same as the abundances in the interstellar gas from which the star formed. Thus, the abundance of massive elements in the interstellar medium provides a record of the cumulative amount of star formation that has taken place up to the present time. Gas that shows large abundances of massive elements has gone through a great deal of stellar processing—so it contains more "recycled" material. Gas with low abundances of massive elements is more pristine.

The chemical composition of a star's atmosphere reflects the cumulative amount of star formation that has occurred up to that moment. Stars in globular clusters contain only very small amounts of massive elements; some globular-cluster stars contain only 0.5 percent as much of these massive elements as the Sun does, indicating that they were among the earliest stars to form. Open clusters are younger and contain stars that formed from a more enriched interstellar medium, and therefore they have higher amounts of the more massive elements.

Even the very oldest globular-cluster stars contain *some* amount of massive chemical elements. There must have been at least one generation of massive stars that lived and died, ejecting newly synthesized massive elements into space, before even the oldest globular clusters formed. Further, every red dwarf star less massive than about 0.5 M_{Sun} that ever formed is still around today. Note, though, that even a chemically rich star like the Sun, which is made of gas processed through approximately 9 billion years of previous generations of stars, is still composed of less than 2 percent massive elements. Luminous matter in the universe is still dominated by hydrogen and helium formed before the first stars. In upcoming chapters, you will learn that these variations in the chemical content of stars indicate a lot about the chemical evolution of galaxies.

CHECK YOUR UNDERSTANDING 17.4

If the main-sequence turnoff of a globular cluster occurs near the very top of the main sequence, then the cluster is: (a) very old; (b) very young; (c) very hot; (d) very dense.

Origins

Seeding the Universe with New Chemical Elements

On Earth, massive elements are everywhere. The surfaces of rocky planets contain silicon, oxygen, magnesium, and sodium. The iron-and-nickel solid inner core and liquid outer core of Earth are responsible for Earth's magnetic field. The most common chemical elements in biological molecules are carbon (C), hydrogen (H), nitrogen (N), oxygen (O), phosphorus (P), and sulfur (S)—all but hydrogen created in a dying star. The fact that these elements are here means that the Sun is not a first-generation star: it formed from material in the interstellar medium enriched by material from dying massive stars.

In supergiant, giant, and AGB stars, more massive chemical elements that formed from nuclear fusion deep within their interiors are carried upward and mixed with material in the outer parts of the star. As a star ages, its core grows hotter and hotter, and the temperature gradient within the star grows steeper. Under certain circumstances, convection can spread so deep into a star that chemical elements formed by nuclear burning within the star are dredged up and carried to the star's surface. For example, some AGB stars show an overabundance of carbon and other by-products of nuclear burning in their spectra. This extra carbon originated in each star's helium-burning shell and was carried to the surface by convection. For stars with lower masses, stellar winds and planetary nebulae carry the enriched outer layers off into interstellar space. The nuclear burning that occurs in supergiant stars goes well beyond the formation of elements such as carbon. Supernova explosions seed the universe with much more massive atoms, from iron and nickel up to uranium.

The oxygen atoms in the air you breathe and the water you drink were created by nucleosynthesis in dying stars. The iron atoms that are a key element of hemoglobin, which makes up the red blood cells that carry oxygen from your lungs to the rest of your body, formed in the explosions of massive stars. The nickel, copper, and zinc atoms in the coins in your pocket and the rare-earth atoms in your electronics were created in exploding massive stars. The Sun, the planets (including Earth), and all life on Earth are made of recycled stars. Supernovae are in you.

Astronomers report observations of a nearby series of supernovae. (X)

We Are Swimming in a Superhot Supernova Soup

By **IAN O'NEILL**, News.discovery.com

Approximately 10 million years ago, a nearby cluster of stars erupted as a violent series of supernovae and, according to new observations, the million-degree plasma from these powerful detonations surround the Solar System today.

Although astronomers have known about this tenuous "Local Bubble" of gas for some time, a suborbital NASA sounding rocket launched by scientists in 2012 has removed any doubt about the origins of this 300 light-year wide feature we are cocooned within.

At a temperature in excess of a million degrees kelvin, the supernova "soup" is actually very tenuous—around 0.001 atoms per cubic centimeter—and very different from other particles of matter that occupy the interstellar medium.

The idea is that several neighboring massive stars went supernova after reaching the ends of their lives. Although supernovae are very powerful, apparently they weren't close enough to Earth to sterilize our planet of life. The supernovae occurred during the early stages of human evolution.

Evidence for the Local Bubble has been brewing for several decades and its presence was inferred from X-ray observations of the local galaxy—a background glow of X-ray radiation was detected in all directions. Although the evidence seemed strong for an ancient supernova soup, there was other possible interpretation.

"Within the last decade, some scientists have been challenging the (supernova) interpretation, suggesting that much or all of the soft X-ray diffuse background is instead a result of charge exchange," said F. Scott Porter of the Goddard Space Flight Center, Greenbelt, Maryland.

Charge exchange can occur between solar wind ions (charged particles that lack electrons) and neutral gases. When the two gases come into contact within the Solar System—between the solar wind and a comet's coma, for example—electrons can be stripped from the neutral particles, generating X-ray emissions.

Therefore, many astronomers argued that this diffuse X-ray glow observed in all directions may be a phenomenon *inside* the Solar System and not superhot particles from 10 million year–old supernovae *outside* the Solar System.

To put this debate to bed, an international team of scientists constructed an instrument called the Diffuse X-ray emission from the Local Galaxy (or DXL), which could distinguish between the two scenarios. After launching the instrument 160 miles in altitude atop a sounding rocket, well above the Earth's atmosphere, the instrument specifically detected the amount of charge exchange that is occurring within interplanetary space.

After 5 minutes of observations on December 12, 2012, the DXL was able to determine that only 40 percent of the diffuse soft-X-ray emissions were generated by charge exchange. The rest must be coming from the Local Cloud from outside the Solar System—so therefore the Local Bubble is real and we did indeed get rumbled by a series of supernovae 10 million years ago.

"This is a significant discovery," said Massimiliano Galeazzi, of the University of Miami in Coral Gables, who led the team. "(It) affects our understanding of the area of the galaxy close to the Sun, and can, therefore, be used as a foundation for future models of the galaxy structure."

ARTICLES | **QUESTIONS**

1. Why would a group of stars in a cluster go supernovae at nearly the same time?
2. Why was a rocket needed to observe these X-rays instead of observing them from the ground on Earth?
3. What is the evidence that the X-rays came from outside of the Solar System?
4. How hot is the X-ray gas? Could you safely stick your hand in it—explain.
5. Go to NASA News (http://science.nasa.gov/science-news/science-at-nasa/2014/26aug_localbubble/) and watch the 4-minute ScienceCast video on this discovery. Which part of this story is explained better with video? Search at Nasa.gov—was the next DXL rocket launched in December 2015 as planned?

Summary

As high-mass stars evolve, their interiors form concentric shells of progressive nuclear fusion. Once they leave the main sequence, they may pass through the instability strip and become pulsating variable stars. High-mass stars eventually explode as Type II supernovae, which eject newly formed massive elements into interstellar space. The supernova explosion that ends the life of a massive star leaves behind a neutron star that contains between 1.4 and 3 M_{Sun} of neutron-degenerate matter packed into a sphere 10–20 km in diameter. Accretion of mass onto neutron stars produces X-rays in some binary systems. Pulsars are rapidly spinning, magnetized neutron stars. The Sun, the Solar System, Earth, and all life on Earth contain heavy elements created inside of earlier generations of short-lived massive stars.

LG 1 **Describe how the death of high-mass stars differs from that of low-mass stars.** The larger masses of high-mass stars allow them to fuse heavier elements than those produced in low-mass stars. This leads to a more violent death that leaves massive cores behind.

LG 2 **List the sequence of stages for evolving high-mass stars.** Evolving high-mass stars leave the main sequence as they burn heavier elements. Once an iron core is produced, the star becomes unstable and the core collapses, heating the material to cause photodisintegration of the iron nuclei and the merging of protons and electrons into neutrons. The outer layers bounce off of the dense core and produce a shock wave that travels outward. This shock wave causes neutrons to fuse with atomic nuclei and form more massive elements.

LG 3 **Explain the origin of chemical elements up to and heavier than iron.** The chain of nuclear fusion reactions consists of increasingly shorter stages of burning, resulting in more massive elements up to iron. However, these atoms are destroyed in the set of processes that define the core collapse. Heavy elements in the universe today were formed during the rebound explosion of massive stars as high-energy neutrons penetrated atomic nuclei. These neutrons then decay to protons, creating new elements with higher atomic numbers.

LG 4 **Identify how Hertzsprung-Russell (H-R) diagrams of clusters enable astronomers to measure the ages of stars and test theories of stellar evolution.** Clusters are groups of stars that were born together and are all at about the same distance from Earth. H-R diagrams of clusters show stars leaving the main sequence in a progression from the highest-mass stars to the lowest-mass stars, confirming theories of stellar evolution. The location of the main-sequence turnoff indicates the age of the cluster.

? UNANSWERED QUESTIONS

- Blue straggler stars, found in clusters, are bluer and brighter than the stars at the main-sequence turnoff point. How do they fit into the picture of stellar evolution? They may have resulted from mass transfer in a binary pair or from the merger of two single or two binary stars, either of which could have resulted in a more massive star than what might be expected from the age of the cluster. Astronomers study the environments of these stars by estimating the likelihood of collisions and the number of binary systems, which may be different in clusters where the density of stars is high.

- What creates magnetars? There is a class of pulsars called magnetars, which are characterized by extremely large magnetic fields. These objects are observed to produce bursts of lower energy gamma rays. The origin of their huge magnetic fields is not well understood. These fields may originate from a dynamo in the interior of a superconducting region of the neutron star, but we do not know whether ordinary pulsars go through a magnetar phase.

Questions and Problems

Test Your Understanding

1. Why does the interior of an evolved high-mass star have layers like an onion?
   a. Heavier atoms sink to the bottom because stars are not solid.
   b. Before the star formed, heavier atoms accumulated in the centers of clouds because of gravity.
   c. Heavier atoms fuse closer to the center because the temperature and pressure are higher there.
   d. Different energy transport mechanisms occur at different densities.

2. Arrange the following elements in the order they burn inside the nucleus of a high-mass star during the star's evolution.
   a. helium
   b. neon
   c. oxygen
   d. silicon
   e. hydrogen
   f. carbon

3. Elements heavier than iron originated
   a. in the Big Bang.
   b. in the cores of low-mass stars.
   c. in the cores of high-mass stars.
   d. in the supernova explosions of high-mass stars.

4. A pulsar pulses because
   a. its spin axis crosses Earth's line of sight.
   b. it spins.
   c. it has a strong magnetic field.
   d. its magnetic axis crosses Earth's line of sight.

5. Study Figure 17.5. If it were possible to watch a high-mass star move to the right, along one of these post-main-sequence lines, what would you observe happening to the star's color?
   a. It would become redder.
   b. It would become bluer.
   c. It would remain the same.

6. Study Figure 17.5. If it were possible to watch a high-mass star move to the right, along the topmost of these post-main-sequence lines, what would you observe happening to the star's size?
   a. It would become much larger.
   b. It would become much smaller.
   c. It would remain the same size.

7. Study Figure 17.18. If the Sun were a member of a globular cluster, that cluster's H-R diagram would fall between
   a. (a) and (b)
   b. (b) and (c)
   c. (c) and (d)
   d. (d) and (e)
   e. (e) and (f)

8. In a high-mass star, hydrogen fusion occurs via
   a. the proton-proton chain.
   b. the CNO cycle.
   c. gravitational collapse.
   d. spin-spin interaction.

9. The layers in a high-mass star occur roughly in order of
   a. atomic number.
   b. decay rate.
   c. magnetic field strength.
   d. spin state.

10. Eta Carinae is an extreme example of
    a. a massive star.
    b. a planetary nebula.
    c. a supernova remnant.
    d. an ancient star.

11. Iron fusion cannot support a star because
    a. iron oxidizes too quickly.
    b. iron absorbs energy when it fuses.
    c. iron emits energy when it fuses.
    d. iron is not dense enough to hold up the layers.

12. The start of photodisintegration of iron in a star sets off a process that *always* results in a
    a. supernova.
    b. neutron star.
    c. supergiant.
    d. pulsar.

13. The Crab Nebula is a test of our ideas about supernova explosions because
    a. the system contains an x-ray binary.
    b. the nebula is slowly expanding.
    c. the supernova was observed in 1054 and now astronomers see a pulsar in the nebula.
    d. the original star was like the Sun before exploding.

14. What mechanism provides the internal pressure inside a neutron star?
    a. ordinary pressure from hydrogen and helium gas
    b. degeneracy pressure from neutrons
    c. degeneracy pressure from electrons
    d. rapid rotation

15. Very young star clusters have main-sequence turnoffs
    a. that drop below the main sequence.
    b. at the top left of the main sequence.
    c. at the bottom right of the main sequence.
    d. in the middle of the main sequence.

Thinking about the Concepts

16. Explain the differences between the ways that hydrogen is converted to helium in a low-mass star (proton-proton chain) and in a high-mass star (CNO cycle). What is the catalyst in the CNO cycle, and how does it take part in the reaction?

17. How does a high-mass star begin fusing helium in its core? How is this process different from what happens in low-mass stars?

18. Why does the core of a high-mass star not become degenerate, as the core of a low-mass star does?

19. List the two reasons why each post-helium-fusion cycle for high-mass stars (carbon, neon, oxygen, silicon, and sulfur) becomes shorter than the preceding cycle.

20. Cepheids are highly luminous, variable stars in which the period of variability is directly related to luminosity. Why are Cepheids good indicators for determining stellar distances?

21. Identify and explain two important ways in which supernovae influence the formation and evolution of new stars.

22. Study the Process of Science Figure. Why is this pulsar explanation "simpler"?

23. Describe what an observer on Earth will witness when Eta Carinae explodes.

24. Recordings show that neutrinos from SN 1987A were detected on February 23, 1987. About 3 hours later it was detected in optical light. What was the reason for the time delay?

25. Why can the accretion disk around a neutron star release so much more energy than the accretion disk around a white dwarf, even though the two stars have approximately the same mass?

26. In Section 17.2, you learned that Type II supernovae blast material outward at 30,000 km/s. The material in the Crab Nebula described in Section 17.3 is expanding at only 1,500 km/s. What explains the difference?

27. An experienced astronomer can take one look at the H-R diagram of a star cluster and immediately estimate its age. How is this possible?

28. Explain how astronomers know that there was an even earlier generation of stars before the oldest observed stars.

29. What is the binding energy of an atomic nucleus? How does this quantity help astronomers calculate the energy given off in nuclear fusion reactions?

30. Explain how Earth is made up of material from supernova.

Applying the Concepts

31. Study Figure 17.2. What fraction of the star is helium at time $t = 0$ and at time $t = 7$ million years?

32. Study Figure 17.5. How much hotter, larger, and more luminous than the Sun is the uppermost main-sequence star on this H-R diagram?

33. Study Figure 17.4. Are the radius of the core and the radius of the star represented to scale in this figure? What fraction of the star's radius is the core's radius?

34. If the Crab Nebula has been expanding at an average velocity of 1500 km/s since the year 1054, what was its average radius in the year 2014? (Note: There are approximately 3×10^7 seconds in a year.)

35. Suppose you observe a Classical Cepheid variable with a period of 10 days. What is the luminosity of this star? What other piece of information would you need to find out how far away this star is?

36. Figure 17.13 shows the relative abundance of the elements. Is this a log or a linear plot? Explain what it means that oxygen lies on the y-axis at 10^{-3}.

37. Use Working It Out 17.2 to find the surface gravity on a neutron star with radius 10 km and mass 2.8 M_{Sun}.

38. Use Working It Out 17.2 to find the escape velocity from a neutron star with radius 10 km and mass 2.8 M_{Sun}.

39. The Milky Way has about 50,000 stars of average mass (0.5 M_{Sun}) for every main-sequence star of 20 M_{Sun}. But 20-M_{Sun} stars are about 10,000 times as luminous as the Sun, and 0.5-M_{Sun} stars are only 0.08 times as luminous as the Sun.
    a. How much more luminous is a single massive star than the total luminosity of the 50,000 less massive stars?
    b. How much mass is in the lower-mass stars compared to the single high-mass star?
    c. Which stars—lower-mass or higher-mass stars—contain more mass in the galaxy, and which produce more light?

40. In a large outburst in 1841, the 120-M_{Sun} star Eta Carinae was losing mass at the rate of 0.1 M_{Sun} per year.
    a. The mass of the Sun is 2×10^{30} kg. How much mass (in kilograms) was Eta Carinae losing each minute?
    b. The mass of the Moon is 7.35×10^{22} kg. How does Eta Carinae's mass loss per minute compare with the mass of the Moon?

41. An O star can lose 20 percent of its mass during its main-sequence lifetime. Estimate the average mass loss rate (in solar masses per year) of a 25-M_{Sun} O star with a main sequence lifetime of 7 million years.

42. The approximate relationship between the luminosity and the period of Cepheid variables is L_{star} (in L_{Sun}) = 335 P (in days). Delta Cephei has a cycle period of 5.4 days and a parallax of 0.0033 arcsecond (arcsec). A more distant Cepheid variable appears 1/1,000 as bright as Delta Cephei and has a period of 54 days.
    a. How far away (in parsecs) is the more distant Cepheid variable?
    b. Could the distance of the more distant Cepheid variable be measured by parallax? Explain.

43. For a pulsar that rotates 30 times per second, at what radius in the pulsar's equatorial plane would a co-rotating satellite (rotating about the pulsar 30 times per second) have to be positioned to be moving at the speed of light? Compare this to the pulsar radius of 1 km.

44. Verify the claim in Section 17.3 that Earth would be roughly the size of a football stadium if it were as dense as a neutron star.

45. Estimate the size of a neutron star with the mass of the Sun.

USING THE WEB

46. Go to the Chandra X-ray Observatory's "Variable Stars" Web page (http://chandra.harvard.edu/edu/formal/variable_stars /index.html). Do the two exercises on Cepheid variable stars, which ask you to estimate their changes in brightness. You might want to look at Appendix 7 to review apparent magnitudes before you do the projects.

47. The International Astronomical Union's "List of Recent Super-novae" (http://cbat.eps.harvard.edu/lists/RecentSupernovae .html) includes all recently discovered supernovae. Pick a few of the most recent ones. What type of supernova is each one? How bright is it? Why are these so much fainter than the no-vae you looked at in Chapter 16? Are Type Ia or Type II super-novae more common?

48. What method is used by the Intermediate Palomar Transient Factory (iPTF) survey (http://www.ptf.caltech.edu/iptf) to find supernova? What kinds of supernovae has this study found? How will the replacement camera, the Zwicky Tran-sient Factory (ZTF), improve the search for supernova? Has ZTF found any yet?

49. Go to the website for the Gaia mission (http://esa.int/science /gaia). How is this mission contributing to the study of vari-able stars? How is it contributing to the study of novae and supernovae?

50. Go to the "Einstein@Home" website (http://einsteinathome .org). In this distributed computing project, volunteers use their spare computer processing power to help search for new pulsars. Look over the "News" section on the right. Have any pulsars been found lately? Join the project, create an account, download BOINC, and follow directions to look for pulsars.

smartwork

If your instructor assigns homework in SmartWork, access your assignments at digital.wwnorton.com/astro5.

Nuclear reactions usually involve many steps. In the Exploration for Chapter 14, you investigated the proton-proton chain. In this Exploration, you will study the CNO cycle, which is even more complex. Visit the Student Site at the Digital Landing Page, and open the "CNO Cycle" interactive simulation in Chapter 17.

First, press "Play Animation" and watch the animation all the way through. Press "Reset Animation" to clear the screen, and then press "Play Animation" again, allowing the animation to proceed past the first collision before pressing "Pause."

1 Which atomic nuclei are involved in this first collision?

_____

2 What color is used to represent the proton (hydrogen nucleus)?

_____

3 What does the blue squiggle represent?

_____

4 What atomic nucleus is created in the collision?

_____

5 The resulting nucleus is not the same type of element as either of the two that entered the collision. Why not?

_____

Press "Play Animation" again, and then pause as soon as the yellow ball and the dashed line appear.

6 Is this a collision or a spontaneous decay?

_____

7 What does the yellow ball represent?

_____

8 What does the dashed line represent?

_____

9 The resulting nucleus has the same number of nucleons (13), but it is a different element. What happened to the proton that was in the nitrogen nucleus but is not in the carbon nucleus?

_____

Proceed past the next two collisions, to "^{15}O."

10 Study the pattern that is forming. When a blue ball comes in, what happens to the number of nucleons and the type of the nucleus (that is, what happens to the "12" and the "C," or the "14" and the "N")?

_____

11 What is emitted in these collisions?

_____

Proceed until "^{15}N" appears.

12 Is this a collision or a spontaneous decay?

_____

13 Which previous reaction is this most like?

_____

Now proceed to the end of the animation.

14 After the final collision, a line is drawn back to the beginning, telling you what type of nucleus the upper red ball represents. What is this nucleus?

_____

15 How many nucleons are not accounted for by the upper red ball? (Hint: Don't forget the ^{1}H that came into the collision.) These nucleons must be in the nucleus represented by the bottom red ball.

_____

16 Carbon has six protons. Nitrogen has seven. How many protons are in the nucleus represented by the bottom red ball?

_____

17 How many neutrons are in the nucleus represented by the bottom red ball?

_____

18 What element does the bottom red ball represent?

_____

19 What is the net reaction of the CNO cycle? That is, which nuclei are combined and turned into the resulting nucleus?

_____

20 Why is ^{12}C not considered to be part of the net reaction?

_____

18

Relativity and Black Holes

Some stars leave behind a black hole at the end of their lives. Black holes have such extreme conditions that the laws of Newtonian physics are inadequate to describe them. To discuss black holes, we must understand how Albert Einstein changed the way physicists thought about the nature of space and time in the early 20th century. Einstein's special theory of relativity shows that matter behaves differently when it is traveling near the speed of light. Einstein's general theory of relativity shows that space itself is warped near very massive objects. This warping of space is so extreme at a black hole, it is as if there is a hole in space. In this chapter, we will move beyond Newtonian ideas of space and time in order to understand black holes.

LEARNING GOALS

By the conclusion of this chapter, you should be able to:

LG 1 Describe how the motion of the observer affects the observed velocity of objects.

LG 2 Discuss the observable consequences of the relationship between space and time.

LG 3 Explain how gravity is a consequence of the way mass distorts the very shape of spacetime.

LG 4 Explain why the most massive stars end as black holes, and describe the key properties of these stellar black holes.

This artist's rendition of the black hole Cygnus X-1 shows the accretion disk and the massive star that feeds it. ▶ ▶ ▶

How are
space and
time changed
near a black
hole?

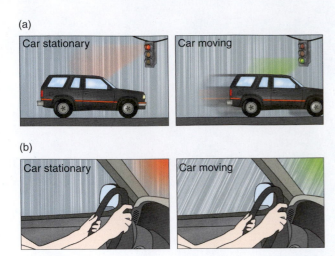

(a)

Car stationary | Car moving

(b)

Car stationary | Car moving

Figure 18.1 On a windless day, the direction in which rain falls depends on the frame of reference in which it is viewed. (a) From outside the car, the rain is seen to fall vertically downward whether the car is stationary or moving. (b) From inside the car, the rain is seen to fall vertically downward if the car is stationary; but if the car is moving, the rain is seen to fall at an angle determined by the speed and direction of the car's motion.

Light from a distant star is like the vertical rain in Figure 18.1.

To an observer on the moving Earth, the starlight is tipped in the direction of Earth's motion.

Apparent motion of star

Over the year, the direction of incoming starlight traces out a loop.

VISUAL ANALOGY

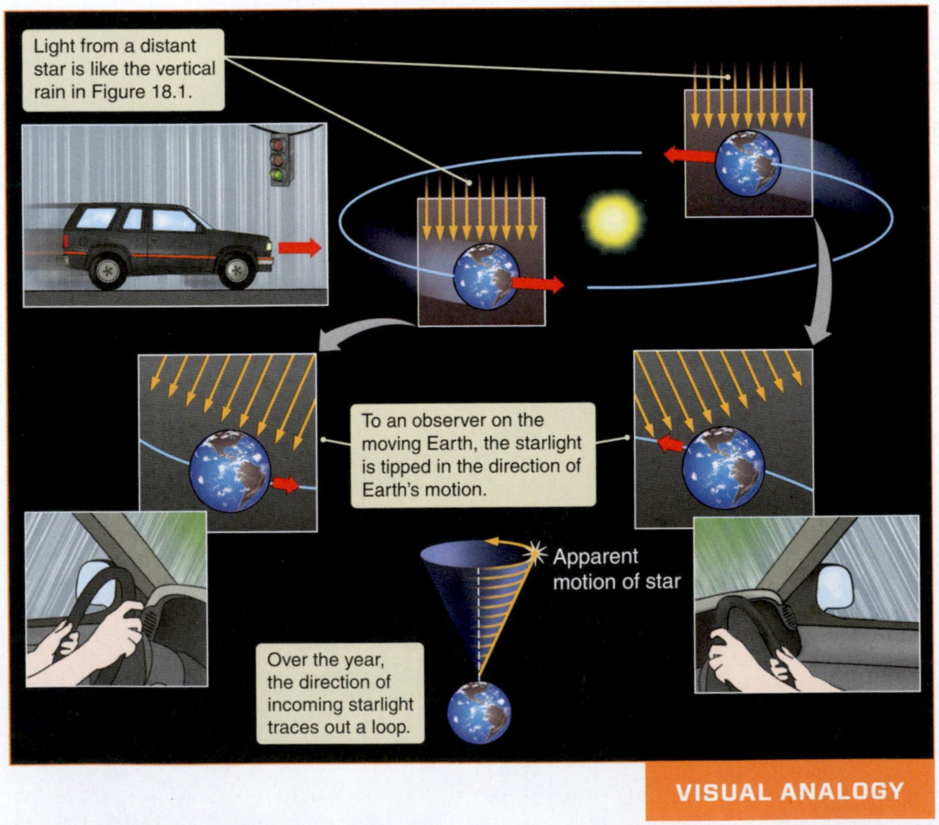

Figure 18.2 When we observe stars, their apparent positions are deflected slightly toward the direction in which Earth is moving. As Earth orbits the Sun, stars appear to trace out small loops in the sky. This effect is called *aberration of starlight*.

18.1 Relative Motion Affects Measured Velocities

All observers, whatever their motion, measure the same speed of light. This observed fact has profound implications for relative motion, space, and time. In this section, we will lay the groundwork for our discussion of relativity by considering how relative motion affects measurements and how these effects change at very high speeds.

Aberration of Starlight

The first direct measurement of the effect of Earth's motion around the Sun was made in the 18th century. Imagine that you are sitting in a car in a windless rainstorm, as shown in **Figure 18.1**. If the car is stationary and the rain is falling vertically, when you look out your side window you see raindrops falling straight down. When the car is moving forward, however, the situation is different. Between the time a raindrop appears at the top of your window and the time it disappears beneath the bottom of your window, the car has moved forward. The raindrop disappears beneath the window behind the point at which it appeared at the top of the window, which means the raindrop looks as if it is falling at an angle, even though in reality it is falling straight down. As you go faster, the apparent front-to-back slant of the raindrops increases, and their apparent paths become more slanted. An observer by the side of the road would say the raindrops are coming from directly overhead, but to you in the moving car they are coming from a direction in front of the car. You are observing this apparent motion of the raindrops from within your own unique frame of reference.

The light from a distant star arrives at Earth from the direction of the star, as shown in **Figure 18.2**. However, just as the raindrops appeared to be coming from in front of the moving car in Figure 18.1b, an observer on the moving Earth sees the starlight coming from a slightly different direction. Because the direction of Earth's motion around the Sun continually changes during the year, the apparent position of a star in the sky moves in a small loop, a phenomenon known as the **aberration of starlight**. This shift in apparent position was first detected in the 1720s by two astronomers—Samuel Molyneux and James Bradley. Measurement of the aberration of starlight shows that Earth moves on its path about the Sun with an average speed of just under 30 kilometers per second (km/s). Because distance equals speed multiplied by time, astronomers were able to use this measurement to determine the circumference, and therefore the radius, of Earth's orbit. The speed of Earth (29.8 km/s) multiplied by the number of seconds in 1 year (3.16×10^7 seconds) gives a circumference of 9.42×10^8 km. Astronomers used this circumference to estimate the radius of Earth's orbit (1.5×10^8 km). The aberration of starlight is an astronomical example of the effects of relative motion on a measurement.

Relative Speeds Close to the Speed of Light

Every observer inhabits a reference frame, a set of coordinates in which the observer measures distances and speeds. A nonaccelerating reference frame, in which no net force is acting, is known as an **inertial reference frame**. Newton's laws of motion predict how the motion of an object will be measured by observers in different inertial reference frames. For example, the observed motion of a ball thrown from a moving car depends on the reference frame of the observer (recall the analogy for the Coriolis effect from Figure 2.11). If light behaved like other objects, the speed of light should differ from one observer to the next as a result of the observer's motion, just like the speed of a ball thrown from a moving car. However, the results of laboratory experiments with light in the late 19th and early 20th centuries conflicted with these predictions. Scientists found instead that all observers measure exactly the same value for the speed of light, regardless of the frame of reference of each observer.

As shown in **Figure 18.3a**, imagine that you are in a red car traveling at 100 kilometers per hour (km/h) on a highway, and you throw a ball at a speed of 50 km/h out the window at an oncoming green car, also traveling at 100 km/h. An observer standing by the side of the road watches the entire event. In *your* reference frame (Figure 18.3a, top panel), the red car is stationary; that is, you and the car are moving together, so the car does not move relative to you. So in your reference frame, you measure the ball traveling at 50 km/h. However, to the observer standing by the road (Figure 18.3a, middle panel), the ball is moving at 150 km/h— 50 km/h from throwing it plus 100 km/h from the motion of the car. To passengers in the oncoming green car moving at 100 km/h (Figure 18.3a, bottom panel), the speed of the ball is 250 km/h, because the ball approaches them at 150 km/h, and they approach the ball at 100 km/h. This example shows that the speed of the ball depends on how the observer, the cars, and the ball are moving relative to one another. The velocities are added together to find the velocity of one object relative to another. This is Galilean relativity, which you use in everyday life.

Figure 18.3b demonstrates how light differs from the ball in the previous example. Imagine that you are riding in the yellow spaceship at half the speed of light (0.5*c*) and you shine a beam of laser light forward (Figure 18.3b, top panel). You measure the speed of the beam of light to

Figure 18.3 (a) The Newtonian rules of motion apply in daily life; however, these rules break down when speeds approach the speed of light. (b) The fact that light itself always travels at the same speed for any observer is the basis of special relativity, and it implies that velocities don't simply add together, as they do in the Newtonian world.

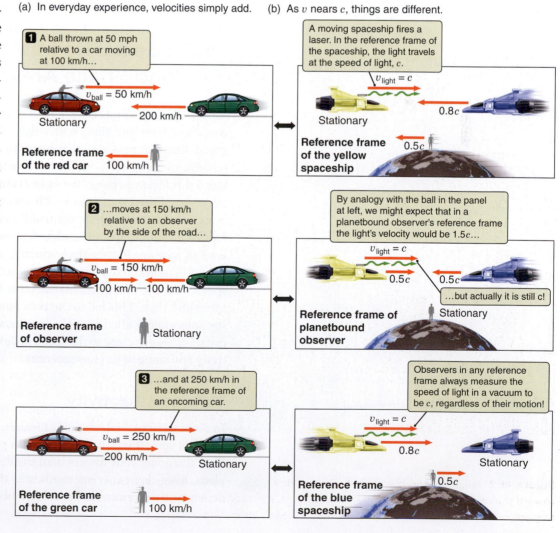

(a) In everyday experience, velocities simply add.

1 A ball thrown at 50 mph relative to a car moving at 100 km/h…

$v_{ball} = 50$ km/h
200 km/h
Stationary

Reference frame of the red car 100 km/h

2 …moves at 150 km/h relative to an observer by the side of the road…

$v_{ball} = 150$ km/h
100 km/h 100 km/h

Reference frame of observer Stationary

3 …and at 250 km/h in the reference frame of an oncoming car.

$v_{ball} = 250$ km/h
200 km/h
Stationary

Reference frame of the green car 100 km/h

(b) As *v* nears *c*, things are different.

A moving spaceship fires a laser. In the reference frame of the spaceship, the light travels at the speed of light, *c*.

$v_{light} = c$
0.8*c*
Stationary

Reference frame of the yellow spaceship 0.5*c*

By analogy with the ball in the panel at left, we might expect that in a planetbound observer's reference frame the light's velocity would be 1.5*c*…

$v_{light} = c$
0.5*c* 0.5*c*

…but actually it is still *c*!

Reference frame of planetbound observer Stationary

Observers in any reference frame always measure the speed of light in a vacuum to be *c*, regardless of their motion!

$v_{light} = c$
0.8*c*
Stationary

Reference frame of the blue spaceship 0.5*c*

be c, or 3×10^8 meters per second (m/s), as expected because you are holding the source of the light. But the observer on the planet also measures the speed of the passing beam of light to be 3×10^8 m/s, not the sum of the speed of light plus the speed of the spaceship, $1.5c$ (Figure 18.3b, middle panel). Even a passenger in an oncoming blue spacecraft traveling at $0.5c$, as shown in the bottom panel of Figure 18.3b, finds that the beam from your light is traveling at exactly c in her own reference frame, and not the $2.0c$ sum of the speeds of the spacecraft and the laser light. At speeds close to the speed of light, speeds do not simply add. This is true not just for light, but for all ordinary objects moving at nearly the speed of light. Using Einstein's relativistic formulas, the relative speed between the two spacecraft in the top and bottom panels of Figure 18.3b ($0.5c + 0.5c$) adds to $0.8c$, not $1.0c$! At **relativistic speeds** (speeds close to the speed of light), everyday experience no longer holds true. Every observer always finds that light in a vacuum travels at exactly the same speed, c, regardless of his or her own motion or the motion of the light source. This directly conflicts with Newtonian theory and Galilean relativity.

CHECK YOUR UNDERSTANDING 18.1

Which beam of light is moving faster: (a) that from the headlight of a parked car; (b) that from the headlight of a moving car; (c) that from the headlight of a moving spaceship; (d) all the beams are moving at c regardless of reference frame.

18.2 Special Relativity Explains How Time and Space Are Related

Albert Einstein's first scientific paper, written when he was a 16-year-old student, was about traveling along with a light wave, moving in a straight line at constant speed. Einstein reasoned that according to Newton's laws of motion, it should be possible to "keep up" with light so that you are moving right along with it. In this inertial reference frame, the light is stationary: an oscillating electric and magnetic wave that does not move. This was impossible according to Maxwell's equations for electromagnetic waves (discussed in Chapter 5). Einstein later took a radical new approach to resolve the conflict between experiments and Newton's laws of motion. Rather than starting with preconceived ideas about space and time, Einstein started with the observed fact that light always travels at the same speed, and then he reasoned backward to find out what that must imply about space and time. This led to the 1905 publication of his **special theory of relativity**, sometimes called *special relativity*, which describes the effects of traveling at constant speeds close to the speed of light. In this section, we explore special relativity and some of its consequences.

Time and Relativity

In developing special relativity, Einstein focused his thinking on pairs of *events*. An **event** is something that happens at a particular location in space at a particular time. Snapping your fingers is an event, because that action has both a time and a place. Everyday experience indicates that the distance between any two events depends on the reference frame of the observer. Imagine you are sitting in a car that is traveling on the highway in a straight line at a constant 60 km/h. You snap your fingers (event 1), and a minute later you snap your fingers again (event 2). In your reference frame you are stationary, and the two events happened at exactly the same place—in the car. The events were, however, separated by a minute in time. This is very different from what happens in the reference frame of an observer sitting by the road. This observer agrees that the second snap of your fingers (event 2) occurred a minute after the first snap of your fingers (event 1), but to this observer the two events were separated from each other in space by a kilometer, the distance your car traveled in the minute. In this "Newtonian" view, the distance between two events depends on the motion of the observer, but the *time* between the two events does not. Special relativity instead finds that both the distance and the time between events varies depending on the motion of the observer.

The notion that different observers will measure time differently is a *very* counterintuitive idea, but it is central to special relativity and therefore to our scientific understanding of the universe as well. To see how Einstein arrived at the concept of relative time, consider his thought experiment known as the boxcar experiment. In this experiment, observer 1 is in a boxcar of a train moving to the right. Observer 1 has a lamp, a mirror (mounted to the roof of the boxcar), and a clock. Observer 2 is standing on the ground outside. The clock is based on a value that everyone can agree on—such as the speed of light.

Figure 18.4a shows the experimental setup as seen by observer 1, who is stationary with respect to the clock. At time t_1, event 1 happens: the lamp gives off a pulse of light. The light bounces off a mirror at a distance l meters away and then heads back toward its source. At time t_2, event 2 happens: the light arrives at the clock and is recorded by a photon detector. (Note that the light beam is shown leaving and arriving at two different locations. This was done by the artist so that you can see both events in the figure. Both events occur at the same location.) The time between events 1 and 2 is just the distance the light travels ($2l$ meters) divided by the speed of light: $t_2 - t_1 = 2l/c$.

Figure 18.4b shows the experiment as seen by observer 2, stationary on the ground outside the train, which is moving at speed v. In observer 2's reference frame, the clock moves to the right between the two events, so the light has farther to go because of the horizontal distance. The time between the two events is still the distance traveled divided by the speed of light, but now that distance is *longer* than $2l$ meters. Because the speed of light is the same for all observers, the time between the two events must be longer as well.

The two events are the same two events, regardless of the reference frame from which they are observed. Because the speed of light is the same for all observers, there must be more time between the two events when viewed from a reference frame in which the clock is moving (observer 2). The seconds of a moving clock are stretched. That is, it takes a moving clock more time than a stationary clock to complete one "tick." Therefore, the passage of time must depend on an observer's frame of reference. Because both frames of reference are equally good places to do physics, both time measurements are valid in their own frames, even though they differ from one another.

In this experiment, light travels farther between events in a moving boxcar than between events in a stationary boxcar and consequently takes a longer time to travel between the events in the moving boxcar. Einstein realized that the *only* way the speed of light can be the same for all observers is *if the passage of time is different from one observer to the next*. For moving observers, the time is stretched out, so that each second is longer, a phenomenon known as **time dilation**.

(a) In observer 1's reference frame, the clock is stationary.

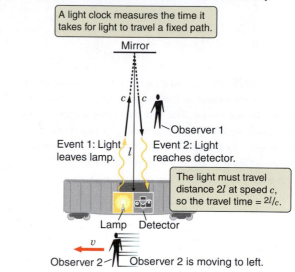

A light clock measures the time it takes for light to travel a fixed path.

Mirror

Observer 1

Event 1: Light leaves lamp.

Event 2: Light reaches detector.

The light must travel distance $2l$ at speed c, so the travel time = $2l/c$.

Lamp   Detector

v

Observer 2   Observer 2 is moving to left.

(b) The clock is moving in observer 2's reference frame.

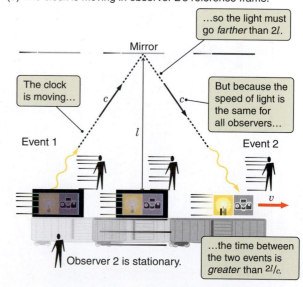

…so the light must go *farther* than $2l$.

Mirror

The clock is moving…

But because the speed of light is the same for all observers…

Event 1   Event 2

Observer 2 is stationary.

…the time between the two events is *greater* than $2l/c$.

Figure 18.4 The "tick" of a light clock is different when seen in two different reference frames: (a) stationary as in the reference frame of observer 1 in the boxcar, and (b) moving as in the reference frame of observer 2 on the tracks. As Einstein's thought experiment demonstrates, if the speed of light is the same for every observer, then moving clocks *must* run slowly compared to stationary clocks.

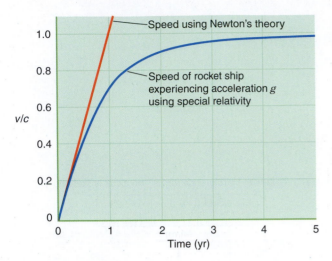

Figure 18.5 The speed of a rocket ship experiencing an acceleration equal to Earth's acceleration of gravity (*g*) increases with time. The red curve shows the speed using Newtonian theory. The blue curve shows the effects of special relativity, where the rocket ship approaches the speed of light but never gets there. The blue line correctly accounts for relativistic effects; the red line does not.

The Newtonian view of the world describes a three-dimensional space through which time marches steadily onward. But Einstein discovered that time flows differently for different observers. He reshaped the three dimensions of space and the one dimension of time into a four-dimensional combination called **spacetime**. Events occur at specific locations within this four-dimensional spacetime, but how much of a spacetime distance is measured in space and how much is measured in time depends on the observer's reference frame.

Einstein did not "disprove" Newtonian physics. At speeds much less than the speed of light, Einstein's equations become identical to the equations of Newtonian physics, so that Newtonian physics is contained within special relativity. Only when objects approach the speed of light do our observations begin to depart measurably from the predictions of Newtonian physics. These departures are called **relativistic** effects. In our everyday lives, we never encounter relativistic effects because we never travel at speeds that approach the speed of light. Even the fastest object ever made by humans, the *Helios II* spacecraft, traveled at only about 0.00023*c*.

Einstein's ideas remained controversial well into the 20th century, and his 1921 Nobel Prize in Physics was awarded for his work on the photoelectric effect (see Chapter 5) and not for his work on relativity because not all physicists were yet convinced that his theory was correct. But as one experiment after another confirmed the strange and counterintuitive predictions of relativity, scientists came to accept its validity.

The Implications of Relativity

Today, special relativity shapes our thinking about the motions of both the tiniest subatomic particles and the most distant galaxies. In this subsection, we discuss only a few of the essential insights that come from Einstein's work.

Mass and Energy What we think of as "mass" and what we think of as "energy" are actually closely related. The energy of an object depends on its speed: the faster it moves, the more energy it has. But Einstein's famous equation, $E = mc^2$, says that even a stationary object has an intrinsic "rest" energy that equals the mass (*m*) of the object multiplied by the speed of light (*c*) squared. The speed of light is a very large number, so a small mass has a very large rest energy. A single tablespoon of water has a rest energy equal to the energy released in the explosion of more than 300,000 tons of TNT. We used this relationship between mass and energy in Chapter 14 when discussing the nuclear fusion that makes stars shine.

In Chapter 3, we connected the *mass* of an object to its inertia—its resistance to changes in motion. At relativistic speeds, it becomes clear that adding to the energy of motion of an object increases its inertia. For example, a proton in a high-energy particle accelerator may travel so close to the speed of light that its total energy is 1,000 times greater than its rest energy. Such an energetic proton is harder to "push around" (in other words, it has more inertia) than a proton at rest. It also more strongly attracts other masses through gravity.

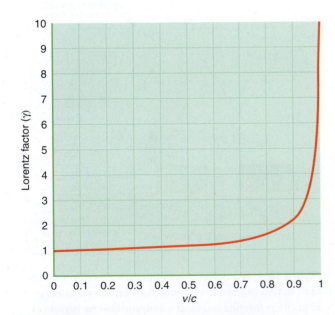

Figure 18.6 In this graph, the Lorentz factor γ is plotted against *v/c*. This factor doesn't become significant until velocities are about 50 percent the speed of light.

The Ultimate Speed Limit We already discussed Einstein's insight that if it were possible to travel at the speed of light, then in that reference frame light would cease to be a traveling wave, and the laws of electromagnetism wouldn't be valid. You can also think about this limit in terms of the equivalence of mass and energy just discussed. As the speed of an object gets closer and closer to the speed of light, the energy of that object, and therefore its mass, become greater and greater, so it becomes increasingly resistant to further changes in its motion. Only a photon or other massless particle can travel at the speed of light. We rely on the fact that light travels at a constant speed in a vacuum whenever we use the travel time of light to describe astronomical distances.

Adding energy to an object will cause its velocity to get closer and closer to the speed of light, but it will never actually reach the speed of light. It would take an *infinite* amount of energy to accelerate an object with a nonzero rest mass to the speed of light. There is not enough energy in the entire universe to accelerate even one electron to the speed of light. The electron can get arbitrarily close to that number—0.99999999999999999999 . . . × c is possible (at least in principle)—but there is not enough energy available to accelerate the electron beyond that to the speed of light. **Figure 18.5** shows how a rocket ship, which experiences a constant acceleration equal to that of gravity on Earth (so that its occupants will feel "normal" gravity), moves faster and faster but never reaches the speed of light. In the Newtonian view shown by the red curve, this limit is not present, and the speed of the spaceship would continually increase at the same rate.

Time Time passes more slowly in a moving reference frame: for moving objects, the seconds are stretched out in time dilation. No inertial reference frame is special. If you compared clocks with an observer moving at nine-tenths the speed of light (0.9c) relative to you, you would find that the other observer's clock was running 2.29 times slower than your clock. The other observer would find instead that *your* clock was running 2.29 times slower. To you, the other observer may be moving at 0.9c, but to the other observer, *you* are the one who is moving. Either frame of reference is equally valid, so you would each find the other's clock to be slow compared to your own. This time dilation effect increases with speed, and this symmetry holds as long as neither frame accelerates. **Figure 18.6** and **Table 18.1** show that how much the time is stretched depends on the object's speed. The factor of 2.29 by which time is stretched in the above example is called the Lorentz factor and is usually denoted by the symbol γ.

A scientific observation demonstrates time dilation in nature. As illustrated in **Figure 18.7**, fast particles called cosmic-ray muons are produced 15 km up in Earth's atmosphere when high-energy **cosmic rays**—elementary particles moving at nearly the speed of light—strike atmospheric atoms or molecules. Muons at rest decay very rapidly into other particles. This decay happens so quickly that even if they could move at the speed of light, virtually all muons would have decayed long before traveling the 15 km to reach Earth's surface. However, time dilation slows the muons' clocks, so the particles live longer and can travel farther and reach the ground. The faster muons move, the slower their clocks run and the more of them that can reach the ground. The same general principle is observed in particle accelerators, where particles that are traveling at speeds near the speed of light live longer before decaying. **Working It Out 18.1** shows some examples of time dilation.

Length An object appears shorter in motion than it is at rest. Moving objects are compressed in the direction of their motion by a factor of $1/\gamma$, where γ is the same Lorentz factor introduced in the discussion of time dilation. This phenomenon is called length contraction. A meter stick moving at 0.9c

TABLE 18.1	Lorentz Factor
v/c	γ
0.10	1.005
0.20	1.02
0.30	1.05
0.40	1.09
0.50	1.15
0.60	1.25
0.70	1.40
0.80	1.67
0.90	2.29
0.95	3.20
0.99	7.09
0.995	10.01
0.999	22.37
0.9999	70.71

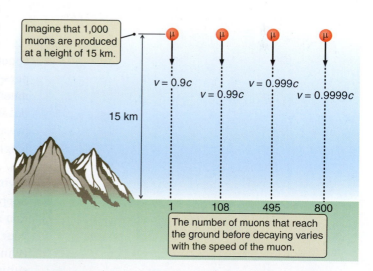

Figure 18.7 Muons created by cosmic rays high in Earth's atmosphere decay long before reaching the ground if they are not traveling at nearly the speed of light. Here, we show what happens to 1,000 muons produced at an altitude of 15 km for a variety of speeds. Faster muons have slower clocks, so more of them survive long enough to reach the ground—many more than would be expected simply due to the faster speed.

18.1 Working It Out Time Dilation

Physicist Hendrik Lorentz (1853–1928) derived the equation for how much time is dilated and how much space is contracted when something is traveling at velocities near the speed of light. This *Lorentz factor* (abbreviated γ) is given by

$$\gamma = \frac{1}{\sqrt{1 - \dfrac{v^2}{c^2}}}$$

Figure 18.6 shows the Lorentz factor plotted against velocity, and Table 18.1 gives the calculated value of γ for different values of velocities v/c. You can see that for something moving at half the speed of light, the Lorentz factor is 1.15. But for something moving at 90 percent of c, the factor is 2.29, and it goes up quickly from there, becoming arbitrarily large as the velocity approaches but never quite reaches the speed of light.

Let's look at a few examples. The first one is a thought experiment. Suppose you take a trip to a star 20 light-years away in order to study the "super-Earth" planets in orbit around it. You travel at 0.99c, while your twin stays behind on Earth. Our common Newtonian experience on Earth would tell you how long the trip takes using the familiar equation that time equals distance divided by speed:

$$\text{Time passed} = \frac{20\,\text{light-years}}{0.99c} = 20.2\,\text{years}$$

The round trip would take 40.4 years. When you return, you will find that 40.4 years have passed on Earth.

To you on the spaceship moving at 0.99c, special relativity says time will pass slower, by the Lorentz factor of 7.09. Your trip to the star would take $1/7 \times 20.2$ years = 2.8 years and another 2.8 years for the return trip. So as you traveled to this star and back, 5.6 years would have passed for you in the spaceship, but *40.4 years* would have passed on Earth. Your twin on Earth would be almost 35 years older than you are!

For a more practical experiment that can be performed on Earth, we've noted that time dilation can be seen with subatomic particles. Suppose one type of particle, called a pion, can "live" for 20 nanoseconds (ns) before it decays into different particles. If pions are produced in a particle accelerator at a speed of 0.999c, how long will physicists see the pions before they decay? Here $v = 0.999c$, so from Table 18.1, $\gamma = 22.37$. In the reference frame of the moving pions, they are still living for 20 ns: they are like the twin that traveled into space at high speed. But in the reference frame of the physicists, special relativity predicts that the particle will last longer, 20 ns × 22.37, or 447 ns. Indeed, physicists observe that pions moving at nearly the speed of light "live" longer and travel farther before decay (like the muons in Figure 18.7).

The same factor γ applies for mass and length. If physicists measured a ruler moving at 0.999c, the length of the ruler would be 22.37 times shorter compared to its length when at rest. In the earlier example, the pions traveling at 0.999c will behave as if their mass were 22.37 times larger: if the high-speed pions collided with other particles, the energy of the collision would be as if they had the higher mass. Particles at high speeds have provided the experimental evidence for special relativity.

appears to be 0.44 meters long. This also explains our muon experiment from the perspective of the muons themselves. In the reference frame of the fast-moving muon produced at a height of 15 km, Earth's atmosphere is moving fast and appears to be much shorter than 15 km; indeed, it is so compressed from the muon's perspective that the muon may be able to reach the ground before decaying. This length contraction effect also increases with speed.

Twin Paradox One interesting consequence of relativity is the *twin paradox*. Suppose you take a trip into space and leave your identical twin back on Earth. You accelerate to nearly the speed of light as you leave Earth. After you arrive at your destination, you return to Earth, again traveling at a speed close to c. To your twin on Earth, you were the one in a moving reference frame, so your twin measures your time as running much slower (recall Working It Out 18.1), and you should return younger than your twin. However, from your perspective, the spaceship didn't move. Instead, Earth receded from you, stopped, and returned. Your twin is the one who moved at just under the speed of light, and your twin's time ran slowly compared to yours, so your twin has aged much less. Both you and your twin cannot be correct, and that is the nature of the paradox.

To resolve the paradox, you must realize that you experienced acceleration during your trip, while your twin did not. Accelerated motion is not uniform motion.

As a result, you changed reference frames during your trip. You changed reference frames when you left Earth, changed again when you stopped at your destination, changed a third time when you left your destination to return home, and changed reference frames one final time when you arrived back at Earth. Your twin, however, remained in Earth's reference frame. Upon your return, you would find that in fact more time has passed for your twin on Earth than for you, and your twin has aged more than you have.

Space Travel What is the actual reality of human travel in space? Some American astronauts have been to the Moon and back, and a number of robotic spacecraft have been sent to explore objects throughout the Solar System. (Only one robotic spacecraft has actually left the Solar System.) But it would be very difficult for humans to visit and explore other planetary systems in the Milky Way Galaxy because of the constraints of energy and the ultimate speed limit of light. With current technology, engineers can construct rockets able to travel at speeds of up to 20,000 m/s. At such a speed, a one-way trip to Earth's nearest neighbor star, Proxima Centauri, at a distance of 4.2 light-years, would take well over 50,000 years. Travel to more distant stars would take even longer.

In principle, travel just under the speed limit c is possible, so one could take advantage of relativistic time dilation to make such adventures well within the lifetime of a space traveler. In the example in Working It Out 18.1, an astronaut experiences a round-trip travel time to planets 20 light-years away in just 5.8 years at a speed of $0.99c$. Or an astronaut could travel to the center of the Milky Way Galaxy and back in just 2 years by traveling at $0.9999999992c$.

Although theoretically possible, travel at these speeds in practice would require an impractical amount of energy. If M is the mass of the astronauts, rocket ship, and fuel, it would take γMc^2 of energy just to accelerate the rocket up to such a high speed, or $10Mc^2$ in the first example and $25,000Mc^2$ in the second example to the center of the Milky Way. For this second example, the energy to accelerate just the astronaut (not even including the spaceship) to such energies is more than that contained in 10 billion nuclear weapons. So, while not theoretically impossible, visits to other stars in our galaxy will not take place anytime soon.

CHECK YOUR UNDERSTANDING 18.2

Suppose that your friend flies past you in a spaceship, and both of you measure the time it takes the spaceship to pass your location. Which of the following is true? (a) The time you measure is longer than the time your friend measures. (b) The time you measure is shorter than the time your friend measures. (c) You both measure the same amount of time.

....

18.3 Gravity Is a Distortion of Spacetime

Our exploration of special relativity began with the observation that the speed of light is always the same regardless of the motion of an observer or the motion of the source of the light. We have seen that three-dimensional space and time are actually just the result of a particular, limited perspective on a four-dimensional spacetime that is different for each observer. This four-dimensional spacetime is itself warped and distorted by the masses it contains. As we discuss the properties of black holes—indeed, of all massive objects in the universe—the concepts

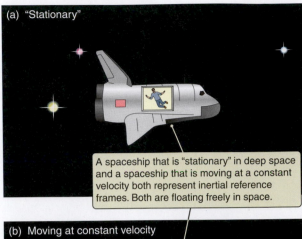

(a) "Stationary"

A spaceship that is "stationary" in deep space and a spaceship that is moving at a constant velocity both represent inertial reference frames. Both are floating freely in space.

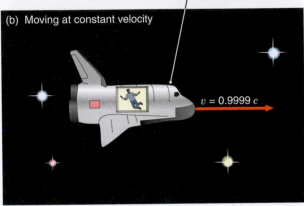

(b) Moving at constant velocity

$v = 0.9999\,c$

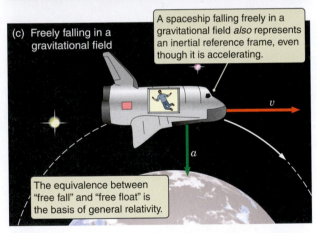

(c) Freely falling in a gravitational field

A spaceship falling freely in a gravitational field *also* represents an inertial reference frame, even though it is accelerating.

v

a

The equivalence between "free fall" and "free float" is the basis of general relativity.

Figure 18.8 Special relativity says that there is no difference between (a) a reference frame that is floating stationary in space and (b) one that is moving through the galaxy at constant velocity. General relativity adds that there is no difference between these inertial reference frames and (c) an inertial reference frame that is falling freely in a gravitational field. Free fall is the same as free float, as far as the laws of physics are concerned.

of space and time will diverge even further from the absolutes of Newtonian physics. In this section, we will explore the general theory of relativity, which describes how mass affects space and time.

The Equivalence Principle

We have already discussed one fundamental connection between gravity and spacetime in Chapter 4 and showed that the *inertial mass* of an object—the mass appearing in Newton's equation $F = ma$—is *exactly* the same as the object's gravitational mass. In addition, any two objects at the same location and moving with the same velocity will follow the same path through spacetime, regardless of their masses. The astronaut in an orbiting spaceship falls around Earth, moving in lockstep with the spaceship itself. A feather dropped by an *Apollo* astronaut standing on the Moon falls toward the surface of the Moon at exactly the same rate as a dropped hammer does. Rather than thinking of gravity as a force that acts on objects, it is more accurate to think of gravity as a consequence of the warping of spacetime in the presence of a mass. *Gravitation is the result of the shape of spacetime that objects move through.* This is one of the key insights of the **general theory of relativity**, Einstein's theory of gravity.

Special relativity tells us that any inertial reference frame is as good as any other. No experiment can distinguish between sitting in an enclosed spaceship floating stationary in deep space, as seen in **Figure 18.8a**, and sitting in an enclosed spaceship traveling at 0.9999 times the speed of light (Figure 18.8b). These two situations feel the same because neither observer feels an acceleration. Each is an equally valid inertial reference frame. As long as nothing accelerates either spaceship, neither observer can distinguish between these two spacecraft.

But what if there is an acceleration? Recall that an acceleration can change either the speed or the direction of an object. Consider an astronaut inside a spaceship orbiting Earth, as shown in Figure 18.8c. This spaceship is accelerating, as the direction of its velocity is constantly changing as it orbits Earth. The astronaut is also accelerating. Because he feels weightless, the astronaut has no way to tell the difference between being inside the spaceship as it falls around Earth and being inside a spaceship floating through interstellar space. Even though its velocity is constantly changing as it falls, the inside of a spaceship orbiting Earth is an inertial frame of reference just as an object drifting along a straight line through interstellar space is an inertial frame of reference.

The idea that a freely falling reference frame is equivalent to a freely floating reference frame is called the **equivalence principle**. If you close your eyes and jump off a diving board, for the brief time that you are falling freely through Earth's gravitational field, the sensation you feel is exactly the same as the sensation you would feel floating in interstellar space.

The natural path that an object will follow through spacetime in the absence of other forces is called the object's **geodesic**. In the absence of a gravitational field, the geodesic of an object is a straight line, in accordance with Newton's first law: an object will move at a constant speed in a constant direction unless acted on by a net external force. However, the shape of spacetime becomes distorted in the presence of mass, so an object's geodesic becomes curved.

Figure 18.9a shows two examples of inertial frames: an astronaut in a spaceship coasting through space, and a person in a box falling toward the ground

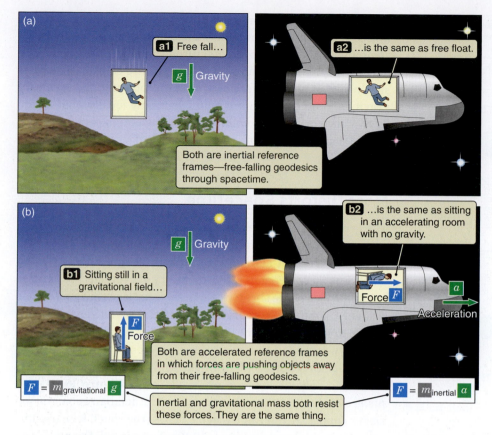

Figure 18.9 According to the equivalence principle, (a) an object falling freely in a gravitational field is in an inertial reference frame, and (b) an object at rest in a gravitational field is in an accelerated reference frame. According to the equivalence principle, sitting in a spaceship accelerating at 9.8 m/s^2 feels the same as sitting still on Earth.

with an acceleration *g*. Both of these people are following their geodesics, so these two inertial reference frames are equivalent, and neither observer can distinguish between them. The equivalence principle also applies in cases of accelerations that result in a change in speed. Imagine an observer sitting in a closed box on the surface of Earth, as in Figure 18.9b. The floor of the box pushes on him to keep him from following his geodesic, and he feels that force. Now imagine the box is inside a spaceship that is accelerating through deep space at a rate of 9.8 meters per second per second (m/s^2) in the direction of the arrow shown in Figure 18.9b. The floor of the box pushes on the observer to overcome his inertia and cause him to accelerate at 9.8 m/s^2, so he feels as though he is being pushed into the floor of the box. In both of these cases, a force acts on the observer so that he does not follow the same free-falling geodesic. The observer feels as though he is being pushed into the floor of the box, so he feels the acceleration, and his frame of reference is not inertial. According to the equivalence principle, sitting in an armchair in a spaceship traveling with an acceleration of 9.8 m/s^2 is equivalent to sitting in an armchair on the surface of Earth reading this book. In each case, it is the same mass—the mass that gives an object inertia—that resists the change. Gravitational mass and inertial mass are the same thing.

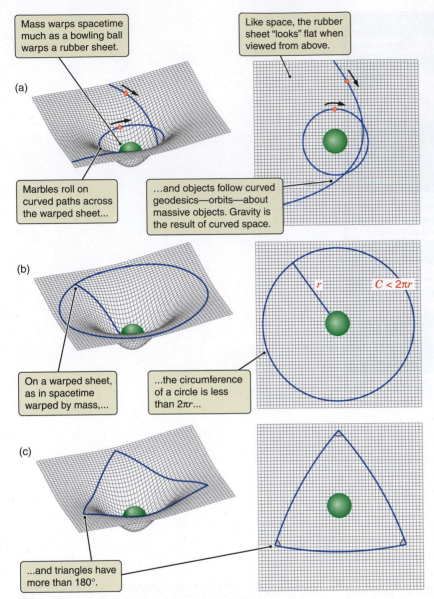

Mass warps spacetime much as a bowling ball warps a rubber sheet.

Like space, the rubber sheet "looks" flat when viewed from above.

(a)

Marbles roll on curved paths across the warped sheet...

...and objects follow curved geodesics—orbits—about massive objects. Gravity is the result of curved space.

(b)

r $C < 2\pi r$

On a warped sheet, as in spacetime warped by mass,...

...the circumference of a circle is less than $2\pi r$...

(c)

...and triangles have more than 180°.

Figure 18.10 Mass warps the geometry of spacetime in much the same way that a bowling ball warps the surface of a stretched rubber sheet. This distortion of spacetime has many consequences; for example, (a) objects follow curved paths or geodesics through curved spacetime, (b) the circumference of a circle around a massive object is less than 2π times the radius of the circle, and (c) angles in triangles need not add to exactly 180 degrees.

There is an important caveat to the equivalence principle. In an accelerated reference frame such as an accelerating spaceship, the same acceleration (both magnitude and direction) is experienced everywhere. In contrast, the curvature of space by a massive object is weaker farther from the object. The effects of gravity and acceleration are equivalent only locally; that is, the equivalence principle is valid as long as attention is restricted to small-enough volumes of space so that changes in gravity can be ignored.

Mass Distorts Spacetime

The general theory of relativity describes how mass distorts the geometry of spacetime. Imagine the surface of a tightly stretched, flat rubber sheet. A marble will roll in a straight line across the sheet. Euclidean geometry, the geometry of everyday life, applies on the surface of the sheet: if you draw a circle, its circumference is equal to 2π times its radius, r; if you draw a triangle, the angles add up to 180 degrees; lines that are parallel anywhere are parallel everywhere.

Now place a bowling ball in the middle of the rubber sheet, creating a deep depression, or "well," as in **Figure 18.10**. The surface of the sheet is no longer flat, and Euclidian geometry no longer applies. If you roll a marble across the sheet, its path dips and curves, as shown in Figure 18.10a. You can roll the marble so that it moves around and around the bowling ball, like a planet orbiting about the Sun. Figure 18.10b shows that if you draw a circle around the bowling ball, the circumference of the circle is less than $2\pi r$. If you draw a triangle, the angles add up to more than 180 degrees (Figure 18.10c).

Mass has an effect on the fabric of spacetime that is like the effect of the bowling ball on the fabric of the rubber sheet. The bowling ball stretches the sheet, changing the distances between any two points on the surface of the sheet. Similarly, mass distorts spacetime, changing the distance between any two locations or events. Larger masses produce larger spacetime distortions. We can visualize how a rubber sheet with a bowling ball on it is stretched through a third spatial dimension, but it is impossible for most people to visualize what a curved four-dimensional spacetime would "look like." Yet experiments verify that the geometry of four-dimensional spacetime is distorted much like the rubber sheet, whether or not it can be easily pictured.

When One Physical Law Supplants Another

Earlier in this book, we described gravity as a force that obeys Newton's universal law of gravitation: $F = Gm_1m_2/r^2$. In this chapter, we have introduced the ideas of general relativity and asked you to change the way you think about gravity. If general relativity is correct, does that mean Newton's formulation of gravity is wrong? If so, then why does Newton's law continue to be used?

These questions go to the heart of how science progresses. As long as a gravitational field is not too strong, Newton's law of gravitation is a very close *approximation* to the results of a calculation using general relativity. In this context, even the gravitational field near the core of a massive main-sequence star would be considered "weak." An astronomer will obtain the same results if she uses a general relativistic formulation of gravity rather than Newton's laws to calculate the structure of a main-sequence star. Similarly, even though spacetime is curved by the presence of mass, this curvature near Earth is so slight that over small regions it can be ignored entirely, and flat Euclidean geometry can be used. This is exactly the kind of approximation people use when they navigate with a flat road map, despite the curvature of Earth.

Similarly, Newton's laws of motion are *approximations* of special relativity and quantum mechanics. In fact, Newton's laws can be mathematically derived from special relativity and quantum mechanics using the assumptions that speeds of objects are much less than the speed of light and that objects are much larger than the particles from which atoms are made. Newton's laws of motion and gravitation are used most of the time because they are far easier to apply than general relativity and because any inaccuracies introduced by using Newtonian approximations are usually far too tiny to measure. This was illustrated in Figure 18.5, where the behavior of an accelerated object moving at a velocity much less than c is the same whether or not relativity is used. Calculations based on general relativity are required only when conditions are very different from those of everyday life (for example, the behavior of an electron in an atom or the behavior of the gravitational field of a black hole) or in special cases when very high accuracy is needed (for example, the precise timing used by the GPS satellite network).

If a new theory is to replace an earlier, highly successful scientific theory, the new theory must hold the old theory within it—it must be able to reproduce the successes of the earlier theory. Special relativity contains Newton's laws of motion, and general relativity holds within it the successful Newtonian description of gravity that we have relied on throughout this book (**Process of Science Figure**).

The Observable Consequences of General Relativity

The curved spacetime of general relativity does have observable consequences. Indeed, there are observations that distinguish general relativity from Newtonian physics in our own Solar System. In Newton's theory, orbits are elliptical and fixed in space. In contrast, general relativity predicts that the long axis of an elliptical orbit slowly rotates, or precesses. **Figure 18.11** illustrates the difference between an orbit in Newton's theory, which remains stationary (left panel), versus one in general relativity, which precesses (right panel). In our Solar System, even after accounting for the effects of other planets on Mercury, there remains a very small shift in its axis equal to 43 arcseconds (arcsec) per century, which cannot be explained by Newton's laws alone. General relativity predicts exactly that precession for Mercury.

There are other unique implications of general relativity. A beam of light moving through empty space travels in a straight line, but a beam of light moving through the distorted spacetime around a massive object is bent by gravity, just as the lines in Figure 18.10 are bent by the curvature of the sheet. This bending of the light path by curved spacetime is called **gravitational lensing** because optical lenses also bend light paths. The first measurement of gravitational lensing came during the total solar eclipse of 1919. Several months before the eclipse, astrophysicist Sir Arthur Stanley Eddington (1882–1944) measured the positions of a number of stars in the direction of the sky where the eclipse would occur. He then

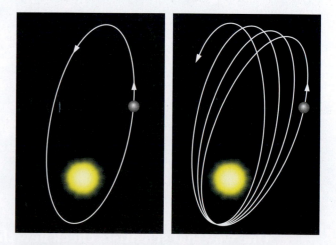

Figure 18.11 The left panel denotes an elliptical orbit about the Sun. In the Newtonian view, this elliptical orbit remains stationary. In the right panel, Mercury's orbit precesses due to the warped spacetime near the Sun.

General relativity is more accurate than Newton's laws: it also explains why gravity acts as it does. Still, for objects like Earth or the Solar System, the two calculations agree.

General relativity is needed when masses are large and distances are small, so the pull of gravity is large.

But far from a mass, where gravity is weak and spacetime relatively flat, general relativity gives the same result that Newton found.

One way that scientists check new theories is by considering the limits. What happens at great distances? What happens if the mass is very small? In these limits, new, more complete theories must be compatible with old ones.

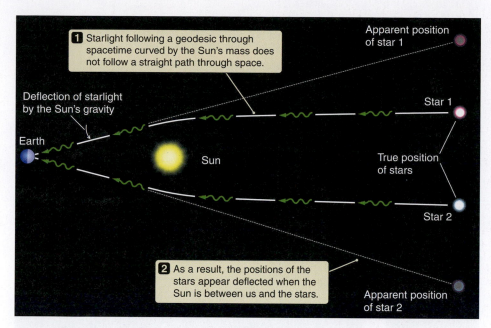

1. Starlight following a geodesic through spacetime curved by the Sun's mass does not follow a straight path through space.

Apparent position of star 1

Deflection of starlight by the Sun's gravity

Earth

Sun

Star 1

True position of stars

Star 2

2. As a result, the positions of the stars appear deflected when the Sun is between us and the stars.

Apparent position of star 2

Figure 18.12 Measurements during the total solar eclipse of 1919 found that the gravity of the Sun bends the light from distant stars by the amount predicted by Einstein's general theory of relativity. This is an example of gravitational lensing. Note that the "triangle" formed by Earth and the two stars contains more than 180 degrees, just like the triangle in Figure 18.10c.

Figure 18.13 This Hubble Space Telescope image shows four images of the same distant supernova (arrows), where the images arise from the gravitational lensing due to a massive foreground galaxy.

repeated the measurements during the eclipse. **Figure 18.12** shows how the light from distant stars curved as it passed the Sun, causing the measured positions of the stars to shift outward. The stars appeared farther apart in Eddington's second measurement than in his first by the amount predicted by general relativity. This is another example of the effects of general relativity—in curved space, parallel lines actually can intersect.

Images of galaxies are sometimes distorted by gravitational lensing by other galaxies or clusters of galaxies. In the Hubble Space Telescope image shown in **Figure 18.13**, a supernova in a distant galaxy is lensed into four images by a nearby massive galaxy. What makes this case particularly interesting is that these four images appeared at different times because the light for each image took a different path, and some were longer than others. In an extreme case, a lensed galaxy image can be distorted into a ring called an **Einstein ring**, as illustrated in **Figure 18.14**.

Mass distorts not only the geometry of space but also the geometry of time. The deeper one descends into the gravitational field of a massive object, the more slowly clocks appear to run from the perspective of a distant observer, an effect called **general relativistic time dilation**. Suppose a light is attached to a clock sitting on the surface of a neutron star. The light is timed so that it flashes once a second. Because time near the surface of the star is dilated due to high gravity, an observer far from the neutron star perceives the light to be pulsing with a lower frequency—less than once a second. Now suppose there is an emission line source on the surface of the neutron star. Because time is running slowly on the surface of the neutron star, the light that reaches the distant observer will have a lower frequency than that when it was emitted. Recall that a lower frequency means a longer wavelength, so the light from the source will be seen at a longer, redder wavelength than the wavelength at which it was emitted. This shift in the wavelengths of light from objects deep within a gravitational well is called the

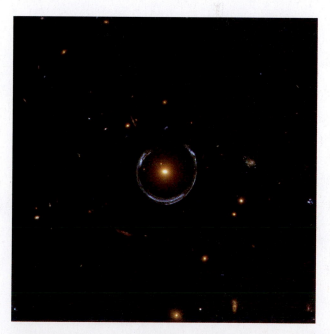

Figure 18.14 This photograph shows an Einstein ring created by the gravity of a luminous red galaxy gravitationally lensing the light from a much more distant blue galaxy.

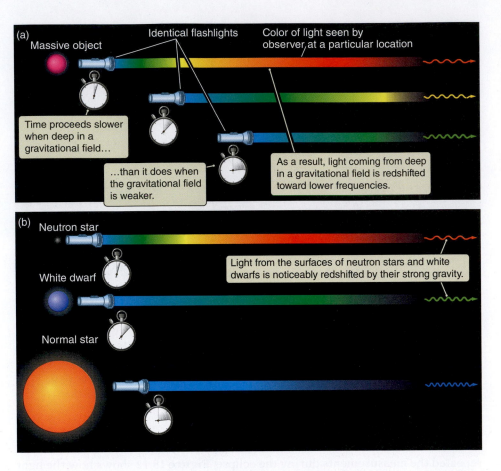

Figure 18.15 Time passes more slowly near massive objects because of the curvature of spacetime. As a result, to a distant observer light from near a massive object will have a lower frequency and longer wavelength. (a) The closer the source of radiation is to the object or (b) the more massive and compact the object is, the greater the gravitational redshift will be.

gravitational redshift, as shown in **Figure 18.15**. The effect of gravitational redshift is similar to the Doppler redshift discussed in Chapter 5. In fact, there is no way to tell the difference between light that has been redshifted by gravity and light from an object moving away from you that has been Doppler shifted.

Bringing the phenomenon of general relativistic time dilation a bit closer to home, a clock on the top of Mount Everest runs faster, gaining about 80 ns a day compared with a clock at sea level. The difference between an object on the surface of Earth and an object in orbit is even greater. A GPS receiver uses the results of sophisticated calculations of the effects of general relativistic gravitational redshift to help you accurately find your position on the surface of Earth. Satellites in orbit travel quickly enough that special relativistic effects are measurable. Even after allowing for slowing due to special relativity, the clocks on the satellites that make up the GPS run faster than clocks on the surface of Earth. If the satellite clocks *and* your GPS receiver did not correct for this and other effects of general relativity, then the position your GPS receiver reported would be in error by up to half a kilometer. The fact that the GPS can be accurate to a few meters provides strong experimental confirmation of two predictions of general relativity—gravitational redshift and general relativistic time dilation.

Gravitational Waves

General relativity also makes predictions that have not yet been confirmed. If you exert a force on the surface of a rubber sheet, accelerating it downward, waves will move away from where you struck it, like ripples spreading out over the surface of a pond. Similarly, the equations of general relativity predict that if you accelerate the fabric of spacetime (for example, with the catastrophic asymmetrical collapse of a high-mass star), then ripples in spacetime, or **gravitational waves**, will move outward at the speed of light. These gravitational waves are like electromagnetic waves in some respects. Accelerating an electrically charged particle gives rise to an electromagnetic wave. Deforming a massive object gives rise to gravitational waves.

Gravitational waves have not yet been directly observed, but there is strong circumstantial evidence for their existence. General relativity predicts that the orbit of binary neutron stars should lose energy, which will be carried away as gravitational waves. In 1974, astronomers discovered a binary system of two neutron stars, one of which is an observable pulsar. Using the pulsar as a precise clock, astronomers accurately measured the orbits of both stars. The orbits are gradually losing energy at the rate predicted by general relativity. Other similar binary pairs have been found with an orbital energy loss consistent with the radiation of gravitational waves. These systems provide indirect evidence for the existence of gravitational waves. As we discussed in Chapter 6, astrophysicists have a new kind of "telescope," called the Laser Interferometer Gravitational-Wave Observatory (LIGO), that might be able to detect the predicted gravitational waves emanating from such events.

CHECK YOUR UNDERSTANDING 18.3

What does gravity mean in relativity? (a) It is a result of mass and energy being the same thing. (b) It is a consequence of length contraction. (c) It is the result of masses acting larger when they move at high speeds. (d) It is the result of the distortion in spacetime around a massive object.

18.4 Black Holes

General relativity also predicts the existence of black holes. There is no well-formulated theory of black holes in the Newtonian world. When a mass is placed on the surface of a rubber sheet, it causes a funnel-shaped distortion that is analogous to the distortion of spacetime by a mass. Now imagine the limit in which the funnel is *infinitely* deep—it gets narrower as it goes deeper but has no bottom. This is the rubber-sheet analog to a black hole. The mathematics describing a black hole approaches infinity in the same way that the mathematical expression $1/x$ does when x approaches zero. Such a mathematical anomaly is called a **singularity**. Black holes contain singularities in spacetime, and this mathematical complication indicates that extreme conditions exist near (and inside) a black hole.

How Black Holes Form

We have seen how stellar evolution can lead to the formation of compact stellar remnants such as white dwarfs and neutron stars. However, there is an even more extreme fate that awaits some massive stars at the end of their evolution. Recall from Chapter 16 that a white dwarf can have a mass of no more than the Chandrasekhar limit, about 1.4 solar masses. If the mass of the object exceeds this limit, then gravity is able to overcome electron degeneracy pressure, and the white dwarf will begin collapsing again.

(a)

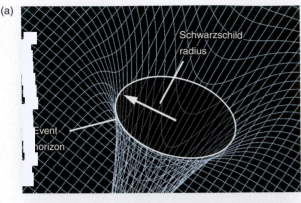

(b)

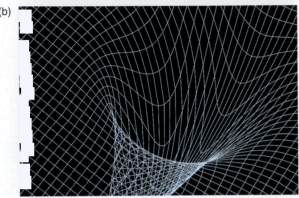

Figure 18.16 (a) A black hole's size is determined by the Schwarzschild radius and the corresponding event horizon. This image is a two-dimensional analogy for a black hole. In reality, the event horizon is a sphere. (b) If the object that formed the black hole was spinning, its angular momentum is conserved, and the black hole twists the spacetime around it.

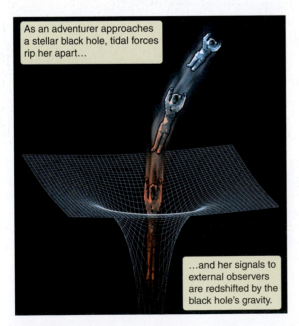

As an adventurer approaches a stellar black hole, tidal forces rip her apart…

…and her signals to external observers are redshifted by the black hole's gravity.

Figure 18.17 An adventurer falling into a black hole would be torn apart by the extreme tidal forces.

The physics of a neutron star is much like the physics of a white dwarf, except that neutrons rather than electrons are what cause a neutron star to be degenerate. Analogous to a white dwarf, if the mass of a neutron star exceeds about 3 M_{Sun}, gravity begins to win out over pressure once again. The neutron star grows smaller, and gravity at the star's surface becomes stronger and stronger at an ever-accelerating pace. Recall from Chapter 4 that the escape velocity from a planet or moon depends on its surface gravity. Now imagine surface gravity so strong that the escape velocity approaches the speed of light. But nothing, not even light, can travel faster than c. So when the neutron star crosses the threshold where the escape velocity from its surface exceeds the speed of light, not even light can escape its gravity. A region of space where neither matter nor radiation can escape the pull of gravity is called a **black hole**.

A neutron star in a binary system can collapse to become a black hole if it accretes enough matter from its companion to push it over the 3-M_{Sun} limit, similar to the way some Type Ia supernovae accrete matter from a companion and approach the Chandrasekhar limit of 1.4 M_{Sun}. Regardless of how it forms, any stellar remnant with a mass greater than 3 M_{Sun} must be a black hole.

Properties of Black Holes

From the outside, you can never actually "see" a black hole. The closer an object is to a black hole, the greater is its escape velocity (the speed that it would need to escape from the gravity of the black hole). There is a distance from the black hole at which the escape velocity reaches the speed of light. The radius where the escape velocity equals the speed of light is called the **Schwarzschild radius**, named for the physicist Karl Schwarzschild (1873–1916), and it is proportional to the mass of the black hole:

$$R_S = \frac{2GM_{BH}}{c^2} = 3\,\text{km} \times \frac{M_{BH}}{M_{Sun}}$$

where R_S is the Schwarzschild radius, G is the universal gravitational constant, M_{BH} is the mass of the black hole, and c is the speed of light. The sphere around the black hole at this distance is called its **event horizon**, a surface through which nothing, not even light, can escape. A black hole with a mass of 1 M_{Sun} has a Schwarzschild radius of about 3 km. A black hole with a mass of 5 M_{Sun} has a Schwarzschild radius 5 times that, or about 15 km. If Earth were squeezed into a black hole, it would have a Schwarzschild radius of only about a centimeter. All the mass of a black hole is concentrated at its very center, but this fact is unobservable from outside the black hole. **Figure 18.16a** shows a rubber-sheet analog to a black hole, with the Schwarzschild radius and the event horizon.

A black hole has only three observable properties: mass, electric charge, and angular momentum. The mass of a black hole determines the Schwarzschild radius. The electric charge of a black hole is the net electric charge of the matter that fell into it. The angular momentum of a rotating black hole twists the spacetime around it, as shown in Figure 18.16b. Apart from these three properties, all information about the material that fell into the black hole is lost. Nothing of its former composition, structure, or history survives.

Imagine that an adventurer falls into a black hole, as illustrated in **Figure 18.17**. From our perspective outside the black hole, the adventurer would appear to fall toward the event horizon. As she fell, her watch would appear to run more and more slowly, and her progress toward the event horizon would slow as well. At the event horizon, the gravitational redshift becomes infinite, and clocks appear to stop altogether. She would approach the event horizon, but from our perspective she would never quite make it. Yet the adventurer's experience would be very different. From her perspective, there would be nothing special about the event horizon. She would fall past the event horizon and continue deeper into the black hole's gravitational well. She would now have entered a region of spacetime cut off from the rest of the universe. The event horizon is like a one-way door: after the adventurer has passed through, she can never again pass back into the larger universe she once belonged to.

Actually, we have overlooked a rather important detail—the adventurer would have been torn to shreds long before she reached the black hole. Near the event horizon of a 3-M_{Sun} black hole, the difference in gravitational acceleration between the adventurer's feet and her head would be about a billion times her gravitational acceleration on the surface of Earth. In other words, her feet would be accelerating a billion times faster than her head. This is not an adventure that anyone would ever find appealing. Although scientific theories must produce testable predictions, not all individual predictions have to be tested directly.

"Seeing" Black Holes

In 1974, the physicist Stephen Hawking realized that black holes should be sources of radiation. In the ordinary vacuum of empty space, particles and their antiparticles spontaneously appear and then, within about 10^{-21} second, annihilate each other and disappear. Such particles are called **virtual particles** because they exist for only a very short time. If a pair of virtual particles comes into existence near the event horizon of a very small black hole, one of the particles might fall into the black hole while the other particle escapes, as illustrated in **Figure 18.18**. Some of the gravitational energy of the black hole will have been used up in making one of the pair of virtual particles real. Hawking showed that through this process, a black hole should emit a Planck blackbody spectrum and that the effective temperature of this spectrum would increase as the black hole became smaller through this "evaporation" process. After a very, very long time (of the order 10^{61} years for a black hole with a mass of the Sun), the black hole would become small enough that it would become unstable and explode. Although the light that emerges, called **Hawking radiation**, is of considerable interest to physicists, in a practical sense the low intensity of Hawking radiation means it is not a likely way to see a black hole.

The strongest direct evidence for black holes that result from supernovae comes from X-ray binary stars. The radio emission from Cygnus X-1 (an object originally identified in X-rays) flickers rapidly, changing in as little as 0.01 second. This means that the source of the X-rays must be smaller than the distance that light travels in 0.01 second, or 3,000 km—smaller than Earth. Cygnus X-1 was also identified both with a radio source and with an already cataloged star called HD 226868. The spectrum of HD 226868 shows that it is a normal O9.7 I supergiant star with a mass of about 19 M_{Sun}, far too cool to produce X-ray emission in the quantity observed. The wavelengths of absorption lines in the spectrum of HD 226868 are Doppler-shifted back and forth with a period of 5.6 days, indicating that HD 226868 is part of a binary system. Using the same techniques we discussed to measure the masses of binary stars in Chapter 13, astronomers

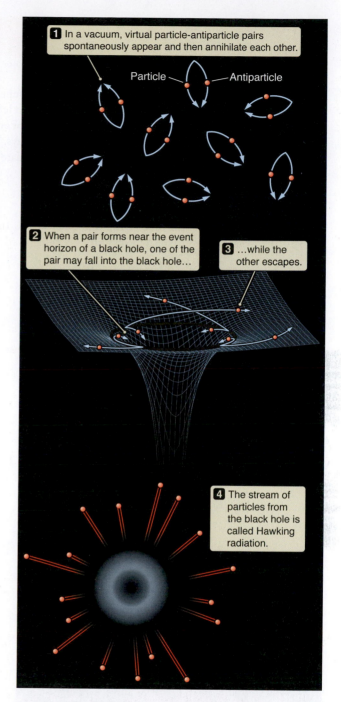

Figure 18.18 In the vacuum of empty space, particles and antiparticles are constantly being created and then annihilating each other. Near the event horizon of a black hole, however, one particle may cross the horizon and fall into the black hole before it recombines with its partner. The remaining particle leaves the vicinity of the event horizon, leading to the production of Hawking radiation.

18.2 Working It Out Masses in X-ray Binaries

Cygnus X-1 is part of a binary system with a blue supergiant star O9.7 I (very close to B0 I) and an unseen compact object located about 0.2 astronomical unit away from it. The blue supergiant and the compact object orbit a common center of mass every 5.6 days. We can use the simple formula from Working It Out 13.4 to calculate the sum of the masses:

$$\frac{M_{blue}}{M_{Sun}} + \frac{M_{compact}}{M_{Sun}} = \frac{A^3}{P^2}$$

with A in astronomical units (AU) and P in years. In this case, $A = 0.2$ AU and $P = 5.6/365.24$ years, so

$$\frac{M_{blue}}{M_{Sun}} + \frac{M_{compact}}{M_{Sun}} = \frac{0.2^3}{\left(\frac{5.6}{365.24}\right)^2} = 34$$

The sum of the masses of the two stars is 34 M_{Sun}.

To find the values of the two individual masses from their orbits, we need to know the velocities of the two stars or the distance of each star to the center of mass and the orbital inclination of the system. Obtaining such information is difficult when one star is compact and not observed separately. However, the mass of the blue supergiant star can be estimated from spectroscopic and photometric data at many wavelengths, assuming the distance to the system is known. When this is done, the mass of the supergiant is estimated at 19 M_{Sun}. Subtracting from 34, the mass of the compact object is 15 M_{Sun}—well over the mass limit for a neutron star. Therefore, Cygnus X-1 is assumed to be a black hole. A recent study with data from several X-ray telescopes concluded that the black-hole mass of Cygnus X-1 is 14.8 M_{Sun}.

found that the mass of the unseen compact companion of HD 226868 must be about 15 M_{Sun} (**Working It Out 18.2**). The companion to HD 226868 is too compact to be a normal star, yet it is much more massive than the Chandrasekhar limit for a white dwarf or the upper mass limit of a neutron star. Such an object can only be a black hole. The X-ray emission from Cygnus X-1 arises when material from the O9.7 I supergiant falls onto an accretion disk surrounding the black hole as illustrated in the chapter-opening figure.

In some similar systems, winds have been observed blowing off the disk around the black hole. These winds are probably caused by magnetic fields in the disk. The fastest winds observed are in the binary system IGR J17091, where the winds are as high as 32 million km/h (about 3 percent of the speed of light). This wind is blowing in many directions, and it may be carrying away more mass than is being captured by the black hole, as illustrated in **Figure 18.19**.

Astronomers have modeled the observational data of dozens of good candidates for stellar-mass black holes in X-ray binary systems in the Milky Way. They have found that the black-hole masses are greater than 4.5–5 M_{Sun}; that is, not very close to the limit of 2.5–3 M_{Sun} for a neutron star. This gap in mass between the most massive neutron stars and the least massive stellar black holes is not yet understood, and it is assumed to be a result of the mass transfer processes between the stars.

The black holes we discussed in this chapter came from collapsing massive stars, but this is not the only type of black hole. In Chapters 19 and 20, you will learn that supermassive black holes can be found at the centers of galaxies, including the Milky Way.

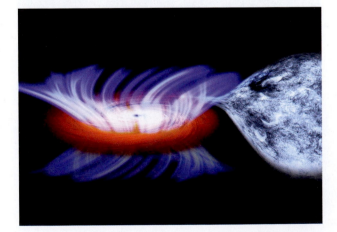

Figure 18.19 This artist's model shows strong winds being emitted from the disk around a stellar black hole. These winds can remove more material than the amount that actually falls into the hole.

CHECK YOUR UNDERSTANDING 18.4

If a black hole suddenly doubled in mass, the event horizon would become _____ its original radius. (a) one-quarter; (b) one-half; (c) twice; (d) 3 times; (e) 4 times

Origins

Gamma-Ray Bursts

The most energetic explosions in the universe are probably related to stellar black holes. **Gamma-ray bursts**, or **GRBs**, are intense bursts of gamma rays. The bursts are followed by a weaker "afterglow" that is observed at many wavelengths. GRBs were first observed in the 1960s by satellites designed to look for radiation from nuclear weapons being tested in space after such tests were banned on Earth. In the 1990s, gamma-ray astronomy satellites discovered that these bursts were coming from all directions in the sky and that they might be associated with supernova explosions in distant galaxies. Short-duration GRBs that last less than 2 seconds probably originate from the merging of two neutron stars or a neutron star and a black hole in a close binary system that collapses into a single black hole. The more common long-duration GRBs are easier to study because they have a longer afterglow. Astronomers think that they originate in the collapse of a very high-mass, rapidly spinning star to a black hole or a neutron star after a supernova explosion. (Supernovae from very high-mass stars are sometimes called **hypernovae**, or Type Ib or Ic supernovae.)

Unlike regular supernovae that radiate equally in all directions, GRBs are beamed events, so most of their enormous energies are concentrated into two opposite jets of emission as illustrated in **Figure 18.20a**. In addition to the electromagnetic radiation, there are relativistic jets of cosmic rays. Astronomers have not observed any GRBs in the Milky Way: there has not been a massive supernova in the Milky Way for at least a century. But the energy of GRBs is so intense that people have wondered what might happen to Earth if one went off nearby with its radiation beamed in Earth's direction,

a possibility imagined in Figure 18.20b. A leading candidate for a future GRB in our galaxy is the massive star Eta Carinae, discussed in Chapter 17. This star is 7,500 light-years away—a neighbor, astronomically speaking. However, its rotation axis is such that is unlikely to form a GRB that beams toward Earth.

Some scientists wonder whether past supernova and GRB events could have affected the history of life on Earth. Supernova "archaeologists" may have found evidence on Earth of past supernovae. In one study, rocks deep in the Pacific Ocean were found to have amounts of a radioactive isotope of iron that is too short-lived to be left over from the formation of Earth. This iron-60 could have been deposited 2.8 million years ago on Earth after a supernova explosion. In another study, high concentrations of nitrates were found in some layers in Antarctic ice cores. Gamma radiation from supernovae can produce excess nitrogen oxides in the atmosphere, which then become converted to nitrates that are trapped in snowfall. Nitrate spikes correlate to 1006 and 1054 CE—two years when bright supernovae are known to have appeared in the Milky Way.

What about more drastic effects on Earth from a nearby supernova or a more distant but beamed GRB? Normally, Earth is protected from cosmic radiation and cosmic-ray particles by its ozone layer and its magnetic field. Cosmic-ray particles might not be a major problem if they arose very close to Earth, but the high-energy gamma-ray radiation could have a more serious effect on Earth. The excess nitrogen oxides they produce in the atmosphere can absorb sunlight, which would cool Earth. The gamma radiation could ionize Earth's atmosphere, reducing or

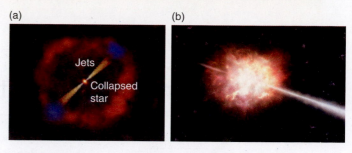

Figure 18.20 An artist's model of a gamma-ray burst. (a) Narrow beams of intense energy are sent in two opposite directions. (b) If one beam is pointed toward the observer, the GRB will appear bright.

destroying the ozone layer that protects life from ultraviolet radiation. Even a short burst of a few seconds could lead to ozone damage lasting for decades. The gamma rays could trigger a burst of solar UV radiation at Earth's surface, which could damage the DNA of phytoplankton a few hundred meters deep in the ocean, affecting their ability to photosynthesize. Phytoplankton are the base of Earth's food chain, so a drastic reduction in phytoplankton could upset the entire biosphere. It has been hypothesized that a GRB may have been responsible for the Ordovician mass extinction event 450 million years ago.

Statistically speaking, GRBs that beam to Earth may be quite rare, and some astronomers have argued that they are less likely to be produced in a Milky Way–type galaxy than in other types of galaxies. There is a lot of uncertainty in any estimate of how close the supernova or GRB must be and how often these explosive events must occur to have a serious effect on Earth. According to one estimate, there is a supernova or GRB explosion close enough to alter Earth's biosphere a few times every billion years, possibly leading to mass extinction events. In Chapter 17, we noted that the chemical elements that make up life were created in supernova explosions. The discussion here suggests that supernovae may have had some effect on the *evolution* of life on Earth as well.

Astronomers simulate the formation of a black hole.

After Neutron Star Death-Match, a Black Hole Is Born

By **IAN O'NEILL**, Discovery News

What happens when two neutron stars collide? Using a sophisticated computer simulation, NASA scientists have visualized this violent scenario in awesome degenerate-matter-crushing detail.

Neutron stars are the result of supernovae spawned by stars 8–30 times the mass of our Sun. Occasionally, however, two neutron stars may meet, becoming entangled in a deep gravitational embrace. Should this scenario play out, one of the most powerful known explosions in the universe may be sparked—a fast gamma-ray burst (GRB).

But before two neutron stars collide, what happens to their structures? What kind of insanely powerful tidal forces are at play?

In a simulation released today (May 13) by NASA Goddard Space Flight Center scientists, two neutron stars are placed a mere 11 miles apart. Keep in mind that although both neutron stars are 1.5 and 1.7 times the mass of our Sun, all of that matter is packed into a tiny sphere only *12 miles wide*. As a result, their densities and gravitational fields are immense—a teaspoonful of neutron star material would weigh as much as Mount Everest. The crushing gravitational forces ensure that atomic structures cannot be sustained, collapsing the material into a neutron degenerate state—only the structure of the neutrons themselves prevent the neutron star's gravity from collapsing it into a point, forming a black hole.

Should more mass be added to the neutron star, a mass threshold may be reached when the gravitational forces overwhelm even the neutron degenerate pressure, causing it to collapse.

As the simulation unfolds, the neutron stars' savage tidal forces rip each other to shreds, cracking open their thin crusts and shedding huge quantities of material into space. As they are so close to one another, the two neutron stars rapidly spin, merging in a fraction of a second. The progression of this simulated neutron star merger produces a ring doughnut–shaped material that forms about the genesis of the newborn black hole in its center.

Simulating these violent events is important for our understanding of not only how some black holes are created: it can develop the science behind GRBs. As an extension, the mysterious source of the heaviest elements in the universe may be found in these events where the rapid cohesion of neutron stars—and the resulting explosions—could forge rapid-neutron capture (or "r-process") elements.

The majority of r-process elements—heavier than iron—found in the universe are thought to be generated inside core-collapse supernovae. However, there's a growing body of evidence that suggests merging neutron stars may be a fertile environment for the largest atomic nuclei to form.

So neutron star mergers are not only a fascinating field of astrophysical curiosity, they may also be the driving production mechanism responsible for the heaviest elements and complex chemistry throughout the cosmos.

ARTICLES   QUESTIONS

1. Explain why the merger of two neutron stars will make a black hole.
2. How do the two stars get so close together?
3. What is space like near the merging neutron stars?
4. R-process elements form from nucleosynthesis after rapid neutron capture. How would capturing a neutron lead to nucleosynthesis?
5. Watch the video at http://svs.gsfc.nasa.gov/vis/a010000/a011500/a011530/index.html. Why do such mergers have to be studied on a computer?

"After neutron star death-match, a black hole is born," by Ian O'Neill. Discovery News, May 13, 2014. Reprinted by permission of Discovery Access.

Summary

The highest-mass stars leave behind black holes. In the environment surrounding black holes, relativistic effects become important. A black hole's mass determines its Schwarzschild radius: the boundary from which light cannot escape. Gamma-ray bursts are beamed high-energy explosions that result from the merger of two compact objects or the rapid collapse of a high-mass star to a black hole. The radiation from these bursts could affect life on Earth.

LG 1 **Describe how the motion of the observer affects the observed velocity of objects.** Even at relatively low relative speeds, the motion of the observer can affect the direction of the measured velocity, as in the aberration of starlight. At higher speeds, the magnitude of the measured velocity is also affected. The speed of light in a vacuum, c, is the ultimate speed limit. Observers in all inertial reference frames will measure the same speed of light.

LG 2 **Discuss the observable consequences of the relationship between space and time.** Special relativity connects space and time into four-dimensional spacetime. How the spacetime distance between events is divided between space and time depends on the observer's motion. As observers approach the speed of light, they observe moving clocks appearing to run more slowly than their own clocks, moving objects to be contracted in length, and moving objects to behave as if they were more massive.

LG 3 **Recognize that gravity is a consequence of the way mass distorts the very shape of spacetime.** Inertial mass and gravitational mass are the same, leading to the principle of equivalence, in which acceleration cannot be distinguished from gravity if the acceleration is small enough. In general relativity, mass warps the fabric of spacetime so that objects move on the shortest path in this warped geometry. Gravity is a consequence of the spacetime warping. Time runs more slowly near massive objects, and radiation from any light source near a black hole's event horizon is redshifted.

LG 4 **Explain why the most massive stars end as black holes, and describe the key properties of these stellar black holes.** There is an upper limit to the mass for both white dwarfs and neutron stars. Dense stellar remnants more massive than about 3 M_{Sun} collapse to form black holes. The supernova explosion that ends the life of a massive star leaves behind a neutron star or a black hole. The mathematical singularity at the center of a black hole is still a mystery to science. However, observational evidence for black holes is very strong: scientists have identified many objects that have strong gravity but are too small to be normal matter. Black holes may, after a very, very long time, be destroyed by evaporation through Hawking radiation.

? UNANSWERED QUESTIONS

- What happens to the information that falls into a black hole? We said earlier that a black hole is characterized by only three properties: mass, angular momentum, and electric charge. Where did all the other information go? To a distant observer, it takes an infinitely long time for material to fall into a black hole, so although the observer sees less and less radiation from the material, the properties of the material seem to be the same for all time. But from the perspective of the infalling material, it takes a finite time to cross the event horizon, so information other than mass, angular momentum, and charge can no longer be shared with the outside world.

- Do wormholes exist in spacetime, connecting one region with another, perhaps through black holes? Wormholes are a mathematical solution to the equations of general relativity. The idea is that when something goes into a black hole, it travels through a wormhole and emerges in a different part of the universe. In this way a wormhole acts as a shortcut through spacetime. In science fiction, wormholes are a popular means of traveling large distances by exploiting the strange geometry of spacetime. But many scientists doubt that wormholes can exist in nature, and even if they do exist, strong tidal forces would pull apart anything that falls into a black hole before it emerged from a wormhole.

Questions and Problems

Test Your Understanding

1. Rank the following in terms of the mass of the star that produces each from least to most massive.
   a. neutron star
   b. black hole
   c. white dwarf

2. A car approaches you at 50 km/h. A fly inside the car is flying toward the back of the car at 7 km/h. From your point of view by the side of the road, the fly is moving at _____ km/h.
   a. 7
   b. 28.5
   c. 43
   d. 57

3. A car approaches you at 50 km/h. The driver turns on the headlights. From your point of view, the light from the headlights is moving at
   a. $c + 50$ km/h.
   b. $c - 50$ km/h.
   c. $(c + 50$ km/h$)/2$.
   d. c.

4. Imagine that you are on a spaceship. A second spaceship rockets past yours at $0.5c$. You start a stopwatch and stop it 10 seconds later. For an astronaut in the other spaceship, the number of seconds that have ticked by during the 10 seconds on your stopwatch is
   a. more than 10 seconds.
   b. equal to 10 seconds.
   c. less than 10 seconds.

5. The International Space Station flies overhead. Using a telescope, you take a picture and measure its length to be _____ than its length as it would be measured if it were sitting on the ground.
   a. much greater
   b. slightly greater
   c. slightly less
   d. much less

6. Astronauts in the International Space Station
   a. have no mass.
   b. have no energy.
   c. are outside of Earth's gravitational field.
   d. are in free fall.

7. Einstein's formulation of gravity
   a. is approximately equal to Newton's universal law of gravitation for small gravitation fields.
   b. is always used to calculate gravitational effects in modern times.
   c. explained why Newton's universal law of gravitation describes the motions of masses.
   d. both a and c

8. As the mass of a black hole increases, its Schwarzschild radius
   a. increases as the square of the mass.
   b. increases proportionately.
   c. stays the same.
   d. decreases proportionately.
   e. decreases as the square of the mass.

9. If a neutron star is more than 3 times as massive as the Sun, it collapses because
   a. the force of electron degeneracy is stronger than gravity.
   b. gravity overpowers the force of electron degeneracy.
   c. gravity overpowers the force of neutron degeneracy.
   d. the force of neutron degeneracy is stronger than gravity.

10. Relative motion between two objects is apparent
    a. even at everyday speeds, such as 10 km/h.
    b. only at very large speeds, such as $0.8c$.
    c. only near very large masses.
    d. only when both objects are in the same reference frame.

11. If a spaceship approaches you at $0.5c$, and a light on the spaceship is turned on pointing in your direction, how fast will the light be traveling when it reaches you?
    a. $1.5c$
    b. between $1.0c$ and $1.5c$
    c. exactly c
    d. between $0.5c$ and $1.0c$

12. Imagine two protons traveling past each other at a distance d, with relative speed $0.9c$. Compared with two *stationary* protons a distance d apart, the gravitational force between these two protons will be
    a. smaller, because they interact for less time.
    b. smaller, because the moving proton acts as if it has less mass.
    c. the same, because the particles have the same mass.
    d. larger, because the moving proton acts as if it has more mass.

13. If two spaceships approach each other, each traveling at $0.5c$ relative to an outside observer, spaceship 1 will measure spaceship 2 to be traveling
    a. much faster than c.
    b. slightly faster than c.
    c. at c.
    d. more slowly than c.

14. The strongest evidence for black holes comes from rapidly flickering X-ray sources. This observation indicates that
    a. X-rays are not light, because even light cannot escape from a black hole.
    b. the X-rays are coming from a very small source—so small that it could be a black hole.
    c. black holes are (at least sometimes) surrounded by hot gas.
    d. black holes have a very high temperature.

15. The current model of long-duration GRBs includes jets from the collapsed star. You have seen jets like this before, when studying (choose all that apply)
    a. giant planets.
    b. white dwarfs.
    c. pulsars.
    d. star formation.
    e. supernovae.
    f. planetary nebulae.

Thinking about the Concepts

16. An astronomer sees a redshift in the spectrum of an object. Without any other information, can she determine whether this is an extremely dense object (exhibiting gravitational redshift) or one that is receding from her (exhibiting Doppler redshift)? Explain your answer.

17. Imagine you are traveling in a spacecraft at 0.9999999c. You point your laser pointer out the back window of the space-craft. At what speed does the light from the laser pointer travel away from the spacecraft? What speed would be observed by someone on a planet traveling at 0.000001c?

18. Einstein's special theory of relativity says that no object can travel faster than, or even at, the speed of light. Recall that light is both an electromagnetic wave and a particle called a photon. If it acts as a particle, how can a photon travel at the speed of light?

19. Twin A takes a long trip in a spacecraft and returns younger than twin B, who stayed behind. Could twin A ever return before twin B was born? Explain.

20. In one frame of reference, event A occurs before event B. Is it possible, in another frame of reference, for the two events to be reversed, so that B occurs before A? Explain.

21. Imagine you are watching someone whizzing by at very high speed in a spacecraft. Will that person's pulse rate appear to be extremely fast or extremely slow?

22. Suppose you had a density meter that could instantly measure the density of an object. You point the meter at a person in a spacecraft zipping by at very high speed. Is that person's mass density larger than an average person's or smaller? Explain.

23. Imagine a future astronaut traveling in a spaceship at 0.866 times the speed of light. Special relativity says that the length of the spaceship along the direction of flight is only half of what it was when it was at rest on Earth. The astronaut checks this prediction with a meter stick that he brought with him. Will his measurement confirm the contracted length of his spaceship? Explain your answer.

24. Looking into a speeding spaceship, you observe that the travelers are playing soccer with a perfectly round soccer ball. What is the shape of the ball according to observers on the spacecraft?

25. You observe a meter stick traveling past you at 0.9999c. You measure the meter stick to be 1 meter long. How is the meter stick oriented relative to you?

26. Suppose astronomers discover a 3-M_{Sun} black hole located a few light-years from Earth. Should they be concerned that its tremendous gravitational pull will lead to Earth's untimely demise?

27. If you could watch a star falling into a black hole, how would the color of the star change as it approached the event horizon?

28. Why don't people detect the effects of special and general relativity in their everyday lives here on Earth?

29. Many movies and television programs (like *Star Wars*, *Star Trek*, and *Battlestar Galactica*) are premised on faster-than-light travel. How likely is it that such technology will be developed in the near future?

30. How could a gamma-ray burst in our galaxy potentially affect life on Earth?

Applying the Concepts

31. Study Figure 18.1b. Does the angle of the rain falling outside the car depend on the speed of the car? Knowing only this angle and the information on your speedometer, how could you determine the speed of the falling rain?

32. Compare Figure 18.1 with Figure 18.2. If you knew the speed of Earth in its orbit from a prior experiment, how could you determine the speed of light from the angle of the aberration of starlight?

33. According to Einstein, mass and energy are equivalent. So, which weighs more on Earth: a cup of hot coffee or a cup of iced coffee? Why? Do you think the difference is measurable?

34. As explained in the Process of Science Figure, a new theory should contain the old theory within it. Study Figure 18.5, which compares two imaginary rocket ships experiencing the same acceleration, g.
    a. Approximately how fast (as a fraction of the speed of light) are the two spaceships going when the effects of relativity begin to be significant?
    b. Convert this speed to kilometers per hour. How does this speed compare to the speeds at which you usually travel? Why do you not usually see relativistic effects in your life?

35. Figure 18.6 shows how the Lorentz factor depends on speed. At about what speed (in terms of *c*) does the Lorentz factor begin to differ noticeably from 1? What happens to the Lorentz factor as the speed of an object approaches the speed of light?

36. The perihelion of Mercury advances 2 degrees per century. How many arcseconds does the perihelion advance in a year? (Recall that there are 60 arcseconds in an arcminute and 60 arcminutes in a degree.) Is it possible to measure Mercury's position well enough to measure the advance of perihelion in 1 year?

37. Study Figure 18.12. If the Sun were twice as massive, would the distance between the apparent positions of stars 1 and 2 increase or decrease during an eclipse?

38. Follow Working It Out 18.1 to find out how much younger you would be than your twin if you made the journey described there at 0.5*c*.

39. Use Working It Out 18.1 to predict how much younger you would be than your twin if you traveled at 0.999*c* instead of 0.5*c* as in question 38. Then calculate the difference in ages and compare the calculated result to your prediction.

40. What is the Schwarzschild radius of a black hole that has a mass equal to the average mass of a person (~70 kilograms)?

41. What is the mass of a black hole with a Schwarzschild radius of 1.5 km?

42. The Moon has a mass equal to $3.7 \times 10^{-8}\ M_{\text{Sun}}$. Suppose the Moon suddenly collapsed into a black hole.
    a. What would be the Schwarzschild radius of the black-hole Moon?
    b. What effect would this collapse have on tides raised by the Moon on Earth? Explain.
    c. Do you think this event would generate gravitational waves? Explain.

43. If a spaceship approaching Earth at 0.9 times the speed of light shines a laser beam at Earth, how fast will the photons in the beam be moving when they arrive at Earth?

44. Suppose you discover signals from an alien civilization coming from a star that is 25 light-years away, and you go to visit it using the spaceship described in the discussion of the twin paradox in Working It Out 18.1.
    a. How long will it take you to reach that planet, according to your clock? According to a clock on Earth? According to the aliens on the other planet?
    b. How likely is it that someone you know will be here to greet you when you return to Earth?

45. Working It Out 18.2 relates the mass of a binary pair to the period and the size of the orbit. Suppose that a spaceship orbited a black hole at a distance of 1 AU, with a period of 0.5 year. What assumptions could you make that would allow you to calculate the mass of the black hole from this information? Make those assumptions, and calculate the mass.

USING THE WEB

46. Go to the "Through Einstein's Eyes" website (http://anu.edu.au/Physics/Savage/TEE/site/tee/home.html), click on "Start Here," and "Take the tour" and "Movie explained." Take the ride on the "relativistic roller coaster." Why do colors look different on the relativistic roller coaster? Click on "Continue tour" and view the cube, tram, and desert road. How do things look different? Why do you get rainbow when driving down the desert at high gamma? Continue for the Solar System tours (or return to the home page). What do you see when you approach the Sun or a planet at a relativistic speed?

47. NASA missions:
    a. Go to NASA's Swift Gamma-Ray Burst Mission website (http://swift.gsfc.nasa.gov/). Locate a recent result related to supernovae, gamma-ray bursts, or stellar black holes. Why would two merging neutron stars likely form a black hole?
    b. NASA's Fermi Gamma-ray Space Telescope (http://www.nasa.gov/content/fermi-gamma-ray-space-telescope and http://fermi.gsfc.nasa.gov) is exploring the gamma-ray universe. What objectives of this mission relate to the study of black holes? What is a recent news story related to black holes?

48. Go to the LIGO website (http://ligo.org/science.php) and read about gravitational waves. Click on "Sources of Gravitational Waves" and listen to the example. What are the differences among the four listed sources of gravitational waves? Click on "Advanced LIGO." What's new with the project?

49. The newest NASA mission to study black holes, gamma-ray bursts, and neutron stars is named NuSTAR (Nuclear Spectroscopic Telescopic Array). Go to the NuSTAR website (http://www.nustar.caltech.edu/news). What wavelengths and energies does this telescope observe? What has been observed? What new science has been learned?

50. Go to the "Inside Black Holes" website (http://jila.colorado.edu/~ajsh/insidebh), enter, and click on "Schwarzschild." Work your way down the page, watching the videos. What does it look like when you go into a black hole? Why is there gravitational lensing when Earth is in orbit? What happens when you fall through the horizon: is everything black? Click on "Reissner-Nordström" to see an electrically charged black hole. What is a wormhole? Why is there a warning at the top of the page? Click on "4D perspective." What does it look like if you move toward the Sun at the speed of light?

smartwork

If your instructor assigns homework in SmartWork, access your assignments at digital.wwnorton.com/astro5.

Because it's not possible to go grab a black hole and bring it into the lab, and because Earth has never actually been close to one, astronomers can only conduct thought experiments to explore the properties of black holes. Following are a few thought experiments to help you think about what's happening near and around a black hole.

Imagine a big rubber sheet. It is very stiff and not easily stretched, but it does have some "give" to it. At the moment, it is perfectly flat. Imagine rolling some golf balls across it.

1 Describe the path of the golf balls across the sheet.

_____

_____

Now imagine putting a bowling ball (very much heavier than a golf ball) in the middle of the sheet, so that it makes a big, slope-sided pit. Roll some more golf balls.

2 What happens to the path of the golf balls when they are very far from the bowling ball?

_____

_____

3 What happens to the path of the golf balls when they come just inside the edge of the dip?

_____

_____

4 What happens to the path of the golf balls when they go directly toward the bowling ball?

_____

_____

5 How do each of the three cases in questions 2–4 change if the golf balls are moving very, very fast? What if they are moving very slowly?

_____

_____

6 What happens to the depth and width of the pit as the golf balls fall into the center near the bowling ball? (Imagine putting *lots* of golf balls in.)

_____

_____

All of the preceding thought experiments relate to ordinary stuff. Stars, people, planets—everything interacts in this way because of gravity. With black holes, things are a bit different. In this case, it is more accurate to think of the bowling ball as a hole in the sheet that pulls it down, rather than as an object that sits on it. But the bowling ball still affects the sheet in the same way. The hole is a good analogy for the event horizon of a black hole. Objects outside the event horizon will know that the black hole is there, because the sheet is sloping, but they won't be captured unless they come within the event horizon. Think about light for a moment as though it were, say, grains of sand rolling across the sheet.

7 What happens to the light as it passes far from the pit? What happens if it reaches the hole?

_____

_____

Now suppose you roll another bowling ball across the sheet.

8 What happens to the sheet when the second bowling ball falls in after the first? Would this change affect your golf balls and grains of sand? How? What happens to the hole? What happens to the size of the pit?

_____

_____

_____

_____

_____

None of these thought experiments take into account relativistic effects (length contraction and time dilation). Imagine for a moment that you are traveling close to the black hole.

9 Look out into the galaxy and describe what you see. Consider the lifetimes of stars, the distances between them, their motions in your sky, and how they die. Add anything else that occurs to you.

_____

_____

_____

_____

19 Galaxies

It has been less than a century since astronomers realized that the universe is filled with huge collections of stars, gas, and dust called galaxies. Just as stars vary in their mass or their stage of evolution, galaxies come in many forms. This chapter begins our discussion of galaxies with a survey of the types of galaxies and their basic properties to understand better the differences among them. The next chapter looks in detail at our galaxy, the Milky Way, and in subsequent chapters we will look at the evolution of galaxies and of the universe itself.

LEARNING GOALS

By the conclusion of this chapter, you should be able to:

LG 1 Determine a galaxy's type from its appearance, and describe the motions of its stars.

LG 2 Explain the distance ladder and how distances to galaxies are measured.

LG 3 Describe the evidence suggesting that galaxies are composed mostly of dark matter.

LG 4 Discuss the evidence indicating that most—perhaps all—large galaxies have supermassive black holes at their centers.

The large, barred spiral galaxy NGC 1300 is about 20 million parsecs away. ▶ ▶ ▶

How do galaxies get their shapes?

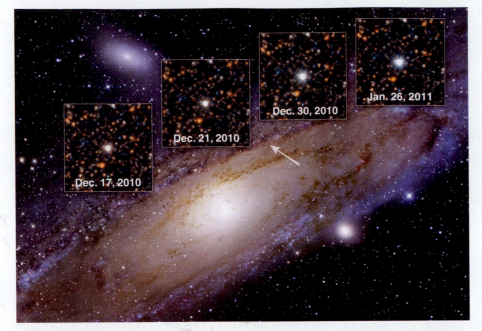

Figure 19.1 The Andromeda Galaxy, the nearest large galactic neighbor to the Milky Way, is about 2.5 million light-years (780,000 pc) away. The arrow points to the Cepheid variable star V1, a standard candle that Hubble used to estimate the distance to Andromeda. The insets show V1's variability. Hubble's measurement provided the first observational evidence of the vastness of the universe.

19.1 Galaxies Come in Different Shapes and Sizes

A **galaxy** is a gravitationally bound collection of dust, gas, and a million to hundreds of billions of stars. The universe contains more galaxies than there are stars in the Milky Way Galaxy. Most of these galaxies are located at such astonishing distances that they appear too small and faint to detect with any but the most powerful telescopes. In this section, we will discuss how astronomers concluded that galaxies were separate from our Milky Way and will differentiate the various types of galaxies.

The Discovery of Galaxies in the 20th Century

Observers have long known that the sky contains faint, misty patches of light. These objects were originally called nebulae (singular: *nebula*) because of their nebulous (fuzzy) appearance. In 1784, astronomer Charles Messier (1730–1817) published a catalog of 103 nebulous objects. Twenty years later, because of the observations by the astronomers William Herschel and his sister Caroline Herschel, that number jumped to 2,500. Although some of the nebulae looked diffuse and amorphous, most were round or elliptical or resembled spiraling whirlpools. These distinctions were the basis for the original three categories—diffuse, elliptical, and spiral—used to classify nebulae.

Over the next 140 years, it was suggested that spiral nebulae might be relatively nearby planetary systems in various stages of formation. Alternatively, the influential 18th century philosopher Immanuel Kant (1724–1804) speculated that spiral nebulae were instead "island universes"—separate from the Milky

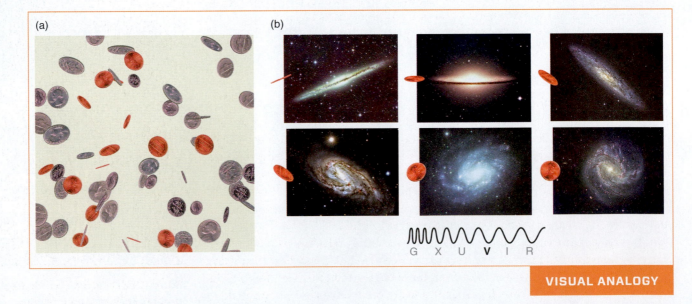

VISUAL ANALOGY

Figure 19.2 (a) A handful of coins thrown in the air provides an analogy for the difficulties in identifying the shapes of certain types of galaxies. Like the coins in this picture, galaxies are seen in various orientations—some face on, some edge on, and most somewhere in between. (b) These disk-shaped galaxies are seen from various perspectives or angles, corresponding to the range of perspectives for coins.

Way Galaxy. The presence of interstellar dust complicated early attempts to understand the size of the Milky Way. Early astronomers did not know of the existence or consequences of the dust that blocks the passage of visible light through the Milky Way. Unable to see past this obscuring shroud, they concluded that the Milky Way is a system of stars some 1,800 parsecs (pc) across (recall from Chapter 13 that 1 parsec = 3.26 light-years). In the beginning of the 20th century, astronomer Harlow Shapley (1885–1972) used observations of globular clusters to estimate that instead, the Milky Way is more than 50 times larger than the earlier estimates—300,000 light-years, or 92,000 pc, in size.

Shapley thought his far larger estimate for the size of the Milky Way meant that it was big enough to encompass everything in the universe, and therefore he thought that the spiral and elliptical nebulae were inside the Milky Way. Astronomer Heber D. Curtis (1872–1942) preferred the earlier, smaller model of the Milky Way. He also favored the idea that the spiral nebulae were in fact galaxies separate from the Milky Way and that therefore the whole universe was larger than the Milky Way. In 1920, Shapley and Curtis met in Washington, D.C., to debate publicly their interpretations of the nature of spiral nebulae. Historians call this meeting astronomy's *Great Debate*. While this debate did not resolve the issue at the time, it set the stage and gave direction to the subsequent work of Edwin P. Hubble (1889–1953), whose discoveries fundamentally changed the modern understanding of the universe.

Using the newly finished 100-inch telescope on Mount Wilson, high above the then small city of Los Angeles, Hubble was able to find some variable stars in the spiral nebula Andromeda, as shown in **Figure 19.1**. He recognized that these stars were very similar to but appeared fainter than other known Cepheid variable stars. Using the period-luminosity relation for Cepheid variable stars discussed in Chapter 17, Hubble turned his observations of these stars into measurements of the distances to these objects. The results showed that the distances to these nebulae are far greater than Shapley's size of the Milky Way. Kant was correct: the Milky Way is one of many island universes. Most diffuse nebulae are clouds of gas and dust near Earth in the Milky Way Galaxy, but what Messier and others thought of as elliptical and spiral nebulae are instead galaxies similar in size to the Milky Way but located at truly immense distances.

Types of Galaxies

Imagine taking a handful of coins and throwing them in the air, as shown in **Figure 19.2a**. You know that all of these objects are flat and circular. When you look at the objects falling through the air, however, they do not appear all the same. Some coins appear face on and look circular. Some coins are seen edge on and look like thin lines. Most coins are seen from an angle between these two extremes and appear with various degrees of ellipticity, or flattening. Even if this image of many coins was the only information you had, you could use it to figure out the three-dimensional shape of a coin—flat and circular.

Astronomers use a similar method to discover the true three-dimensional shapes of galaxies. Figure 19.2b shows a set of galaxies seen from various viewing angles, from face on to edge on. You can infer from these images that, just like the coins in Figure 19.2a, some galaxies are disk-shaped and are randomly oriented on the sky.

The classifications for galaxies used today date back to the 1930s, when Edwin Hubble sorted the different shapes into categories like those shown in **Figure 19.3**. Hubble grouped all galaxies according to appearance and positioned them on a diagram that resembles the tuning fork used in the tuning of a musical in-

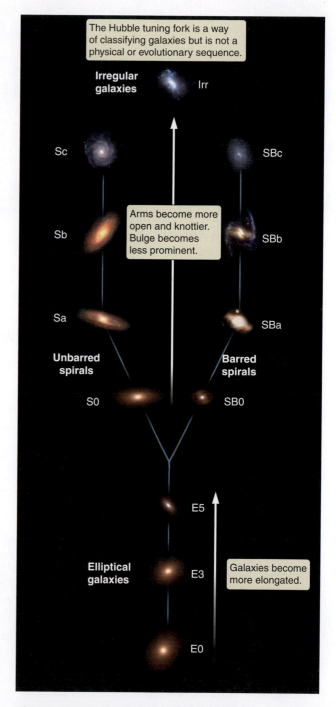

Figure 19.3 This tuning fork diagram illustrates Edwin Hubble's scheme for classifying elliptical (E), spiral (S, SB), and S0 galaxies according to their appearance only. Irregular (Irr) galaxies were not placed on the tuning fork. Hubble's scheme does not indicate an evolutionary sequence of galaxies.

strument. Galaxies come in three basic types: spirals, ellipticals, and irregulars—the latter including all other shapes. Originally, Hubble thought that his tuning fork diagram might indicate an evolutionary sequence for galaxies similar to what the Hertzsprung-Russell (H-R) diagram had done for stars. Even though it is *not* an evolutionary sequence, Hubble's classification scheme is still used for sorting galaxy types by their appearance in visible light.

At the bottom of Figure 19.3 are objects that are generally elliptical in three dimensions. These **elliptical galaxies** (labeled "E" on the diagram) can have either spherical or ellipsoidal shapes, and they show little evidence of the flat, round disk seen in other types of galaxies. They have numbered subtypes ranging from nearly spherical (E0) to quite flattened (E7). As with the coins tossed in the air, the appearance of an elliptical galaxy in the sky does not necessarily tell us its true shape. For example, a galaxy might actually be shaped like a rugby ball (which has rounded ends), but if viewed end on, it looks round like a soccer ball instead.

Spiral galaxies, labeled "S" on the left side of Figure 19.3, are characterized by spiral arms that lie in a flattened, rotating disk. In addition to disks and arms, a spiral galaxy has a central **bulge**, which extends above and below the disk. Hubble noticed that the bulges of about half of the spiral galaxies are bar-shaped: these galaxies are called **barred spirals**, labeled "SB" on the right side of Figure 19.3. Both spirals and barred spirals are subdivided into types a, b, and c according to the prominence of the central bulge and how tightly the spiral arms are wound. For example, Sa and SBa galaxies have the largest bulges and display tightly wound and smooth spiral arms. Sc and SBc galaxies have small central bulges and more loosely woven spiral arms that are often very knotty in appearance. The Milky Way Galaxy is a barred spiral (SBbc).

The distinction between spiral and elliptical galaxies is not always clear. Some galaxies, known as **S0 galaxies**, are a combination of the two types, having stellar disks but no spiral arms, so that the disk is smooth in appearance, like an elliptical galaxy. Hubble differentiated S0 galaxies as either barred (SB0) or unbarred (S0). Modern telescopic observations have revealed that many, if not most, elliptical galaxies contain small rotating disks at their centers, and both elliptical and S0 galaxies have little star formations. Galaxies that fall into none of these classes are **irregular galaxies** (labeled "Irr" in Figure 19.3). As their name implies, irregular galaxies are often without symmetry in shape or structure. About 25 percent of galaxies are irregular, and now astronomers think that most of them once were spirals or ellipticals that became distorted by the gravity of another galaxy. **Table 19.1** summarizes the criteria that Hubble used to classify galaxies.

Astronomy in Action: Galaxy Shapes and Orientation

TABLE 19.1	The Hubble Classification of Galaxies

A Classification Scheme Based on the Properties of Galaxies

Category	Criteria	Abbreviation	Range of Features		
Ellipticals	Mostly bulge Old, red stellar population Smooth-appearing	E0 ↕ E7	More spherical ↕ More elongated		
S0 (unbarred/barred)	Bulge and disk with no arms and with mostly old, red stars	S0/SB0	Smooth disk and bulge		
Spirals (unbarred/barred)	Bulge and disk with arms Bulge has old, red stars Disk has both old, red stars and young, blue stars Spirals (S) have roundish bulges Barred spirals (SB) have elongated or barred bulges	Sa/SBa Sb/SBb Sc/SBc	More bulge ↕ Little bulge	Tightly wound arms ↕ Open arms	Smooth arms ↕ Knotty arms
Irregulars	No arms, no bulge Some old stars, but mostly young stars, gas and dust, giving a knotty appearance	Irr			

Stellar Motions and Galaxy Shape

Stellar motions determine galaxy shapes. A galaxy is not a solid object like a coin, but a collection of stars, gas, and dust. In an elliptical galaxy, stars move in all possible directions, following orbits with a wide range

Figure 19.4 Elliptical galaxies take their shape from the orbits of the stars they contain. The colored lines superimposed on the galaxy represent the complex orbits of its stars

of shapes, as shown in **Figure 19.4**. These orbits are more complex than the orbits of planets about a star because the gravitational field within an elliptical galaxy does not come from a single central object. Taken together, all of these stellar orbits give an elliptical galaxy its shape.

Orbital speeds are also a factor. The faster the stars are moving, the more spread out the galaxy is. If the stars in an elliptical galaxy are moving in truly random directions, the galaxy will have a spherical shape. However, if stars are more likely to have certain directions of motion than others at each location, the galaxy will be more spread out in that direction, giving it an elliptical shape.

The orbits of stars in the disks of spiral galaxies are quite different from those of stars in elliptical galaxies. The components of a barred spiral galaxy are shown in **Figure 19.5**. The defining feature of a spiral galaxy is that it has a flattened, rotating disk. Like the planets of the Solar System, most of the stars in the disk of a spiral galaxy follow nearly circular orbits and travel in the same direction around a concentration of mass at the center of the galaxy. But the stellar orbits in a spiral galaxy's central bulge are quite different from those in the galaxy's disk. As with elliptical galaxies, the gravitational field within the bulge does not come from a single object, and stars therefore follow random orbits. The bulges of unbarred spiral galaxies are thus roughly spherical in shape.

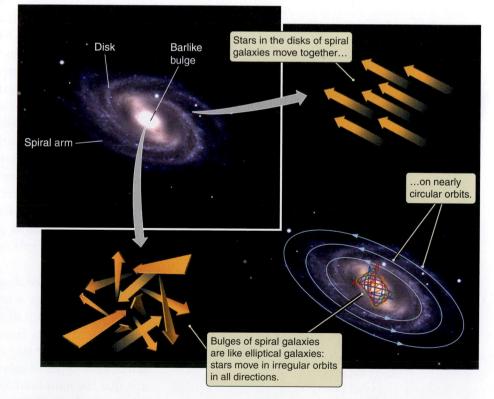

Figure 19.5 The components of a barred spiral galaxy include a barlike bulge, a disk, and spiral arms. The orbits of stars in the rotating disk are different than the orbits of stars in the elliptically shaped bulge.

CHECK YOUR UNDERSTANDING 19.1

Galaxies are classified according to: (a) mass; (b) color; (c) density; (d) shape.

G X U V I R

Figure 19.6 This Hubble Space Telescope (HST) image shows the nearly edge-on spiral galaxy M104 (the Sombrero Galaxy; type Sa). The dust in the plane is seen as a dark, obscuring band in the midplane of the galaxy. Note the bright halo made up of stars and globular clusters. Compare this image with Figure 15.1, which shows the dust in the plane of the Milky Way.

Other Differences among Galaxies

In addition to the differences in their stellar orbits, there are other important distinctions between spiral and elliptical galaxies. These distinctions carry information about the way they have evolved in the past and the way they will evolve in the future.

Gas and Dust   Most spiral galaxies contain large amounts of dust and cold, dense molecular gas concentrated in the midplanes of their disks. Just as the dust in the disk of the Milky Way can be seen on a clear summer night as a dark band slicing the galaxy in two (see Figure 15.1), the dust in an edge-on spiral galaxy appears as a dark, obscuring band running down the midplane of the disk (**Figure 19.6**). The cold molecular gas that accompanies the dust can also be seen in radio observations of spiral galaxies. In contrast, giant elliptical galaxies contain large amounts of very hot gas that astronomers see primarily by observing the X-rays that the gas emits.

The difference in shape between elliptical and spiral galaxies offers some insight into why the gas in giant ellipticals is hot, while in spirals it is cold and dense. Conservation of angular momentum causes cold gas to settle into the disk of a spiral galaxy, just as gas settles into a disk around a forming star. In contrast, elliptical galaxies do not have a net rotation, so the gas does not settle into a disk. The only place in an elliptical galaxy where cold gas could collect is at the center. However, the density of stars in elliptical galaxies is so high that evolving stars and Type Ia supernovae continually reheat this gas, preventing most of it from cooling off and forming cold interstellar clouds.

Color   The colors of spiral and elliptical galaxies reveal a great deal about their star formation histories. Recall from Chapter 15 that stars form from dense clouds of cold molecular gas. Because the gas seen in elliptical galaxies is very hot, astronomers know that active star formation is not taking place in those galaxies today. The reddish colors of elliptical and S0 galaxies confirm that little or no star formation has occurred there for quite some time. The stars in these galaxies are an older population of lower-mass stars. In contrast, the bluish colors of the disks of spiral galaxies indicate that massive, young, hot stars are forming in the cold molecular clouds contained within the disk. Even though most of the stars in a spiral disk are old, the massive, young stars are so luminous that their blue light dominates. Star formation in most irregular galaxies is like that in spiral galaxies. Some irregular galaxies form stars at prodigious rates, given their relatively small sizes. Irregular and disk galaxies that undergo intense bursts of star formation are called starburst galaxies.

Luminosity   The relationship between luminosity and radius among the different types of galaxies is not straightforward. Galaxies range in luminosity from tens of thousands up to a trillion (a million million) solar luminosities (10^4 to 10^{12} L_{Sun}) and in size from a few hundred to hundreds of thousands of parsecs. There is no distinct size difference between elliptical and spiral galaxies: about half of both types of galaxies fall within a similar range of sizes. Although it is true that the most luminous elliptical galaxies are more luminous than the most luminous spiral galaxies, there is considerable overlap in the range of luminosities among all Hubble types.

(a)

(b)

Figure 19.7 The mass or size of a spiral galaxy does not determine its appearance. Even though these galaxies appear to be similar in size and luminosity, the larger galaxy (a) is 4 times more distant and 10 times more luminous than the smaller galaxy (b).

Mass Mass is the single most important parameter in determining the properties and evolution of a star. In contrast, differences in mass and size do not lead to obvious differences among galaxies. Only subtle differences in color and concentration exist between large and small galaxies, making it difficult to distinguish which are large and which are small. Even when a larger, more distant spiral galaxy (**Figure 19.7a**) is seen next to a smaller, nearby spiral galaxy (Figure 19.7b), it can be hard to tell which is which by appearance alone. Galaxies that have relatively low luminosity (less than 1 billion L_{Sun}) are called dwarf galaxies, and those that are more than 1 billion L_{Sun} are called giant galaxies, because the luminosity indicates the number of stars and therefore the total amount of stellar mass. Only elliptical and irregular galaxies come in both types: among spiral and S0 galaxies, there are only giants. It is relatively easy to tell the difference between a dwarf elliptical galaxy and a giant elliptical galaxy (**Figure 19.8**). Giant elliptical galaxies have a much higher density of stars, and these are more centrally concentrated than stars in dwarf ellipticals.

(a) Dwarf elliptical galaxy

(b) Giant elliptical galaxy

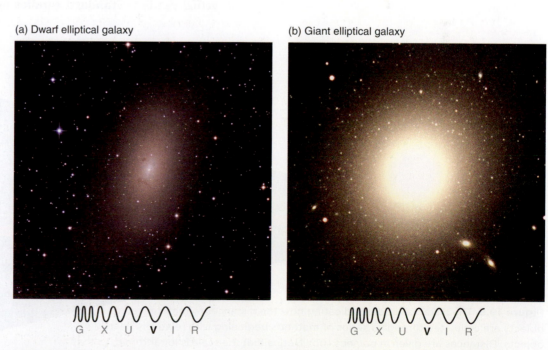

CHECK YOUR UNDERSTANDING 19.2

Currently, star formation rates are highest in (a) elliptical galaxies; (b) S0 galaxies; (c) spiral galaxies; (d) all of the above

Figure 19.8 Dwarf elliptical galaxies (a) differ in appearance from giant elliptical galaxies (b). Stars in giant elliptical galaxies are more centrally concentrated than those in dwarf elliptical galaxies.

19.2 Astronomers Use Several Methods to Find Distances to Galaxies

To determine the distances to galaxies, astronomers started with the closest objects and looked for patterns that would help them to find distances to the farthest objects—just as they did for stars. Distances are measured in a series of different methods called the **distance ladder**, which relates distances on a variety of overlapping scales, each method building on the previous one. In this section, we will discuss these distance methods and how they led to the surprising discovery that the galaxies are moving away from us.

The Distance Ladder

 Nebraska Simulation: Spectroscopic Parallax Simulator

The distance ladder is summarized in **Figure 19.9**. In the 18th and 19th centuries, astronomers used several methods to estimate the astronomical unit (AU; the average distance from Earth to the Sun), and thus the distances to the planets. (Since the 1960s, distances within the Solar System are found using radar and signals from space probes.) Once the value of the AU was known, astronomers used trigonometric parallax, as discussed in Chapter 13, to measure distances to nearby stars and thereby to build up the H-R diagram. For more distant stars, astronomers use the spectral and luminosity classification of a star to determine its position on the H-R diagram. That position provides a star's luminosity, which in turn enables astronomers to estimate its distance by comparing its apparent brightness with its luminosity using spectroscopic parallax (as described in Chapter 13 and Appendix 7).

 Nebraska Simulation: Supernova Light Curve Fitting Explorer

Astronomers measure the distance to relatively nearby galaxies using *standard candles*, a term borrowed from an old unit of light intensity that was based on actual candles. **Standard candles** are objects that have a known luminosity,

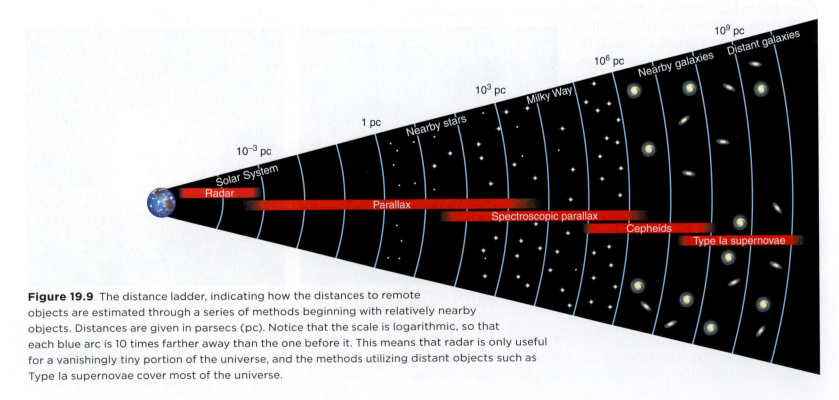

Figure 19.9 The distance ladder, indicating how the distances to remote objects are estimated through a series of methods beginning with relatively nearby objects. Distances are given in parsecs (pc). Notice that the scale is logarithmic, so that each blue arc is 10 times farther away than the one before it. This means that radar is only useful for a vanishingly tiny portion of the universe, and the methods utilizing distant objects such as Type Ia supernovae cover most of the universe.

usually because they have been observed within the Milky Way. The objects must also be bright enough to be recognizable in the distant galaxy. Astronomers assume that the luminosity of each object is the same as that for a similar object of that type in the Milky Way, and then they compare the luminosity and apparent brightness of the standard candle to find its distance.

Objects that can be used as standard candles include main-sequence O stars, globular clusters, planetary nebulae, novae, variable stars such as RR Lyraes and Cepheids, and supernovae. For example, Hubble Space Telescope (HST) observations of Cepheid variables enable astronomers to measure distances accurately to galaxies as far away as 30 million parsecs (also called 30 megaparsecs [Mpc]). Even more luminous than the Cepheids, and thus detectable at greater distances, are Type Ia supernovae.

Recall from Chapter 16 that Type Ia supernovae can occur when gas flows from an evolved star onto its white dwarf companion, pushing the white dwarf up toward the Chandrasekhar limit for the mass of an electron-degenerate object: $1.4\ M_{Sun}$. When this happens, the white dwarf burns carbon, collapses, and then explodes. At first, astronomers thought that these Type Ia supernovae all occurred in white dwarfs of just below $1.4\ M_{Sun}$. In that situation, all such explosions would occur at the same mass and have similar luminosity, with some calibration adjustment for the rate at which the brightness declines after it peaks. But now astronomers estimate that 80 percent of Type Ia supernova come from double-degenerate systems; for example, two white dwarfs that merge and then explode. The double white dwarf supernova could have up to twice the mass of a single white dwarf supernova. In addition, there may be other variations in Type Ia supernovae based on their color. To test whether they all have about the same luminosity, astronomers observe nearby Type Ia supernovae in galaxies with distances determined from other methods, such as by Cepheid variables. But more distant supernovae might have different luminosity. With a peak luminosity that can outshine a billion Suns (**Figure 19.10**), Type Ia supernovae can be seen and measured with modern telescopes at very large distances (**Working It Out 19.1**).

The motions of the stars in galaxies give rise to two "secondary" methods for estimating the distance to a galaxy. In rotating spiral galaxies, some of the light is approaching Earth and thus blueshifted, and some light is moving away from Earth and thus redshifted. These redshifts and blueshifts together broaden a spectral line in the galaxy, and the amount of broadening indicates how fast the galaxy is rotating. Astronomers can measure the broadening from the Doppler shifts of the 21-centimeter (21-cm) radio emission line of hydrogen, which tells them the speed of rotation, which then relates to the galaxy's mass by Newton's version of Kepler's third law. The more massive galaxies have more stars and are therefore more luminous. This empirical relation between the measured width of the 21-cm line and the luminosity of the spiral galaxy is called the Tully-Fisher relation. Once the luminosity of the galaxy is known, it can be compared to the galaxy's observed apparent brightness to estimate its distance. This method is thought to work out to about 100 Mpc.

Elliptical galaxies and the bulges of S0 galaxies do not rotate, so instead, astronomers look at the distribution of the surface brightness of a galaxy. Closer galaxies show more variations in the surface brightness because the distribution of stars throughout the galaxy isn't perfectly uniform. For more distant galaxies, these variations are less noticeable, and the surface brightness appears more uniform across the galaxy. This method is less precise than the Tully-Fisher method for spirals, but generally it also is thought to work out to about 100 Mpc.

(a)

G   X   U   V   I   R

(b)

SN 2011fe

G   X   U   V   I   R

Figure 19.10 Type Ia supernovae are extremely luminous standard candles. The Pinwheel Galaxy is 6.4 Mpc away, shown before (a) and after (b) Supernova 2011fe.

Let's see how astronomers use a standard candle to estimate distance. Figure 19.10 shows the Pinwheel Galaxy (M101), with a supernova that was observed in 2011. Astronomers can compare the peak observed brightness of this supernova with the peak luminosity for this type of supernova to compute the distance.

In Section 13.1, we gave the equation relating brightness, luminosity, and distance:

$$\text{Brightness} = \frac{\text{Luminosity}}{4\pi d^2}$$

Rearranging to solve for distance gives:

$$d = \sqrt{\frac{\text{Luminosity}}{4\pi \times \text{Brightness}}}$$

The maximum observed brightness of this supernova is 7.5×10^{-12} watts per square meter (W/m²). The graph in Figure 17.11 shows that the typical maximum luminosity L of a Type Ia supernova is 9.5×10^9 times the luminosity of the Sun:

$$L = 9.5 \times 10^9 \times L_{\text{Sun}}$$

$$L = 9.5 \times 10^9 \times (3.9 \times 10^{26}\,\text{W}) = 3.7 \times 10^{36}\,\text{W}$$

Thus, we can solve the equation:

$$d = \sqrt{\frac{3.7 \times 10^{36}\,\text{W}}{4\pi \times 7.5 \times 10^{-12}\,\text{W/m}^2}} = 2.0 \times 10^{23}\,\text{m}$$

To put this into megaparsecs:

$$d = \frac{2.0 \times 10^{23}\,\text{m}}{3.1 \times 10^{22}\,\text{m/Mpc}} = 6.4\,\text{Mpc}$$

The distance is 6.4 Mpc. This supernova was detected early, while it was in the earliest stages of the explosion. Because the Pinwheel Galaxy is relatively close, other standard candles can be observed in this galaxy to help calibrate the distance. Also note that 6.4 Mpc = 21 million light-years, so this supernova explosion took place 21 million years ago.

The Discovery of Hubble's Law

In the 1920s, Hubble and his coworkers were studying the properties of a large collection of galaxies. Another astronomer, Vesto Slipher (1875–1969), was obtaining spectra of these galaxies at Lowell Observatory in Flagstaff, Arizona. Slipher's galaxy spectra looked like the spectra of ensembles of stars with a bit of glowing interstellar gas mixed in. But he was surprised to find that the emission and absorption lines in the spectra of these galaxies were seldom seen at the same wavelengths as in the spectra of stars observed in the Milky Way Galaxy. The lines were almost always shifted to longer wavelengths as seen in **Figure 19.11**.

Slipher characterized most of the observed shifts in galaxy spectra as redshifts because the light from these galaxies is shifted to longer, or redder, wavelengths. Hubble interpreted Slipher's redshifts as Doppler shifts, and he concluded that almost all of the galaxies in the universe are moving away from the Milky Way. Recall from Chapter 5 that objects with larger Doppler redshifts are moving away more quickly than those with smaller redshifts. When Hubble combined the measurements of galaxy velocities with his own estimates of the distances to these galaxies, he found that distant galaxies are moving away from Earth more rapidly than are nearby galaxies. Specifically, *the velocity at which a galaxy is moving away from an observer is proportional to the distance of that galaxy*. This relationship between distance and recession velocity has become known as **Hubble's law**.

Figure 19.12 plots the measured recession velocities of galaxies against their measured distances. Because the velocity and distance are proportional to each other, the points lie along a line on the graph with a slope equal to the proportionality constant H_0, called the **Hubble constant**. Notice how well the data line up along the line. This strong correlation indicates that the universe follows Hubble's law; for example, a galaxy at a distance of 30 Mpc from Earth moves away twice as fast as a galaxy that is 15 Mpc distant. The original value for the

Nebraska Simulation: Galactic Redshift Simulator

▶❚❚ **AstroTour:** Hubble's Law

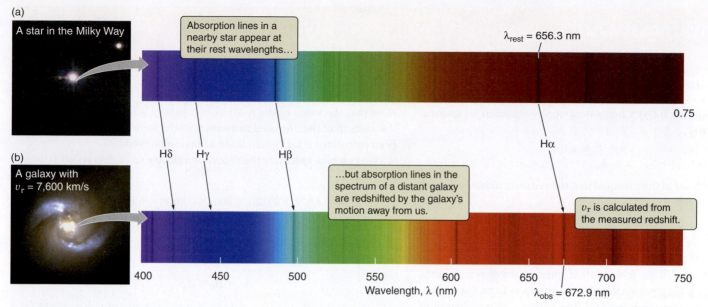

Figure 19.11 (a) The spectrum of a star in our galaxy shows absorption lines, which in this case lie at the rest wavelength. (b) A distant galaxy, shown with its spectrum at the same scale as that of the star, has lines that are redshifted to longer wavelengths. v_r is recession velocity, or radial velocity.

Hubble constant was 8 times too large, which led to inconsistencies, but the problem was resolved when astronomers realized there that there are two types of Cepheids with slightly different period-luminosity relationships. Today, astronomers have measured the Hubble constant to an accuracy of a few percent by several different methods, using observations from the Wilkinson Microwave Anisotropy Probe (*WMAP*), HST, Spitzer Space Telescope, and the Planck space observatory. The value is likely to be further refined in the years to come. In this text, we use a value of 70 km/s/Mpc as an approximation to the best current measured values of 67 to 74 km/s/Mpc, with uncertainties of 1–3 km/s/Mpc depending on the measurement.

Using this constant, Hubble's law is written as $v_r = H_0 \times d_G$, where d_G is the distance to the galaxy, and v_r is the galaxy's recession velocity (**Working It Out 19.2**). Hubble's law gives astronomers a practical tool for measuring distances to remote galaxies. Once they know the value of H_0, they can use a straightforward measurement of the redshift of a galaxy to find its distance. In other words, once H_0 is known, Hubble's law makes the once-difficult task of measuring distances in the universe relatively easy, providing astronomers with a tool to map the structure of the observable universe. We will return to Hubble's law and its implications for understanding the universe as a whole in Chapter 21.

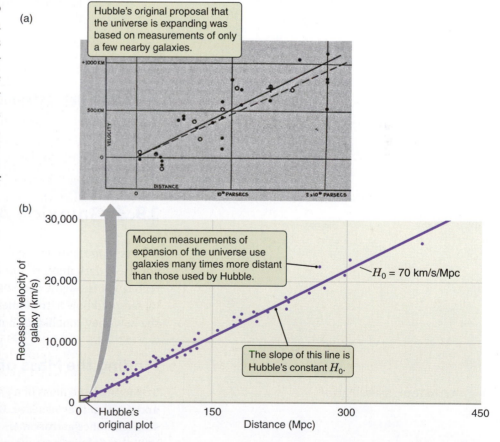

Figure 19.12 (a) Hubble's original graph shows that more distant galaxies are receding faster than less distant galaxies. (b) Modern data on galaxies many times farther away than those studied by Hubble show that recession velocity is proportional to distance.

> ## 19.2 Working It Out Redshift—Calculating the Recession Velocity and Distance of Galaxies
>
> Recall from Working It Out 5.2 that the Doppler equation for spectral lines showed that
>
> $$v_r = \frac{\lambda_{obs} - \lambda_{rest}}{\lambda_{rest}} \times c$$
>
> The fraction in front of the c is equal to z, the redshift. Substituting for the fraction, we get
>
> $$v_r = z \times c$$
>
> (Note: This correspondence requires a correction as velocities approach the speed of light.)
>
> Because spectral lines from distant galaxies have wavelengths shifted to the red, the galaxies must be moving away from Earth. Suppose astronomers observe a spectral line with a rest wavelength of 373 nanometers (nm) in the spectrum of a distant galaxy. If the observed wavelength of the spectral line is 379 nm, then its redshift (z) is
>
> $$z = \frac{\lambda_{obs} - \lambda_{rest}}{\lambda_{rest}}$$
>
> $$z = \frac{379 \, \text{nm} - 373 \, \text{nm}}{373 \, \text{nm}} = 0.0161$$
>
> Note that the value of the redshift of a galaxy is independent of the wavelength of the line used to measure it: the same result would have been calculated if a different line had been observed.
>
> We can now calculate the recession velocity from this redshift as follows:
>
> $$v_r = z \times c = 0.0161 \times 300{,}000 \, \text{km/s} = 4{,}830 \, \text{km/s}$$
>
> How far away is the distant galaxy? This is where *Hubble's law* and the *Hubble constant* apply. Hubble's law relates a galaxy's recession velocity to its distance as
>
> $$v_r = H_0 \times d_G$$
>
> where d_G is the distance to a galaxy measured in megaparsecs. Dividing through by $H_0 = 70$ km/s/Mpc yields
>
> $$d_G = \frac{v_r}{H_0} = \frac{4{,}830 \, \text{km/s}}{70 \, \text{km/s/Mpc}} = 69 \, \text{Mpc}$$
>
> From a measurement of the wavelength of a spectral line, we see that the distant galaxy is approximately 69 Mpc away.

CHECK YOUR UNDERSTANDING 19.3

Hubble's law was discovered using measurements of two properties of a galaxy: _____ and _____. (a) size; mass (b) distance; rotation speed (c) distance; recession velocity (d) size; recession velocity

19.3 Galaxies Are Mostly Dark Matter

Efforts to measure the masses of galaxies during the 20th century led to the discovery of dark matter—mass that does not interact with light and cannot be detected via the light they emit. To understand this discovery, we first need to understand how astronomers go about measuring the mass of a galaxy, and then see how they concluded that much of the mass in a galaxy is dark matter.

Finding the Mass of a Galaxy

▶|| **AstroTour:** Dark Matter

To measure the mass of a galaxy, astronomers add up the mass of the stars, dust, and gas that they observe. Because a galaxy's spectrum is composed primarily of starlight, once astronomers know what types of stars are in the galaxy, they can use what is known about stellar evolution to estimate the total stellar mass from the galaxy's luminosity. Astronomers then estimate the mass of the dust and gas by using the physics of radiation from interstellar gas at X-ray, infrared, and radio wavelengths. Together, the stars, gas, and dust in a galaxy are called **luminous**

matter (or simply **normal matter**) because this matter emits or scatters electromagnetic radiation.

There is an independent method for determining mass that does not involve luminosity: the effect of gravity on an object's motion can be used to determine its mass. Stars in disks follow orbits that are much like the Keplerian orbits of planets around their parent stars and binary stars around each other (see Working It Out 13.4). To measure the mass of a spiral galaxy, astronomers apply Kepler's laws, just as they do for those other systems.

Detections of Dark Matter

Astronomers originally hypothesized that the mass and the light in a galaxy are distributed in the same way; that is, they assumed that all the mass in a galaxy is luminous matter. They observed that the light of all galaxies, including spiral galaxies, is highly concentrated toward the center (**Figure 19.13a**). On the basis of the observed location of light, astronomers predicted that nearly all mass in a spiral galaxy is concentrated toward its center (Figure 19.13b). This situation is much like the Solar System, where nearly all the mass is at the Solar System's center—in the Sun. Therefore, they predicted faster orbital velocities near the center of the spiral galaxy and slower orbital velocities farther out (Figure 19.13c).

To test this prediction, astronomers used the Doppler effect to measure orbital motions of stars, gas, and dust at various distances from a galaxy's center. The velocities of stars are obtained from observations of absorption lines in their spectra. The velocities of interstellar gas are obtained using emission lines such as those produced by hydrogen alpha (Hα) emission or 21-cm emission from neutral hydrogen (see Chapter 15). Once the velocities have been found, astronomers create a graph—called a **rotation curve**—that shows how orbital velocity in a galaxy varies with distance from the galaxy's center. The rotation curve of a spiral galaxy enables astronomers to determine directly how the mass in that galaxy is distributed by applying Kepler's laws to the rotation curves.

Vera Rubin pioneered work on galaxy rotation rates in the 1970s. She discovered that, contrary to earlier prediction (Figure 19.13c), the rotation velocities of spiral galaxies remain about the same out to the most distant measured parts of the galaxies (Figure 19.13d). Observations of 21-cm radiation from neutral hydrogen show that the rotation curves appear level, or "flat," in their outer parts even well outside the extent of the visible disks. These observations indicated that mass and light are distributed differently.

What mass distribution would cause this unexpected rotation curve? Recall from Chapter 4 that only the mass inside a given radius contributes to the net gravitational force felt by an orbiting object. From the rotation velocity, you can calculate the mass within the orbit of the object. **Figure 19.14** shows the result of such a calculation for the spiral galaxy NGC 3198. The black line shows the speed of rotation of this galaxy at a particular radius. The red line shows how much luminous mass is inside that radius. To produce a rotation curve like the one shown in black, this galaxy must have a second component consisting of matter that does not show up in the census of stars, gas, and dust. This material, which does not interact with light, and therefore reveals itself only by the influence of its gravity, is called

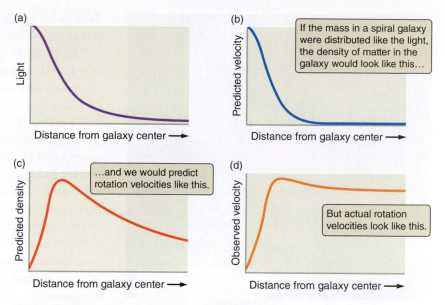

Figure 19.13 (a) The profile of visible light in a typical spiral galaxy drops off with distance from the center. (b) The predicted mass density of stars and gas located at a given distance from the galaxy's center follows the light profile. If stars and gas accounted for all of the mass of the galaxy, then the galaxy's rotation curve would be as shown in (c). However, observed galaxy rotation curves look more like the curve shown in (d).

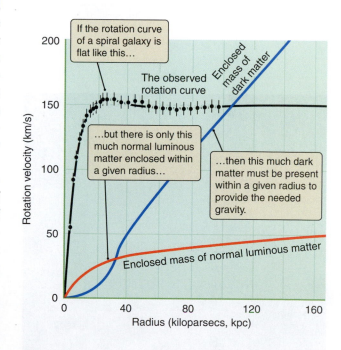

Figure 19.14 The flat rotation curve of the spiral galaxy NGC 3198 can be used to determine the total mass within a given radius. Notice that the normal mass that can be accounted for by stars and gas provides only part of the needed gravity. Extra dark matter is needed to explain the rotation curve. (Note: 1 kiloparsec [kpc] = 1,000 parsecs.)

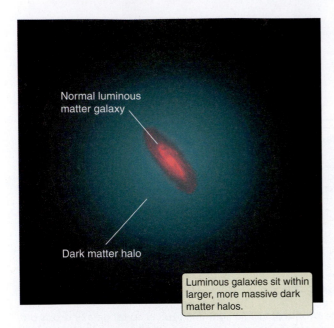

Normal luminous
matter galaxy

Dark matter halo

Luminous galaxies sit within
larger, more massive dark
matter halos.

Figure 19.15 In addition to the matter that is visible, galaxies are surrounded by halos containing a large amount of dark matter.

G X U V I R

Figure 19.16 In this combined visible-light and X-ray image of elliptical galaxy NGC 1132, the false-color blue and purple halo is X-ray emission from hot gas surrounding the galaxy. The hot gas extends well beyond the visible light from stars.

dark matter. The blue line shows how much dark matter must be inside a particular radius in order to provide enough mass to make the galaxy rotate as it does.

The rotation curves of the inner parts of spiral galaxies match predictions based on their luminous matter, indicating that the inner parts of spiral galaxies are mostly luminous matter. Within the entire *visual* image of a galaxy, the mix of dark and luminous matter is about half and half. However, rotation curves measured with 21-cm radiation from neutral hydrogen indicate that the outer parts of spiral galaxies are mostly dark matter. Astronomers currently estimate that as much as 95 percent of the total mass in some spiral galaxies consists of a **dark matter halo**, shown in **Figure 19.15**, which can extend up to 10 times farther than the visible spiral portion of the galaxy located at the galaxy's center. This is a startling statement. The luminous part of a spiral galaxy is only a part of a much larger distribution of mass that is dominated by some type of invisible dark matter.

What about elliptical galaxies? Again, astronomers need to compare the luminous mass measured from the light they can see with the gravitational mass measured from the effects of gravity. Because elliptical galaxies do not rotate, astronomers cannot use Kepler's laws to measure the gravitational mass from the rotation of a disk. Instead, they noticed that an elliptical galaxy's ability to hold on to its hot, X-ray-emitting gas depends on its mass. If the galaxy is not massive enough, the hot atoms and molecules will escape into intergalactic space. To find the mass of an elliptical galaxy, astronomers first infer the total amount of gas from X-ray images, such as the (false-color) blue and purple halo seen in **Figure 19.16**. Then they calculate the mass that is needed to hold on to the gas and compare that gravitational mass with the luminous mass. The amount of dark matter is the difference between what is needed to hold on to the inferred amount of gas and the observed amount of luminous matter.

Some elliptical galaxies contain up to 20 times as much mass as can be accounted for by their stars and gas alone, so they must be dominated by dark matter, just like spiral galaxies. As with spirals, the luminous matter in ellipticals is more centrally concentrated than the dark matter. The transition from the inner parts of galaxies (where luminous matter dominates) to the outer parts (where dark matter dominates) is remarkably smooth. Some galaxies may contain less dark matter than others, but about 90–95 percent of the total mass in a typical galaxy is in the form of dark matter. The high percentage of dark matter distinguishes smaller dwarf galaxies from globular clusters, which do not have dark matter. This difference is an important observation that will need to be explained in the context of the evolution of galaxies.

The Composition of Dark Matter?

What is the dark matter that makes up most of a galaxy? A number of suggestions are under investigation. Some candidates are objects such as large planets, compact stars, black holes, and exotic unknown elementary particles. These candidates have been lumped into two groups: MaCHOs and WIMPs.

Dark matter candidates such as small main-sequence M stars, Jupiter-sized planets, white dwarfs, neutron stars, or black holes are collectively referred to as **MaCHOs**, which stands for *massive compact halo objects*. If the dark matter in a galaxy's halo consists of MaCHOs, there must be a lot of these objects, and they must each exert gravitational force but not emit much light. Because they have mass, MaCHOs gravitationally deflect light according to Einstein's general theory of relativity—a phenomenon called gravitational lensing (see Chapter 18). If astronomers were observing a distant star and a MaCHO passed between Earth and the star, the star's light would be deflected and, if the geometry were just right, focused by the intervening MaCHO as it passed across their line of sight. Because gravity affects all wavelengths equally, such lensing events should look the same in all colors, ruling out other causes of variability.

Astronomers would be remarkably lucky if such an event occurred just as they were observing a single distant star. When they monitored the stars in two small companion galaxies of the Milky Way, observing tens of millions of stars for several years, they saw some lensing events but not nearly enough to account for the amount of dark matter in the halo of our galaxy. The implication of this result is that dark matter in the Milky Way (and therefore other galaxies too) must be composed of something other than MaCHOs.

The other dark matter candidates are the exotic unknown elementary particles commonly known as **WIMPs**, which stands for *weakly interacting massive particles*. WIMPs are predicted to be similar to neutrinos in that they would barely interact with ordinary matter, yet would be more massive and would move more slowly. WIMPs are currently the favored candidates for dark matter because there are not enough MaCHOs to account for the observed effects. Experiments are under way at the Large Hadron Collider and on the International Space Station to detect the existence of such particles, and additional experiments are being done to detect such particles from the dark matter halo of the Milky Way as they pass through Earth.

CHECK YOUR UNDERSTANDING 19.4

Astronomers detect dark matter: (a) by comparing luminous mass to gravitational mass; (b) because it blocks background light; (c) because more distant galaxies move away faster; (d) because it emits lots of X-rays.

19.4 Most Galaxies Have a Supermassive Black Hole at the Center

Studying the centers of galaxies is difficult because there are so many stars and so much dust and gas in the way that astronomers cannot get a clear picture of the center, even for nearby galaxies. Instead, it was observations of the most distant objects in the universe that provided the clues to understanding what lies in the centers of massive galaxies.

The Discovery of Quasars

In the late 1950s, radio surveys detected a number of bright, compact objects that at first seemed to have no optical counterparts. Improved radio positions revealed that the radio sources coincided with faint, very blue, starlike objects. Unaware of the true nature of these objects, astronomers called them "radio stars." Obtaining spectra of the first two radio stars was a laborious task, requiring 10-hour exposures. Astronomers were greatly puzzled by the results because these spectra did not display the expected absorption lines characteristic of blue stars. Instead, the spectra showed only a single pair of emission lines that were broad—indicating very rapid motions within these objects—and that did not seem to correspond to the lines of any known substances.

For several years, astronomers believed they had discovered a new type of star, until astronomer Maarten Schmidt realized that these broad spectral lines, shown in **Figure 19.17**, were the highly redshifted lines of ordinary hydrogen.

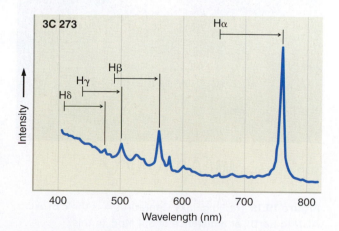

Figure 19.17 This figure shows the spectrum of quasar 3C 273, one of the closest and most luminous known quasars. The emission lines are redshifted by $z = 0.16$, from marked rest wavelengths, indicating that the quasar is at a distance of about 750 Mpc.

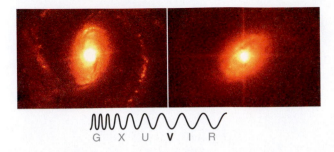

G X U V I R

Figure 19.18 These HST images show quasars embedded in the centers of galaxies.

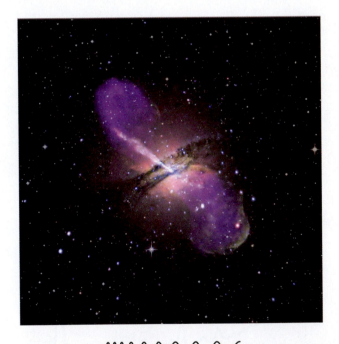

G X U V I R

Figure 19.19 Radio galaxy Centaurus A is the closest AGN to Earth, at a distance of 3.4 Mpc. In this composite image, the visible-light image shows the galaxy, the X-ray image (pink) shows the hot gas and an energetic jet blasting from the AGN, and the radio image (purple) shows the jets and lobes.

The implications were surprising: these "stars" were not stars. They were extraordinarily luminous objects at enormous distances. These "quasi-stellar radio sources" were named **quasars**. Other quasars were soon found by the same techniques. As still more were found, astronomers began cataloging them.

Quasars are phenomenally powerful, shining with the luminosity of a trillion to a thousand trillion (10^{12} to 10^{15}) Suns. They are also very distant from Earth: hundreds or thousands of megaparsecs. Billions of galaxies are closer to Earth than is the nearest quasar. Recall that the distance to an object also indicates the amount of time that has passed since the light from that object left its source. The fact that quasars are seen only at great distances implies that they are quite rare in the universe at this time but were once much more common. The discovery that quasars existed in the distant and therefore earlier universe provided one of the first pieces of evidence demonstrating that the universe has evolved over time.

Quasars are now recognized as the result of the most extreme form of activity that can occur in the nuclei of galaxies (**Figure 19.18**), often resulting from interactions with other galaxies. Quasars are a type of **active galactic nuclei**, or simply **AGNs**. The distinct types of active nuclei are identified from the spectrum of the galaxy. A "normal" galaxy has an absorption spectrum that is a composite of the light from its billions of stars. A galaxy with an AGN exhibits emission lines in addition to the stellar absorption spectrum. Active galactic nuclei are identified by the emission lines in their spectra, which distinguishes them from normal galaxies, which mostly show only absorption lines.

AGNs come in several types and can occur in spiral or elliptical galaxies. **Seyfert galaxies**, named after Carl Seyfert (1911–1960), who discovered them in 1943, are spiral galaxies whose centers contain AGNs. The luminosity of a typical Seyfert nucleus can be 10 billion to 100 billion L_{Sun}, comparable to the luminosity of the rest of the host galaxy as a whole. Similarly, **radio galaxies** are elliptical galaxies whose centers contain AGNs; their emission is usually most prominent in radio wavelengths. Radio galaxies and the more distant and luminous quasars are often the sources of slender jets that extend outward millions of light-years from the galaxy, powering twin lobes of radio emission (**Figure 19.19**).

Much of the light from AGNs is synchrotron radiation. This is the same type of radiation that comes from extreme environments such as the Crab Nebula supernova remnant. Synchrotron radiation comes from relativistic charged particles spiraling around the direction of a magnetic field. The fact that AGNs accelerate large amounts of material to nearly the speed of light indicates that they are very violent objects. In addition to the continuous spectrum of synchrotron emission, the spectra of many quasars and Seyfert nuclei also show emission lines that are smeared out by the Doppler effect across a wide range of wavelengths. This observation implies that gas in AGNs is swirling around the centers of these galaxies at speeds of thousands or even tens of thousands of kilometers per second.

AGNs Are the Size of the Solar System

The enormous radiated power and mechanical energy of active galactic nuclei are made even more spectacular by the fact that all of this power emerges from a region that can be no larger than a light-day or so across—comparable in size to the Solar System. Although the HST and large, ground-based telescopes show faint fuzz—light from the surrounding galaxy—around the images of some quasars and other AGNs, the objects themselves remain as unresolved points of light.

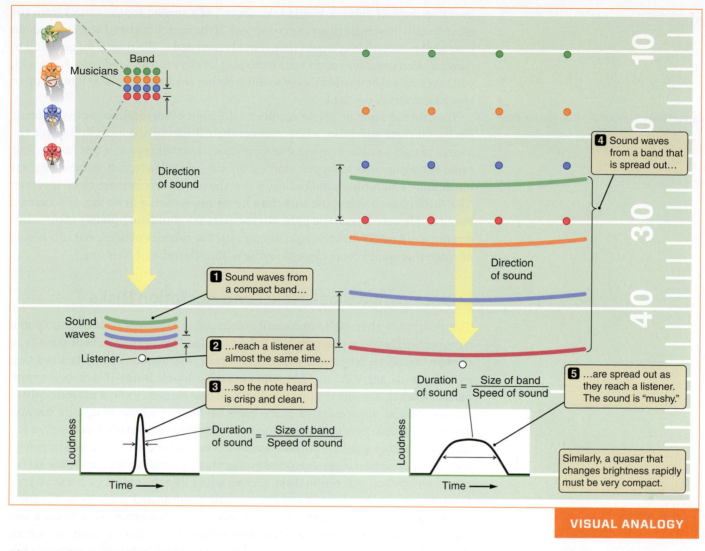

Figure 19.20 shows:
- Band / Musicians
- Direction of sound
- **1** Sound waves from a compact band…
- Sound waves / Listener
- **2** …reach a listener at almost the same time…
- **3** …so the note heard is crisp and clean.
- Loudness / Time →
- $\text{Duration of sound} = \dfrac{\text{Size of band}}{\text{Speed of sound}}$
- **4** Sound waves from a band that is spread out…
- Direction of sound
- $\text{Duration of sound} = \dfrac{\text{Size of band}}{\text{Speed of sound}}$
- **5** …are spread out as they reach a listener. The sound is "mushy."
- Similarly, a quasar that changes brightness rapidly must be very compact.
- Loudness / Time →

VISUAL ANALOGY

Figure 19.20 A marching band spread out across a field cannot play a clean note. Similarly, AGNs must be very compact to explain their rapid variability.

To understand why astronomers think that AGNs are compact objects, think about the halftime show at a local football game. **Figure 19.20** illustrates a problem faced by every director of a marching band. When a band is all together in a tight formation at the center of the field, the notes you hear in the stands are clear and crisp; the band plays together beautifully. But as the band spreads out across the field, its sound begins to get mushy. This is not because the marchers are poor musicians. Instead, it is because sound travels at a finite speed. On a cold, dry December day, sound travels at a speed of about 330 meters per second (m/s). At this speed, it takes sound approximately $\frac{1}{3}$ of a second to travel from one end of the football field to the other. Even if every musician on the field plays a note at exactly the same instant in response to the director's cue, in the stands you hear the instruments close to you first but have to wait longer for the sound from the far end of the field to arrive.

If the band is spread from one end of the field to the other, then the beginning of a note will be smeared out over about $\frac{1}{3}$ of a second, which is the time it takes for sound to travel from one end of the field to the other. If the band were spread out

▶Ⅱ **AstroTour:** Active Galactic Nuclei

over two football fields, it would take about $\frac{2}{3}$ of a second for the sound from the most distant musicians to arrive at your ear. If the marching band were spread out over a kilometer, then it would take roughly 3 seconds—the time it takes sound to travel a kilometer—for you to hear a crisply played note start and stop. Even with your eyes closed, it would be easy to tell whether the band was in a tight group or spread out across the field.

Astronomy in Action: Size of Active Galactic Nuclei

Exactly the same principle applies to the light observed from active galactic nuclei. Quasars and other AGNs change their brightness dramatically over the course of only a day or two—and in some cases as quickly as in a few hours. This rapid variability sets an upper limit on the size of the AGN, just as hearing clear music from a marching band indicates that the band musicians are close together. The AGN powerhouse must therefore be no more than a light-day or so across because if it were larger, the light astronomers see could not possibly change in a day or two. An AGN has the light of up to 10,000 galaxies pouring out of a region of space that would come close to fitting within the orbit of Neptune.

Supermassive Black Holes and Accretion Disks

The conclusion that AGNs are small compared to their entire host galaxy and have incredible energy outputs had to be explained. In thinking about what type of object could be very small yet very energetic, astronomers hypothesized that galaxies with an AGN contain **supermassive black holes**—black holes with masses from thousands to tens of billions of solar masses. Violent accretion disks surround these supermassive black holes. As matter in the disk falls inward, gravitational energy is converted into heat, providing the luminous energy of the AGN. Recall that you have already encountered accretion disks several times in this book. Accretion disks surround young stars, providing the raw material for planetary systems. Accretion disks around white dwarfs, fueled by material torn from their bloated evolving companions, lead to novae and some Type Ia supernovae. Accretion disks around neutron stars and stellar-mass black holes a few kilometers across are seen as X-ray binary stars. Now take the neutron star or stellar black hole examples and scale them up to a black hole with a mass of a billion solar masses and a radius comparable in size to the orbit of Neptune. To attain such a high luminosity, an AGN has an accretion disk that is fed by several solar masses every year rather than by the small amounts of material siphoned off a star (**Working It Out 19.3**).

In our discussion of star formation in Chapter 15, we showed that gravitational energy is converted to thermal energy as material moves inward toward the growing protostar. Here, as material moves inward toward the supermassive black hole, conversion of gravitational energy heats the accretion disk to hundreds of thousands of kelvins, causing it to glow brightly in visible, ultraviolet, and X-ray light. Conversion of gravitational energy to thermal energy as material falls onto the accretion disk is also a source of energetic emission. As much as 20 percent of the mass of infalling material around a supermassive black hole is converted to luminous energy. The rest of that mass is pulled into the black hole itself, causing it to grow even more massive.

The interaction of the accretion disk with the black hole creates powerful radio jets that emerge perpendicular to the disk (as in the jet in the upper left of Figure 19.19). Throughout, twisted magnetic fields accelerate charged particles such as electrons and protons to relativistic speeds, producing synchrotron emission. Gas in the accretion disk or in nearby clouds orbiting the central black hole

19.3 ▶ Working It Out The Size, Density, and Power of a Supermassive Black Hole

Size. What are the sizes of supermassive black holes? Recall from Chapter 18 our discussion of the Schwarzschild radius, where you saw that stellar-mass black holes are kilometers in size. The formula for the Schwarzschild radius is given by

$$R_S = \frac{2GM_{BH}}{c^2}$$

where G is the gravitational constant, and c is the speed of light.

The largest supermassive black holes observed have about 10 billion solar masses (M_{Sun}). For example, the black hole at the center of the galaxy M87 is 6.6 billion M_{Sun}. To compute its size, recall that $M_{Sun} = 1.99 \times 10^{30}$ kilograms (kg), $c = 3 \times 10^5$ km/s, and $G = 6.67 \times 10^{-20}$ km³/(kg s²). Then, a 6.6-billion-M_{Sun} black hole has a Schwarzschild radius of

$$R_S = \frac{2 \times [6.67 \times 10^{-20}\,\text{km}^3/(\text{kg}\,\text{s}^2)] \times (6.6 \times 10^9 \times 1.99 \times 10^{30}\,\text{kg})}{(3 \times 10^5\,\text{km/s})^2}$$

$$R_S = 2.0 \times 10^{10}\,\text{km}$$

We can convert this value into astronomical units. Recall that 1 astronomical unit (AU) = 1.5×10^8 km. Therefore, this supermassive black hole has a radius of 130 AU—somewhat larger than the Solar System. We know that light takes $8\frac{1}{3}$ minutes to reach Earth from the Sun at a distance of 1 AU, so this 130 AU corresponds to a distance of 1,080 light-minutes, or 18 light-hours.

Density. What is the average density of this object inside of the event horizon? The mass of the black hole divided by the volume within the Schwarzschild radius is

$$\text{Density} = \frac{\text{Mass}}{\text{Volume}} = \frac{(6.6 \times 10^9) \times (1.99 \times 10^{30}\,\text{kg})}{\frac{4}{3} \times \pi \times (2.0 \times 10^{10}\,\text{km})^3}$$

$$\text{Density} = 3.9 \times 10^8\,\text{kg/km}^3 = 0.39\,\text{kg/m}^3$$

This is 2,500 times less dense than water. Supermassive black holes do not have the extremely high mean densities of stellar-mass black holes.

Feeding an AGN. Power for an AGN is produced when matter falls onto the accretion disk around the central supermassive black hole. Some of this high-velocity mass is radiated away according to Einstein's mass-energy equation: $E = mc^2$. About how much material has to be accreted to produce the observed luminosities? Astronomers estimate the efficiency of the accretion to be about 10–20 percent. Here, we'll assume that 15 percent of the infalling matter is radiated away as energy, or

$$E = 0.15mc^2$$

Astronomers can measure how much energy is produced by the infalling material and radiated to space. For a relatively weak AGN like that of the earlier example of the galaxy M87, $L = 5 \times 10^{35}$ joules per second (J/s), or 5×10^{35} kg m²/s² each second. Here, we'll use $c = 3 \times 10^8$ m/s. Dividing both sides of Einstein's equation by $0.15c^2$ gives us the mass consumed each second:

$$m = \frac{E}{0.15c^2} = \frac{5 \times 10^{35}\,\text{kg}\,\text{m}^2/\text{s}^2}{0.15 \times (3 \times 10^8\,\text{m/s})^2}$$

$$m = 3.7 \times 10^{19}\,\text{kg}$$

Multiplying this result (the mass consumed each second) by 3.2×10^7 seconds per year shows that this AGN accretes 10^{27} kg, or about half the mass of Jupiter, each year, which is then radiated away as energy.

If we consider a quasar with a luminosity (L) of 10^{39} J/s = 10^{39} kg m²/s² each second (= 2.5 trillion L_{Sun}), then the mass accreted each second is given by

$$m = \frac{10^{39}\,\text{kg}\,\text{m}^2/\text{s}^2}{0.15 \times (3 \times 10^8\,\text{m/s})^2}$$

$$m = 7.4 \times 10^{22}\,\text{kg}$$

Multiplying by 3.2×10^7 seconds per year yields a mass of 2.4×10^{30} kg per year. Recall that the mass of the Sun is 1.99×10^{30} kg. Therefore, this quasar supermassive black hole is accreting about 1.2 M_{Sun} each year to radiate this much energy. A quasar with 10 times this luminosity would be accreting 10 times the mass.

at high speeds produces emission lines that are smeared out by the Doppler effect into the broad lines seen in AGN spectra. This accretion disk surrounding a supermassive black hole is the "central engine" that powers AGNs.

The Unified Model of AGN

Astronomers have developed the basic picture of a supermassive black hole surrounded by an accretion disk into a more complete AGN model. The **unified model of AGN** attempts to explain all types of AGNs—quasars, Seyfert galaxies,

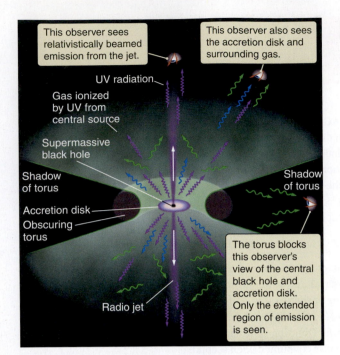

This observer sees relativistically beamed emission from the jet.

This observer also sees the accretion disk and surrounding gas.

UV radiation

Gas ionized by UV from central source

Supermassive black hole

Shadow of torus

Shadow of torus

Accretion disk

Obscuring torus

The torus blocks this observer's view of the central black hole and accretion disk. Only the extended region of emission is seen.

Radio jet

Figure 19.21 This figure illustrates the basic model of an active galactic nucleus, with a supermassive black hole surrounded by an accretion disk at the center. A larger, dusty torus sometimes blocks the view of the black hole. The mass of the central black hole, the rate at which it is being fed, and the viewing angle determine the observational properties of an AGN.

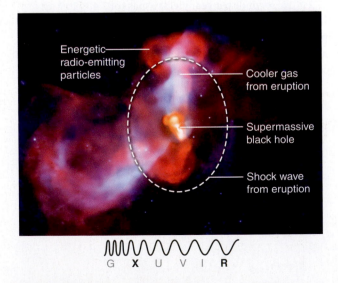

Energetic radio-emitting particles

Cooler gas from eruption

Supermassive black hole

Shock wave from eruption

G   X   U   V   I   R

Figure 19.22 This image of M87 in radio and X-rays shows the location of the supermassive black hole.

and radio galaxies. **Figure 19.21** shows the various components of this AGN model, in which an accretion disk surrounds a supermassive black hole. Much farther out from the accretion disk lies a large torus, or "doughnut" of gas and dust consisting of material that is feeding the central engine. Located far from the inner turmoil of the accretion disk, and far larger than the central engine, some of this torus is ionized by UV light from the AGN (**Process of Science Figure**).

In the unified model of AGN, the different types of AGNs observed from Earth are partly explained by astronomers' view of the central engine. The torus of gas and dust obscures this view in different ways, depending on the viewing angle. Variation in this angle, in the mass of the black hole, and in the rate at which it is being fed accounts for a wide range of AGN properties. When the AGN is viewed edge on, astronomers see emission lines from the surrounding torus and other surrounding gas. They can also sometimes see the torus in absorption against the background of the galaxy. From this nearly edge-on orientation, they cannot see the accretion disk itself, so they do not see the Doppler-smeared lines that originate closer to the supermassive black hole. If jets are present in the AGN, however, these should be visible emerging from the center of the galaxy.

If astronomers observe the accretion disk somewhat more face on, they can see over the edge of the torus and thus get a more direct look at the accretion disk and the location of the black hole. In this case, they see more of the synchrotron emission from the region around the black hole and the Doppler-broadened lines produced in and around the accretion disk. **Figure 19.22** shows an image of one such object, the galaxy M87, at an intermediate inclination. M87 is a source of powerful jets that continue outward for 100,000 light-years but originate in the tiny engine at the heart of the galaxy. Spectra of the disk at the center of this galaxy show the rapid rotation of material around a central black hole that has a mass of 3 billion (3×10^9) M_{Sun}.

The material in an AGN jet travels very close to the speed of light. As a result, what astronomers see is strongly influenced by relativistic effects. One of these is called **relativistic beaming**: matter traveling at close to the speed of light concentrates any radiation it emits into a tight beam pointed in the direction in which it is moving. So astronomers often observe only one side of the jets from AGNs, even though the radio lobes of radio galaxies are usually two-sided. The jet moving away is just too faint to observe.

In rare instances when the accretion disk in a quasar or radio galaxy is viewed almost directly face on, relativistic beaming dominates the observations. In these *blazars*, emission lines and other light coming from hot gas in the accretion disk are overwhelmed by the bright glare of jet emission beamed directly at Earth.

Normal Galaxies and AGNs

The essential elements of an AGN are a central engine (an accretion disk surrounding a supermassive black hole) and a source of fuel (gas and stars flowing onto the accretion disk). Without a source of matter falling onto the black hole, an AGN would no longer be an active nucleus. Astronomers looking at such an object would observe a normal galaxy with a supermassive black hole sitting in its center.

Only a few percent of present-day galaxies contain AGNs as luminous as the host galaxy. But when astronomers look at more distant galaxies (and therefore look further back in time), the percentage of galaxies with AGNs is much larger. These observations show that when the universe was younger, there were many more AGNs than there are today. If astronomers' understanding of AGNs is correct, then all the supermassive black holes that powered those dead AGNs should

Process of Science

FINDING THE COMMON THREAD

When first discovered, active galactic nuclei seemed to come in many different types, with dramatically different spectra. Later, it was realized that this could be an orientation effect: the many different types of objects could all be explained with one type of object, viewed from different angles.

"Blazar"
Viewing
down the jet

"Quasar/Seyfert 1"
Viewing at an
angle to the jet

"Radio Galaxy/
Seyfert 2"
Viewing at 90°
from the jet

Black hole

Accretion disk

Torus of neutral
gas and dust

Radio jet

Scientists seek underlying principles that explain more than one phenomenon. In the case of AGNs, scientists developed one model to explain many types of objects. This model encompasses the idea that our classification can be affected by our viewing angle, as in the picture of the coins at the beginning of the chapter. A unified model is much simpler than having a different model for every type of object.

still be around. If astronomers combine what they know of the number of AGNs in the past with ideas about how long a given galaxy remains in an active AGN phase, they are led to predict that many—perhaps even *most*—normal galaxies today contain supermassive black holes.

If supermassive black holes are present in the centers of normal galaxies, these black holes should reveal themselves in a number of ways. For one thing, such a concentration of mass at the center of a galaxy should draw surrounding stars close to it. The central region of such a galaxy should be much brighter than could be explained if stars alone were responsible for the gravitational field in the inner part of the galaxy. Stars feeling the gravitational pull of a supermassive black hole in the center of a galaxy should also orbit at very high velocities and therefore show large Doppler shifts. Astronomers have found these large Doppler shifts in every normal galaxy with a substantial bulge in which a careful search has been conducted. The masses inferred for these black holes range from 10,000 M_{Sun} to 20 billion M_{Sun}. The mass of the supermassive black hole seems to be related to the mass of the bulge in which it is found. Most large galaxies, whether elliptical or spiral, probably contain supermassive black holes. These observations reveal something remarkable about the structure and history of normal galaxies.

Apparently the only difference between a normal galaxy and an active galaxy is whether the supermassive black hole at its center is being fed at the time we see that galaxy. The rarity of present-day galaxies with very luminous AGNs does not indicate which galaxies have the potential for AGN activity. Rather, it indicates which galaxy centers are being lit up at the moment. If a large amount of gas and dust were dropped directly into the center of any large galaxy, this material would fall inward toward the central black hole, forming an accretion disk and a surrounding torus. This process would change the nucleus of this galaxy into an AGN.

In Chapter 23, we will discuss galaxy evolution and note that many of the observed properties of galaxies discussed in this chapter, including the formation of galaxy type, spiral structure, star formation, and AGN, depend on the interactions and mergers between galaxies. To account for the many large galaxies visible today, interactions and mergers must have been much more prevalent in the past when the universe was younger; this is one explanation for the larger number of AGNs that existed in the past. Computer models show that galaxy-galaxy interactions can cause gas located thousands of parsecs from the center of a galaxy to fall inward toward the galaxy's center, where it can provide fuel for an AGN. During mergers, a significant fraction of a galaxy might wind up being cannibalized. HST images of quasars often show that quasar host galaxies are tidally distorted or are surrounded by other visible matter that is probably still falling into the galaxies. Galaxies that show evidence of recent interactions with other galaxies are more likely to house AGNs in their centers (**Figure 19.23**). Any large galaxy might be only an encounter away from becoming an AGN.

Figure 19.23 The Swift Gamma-Ray observatory has detected active black holes (circles) in these merging galaxies.

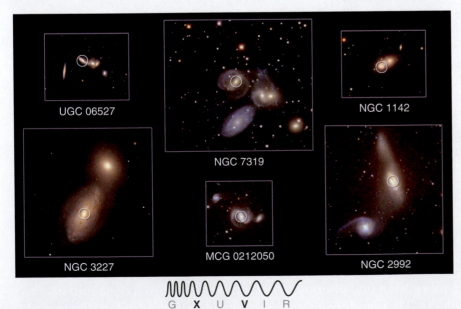

UGC 06527

NGC 7319

NGC 1142

NGC 3227

MCG 0212050

NGC 2992

G X U V I R

CHECK YOUR UNDERSTANDING 19.5

Supermassive black holes: (a) are extremely rare—there are only a handful in the universe; (b) are completely hypothetical; (c) occur in most, perhaps all, large galaxies; (d) occur only in the space between galaxies.

Origins

Habitability in Galaxies

In this chapter, we discussed the different types of galaxies that have been observed. Can we say anything about their potential for life? The short answer is that there is no solid information. These galaxies are too far away for astronomers to have detected any planets around their stars. So all we can do is speculate about the habitability of other galaxies. Two key requirements are the presence of heavy elements to form planets (and life) and an environment without too much radiation that might be damaging to life.

A study of the host stars of exoplanet candidates discovered by the Kepler telescope suggests that stars with a higher percentage of heavier elements may be more likely to have planets. (This finding fits with the core accretion models of planet formation discussed in Chapter 7.) The first generation of stars made from the hydrogen and helium of Big Bang nucleosynthesis do not have heavy elements. Recall from the chapters on stellar evolution that elements heavier than helium are created in the cores of dying stars and then are scattered into the galactic environment through planetary nebulae, stellar winds, and supernova explosions. So the amount of heavier elements in a star depends on the cosmic history of the material from which the star formed. Therefore, astronomers must consider the galactic environment of the star, which varies among different types of galaxies and different locations within the galaxies.

Spiral galaxies have had more continual star formation in their disks throughout their history. They contain more stars born from recycled material, and therefore more stars with a higher fraction of heavy elements. Elliptical (and S0) galaxies have older, redder populations of stars and very little current star formation. Old, massive ellipticals have a larger percentage of lower-mass stars than that of smaller ellipticals or spirals. Astronomers had previously thought this difference meant that large elliptical galaxies would not be good environments for planet formation. But the Kepler telescope has found many planets around small, red, main-sequence stars like the ones that populate elliptical galaxies. One study of two elliptical galaxies showed that both had some fraction of stars with a heavy-element fraction similar to that of the stars hosting Kepler exoplanets in the Milky Way.

Another issue is the presence of radiation that might be hazardous to life. This radiation is most likely to come from the center of the galaxy. Galaxies that are in an active AGN state might have too much radiation in regions close to their centers to be conducive to life. Stars whose orbits cross spiral arms many times might also be exposed to higher-than-average levels of radiation, but this is not the case for the majority of stars in a galaxy.

The conditions in these galaxies may also change as the galaxies evolve. Galaxy mergers can shake up stellar orbits and relocate stars and their planets to different locations. Mergers may also affect the growth and activity level of supermassive black holes and thus the presence of radiation. Some galactic environments just may not remain habitable for the length of time—billions of years—that it took life to evolve from bacteria to intelligence on Earth.

READING ASTRONOMY NEWS

ARTICLES   QUESTIONS

In this news release story, astronomers used the Hubble and Gemini telescopes to find a massive black hole in a small galaxy.

Hubble Helps Find Smallest Known Galaxy with a Supermassive Black Hole

Hubblesite.org

Astronomers have found an unlikely object in an improbable place: a monster black hole lurking inside one of the tiniest galaxies known.

Though the black hole is five times the mass of the black hole at the center of our Milky Way, it is inside a galaxy that crams 140 million stars within a diameter of about 300 light-years, only 1/500th of our galaxy's diameter.

The dwarf galaxy containing the black hole, called M60-UCD1, is the densest galaxy ever seen (**Figure 19.24**). If you lived inside of it,

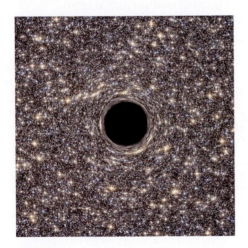

Figure 19.24 An artist's concept of a giant black hole in the center of the ultracompact galaxy M60-UCD1.

the night sky would dazzle with at least 1 million stars visible to the naked eye (as opposed to 4,000 stars in our nighttime sky, as seen from Earth's surface).

The finding implies that there are many other very compact galaxies in the universe that contain supermassive black holes. The observation also suggests that dwarf galaxies may actually be the stripped remnants of larger galaxies that were torn apart during collisions with yet other galaxies—rather than small islands of stars born in isolation.

"We don't know of any other way you could make a black hole so big in an object this small," said University of Utah astronomer Anil Seth, lead author of an international study of the dwarf galaxy published in the journal *Nature*.

His team of astronomers used the Hubble Space Telescope and the Gemini North 8-meter optical and infrared telescope on Hawaii's Mauna Kea to observe M60-UCD1 and measure the black hole's mass. The sharp Hubble images provide information about the galaxy's diameter and stellar density. Spectroscopy with Gemini measures the stellar motions as affected by the black hole's pull. These data are used to calculate the mass of the unseen black hole.

Black holes are gravitationally collapsed, ultracompact objects that have a gravitational pull so strong that even light cannot escape.

Supermassive black holes—those with the mass of at least 1 million stars like our Sun—are thought to be at the centers of many galaxies.

The black hole at the center of our Milky Way galaxy has the mass of 4 million Suns, but as heavy as that is, it is less than 0.01 percent of the Milky Way's total mass. By comparison, the supermassive black hole at the center of M60-UCD1 is a stunning 15 percent of the small galaxy's total mass. "That is pretty amazing, given that the Milky Way is 500 times larger and more than 1,000 times heavier than the dwarf galaxy M60-UCD1," Seth said.

One explanation is that M60-UCD1 was once a large galaxy containing 10 billion stars, but then it passed very close to the center of an even larger galaxy, M60, and in that process all the stars and dark matter in the outer part of the galaxy got torn away and became part of M60.

The team believes that M60-UCD1 may eventually be pulled back to merge with the center of M60, which has its own monster black hole, weighing a whopping 4.5 billion solar masses (more than 1,000 times bigger than the black hole in our galaxy). When that happens, the black hole in M60-UCD1 will merge with the far more massive black hole in M60. The galaxies are 50 million light-years away.

ARTICLES   QUESTIONS

1. How far away is M60-UCD1 in parsecs and kiloparsecs?
2. On average, how close are stars to each other in this galaxy?
3. Explain how the black hole's mass was estimated.
4. How is this merger different from those shown in Figure 19.23?
5. Watch the short video of a computer simulation of M60-UCD1 (http://www.spacetelescope.org/videos/heic1419a/). How do astronomers think this galaxy formed?

Summary

Galaxies are classified on the basis of their shape and the types of orbits of their stars. Most of the mass of a galaxy is dark matter, which interacts with light very weakly, if at all. The form of this matter is not yet known. Most galaxies have a supermassive black hole at the center, which may become an AGN if gas accretes onto it. In thinking about the potential habitability of galaxies, astronomers consider the activity state of the galaxy, including the presence or absence of an AGN and mergers, and the amount of heavy elements in the stars in the galaxy, which is related to the star formation rate and galaxy type.

LG 1 **Determine a galaxy's type from its appearance, and describe the motions of its stars.** Spiral galaxies are distinguished by their flat disk and spiral arms. The stars in this disk all orbit the center of the galaxy in the same direction. Elliptical galaxies are roughly egg-shaped, and the stars orbit in all directions. Irregular galaxies are galaxies that fit neither of these classifications, usually because they are interacting with another galaxy.

LG 2 **Explain the distance ladder and how distances to galaxies are estimated.** Astronomers build a distance ladder to galaxies by observing objects of known luminosity, such as Cepheid variable stars and Type Ia supernovae, in distant galaxies. Observations of the distance and the velocity of galaxies show that the two factors are related: more distant galaxies move away from us faster. Hubble's law $v = H_0 d$ provides a method to find the distances of the most remote objects.

LG 3 **Describe the evidence suggesting that galaxies are composed mostly of dark matter.** Most of the mass in galaxies does not reside in gas, dust, or stars; instead, galaxy rotation curves indicate that about 90 percent of a galaxy's mass is in the form of dark matter, which does not emit or absorb light to any significant degree. Dark matter is identified by its gravitational interaction with ordinary matter. The two main groups of candidates for the composition of dark matter are MaCHOs (astronomical objects such as planets, dead stars, and black holes) and WIMPs (massive, weakly interacting elementary particles).

LG 4 **Discuss the evidence indicating that most—perhaps all—large galaxies have supermassive black holes at their centers.** Observations of distant quasars reveal extremely luminous, compact sources near the centers of galaxies. These active galactic nuclei (AGNs) are best explained as supermassive black holes, surrounded by an accretion disk and a torus of dust and gas. AGNs can emit as much as 1,000 times the light of the whole galaxy, all coming from a region the size of the Solar System. Surveys of the centers of galaxies show that many have AGNs at their centers.

? UNANSWERED QUESTIONS

- How standard are "standard candles"? For example, Cepheid variable light curves are slightly different, depending on the amount of heavy elements in the stars, so this variation must be calibrated. A majority of Type Ia supernovae may originate from the merging of two compact objects rather than from one white dwarf accreting mass and exploding when it gets close to 1.4 M_{Sun}. Astronomers try to address these difficulties by using multiple methods to find distances to galaxies—for example, observing numerous Cepheids, bright O stars, and Type Ia supernovae in the same galaxy—to check that their calculated distances agree. In one recent study, astronomers used HST to observe more than 600 Cepheid variable stars in eight galaxies in which there were Type Ia supernova detections, and they were able to reduce the uncertainty in their distances (and thus their calculated uncertainty in their value of the Hubble constant). The Type Ia supernovae calibrated in this way are consistent with earlier-calibrated maximum luminosities. But it is possible that those supernovae in galaxies far enough away that there are no other standard candles may have slightly different maximum luminosities, leading to a somewhat incorrect distance estimate (and value of H_0).

- Where do supermassive black holes come from? To explore this question, astronomers are studying computer models in which the supermassive black hole grew from a large number of stellar-mass black holes, or grew along with the galaxy by swallowing large amounts of central gas, or increased after the merger of two or more galaxies. We will return to this question when we discuss galaxy evolution in Chapter 23.

Questions and Problems

Test Your Understanding

1. Which of the following contributes the largest percentage to the total mass of a spiral galaxy?
   a. dark matter
   b. central black hole
   c. stars
   d. dust and gas

2. In the context of spiral galaxies, Kepler's laws could be used to estimate
   a. P, the period.
   b. A, the semimajor axis.
   c. M, the mass of the galaxy.
   d. v, the rotation speed of the galaxy.

3. Astronomers determine the radius of an AGN by measuring
   a. how much light comes from it.
   b. how hard it pulls on stars nearby.
   c. how quickly its light varies.
   d. how quickly it rotates.

4. If you observed a galaxy with an Hα emission line that had a wavelength of 756.3 nm, what would be the galaxy's redshift? Note that the rest wavelength of the Hα emission line is 656.3 nm.
   a. 0.01
   b. 0.05
   c. 0.10
   d. 0.15

5. As astronomers extend their distance ladder beyond 30 Mpc, they change their measuring standard from Cepheid variable stars to Type Ia supernovae. Why is this change necessary?
   a. Type Ia supernovae are more luminous than Cepheid variables.
   b. Type Ia supernovae are less luminous than Cepheid variables.
   c. Type Ia supernovae vary more slowly than do Cepheid variables.
   d. Type Ia supernovae vary more quickly than do Cepheid variables.

6. Which galaxy type has a spherical bulge and a well-defined disk?
   a. spiral
   b. barred spiral
   c. elliptical
   d. irregular

7. Which galaxy type is shaped like a rugby ball?
   a. Sb
   b. SBb
   c. E0
   d. E5

8. For a galaxy, the term *morphology* refers to
   a. its shape.
   b. its evolution over time.
   c. the motion of its stars.
   d. its overall density.

9. If all the stars in an elliptical galaxy traveled in random directions in their orbits, the elliptical galaxy would be type
   a. E0.
   b. E2.
   c. E5.
   d. E7.

10. The flat rotation curves of spiral galaxies imply that the distribution of mass resembles
    a. the Solar System; most mass is concentrated in the center.
    b. a wheel; the density remains the same as the radius increases.
    c. the light distribution of the galaxy; a large concentration occurs in the middle, but significant mass exists quite far out.
    d. an invisible sphere much larger than the visible galaxy.

11. Astronomers observe two galaxies, A and B. Galaxy A has a recession velocity of 2,500 km/s, while galaxy B has a recession velocity of 5,000 km/s. According to these data,
    a. galaxy A is 4 times as far away as galaxy B.
    b. galaxy A is twice as far away as galaxy B.
    c. galaxy B is twice as far away as galaxy A.
    d. galaxy B is 4 times as far away as galaxy A.

12. The Hubble constant is found from the
    a. slope of the line fit to the data in Hubble's law.
    b. y-intercept of the line fit to the data in Hubble's law.
    c. spread in the data in Hubble's law.
    d. inverse of the slope of the line fit to the data in Hubble's law.

13. Astronomers know that dark matter is present in galactic halos because the speeds of orbiting stars _____ far from the center of the galaxy.
    a. decrease
    b. increase
    c. remain about constant
    d. fluctuate dramatically

14. What accounts for the differences among various types of AGNs?
    a. the type of the host galaxy
    b. the size of the central black hole
    c. the amount of dark matter in the galaxy's halo
    d. our viewing angle

15. If a Seyfert galaxy's nucleus varies in brightness on the time-scale of 10 hours, then approximately what is the size of the emitting region?
    a. 20 AU
    b. 70 AU
    c. 90 AU
    d. 140 AU

Thinking about the Concepts

16. What was the subject of the Great Debate, and why was it important to astronomers' understanding of the scale of the universe?

17. How did observations of Cepheid variable stars finally settle the Great Debate?

18. Why is it better to observe more than one type of standard candle in a distant galaxy?

19. Explain what astronomers mean by *distance ladder*.

20. Why is it important to know the type of progenitor of a Type Ia supernova in a distant galaxy?

21. Some galaxies have regions that are relatively blue; other regions appear redder. What does this variation indicate about the differences between these regions?

22. Describe how elliptical galaxies and spiral bulges are similar.

23. Which is more luminous: a quasar or a galaxy with 100 billion solar-type stars? Explain your answer.

24. The nearest observed quasar is about 750 Mpc away. Why don't astronomers observe any that are closer?

25. What distinguishes a normal galaxy from one that contains an AGN? It is a principle of science, often attributed to Einstein, that one should make things as simple as possible, but not simpler. Explain how this principle is at work in the distinction between normal galaxies and AGNs.

26. Contrast the size of a typical AGN with the size of the Solar System. How do astronomers know the size of an AGN?

27. Describe what astronomers think is happening at the center of a galaxy that contains an AGN.

28. It is likely that most galaxies contain supermassive black holes, yet many galaxies display no obvious evidence for their existence. Why do some black holes reveal their presence while others do not?

29. Study Figure 19.9.
    a. Why is it important that the different "rungs" of the distance ladder overlap in the distances that they measure?
    b. Does the figure end at the right edge because there are no more ways to measure distance or because there is no more universe to measure? How do you know?

30. Why do astronomers think that planets around stars near the centers of galaxies would not be good locations for the formation of life?

Applying the Concepts

31. Study Figure 19.14. The small vertical bars (known as error bars) on the data points indicate the size of the measurement error.
    a. At a radius of 25,000 parsecs (pc), what is the approximate measurement error in the rotation velocity?
    b. What is this value as a percentage of the measured velocity?
    c. Error bars are important because they show how wrong the measurement could possibly be. One way to think about this is that the black line could be as high as the top of the error bars or as low as the bottom of the error bars. In either case, would shifting the black line change the overall conclusion about redshift and distance? Why or why not?

32. Suppose the number density of galaxies in the universe is, on average, 3×10^{-68} galaxy/m³. If astronomers could observe all galaxies out to a distance of 10^{10} pc, how many galaxies would they find?

33. The spectrum of a distant galaxy shows the Hα line of hydrogen ($\lambda_{rest} = 656.28$ nm) at a wavelength of 750 nm. Assume that $H_0 = 70$ km/s/Mpc.
    a. What is the redshift (z) of this galaxy?
    b. What is its recession velocity (v_r) in kilometers per second?
    c. What is the distance of the galaxy in megaparsecs?

34. The nearest known quasar is 3C 273. It is located in the constellation Virgo and is bright enough to be seen in a medium-sized amateur telescope. With a redshift of 0.158, what is the distance to 3C 273 in parsecs?

35. The quasar 3C 273 has a luminosity of $10^{12} L_{Sun}$. Assuming that the total luminosity of a large galaxy, such as the Andromeda Galaxy, is 10 billion times that of the Sun, compare the luminosity of 3C 273 with that of the entire Andromeda Galaxy.

36. A quasar has the same brightness as a galaxy that is seen in the foreground 2 Mpc distant. If the quasar is 1 million times more luminous than the galaxy, what is the distance of the quasar?

37. Estimate the Schwarzschild radius for a supermassive black hole with a mass of 26 billion M_{Sun}.

38. You read in the newspaper that astronomers have discovered a "new" cosmological object that appears to be flickering with a period of 83 minutes. Because you have read *21st Century Astronomy*, you are able to estimate quickly the maximum size of this object. How large can it be?

39. A quasar has a luminosity of 10^{41} W, or J/s, and 10^8 M_{Sun} to feed it. Assuming constant luminosity and 20 percent conversion efficiency, what is your estimate of the quasar's lifetime?

40. A solar-type star ($M = 2 \times 10^{30}$ kg) approaches a supermassive black hole. As it crosses the event horizon, half of its mass falls into the black hole while the other half is completely converted to energy in the form of light. How much energy does this dying star send out to the rest of the universe?

41. Suppose a Type Ia supernova is found in a distant galaxy. The measured supernova brightness is 10^{-17} W/m². What is the distance of the galaxy?

42. Suppose that an object with the mass of Earth ($M_{Earth} = 5.97 \times 10^{24}$ kg) fell into a supermassive black hole with a 10 percent energy conversion.
    a. How much energy (in joules) would be radiated by the black hole?
    b. Compare your answer with the energy radiated by the Sun each second: 3.85×10^{26} J.

43. If a luminous quasar has a luminosity of 2×10^{41} W, or J/s, how many solar masses ($M_{Sun} = 2 \times 10^{30}$ kg) per year does this quasar consume to maintain its average energy output?

44. Material ejected from the supermassive black hole at the center of galaxy M87 extends outward from the galaxy to a distance of approximately 30,000 pc. M87 is approximately 17 Mpc away.
    a. If this material were visible to the naked eye, how large would it appear in the nighttime sky? Give your answer in degrees (1 radian = 57.3°).
    b. Compare this size with the angular size of the Moon.

45. A lobe in a visible jet from galaxy M87 is observed at a distance of 1,530 pc (5,000 light-years) from the galaxy's center moving outward at a speed of 0.99 times the speed of light (0.99c). Assuming constant speed, how long ago was the lobe expelled from the supermassive black hole at the galaxy's center?

USING THE WEB

46. Go to the Goddard Media Studios website and view the animation of a Cepheid variable star in a spiral galaxy (http://svs.gsfc.nasa.gov/goto?10145). Explain how astronomers use data like these to estimate the distance to the galaxy. What is actually observed, what is assumed, and what is calculated? (Review the discussion in Chapter 16 as needed.)

47. Go to the Astronomy Picture of the Day app or website (http://apod.nasa.gov/apod) and look at some recent pictures of galaxies. In each case, consider the following questions: Was the picture taken from a large or small telescope; from the ground or from space? Are galaxies in the image face on, edge on, or at an angle? What wavelengths were used for making the image? Are any of the colors "false colors"? If the picture is a combination of images from several telescopes, what do the different colors indicate?

48. Citizen science:
    a. Go to the website for Galaxy Zoo (http://galaxyzoo.org), the original Zooniverse citizen science project. (Log in with your Zooniverse password.) The specific project in action at any given time depends on the real data that need to be examined. One of the projects is likely a classification project. Click on and read, "Story," "Science," and "Classify," and then classify some galaxies. Save a copy of your classifications for your homework if necessary.
    b. Go to the website for Radio Galaxy Zoo (http://radio.galaxyzoo.org/). This project asks people to help locate supermassive black holes and their associated jets. Click on "Science" to read about the project, "Begin Hunting" to see some data, and then "Classify." Work through the given example, and then classify a few more.

49. a. Go to the website for the Fermi Gamma-ray Space Telescope (http://fermi.gsfc.nasa.gov). Scroll down to click on "Full News Archive" and look for a story about dark matter. What has this telescope discovered about dark matter?
    b. Go to the website for the Alpha Magnetic Spectrometer (http://ams02.org), a particle physics detector located on the International Space Station to search for dark matter, including WIMPs. What are the latest results?

50. a. Go to the website for the NASA Swift Gamma-Ray observatory (swift.gsfc.nasa.gov), which studies gamma-ray bursts. Click on "Latest Swift News" and look for a story about supermassive black holes. What has been discovered?
    b. Go to the website for NuSTAR (Nuclear Spectroscopic Telescope Array—http://www.nustar.caltech.edu), a space telescope launched by NASA in 2012. This mission is studying active galaxies hosting supermassive black holes. What type of telescope is this (wavelengths observed, general design)? What has been discovered?

smartwork

If your instructor assigns homework in SmartWork, access your assignments at digital.wwnorton.com/astro5.

digital.wwnorton.com/astro5

Galaxy classification sounds simple, but it can become complicated when you actually attempt it. **Figure 19.25**, taken by the Hubble Space Telescope, shows a small portion of the Coma Cluster of galaxies. The Coma Cluster contains thousands of galaxies, each containing billions of stars. Some of the objects in this image (the ones with a bright cross) are foreground stars in the Milky Way. Some of the galaxies in this image are far behind the Coma Cluster. Working with a partner, in this Exploration you will classify the 20 or so brightest galaxies in this cluster.

First, make a map by laying a piece of paper over the image and numbering the 20 or so brightest (or largest) galaxies in the image (label them "galaxy 1," "galaxy 2," and so on). Copy this map so that you and your partner each have a list of the same galaxies.

Separately, classify each galaxy by type. If it is a spiral galaxy, what is its subtype: a, b, or c? If it is an elliptical, how elliptical is it? Make a table that contains the galaxy number, the type you have assigned it, and any comments that will help you remember why you made that choice. When you are done classifying, compare your list with your partner's. Now comes the fun part! Argue about the classifications until you agree—or until you agree to disagree.

Figure 19.25 This Hubble Space Telescope image of the Coma Cluster shows a diversity of shapes.

1 Which galaxy type was easiest to classify?

_____

_____

2 Which galaxy type was hardest to classify?

_____

_____

3 What makes it hard to classify some of the galaxies?

_____

_____

4 Which galaxy type did you and your partner agree about most often?

_____

_____

5 Which galaxy type did you and your partner disagree about most often?

_____

_____

6 How might you improve your classification technique?

_____

_____

If you found this activity interesting and rewarding, astronomers can use your help: go to http://galaxyzoo.org to get involved in a citizen science project to classify galaxies, some of which have never been viewed before by human eyes.

20

The Milky Way—A Normal Spiral Galaxy

Of the hundreds of billions of galaxies in the universe, the Milky Way is the only one that astronomers can study at close range. In this chapter, we focus our attention on the Milky Way and how it offers clues to understanding all galaxies.

LEARNING GOALS

By the conclusion of this chapter, you should be able to:

LG 1 Explain how astronomers discovered the size and spiral structure of the Milky Way.

LG 2 List the clues of galaxy formation that can be found from the components of the Milky Way.

LG 3 Explain the evidence for a dark matter halo and for the supermassive black hole at the center of the Milky Way.

LG 4 Describe the Local Group of galaxies and how it provides clues about the evolution of the Milky Way.

The Milky Way in Earth's night sky has both bright patches of starlight and dark patches where stars are hidden by dust and gas. ▶▶▶

How do we know what the Milky Way Galaxy looks like from the outside?

20.1 Astronomers Have Measured the Size and Structure of the Milky Way

As you saw in Chapter 19, the universe is full of galaxies of many sizes and types. Because Earth is embedded within the Milky Way, the details of the shape and structure of the Milky Way are actually more difficult to determine than for other galaxies. Comparing observations of the Milky Way with observations of more distant galaxies improves our understanding of the Milky Way. In this section, we explore how astronomers infer the size and structure of the Milky Way from observations both within and without it.

Spiral Structure in the Milky Way

Figure 20.1a shows the Milky Way Galaxy in Earth's night sky. From a dark location at night, you can see dark bands of interstellar gas and dust that obscure much of the central plane of the Milky Way. This view of the Milky Way from inside offers a different and much closer perspective of a galaxy than can be obtained by viewing external galaxies. Compare this image of the Milky Way to the image in Figure 20.1b, which shows an edge-on spiral galaxy. The similarities between these two images suggest that the Milky Way is a spiral galaxy and that we are viewing it edge on, from a location inside the disk.

Finding further information about the size and shape of the Milky Way requires more extensive observations in the visible, infrared, and radio regions of the electromagnetic spectrum. Recall from Chapter 15 that neutral hydrogen

(a)

(b)

Figure 20.1 (a) We see the Milky Way as a luminous band across the night sky. Prominent dark lanes caused by interstellar dust obscure the light from more distant stars. (b) The edge-on spiral galaxy NGC 891, whose disk greatly resembles the Milky Way.

emits radiation at a wavelength of 21 centimeters (cm), in the radio region of the spectrum. Maps of this radiation show spiral structure in other galaxies and suggest spiral structure in the Milky Way. In addition, observations of ionized hydrogen gas in visible light show two spiral arms with concentrations of young, hot O and B stars. These observations confirm that the Milky Way is a spiral galaxy.

In 2005, Spitzer Space Telescope observations of the distribution and motions of stars toward the center of the galaxy confirmed that the Milky Way has a substantial bar with a modest bulge at its center. **Figure 20.2** shows an artist's rendering of the major features of the Milky Way. Two major spiral arms— Scutum-Centaurus and Perseus—connect to the ends of the central bar and sweep through the galaxy's disk, just like the arms observed in external spiral galaxies. There are several smaller arm segments, including the Orion Spur, which contains the Sun and Solar System. Astronomers conclude that the Milky Way is a giant barred spiral that is more luminous than an average spiral. If viewed from the outside, the Milky Way would look much like the barred spiral galaxy M109, shown in **Figure 20.3**.

Spiral Arms and Star Formation

In pictures of other spiral galaxies, the arms are often the most prominent feature, as in the Andromeda Galaxy, shown in **Figure 20.4**. The spiral arms are quite prominent in the ultraviolet image (Figure 20.4a), and while they are less prominent in visible light (Figure 20.4b), they are still clearly defined. From this, you might conclude that most stars in the disk of a spiral galaxy are concentrated in the spiral arms. This turns out not to be the case: although stars are slightly concentrated in spiral arms, this concentration is not strong enough to account for their prominence. However, structures associated with star formation, such as molecular clouds and associations of luminous O and B stars, are all concentrated in spiral arms. Spiral arms are prominent because star formation is occurring there, so the arms contain significant concentrations of young, massive, luminous stars.

Recall from Chapter 15 that stars form when dense interstellar clouds become so dense that they begin to collapse under the force of their own gravity. If stars form in spiral arms, then spiral arms must be places where clouds of interstellar gas and dust pile up and are compressed. These clouds can be observed where the dust

Figure 20.2 Infrared and radio observations contribute to an artist's model of the Milky Way Galaxy. The galaxy's two major arms (Scutum-Centaurus and Perseus) are seen attached to the ends of a thick central bar.

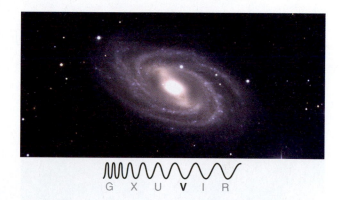

G X U V I R

Figure 20.3 From the outside, the Milky Way would look much like this barred spiral galaxy, M109.

(a) Ultraviolet light

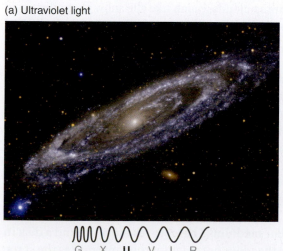

G X U V I R

(b) Visible light

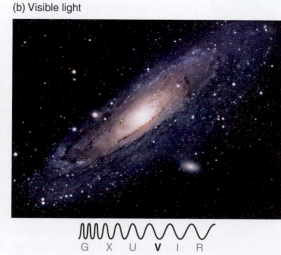

G X U V I R

Figure 20.4 These photos show the Andromeda Galaxy in ultraviolet light (a) and visible light (b). Note that the spiral arms, which are dominated by hot young stars, are most prominent in ultraviolet light. The spiral arms are less prominent in visible light.

(a)

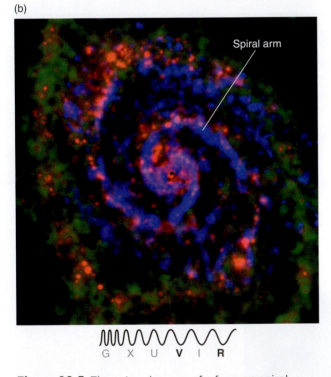

Dust lane in spiral arm

G X U V I R

(b)

Spiral arm

G X U V I R

Figure 20.5 These two images of a face-on spiral galaxy show the spiral arms. (a) This visible-light image also shows dust absorption. (b) This image shows the distribution of neutral interstellar hydrogen (green), carbon monoxide (CO) emission from cold molecular clouds (blue), and hydrogen alpha (Hα) emission from ionized gas (red).

and gas block starlight, as in **Figure 20.5a**, or where gases such as neutral hydrogen or carbon monoxide produce emission at various wavelengths (Figure 20.5b).

Why do spiral arms exist at all? Part of the answer is that disks do not rotate like a solid body, such as a wheel. Instead, material that is closer to the center takes less time to complete a revolution around the galaxy than material farther out in the galaxy, and so the inner part of the disk gets ahead of the outer part. This means that any disturbance in the disk of a spiral galaxy will produce a spiral pattern as the disk rotates. **Figure 20.6** illustrates the point. The first frame shows a single linear disturbance through the center of a model galaxy. In the second frame, the outer part of the line is trailing behind the inner part. As the galaxy rotates, a straight line through the center becomes a spiral. In the time it takes for objects in the inner part of the galaxy to complete several rotations, objects in the outer part of the galaxy may not have completed even a single revolution. In the process, the originally straight arms are slowly made into the spiral structure shown.

A spiral galaxy can be disturbed, for example, by gravitational interactions with other galaxies or by a burst of star formation. However, a single disturbance will not produce a stable spiral-arm pattern. Spiral arms from a single disturbance will wind tightly around the center in two or three rotations and disappear. Disturbances that are repetitive, however, can sustain spiral structure indefinitely. For example, a bar in the center of a spiral galaxy gravitationally disturbs the disk. As the disk rotates through this disturbance, the disturbance is repeated. Repeated episodes of star formation occur and maintain stable spiral arms.

Many galaxies show clear evidence of a relationship between the shapes of their bulges and the structure of their spiral arms. Barred spirals, for example, have a characteristic two-armed spiral pattern that is connected to the elongated bulge, as seen in Figure 20.3. Even the bulges of galaxies that are not obviously barred may be elongated enough to contribute to the formation of a two-armed spiral structure. Smaller galaxies in orbit about larger galaxies can also give rise to a periodic gravitational disturbance, triggering the same sort of two-armed structure.

The process of star formation itself can also create spiral structure. Regions of star formation release considerable energy into their surroundings through UV radiation, stellar winds, and supernova explosions. This energy compresses clouds of gas and triggers more star formation. Typically, many massive stars form in the same region at about the same time, and their combined mass outflows and supernova explosions occur one after another in the same region of space over the course of only a few million years. The result can be large, expanding bubbles of hot gas that sweep out cavities in the interstellar medium and concentrate the swept-up gas into dense, star-forming clouds, much like the snow that piles up in front of a snowplow. In this way, star formation can propagate through the disk of a galaxy. Rotation bends the resulting strings of star-forming regions into spiral structures.

Stars move in and out of arms as they orbit the center of a galaxy. Consequently, the stars in an arm today are not the same stars that were in the arm 20 million years ago. This is roughly analogous to a traffic jam on a busy highway. The cars in the jam are changing all the time, yet the traffic jam persists as a place of higher density—where there are more cars than usual. The traffic jam itself moves slowly backward, even as the cars move forward and pass through it. Just like the traffic jam, the disturbance of the spiral arm also moves at a different speed than the individual stars. These disturbances in the disks of spiral galaxies are called **spiral density waves** because they are waves of greater mass density and increased pressure in the galaxy's interstellar medium. These waves move around a disk in the pattern of a two-armed spiral that does not rotate at the same rate as the stars,

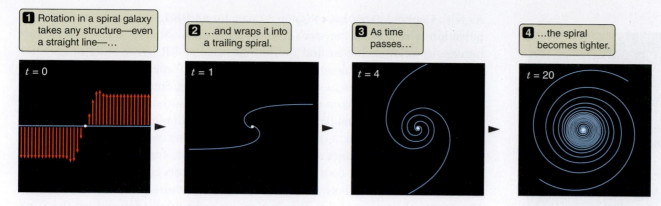

1 Rotation in a spiral galaxy takes any structure—even a straight line—...

$t = 0$

2 ...and wraps it into a trailing spiral.

$t = 1$

3 As time passes...

$t = 4$

4 ...the spiral becomes tighter.

$t = 20$

Figure 20.6 The rotation of a spiral galaxy will naturally take even an originally linear structure ($t = 0$) and wrap it into a progressively tighter spiral as time (t) goes by.

gas, or dust. As material in the disk orbits the center of the galaxy, it passes through these spiral density waves.

A spiral density wave has very little effect on the motions of stars as they pass through it, but it does compress the gas that flows through it. As an analogy for this process, consider what happens when you turn on the tap in your kitchen sink (**Figure 20.7**). The water hits the bottom of the sink and spreads out in a thin, rapidly moving layer. A few centimeters out, depending on the rate at which water is flowing, there is a sudden increase in the depth of the water. Spiral arms in galaxies work in much the same way. Gas flows into the spiral density wave and piles up like the water in the sink. Stars form in the resulting compressed gas. Massive stars are concentrated in the arms because they have such short lives (typically 10 million years or so) that they never have the chance to drift far from the spiral arms where they were born. Less massive stars, however, have plenty of time to move away from their places of birth, so they form a smooth underlying disk.

Nebraska Simulation: Traffic Density Analogy

The Size of the Milky Way Galaxy

Because the Solar System is inside the dusty disk of the galaxy itself, the visible-light view of the Milky Way is badly obscured. If you go out on a dark night, away from any streetlights, and look in the direction of the center of the Milky Way—located in the constellation Sagittarius—you will see the dark lane of dusty clouds shown in Figure 20.1a. To probe the structure of the Milky Way, modern astronomers use long-wavelength infrared and radio radiation that can penetrate the dust in the disk. The most powerful tool for this work is the same 21-cm line from neutral interstellar hydrogen, which (as we described in Chapter 19) is used to measure the rotation of other galaxies. The distance to the center of the galaxy, however, still cannot be measured directly. In the 1920s, Harlow Shapley made a three-dimensional map of globular clusters in the Milky Way, which led to the first determination of the size of the Milky Way and the Sun's offset from the center.

Recall from Chapter 17 that globular clusters are large, spheroidal groups of stars held together by gravity. The Milky Way contains more than 150 cataloged globular clusters, and many more are hidden by dust in the disk. Globular clusters are very luminous (as much as 1 million L_{Sun}), so the ones that lie outside of the dusty disk can be easily seen as round, fuzzy blobs even through small telescopes, and even at great distances.

In a Hertzsprung-Russell (H-R) diagram of an old cluster, the horizontal branch crosses the instability strip, which contains pulsating stars such as RR Lyrae stars and Cepheid variables. RR Lyrae stars are easy to spot in globular clusters because they are relatively luminous and have a distinctive light curve.

Figure 20.7 Water from a tap flows in a thin layer along the bottom of a kitchen sink. There is a sudden increase in the depth of the water away from the drain.

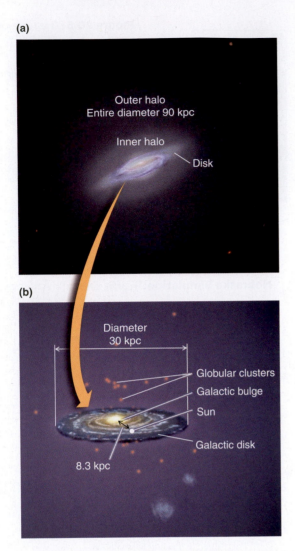

Figure 20.8 Parts of the Milky Way include (a) the disk and the inner and outer halos; (b) the galactic bulge and disk. Globular clusters are located in the halo, and the Sun is located in the disk, about 8.3 kpc from the center. (1 kpc = 1,000 pc.)

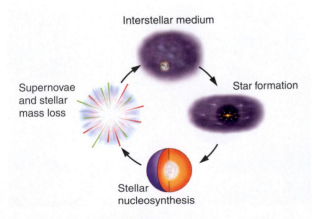

Figure 20.9 Matter moves from the interstellar medium into stars and back again in a progressive cycle that has enriched today's universe with massive elements.

As with Cepheid variables, the time it takes for an RR Lyrae star to undergo one pulsation is related to the star's luminosity. Harlow Shapley used this period-luminosity relationship to find the luminosities of RR Lyrae stars in globular clusters. He then combined these luminosities with measured brightnesses to determine the distances to globular clusters. Finally, Shapley cross-checked his results by noting that more distant clusters (as measured by the RR Lyrae stars) also tended to appear smaller in the sky, as expected.

These globular clusters trace out the luminous part of the galactic halo of the Milky Way Galaxy, a large spherical volume of space surrounding the disk and bulge. The center of the distribution of globular clusters coincides with the gravitational center of the galaxy. Shapley realized that because he could determine the distance to the center of this distribution, he had actually determined the Sun's distance from the center of the Milky Way, as well as the size of the galaxy itself. His map showed that globular clusters occupy a roughly spherical region of space with a diameter of about 90 kiloparsecs (kpc), or 90,000 parsecs (pc). Shapley, however, did not know about gas and dust, so he overestimated the distance to the globular clusters (**Process of Science Figure**). A modern determination indicates that the Sun is located about 8,300 pc (27,000 light-years) from the center of the galaxy, or roughly halfway out toward the edge of the disk.

CHECK YOUR UNDERSTANDING 20.1

Name three pieces of evidence in support of the theory that the Milky Way is a spiral galaxy.

20.2 The Components of the Milky Way Provide Clues about the Formation of Spiral Galaxies

Figure 20.8 illustrates the Sun's position and relationship to the rest of the Milky Way. The Sun is a middle-aged disk star located among other middle-aged stars that orbit around the galaxy within the galactic disk, as do the gas and dust in the disk. The stars in the halo move in random orbits similar to those of stars in elliptical galaxies, sometimes at high velocities. Some of these stars can be observed near the Sun, as their orbits carry them swiftly through the disk. Most of these stars are much older than the Sun. The bar in the galactic bulge of the Milky Way is shaped primarily by stars and gas moving both in highly elongated orbits up and down the long axis of the bar and in short orbits aligned perpendicular to the bar. All these stellar orbits determine the shapes of the different parts of the galaxy, and it is much easier to measure stellar orbits in the Milky Way than in other galaxies. Using the ages, chemical abundances, and motions of nearby stars, astronomers can differentiate between disk and halo stars to learn more about the galaxy's structure. In this section, you will learn how astronomers study the constituents of the Milky Way to find direct clues about how spiral galaxies form.

Age and Chemical Compositions of Stars

Over time, the chemical content of a galaxy changes as stars are born, live and die, and progressively enrich the interstellar medium, as shown in **Figure 20.9**. The interstellar medium therefore reflects all the stellar evolution that has taken place

Process of Science

Shapley's initial efforts to measure the Milky Way did not include dust and gas, because he did not know about them.

Shapley measures the disk of the Milky Way to be roughly 90 kiloparsecs (kpc) across, with the Sun 50 kpc from the center.

Dust and gas are discovered in the disk. Reddening caused Shapley to overestimate the distance.

Shapley's original measurement

Modern value

0 kpc   30 kpc   60 kpc   90 kpc

Today, astronomers routinely account for reddening. The Milky Way's disk is about 30 kpc in diameter and the Sun is 8.3 kpc from the center.

Often, scientists are very aware of what they don't know—for example, the composition of dark matter. Other times, an "unknown unknown" is later discovered, and prior results must be modified to incorporate the new knowledge.

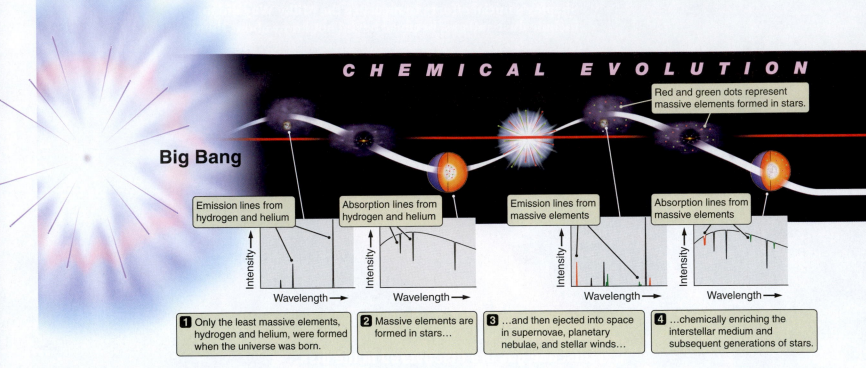

Figure 20.10 As subsequent generations of stars form, live, and die, they enrich the interstellar medium with massive elements—the products of stellar nucleosynthesis. The chemical evolution of the Milky Way and other galaxies can be traced in many ways, including by the strength of interstellar emission lines and stellar absorption lines.

up to the present time. Gas that is rich in massive elements has gone through a great deal of stellar processing. This evolution is similarly reflected in the composition of stars within the galaxy. As illustrated in **Figure 20.10**, the chemical composition of a star's atmosphere reflects the cumulative amount of star formation that occurred before that star formed. Although the details are complex, several clear and important lessons can be learned from the observed patterns in the amounts of massive elements in the galaxy.

Stars in globular clusters, being among the earliest stars to form, should contain only very small amounts of massive elements. Some globular-cluster stars contain only 0.5 percent as much of these massive elements as the Sun. This relationship between age and abundances of massive elements is evident throughout much of the galaxy. Within the disk, younger stars typically have higher abundances of massive elements than those of older stars. Similarly, older stars in the outer parts of the galaxy's bulge have lower massive-element abundances than those of young stars in the disk. Such lower amounts of heavy elements characterize not only globular-cluster stars but also all of the stars in the galaxy's halo, where globular-cluster stars constitute only a minority among the total number of stars in the galactic halo.

In addition to the globular clusters in the halo, younger open clusters orbit in the disk of the Milky Way. As with globular clusters, the stars in an open cluster all formed in the same region at about the same time. Open clusters have a wider range of ages. Some open clusters contain the very youngest stars known; others contain stars that are somewhat older than the Sun. The oldest open clusters in the disk are several billion years younger than the youngest globular clusters in the halo. The differences in ages between globular and open clusters indicate that stars in the halo formed first, but this epoch of star formation did not last long. No young globular clusters are seen. Star formation in the disk started later and has been continuing ever since.

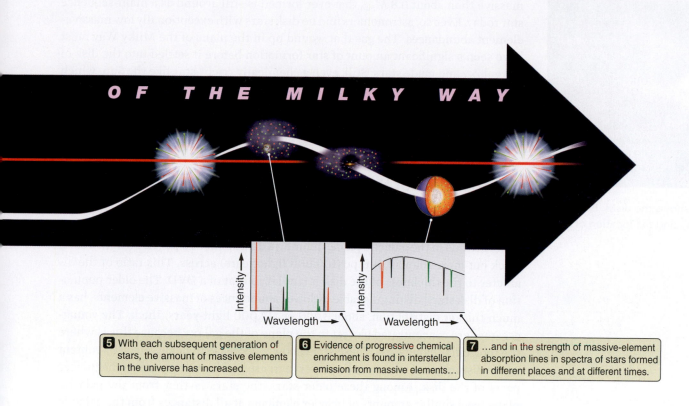

OF THE MILKY WAY

Intensity →

Wavelength →

Intensity →

Wavelength →

5 With each subsequent generation of stars, the amount of massive elements in the universe has increased.

6 Evidence of progressive chemical enrichment is found in interstellar emission from massive elements…

7 …and in the strength of massive-element absorption lines in spectra of stars formed in different places and at different times.

Within the galaxy's disk, astronomers observe differences in abundances of massive elements from place to place, related to the rate of star formation in different regions. Star formation is generally more active in the denser, inner part of the Milky Way than in the outer parts. Observations of chemical abundances in the interstellar medium, based both on interstellar absorption lines in the spectra of stars and on emission lines in glowing clouds of gas known as H II regions, confirm this prediction by showing that there is a smooth decline in abundances of massive elements from the inner to the outer parts of the disk. Astronomers have observed similar trends in other galaxies. Within a galactic disk, relatively old stars near the center of a galaxy often have greater massive-element abundances than those of young stars in the outer parts of the disk.

The basic idea that higher massive-element abundances should follow from the more prodigious star formation in the inner galaxy seems correct, but the full picture is not this simple. New material falling into the galaxy might affect the amounts of heavy elements in the interstellar medium. Chemical elements produced in the inner disk might be blasted into the halo in great "fountains" powered by the energy of massive stars, only to fall back onto the disk elsewhere. Past interactions with other galaxies might have stirred the Milky Way's interstellar medium, mixing gas from those other galaxies with gas already there. The variations of chemical abundances within the Milky Way and other galaxies—and what these variations tell us about the history of star formation and the formation of elements—remain active topics of research.

Even the very oldest globular-cluster stars contain some chemical elements fused in previous generations of more massive stars. This observation implies that globular-cluster stars and other halo stars were not the first stars in the Milky Way to form. At least one generation of massive stars lived and died, ejecting newly synthesized massive elements into space, before even the oldest globular clusters formed. (We will return to these first stars in Chapter 23.) Every star less

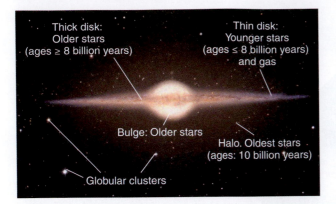

Figure 20.11 This illustration shows the disks, bulge, and inner halo of the Milky Way and the location of globular clusters.

massive than about 0.8 M_{Sun} that ever formed is still around as a main-sequence star today. Even so, astronomers find no disk stars with exceptionally low massive-element abundances. The gas that wound up in the plane of the Milky Way must have seen a significant amount of star formation before it settled into the disk of the galaxy and made stars. Still, even a chemically "rich" star like the Sun, which is made of gas processed through approximately 9 billion years of previous generations of stars, is composed of less than 2 percent massive elements. Luminous matter in the universe is still dominated by hydrogen and helium formed just after the Big Bang, long before the first stars.

Components of the Disk

The disk of the Milky Way has a thin component and a thick component, as illustrated in **Figure 20.11**. The youngest stars in the galaxy are most strongly concentrated in the galactic plane, defining a disk about 300 pc (1,000 light-years) thick but more than 30,000 pc (100,000 light-years) across. This ratio of the diameter to the thickness of the disk is similar to that of a DVD. The older population of disk stars, distinguishable by lower abundances of massive elements, has a much thicker distribution, about 3,700 pc (12,000 light-years) thick. The youngest stars are concentrated closest to the plane of the galaxy because this is where the molecular clouds are. These stars show a decrease in heavy-element content as the distance from the galactic center increases. Older stars make up the thicker parts of the disk. Among these older stars, the stars farther from the galactic plane have similar amounts of heavier elements at all distances from the galactic center.

There are two hypotheses for the origin of this thicker disk. One suggests that these stars formed in the midplane of the disk long ago but were affected by gravitational interactions with massive molecular clouds in the spiral arms that kicked them up out of the plane of the galaxy. The other hypothesis suggests that these stars were acquired from the merging processes that formed the Milky Way Galaxy. As galaxies merge, their clouds of gas cannot pass through each other, so the colliding gas clouds instead settle into a disk at the midplane of the newly formed spiral galaxy. These clouds appear as the concentrated dust lanes that slice the disks of spiral galaxies (as shown in Figure 20.1). This thin lane slicing through the middle of the disk is the place where new stars form and where new stars are found. However, stars are free to move back and forth from one side of the disk to the other, so the older stars can be found above and below this thin disk.

The interstellar medium is a dynamic place—energy from star-forming regions can shape it into impressively large structures. As we mentioned earlier, energy from regions of star formation can form interesting structures in the interstellar medium, clearing out large regions of gas in the disk of a galaxy. Many massive stars forming in the same region can blow "fountains" of hot gas out through the disk of the galaxy via a combination of supernova explosions and strong stellar winds. In the process, dense interstellar gas can be thrown high above the plane of the galaxy as illustrated in the artist's depiction of **Figure 20.12**. Once the gas is a few kiloparsecs above or below the disk, it radiates and cools, falling back to the disk. Maps of the 21-cm emission from neutral hydrogen in the galaxy, X-ray observations, and visible-light images of hydrogen emission from some edge-on external galaxies show numerous vertical structures in the interstellar medium of disk galaxies. These vertical structures are often interpreted as the "walls" of the fountains. If enough massive stars are formed together, sufficient energy may be deposited to blast holes all the way through the plane of the galaxy.

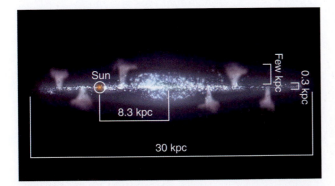

Figure 20.12 This artist's sketch illustrates the concept of a "galactic fountain" in which gas is pushed away from the plane of the galaxy by energy released by young stars and supernovae and then falls back onto the disk. The distance of the Sun from the center of the galaxy and the diameter of the disk are shown here for scale.

Components of the Halo

The globular clusters in the galactic halo indicate that star formation there was early and brief. Yet globular clusters account for only about 1 percent of the total mass of stars in the halo. As halo stars fall through the disk of the Milky Way, some pass close to the Sun, providing a sample of the halo that can be studied at closer range. Most of the stars near the Sun are disk stars like the Sun, but astronomers can distinguish nearby halo stars in two ways. First, most halo stars have much lower amounts of massive elements than those of disk stars. Second, halo stars appear to be moving at higher relative velocities than disk stars. The disk stars near the Sun orbit the center of the galaxy at nearly the same speed, in roughly the same direction. In contrast, halo stars orbit the center of the galaxy in random directions. So the relative velocity between the halo stars and the Sun tends to be high. These stars are known as high-velocity stars.

By studying the orbits of high-velocity stars, astronomers have determined that the halo has two separate components: an inner halo that includes stars up to about 15 kpc (50,000 light-years) from the center, and an outer halo that extends far beyond that (see Figure 20.8a). The stars in the outer halo have lower fractions of heavier elements, suggesting that they formed very early. Many of them are moving in a direction opposite to the rotation of the galaxy. This suggests that the outer halo may have its origins in a merger with a small dwarf galaxy long ago. The orbits of halo stars suggest that these stars fill a volume of space similar to that occupied by the globular clusters in the halo.

X-ray observations indicate that there may be a halo of hot gas surrounding the Milky Way, as shown in the artist's concept in **Figure 20.13**. This gas halo may extend for about 100 kpc (300,000 light-years) from the galactic center, encompassing two nearby small galaxies, and containing as much mass as that of all the stars in the galaxy. Its temperature is estimated to be about 2 million kelvins (K), so the gas particles are moving very quickly. But the gas is extremely diffuse, so the particles are not colliding with each other and transferring energy. The gas wouldn't feel hot, much like the solar corona that we discussed in Chapter 14.

Figure 20.13 This artist's concept shows a hot gas halo surrounding the Milky Way, which may contain as much mass as that of all of the stars in the galaxy combined. The Large Magellanic Cloud (LMC) and Small Magellanic Cloud (SMC) are nearby dwarf galaxies.

Magnetic Fields and Cosmic Rays Fill the Galaxy

The interstellar medium of the Milky Way is laced with magnetic fields that are wound up and compressed by the rotation of the galaxy's disk. The total interstellar magnetic field, however, is about a hundred thousand times weaker than Earth's magnetic field. Charged particles and magnetic fields interact strongly; the particles spiral around magnetic fields, moving along the field rather than across it. Conversely, magnetic fields cannot freely escape from a cloud of gas containing even a small amount of charged particles. The dense clouds of interstellar gas in the midplane of the Milky Way (**Figure 20.14**), anchor the galaxy's magnetic field to the disk, in turn anchoring high-energy charged particles, known as cosmic rays, to the galaxy.

Cosmic rays are charged particles that originate in space and travel close to the speed of light. Despite their name, cosmic rays are not a form of electromagnetic radiation: they were named before their true nature was known. Most cosmic-ray particles are protons, but some are nuclei of helium, carbon, and other elements produced by nucleosynthesis. A few are high-energy electrons and other subatomic particles. Cosmic rays span an enormous range in particle

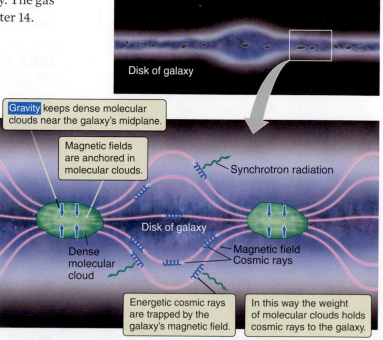

Figure 20.14 The weight of interstellar clouds anchors the magnetic field of the Milky Way to the disk of the galaxy. The magnetic field, in turn, traps the galaxy's cosmic rays.

energy. Astronomers can observe the lowest-energy cosmic rays with interplanetary spacecraft. These cosmic rays have energies as low as about 10^{-11} joule (J), which corresponds to the energy of a proton moving at a velocity of a few tenths the speed of light.

In contrast, the most energetic cosmic rays are 10 trillion (10^{13}) times as energetic as the lowest-energy cosmic rays, and they move very close to the speed of light, $0.999999c$. These high-energy cosmic rays are continually hitting Earth and are detected from the showers of elementary particles that they cause when crashing through Earth's atmosphere. Astronomers hypothesize that most of these cosmic rays are accelerated to these incredible energies by the shock waves produced in supernova explosions. The very highest-energy cosmic rays are as much as a hundred million times more energetic than any particle ever produced in a particle accelerator on Earth. These extremely high energies make these cosmic rays much more difficult to explain than those with lower energies.

The disk of the galaxy glows from synchrotron radiation (see Chapter 17) produced by cosmic rays, mostly electrons, spiraling around the galaxy's magnetic field. Such synchrotron emission is seen in the disks of other spiral galaxies as well, indicating that they, too, have magnetic fields and populations of energetic cosmic rays. The very highest-energy cosmic rays are moving much too fast to be confined by the gravitational force of their originating galaxy. Any such cosmic rays that formed in the Milky Way would soon stream away from the galaxy into intergalactic space. Thus, it is likely that some of the energetic cosmic rays reaching Earth originated in energetic events outside of the Milky Way Galaxy.

The total energy of all the cosmic rays in the galactic disk can be estimated from the energy of the cosmic rays reaching Earth. The strength of the interstellar magnetic field can be measured by observing the effect that it has on the properties of radio waves passing through the interstellar medium. These measurements indicate that in the Milky Way Galaxy, the magnetic-field energy and the cosmic-ray energy are about equal to each other. Both are comparable to the energy present in other energetic components of the galaxy, including the motions of interstellar gas and the total energy of electromagnetic radiation within the galaxy.

CHECK YOUR UNDERSTANDING 20.2

What parts of the Milky Way contain old stars, and what parts contain young stars?

20.3 Most of the Milky Way Is Unseen

Nebraska Simulation: Milky Way Rotational Velocity

As in other galaxies, the most interesting parts of the Milky Way may be the parts that can't be seen directly but only detected by their gravitational influence on the stars around them. Dark matter accounts for the vast majority of the mass in a galaxy and extends far beyond a galaxy's visible boundary. As in all other spiral galaxies, there is compelling evidence that dark matter dominates the Milky Way. From radio and infrared observations, astronomers can figure out how the disk of the Milky Way moves, and from that motion, in turn, they can determine its mass. The supermassive black hole at the center of the Milky Way provides a different kind of observation problem. This object is also detected by its gravitational effects on the stars nearby but cannot be seen directly. In this section, we will explore what can be inferred about these two components of the Milky Way from the effect of their gravity on other objects.

Dark Matter in the Milky Way

The rotation of the disk of the Milky Way can be determined from observations of the relative velocities of interstellar hydrogen measured from 21-cm radiation. **Figure 20.15** shows how these velocities vary with viewing direction from the Sun. Looking toward the center of the galaxy, the gas is stationary relative to the Sun. Looking in the direction of the Sun's motion around the galactic center, hydrogen clouds appear to be moving toward Earth, while in the opposite direction, clouds are moving away from Earth. In other directions, the measured velocities are complicated by Earth's moving vantage point within the disk and so are more difficult to interpret at a glance. This is a pattern of the rotation velocity of gas in a disk like those you learned about in the past chapter. The only difference is that instead of looking at it from outside, we see the Milky Way Galaxy's rotation curve from a vantage point located within—and rotating with—the galaxy. Even so, observed velocities of neutral hydrogen enable astronomers to measure the Milky Way Galaxy's rotation curve and even determine the structure present throughout its disk.

Recall from the previous chapter that observations of rotation curves led astronomers to conclude that the masses of spiral galaxies consist mostly of dark matter. **Figure 20.16** shows the rotation curve of the Milky Way as inferred primarily from 21-cm observations. The orbital motion of the nearby dwarf galaxy called the Large Magellanic Cloud provides data for the outermost point in the rotation curve, at a distance of roughly 50,000 pc (160,000 light-years) from the center of the galaxy. Like other spiral galaxies, the Milky Way has a fairly flat rotation curve. The mass outside of the Sun's orbit does not greatly affect the Sun's orbit.

The total mass of the Milky Way Galaxy is currently estimated to be about 1.0 trillion to 1.5 trillion times the mass of the Sun. However, the luminous mass, estimated by adding the masses of stars, dust, and gas, is only about one-tenth

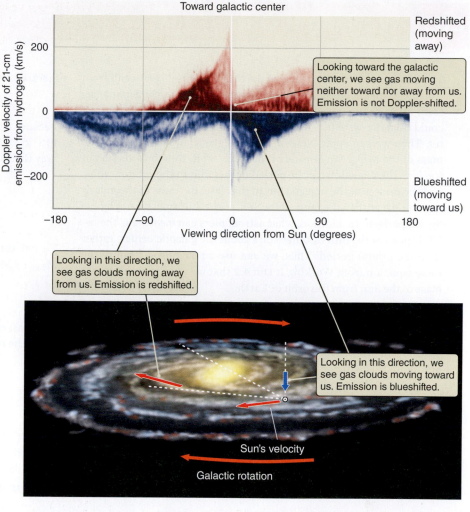

Toward galactic center

Redshifted (moving away)

Looking toward the galactic center, we see gas moving neither toward nor away from us. Emission is not Doppler-shifted.

Blueshifted (moving toward us)

Looking in this direction, we see gas clouds moving away from us. Emission is redshifted.

Looking in this direction, we see gas clouds moving toward us. Emission is blueshifted.

Sun's velocity

Galactic rotation

Figure 20.15 The graph shows Doppler velocities measured from observations of 21-cm emission from interstellar clouds of neutral hydrogen. These velocities vary between redshift and blueshift as an observer looks around in the plane of the Milky Way, along the sight lines shown as dashed white lines. From Earth's perspective within the Solar System, the signature of a rotating disk is clear from views of either side of the galactic center.

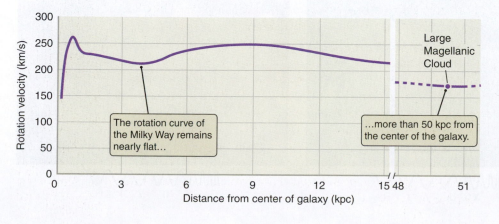

Large Magellanic Cloud

The rotation curve of the Milky Way remains nearly flat…

…more than 50 kpc from the center of the galaxy.

Figure 20.16 Rotation velocity is plotted against distance from the center of the Milky Way. The most distant point comes from measurements of the orbit of the Large Magellanic Cloud. The nearly flat rotation curve indicates that dark matter dominates the outer parts of the Milky Way. Part of this graph has been hidden in order to fit the Large Magellanic Cloud into the figure. This is indicated by the broken x-axis, marked by the diagonal hatch marks.

20.1   Working It Out   The Mass of the Milky Way inside the Sun's Orbit

The Sun orbits about the center of the Milky Way Galaxy. Recall from Chapter 4 that even though the gravitational pull on the Sun comes from all of the material inside this orbit, Newton showed that we could treat the system as if all the mass were concentrated at the center. Then we can apply Newton's and Kepler's laws to calculate the mass of the Milky Way inside of the Sun's orbit. Newton's version of Kepler's third law relates the period of the orbit to the orbital radius and the masses of the objects. But in this case, the mass of the galaxy is much larger than the mass of the Sun, so the Sun's mass is negligible by comparison. In addition, what astronomers can *measure* is the orbital speed of the Sun or other stars about the galactic center, rather than the orbital period. Thus, we can use a rearranged form of the same equation from Working It Out 4.2 that we used to estimate the mass of the Sun from the orbit of Earth:

$$M = \frac{r v_{circ}^2}{G}$$

The Sun orbits the center of the galaxy at 220 kilometers per second (km/s). The distance of the Sun from the center of the galaxy is 8,300 pc, which converts to kilometers as

$$8{,}300 \text{ pc} \times (3.09 \times 10^{13} \text{ km/pc}) = 2.56 \times 10^{17} \text{ km}$$

Because we know that the gravitational constant $G = 6.67 \times 10^{-20} \text{ km}^3/(\text{kg s}^2)$, we can calculate the mass of the portion of the Milky Way that is inside of the Sun's orbit:

$$M = \frac{(2.56 \times 10^{17} \text{ km}) \times (220 \text{ km/s})^2}{6.67 \times 10^{-20} \text{ km}^3/(\text{kg s}^2)}$$

$$M = 1.86 \times 10^{41} \text{ kg}$$

To put this in units of the Sun's mass, we divide this answer by $M_{Sun} = 1.99 \times 10^{30}$ kg, yielding

$$M = \frac{1.86 \times 10^{41} \text{ kg}}{1.99 \times 10^{30} \text{ kg}/M_{Sun}} = 9.35 \times 10^{10} M_{Sun}$$

The mass of the Milky Way inside of the Sun's orbit is about 94 billion times the mass of the Sun.

as much. Like other spiral galaxies, the Milky Way's mass consists mainly of dark matter. The spatial distribution of dark and normal matter within the Milky Way is also much like that of other galaxies, with dark matter dominating its outer parts (**Working it Out 20.1**).

The Supermassive Black Hole

Figure 20.17 shows images of the Milky Way's center taken with the Chandra X-ray Observatory and the Spitzer Space Telescope. The X-ray view (Figure 20.17a) shows the location of a strong radio source called Sagittarius A* (abbreviated

(a)

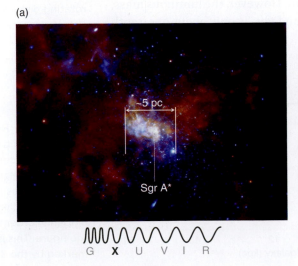

(b)

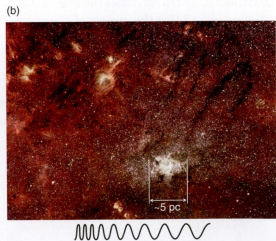

Figure 20.17 (a) This X-ray view of the Milky Way's central region shows the active source, Sagittarius A* (Sgr A*), as the brightest spot at the middle of the image. Lobes of superheated gas (shown in red) are evidence of recent, violent explosions happening near Sgr A*. (b) This infrared view of the central core of the Milky Way shows hundreds of thousands of stars. The bright white spot at the lower right marks Sgr A*, the location of the supermassive black hole.

Sgr A*), which lies at the center of the Milky Way. The infrared image (Figure 20.17b) cuts through the dust to reveal the galaxy's crowded, dense core containing hundreds of thousands of stars.

Studies of the motions of stars closest to the Sgr A* source suggest a central mass very much greater than that of the few hundred stars orbiting there. Furthermore, observations of the galaxy's rotation curve show rapid rotation velocities very close to the galactic center. Stars closer than 0.1 light-year from the galactic center follow Kepler's laws, indicating that their motion is dominated by mass within their orbit. The closest stars studied are only about 0.01 light-year from the center of the galaxy—so close that their orbital periods are only about a dozen years. The positions of these stars change noticeably over time, and astronomers can see them speed up as they whip around what can only be a supermassive black hole at the focus of their elliptical orbits, as shown in **Figure 20.18**. Using Newton's version of Kepler's third law, we can then estimate that the black hole at the center of the Milky Way Galaxy is a relative lightweight, having a mass of "only" 4 million times the mass of the Sun (**Working It Out 20.2**).

Clouds of interstellar gas at the galaxy's center are heated to millions of degrees by shock waves from supernova explosions and colliding stellar winds blown outward by young, massive stars. Superheated gas produces X-rays, and the Chandra X-ray Observatory has detected more than 9,000 X-ray sources within the central region of the galaxy. These include frequent, short-lived X-ray flares near Sgr A* (see Figure 20.17a), which provide direct evidence that matter falling toward the supermassive black hole fuels the energetic activity at the galaxy's center.

The Fermi Gamma-ray Space Telescope has observed gamma-ray-emitting bubbles that extend 8 kpc (25,000 light-years) above and below the galactic plane. The bubbles may have formed after a burst of star formation a few million years ago produced massive star clusters near the center of the galaxy. If some of the gas formed stars and about 2,000 M_{Sun} of material fell into the supermassive black hole, enough energy could have been released to power the bubbles. More recently, faint gamma-ray signals were observed that look like jets coming from the

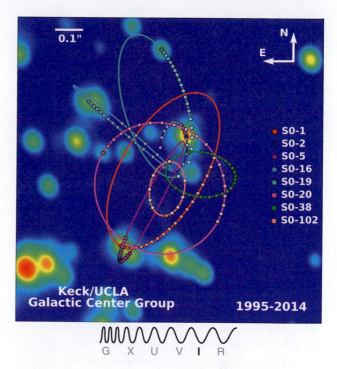

Figure 20.18 This figure shows orbits of seven stars within 0.03 pc (0.1 light-year, or about 6,000 astronomical units [AU]) of the Milky Way's center. The Keplerian motions of these stars reveal the presence of a 4-million-M_{Sun} supermassive black hole at the galaxy's center. Colored dots show the measured positions of each of the stars over many years: the dots progress from lighter in 1995 to darker in 2014.

20.2 Working It Out   The Mass of the Milky Way's Central Black Hole

Figure 20.18 illustrates data points for the stars in the central region orbiting closely to the central black hole of the Milky Way Galaxy. These stars have highly elliptical orbits with changing speeds, but the orbital periods are short enough that they can be observed and measured. Star S0-2 in the figure has a measured orbital period of 15.8 years. The semimajor axis of its orbit is estimated to be 1.5×10^{11} km = 1,000 AU. With this information, we can use Newton's version of Kepler's third law to estimate the mass inside of S0-2's orbit. Setting up the equation as we did in Working It Out 13.4:

$$\frac{m_{BH}}{M_{Sun}} + \frac{m_{S0\text{-}2}}{M_{Sun}} = \frac{A_{AU}^3}{P_{years}^2}$$

The mass of star S0-2 is much less than the mass of the black hole, so the sum of the two is very close to the mass of the black hole. Therefore, we can write

$$\frac{m_{BH}}{M_{Sun}} = \frac{A_{AU}^3}{P_{years}^2} = \frac{1{,}000^3}{15.8^2} = 4.0 \times 10^6$$

$$m_{BH} = 4.0 \times 10^6 \, M_{Sun}$$

The supermassive black hole at the center of the Milky Way has a mass 4 million times that of the Sun. This is quite a bit less than the billion-solar-mass black holes in some active galactic nuclei (AGNs) that we discussed in Chapter 19.

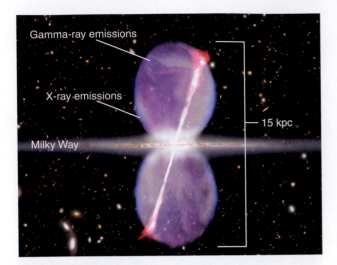

Figure 20.19 The Fermi Gamma-ray Space Telescope observed gamma-ray bubbles (purple) extending 8 kpc above and below the galactic plane. Hints of the edges of the bubbles were first observed in X-rays (blue) in the 1990s. In this artist's conceptual view from outside of the galaxy, the gamma-ray jets (pink) are tilted with respect to the bubbles, which might suggest that the accretion disk around the black hole is tilted as well.

center, within the bubbles, as shown in the artist's depiction in **Figure 20.19**. If these jets are originating from material falling into the supermassive black hole, activity might be even more recent—perhaps 20,000 years ago. Some astronomers predict that gas clouds are heading toward the center and will soon be accreted by the black hole. Currently, the observed activity is not as intense as that seen in active galaxies with central, supermassive black holes. The inner Milky Way is a reminder that it was almost certainly "active" in the past and could become active once again.

CHECK YOUR UNDERSTANDING 20.3

Which property is detectable for both dark matter and the supermassive black hole at the center of the Milky Way? (a) their luminosities; (b) their temperatures; (c) their gravity; (d) their composition

20.4 The History and Future of the Milky Way

One of the fundamental goals of stellar astronomy is to understand the life cycle of stars, including how stars form from clouds of interstellar gas. In Chapter 15, we told a fairly complete story of this process, at least as it occurs today, and tied this story strongly to observations of Earth's galactic neighborhood. Galactic astronomy has a similar basic goal. Astronomers would like to have a complete and well-tested theory of how the Milky Way formed and to be able to make predictions about its future. The distribution of stars of different ages with different amounts of heavy elements is one clue. Additional clues come from studying other galaxies at different distances (and therefore of different ages), their supermassive black holes, and their merger history. In this section, we explore the history and the future of the Milky Way.

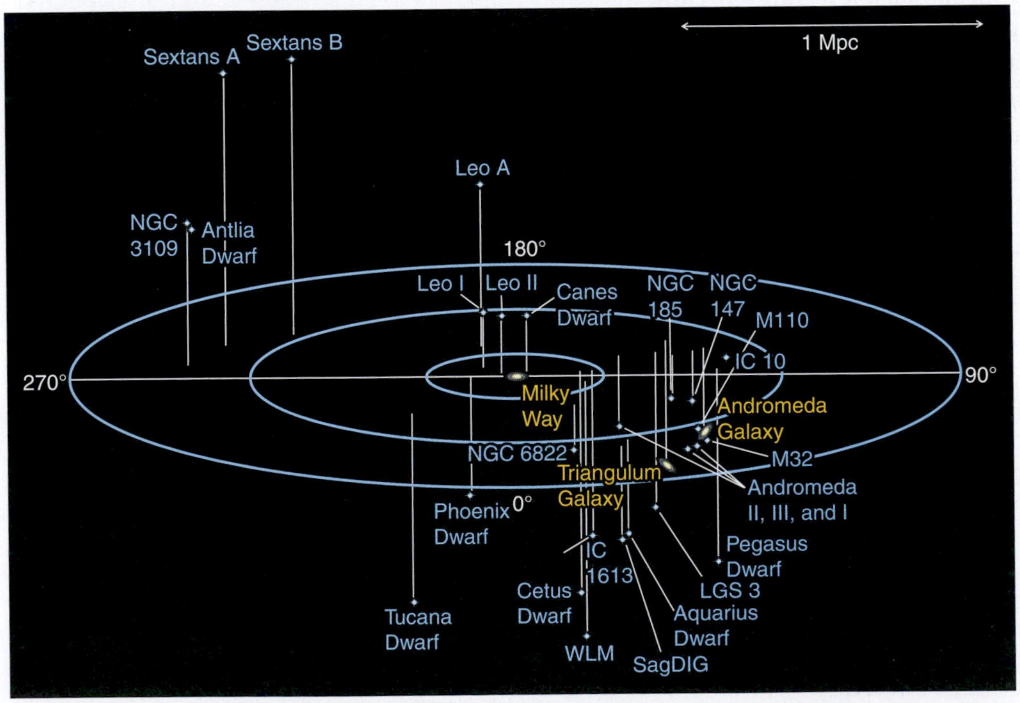

The Local Group

Galaxies do not exist in isolation. The vast majority of galaxies are parts of gravitationally bound collections of galaxies. The smallest and most common of these are called **galaxy groups**. A galaxy group contains as many as several dozen galaxies, most of them dwarf galaxies. As we saw in Chapter 1, the Milky Way is a member of the Local Group, first identified by Edwin Hubble in 1936. Hubble labeled 12 galaxies as part of the Local Group, but now astronomers count at least 50. As shown in **Figure 20.20**, the Local Group includes the two giant barred spirals—the Milky Way Galaxy and the Andromeda Galaxy—along with a few ellipticals and irregulars and at least 30 smaller dwarf galaxies in a volume of space

Figure 20.20 This graphical map shows some of the members of the Local Group of galaxies. Most are dwarf galaxies. Spiral galaxies are shown in yellow. The closest galaxies to the Milky Way are not seen on this scale. (1 Mpc = 1 megaparsec = 1 million parsecs.)

about 3 million pc (10 million light-years) in diameter. Almost 98 percent of all the galaxy mass in the Local Group resides in just its two giant galaxies. The third largest galaxy, Triangulum, is an unbarred spiral with about one-tenth the mass of the Milky Way or Andromeda. Most but not all of the dwarf elliptical and dwarf spheroidal galaxies in the group are satellites of the Milky Way or Andromeda. The Local Group interacts with a few nearby groups, which will be discussed further in Chapter 23.

There are more than 20 of these satellite dwarf galaxies, although it's not certain that all are gravitationally bound to the Milky Way. Some of the fainter dwarf galaxies were discovered only very recently because of their low luminosity. The dwarf galaxies are the lowest-mass galaxies observed, and they are dominated by an even greater percentage of invisible dark matter than are other known galaxies. They also contain stars very low in elements more massive than helium. These ultrafaint dwarf galaxies offer clues to the formation of the Local Group. In addition, observations of the motions and speeds of the dwarf galaxies about the Milky Way may lead to new estimates of the dark matter mass within the Milky Way itself.

The Formation of the Milky Way

We have seen that globular clusters and high-velocity stars must have been among the first stars formed in the Milky Way that still exist. The fact that they are not concentrated in the disk or bulge of the galaxy indicates that they formed from clouds of gas well before those clouds settled into the galaxy's disk. This hypothesis is supported by observations that globular clusters are very old and that the youngest globular cluster is older than the oldest disk stars. The presence of extremely small amounts of massive elements in the atmospheres of halo stars also indicates that at least one generation of stars must have lived and died *before* the formation of the halo stars visible today. This line of reasoning implies that the Milky Way formed from the merger of a number of smaller clumps of matter, which included both stars and clouds of dust and gas.

Combining that conclusion with the presence of the central, supermassive black hole and the number of nearby dwarf galaxies, astronomers conclude that the Milky Way must have formed when the gas within a huge "clump" of dark matter collapsed into a large number of small protogalaxies. Some of these protogalaxies then merged to form the large, barred spiral galaxies in the Local Group, but some of these smaller protogalaxies are still around today in the form of the small, satellite dwarf galaxies near the Milky Way. The largest among them are the Large Magellanic Cloud and the Small Magellanic Cloud (**Figure 20.21**), which are easily seen by the naked eye in the Southern Hemisphere and appear much like detached pieces of the Milky Way. The Magellanic Clouds were named for Ferdinand Magellan (circa 1480–1521), who headed the first European expedition that ventured far enough into the Southern Hemisphere to see them.

The Future of the Milky Way

Mergers and collisions of Local Group galaxies continue today. Among the closest companions to the Milky Way is the Sagittarius Dwarf Galaxy, which is plowing through the disk of the Milky Way on the other side of the bulge. Astronomers have observed streams of stars, as sketched in **Figure 20.22**, from Sagittarius

Large Magellanic Cloud

Small Magellanic Cloud

G X U V I R

Figure 20.21 The Milky Way is surrounded by more than 20 dwarf companion galaxies: the largest among them are the Large Magellanic Cloud (LMC) and Small Magellanic Cloud (SMC) (see Figure 20.13).

Figure 20.22 This artist's impression shows tidal tails of stars from the Sagittarius Dwarf elliptical galaxy (reddish-orange). These stars have been stripped from the dwarf galaxy by the much more massive Milky Way, and the two galaxies will eventually merge in billions of years.

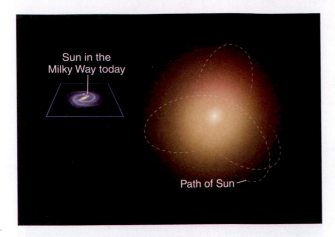

Figure 20.23 A computer simulation of the orbit of the Sun within the Milky Way–Andromeda merger remnant 10 billion years from now.

Dwarf and some of the other dwarf galaxies that are being tidally disrupted by the Milky Way. These dwarf galaxies will become incorporated into the Milky Way—an indication that the galaxy is still growing. Computer simulations suggest that such mergers could cause the spiral-arm structure.

The Andromeda Galaxy appears to violate Hubble's law because its spectrum shows blueshifts, indicating that the galaxy is moving toward, not away from, the Milky Way at 110 km/s (400,000 km/h). Andromeda and the Milky Way are the two largest galaxies in the Local Group, about 770 kpc (2.5 million light-years) apart. (If each galaxy were the size of a quarter, they would be about an arm's length apart.) Since Edwin Hubble noticed nearly 100 years ago that Andromeda's spectrum was blueshifted, indicating that it was approaching the Milky Way, astronomers have wondered if and when Andromeda and the Milky Way would collide and form a merged galaxy.

Astronomers used the Hubble Space Telescope to measure the perpendicular motion of Andromeda and determine whether a collision will be head-on, partial, or a total miss. They concluded that the two galaxies will "collide" head-on in about 4 billion years, although because most of a galaxy is space between the stars, actual collisions between the stars themselves are quite unlikely: while diffuse gas in the interstellar medium will collide and cause hot shocks, most of the material of one galaxy will pass through the other. It will take another 2 billion years for the two galaxies to merge completely and form one giant elliptical galaxy. The third spiral in the Local Group, the Triangulum Galaxy, may merge first with the Milky Way, or with Andromeda, or be ejected from the Local Group.

Note the timing here. This first close encounter with Andromeda in 4 billion years will occur *before* the Sun runs out of hydrogen in its core, although it is likely that the Sun will have increased its luminosity enough by then that Earth's habitability will have been affected. In 6 billion years, the Sun and its planets could end up near the center of the merged galaxy, or more likely they will have a new location farther from the center with a different orbit and in a different stellar neighborhood (**Figure 20.23**). If by chance the encounter leads to a star passing close to the Sun, the orbits of the Solar System planets could be disrupted, causing them to be at different distances from the Sun. **Figure 20.24**

Figure 20.24 This computer simulation depicts the view in the sky on Earth when the Andromeda and Milky Way galaxies collide. The spiral Andromeda appears larger in the sky as it gets closer, and then the sky becomes increasingly bright as the two collide.

shows what this merged galaxy might look like in Earth's sky—if anyone is still around to see it!

Why is Andromeda now moving toward us?

Origins

The Galactic Habitable Zone

In Chapter 19, we discussed the concept of galactic habitable zones in general; in this Origins, we will focus more specifically on ideas about the habitable zone of the Milky Way. Stars that are situated too far from the galactic center may have protoplanetary disks with insufficient quantities of heavy elements—such as oxygen, silicon, iron, and nickel (to make up rocky planets like Earth) or carbon, nitrogen, and oxygen (to make up the molecules of life). Stars that are too close to the galactic center may have planets that are too strongly affected by its high-energy radiation environment (X-rays and gamma rays from the supermassive black hole) and by supernova explosions and gamma-ray bursts (GRBs). The bulge has a higher density of stars creating a strong radiation field, and the halo and thick disk have stars with lower amounts of heavy elements, so perhaps only stars in the thin disk of the galaxy are candidates for residence in a galactic habitable zone.

Astronomers must also consider stellar lifetimes versus the 4 billion years after the formation of Earth that it took for life to evolve into land animals—so only stars with masses low enough that they will live at least 4 billion years on the main sequence are considered potential hosts of more complex life. In this simple model based on the evolution times for life on Earth, the galactic habitable zone could be a doughnut-shaped region around the galactic center. In one version of the model, this zone is estimated to contain stars born 4 billion to 8 billion years ago located between 7 and 9 kpc from the galactic center, with the Sun exactly in the middle of the doughnut. The doughnut would grow larger over time as heavier-element formation spread outward from the galactic center.

Some have conducted additional studies and proposed more complex models. For example, one group initiated a search for the molecule formaldehyde (H_2CO)—a key "prebiotic" molecule—by observing molecular clouds in the outer parts of the Milky Way, 12–23.5 kpc from the galactic center. Formaldehyde was detected in two-thirds of the group's sample of 69 molecular clouds, suggesting that at least one important prebiotic molecule is available far from the galactic center.

Another computer model took into account details of the evolution of individual stars within the Milky Way, including birth rates, locations, distribution within the galaxy, abundances of heavy elements, stellar masses, main-sequence lifetimes, and the likelihood that stars became or will become supernovae. The model also assumed 4 billion years to the development of complex life. One result of this model is that stars in the inner part of the Milky Way are more likely to be affected by supernova explosions, but these stars are even more likely to have the heavier elements for the formation of planets. So in this model, the inner part of the galaxy, about 2.5–4 kpc from the center, in and near the midplane of the thin disk, is the most likely place for habitable planets. In this case, the Sun is *not* in the middle of the most probable zone for habitable planets. As more extrasolar planets are discovered, astronomers will have a better idea of their distribution throughout the Milky Way.

 Nebraska Simulation: Milky Way Habitability Explorer

As noted, mergers with other galaxies could cause stars to migrate into or out of the galactic habitable zone. The uncertainties increase as the assumptions in these models move from the astronomical to the biological. For example, maybe life evolved faster on other planets, so stars of higher mass and shorter lifetimes should be included. Or maybe life evolved slower elsewhere, in which case the older stars would be the best candidates. It is unknown whether intense radiation from a supernova or a gamma-ray burst would permanently sterilize a planet or only affect evolution for a while. For example, if Earth's ozone layer was temporarily destroyed, life on land might die out, but life in the oceans would continue. Habitability in the Milky Way Galaxy is complicated, with many unanswered questions.

ARTICLES   QUESTIONS

In this article, astronomers report on their estimate of the amount of dark matter in the Milky Way Galaxy.

Dark Matter Half What We Thought, Say Scientists

International Centre for Radio Astronomy Research

A new measurement of dark matter in the Milky Way has revealed there is half as much of the mysterious substance as previously thought.

Australian astronomers used a method developed almost 100 years ago to discover that the weight of dark matter in our own galaxy is 800,000,000,000 (or 8×10^{11}) times the mass of the Sun.

They probed the edge of the Milky Way, looking closely, for the first time, at the fringes of the galaxy about 5 million trillion kilometers from Earth.

Astrophysicist Dr. Prajwal Kafle, from The University of Western Australia node of the International Centre for Radio Astronomy Research, said we have known for a while that most of the universe is hidden.

"Stars, dust, you and me, all the things that we see, only make up about 4 percent of the entire universe," he said.

"About 25 percent is dark matter and the rest is dark energy."

Dr. Kafle, who is originally from Nepal, was able to measure the mass of the dark matter in the Milky Way by studying the speed of stars throughout the galaxy, including the edges, which had never been studied to this detail before.

He used a robust technique developed by British astronomer James Jeans in 1915—decades before the discovery of dark matter.

Dr. Kafle's measurement helps to solve a mystery that has been haunting theorists for almost two decades.

"The current idea of galaxy formation and evolution, called the Lambda Cold Dark Matter theory, predicts that there should be a handful of big satellite galaxies around the Milky Way that are visible with the naked eye, but we don't see that," Dr. Kafle said.

"When you use our measurement of the mass of the dark matter the theory predicts that there should only be three satellite galaxies out there, which is exactly what we see; the Large Magellanic Cloud, the Small Magellanic Cloud, and the Sagittarius Dwarf Galaxy."

University of Sydney astrophysicist Professor Geraint Lewis, who was also involved in the research, said the missing satellite problem had been "a thorn in the cosmological side for almost 15 years."

"Dr. Kafle's work has shown that it might not be as bad as everyone thought, although there are still problems to overcome," he said.

The study also presented a holistic model of the Milky Way, which allowed the scientists to measure several interesting things such as the speed required to leave the galaxy.

"Be prepared to hit 550 kilometers per second if you want to escape the gravitational clutches of our galaxy," Dr. Kafle said.

"A rocket launched from Earth needs just 11 kilometers per second to leave its surface, which is already about 300 times faster than the maximum Australian speed limit in a car!"

ARTICLES   QUESTIONS

1. How far from the center of the galaxy in parsecs were the measured stars?
2. Where would this measurement fit on Figure 20.16?
3. How does this mass of dark matter compare with the estimate of mass inside of the Sun's orbit?
4. Where is this additional mass located?
5. Explain how a new estimate of the mass of the Milky Way affects the calculation of the escape velocity from the Milky Way.

"Dark matter half what we thought, say scientists." ICRAR Media Release, October 10, 2014. Reprinted by permission of the International Centre for Radio Astronomy Research.

Summary

Astronomers compare images and spectra of other galaxies to observations of the Milky Way. From this, they determine that the Milky Way is a barred spiral of type SBbc. The Milky Way formed from a collection of smaller protogalaxies that collapsed out of a halo of dark matter. The idea of a galactic habitable zone is that certain parts of the Milky Way may be more suitable for the existence of habitable planets. This zone would have enough heavy elements for the formation of rocky planets and organic molecules, but not too much radiation that it would damage any life.

LG 1 **Explain how astronomers discovered the size and spiral structure of the Milky Way.** The distances to globular clusters can be found from the luminosity of variable stars within them. Because these globular clusters are symmetrically distributed around the center of the Milky Way, the center of the distribution is located at the center of the galaxy. The Sun is located about 8,300 pc (27,000 light-years) from the Milky Way's center, and the Milky Way is 30,000 pc (100,000 light-years) across.

LG 2 **List the clues of galaxy formation that can be found from the components of the Milky Way.** The chemical composition of the Milky Way has evolved with time, as material cycles between stars and the interstellar medium. There must have been a generation of stars before the oldest halo and globular-cluster stars we see today formed. The Milky Way has a disk consisting of two parts, the thick disk of old stars and the thin disk of young stars, implying that gas and dust from merging galaxies settled onto the disk, while the stars passed through. The galactic halo consists of an inner halo and outer halo of stars and globular clusters, as well as a large, hot gas halo. The abundance of heavy elements in the Milky Way has increased with time as each generation of stars has produced more of these elements during the final phases of the stars' lives.

LG 3 **Explain the evidence for the dark matter halo and for the supermassive black hole at the center of the Milky Way.** The Doppler velocities of radio spectral lines show that the rotation curve of the Milky Way is flat, like those of other galaxies. But the inferred mass cannot be accounted for by the mass that is observed directly. This indicates that the Milky Way's mass is mostly in the form of dark matter. Evidence for the black hole at the center of the galaxy includes rapid orbital velocities of nearby stars and symmetric X-ray and gamma-ray outflows of material.

LG 4 **Describe the Local Group of galaxies and how it provides clues about the evolution of the Milky Way.** The Milky Way is part of the Local Group of galaxies, which consists of two large, barred spirals and several dozen smaller galaxies. Collisions and mergers between these galaxies likely happened in the past, and a merger with the Andromeda Galaxy may be part of the Milky Way's future. The dwarf satellites and other neighbors in the Local Group are evidence that the Milky Way is growing through accretion.

? UNANSWERED QUESTIONS

- Are there many more ultrafaint dwarf galaxies than have so far been detected? These types of dwarf galaxies are so faint that they are hard to detect even when close to the Milky Way. As we will discuss in Chapter 23, the giant spirals such as the Milky Way were built up by mergers of these small, faint galaxies, so models predict that there should be hundreds or even thousands of them. They could be so dominated by dark matter that they are not at all visible, or too small ever to have formed stars, or they might have merged with the Milky Way or other Local Group members long ago.

- Will the merger of the Milky Way and Andromeda form a quasar at the center of the new elliptical galaxy? If the dust and gas at the center of the galaxy did not completely block the view from Earth, such a quasar could be brighter than the full Moon. The supermassive black hole at the center of Andromeda is thought to be 25–50 times larger than the one in the Milky Way, but both black holes combined still would be on the lower end of the masses of the black holes in AGNs discussed in Chapter 19.

Questions and Problems

Test Your Understanding

1. The size of the Milky Way is determined from studying _____ stars in globular clusters.
   a. Cepheid variable
   b. blue supergiant
   c. RR Lyrae
   d. Sun-like

2. Detailed observations of the structure of the Milky Way are difficult because
   a. the Solar System is embedded in the dust and gas of the disk.
   b. the Milky Way is mostly dark matter.
   c. there are too many stars in the way.
   d. the galaxy is rotating too fast (about 200 km/s).

3. Older stars are found farther from the midplane of a galactic disk because
   a. the disk used to be thicker.
   b. the stars have lived long enough to move there.
   c. the younger stars in the thick disk were more massive and have already died.
   d. they passed through the disk when the galaxy formed.

4. The magnetic field of the Milky Way has been detected by
   a. synchrotron radiation from cosmic rays.
   b. direct observation of the field.
   c. its interaction with Earth's magnetic field.
   d. studying molecular clouds.

5. Evidence of a supermassive black hole at the center of the Milky Way comes from
   a. direct observations of stars that orbit it.
   b. visible light from material that is falling in.
   c. strong radio emission from the black hole itself.
   d. streams of cosmic rays from the center of the galaxy.

6. The Large Magellanic Cloud and Small Magellanic Cloud will likely
   a. become part of the Milky Way.
   b. remain orbiting forever.
   c. become attached to another passing galaxy.
   d. escape from the gravity of the Milky Way.

7. Globular clusters are important to understanding the Milky Way because
   a. they are so young that they provide information about current star formation.
   b. they provide information about dwarf ellipticals from which the Milky Way formed.
   c. they reveal the size of the Milky Way and Earth's location in it.
   d. the stars in them are highly enhanced in metals.

8. A globular cluster with no variable stars would have been left out of Shapley's study because
   a. the cluster would be too far away.
   b. the distance to the cluster could not be determined.
   c. the cluster would be too faint to see.
   d. the cluster would be too young to determine its evolutionary state.

9. The best evidence for the presence of dark matter in the Milky Way comes from the observation that the rotation curve
   a. is flat at great distances from the center.
   b. rises swiftly in the interior.
   c. falls off and then rises again.
   d. has a peak at about 2,000 light-years from the center.

10. Cosmic rays are
    a. a form of electromagnetic radiation.
    b. high-energy particles.
    c. high-energy dark matter.
    d. high-energy photons.

11. What kind of galaxy is the Milky Way?
    a. elliptical
    b. spiral
    c. barred spiral
    d. irregular

12. Where are the youngest stars in the Milky Way Galaxy?
    a. in the core
    b. in the bulge
    c. in the disk
    d. in the halo

13. Halo stars are found in the vicinity of the Sun. What observational evidence does not distinguish them from disk stars?
    a. the direction of their motion
    b. their speed
    c. their composition
    d. their temperature

14. Why are most of the Milky Way's satellite galaxies so difficult to detect?
    a. They are very small.
    b. They are very far away.
    c. The halo of the Milky Way blocks the view.
    d. They are very faint.

15. The concept of a galactic habitable zone does not consider
    a. the radiation field.
    b. the ages of stars.
    c. the amount of heavy elements.
    d. the distance of a planet from its central star.

Thinking about the Concepts

16. Scientists can never know everything, especially at the beginning of a set of research programs. For example, Shapley did not know about dust in the Milky Way. Yet, scientists must often set aside this problem of the unknowns they don't know and do the best they can with the knowledge they have. Explain why this is a necessary step toward scientific understanding.

17. Describe the distribution of globular clusters within the Milky Way, and explain what that implies about the size of the galaxy and our distance from its center.

18. In which parts of the Milky Way do astronomers find open clusters? In which parts do they find globular clusters?

19. Old stars in the inner disk of the Milky Way have higher abundances of massive elements than those of young stars in the outer disk. Explain how this difference might have developed.

20. How do 21-cm radio observations reveal the rotation of the Milky Way Galaxy?

21. What does the rotation curve of the Milky Way indicate about the presence of dark matter in the galaxy?

22. Halo stars are found in the vicinity of the Sun. What observational evidence distinguishes them from disk stars?

23. What is one source of synchrotron radiation in the Milky Way, and where is it found?

24. Why must astronomers use X-ray, infrared, and 21-cm radio observations to probe the center of the galaxy?

25. What is Sgr A* and how was it detected?

26. Explain the evidence for a supermassive black hole at the center of the Milky Way. How does the mass of the supermassive black hole at the center of our galaxy compare with that found in most other spiral galaxies?

27. To observers in Earth's Southern Hemisphere, the Large Magellanic Cloud and Small Magellanic Cloud look like detached pieces of the Milky Way. What are these "clouds," and why is it not surprising that they look so much like pieces of the Milky Way?

28. What is the origin of the Milky Way's satellite galaxies? What has been the fate of most of the Milky Way's satellite galaxies? Why are most of the Milky Way's satellite galaxies so difficult to detect?

29. Use your imagination to describe how Earth's skies might appear if the Sun and Solar System were located (a) near the center of the galaxy; (b) near the center of a large globular cluster; (c) near the center of a large, dense molecular cloud.

30. What factors do astronomers consider when thinking about a galactic habitable zone in the Milky Way?

Applying the Concepts

31. From Figure 20.8, estimate the ratio between the radius of the Milky Way's outer halo and the radius of the disk.

32. Study Figure 20.16.
    a. What is the rotation velocity of a disk star located 6,000 pc from the center of the Milky Way?
    b. Assuming a circular orbit, how long does it take that star to orbit once?

33. From the data in Figure 20.16, estimate the time it would take the Large Magellanic Cloud to orbit the Milky Way if the Large Magellanic Cloud were on a circular orbit.

34. The Sun completes one trip around the center of the galaxy in approximately 230 million years. How many times has the Solar System made the circuit since its formation 4.6 billion years ago?

35. The Sun is located about 8,300 pc from the center of the galaxy, and the galaxy's disk probably extends another 9,000 pc farther out from the center. Assume that the Sun's orbit takes 230 million years to complete.
    a. With a truly flat rotation curve, how long would it take a globular cluster located near the edge of the disk to complete one trip around the center of the galaxy?
    b. How many times has that globular cluster made the circuit since its formation about 13 billion years ago?

36. Parallax measurements of the variable star RR Lyrae indicate that it is located 230 pc from the Sun. A similar star observed in a globular cluster located far above the galactic plane appears 160,000 times fainter than RR Lyrae.
    a. How far from the Sun is this globular cluster?
    b. What does your answer to part (a) tell you about the size of the galaxy's halo compared to the size of its disk?

37. How do astronomers know that the center of the Milky Way contains a black hole and not just dark matter?

38. Given what you have learned about the distribution of massive elements in the Milky Way and what you know about the terrestrial planets, where do you think such planets are most likely and least likely to form?

39. A cosmic-ray proton is traveling at nearly the speed of light (3×10^8 m/s).
    a. Using Einstein's familiar relationship between mass and energy ($E = mc^2$), show how much energy (in joules) the cosmic-ray proton would have if m were based only on the proton's rest mass (1.7×10^{-27} kg).
    b. The actual measured energy of the cosmic-ray proton is 100 J. What, then, is the relativistic mass of the cosmic-ray proton?
    c. How much greater is the relativistic mass of this cosmic-ray proton than the mass of a proton at rest?

40. One of the fastest cosmic rays ever observed had a speed of $(1.0 - [1.0 \times 10^{-24}]) \times c$ (very, very close to c). Assume that the cosmic ray and a photon left a source at the same instant. To a stationary observer, how far behind the photon would the cosmic ray be after traveling for 100 million years?

41. Consider a black hole with a mass of 5 million M_{Sun} ($M_{Sun} = 2 \times 10^{30}$ kg). A star's orbit about the black hole has a semimajor axis of 0.02 light-year (1.9×10^{14} meters). Calculate the star's orbital period. (Hint: You may want to refer back to Chapter 4.)

42. A star in a circular orbit about the black hole at the center of the Milky Way (whose mass $M_{BH} = 8 \times 10^{36}$ kg) has an orbital radius of 0.0131 light-year (1.24×10^{14} meters). What is the average speed of this star in its orbit? (Hint: You may want to refer back to Chapter 4.)

43. What is the Schwarzschild radius of the black hole at the center of the Milky Way? What is its density? How does this compare with the density of a stellar black hole?

44. A star is observed in a circular orbit about a black hole with an orbital radius of 1.5×10^{11} km and an average speed of 2,000 km/s. What is the mass of this black hole in solar masses?

45. One model of the galactic habitable zone contains stars in a doughnut-shaped region between 7 and 9 kpc from the center of the galaxy, in the disk. Assuming that this doughnut is as thick as the disk itself, what fraction of the disk of the Milky Way lies in this habitable zone?

USING THE WEB

46. Go to the "Night Sky" Web page of the National Park Service (NPS; http://nature.nps.gov/night). What is a "natural lightscape"? Click on and read "Light Pollution" and "Measuring Lightscapes." Why is it becoming more rare for people to see the Milky Way? Why does the NPS consider viewing the Milky Way an important part of the experience for people visiting the park?

47. a. Go to the Astronomy Picture of the Day (APOD) app or website for July 2, 2012, and watch the video clip "Zoom-ing into the Center of the Milky Way" (http://apod.nasa.gov/apod/ap120702.html). Why is there a shift in wavelengths of the selected pictures? On APOD, run a search for "Milky Way" to look at some of the best photographs of it. Where were the pictures taken from? Can you see the Milky Way from *your* location on a clear night?

b. Why does the galactic center have to be observed in infrared and X-ray wavelengths? Go to the "Milky Way Galaxy" page of the Chandra X-ray telescope website (http://chandra.harvard.edu/photo/category/milkyway.html) and to the Spitzer infrared telescope website (http://spitzer.caltech.edu). Are there any new images of the galactic center? What has been learned?

48. Go to the websites of the two main groups studying the supermassive black hole at the galactic center: the UCLA Galactic Center Group (http://astro.ucla.edu/~ghezgroup/gc) and the Galactic Center Research group at the Max Planck Institute for Extraterrestrial Physics (http://mpe.mpg.de/ir/GC). Watch some of the time-lapse animations of the stars orbiting something unseen. Why is it assumed that the unseen object is a black hole? What new results are these groups reporting?

49. Citizen science: Go to the Milky Way Project website (http://milkywayproject.org). This project examines Spitzer telescope observations of the dusty material in the galaxy. Log in with your Zooniverse account name, and read the information under Menu: "Science," "FAQ," and "Classify." Participants in this project have already discovered some of these bubbles (http://spitzer.caltech.edu/images/4938-sig12-002-Finding-Bubbles-in-the-Milky-Way). Classify some images.

50. Go to the Hubble Space Telescope website and watch the videos about the possible collision of the Milky Way and Andromeda (http://hubblesite.org/newscenter/archive/releases/galaxy/2012/20/video). Read the report under "The Full Story." A newer simulation is here: www.icrar.org/multimedia/videos/video-pages/andromeda-and-the-milky-way-collide! Why will Andromeda "eat" the Milky Way instead of the other way around?

smartwork

If your instructor assigns homework in SmartWork, access your assignments at digital.wwnorton.com/astro5.

EXPLORATION

The Center of the Milky Way

digital.wwnorton.com/astro5

Adapted from Learning Astronomy by Doing Astronomy *by Ana Larson.*

Astronomers once thought that the Sun was at the center of the Milky Way. In this Exploration, you will repeat Harlow Shapley's globular-cluster experiment that led to a more accurate picture of the size and shape of the Milky Way.

Imagine that the disk of the Milky Way is a flat, round plane, like a pizza. Globular clusters are arranged in a rough sphere around this plane. To map globular clusters on **Figure 20.25**, imagine that a line is drawn straight "down" from a globular cluster to the plane of the Milky Way. The "projected distance" in kiloparsecs is the distance from the Sun to the place where the line hits the plane. The galactic longitude indicates the direction toward that point; it is marked around the outside of the graph, along with the several constellations.

Make a dot at the location of each globular cluster by finding the galactic longitude indicated outside the circle and then coming in toward the center to the projected distance. The two globular clusters in boldface in **Table 20.1** have been plotted for you as examples. After plotting all of the globular clusters, estimate the center of their distribution and mark it with an X. This is the center of the Milky Way.

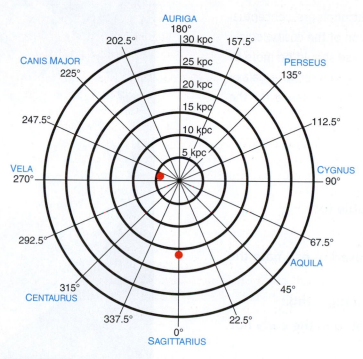

Figure 20.25 This polar graph can be used to plot distance and direction.

1 What is the approximate distance from the Sun to the center of the Milky Way?

_____

2 What is the galactic longitude of the center of the Milky Way?

_____

3 How do astronomers know that the Sun is not at the center of the Milky Way?

_____

TABLE 20.1 Globular Cluster Data

Cluster	Galactic Longitude	Projected Distance (kpc)	Cluster	Galactic Longitude	Projected Distance (kpc)
104	306	3.5	6273	357	7
362	302	6.6	**6287**	**0**	**16.6**
2808	283	8.9	6333	5	12.6
4147	**251**	**4.2**	6356	7	18.8
5024	333	3.4	6397	339	2.8
5139	309	5	6535	27	15.3
5634	342	17.6	6712	27	5.7
Pal 5	1	24.8	6723	0	7
5904	4	5.5	6760	36	8.4
6121	351	4.1	Pal 10	53	8.3
O 1276	22	25	Pal 11	32	27.2
6638	8	15.1	6864	20	31.5
6171	3	15.7	6981	35	17.7
6218	15	6.7	7089	54	9.9
6235	359	18.9	Pal 12	31	25.4
6266	353	11.6	288	147	0.3
6284	358	16.1	1904	228	14.4
6293	357	9.7	Pal 4	202	30.9
6341	68	6.5	4590	299	11.2
6366	18	16.7	5053	335	3.1
6402	21	14.1	5272	42	2.2
6656	9	3	5694	331	27.4
6717	13	14.4	5897	343	12.6
6752	337	4.8	6093	353	11.9
6779	62	10.4	6541	349	3.9
6809	9	5.5	6626	7	4.8
6838	56	2.6	6144	352	16.3
6934	52	17.3	6205	59	4.8
7078	65	9.4	6229	73	18.9
7099	27	9.1	6254	15	5.7

21 The Expanding Universe

The cosmological principle has been at the center of astronomers' conceptual understanding of the universe. An important prediction of the cosmological principle is that the conclusions we reach about our universe should be more or less the same, regardless of whether we live in the Milky Way or in a galaxy billions of light-years away. In this chapter, we look at the evidence for the cosmological principle and for the Big Bang.

LEARNING GOALS

By the conclusion of this chapter, you should be able to:

LG 1 Explain in detail the cosmological principle.

LG 2 Describe how the Hubble constant can be used to estimate the age of the universe.

LG 3 Describe the observational evidence for the Big Bang.

LG 4 Explain which chemical elements were created in the early hot universe.

This image shows the oldest light in the universe, the cosmic microwave background radiation, detected by the Planck space observatory. ▶ ▶ ▶

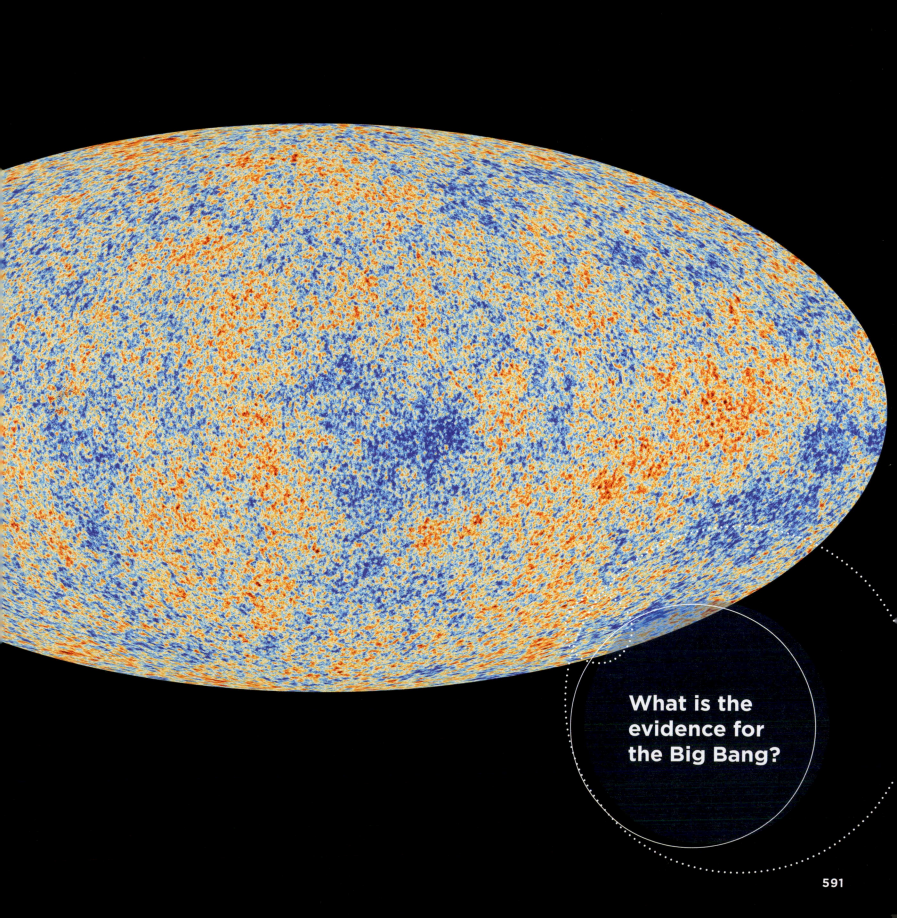

What is the
evidence for
the Big Bang?

21.1 The Cosmological Principle

In Chapter 19, we described observations of galaxy motions and distances. From these observations, Edwin Hubble created a graph that showed the galaxies moving apart, with the more distant ones moving faster—a result known as Hubble's law. In this section, we will further explore the implications of Hubble's law.

The Homogeneous and Isotropic Universe

Cosmology is the study of space and time and the dynamics of the universe as a whole. In the 1920s, around the same time that astronomers were first measuring distances to galaxies, theoretical physicists were applying Einstein's general theory of relativity to cosmology. The cosmologist Alexander Friedmann (1888–1925) produced mathematical models of the universe that assumed the **cosmological principle**, which is stated as follows: *the physical laws that apply to one part of the universe apply everywhere.* The cosmological principle forms the basis of our study of distant galaxies and of the universe itself and is now a fundamental tenet of modern cosmology.

We have used this principle throughout this book when applying laws of physics and chemistry to objects near and far in the universe. For example, according to the cosmological principle, gravity works the same way in distant galaxies as it does here on Earth. The cosmological principle is a testable scientific theory. An important prediction of the principle is that the conclusions that scientists reach about the universe should be the same whether we observe the universe from the Milky Way or from a galaxy billions of parsecs away. In other words, if the cosmological principle is correct, then the universe is **homogeneous**, having the same composition and properties at all places.

Clearly, the universe is not homogeneous in an absolute sense, as the conditions on Earth are very different from those in deep space or in the center of the Sun. In cosmology, homogeneity of the universe means that stars and galaxies in Earth's part of the universe are much the same, and behave in the same manner, as stars and galaxies in remote corners of the universe. It also means that stars and galaxies everywhere are distributed in space in much the same way that they are distributed in Earth's cosmic neighborhood, and that observers in those galaxies would see the same properties for the universe that astronomers see from here.

It is not easy to verify the prediction of homogeneity directly. Scientists cannot travel from the Milky Way to a galaxy in the remote universe to see whether conditions are the same. However, they can compare light arriving from closer and farther locations in the distant universe and see the ways in which features look the same or different. For example, they can look at the way galaxies are distributed in distant space and see whether that distribution is similar (that is, homogeneous) to the distribution nearby.

In addition to predicting that the universe is homogeneous, the cosmological principle requires that all observers measure the same properties of the universe, regardless of the direction in which they are looking. If something is the same in all directions, then it is **isotropic**. This prediction of the cosmological principle is much easier to test directly than is homogeneity. For example, if galaxies were lined up in rows, astronomers would measure very different properties of the universe, depending on the direction they looked. The universe would still be homogeneous, but not isotropic, so it would not satisfy the cosmological principle.

In most instances, isotropy goes together with homogeneity, and the cosmological principle requires both. **Figure 21.1** shows examples of how the universe could have violated the cosmological principle by not being homogeneous or isotropic, as well as examples of how the universe might satisfy the cosmological principle. All observations show that the properties of the universe are basically the same, regardless of the direction in which observers are looking. When averaged over very large scales, thousands of millions of parsecs, the universe appears homogeneous as well.

The Hubble Expansion

Recall from Chapter 19 our discussion of Hubble's law—that the distance of a galaxy is proportional to its recession velocity. Hubble's law helps astronomers investigate whether the universe is homogeneous and isotropic. They can confirm its isotropy by observing that galaxies in one direction in the sky obey the same Hubble law as galaxies in other directions in the sky. Hubble's law says that Earth is located in an expanding universe and that the expansion looks the same,

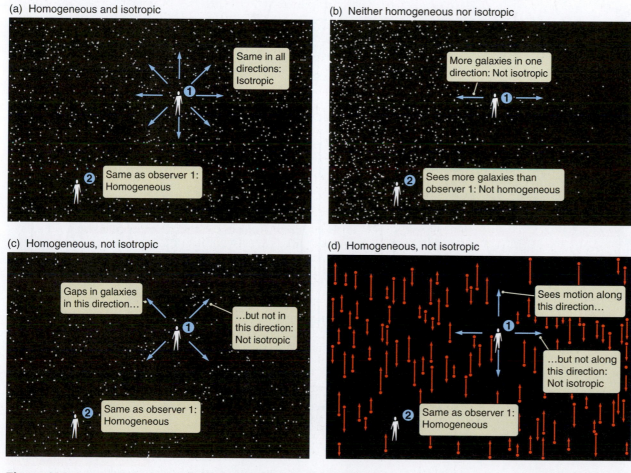

Figure 21.1 Homogeneity and isotropy in four different theoretical models of a universe. Blue arrows indicate the direction of view. (a) The distribution of galaxies is uniform, so this universe is both homogeneous and isotropic. (b) The density of galaxies is decreasing in one direction, so this universe is neither homogeneous nor isotropic. (c) The bands of galaxies lie along a unique axis, making this universe not isotropic. (d) The distribution of galaxies is uniform, but galaxies move along only one direction, so this universe also is not isotropic.

regardless of the location of the observer. To help you visualize this, we now turn to a useful model that you can build for yourself with materials you probably have in your desk.

Figure 21.2 shows a long rubber band with paper clips attached along its length. If you stretch the rubber band, the paper clips—which represent galaxies in an expanding universe—get farther and farther apart. To get a sense of what this expansion would look like up close, imagine yourself an ant riding on paper clip A. As the rubber band is stretched, you notice that all of the paper clips are moving away from you. Clip B, the closest one, is moving away slowly. Clip C,

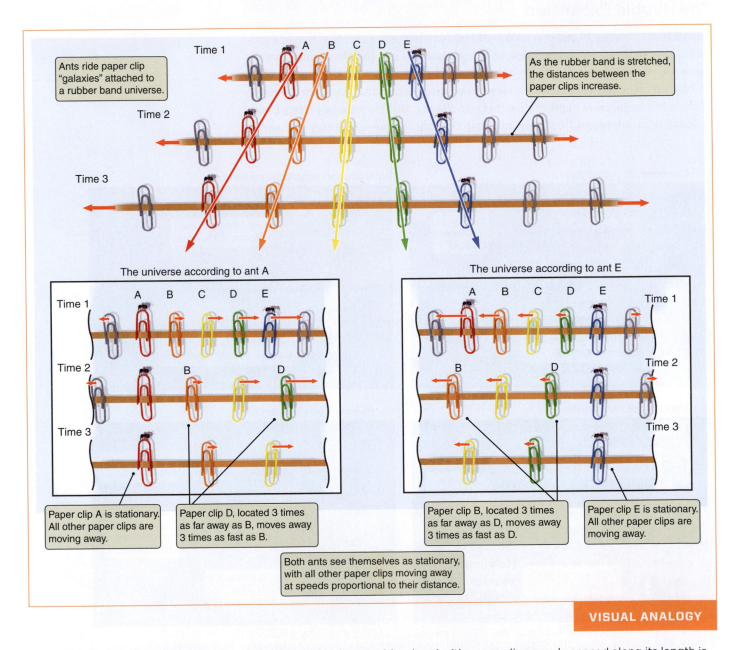

Figure 21.2 In this one-dimensional analogy of Hubble's law, a rubber band with paper clips evenly spaced along its length is stretched. As the rubber band stretches, an ant riding on clip A observes clip C moving away twice as fast as clip B. Similarly, an ant riding on clip E sees clip C moving away twice as fast as clip D. Any ant sees itself as stationary, regardless of which paper clip it is riding, and it sees the other clips moving away with speed proportional to distance.

located twice as far from you as clip B, is moving away twice as fast as clip B. Clip E, located 4 times as far away as clip B, is moving away 4 times as fast as clip B. From your perspective as an ant riding on clip A, all of the other paper clips on the rubber band are moving away with a velocity proportional to their distance. The paper clips located along the rubber band obey a Hubble-like law.

The key insight to the analogy comes from realizing that there is nothing special about the perspective of paper clip A. If, instead, the ant were riding on clip E, clip D would be the one moving away slowly and clip A would be moving away 4 times as fast. Repeat this experiment for any paper clip along the rubber band, and you will arrive at the same result. For an ant on any paper clip along the rubber band, the speed at which other clips are moving away from the ant is proportional to their distance. The stretching rubber band, like the universe, is "homogeneous." The same Hubble-like law applies, regardless of where the observer ant is located.

The observation that nearby paper clips move away slowly and distant paper clips move away more rapidly does not say that the paper clip selected as a vantage point is at the center of anything. Instead, it says that the rubber band is being stretched uniformly along its length. Similarly, Hubble's law for galaxies means that nearby galaxies are carried away slowly by expanding space, and distant galaxies are carried away more rapidly (**Process of Science Figure**). Our galaxy is not at the center of an expanding universe; the universe is expanding uniformly. Any observer in any galaxy sees nearby galaxies moving away slowly and more distant galaxies moving away more rapidly. The same Hubble law applies from their vantage point as applies from our vantage point on Earth. The expansion of the universe is homogeneous.

The only exception to this rule is the case of galaxies that are close together, in which case gravitational attraction dominates over the expansion of space. As you saw in the previous chapter, the Andromeda Galaxy and the Milky Way are being pulled together by their gravity. The Andromeda Galaxy is approaching the Milky Way at about 110 kilometers per second (km/s), so the light from the Andromeda Galaxy is blueshifted, not redshifted. This velocity is caused by the gravitational interaction between the two nearby galaxies, rather than the expansion of space. The fact that gravitational or electromagnetic forces can overwhelm the expansion of space also explains why the Solar System is not expanding, and neither are you.

The Universe in Space and Time

Hubble's law gives astronomers a practical tool for measuring distances to remote objects. Once they know the value of the Hubble constant, H_0, they can use a straightforward measurement of the redshift of a galaxy to find its distance. In other words, once H_0 is known, Hubble's law makes the once-difficult task of measuring distances in the universe relatively easy, providing astronomers with a tool to map the structure of the observable universe. This may seem like a logical impossibility, because they are using redshifts and distances to find H_0 and then using H_0 to find the distances. But astronomers find H_0 from nearby galaxies using standard candles, and then use that value to find the distance to a different, more distant set of galaxies.

Hubble's law does more than place galaxies in space. It also places galaxies in time. Light travels at a huge but finite speed. Recall from Chapter 1 that when you look

DATA ARE THE ULTIMATE AUTHORITY

Einstein is one of the most famous scientists of all time. His genius is recognized by everyone. But even Einstein had to change his mind in the face of new data.

Einstein develops general relativity, which predicts the universe is expanding or contracting.

He inserts an extra constant into his equations to make the universe stationary because he believes it should be.

Hubble presents his data to Einstein. The universe is expanding!

Einstein changes his mind, and describes his reluctance to accept the prediction of his original theory as "his biggest blunder."

Even the most brilliant scientists must yield when the data contradict their conclusions.

at the Sun, you see it as it existed $8\frac{1}{3}$ minutes ago. When you look at Alpha Centauri, the nearest stellar system beyond the Sun, you see it as it existed 4.3 years ago. If you look at the center of the Milky Way, the light you see is 27,000 years old. The **look-back time** of a distant object is the time it has taken for the light from that object to reach a telescope on Earth. As astronomers look into the distant universe, look-back times become very large. The distance to a galaxy whose redshift $z = 0.1$ is 1.4 billion light-years (assuming $H_0 = 70$ km/s/Mpc), so the look-back time to that galaxy is 1.4 billion years. The look-back time to a galaxy where $z = 0.2$ is 2.7 billion years. The look-back time of the most distant galaxies observed, at about $z = 10$, is 13.2 billion years. As astronomers observe objects with greater and greater redshifts, they are seeing increasingly younger stages of the universe.

CHECK YOUR UNDERSTANDING 21.1

In astronomy, *isotropy* means that the universe is the same _____, and *homogeneity* means that the universe is the same _____. (a) in all locations; (b) in all directions; (c) at all times; (d) at all size scales.

21.2 The Universe Began in the Big Bang

Imagine watching a video of the universe, with the galaxies moving apart. Now, reverse the video, and run it backward in time. The galaxies become closer and closer together as the universe becomes younger and younger. In this way, the observation of the expansion of the universe leads to the idea of a beginning to the universe at a time that can be estimated from Hubble's law. In this section, we explore this implication of Hubble's law.

Expansion and the Age of the Universe

Hubble's law gives an estimate of the age of the universe. If we assume the speed of the expansion has always been constant, then the age of the universe can be estimated from the slope of the line in a graph of the velocity of galaxies plotted against their distance. This slope has units that reduce to 1/time, so its inverse has units of time. This slope is the Hubble constant, H_0, and its inverse (1 divided by H_0) is the **Hubble time**. The Hubble time is an estimate of the universe's age: 13.8 billion years (**Working It Out 21.1**). If the expansion were faster, the Hubble constant would be larger, and the universe would be younger; similarly, a slower expansion would yield a smaller Hubble constant and an older universe.

If the universe expanded uniformly, then when it was about half its current age, about 6.9 billion years ago, all of the galaxies in the universe were half as far apart as they are now, and 12.4 billion years ago, all of the galaxies in the universe were about a tenth as far apart. Assuming that galaxies have been moving apart all that time at the same speed as they do today, then a little less than one Hubble time ago—13.8 billion years ago—there was almost no space between the particles that constitute today's universe. All such matter as well as energy in the universe then must have been unimaginably dense. Because

▶❚❚ **AstroTour:** Hubble's Law

Astronomy in Action: Expanding Balloon Universe

21.1 | Working It Out Expansion and the Age of the Universe

We can use Hubble's law to estimate the age of the universe. Consider two galaxies located 30 Mpc ($d_G = 9.3 \times 10^{20}$ km) away from each other, as shown in **Figure 21.3**. If these two galaxies are moving apart, then at some time in the past they must have been together in the same place at the same time. According to Hubble's law, and assuming that $H_0 = 70$ km/s/Mpc, the distance (d_G) between these two galaxies is increasing at the following rate:

$$v_r = H_0 \times d_G$$

$$v_r = 70 \text{ km/s/Mpc} \times 30 \text{ Mpc}$$

$$v_r = 2{,}100 \text{ km/s}$$

Knowing the velocity (v_r) at which they are traveling, we can calculate the time it took for the two galaxies to become separated by 30 Mpc:

$$\text{Time} = \frac{\text{Distance}}{\text{Velocity}} = \frac{9.3 \times 10^{20} \text{ km}}{2{,}100 \text{ km/s}} = 4.4 \times 10^{17} \text{ s}$$

Dividing by the number of seconds in a year (about 3.16×10^7 s/yr) gives

$$\text{Time} = 1.4 \times 10^{10} \text{ yr} = 14 \text{ billion yr}$$

In other words, *if* expansion of the universe has been constant, two galaxies that today are 30 Mpc apart started out at the same place 14 billion years ago.

Now let's do the same calculation with two galaxies that are 60 Mpc, or 18.6×10^{20} km, apart. These two galaxies are twice as far apart, but the distance between them is increasing twice as rapidly:

$$v_r = H_0 \times d_G = 70 \text{ km/s/Mpc} \times 60 \text{ Mpc} = 4{,}200 \text{ km/s}$$

Therefore,

$$\text{Time} = \frac{18.6 \times 10^{20} \text{ km}}{4{,}200 \text{ km/s}} = 4.4 \times 10^{17} \text{ s} = 1.4 \times 10^{10} \text{ yr}$$

Again, we calculate time as distance divided by velocity (twice the distance divided by twice the velocity) to find that these galaxies also took about 14 billion years to reach their current locations. We can do this calculation again and again for any pair of galaxies in the universe today. The farther apart the two galaxies are, the faster they are moving. But all galaxies took the same amount of time to get to where they are today.

Working out the example using words instead of numbers makes it clear why the answer is always the same. Because the velocity we are calculating comes from Hubble's law, velocity equals the Hubble constant multiplied by distance. Writing this out as an equation, we get

$$\text{Time} = \frac{\text{Distance}}{\text{Velocity}} = \frac{\text{Distance}}{H_0 \times \text{Distance}}$$

Distance divides out to give

$$\text{Time} = \frac{1}{H_0}$$

where $1/H_0$ is the Hubble time. This is one way of estimating the age of the universe.

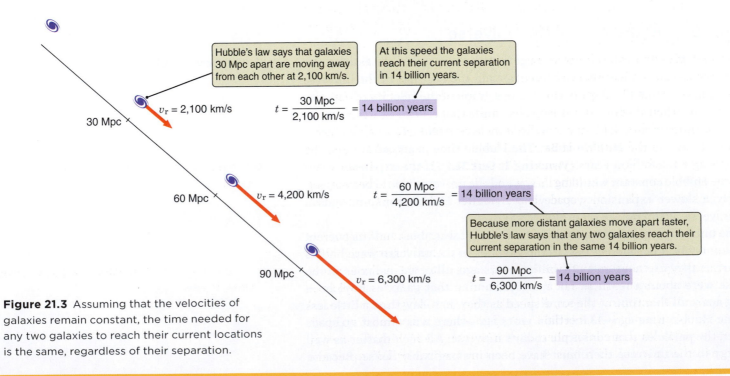

Figure 21.3 Assuming that the velocities of galaxies remain constant, the time needed for any two galaxies to reach their current locations is the same, regardless of their separation.

expanding gases cool down, the universe then must have been much hotter than it is today in its expanded state. This hot, dense beginning, 13.8 billion years ago, is called the **Big Bang (Figure 21.4)**.

Georges Lemaître (1894–1966) was the first to propose the theory of the Big Bang. This idea greatly troubled many astronomers in the early and middle years of the 20th century. Several different suggestions were put forward to explain the observed fact of Hubble expansion without resorting to the idea that the universe came into existence in an extraordinarily dense "fireball" billions of years ago. However, as more and more distant galaxies have been observed, and as more discoveries about the structure of the universe have been made, the Big Bang theory has grown stronger. The major predictions of the Big Bang theory have proven to be correct.

The implications of Hubble's law forever changed the scientific concepts of the origin, history, and possible future of the universe. At the same time, Hubble's law has pointed to many new questions about the universe. To address them, we next need to consider exactly what is meant by the term *expanding universe*.

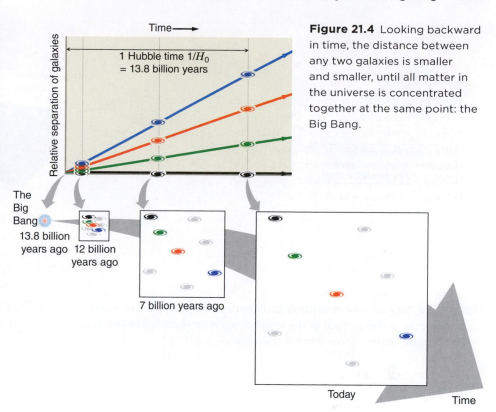

Figure 21.4 Looking backward in time, the distance between any two galaxies is smaller and smaller, until all matter in the universe is concentrated together at the same point: the Big Bang.

Galaxies and the Expansion

At this point in our discussion, you may be picturing the expanding universe as a cloud of debris from an explosion flying outward through surrounding space. This is a common depiction of the Big Bang in movies and television shows, where they portray a tiny bright spot that explodes to fill the screen. However, the Big Bang is not an explosion in the usual sense of the word, and in fact there is no surrounding space into which the universe expands.

We must draw a distinction between the universe and the observable universe. The observable universe is the part of the universe that we can see. The observable universe extends 13.8 billion light-years in every direction. This limit exists because that is the length of time the universe has been around. The light from more distant regions has not yet had time to travel to us, and so we cannot see it yet.

A common question about the Big Bang is, Where did it take place? The answer is that the Big Bang took place *everywhere*. Wherever anything is in the universe today, it is at the site of the Big Bang. The reason is that galaxies are not flying apart through space at all. Rather, space itself is expanding, carrying the stars and galaxies that populate the universe along with it.

We have already dealt with the basic ideas that explain the expansion of space. In our discussion of black holes in Chapter 18, you encountered Einstein's general theory of relativity. General relativity says that space is distorted by the presence of mass, and that the consequence of this distortion is gravity. For example, the mass of the Sun, like any other object, distorts the geometry of spacetime around it; so Earth, coasting along in its inertial frame of reference, follows a curved path around the Sun. We illustrated this phenomenon in Figure 18.10 with the analogy of a ball placed on a stretched rubber sheet, showing how the ball distorted the surface of the sheet.

Astronomy in Action: Observable vs. Actual Universe

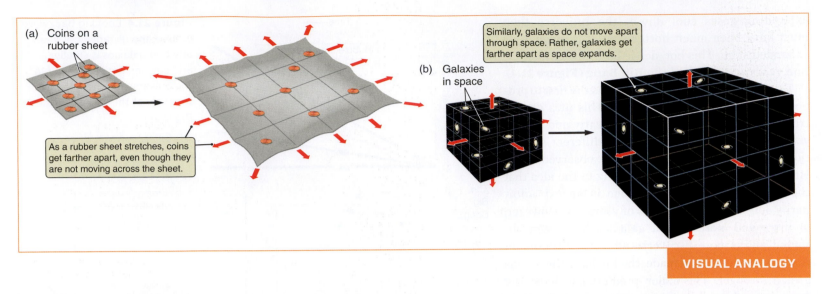

Figure 21.5 (a) As a rubber sheet is stretched, coins on its surface move farther apart, even though they are not moving with respect to the sheet itself. Any coin on the surface of the sheet observes a Hubble-like law in every direction. (b) Galaxies in an expanding universe are not flying apart through space. Rather, space itself is stretching.

The surface of a rubber sheet can be distorted in other ways as well. Imagine a number of coins placed on a rubber sheet, illustrated in **Figure 21.5**. Then imagine grabbing the edges of the sheet and beginning to pull them outward. As the rubber sheet stretches, each coin remains at the same location on the surface of the sheet, but the distances between the coins increase. Two coins sitting close to each other move apart only slowly, while coins farther apart move away from each other more rapidly. The distances and relative motions of the coins on the surface of the rubber sheet obey a Hubble-like relationship as the sheet is stretched.

This movement is analogous to what is happening in the universe, with galaxies taking the place of the coins and space itself taking the place of the rubber sheet. In the case of coins on a rubber sheet, there is a limit to how far the sheet can be stretched before it breaks. With space and the real universe, there is no such limit. The fabric of space can, in principle, go on expanding forever. Hubble's law is the observational consequence of the fact that the space making up the universe is expanding.

How will this expansion of the universe behave in the future? You have learned that most of the mass in galaxies consists of dark matter, and the gravity of the dark matter has the effect of slowing down the expansion of the universe. In the next chapter, you will also learn that there is another unseen constituent of the universe, called *dark energy*, and this constituent causes the expansion of the universe to *accelerate*. At the current stage in the expansion of the universe, the accelerating effect of dark energy dominates over the slowing effect of dark matter, and therefore the universe will continue to expand at an ever-faster rate.

CHECK YOUR UNDERSTANDING 21.2

Where in the universe did the Big Bang take place? (a) near the Milky Way Galaxy; (b) near the center of the universe; (c) near some unknown location on the other side of the universe; (d) everywhere in the universe

21.3 Expansion Is Described with a Scale Factor

As the universe expands, the distance between any two objects increases because of the stretching of space. Astronomers find it useful to discuss this expansion in terms of the *scale factor* of the universe.

Scale Factor

Let's return to the analogy of the rubber sheet. Suppose you place a ruler on the surface of the rubber sheet and draw a tick mark every centimeter, as illustrated in **Figure 21.6a**. To measure the distance between two points on the sheet, you can count the marks between the two points and multiply by 1 centimeter (cm) per tick mark.

As the sheet is stretched, however, the distance between the tick marks does not remain 1 cm. When the sheet is stretched to 150 percent of the size it had when the tick marks were drawn, each tick mark is separated from its neighbors by $1\frac{1}{2}$ times the original distance, or 1.5 cm. The distance between two points can still come from counting the marks, but you need to scale up the distance between tick marks by 1.5 to find the distance in centimeters. If the sheet were twice the size it was when the tick marks were drawn (Figure 21.6b), each mark would correspond to 2 cm of actual distance. Astronomers use the term **scale factor** to indicate the size of the sheet relative to its size at the time when the tick marks were drawn. The scale factor also indicates how much the distance between points on the sheet has changed. In the first example, the scale factor of the sheet is 1.5; in the second, the scale factor of the sheet is 2.

Suppose astronomers choose today to lay out a "cosmic ruler" on the fabric of space, placing an imaginary tick mark every 10 Mpc. The scale factor of the universe at this time is defined to be 1. In the past, when the universe was smaller, distances between the points in space marked by this cosmic ruler would have been less than 10 Mpc. The scale factor of that younger, smaller universe would have been less than 1 compared to the scale factor today. In the future, as the universe continues to expand, the distances between the tick marks on this cosmic ruler will grow to more than 10 Mpc, and the scale factor of the universe will be greater than 1. Astronomers use the scale factor, usually written as R_U, to keep track of the changing scale of the universe.

It is important to remember that the laws of physics are themselves unchanged by the expansion of the universe, just as stretching a rubber sheet does not change the properties of the coins on its surface. At noncosmological scales, the nuclear and electromagnetic forces within and between atoms, as well as the gravitational forces between relatively nearby objects, dominate over the expansion. As the universe expands, the sizes and other physical properties of atoms, stars, and galaxies also remain unchanged.

Looking back in time, the scale factor of the universe gets smaller and smaller, approaching zero as it comes closer and closer to the Big Bang. The fabric of space that today spans billions of parsecs spanned much smaller distances when the universe was young. When the universe was only a day old, all of the space visible today amounted to a region only a few times the size of the Solar System. When the universe was 1/50 of a second old, the vast expanse of space that

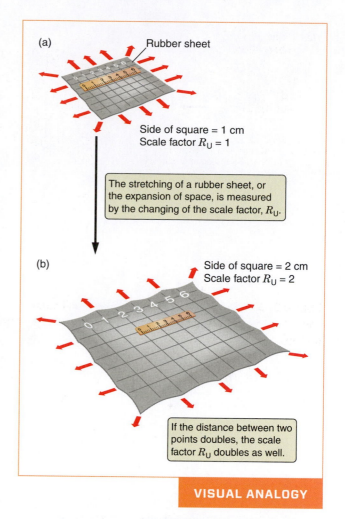

(a) Rubber sheet

Side of square = 1 cm
Scale factor $R_U = 1$

The stretching of a rubber sheet, or the expansion of space, is measured by the changing of the scale factor, R_U.

(b) Side of square = 2 cm
Scale factor $R_U = 2$

If the distance between two points doubles, the scale factor R_U doubles as well.

VISUAL ANALOGY

Figure 21.6 (a) On a rubber sheet, tick marks are drawn 1 cm apart. As the sheet is stretched, the tick marks move farther apart. (b) When the spacing between the tick marks is 2 cm, or twice the original value, the scale factor of the sheet, R_U, is said to have doubled. A similar scale factor, R_U, is used to describe the expansion of the universe.

makes up today's observable universe (and all the matter in it) occupied a volume only the size of today's Earth. Going backward in time approaching the Big Bang itself, the space that makes up today's observable universe becomes smaller and smaller—the size of a grapefruit, a marble, an atom, a proton. Every point in the fabric of space that makes up today's universe was right there at the beginning, a part of that unimaginably tiny, dense universe that emerged from the Big Bang.

The Big Bang did not occur at a specific point in space, because space itself came into existence with the Big Bang. The Big Bang happened everywhere; there is no particular point in today's universe that marked the site of the Big Bang. A Big Bang universe is homogeneous and isotropic, consistent with the cosmological principle.

CHECK YOUR UNDERSTANDING 21.3

The scale factor keeps track of: (a) the movement of galaxies through space; (b) the current distances between many galaxies; (c) the changing distance between any two galaxies; (d) the location of the center of the universe.

Redshift Is Due to the Changing Scale Factor of the Universe

The ideas of general relativity discussed in Chapter 18 are powerful tools for interpreting Hubble's great discovery of the relationship between the velocity and distance of galaxies and of the expanding universe. These velocities were determined from the redshift of the galaxies using the Doppler effect. Although it is true that the distance between galaxies is increasing as a result of the expansion of the universe, and that we can use the equation for Doppler shifts to measure the redshifts of galaxies, these redshifts are not due to Doppler shifts in the same way that we described for a moving star. Light that comes from very distant objects was emitted at a time when the universe was younger and therefore smaller. As this light comes toward Earth from distant galaxies, the scale factor of the space through which the light travels is constantly increasing, and as it does, the distance between adjacent light-wave crests increases as well. The light is "stretched out" as the space it travels through expands.

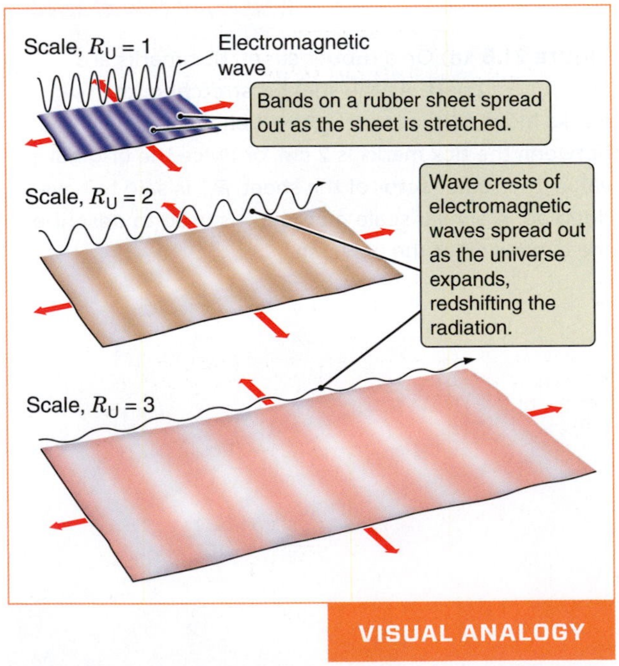

Recall the rubber sheet analogy that we used when discussing black holes in Chapter 18. In that case, we imagined space as a two-dimensional rubber sheet. **Figure 21.7** uses this rubber-sheet analogy to explain why the increasing distance between galaxies causes the light to be redshifted. If you draw a series of bands on the rubber sheet to represent the crests of an electromagnetic wave, you can watch what happens to the wave as the sheet is stretched out. By the time the sheet is stretched to twice its original size—that is, by the time the scale factor of the sheet is 2—the distance between wave crests has doubled. When the sheet has been stretched to 3 times its original size (a scale factor of 3), the wavelength of the wave will be 3 times what it was originally.

Consider the light coming from a distant galaxy. When the light left the distant galaxy, the scale factor of the universe was smaller than it is today. As light comes toward us from distant galaxies, the space through which the light travels is stretching, and the light is also "stretched out" as the space through which it travels expands. The universe expanded while the light was in transit, and as it did so, the wavelength of the light grew longer in proportion to the increasing scale factor of the universe. The redshift of light from distant galaxies is therefore a direct

Scale, $R_U = 1$

Electromagnetic wave

Bands on a rubber sheet spread out as the sheet is stretched.

Scale, $R_U = 2$

Wave crests of electromagnetic waves spread out as the universe expands, redshifting the radiation.

Scale, $R_U = 3$

VISUAL ANALOGY

Figure 21.7 Bands drawn on a rubber sheet represent the positions of the crests of an electromagnetic wave in space. As the rubber sheet is stretched—that is, as the universe expands—the wave crests get farther apart. The light is redshifted.

measure of how much the universe has expanded since the time when the radiation left its source. Redshift measures how much the scale factor of the universe, R_U, has changed since the light was emitted. This redshift can be greater than 1 if the galaxy is far enough away that the recession velocity is relativistic, as described in **Working It Out 21.2**.

21.2 Working It Out   When Redshift Exceeds One

In our discussion of the Doppler shift in Chapter 5, we noted that $(\lambda_{obs} - \lambda_{rest})/\lambda_{rest}$ is equal to the velocity of an object moving away divided by the speed of light. Edwin Hubble used this result to interpret the observed redshifts of galaxies as evidence that galaxies throughout the universe are moving away from the Milky Way. Einstein's special theory of relativity says that nothing can move faster than the speed of light. Hubble initially assumed that redshifts are due to the Doppler effect. The resulting relation, $z = v_r/c$, would then seem to imply that no object can have a redshift (z) greater than 1. Yet that is not the case. Astronomers routinely observe redshifts significantly in excess of 1. As of this writing, the most distant objects known have redshifts as large as 9 or 10. How can redshifts exceed 1?

To arrive at the expression for the Doppler effect, $v_r/c = (\lambda_{obs} - \lambda_{rest})/\lambda_{rest}$, we have to *assume* that v_r is much less than c. If v_r was close to c, we would have to consider more than just the fact that the waves from an object are stretched out by the object's motion away. We would also have to consider relativistic effects, including the fact that moving clocks run slowly (see Working It Out 18.1). When combining these effects, we would find that as the speed of an object approaches the speed of light, its redshift approaches infinity, as shown in **Figure 21.8**.

Doppler's original formula is essentially correct—for objects near enough that their measured velocities are far less than the speed of light. When astronomers look at the motions of orbiting binary stars or the peculiar velocities of galaxies relative to the Hubble flow, this equation works just fine. But anytime there is a redshift of 0.4 or greater, relativity must be taken into account.

Another source of redshift is the gravitational redshift discussed in Chapter 18. As light escapes from deep within a gravitational well, it loses energy, so photons are shifted to longer and longer wavelengths. If the gravitational well is deep enough, then the observed redshift of this radiation can be boundlessly large. In fact, the event horizon of a black hole—that is, the surface around the black hole from which not even light can escape—is where the gravitational redshift becomes infinite.

Cosmological redshift, which is most relevant to this chapter, results from the amount of "stretching" that space has undergone during the time the light from its original source has been en route to Earth. The amount of stretching is given by the factor $1 + z$. When astronomers observe light from a distant galaxy whose redshift $z = 1$, then the wavelength of this light is twice as long as when it left the galaxy. When the light left its source, the universe was half the size that it is today. When they see light from a galaxy with $z = 2$, the

wavelength of the radiation is 3 times its original wavelength, and they are seeing the universe when it was one-third its current size. This direct relationship enables astronomers to use the observed redshift of the galaxy to calculate the size of the universe at the look-back time to that galaxy. Nearby, this means that distance and look-back time are proportional to z. As they look back closer and closer to the Big Bang, however, redshift climbs more and more rapidly.

Written as an equation, the scale factor of the universe that astronomers see when looking at a distant galaxy is equal to 1 divided by 1 plus the redshift of the galaxy:

$$R_U = \frac{1}{1 + z}$$

For example, when astronomers report they have observed a galaxy with a redshift of 9, the scale factor at the time the light was emitted was $R_U = 1/(1 + 9)$, or 1/10.

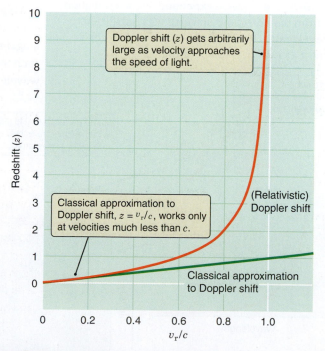

Figure 21.8 This graph shows the plot of the redshift (z) of an object versus its recession velocity (v_r) as a fraction of the speed of light. According to special relativity, as v_r approaches c, the redshift becomes large without limit.

21.4 Astronomers Observe Cosmic Microwave Background Radiation

What is the evidence that the Big Bang actually took place? One piece of evidence comes from observations of the early universe across the entire sky. In this section, you will learn about the cosmic microwave background radiation, one of the major confirming observations of the theory that the universe had a beginning.

Radiation from the Big Bang

In the late 1940s, cosmologists Ralph Alpher (1921–2007), Robert Herman (1914–1997), and George Gamow (1904–1968) reasoned that because a compressed gas cools as it expands, the universe should also be cooling as it expands. When the universe was very young and small, it must have consisted of an extraordinarily hot, dense gas. As with any hot, dense gas, this early universe would have been awash in blackbody radiation that exhibits a Planck blackbody spectrum (see Chapter 5).

Gamow and Alpher took this idea a step further, noting that as the universe expanded, this radiation would have been redshifted to longer and longer wavelengths. Recall Wien's law from Chapter 5, which states that the temperature associated with Planck blackbody radiation is inversely proportional to the peak wavelength: $T = (2{,}900{,}000 \text{ nm K})/\lambda_{peak}$. Shifting the Planck radiation to longer and longer wavelengths is therefore the equivalent of shifting the temperature of the radiation to lower and lower values. As illustrated in **Figure 21.9**, doubling the wavelength of the photons in a Planck blackbody spectrum by stretching space and doubling the scale factor of the universe is equivalent to cutting the temperature of the Planck spectrum in half. The conclusion was that this radiation should still be detectable today and should have a Planck blackbody spectrum with a temperature of about 5–50 kelvins (K). Alpher searched for the signal, but the technology of the late 1940s and early 1950s was not sufficiently advanced. A decade later, physicist Robert Dicke (1916–1997) and his colleagues at Princeton University also predicted a hot early universe, arriving independently at the same basic conclusions that Alpher and Gamow had reached earlier.

Figure 21.9 As the universe expanded, Planck radiation left over from the hot young universe was redshifted to longer wavelengths. Redshifting a Planck spectrum is equivalent to lowering its temperature.

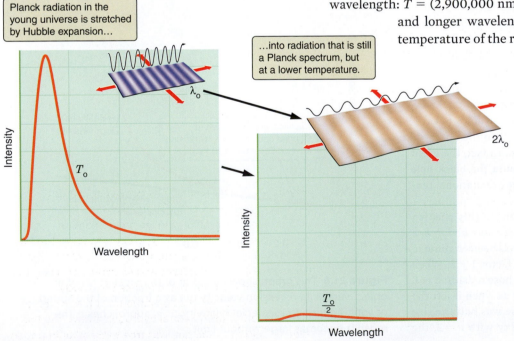

Planck radiation in the young universe is stretched by Hubble expansion…

…into radiation that is still a Planck spectrum, but at a lower temperature.

Measuring the Temperature of the Cosmic Microwave Background Radiation

As we noted at the end of Chapter 6, in the early 1960s two physicists at Bell Laboratories—Arno Penzias and Robert Wilson (**Figure 21.10**)—detected a faint microwave signal in all parts of the sky. When Penzias and Wilson had found the signal, they interpreted it as the radiation left behind by the hot early universe. The strength of the detected signal was consistent with the glow from a blackbody with a temperature of about 3 K, very close to the predicted value. Their results, published in 1965, reported the discovery of the "glow" left behind by the Big Bang.

This radiation left over from the early universe is called the **cosmic microwave background radiation (CMB)**. When the universe was young, it was hot enough that all atoms were ionized, so that the electrons were separate from the atomic nuclei. Recall from our discussion of the structure of the Sun and stars in Chapter 14 that free electrons in such conditions interact strongly with radiation, blocking its progress so radiation does not travel well through ionized plasma.

As illustrated in **Figure 21.11a**, the conditions within the early universe were much like the conditions within a star: hot, dense, and opaque. As the universe

Figure 21.10 Arno Penzias (left) and Robert Wilson next to the Bell Labs radio telescope antenna with which they discovered the cosmic background radiation. This antenna is now a U.S. National Historic Landmark.

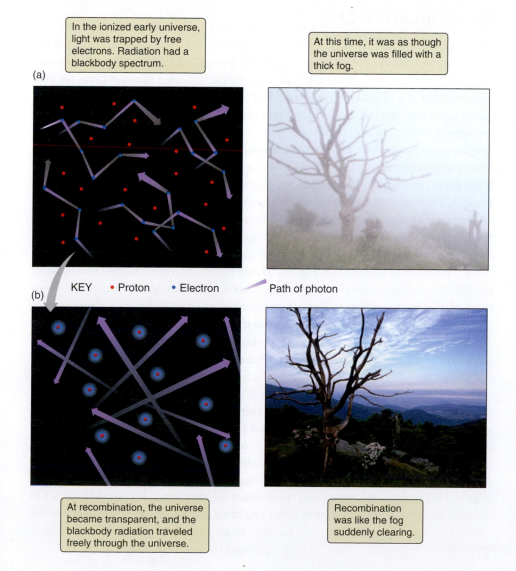

(a)

In the ionized early universe, light was trapped by free electrons. Radiation had a blackbody spectrum.

At this time, it was as though the universe was filled with a thick fog.

KEY • Proton • Electron → Path of photon

(b)

At recombination, the universe became transparent, and the blackbody radiation traveled freely through the universe.

Recombination was like the fog suddenly clearing.

Figure 21.11 The cosmic microwave background radiation originated at the moment the universe became transparent. (a) Before recombination, the universe was like a foggy day, except that the "fog" was a sea of electrons and protons. Radiation interacted strongly with free electrons and so could not travel far. The trapped radiation had a Planck blackbody spectrum. (b) When the constituents of the universe recombined to form neutral hydrogen atoms, the fog cleared and this radiation was free to travel unimpeded.

expanded, the gas filling it cooled. By the time the universe was several hundred thousand years old, and about a thousandth of its current size, the temperature had dropped to a few thousand kelvins. Hydrogen and helium nuclei combined with electrons to form neutral atoms, an event called the **recombination** of the universe, illustrated in Figure 21.11b.

Hydrogen atoms block radiation much less effectively than do free electrons, so when recombination occurred, the universe suddenly became transparent to radiation. Since that time, the radiation left behind from the Big Bang has been able to travel largely unimpeded throughout the universe. At the time of recombination, when the temperature of the universe was about 3000 K, the wavelength of this radiation peaked at about 1 micron (μm), according to Wien's law (see Chapter 5). As the universe expanded, this radiation was redshifted to longer and longer wavelengths—and therefore cooler temperatures as in Figure 21.9. Today, the scale of the universe has increased a thousandfold since recombination, and the peak wavelength of the cosmic background radiation has increased by a thousandfold as well, to a value close to 1 millimeter (mm). The spectrum of the CMB still has the shape of a Planck blackbody spectrum, but with a characteristic temperature a thousandth what it was at the time of recombination.

Variations in the CMB

The presence of cosmic background radiation with a Planck blackbody spectrum is a strong prediction of the Big Bang theory. Penzias and Wilson had confirmed that a signal with the correct strength was there, but they could not say for certain whether the signal they saw had the spectral shape of a Planck blackbody spectrum. From the late 1960s to the 1980s, experiments at different wavelengths supported these same conclusions. The Cosmic Background Explorer (COBE) satellite made extremely precise measurements of the CMB at many wavelengths, from a few microns out to 1 cm. As you can see in **Figure 21.12**, the COBE spectrum of the CMB is a Planck blackbody spectrum with a temperature of 2.73 K. The observed spectrum so perfectly matches the one predicted by Big Bang cosmology that there can be no real doubt that this is the residual radiation left behind from the primordial fireball of the early universe.

COBE data included much more than a measurement of the spectrum of the cosmic background radiation. **Figure 21.13a** shows a map obtained by COBE of the CMB from the entire sky. The different colors in the map correspond to variations of about 0.1 percent in the temperature of the CMB. Most of this range of temperature is present because one side of the sky looks slightly warmer than the opposite side of the sky. This difference has nothing to do with the large-scale structure of the universe itself, but rather is the result of the motion of Earth with respect to the CMB.

We have emphasized that there is no preferred frame of reference. The laws of physics are the same in any inertial reference frame, so none is better than any other. Yet in a certain sense there is a preferred frame of reference at every point in the universe. This is the frame of reference that is at rest with respect to the expansion of the universe and in which the CMB is isotropic, or the same in all directions. The COBE map shows that one side of the sky is slightly hotter than the other because Earth and the Sun are moving at a velocity of 368 km/s in the direction of the constellation Crater relative to this cosmic reference frame.

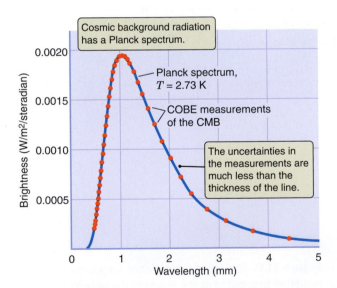

Figure 21.12 The spectrum of the CMB as measured by the Cosmic Background Explorer (COBE) satellite is shown by the red dots. The uncertainty in the measurement at each wavelength is much less than the size of a dot. The line running through the data is a blackbody spectrum with a temperature of 2.73 K.

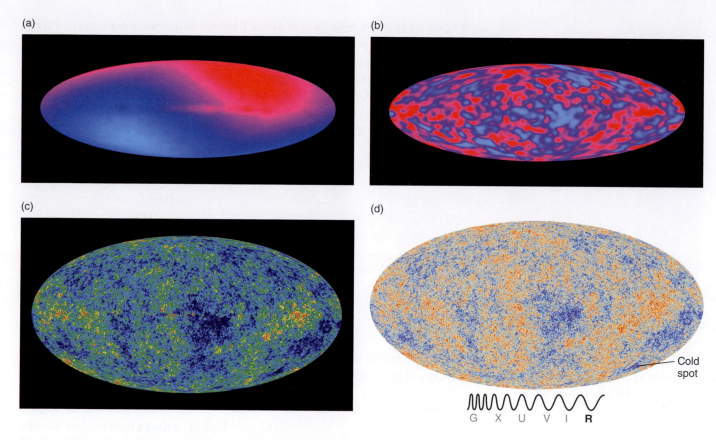

Figure 21.13 (a) The COBE satellite mapped the temperature of the CMB. The CMB is slightly hotter (by about 0.003 K) in one direction in the sky than in the other direction. This difference is due to Earth's motion relative to the CMB. (b) The COBE map with Earth's motion removed, showing tiny ripples remaining in the CMB. (c) *WMAP* (Wilkinson Microwave Anisotropy Probe) confirmed the fundamentals of cosmological theory at small and intermediate scales. (d) The Planck space observatory has provided the highest resolution yet of the CMB and has detected some surprises, such as the "cold spot." The radiation seen here was emitted less than 400,000 years after the Big Bang. Blue spots are cooler, red spots are warmer.

Radiation coming from the direction in which Earth is moving is slightly blue-shifted (and thus shifted to a higher temperature) by this motion, whereas radiation coming from the opposite direction is Doppler-shifted toward the red (or cooler temperatures). Earth's motion is due to a combination of factors, including the motion of the Sun around the center of the Milky Way Galaxy and the motion of the Milky Way relative to the CMB.

When this asymmetry in the CMB caused by the motion of Earth is subtracted from the COBE map, only slight variations in the CMB remain, as shown in Figure 21.13b. The brighter parts of this image are only about 0.001 percent brighter than the fainter parts. These slight variations might not seem like much, but they are actually of crucial importance in the history of the universe. Recall from Chapter 18 that gravity itself can create a redshift. These tiny fluctuations in the cosmic background radiation are the result of gravitational redshifts caused by concentrations of mass that existed in the early universe. These concentrations later gave rise to galaxies and the rest of the structure that is evident in the universe today.

Beginning in 2001, NASA's *WMAP* (Wilkinson Microwave Anisotropy Probe) satellite made more precise measurements of the variations of the CMB. Figure 21.13c shows the ripples measured by *WMAP* with much higher resolution than could be detected by COBE. The European Space Agency's Planck space observatory collected data of even higher resolution from 2009 to 2012. A "cold spot" first detected in the *WMAP* data was confirmed to be real (Figure 21.13d). The much higher-resolution maps obtained by *WMAP* and Planck enable astronomers to

Origins

Big Bang Nucleosynthesis

The expansion of the universe and the cosmic microwave background radiation are two of the key pieces of observational evidence supporting the Big Bang theory. The third major piece of supporting evidence comes from observations of the number and types of chemical elements in the universe. For a short time after the Big Bang, the temperature and density of the universe was high enough for nuclear reactions to take place. Collisions between protons in the early universe built up low-mass nuclei, including deuterium (heavy hydrogen) and isotopes of helium, lithium, and beryllium. This process of element creation, called **Big Bang nucleosynthesis**, determined the final chemical composition of the matter that emerged from the hot phase of the Big Bang. Because the universe was rapidly expanding and cooling, the density and temperature of the universe fell too low for fusion to heavier elements such as carbon to occur. Therefore, all elements more massive than beryllium, including most of the atoms that make

up Earth and its life, must have formed in subsequent generations of stars.

▶❚❚ **AstroTour:** Big Bang Nucleosynthesis

Figure 21.14 shows the observed and calculated predictions of the amounts of deuterium, helium, and lithium from Big Bang nucleosynthesis, plotted as a function of the observed present-day density of normal (luminous) matter in the universe. Observations of current abundances are shown as horizontal bands. Theoretical predictions, which depend on the density of the universe, are shown as darker, thick lines. Big Bang nucleosynthesis predicts that about 24 percent of the mass of the normal matter formed in the early universe should have ended up in the form of the very stable isotope ^{4}He, regardless of the total density of matter in the universe. Indeed, this is what is observed—about 24 percent of the mass of normal matter in the universe today is in the form of ^{4}He, in complete agreement with the prediction of Big Bang

nucleosynthesis. This agreement between theoretical predictions and observation provides powerful evidence that the universe began in a Big Bang.

Unlike helium, most other isotope abundances depend on the density of normal matter in the universe, so comparing current abundances with models of isotope formation in the Big Bang helps pin down the density of the early universe. Beginning with the abundances of isotopes such as ^{2}H (deuterium) and ^{3}He found in the universe today (shown as the roughly horizontal, light-colored bands in Figure 21.14) and comparing them to predictions in different models of how abundant isotopes *should* be when formed at different densities (dark-colored curves in Figure 21.14), cosmologists can find the density of normal matter (the vertical yellow band in Figure 21.14). The best current measurements give a value of about 3.9×10^{-28} kg/m^3 for the average density of normal matter in the universe today (at sea level, the density of air is about 1.2 kg/m^3). This value lies refine their ideas about the development of structure in the early universe, which we will discuss in Chapters 22 and 23.

CHECK YOUR UNDERSTANDING 21.5

The existence of the cosmic microwave background radiation tells us that the early universe was: (a) much hotter than it is today; (b) much colder than it is today; (c) about the same temperature as today but was much more dense.

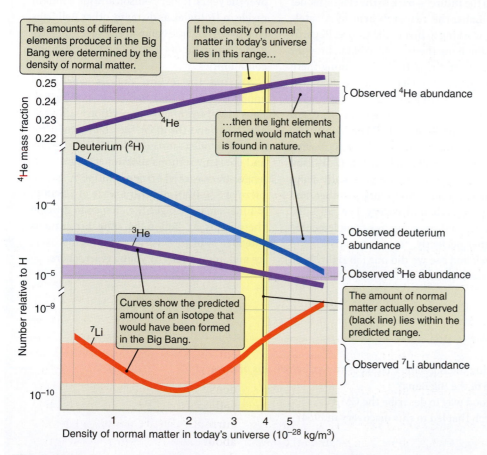

The amounts of different elements produced in the Big Bang were determined by the density of normal matter.

If the density of normal matter in today's universe lies in this range...

...then the light elements formed would match what is found in nature.

Curves show the predicted amount of an isotope that would have been formed in the Big Bang.

The amount of normal matter actually observed (black line) lies within the predicted range.

Observed ^{4}He abundance

Observed deuterium abundance

Observed ^{3}He abundance

Observed ^{7}Li abundance

^{4}He

Deuterium (^{2}H)

^{3}He

^{7}Li

^{4}He mass fraction

Number relative to H

Density of normal matter in today's universe (10^{-28} kg/m^3)

Figure 21.14 Observed and calculated abundances of the products of Big Bang nucleosynthesis, plotted against the density of normal matter in today's universe. Big Bang nucleosynthesis correctly predicts the amounts of these isotopes found in the universe today. (Note the two scale breaks on the y-axis.)

well within the range predicted by the observations shown in Figure 21.14. The agreement is remarkable, and it holds for many different isotopes.

Turning this around, cosmologists can begin with an observation of the amount of normal matter in and around galaxies and then compare that with calculations of what the chemical composition emerging from the Big Bang should have been. The observations agree remarkably well with the amounts of these elements actually found in nature. The idea that light elements originated in the Big Bang is resoundingly confirmed.

This agreement also provides a powerful constraint on the nature of dark matter, which dominates the mass in the universe. Dark matter cannot consist of normal matter made up of neutrons and protons; if it did, the density of neutrons and protons in the early universe would have been much higher, and the resulting abundances of light elements in the universe would have been much different from what is actually observed.

ARTICLES   QUESTIONS

Scientists celebrate 50 years since the detection of the cosmic microwave background radiation.

50th Anniversary of the Big Bang Discovery

By **JOANNE COLELLA, The Journal NJ**

A unique gathering of some of the most brilliant minds in the world was held last month on Holmdel's Crawfords Hill on a beautiful spring day, celebrating the 50th anniversary of a truly stellar discovery: the detection of cosmic microwave background radiation (CMB), the thermal echo of the universe's explosive birth, and the evidence that proved the famed Big Bang theory, which would have taken place about 13.8 billion years ago. On May 20, 1964, American radio astronomers Robert Wilson and Arno Penzias confirmed that discovery, admittedly by accident but only after an exhaustive amount of investigative research to rule out every possible explanation for an odd buzzing sound that came from all parts of the sky at all times of day and night. The hum was detected by the enormous Horn Antenna at the Bell Labs site, now a national landmark. Puzzled by the noise, but initially not suspecting its significance, the pair went to great lengths to determine, or rule out, any possible source—including some pigeons that had nested in the antenna and were determined to return, even after being shipped to a distant location.

In 1978, Dr. Wilson and Dr. Penzias won the esteemed Nobel Prize in Physics for their work. Now 78 and 81 years old, respectively, the two came together again at the Horn Antenna site to celebrate the momentous anniversary with current and former Bell Labs colleagues. The event was headed up by Bell Labs President and Corporate CTO Marcus Weldon, who lauded their achievement and spoke about the company initiative to return "back to the future"—back to the classic model of Bell Labs, the research arm of Alcatel-Lucent, working to invent the future. Regarding the Big Bang theory, Mr. Weldon brought chuckles by stating, "In the beginning there was nothing and then it exploded . . . and then there was Arno and Bob." Subtle humor was a recurring theme by some speakers, who even poked fun at myths about Bell Labs, which has produced 12 Nobel Prize laureates, harboring extraterrestrial staff and facilities. Dr. Wilson and Dr. Penzias each spoke personally about their backgrounds, their work together, and the meaning of their discovery. The pair's distinct styles and personalities complemented each other perfectly. "It is very satisfying to look back and see we did our job right," said Dr. Wilson quietly and humbly. "This was

pretty heady stuff," stated the talkative Dr. Penzias. "This is as close to being religious as I can be."

Throughout the presentations and celebratory luncheon, held under oversized tents on the expansive grounds, the air of excitement, pride, and mutual admiration among the attendees was palpable, as they looked back—and looked ahead—to the incredible legacy and pool of talent shared by Bell Labs personnel over the years. Robert Wilson and his wife still reside in Holmdel, as do many other Bell Labs employees, and the company has been an integral component of Holmdel and surrounding communities.

During the event, details were also announced about the establishment of the Bell Labs Prize, an annual competition to give scientists around the globe the chance to introduce their ideas in the fields of information and communications technology. The challenge offers a grand prize of $100,000, second prize of $50,000, and third prize of $25,000. Winners may also get the chance to develop their ideas at Bell Labs. The program is intended to inspire world-changing discoveries and innovations by young researchers.

ARTICLES   **QUESTIONS**

1. What were the other main pieces of evidence supporting the Big Bang that were known before this discovery?
2. Why was this discovery seen as the final confirmation of the Big Bang?
3. Is "thermal echo of the universe's explosive birth" a good way to describe the CMB?
4. Why might Bell Labs have been supporting the research that led to this discovery in 1964?

"50th anniversary of the Big Bang discovery," by Joanne Colella. The Journal, May 29, 2014. Reprinted by permission of The Holmdel Journal, Shrewsbury, NJ.

Summary

The universe has been expanding since the Big Bang, which occurred nearly 14 billion years ago. The Big Bang happened everywhere: it is not an explosion spreading out from a single point. There are three major pieces of evidence for the Big Bang. Hubble's law states that more distant galaxies have higher redshifts, and these observed redshifts result from the increasing space between them. Observations of the cosmic microwave background radiation independently confirm that the universe had a hot, dense beginning. Lastly, the amounts of helium and trace amounts of other light elements measured today agree with what would be expected from nuclear reactions of normal matter in the hot early universe.

LG 1 **Explain in detail the cosmological principle.** Observations suggest that on the largest scales, the universe is homogeneous (looks the same to all observers) and isotropic (looks the same in all direction), in agreement with the cosmological principle.

LG 2 **Describe how the Hubble constant can be used to estimate the age of the universe.** Hubble found that all galaxies are moving away, with a recession velocity proportional to distance, indicating that the universe is expanding. The local physics and structure of objects are not affected by the expansion. Running the Hubble expansion backward leads to the Hubble time, and the age of the universe, found from the inverse of the Hubble constant.

LG 3 **Describe the observational evidence for the Big Bang.** Hubble's law states that light from distant galaxies is redshifted; this occurs because space itself is expanding. Hubble's law suggests that the universe was once very hot and very dense, a beginning known as the Big Bang. Big Bang theory predicts that we should be able to observe the radiation from a few hundred thousand years after the Big Bang. This radiation, called the cosmic microwave background radiation, has the same spectrum in every direction.

LG 4 **Explain which chemical elements were created in the early hot universe.** During the first few minutes after the Big Bang, the universe was hot enough for nucleosynthesis to take place through the process of fusion. Deuterium, helium, lithium, and beryllium, but not the heavier elements, were created before the universe cooled too much for this fusion to take place.

? UNANSWERED QUESTIONS

- What existed before the Big Bang? The usual (and somewhat unsatisfactory) answer is that the Big Bang was the beginning of space and *time*, so there could be no time before it happened. A more recent answer is that this universe may be just one of many universes, and the Big Bang was the beginning of this universe only. We will return to this topic in Chapter 22.

- If there were evidence that the cosmological principle were not true, then could we deduce anything about the evolution of the universe?

Questions and Problems

Test Your Understanding

1. What do astronomers mean when they say that the universe is homogeneous?
   a. The universe looks exactly the same from every perspective.
   b. Galaxies are generally distributed evenly throughout the universe.
   c. All stars in all galaxies have planetary systems just like ours.
   d. The universe has looked the same at all times in its history.

2. What do astronomers mean when they say that the universe is isotropic?
   a. More distant parts of the universe look just like nearby parts.
   b. Intergalactic gas has the same density everywhere in the universe.
   c. The laws of physics apply everywhere in the universe.
   d. The universe looks the same in every direction.

3. Cosmological redshifts are calculated from observations of spectral lines from
   a. individual stars in distant galaxies.
   b. clouds of dust and gas in distant galaxies.
   c. spectra of entire galaxies.
   d. rotations of the disks of distant galaxies.

4. When they look into the universe, astronomers observe that nearly all galaxies are moving away from the Milky Way. This observation suggests that
   a. the Milky Way is at the center of the universe.
   b. the Milky Way must be at the center of the expansion.
   c. the Big Bang occurred at the location of the Milky Way.
   d. an observer in a distant galaxy would make the same observation.

5. Some galaxies have redshifts z that if equated to v_r/c correspond to velocities greater than the speed of light. Special relativity is not violated in this case
   a. because of relativistic beaming.
   b. because of superluminal motion.
   c. because redshifts carry no information.
   d. because these velocities do not measure motion through space.

6. The Big Bang theory predicted (select all that apply)
   a. the Hubble law.
   b. the cosmic microwave background radiation.
   c. the cosmological principle.
   d. the abundance of helium.
   e. the period-luminosity relationship of Cepheid variables.

7. The simplest way to estimate the age of the universe is from
   a. using the slope of Hubble's law.
   b. the age of Moon rocks.
   c. models of stellar evolution.
   d. measurements of the abundances of elements.

8. The CMB includes information about (select all that apply)
   a. the age of the universe.
   b. the temperature of the early universe.
   c. the density of the early universe.
   d. density fluctuations in the early universe.
   e. the motion of Earth around the center of the Milky Way.

9. Repeated measurements showing that the current helium abundance is much less than the value predicted by the Big Bang would imply that
   a. some part of the Big Bang theory is incorrect or incomplete.
   b. the current helium abundance is wrong.
   c. scientists don't know how to measure helium abundances.

10. The cosmological principle says that
    a. the universe is expanding.
    b. the universe began in the Big Bang.
    c. the rules that govern the universe are the same everywhere.
    d. the early universe was 1,000 times hotter than the characteristic temperature of the CMB.

11. Why is the Milky Way Galaxy not expanding together with the rest of the universe?
    a. It is not expanding because it is at the center of the expansion.
    b. It is expanding, but the expansion is too small to measure.
    c. The Milky Way is a special location in the universe.
    d. Local gravity dominates over the expansion of the universe.

12. The scale factor keeps track of
    a. the movement of galaxies through space.
    b. the current distances between many galaxies.
    c. the changing distance between any two galaxies.
    d. the location of the center of the universe.

13. The Big Bang is
    a. the giant supernova explosion that triggered the formation of the Solar System.
    b. the explosion of a supermassive black hole.
    c. the eventual demise of the Sun.
    d. the beginning of space and time.

14. The CMB is essentially uniform in all directions in the sky. This is an example of
    a. anisotropy.
    b. isotropy.
    c. thermal fluctuations.
    d. Wien's law.

15. Which of the following was *not* created as a result of Big Bang nucleosynthesis?
    a. helium
    b. lithium
    c. hydrogen
    d. deuterium
    e. carbon

Thinking about the Concepts

16. Imagine that you are standing in the middle of a dense fog.
    a. Would you describe your environment as isotropic? Why or why not?
    b. Would you describe it as homogeneous? Why or why not?

17. As the universe expands from the Big Bang, galaxies are not actually flying apart from one another. What is really happening?

18. We see the universe around us expanding, which gives distant galaxies an apparent velocity of 70 km/s/Mpc. If you were an astronomer living today in a galaxy that was located 1 billion light-years away from us, at what rate would you see the galaxies moving away from you?

19. Does Hubble's law imply that our galaxy is sitting at the center of the universe? Explain.

20. What does the value of R_U, the scale factor of the universe, tell us?

21. Does the expansion of the universe make the Sun bigger? What about the Milky Way? Why or why not?

22. Science's greatest strength is the self-correcting nature of scientific inquiry: minds can be changed in the face of new data, as described in the Process of Science Figure. Consider a modern paradigm of science such as the Big Bang, climate change, or the power source of stars. In a short paragraph, describe your current viewpoint, and give a piece of evidence that would make you change your mind if it were true. For example, Einstein believed the universe was static. Measurements of the movement of galaxies caused him to change his mind.

23. Name two predictions of the standard Big Bang theory that have been verified by observations.

24. Knowing that you are studying astronomy, a curious friend asks where the center of the universe is located. You answer, "Right here and everywhere." Explain in detail why you give this answer.

25. The general relationship between recession velocity (v_r) and redshift (z) is $v_r = cz$. This simple relationship fails, however, for very distant galaxies with large redshifts. Explain why.

26. Why is it significant that the CMB displays a Planck black-body spectrum?

27. What is the significance of the tiny brightness variations that are observed in the CMB?

28. What important characteristics of the early universe are revealed by today's observed abundances of various isotopes, such as ^{2}H and ^{3}He?

29. Why were only a few of the chemical elements created at the Big Bang?

30. How do astronomers know that some of the observed helium is left over from the Big Bang?

Applying the Concepts

31. In Figure 21.6, a rubber sheet is shown as an analogy to help you think about the scale factor. Between the moments shown in parts (a) and (b), each square doubles in size on every edge. How does the area of a square change? Imagine that the sheet is now a block of rubber, expanding in three dimensions instead of two. How would the volume of a cube change between the moment shown in part (a) and the moment shown in part (b)?

32. Study Figure 21.12. Error bars have not been plotted in this figure. Why not? Was this a very precise measurement or a very imprecise measurement? How does the precision of the measurement affect your confidence in the conclusions drawn from it?

33. Study Figure 21.14. This figure includes both predictions and observations.
    a. What do the vertical yellow bar and the slanted lines and curves represent: theory or observation?
    b. What do the pastel horizontal lines and the vertical black line represent: predictions or observations?
    c. Do the predictions and observations match? Choose one example, and explain how you know.

34. The Hubble time ($1/H_0$) represents the age of a universe that has been expanding at a constant rate since the Big Bang. Calculate the age of the universe in years if $H_0 = 80$ km/s/Mpc. (Note: 1 year $= 3.16 \times 10^7$ seconds, and 1 Mpc $= 3.09 \times 10^{19}$ km.)

35. Throughout the latter half of the 20th century, estimates of H_0 ranged from 50 to 100 km/s/Mpc. Calculate the age of the universe in years for each of these estimated values of H_0. Do any answers contradict the ages obtained from stellar evolution or geology on Earth?

36. Suppose a galaxy is observed with a redshift equal to 2. How much has the universe expanded since that light was emitted from these galaxies?

37. How much has the universe expanded since light was emitted from a galaxy with a redshift of $z = 8$?

38. You observe a distant quasar in which a spectral line of hydrogen with rest wavelength $\lambda_{rest} = 121.6$ nm is found at a wavelength of 547.2 nm. What is its redshift? When the light from this quasar was emitted, how large was the universe compared to its current size?

39. A distant galaxy has a redshift $z = 5.82$ and a recession velocity $v_r = 287{,}000$ km/s (about 96 percent of the speed of light).
    a. If $H_0 = 70$ km/s/Mpc and if Hubble's law remains valid out to such a large distance, then how far away is this galaxy?
    b. Assuming a Hubble time of 13.8 billion years, how old was the universe at the look-back time of this galaxy?
    c. What was the scale factor of the universe at that time?

40. The spectrum of the CMB is shown as the red dots in Figure 21.12, along with a blackbody spectrum for a blackbody at temperature of 2.73 K. From the graph, determine the peak wavelength of the CMB spectrum. Use Wien's law to find the temperature of the CMB. How does this rough measurement that you just made compare to the accepted temperature of the CMB?

41. COBE observations show that the Solar System is moving in the direction of the constellation Crater at a speed of 368 km/s relative to the cosmic reference frame. What is the blueshift (negative value of z) associated with this motion?

42. The average density of normal matter in the universe is 4×10^{-28} kg/m^3. The mass of a hydrogen atom is 1.66×10^{-27} kg. On average, how many hydrogen atoms are there in each cubic meter in the universe?

43. To get a feeling for the emptiness of the universe, compare its density (4×10^{-28} kg/m³) with that of Earth's atmosphere at sea level (1.2 kg/m³). How much denser is Earth's atmosphere? Write this ratio using standard notation.

44. Assume that the most distant galaxies have a redshift $z = 10$. The average density of normal matter in the universe today is 4×10^{-28} kg/m³. What was its density when light was leaving those distant galaxies? (Hint: Keep in mind that volume is proportional to the cube of the scale factor.)

45. What was the size of the universe (compared to the present) when the CMB was emitted, at $z = 1{,}000$?

USING THE WEB

46. For more details on the history of the discovery of the expanding universe, go to the American Institute of Physics' "Cosmic Journey: A History of Scientific Cosmology" website (www.aip.org/history/cosmology/). Read through the sections titled "Island Universes," "The Expanding Universe," and "Big Bang or Steady State?" Why was Albert Einstein "irritated" by the idea of an expanding universe? What was the contribution of Belgian astrophysicist (and Catholic priest) Georges Lemaître? What is the steady-state theory, and what was the main piece of evidence against it?

47. Go to the American Institute of Physics page on the age of the universe, http://www.aip.org/history/curie/age-of-earth.htm. How does the age of the universe from Hubble's law compare with the age of the Solar System? How does it compare with the age of the oldest globular clusters? Do a search for "age of the universe"—has the value been updated?

48. A series of public lectures at Harvard commemorating the 50th anniversary of the discovery of the CMB can be found here: http://astronomy.fas.harvard.edu/news/public-talk-50-year-anniversary-discovery-cosmic-micrrowave-background. Robert Wilson's talk starts at the 27-minute mark. Why does he say this is the most important thing he ever measured? How did he identify this radiation?

49. Updated observations of the CMB from the Planck space observatory are reported on the European Space Agency's Planck website (http://esa.int/SPECIALS/Planck). What has been learned from this mission? What is their value of the Hubble constant?

50. Go to the University of Washington Astronomy Department's "Hubble's Law: An Introductory Astronomy Lab" Web page (http://www.astro.washington.edu/courses/labs/clearinghouse/labs/HubbleLaw/hubbletitle) and do the lab exercise, which uses real data from galaxies to calculate Hubble's constant. Your instructor will indicate whether you should use the regular or the shorter version.

smartwork

If your instructor assigns homework in SmartWork, access your assignments at digital.wwnorton.com/astro5.

EXPLORATION

digital.wwnorton.com/astro5

The expansion of the universe is extremely difficult to visualize, even for professional astronomers. In this Exploration, you will use the surface of a balloon to get a feel for how an "expansion" changes distances between objects. Throughout this Exploration, remember to think of the surface of the balloon as a two-dimensional object, much as the surface of Earth is a two-dimensional object for most people. The average person can move east or west, or north or south, but into Earth and out to space are not options. For this Exploration you will need a balloon, 11 small stickers, a piece of string, and a ruler. A partner is helpful as well. **Figure 21.15** shows some of the steps involved.

Blow up the balloon partially and hold it closed, but *do not tie it shut*. Stick the 11 stickers on the balloon (these represent galaxies) and number them. Galaxy 1 is the reference galaxy.

Measure the distance between the reference galaxy and each of the galaxies numbered 2–10. The easiest way to do this is to use your piece of string. Lay it along the balloon between the two galaxies and then measure the length of the string. Record these data in the "Distance 1" column of a table like the one shown later.

Simulate the expansion of your balloon universe by *slowly* blowing up the balloon the rest of the way. Have your partner count the number of seconds it takes you to do this, and record this number in the "Time Elapsed" column of the table (each row has the same time elapsed, because the expansion occurred for the same amount of time for each galaxy). Tie the balloon shut. Measure the distance between the reference galaxy and each numbered galaxy again. Record these data under "Distance 2."

Subtract the first measurement from the second. Record the difference in the table.

Divide this difference, which represents the distance traveled by the galaxy, by the time it took to blow up the balloon. Distance divided by time gives an average speed.

Make a graph with velocity on the *y*-axis and distance 2 on the *x*-axis to get "Hubble's law for balloons." You may wish to roughly fit a line to these data to clarify the trend.

Figure 21.15 Measure the distance around a curved balloon using a string.

Galaxy Number	Distance 1	Distance 2	Difference	Time Elapsed	Velocity
1 (reference)	0	0	0		0
2					
3					
4					
5					
6					
7					
8					
9					
10					
11					

1 Describe your data. If you fit a line to them, is it horizontal or does it trend upward or downward?

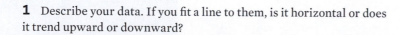

2 Is there anything special about your reference galaxy? Is it different in any way from the others?

3 If you had picked a different reference galaxy, would the trend of your line be different? If you are not sure of the answer, get another balloon and try it.

4 The expansion of the universe behaves similarly to the movement of the galaxies on the balloon. We don't want to carry the analogy too far, but there is one more thing to think about. In your balloon, some areas probably expanded less than others because the material was thicker; there was more "balloon stuff" holding it together. How is this similar to some places in the actual universe?

22 Cosmology

Cosmology is the study of the large-scale universe, including its nature, origin, evolution, and ultimate destiny. In this chapter, we take a closer look at the nature of the universe, how it has evolved over time, and its ultimate fate. We also discuss the physics of the smallest particles, which is necessary for describing the earliest moments of the universe.

LEARNING GOALS

By the conclusion of this chapter, you should be able to:

LG 1 Explain how mass within the universe and the gravitational force it produces affect the history, shape, and fate of the universe.

LG 2 Describe the evidence for the accelerating expansion of the universe.

LG 3 Describe the early period of rapid expansion known as inflation.

LG 4 Explain how the events that occurred in the earliest moments of the universe are related to the forces that operate in the modern universe.

The arrows point to a Type Ia supernova observed in the galaxy NGC 1365. ▶ ▶ ▶

What will be
the fate of
the universe?

22.1 Gravity and the Expansion of the Universe

The fate of the universe is a central question of modern cosmology. The simplest answer depends in part on the average mass in the universe within a fixed volume. A number of factors determine the way mass is distributed on large scales across the universe. In this section, we will look at one of these factors: gravity.

Mass Distribution

To see how gravity affects the expansion of the universe, recall gravity's effects on the motion of projectiles and the discussion of escape velocity from Chapter 4. For example, the fate of a projectile fired straight up from the surface of a planet depends on its speed. As long as the projectile speed is less than the planet's escape velocity, gravity will eventually stop the rise of the projectile and pull it back to the planet's surface. But if the speed of the projectile is greater than the planet's escape velocity, gravity will lose. In this case, although the projectile will slow down, it will never stop, and it will escape from the planet entirely.

Recall that the escape velocity from a planet's surface depends on the mass and radius of the planet. If the planet is massive enough, it will slow the projectile, stop it, and pull it back down. However, if the planet is not massive enough, the projectile will slow down but still escape to space. The size of the planet is also important. A smaller planet's mass is more densely packed: if the projectile is fired from the surface, which is closer to the plant's center, the gravitational pull is stronger, and the planet can pull the projectile down. However, if the planet has lower density, then the projectile at the surface starts farther from the center of the planet, the gravitational pull is weaker, and the projectile will escape. Whether the projectile escapes depends on both the planet's mass and the planet's radius.

Just as the mass of the planet gravitationally pulls on a projectile to slow its climb, the mass distributed across the universe, as characterized by its average density, gravitationally slows the expansion of the universe. If there is enough mass in the universe, then gravity will be strong enough to stop the expansion. However, if the radius of the universe is large, even a lot of mass won't lead to a high escape velocity. Thus, the density is critical in determining the fate of the universe. If the mass is packed closely together, so that the average density is high, the expansion will slow, stop, and reverse. If the mass is very spread out, so that the average density is low, the universe will expand forever.

Critical Density

The "escape velocity" of the universe is also determined by its average density. A faster expansion requires a higher density to stop the expansion. Astronomers define a **critical density** for which the mass in the universe would cause it to just barely stop expanding after a very long time. This critical density determines the dividing line between two possible fates of the universe: expanding forever or collapsing. If the universe is less dense than the critical density, gravity will be too weak, and the universe will expand forever. If the universe is denser on average than the critical density, then gravity will be strong enough to stop and reverse the expansion eventually.

The faster the universe is expanding, the more mass is needed to turn that expansion around. For this reason, the critical density depends on the value of several constants, including the gravitational constant, G, and the Hubble constant, H_0. If gravity is the only factor affecting the expansion, the critical density of the universe is less than the mass of six hydrogen atoms in every cubic meter (**Working It Out 22.1**).

Astronomers use omega, Ω, the very last letter in the Greek alphabet, to discuss densities and the ultimate fate of the universe. The ratio of actual density to critical density is called the *matter density parameter*, written as Ω_m (pronounced "omega matter"). Because it is a ratio of two densities, it has no units.

Figure 22.1 plots the expansion of the universe versus time for different values of Ω_m for a universe in which the expansion is controlled solely by the gravitational attraction of its mass. The colored lines show different possible values of Ω_m. If the universe has a density higher than the critical density, Ω_m is greater than 1, and gravity is strong enough to turn the expansion around. The expansion will slow and eventually stop, and the universe will then fall back in on itself. Conversely, if the universe has a density lower than the critical density, Ω_m is less than 1. The expansion of the universe will be slowed by gravity, but it will still expand forever. The dividing line, where Ω_m equals 1, corresponds to a universe that expands more and more slowly—continuing forever, but never quite stopping. Such a universe is expanding at exactly the "escape velocity." If gravity were not present, the Hubble expansion would not be changing, and the plots in Figure 22.1 would be straight lines instead of curved ones.

Until the closing years of the 20th century, most astronomers thought that this straightforward application of gravity was all that was needed to understand the expansion and fate of the universe. Researchers carefully measured the masses of galaxies and assemblages of galaxies in the expectation that these data would reveal the density and therefore the fate of the universe. The luminous matter

---

22.1 Working It Out Calculating the Critical Density

Using the assumption of the cosmological principle and using a set of equations from Einstein on gravitation, Alexander Friedmann (1888–1925) derived equations for cosmology in an expanding universe and discussed the idea of critical density. When he assumed that the energy of empty space was zero, making gravity from the total mass of the universe the only factor, he derived the following equation for the critical density, ρ_c:

$$\rho_c = \frac{3H_0^2}{8\pi G}$$

where H_0 is the Hubble constant, and G is the gravitational constant. Using the value for H_0 of 70 kilometers per second per megaparsec (km/s/Mpc) and 1 Mpc = 3.1×10^{19} km, we can rewrite H_0 as

$$H_0 = \frac{70 \text{ km/s}}{3.1 \times 10^{19} \text{ km}} = 2.3 \times 10^{-18}/\text{s}$$

Then, using $G = 6.67 \times 10^{-20}$ km³/(kg s²), the current value of the critical density is given by

$$\rho_c = \frac{3 \times (2.3 \times 10^{-18}/\text{s})^2}{8 \times \pi \times [6.67 \times 10^{-20} \text{ km}^3/(\text{kg s}^2)]}$$

$$\rho_c = 9.5 \times 10^{-18} \text{ kg/km}^3 = 9.5 \times 10^{-27} \text{ kg/m}^3$$

Dividing by the mass of a hydrogen atom, 1.67×10^{-27} kg, we find that this equals about 5.7 hydrogen atoms per cubic meter. The observed mass density of the universe is only a few percent of this value of the critical density.

The Hubble constant, which measures the rate of expansion of the universe, changes with time. Consequently, the critical density changes with time as well.

Figure 22.1 There are three possible fates of the universe if gravity is the only important factor.

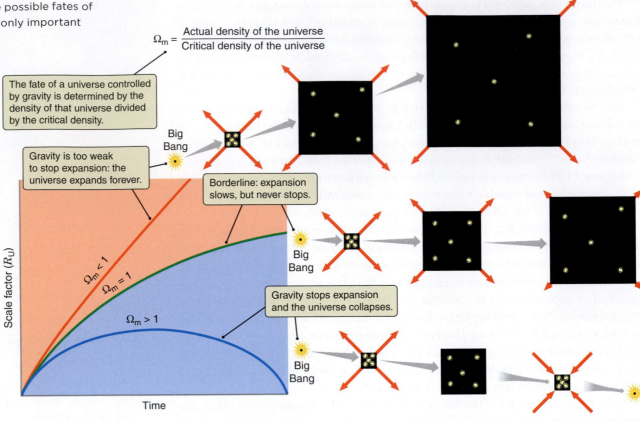

$$\Omega_m = \frac{\text{Actual density of the universe}}{\text{Critical density of the universe}}$$

The fate of a universe controlled by gravity is determined by the density of that universe divided by the critical density.

Gravity is too weak to stop expansion: the universe expands forever.

Borderline: expansion slows, but never stops.

Gravity stops expansion and the universe collapses.

Big Bang

$\Omega_m < 1$

$\Omega_m = 1$

$\Omega_m > 1$

Scale factor (R_U)

Time

that is seen in galaxies and groups of galaxies gives an Ω_m value of about 0.02. But galaxies contain about 10 times as much dark matter as normal luminous matter, so adding in the dark matter in galaxies pushes the value of Ω_m up to about 0.2. When the mass of dark matter *between* galaxies is included (a subject we will return to in the next chapter), Ω_m could increase to 0.3 or higher. By this accounting, there is only at most a third as much mass in the universe as is needed to stop the universe's expansion. Many astronomers were convinced that the expansion of the universe would never be zero, but it would slow down due to gravity. They were wrong.

CHECK YOUR UNDERSTANDING 22.1

If there is enough mass in the universe that the density is higher than the critical density, the universe will: (a) expand forever; (b) expand, but gradually slow down; (c) eventually collapse; (d) neither expand nor contract.

22.2 The Accelerating Universe

Edwin Hubble showed that the universe is expanding. If the expansion of the universe is slowing down due to gravity, then the expansion must have been faster in the past. If this is the case, objects that are very far away—so that we see them as they were long ago—are moving faster. They should therefore have larger redshifts than Hubble's law derived from local galaxies would suggest. However, if objects that are very far away have smaller redshifts than Hubble's law derived from local galaxies indicates, it means the expansion of the universe is speeding up. In this section, we examine the evidence that the expansion of the universe is accelerating.

The Cosmological Constant

During the 1990s, astronomers measured the brightness of Type Ia supernovae in very distant galaxies and compared the brightness of those supernovae with their expected brightness, which was based on the redshift distances of those galaxies. (Recall from Chapter 19 that Type Ia supernovae have a very high peak luminosity that can be calibrated, and therefore they are used as "standardizable" candles for measuring distances to galaxies.) The findings of these studies sent a wave of excitement through the astronomical community. The observational data of the Type Ia supernovae at different distances are plotted in **Figure 22.2**. Rather than showing that the expansion of the universe has slowed down over time, the data indicated that the expansion is *speeding up*—accelerating. An "accelerating universe" does not mean that the universe is zooming through space faster and faster like a car along a road, but instead means that the expansion is happening faster and faster. For this to be true, a previously unknown force stronger than gravity must be pushing the *entire universe* outward in opposition to gravity. Results from the *WMAP* (Wilkinson Microwave Anisotropy Probe) spacecraft early in the 21st century confirmed the result independently, and in 2011, the Nobel Prize in Physics was awarded to Saul Perlmutter, Brian P. Schmidt, and Adam G. Riess for their observations of Type Ia supernovae and the discovery of the accelerating universe.

The idea of a repulsive force that opposes the attractive force of gravity is not new. In the early 20th century, Einstein used his newly formulated equations of general relativity to calculate the structure of spacetime in the universe. The equations clearly indicated that any universe containing mass could not be static, any more than a ball can hang motionless in the air. He found the same result that Figure 22.1 illustrates; namely, that gravity always makes the universe move toward slower expansion or even collapse. However, Einstein's formulation of spacetime came more than a decade before Hubble discovered the expansion of the universe, and the conventional wisdom at the time was that the universe was static—that it neither expands nor collapses.

To force his new general theory of relativity to allow for a static universe, Einstein inserted a "fudge factor" called the **cosmological constant** into his equations. Einstein's cosmological constant acts as a repulsive force that opposes gravity. If it has just the right value, the cosmological constant can lead to a static universe in which galaxies remain stationary despite their mutual gravitational attraction.

When Hubble announced his discovery that the universe is expanding, Einstein realized his mistake. Einstein could have predicted that the universe must either be expanding or contracting with time, but instead forced his equations to comply with conventional wisdom. He called the introduction of his fudge factor, the cosmological constant, the "biggest blunder" of his scientific career. However, the much more recent discovery of an accelerating universe restores the cosmological constant to general relativity. With the results on the brightness of Type Ia supernovae, Einstein's cosmological constant turned out to

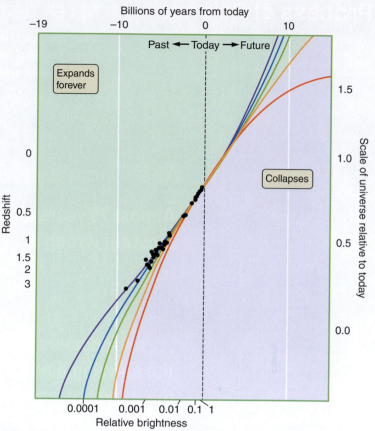

Figure 22.2 This graph plots the observed brightness of Type Ia supernovae as a function of their redshift. The different-colored lines represent different values of the cosmological constant. The observations (data points) indicate that the redshifts are too small for their distances: they best fit the line for an accelerating universe—a universe that is expanding faster today than it did in the past. Note that the colored lines are more separate at higher redshifts, so astronomers look for higher-redshift supernovae to differentiate the data better.

NEVER THROW ANYTHING AWAY

The accelerating universe is a good example of a scientist being right for the wrong reason. Einstein's famous mathematical device has become useful in a way he never would have imagined.

Almost 75 years after Hubble proves Einstein wrong . . .

. . . astronomers were working to measure the slowing down of the expansion of the universe, and to their surprise, they found that the expansion rate is accelerating.

Einstein's "fudge factor" is revived as the cosmological constant, which accounts for the acceleration.

Science has a memory that revisits old ideas in the light of new data, often saving a great deal of effort, even when the interpretation has changed.

be critical to understanding the expansion of the universe. A repulsive force, like the force associated with the cosmological constant in Einstein's equations, is just what is needed to describe a universe that is expanding at an accelerating rate (**Process of Science Figure**).

Today, we write the cosmological constant as Λ (uppercase lambda). The fraction of the critical density provided by the cosmological constant is written as Ω_Λ (pronounced "omega lambda"). In a universe with an accelerating expansion, Ω_Λ is not zero. If Ω_Λ is not zero, the fate of the universe is no longer controlled exclusively by Ω_m, but rather is affected by both Ω_Λ and Ω_m. If something is effectively pushing outward from within the universe, adding to its expansion, then gravity will have a harder time turning the expansion around. In that case, the mass density needed to halt the expansion of the universe will be greater than the critical mass density we already discussed. **Figure 22.3** shows plots of size of the universe versus time that are similar to those in Figure 22.1, but now we have included the effects of Ω_Λ. The fate of the universe depends on the balance between Ω_Λ and Ω_m. If the mass density is sufficiently large (that is, Ω_m is large), gravity will stop the expansion, and the universe will collapse back on itself, regardless of the value of Ω_Λ. In contrast, if Ω_m is less than 1, and therefore not large enough to stop the expansion so that the universe expands forever, then whether the expansion of the universe will accelerate or decelerate depends on Ω_Λ.

When the universe was young and compact, the effect of gravity was strong enough to dominate the effect of the cosmological constant. As the universe expanded, gravity grew weaker and weaker relative to the cosmological constant because the mass was more and more spread out. Unlike gravity, however, the effect of the cosmological constant has become increasingly *greater* as the universe expands. Unless gravity is strong enough to counter the cosmological constant, the cosmological constant will win in the end, causing the expansion to continue accelerating forever. Indeed, even if the density of the universe is greater than the critical density, (Ω_m is greater than 1), so that otherwise it would have collapsed back on itself because of gravity, a large enough cosmological constant could overwhelm gravity and make the universe expand forever (dashed blue line in Figure 22.3).

When Einstein added the cosmological constant to his equations of general relativity, he considered it a new fundamental constant, similar to Newton's universal gravitational constant G. Today, physicists call "empty space" the **vacuum** and recognize that it has some distinct physical properties. For example, the vacuum can have nonzero energy even in the total absence of matter. This energy of empty space produces exactly the kind of repulsive force that could accelerate the expansion of the universe. This repulsive force is called **dark energy**. Einstein's cosmological constant is an example of dark energy that accelerates the expansion of the universe.

But while the cosmological constant does not change, other versions of dark energy can evolve over time. Dark energy is a very active area of study, as scientists work to figure out where this energy comes from, its form and properties, and whether the amount changes over time.

Dark Energy and the Fate of the Universe

What fate, then, actually awaits the universe? Astronomers have data from several different experiments showing the range of possible values for Ω_m and Ω_Λ. The data from Type Ia supernovae, from *WMAP*, and from clusters of galaxies (which will be discussed in Chapter 23) are all consistent with values for Ω_m and

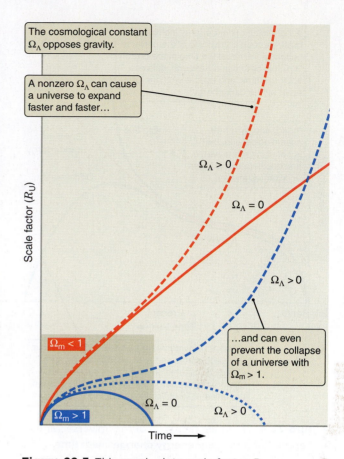

Figure 22.3 This graph plots scale factor R_U versus time for a possible universe with or without a cosmological constant, Ω_Λ. The solid curves for $\Omega_\Lambda = 0$ are the same as in Figure 22.1. The dashed curves show universes with $\Omega_\Lambda > 0$ for different Ω_m. If there is enough mass in a universe, gravity could still overcome the cosmological constant and cause that universe to collapse. In the presence of a cosmological constant, any universe without enough mass to collapse will instead eventually end up expanding at an ever-increasing rate.

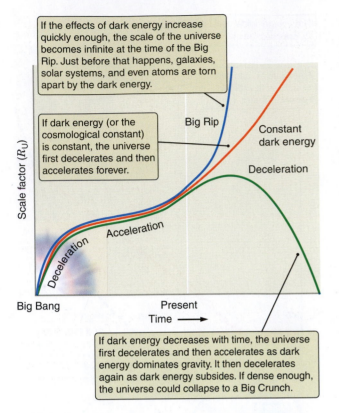

If the effects of dark energy increase quickly enough, the scale of the universe becomes infinite at the time of the Big Rip. Just before that happens, galaxies, solar systems, and even atoms are torn apart by the dark energy.

If dark energy (or the cosmological constant) is constant, the universe first decelerates and then accelerates forever.

Big Rip

Constant dark energy

Deceleration

Acceleration

Deceleration

Big Bang

Present
Time →

Scale factor (R_U)

If dark energy decreases with time, the universe first decelerates and then accelerates as dark energy dominates gravity. It then decelerates again as dark energy subsides. If dense enough, the universe could collapse to a Big Crunch.

Figure 22.4 The scale factor R_U of the universe varies depending on how dark energy changes with time.

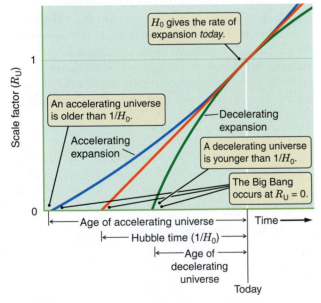

H_0 gives the rate of expansion *today*.

An accelerating universe is older than $1/H_0$.

Decelerating expansion

Accelerating expansion

A decelerating universe is younger than $1/H_0$.

The Big Bang occurs at $R_U = 0$.

Scale factor (R_U)

1

0

← Age of accelerating universe → Time →
← Hubble time ($1/H_0$) →
← Age of decelerating universe →

Today

Figure 22.5 This graph shows plots of the scale factor R_U versus time for three possible universes. If the universe has expanded at a constant rate, then its age is equal to the Hubble time, $1/H_0$. If the expansion of the universe has slowed with time, then the universe is younger than the Hubble time. If the expansion has sped up with time, the universe is older than $1/H_0$.

Ω_Λ of about 0.3 and 0.7, respectively. Thus, it appears that the expansion of the universe is accelerating under the dominant effect of dark energy and has been doing so for 5 billion to 6 billion years.

Because scientists do not yet understand the origin of dark energy, it is possible that it is not really a constant of nature and instead could be either increasing or decreasing with time. A changing cosmological constant would significantly change the future of the universe, as illustrated in **Figure 22.4**. For example, if dark energy were to decrease rapidly enough with time, the accelerating expansion of the universe that is observed now would change to a *deceleration* as the mass once again dominates over dark energy. In fact, if the universe were much denser than measured values indicate, the expansion could reverse, and the universe could collapse to what astronomers call the **Big Crunch**. By contrast, if the effect of dark energy were to increase with time, the universe would accelerate its expansion at an ever-increasing rate. Ultimately, expansion could be so rapid that the scale factor would become infinite within a finite period of time—a phenomenon called the **Big Rip**. In the Big Rip, the repulsive force of dark energy would become so dominant that the entire universe would come apart. First, gravity would no longer keep groups of galaxies together; then, gravity would no longer be able to hold individual galaxies together; and so on. Just before the end, the Solar System would come apart, and even atoms would be ripped into their constituent components. But don't worry too much about the Big Crunch or the Big Rip: the best observational data seem consistent with constant dark energy. In this case, the universe continues expanding, and after 100 trillion years or so, the universe is cold and dark, filled with black holes, dead stars, and dead planets (sometimes called the *Big Freeze* or the *Big Chill*).

The Age of the Universe

The values for Ω_m and Ω_Λ not only affect predictions for the future of the universe but also influence how astronomers interpret the past. **Figure 22.5** shows plots of the scale factor of the universe versus time. Measurement of the Hubble constant (H_0) indicates how fast the universe is expanding *today*. That is, the Hubble constant indicates the slope of the curves in Figure 22.5 at the current time. If the expansion of the universe has not changed with time, then the plot of the scale factor versus time is the straight red line in Figure 22.5. The age of the universe in this case is equal to the Hubble time: $1/H_0$. If the expansion of the universe has been slowing down with time (green line in Figure 22.5), then the universe is actually younger than the Hubble time. If the expansion of the universe has been speeding up with time (blue line), then the true age of the universe is greater than the Hubble time. This is also illustrated in **Figure 22.6**, where looking down the blue timeline from the present to the past shows that the different models take different amounts of time to go from the Big Bang to the present.

Recall that a Hubble constant of $H_0 = 70$ km/s/Mpc corresponds to a Hubble time ($1/H_0$) of about 13.8 billion years. If the expansion of the universe has slowed over time, the universe is actually younger than 13.8 billion years. Having a younger universe is a problem if the measured ages of globular clusters—13 billion years—is correct, because globular clusters clearly cannot be older than the universe that contains them. But if the expansion of the universe has sped up with time, as suggested by the observations of Type Ia supernovae and by *WMAP*, then the universe is at least 13.8 billion years old—comfortably older than globular clusters. The effects from gravity and dark energy have nearly canceled out at the present time.

The Shape of the Universe

We have already discussed such properties of the universe as density, dark energy, and age. The universe also has another key property: its shape in spacetime. Recall the concept of spacetime as described by general relativity. Space is a "rubber sheet" that has stretched outward from the Big Bang. In Chapter 18, you saw that the rubber sheet of space is also curved by the presence of mass. You saw how the shape of space around a massive object is detected through changes in geometric relationships, such as the ratio of the circumference of a circle to its radius or the sum of the angles in a triangle. Recall that the mass of a star, planet, or black hole causes a distortion in the shape of space; similarly, the mass of everything in the universe—including galaxies, dark matter, and dark energy—distorts the shape of the universe *as a whole*.

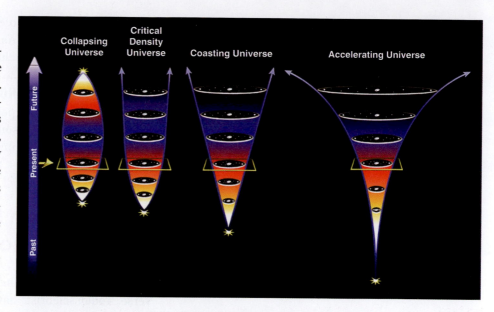

Figure 22.6 Past, present, and future for different scenarios. All models start in the past with a Big Bang (yellow star), expand to the present (yellow square), and then have different possible futures. The age of the universe at the present time is different in each model as in Figure 22.5.

Three basic shapes are possible for the universe, as shown in **Figure 22.7**. Which shape actually describes the universe is determined by the total amount of mass and energy; in other words, the sum of Ω_m and Ω_Λ. Continuing with the rubber-sheet analogy, the first possibility, corresponding to $\Omega_m + \Omega_\Lambda = 1$, is that the universe is a **flat universe**. A flat universe is described overall by the rules of the basic Euclidean geometry. As shown in Figure 22.7a, circles in a flat universe have a circumference of 2π times their radius ($2\pi r$), and triangles contain angles whose sum is 180 degrees. A flat universe stretches on forever.

The second possibility, shown in Figure 22.7b, is that the universe is shaped something like the surface of a saddle. This type of universe, in which $\Omega_m + \Omega_\Lambda < 1$, is also infinite and is called an **open universe**. In an open universe, the circumference of a circle is greater than $2\pi r$, and triangles contain less than 180 degrees.

The third possibility, shown in Figure 22.7c, in which $\Omega_m + \Omega_\Lambda > 1$, is a universe shaped like the surface of a sphere. The geometric relationships on a sphere are similar to those in the vicinity of a massive object, as discussed in Chapter 18. The circumference of a circle on a sphere is less than $2\pi r$, and triangles contain more than 180 degrees. This possibility is called a **closed universe** because space is finite and closes back on itself. Imagine a universe with two spatial dimensions on the surface of a sphere. Then the surface (and universe) is finite with no boundary. The cosmological principle is satisfied. Space is locally flat, and the expansion of the universe maintains both the cosmological principle and the local flatness of space.

Figure 22.7 These two-dimensional representations show possible geometries that space can have in a universe. In a flat universe (a), Euclidean geometry holds: triangles have angles that sum to 180 degrees, and the circumference of a circle equals 2π times the radius. In an open universe (b) or a closed universe (c), these relationships are no longer correct over very large distances.

(a) Flat geometry

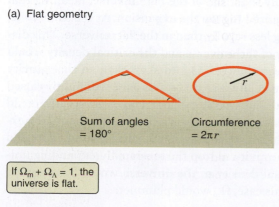

Sum of angles = 180°

Circumference = $2\pi r$

If $\Omega_m + \Omega_\Lambda = 1$, the universe is flat.

(b) Open (saddle) geometry

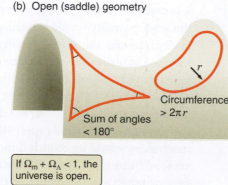

Sum of angles < 180°

Circumference > $2\pi r$

If $\Omega_m + \Omega_\Lambda < 1$, the universe is open.

(c) Closed (spherical) geometry

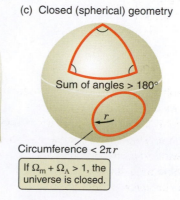

Sum of angles > 180°

Circumference < $2\pi r$

If $\Omega_m + \Omega_\Lambda > 1$, the universe is closed.

The measurements to estimate directly which of these shapes describes the universe are difficult. As noted earlier, the data suggest that $\Omega_m = 0.3$ and $\Omega_\Lambda = 0.7$, so $\Omega_m + \Omega_\Lambda$ is close to 1, meaning that the universe is very nearly flat.

CHECK YOUR UNDERSTANDING 22.2

Dark energy has been hypothesized to solve which problem? (a) The universe is expanding. (b) The cosmic microwave background radiation is too smooth. (c) The expansion of the universe is accelerating. (d) Stars orbit the centers of galaxies too fast.

22.3 Inflation Solves Several Problems in Cosmology

A century ago, astronomers were struggling to understand the size of the universe. Today scientists have a comprehensive theory that ties together many diverse facts about nature: the constancy of the speed of light, the properties of gravity, the motions of galaxies, and even the origins of the atoms that make up planets and life. The case for the Big Bang is compelling. Even so, improved observations of the cosmic background radiation and measurements of the expansion of the universe have raised some questions about how the universe expanded when it was very young. In this section, we will look at these questions and some potential answers.

The Flatness Problem

From the cosmic microwave background radiation, astronomers have found that the universe is flat—too close to being exactly flat for this to have happened by chance. Any deviation from flatness would grow over time, so if the universe originally had a value of $\Omega_m + \Omega_\Lambda$ even slightly different from 1, the value would by now be drastically different and easily detectable. For the present-day value of $\Omega_m + \Omega_\Lambda$ to be as close to 1 as it is, when the universe was 1 second old, $\Omega_m + \Omega_\Lambda$ must have been equal to 1 all the way out to at least the tenth decimal place. At even earlier times, it had to be much flatter still. This is too special a situation to be the result of chance—a fact referred to in cosmology as the flatness problem: the universe is so flat that something about the early universe must have caused $\Omega_m + \Omega_\Lambda$ to have a value incredibly close to 1.

A universe that contains mass and does not start out perfectly flat has a very different fate. If a universe started out with Ω_m even slightly greater than 1, its expansion would slow more rapidly than that of the flat universe, meaning that less and less density would be required to stop the expansion. At the same time, the actual density would be falling less rapidly than in the flat universe. This disparity between the actual density of the universe and the critical density would increase, causing the ratio between the two, Ω_m, to skyrocket, so that gravity quickly overwhelms the expansion. A universe that starts out even slightly closed rapidly becomes obviously closed and would collapse long before stars could form. Conversely, if a universe started with Ω_m even a tiny bit less than 1, the expansion would slow less rapidly than in a flat universe. As time passed, more and more mass would be required for gravity to stop the too-rapidly-expanding universe. At the same time, the actual density of the universe would be dropping faster than in a flat universe. In this case, Ω_m would plummet.

Adding Ω_Λ to the picture makes the math a bit more complex, but it does not change the basic results. Try balancing a razor blade on its edge. If the blade is tipped just a tiny bit in one direction, it quickly falls that way. If the blade is tipped just a tiny bit in the other direction, it quickly falls in the other direction instead. It would seem that the actual universe should be obviously open or obviously closed—analogous to the tipped razor blade. Instead, the universe has $\Omega_m + \Omega_\Lambda$ so close to 1 that it is difficult to tell which way the razor blade is tipped at all. Discovering that $\Omega_m + \Omega_\Lambda$ is extremely close to 1 after more than 13 billion years is like balancing a razor blade on its edge and coming back 10 years later to find that it still has not tipped over.

The Horizon Problem

Another problem faced by cosmological models is that the cosmic microwave background radiation is surprisingly smooth. After the discovery of the cosmic microwave background radiation (CMB) in the 1960s, many observational cosmologists turned their attention to mapping this background radiation. At first, result after result showed that the temperature of the CMB is remarkably constant, with variations of less than one part in 3,000, regardless of where one looks in the sky. Yet over time this strong confirmation of Big Bang cosmology challenged cosmologists' view of the early universe. Once Earth's motion relative to the CMB is removed from the picture, the CMB is not just smooth—it is *too* smooth.

In Chapter 5, we discussed the bizarre world of quantum mechanics that shapes the world of atoms, light, and elementary particles. When the universe was extremely young, it was so small that quantum mechanical effects played a role in shaping the structure of the universe as a whole. The early universe was subject to the quantum mechanical **uncertainty principle**, which says that as a system is studied at extremely small scales, the properties of that system become less and less well determined. This principle applies to the properties of an electron in an atomic orbital or to the entire universe at the time when it was very young and would have fit within the size of an atom.

Consider a simple analogy of how the uncertainty principle applies to the universe. Imagine sitting on the beach looking out across the ocean. Off in the distance, you see more total ocean and average the surface over larger scales; therefore, the surface of the ocean appears smooth and flat. The horizon looks almost like a geometric straight line. Yet the apparent smoothness of the ocean as a whole hides the tumultuous structure present at smaller scales, where waves and ripples fluctuate dramatically from place to place. Similarly, while the universe on large scales seems steady and smooth, quantum mechanics says that at smaller and smaller scales in the universe, conditions must fluctuate in unpredictable ways. In particular, quantum mechanics says that the earlier in the history of the universe, the more dramatic those fluctuations are. When the universe was young, it could not have been smooth. There must have been dramatic variations ("ripples") in the density and temperature of the universe from place to place.

If the universe had expanded slowly, those ripples would have smoothed themselves out. But the universe expanded much too rapidly for such smoothing to be possible. There just wasn't enough time after the Big Bang for a smoothing signal to travel from one region to the other. So when cosmologists look at the universe today, they should see the fingerprint of those early ripples imprinted on the cosmic background radiation—but they do not. The fact that the CMB is so smooth is called the *horizon problem* in cosmology. The **horizon problem** states that different parts of the universe are too much like other parts of the universe that should have been "over their horizon" and beyond the reach of any signals that might have smoothed out the early quantum fluctuations. Basically, the horizon problem is as follows: How can different parts of the universe that underwent different fluctuations and were never able to communicate with one another still show the same temperature in the cosmic background radiation to an accuracy of better than one part in 100,000?

Inflation: Early, Rapid Expansion

In the early 1980s, physicist Alan Guth (1947–) offered a solution to the flatness and horizon problems of cosmology. Guth suggested that the universe has not expanded at a steady pace, but that it started out much more compact than steady expansion would predict. Then, for a brief time, the young universe expanded at a rate *far* in excess of the speed of light. This rapid expansion of the universe is called **inflation**. In the first 10^{-33} second of the universe, the distance between points in space increased by a factor of at least 10^{30} and perhaps very much more, as illustrated in **Figure 22.8**. In that incomprehensibly brief instant, the size of the observable universe grew from ten-trillionths the size of the nucleus of an atom to a region about 1 meter across. That is like a grain of very fine sand growing to the size of today's observable universe—all in a billionth the time it takes light to cross the nucleus of an atom. During inflation, space itself expanded so rapidly that the distances between points in space increased faster than the speed of light. Note that inflation does not violate the rule that no signal can travel through space at greater than the speed of light because the space itself was expanding.

To understand how inflation solves the flatness and horizon problems of cosmology, imagine that you are an ant living in the two-dimensional universe defined by the surface of a golf ball, as illustrated in **Figure 22.9**. This universe is positively curved, like the surface of Earth. If you were to walk around the circumference of a circle in your two-dimensional universe and then measure the radius of the circle, you would find the circumference to be less than $2\pi r$. If you were to draw a triangle in your universe, the sum of its angles would be greater than 180 degrees. Another obvious characteristic would be the dimples, approximately a half-millimeter deep, on the surface of the golf ball.

Now imagine that the golf-ball universe suddenly grew to the size of Earth. The curvature of the universe would no longer be apparent. An ant (or person) walking along the surface of the golf-ball universe would think the universe is flat. The circumference of a circle would be $2\pi R$, and there would be 180 degrees in a triangle. (In fact, it took most of human history to realize that Earth is a sphere). In the case of inflationary cosmology, the universe after inflation would be extraordinarily flat (that is, having $\Omega_m + \Omega_\Lambda$ extraordinarily close to 1) *regardless* of what the geometry of the universe was before inflation. Because the universe was inflated by a factor of at least 10^{30}, $\Omega_m + \Omega_\Lambda$ immediately after inflation must have been equal to 1 within one part in 10^{60}, which is flat enough for $\Omega_m + \Omega_\Lambda$ to remain close to 1 today. If inflation occurred, then today's universe is not flat by chance. It is flat because any universe that underwent inflation would become flat.

What about the horizon problem? When the golf-ball universe inflates to the size of Earth, the dimples that covered the surface of the golf ball stretch out as well. Instead of being a half millimeter or so deep and a few millimeters across, these dimples now are only an atom deep but are hundreds of kilometers across. The ant would not detect any dimples at all. In the case of our universe, inflation took the large fluctuations in conditions caused by quantum uncertainty in the preinflationary universe and stretched them out so much that they are not

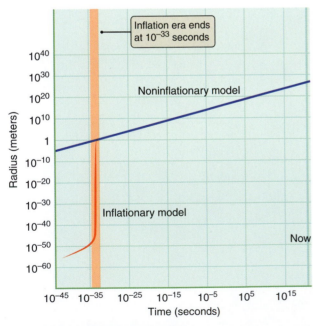

Figure 22.8 This graph illustrates expansion of the observable universe. In an inflationary model, between 10^{-35} and 10^{-33} seconds after the Big Bang, the universe expanded greatly.

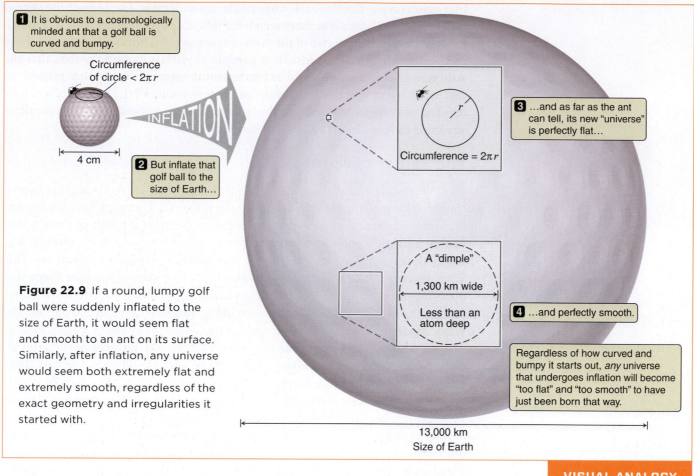

Figure 22.9 If a round, lumpy golf ball were suddenly inflated to the size of Earth, it would seem flat and smooth to an ant on its surface. Similarly, after inflation, any universe would seem both extremely flat and extremely smooth, regardless of the exact geometry and irregularities it started with.

1 It is obvious to a cosmologically minded ant that a golf ball is curved and bumpy.

Circumference of circle < $2\pi r$

4 cm

2 But inflate that golf ball to the size of Earth...

INFLATION

Circumference = $2\pi r$

3 ...and as far as the ant can tell, its new "universe" is perfectly flat...

A "dimple"
1,300 km wide
Less than an atom deep

4 ...and perfectly smooth.

Regardless of how curved and bumpy it starts out, *any* universe that undergoes inflation will become "too flat" and "too smooth" to have just been born that way.

13,000 km
Size of Earth

VISUAL ANALOGY

measurable in today's postinflationary local universe. The slight irregularities observed in the CMB are the faint ghosts of quantum fluctuations that occurred as the universe inflated.

An early era of inflation in the history of the universe offers a way of solving the horizon and flatness problems, but it seems quite remarkable that the universe should have undergone a period during which it expanded at such a high rate. The cause of inflation lies in the fundamental physics that governed the behavior of matter and energy at the earliest moments of the universe. While the existence of an inflationary epoch is difficult to test, it is not impossible, and astronomers are currently devising ways to test whether inflation occurred in the early universe.

CHECK YOUR UNDERSTANDING 22.3

Identify the two problems of cosmology that are solved by inflation.

22.4 The Earliest Moments of the Universe Connect the Very Largest Size Scales to the Very Smallest

At first glance, particle physics and cosmology might seem to have almost nothing in common. Particle physics is the study of subatomic particles, which are smaller than atoms. Whereas particle physics looks at the quantum mechanical world that exists on the tiniest scales imaginable, cosmology is the study of the changing structure of a universe that extends for billions of parsecs and probably much farther. Yet the last quarter of the 20th century saw the boundary between these two fields fade as cosmologists and particle physicists came to realize that the structure of the universe and the fundamental nature of matter are related. In this section, you will learn about the earliest moments of the universe, when atoms had not yet formed, and how the interaction among these subatomic particles governed the conditions and events that took place.

The Forces of Nature

To understand the universe requires an understanding of the forces that govern the behavior of all matter and energy in the universe. There are four fundamental forces in nature, and everything in the universe is a result of their action (**Table 22.1**). The **electromagnetic force**, which includes both electric and magnetic interactions, acts on charged particles like protons and electrons. This force governs not only chemistry but also light. The **strong nuclear force** that binds together the protons and neutrons in the nuclei of atoms governs reactions such as the fusion reactions in the heart of the Sun, discussed in Chapter 14. The **weak nuclear force** governs the radioactive decay of unstable nuclei. Finally, *gravity*, which plays a major role throughout astronomy, governs how matter affects the geometry of spacetime. In models of particle physics, the first three forces combine in a single force, leaving only gravity to stand alone. To understand how these four forces formed and came to govern the universe today, we must explore backward in time, toward the Big Bang itself.

The underlying concept of the standard model is that forces between particles are caused by the exchange of carrier particles. Recall from Chapter 5 the description of light as an electromagnetic wave resulting from electric and magnetic fields and the quantum mechanical description of light as a stream of particles called photons. These two descriptions of electromagnetism coexist. The branch of physics that deals with a quantum description of radiation is called **quantum electrodynamics (QED)**.

Quantum electrodynamics treats charged particles almost as if they were baseball players engaged in an endless game of catch. As baseball players throw and catch baseballs, they experience forces. Similarly, in QED, charged particles "throw" and "catch" an endless stream of "virtual photons" (**Figure 22.10**). Quantum mechanics is a science of probabilities rather than certainties. The QED description of the electromagnetic interaction between two charged particles is an average of all the possible ways that the particles could throw photons back

---

TABLE 22.1 The Four Fundamental Forces of Nature

Force	Relative Strength	Range of Force	Particles That Can Carry the Force	Example of What the Force Does
Strong nuclear	1	10^{-15} m	Gluons	Holds protons and neutrons together in atomic nuclei.
Electromagnetic	10^{-2}	Infinite	Photons	Binds the electrons in an atom to the nucleus.
Weak nuclear	10^{-4}	10^{-16} m	W^+, W^-, and Z^0	Responsible for beta decay.
Gravitational	10^{-38}	Infinite		Holds you to Earth; binds together planetary systems, stars, galaxies, clusters of galaxies, and so forth.

and forth. The resulting force acts, over large scales, like the classical electromagnetic force that we observe in chemistry. Physicists describe this as the electromagnetic force being mediated by the exchange of photons.

The central idea of QED—forces mediated by the exchange of carrier particles—provides a template for understanding two of the other three fundamental forces in nature. The electromagnetic and weak nuclear forces have been combined into a single theory called **electroweak theory**. This theory predicts the existence of three particles—labeled W^+, W^-, and Z^0—that mediate the weak nuclear force. Sheldon Glashow, Abdus Salam, and Steven Weinberg received the 1979 Nobel Prize in Physics for their work on the theory of the unified weak and electromagnetic forces. In the 1980s, physicists identified these particles in laboratory experiments and confirmed the essential predictions of electroweak theory.

The strong nuclear force is described by a distinct theory called **quantum chromodynamics (QCD)**. This theory states that particles such as protons and neutrons are composed of more fundamental building blocks called **quarks**, which are bound together by the exchange of another type of carrier particle, dubbed **gluons**. Together, electroweak theory and QCD compose the **standard model** of particle physics. Excluding gravity, the standard model explains all the currently observed interactions of matter and has made many predictions that were subsequently confirmed by laboratory experiments. However, the standard model leaves many questions unanswered, such as whether neutrinos have mass or why strong interactions are so much stronger than weak interactions.

At high temperatures, the different forces are indistinguishable because the different carrier particles have such high energy. Therefore, our universe started out with all the forces unified, as described by one (as yet unknown) theory of everything. As the universe expanded and cooled, symmetry between the particles was broken, and carrier particles for different forces became distinguishable. You may have heard of the Higgs field, which is responsible for breaking the symmetry between different kinds of carrier particles. This field is manifested as the Higgs boson, an elementary particle.

The Higgs boson is the particle that all other particles must interact with to gain their masses. In the standard model, all particles are created without mass. When the electroweak symmetry breaks (as expansion and cooling continue), this "special" particle is created throughout the universe. All existing particles interact with the Higgs field and gain their mass in this process. The more they interact, the more massive (heavier) they become. The existence of this particle was predicted in 1964 and finally detected at the Large Hadron Collider in Europe in 2012. In 2013, Peter Higgs and François Englert shared the Nobel Prize for their work predicting this particle. As the universe cooled, the carrier particles (and also other particles like electrons) gained mass due to the Higgs field, and so the forces also began to be distinguishable.

A Universe of Particles and Antiparticles

In the standard model, every particle in nature has an **antiparticle** that is identical in mass, but opposite in charge, to the particle. For example, the positron emitted in the proton-proton chain, discussed in Chapter 14, is the antiparticle of an electron. A positron is identical to an electron except that it has a positive charge instead of a negative charge. For the proton there is the antiproton; for the neutron, the antineutron; and so on down the list. Collectively, these antiparticles are called **antimatter**.

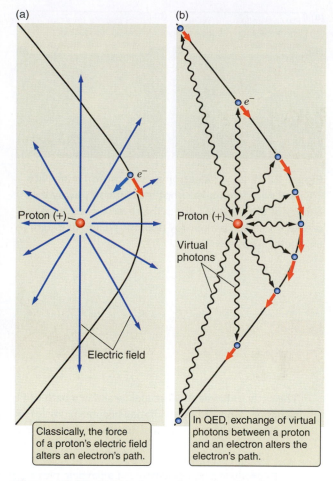

(a)

(b)

Proton (+)

e^-

Electric field

Classically, the force of a proton's electric field alters an electron's path.

Proton (+)

Virtual photons

e^-

In QED, exchange of virtual photons between a proton and an electron alters the electron's path.

Figure 22.10 (a) This figure shows the classical view of an electron being deflected from its course (black line) by the electric field from a proton. (b) According to quantum electrodynamics models, the interaction is viewed as an ongoing exchange of virtual photons between the two particles.

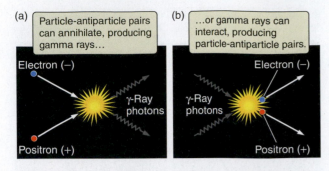

(a) Particle-antiparticle pairs can annihilate, producing gamma rays…

Electron (−)

γ-Ray photons

Positron (+)

(b) …or gamma rays can interact, producing particle-antiparticle pairs.

Electron (−)

γ-Ray photons

Positron (+)

Figure 22.11 (a) An electron and a positron annihilate, creating two gamma-ray photons that carry away the energy of the particles. (b) In the reverse process, pair production, two gamma-ray photons collide to create an electron-positron pair.

One property of particle-antiparticle pairs is that if you bring such a pair together, the two particles annihilate each other. When a particle-antiparticle pair annihilates, the mass of the two particles is converted into energy in accord with Einstein's equation $E = mc^2$. For example, in **Figure 22.11a**, an electron and a positron annihilate each other, and the energy is carried away by a pair of gamma-ray photons. Particle-antiparticle pairs were produced when two high-energy photons collided with each other, as in Figure 22.11b, creating in their place an electron-positron pair. **Pair production**—when an energetic event creates a particle and its corresponding antiparticle—has been observed in particle accelerators.

In principle, *any* type of particle and its antiparticle can be created by pair production. The only limitation comes when there is not enough energy available to supply the mass of the particles being created (**Working It Out 22.2**). If two gamma-ray photons with a combined energy greater than the rest mass energy of an electron-positron pair collide, then the two photons may disappear and leave an electron-positron pair behind in their place. If the photons have more than the

22.2 | Working It Out Pair Production in the Early Universe

The early universe was hot and awash in a bath of Planck blackbody radiation. How did this radiation become particles? Recall Einstein's equation for the conversion of mass to energy and energy to mass:

$$E = mc^2$$

To produce a pair consisting of a particle plus an antiparticle of a certain rest mass, a minimum amount of energy is required. The formula shows that producing a higher-mass particle plus antiparticle requires a greater amount of energy than does producing a lower-mass pair. For example, a proton has 1,836 times the mass of an electron, so it will require 1,836 times as much energy to produce a proton-antiproton pair of particles than to produce an electron-positron pair.

We can relate this amount of energy to the average energy of particles at a given temperature through the following equation:

$$E = 3/2 \, kT$$

where k is Boltzmann's constant: 1.38×10^{-23} kg m²/s²/K (which is joules per kelvin; J/K). Equating the two energies yields

$$mc^2 = 3/2 \, kT$$

Rearranging,

$$T = \frac{2mc^2}{3k}$$

Let's look at some examples. The proton and antiproton each have a mass of 1.67×10^{-27} kg. What temperature would the radiation have to be to produce this proton-antiproton pair?

$$T = \frac{2 \times (2 \times 1.67 \times 10^{-27}\,\text{kg}) \times (3.0 \times 10^8\,\text{m/s})^2}{3 \times (1.38 \times 10^{-23}\,\text{kg m}^2/\text{s}^2/\text{K})}$$

$$T = 1.45 \times 10^{13}\,\text{K}$$

Such high temperatures are thought to have existed only during the first few seconds after the Big Bang.

Similarly, to produce an electron-positron pair, each of which has a mass of 9.11×10^{-31} kg, the temperature of the CMB would have to be

$$T = \frac{2 \times (2 \times 9.11 \times 10^{-31}\,\text{kg}) \times (3.0 \times 10^8\,\text{m/s})^2}{3 \times (1.38 \times 10^{-23}\,\text{kg m}^2/\text{s}^2/\text{K})}$$

$$T = 7.9 \times 10^9\,\text{K}$$

This is still very hot, but less than the temperature required to form a proton-antiproton pair, so electron-positron production lasted longer after the Big Bang than did proton-antiproton production.

We can also think of this in terms of the energy of the photons involved in creating these particles. From Chapter 5, recall that the energy of a photon is related to its wavelength by

$$E = \frac{hc}{\lambda} \quad \text{or} \quad \lambda = \frac{hc}{E}$$

where h is Planck's constant ($h = 6.63 \times 10^{-34}$ kg m²/s). We then use $E = 3/2 \, kT$ with our value of T for the electron-positron production above, yielding

$$\lambda = \frac{2hc}{3kT} = \frac{2 \times (6.63 \times 10^{-34}\,\text{kg m}^2/\text{s}) \times (3.0 \times 10^8\,\text{m/s})}{3 \times (1.38 \times 10^{-23}\,\text{kg m}^2/\text{s}^2/\text{K}) \times (7.92 \times 10^9\,\text{K})}$$

$$\lambda = 1.21 \times 10^{-12}\,\text{m}$$

The electromagnetic spectrum pictured in Figure 5.6 shows that the photons involved in the pair production of electrons and positrons are high-energy gamma-ray photons.

necessary energy, the extra energy goes into the kinetic energy of the two newly formed particles.

Now we apply this idea to a hot universe awash in a bath of blackbody radiation. When the universe was less than about 100 seconds old and had a temperature greater than a billion kelvins, it was filled with energetic photons that were constantly colliding, creating electron-positron pairs, and these electron-positron pairs were constantly annihilating each other, creating pairs of gamma-ray photons. These two processes reached equilibrium, determined strictly by temperature, in which pair creation and pair annihilation exactly balanced each other. Rather than being filled only with a swarm of photons, at this time the universe was filled with a swarm of photons, electrons, and positrons. Earlier, when the universe was even hotter, photons would have produced a swarm of protons and antiprotons. Still earlier, there was a swarm of quarks/antiquarks and gluons called a "quark-gluon plasma," as has been observed in some heavy-nucleus accelerators on Earth.

Grand Unified Theories

In the process of pair production, there is a symmetry between matter and antimatter: for every particle created, its antiparticle is created as well. As the universe cooled, there was no longer enough energy to create particle pairs, so the swarm of particles and antiparticles that filled the early universe annihilated each other and were not replaced. When this cooling happened, first every proton should have been annihilated by an antiproton. Then at still cooler temperatures, every electron should have been annihilated by a positron. This was almost the case, but not quite. For every proton and electron in the universe today, there were 10 billion and one protons and electrons in the early universe, but only 10 billion antiprotons and positrons. This one-part-in-billions excess of electrons over positrons meant that when electron-positron pairs finished annihilating each other, some electrons were left over—enough to account for all the electrons in all the atoms in the universe today illustrated in **Figure 22.12**. Similarly, there was an excess of protons over antiprotons in the early universe, and the protons observed today are all that is left from the annihilation of proton-antiproton pairs.

If the standard model of particle physics were a complete description of nature, then the imbalance of one part in 10 billion between matter and antimatter would not have been present in the early universe. The symmetry between matter and antimatter would have been complete. No matter at all would have survived into today's universe, and galaxies, stars, and planets would not exist. The fact that you are reading this page demonstrates that something more needs to be added to the model.

Several competing ideas seek to explain why the amounts of matter and antimatter were not equal in the early universe. One set of ideas is called **grand unified theories (GUTs)** because they join the electromagnetic force, weak nuclear force, and strong nuclear force together into a single force. Grand unified theories explain why the universe is composed of matter rather than antimatter. When the universe was very young (younger than about 10^{-35} second) and very hot (hotter than about 10^{27} K), enough energy was available for particles associated with a GUT to be freely created. During this time, the distinction among the electromagnetic, weak nuclear, and strong nuclear forces had not yet taken place. There was only the one unified force. During this era of GUTs, the apparent size of the entire observable universe was less than a trillionth the size of a single proton.

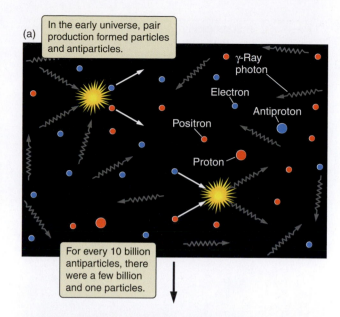

(a) In the early universe, pair production formed particles and antiparticles.

γ-Ray photon
Electron
Antiproton
Positron
Proton

For every 10 billion antiparticles, there were a few billion and one particles.

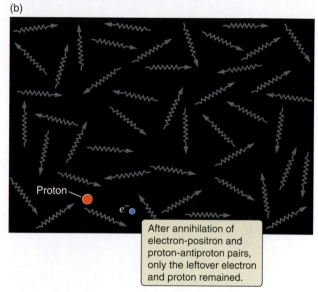

(b)

Proton
e^-

After annihilation of electron-positron and proton-antiproton pairs, only the leftover electron and proton remained.

Figure 22.12 (a) For every 10 billion antiparticles in the early universe, there were 10 billion and one particles. (b) After these particles annihilated, only one electron and one proton remained.

Many possible GUTs exist, and they make many predictions about the universe. Unfortunately, most of those predictions are impossible to test with even the largest of today's particle colliders. The problem is that the particles carrying the forces are so massive that it takes enormous amounts of energy to bring them into existence—roughly a trillion times as much energy as can be achieved in today's particle accelerators. Even so, some predictions of GUTs can be tested with current technology. For example, GUTs predict that protons should be unstable particles that, given enough time, will decay into other types of elementary particles. This is a *very* slow process. Over the course of 100 years, GUTs predict that there may be as much as a 1 percent chance that *one* of the 10^{28} or so protons in your body will decay. Experiments have given a lower limit on the lifetime of a proton of 10^{34} years, but as of this writing, proton decay has yet to be observed.

How does gravity fit into this scheme? General relativity provides a description of gravity that correctly predicts the orbits of planets, describes the ultimate collapse of stars, and enables astronomers to calculate the structure of the universe. Yet general relativity's description of gravity is very different from the theories of the other three forces. Rather than talking about the exchange of photons or gluons or other mediating particles, general relativity talks about the large, smooth, continuous canvas of spacetime that events are painted on. The era of GUTs is described perfectly if gravity is treated as a force separate from the other three forces. There is a time even closer to the Big Bang when a description of the universe requires that gravity be treated as a quantum phenomenon.

Toward a Theory of Everything

Even earlier in the universe than the time when the three forces were unified, when the universe was younger than about 10^{-43} second, its density was incomprehensibly high. The observable universe was so small that 10^{60} universes would have fit into the volume of a single proton. Under these extreme conditions, general relativity can no longer describe spacetime: quantum physics is needed to describe not just particles, but spacetime itself. Rather than a smooth sheet, spacetime was a quantum mechanical "foam." The failure of general relativity to describe this early universe is much like the failure of Newtonian mechanics to describe the structure of atoms. An electron in an atom must be thought of in terms of probabilities rather than certainties. Similarly, there is no deterministic history for the earliest moments after the Big Bang. This era in the history of the universe is called the **Planck era**, signifying that physicists can understand the structure of the universe during this period only by using the ideas of quantum mechanics.

The conflict between general relativity and quantum mechanics is at the current limits of human knowledge. Known physics can explain things back to a time when the universe was a ten-millionth of a trillionth of a trillionth of a trillionth of a second old, but to push back any further, something new is needed. To understand the earliest moments of the universe, physicists need a theory that combines general relativity and quantum mechanics into a single theoretical framework unifying all *four* of the fundamental forces—a **theory of everything (TOE)**.

A successful theory of everything would do more than unify general relativity with quantum mechanics. It would suggest which of the possible GUTs is correct and would tell us the nature of dark matter. A successful theory of everything would also explain the how, when, and why of inflation and the underlying physics of the dark energy that is accelerating the expansion of the universe. Physicists are currently grappling with what a TOE might look like. One leading contender is superstring theory, discussed in Section 22.5.

The Forces Separated in the Cooling Universe

To understand the very earliest moments in the history of the universe, physicists look backward to earlier and earlier times, and consequently to higher and higher energies. Now let's organize the events the other way, beginning at the beginning.

The universe started out with one TOE with all four forces united, and as the universe expanded and cooled, the various forces emerged separately. **Figure 22.13** illustrates how the four fundamental forces emerged in the evolving universe. In the first 10^{-43} second after the Big Bang, as described by the TOE, the physics of elementary particles and the physics of spacetime were one and the same. As the universe expanded and cooled, gravity separated from the forces described by the GUT. Spacetime took on the properties described by general relativity. Inflation may also have been taking place at this time.

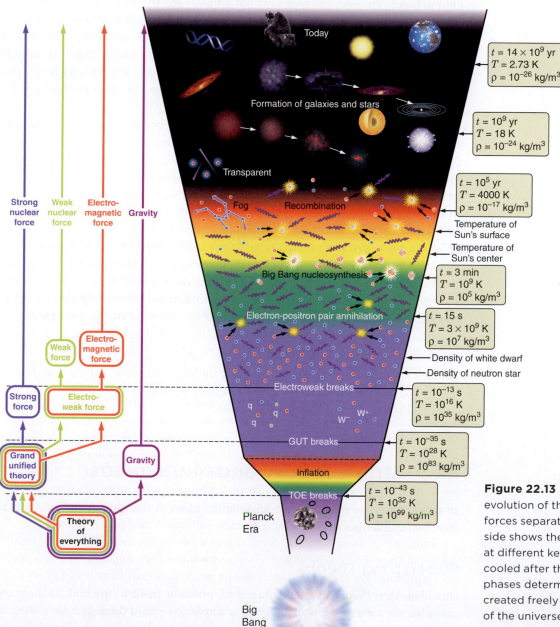

Figure 22.13 This figure conceptualizes eras in the evolution of the universe. The left side shows the four forces separating in stages after the Big Bang. The right side shows the temperature and density of the universe at different key times. As the universe expanded and cooled after the Big Bang, it went through a series of phases determined by what types of particles could be created freely at that temperature. Later, the structure of the universe was set by the gravitational collapse of material to form galaxies and stars and by the chemistry made possible by elements formed in stars.

As the universe continued to expand and its temperature fell further, less and less energy was available for the creation of particle-antiparticle pairs. When the particles responsible for mediating GUT interactions could no longer form, the strong force split off from the others. As the unity of the original TOE was lost, the symmetry between matter and antimatter was broken. As a result, the universe ended up with more matter than antimatter.

The next big change took place when particles responsible for unifying the electromagnetic and weak nuclear forces separated, leaving these two forces independent of each other. All four fundamental forces of nature that govern today's universe were then separate. At one 10-trillionth of a second, the temperature of the universe had fallen to 10^{16} K. It was a full minute or two before the universe cooled to the billion-kelvin mark, below which not even pairs of electrons and positrons could form.

The universe at that point was too cool to form additional particles and their antiparticles. However, it was still hot enough for the fast-moving protons to overcome the electric barriers between them, allowing nuclear reactions to take place. These reactions formed the least massive elements and isotopes, including deuterium, helium, and lithium. Recall from stellar evolution that increasingly high temperatures are needed for the nucleosynthesis of increasingly heavy elements. So as the universe continued to expand, it soon became too cool for the nucleosynthesis of more massive elements.

The Big Bang nucleosynthesis discussed at the end of the past chapter came to an end by the time the universe was about 15–20 minutes old and the temperature of the universe had dropped below about 800 million K. The density of the universe at this point had fallen to only about a tenth that of water. Normal matter consisted of atomic nuclei and electrons, but the universe was still dominated by the radiation from the Big Bang. After several hundred thousand years, the temperature dropped so low that electrons were able to combine with atomic nuclei to form neutral atoms. This was the era of recombination, which is seen directly when astronomers observe the cosmic microwave background radiation. At this stage, the radiation background could no longer dominate over matter: the universe became transparent, and light could move freely through it for the first time. Additionally, gravity began playing its role in forming the vast structure of the universe that is now observed.

CHECK YOUR UNDERSTANDING 22.4

The vast majority of antimatter in the early universe: (a) is still around today, filling the space between galaxies; (b) became dark matter; (c) formed antimatter galaxies and stars; (d) annihilated with matter.

22.5 String Theory and Multiverses

In this section, we discuss some speculative ideas in theoretical cosmology that pose real challenges to test.

Superstring Theory

One idea for a theory of everything is superstring theory, in which elementary particles are viewed not as points but as tiny loops called "strings." According to superstring theory, different types of elementary particles are like different "notes" played by vibrating loops of string. This is analogous to the way in which a guitar string vibrates in one way to play an F, another way to play a G, and yet another way to play an A.

In principle, superstring theory provides a way to reconcile general relativity and quantum mechanics. To make superstring theory work, physicists imagine that these tiny strings are vibrating in a universe with 10 spatial dimensions instead of three. (Adding time to the list would make the universe 11-dimensional.) Whereas the usual three spatial dimensions spread out across the vastness of the universe, the other seven dimensions predicted by superstring theory wrap tightly around themselves, extending no further today than they did a brief instant after the Big Bang.

To visualize this, imagine what it would be like to live in a three-dimensional universe (like the one you experience) in which one of those dimensions extended for only a tiny distance. Living in such a universe would be like living within a thin sheet of paper that extended billions of parsecs in two directions but was far smaller than an atom in the third. In such a universe, you would easily be aware of length and width: you could move in those directions at will. In contrast, you would have no freedom to move in the third dimension at all, and you might not even recognize that the third dimension existed. Perhaps your only inkling of the true nature of space would come from the fact that in order to explain the results of particle physics experiments, you would have to assume that particles extended into a third, unseen dimension. If superstring theory is correct, everyone now sees three spatial dimensions extending possibly forever, but doesn't notice that each point in this three-dimensional space actually has a tiny but finite extent in the seven other dimensions at the same time.

Although called a "theory," superstring theory is not like the well-tested theories that have been discussed throughout this book: it is no more than a promising idea that provided direction to theorists searching for a TOE. Therefore, "string hypothesis" or "string idea" would be more consistent with the definitions of idea, hypothesis, and theory given in Chapter 1. Physicists will probably never be able to build particle accelerators that enable them to search directly for the most fundamental particles predicted by a TOE: the energies required are simply too high. However, some progress may be made by studying the ultimate particle accelerator: the Big Bang itself.

Multiverses

Is our universe the only one? Because we've defined the universe as "everything," what does it mean to say "multiple universes"? Are there parallel universes either separated in space or even occupying exactly the same space as "our" universe? Are there experiments or observations that can test the idea of multiple universes? These ideas are quite speculative, but some cosmologists think seriously about the idea of multiple universes, or **multiverses**—collections of parallel universes.

The simplest example of such parallel universes is illustrated in **Figure 22.14**. The age of our universe—that is, the amount of time that has passed since the Big Bang—is 13.8 billion years. That means that light reaching Earth today can have traveled a maximum distance of about 13.8 billion light-years. Therefore, the observable universe—everything astronomers can possibly observe today—must be within a sphere of radius 13.8 billion light-years. Anything farther than that is outside of the observable universe and cannot be seen. As we discussed earlier in the chapter, the observational evidence suggests that the geometry of space is flat.

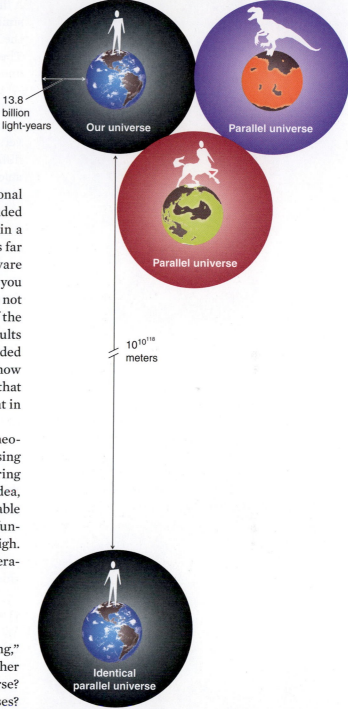

Figure 22.14 The observable universe is a sphere with a radius equal to the distance light has traveled since the Big Bang (13.8 billion light-years). Because the universe is infinite, there must be an infinite number of similar spheres. The rules of probability dictate that some of these are *exactly* like our own.

A flat universe is infinite in size and must therefore contain an infinite number of similar spheres. As dark energy causes the universe to expand faster and faster, the separate observable universes move farther apart and will never overlap. These parallel universes are simply too far away to ever be observed from Earth, and they are moving farther away all the time.

What are these other parallel universes like? Physicists suggest several things on the basis of what has been learned about the observable universe. First, if the cosmological principle holds, then on large scales each of these observable universes should look pretty much like our own, although the details could be very different. In a truly infinite universe there must be an infinite number of observable universes exactly like this one, with exact copies of you reading identical versions of *21st Century Astronomy*. The argument is that if our own observable universe is cooler than about 10^8 K everywhere, there can be no more than 10^{118} particles in the observable universe, and there are only so many ways that those particles can be distributed. If you then ask how far you must go before you will find an observable universe *identical to* our own, the answer is about $10^{10^{118}}$ Mpc. Yes, that's 10 raised to the power 10^{118}. So, in an infinite universe—as enormous as it might be—the nearest identical universe must be at about that distance.

Other types of multiverses include those in which a universe undergoes **eternal inflation**, with no beginning or end to the inflation. If such a universe exists, then small quantum fluctuations may cause some regions to expand more slowly than the rest of the universe. As a result, such a region may form a bubble whose inflating phase ends quickly. In this scenario, Earth is inside such a region, and "our" Big Bang would just be the condensation of our bubble within the eternally inflating universe. This type of multiverse neatly answers the question of what there was before the Big Bang. Because the universe has been inflating and will continue to inflate forever, there is no beginning or end. "Our" own bubble or parallel universe separated from the rest of the universe at a time called the Big Bang, but other bubbles are constantly separating and becoming their own parallel-universe big bangs.

In another type of multiverse derived from quantum physics, which describes a probabilistic universe, each event in the universe spawns multiple universes in which each possible outcome of the event exists. While cosmologists consider this at the level of particle interactions, it is more simply explained at the human scale: you made a decision about having breakfast this morning. That event caused two universes: one in which you did have breakfast and one in which you didn't. Which one are you in while you read this book? Yet another type of multiverse is characterized by parallel universes that have different mathematical structure to describe the different physics within these universes. In this all-encompassing case, almost any behavior for the universe is possible.

Multiverses are common in science fiction and in popular science books and form a type of common mythology about alternative realities that is explored with enthusiasm in popular culture. Many astronomers ask if the idea of parallel universes, or multiverses, is really science. Throughout this book, we have emphasized that any legitimate scientific theory must be testable and ultimately falsifiable. Are there tests capable of proving these multiverse ideas to be wrong? Possibly. For example, the first multiverse we described, involving distinct observable bubbles, is tested when astronomers measure the isotropy of the CMB, the flatness of the universe, or the large-scale distribution of galaxies. The eternal inflation model is difficult to test because it is impossible to observe parallel universes directly. But if physicists obtain a theory of everything that predicts eternal inflation, and if that theory of everything is itself falsifiable, then there is a connection between eternal inflation and observation.

There is considerable debate within the scientific community as to whether the multiverse hypotheses, except for the first type, can be tested and falsified. Even if there are some tests, the concern is that the tests are not truly meaningful. For example, you might test Newton's theory of gravity by releasing an apple and watching whether it falls upward or downward. But this test is not very discerning: it might not always distinguish between two or more sensible theories. Similarly, for the multiverse hypothesis, it is still debatable whether the tests that seem possible would be meaningful tests of the theory.

CHECK YOUR UNDERSTANDING 22.5

Parallel universes are: (a) an observation; (b) a hypothesis; (c) a law; (d) a theory.

Origins

Our Own Universe Must Support Life

In some models of an inflationary universe, each bubble could contain different values of the fundamental constants of physics. In some bubble universes, for example, the strength of the nuclear force might be larger, the electric charge might be smaller, and the gravitational constant G might be much smaller than in others. To address this question, scientists invoke the **anthropic principle**, which states that our universe (or our bubble in the universe) must have physical properties that allow for the development of intelligent life. Because humans exist, are intelligent, and can observe the surrounding universe, our universe must have the properties that would allow intelligent life to evolve. That is, this universe must have had the right physical properties and existed long enough for atoms, stars, galaxies, planets, and life to have formed; otherwise, we wouldn't be here to observe it.

In the case of a multiverse that contains bubbles with different physical constants in each of them, the anthropic principle provides information about the values of those physical constants.

Consider a few examples. In a bubble universe where the gravitational constant G was much bigger than G as measured in our universe, stellar evolution would occur much faster, and there might not be time for intelligent life to evolve on a planet before its star burned itself out. Similarly, the anthropic principle provides a relatively narrow range for the strength of the strong nuclear force that holds nuclei together. If that force were much weaker than what physicists now measure, nuclei would not be able to overcome their electric repulsion in order to fuse. Without nuclear fusion, stars could not shine, heavy elements would not form, and planets and life as we know it could not evolve. Alternatively, if the nuclear force were stronger, it would be easier for two protons to fuse together in the early universe; thus, most hydrogen will fuse to helium, and there would be less hydrogen to form water and organic molecules that are necessary for life as we know it.

As another example of the anthropic principle applied to multiverses, recall the fraction of the critical density provided by the cosmological constant, Ω_Λ. If Ω_Λ were 20 times larger than what is currently observed, the universe would have begun accelerating much earlier, and galaxies would not have had time to form. But without galaxies, there would be little star formation. It follows that stellar nucleosynthesis would not have taken place, and rocky planets and life could not have evolved. Alternatively, if the cosmological constant were negative, then this entire bubble universe would have reached a maximum size and begun to collapse even before galaxies and stars could have evolved. A universe with an intermediate value might last only a few billion years—enough time for stars and galaxies to be established, but perhaps not enough time for sufficient amounts of heavy elements to form or for life to evolve to intelligence. As these examples illustrate, the cosmological constant must be within a particular range of values to allow for the intelligent life that exists on Earth to evolve.

ARTICLES   QUESTIONS

In this article, we see that results from one experiment must be corroborated for the science to be accepted.

Cosmic Inflation: How Progress in Science Is Achieved

By **MARIO LIVIO, HuffingtonPost.com**

You may remember that about six months ago, scientists got all excited about the potential detection of ripples from the Big Bang—direct evidence (if it were to be confirmed) for the event known as "cosmic inflation." What happened then was that a team of scientists analyzing data from the BICEP2 telescope at the South Pole claimed to have detected "B-mode polarization"—an imprint in the cosmic microwave background that could have been created by gravity waves resulting from cosmic inflation.

If that result were to hold, this would have undoubtedly been one of the most dramatic discoveries in decades. However, extraordinary claims require extraordinary proof, and a few scientists were quick to point out that dust in our own Milky Way Galaxy could, at least in principle, produce a polarization signal similar to the one observed by the BICEP2 team.

Here is where the scientific method kicks in big time. Scientists knew that maps of the galactic dust distribution from the Planck satellite, coupled to the determination of the polarization signal from this dust, would shed light on the question of whether what BICEP2 saw could indeed be coming from polarizing dust, rather than from the B-modes produced by inflation. The Planck team has now published its results, and those are not encouraging for the inflationary interpretation. At face value, the Planck results suggest that polarization by dust from the Milky Way could explain the signal detected by BICEP2.

The Plank team is careful to note that their results do not entirely rule out the possibility that BICEP2 did detect something from inflation, and the team's paper highlights the need to reduce the uncertainties through an ongoing, joint analysis of the data sets from both Planck and BICEP2. Moreover, new results are expected soon from the Keck Array experiment, and those should further clarify the picture.

As disappointing as these new results may sound, they provide for a powerful demonstration of how science truly progresses. Advances in science are far from being a direct march to the truth. Rather, they consist of a zigzag path that often results in false starts or blind alleys. The important point, however, is that through continuous checks, testable predictions, and new observations, science is able to self-correct and find the right way.

ARTICLES   **QUESTIONS**

1. How does dust from our Milky Way affect the light we see from the CMB?
2. Why would astronomers put a telescope at the South Pole?
3. What is inflation? When and why do astronomers think inflation took place?
4. Why will scientists be excited about obtaining observational evidence of inflation? Do a news search to find out recent results on evidence for inflation.
5. How does this example illustrate the process of science?

"Cosmic inflation: How progress in science is achieved," by Dr. Mario Livio. Originally published by the Huffington Post, September 24, 2014. Reprinted with permission from Dr. Mario Livio, astrophysicist, author of *Brilliant Blunders*.

Summary

Both gravity and the cosmological constant (or dark energy) determine the fate of the universe. Observations indicate that rather than slowing down, the expansion of the universe is accelerating. The best explanation for this phenomenon is dark energy. The very early universe may have gone through a brief but dramatic period of exceptionally rapid expansion, known as inflation. If this is true, inflation would explain both the flatness and the uniformity of the universe we see today. During the very earliest moments in the universe, the four fundamental forces of nature were unified. Ideas about multiple universes have been proposed, but none are yet directly testable. Our observable universe must be one in which physics can support the formation of life.

LG 1 **Explain how mass within the universe and the gravitational force it produces affect the history, shape, and fate of the universe.** The rate of expansion of the universe is affected by the density (mass/volume) of the universe. Mass can slow and even stop the expansion of the universe. The density also affects the shape of the universe: if density is high, then the universe is shaped more like a sphere; if density is low, the shape is more like a saddle. In between, at the critical density, the universe is flat.

LG 2 **Describe the evidence for the accelerating expansion of the universe.** Recent observations of distant Type Ia supernovae suggest that the expansion of the universe was slower in the past than in the present. This means the expansion is speeding up, implying that there must be a force acting to increase the expansion rate. This force may come from dark energy: the energy of empty space. Because the rate of expansion of the universe is increasing, the universe will likely expand forever.

LG 3 **Describe the early period of rapid expansion known as inflation.** Inflationary models, which state that the universe expanded very rapidly in a very short period of time soon after the Big Bang, were proposed to solve several problems in cosmology. Astronomers are currently looking for an observational signature of inflation.

LG 4 **Explain how the events that occurred in the earliest moments of the universe are related to the forces that operate in the modern universe.** During the earliest moments of the universe, the four forces (gravity, electromagnetism, strong nuclear, and weak nuclear) split off, each becoming separate at a different time. Modern physicists search for the combined theory of all four forces: the theory of everything.

? UNANSWERED QUESTIONS

- What is the origin of dark energy, and why is $\Omega_\Lambda + \Omega_m$ so close to 1 at the present time? Is that a coincidence? We have already said that dark energy is a form of vacuum energy, and it is one of the grand challenges of any successful theory of everything to explain how big Ω_Λ really is. The simplest estimates for the size of Ω_Λ yield results that are much larger than the observed value—by a factor of about 10^{120}. So there must be an as yet undetermined mechanism that affects the size and evolution of Ω_Λ, and that is one of the bigger questions in modern cosmology.

- What mechanism leads to an extra electron for every 10 billion electron-positron pairs in the early universe? In this chapter, we mentioned that grand unified theories predict an asymmetry between particles and antiparticles, as well as proton decay. But physicists do not currently know which GUT is the correct one, and so far they have not actually observed proton decay. Measuring the actual lifetime of a proton would enable physicists to hone in on the correct GUT and therefore the mechanism leading to particle-antiparticle asymmetry. In addition, if the correct TOE was really understood, that theory could predict which GUT describes the universe and therefore the mechanism and amount of asymmetry.

Questions and Problems

Test Your Understanding

1. If astronomers ignored any cosmological constant (or dark energy), the future of the universe could be determined solely from
   a. the mass of the universe.
   b. the volume of the universe.
   c. the amount of light in the universe.
   d. the density of the universe.

2. The cosmological constant accounts for the effects of
   a. dark matter.
   b. the Big Bang.
   c. dark energy.
   d. gravity.

3. The cosmic microwave background radiation indicates that the early universe
   a. was quite uniform.
   b. varied greatly in density from one place to another.
   c. varied greatly in temperature from one place to another.
   d. was shaped differently than the modern universe.

4. According to the definitions of these terms in Chapter 1, superstring theory is
   a. a hypothesis.
   b. a theory.
   c. a law.
   d. a principle.

5. Place in order the following events in the history of the universe.
   a. Planck era
   b. grand unified theory breaks
   c. today
   d. Big Bang nucleosynthesis
   e. electroweak breaks
   f. theory of everything breaks
   g. electron-positron pair annihilation
   h. formation of galaxies and stars
   i. recombination
   j. inflation

6. Astronomers will never directly observe the first few minutes of the universe because
   a. the universe was opaque at that time.
   b. the universe is too large now.
   c. there were no particles or other matter to see.
   d. there were no photons.

7. As applied to the universe, what is the meaning of *critical density*?
   a. Above this density, nebulae collapse to form stars.
   b. Above this density, dark matter becomes important.
   c. Above this density, the universe will eventually collapse.
   d. Above this density, matter becomes degenerate.

8. Of the four fundamental forces in nature, which one depends on electric charge?
   a. gravitational force
   b. electromagnetic force
   c. strong nuclear force
   d. weak nuclear force

9. Suppose you measure the angles of a triangle and find that they add to 185 degrees. From this you can determine that the space the triangle occupies is
   a. flat.
   b. positively curved.
   c. negatively curved.
   d. filled with dark matter.

10. Put the following types of objects in order, starting with the first one to form after the Big Bang and ending with the last one to form.
    a. neutral atoms, protons, nuclei
    b. protons, nuclei, neutral atoms
    c. nuclei, neutral atoms, protons
    d. protons, neutral atoms, nuclei

11. Quarks are
    a. virtual particles.
    b. massless particles.
    c. candidates for dark matter.
    d. building blocks of larger particles.

12. Current understanding indicates that the universe (choose all that apply)
    a. is closed.
    b. is flat.
    c. is open.
    d. is inflating.
    e. has accelerating expansion.

13. When a particle and an antiparticle come together, they
    a. annihilate each other, releasing photons.
    b. create a black hole.
    c. release enormous amounts of energy.
    d. create new particles.

14. Observations of Type I supernovae in distant galaxies have shown that
    a. the star formation rate in galaxies decreases with increasing redshift.
    b. the expansion rate of the universe is increasing.
    c. the cosmological constant is zero.
    d. dark energy is negligible at the present time.

15. The anthropic principle states that
    a. the universe was created so that life exists.
    b. life exists, so the universe must be such that life can exist.
    c. if the universe were otherwise, life would not exist.
    d. life has made the universe the way it is.

Thinking about the Concepts

16. What set of circumstances would cause an expanding universe to reverse its expansion and end up in a "Big Crunch"?

17. Describe the observational evidence suggesting that Einstein's cosmological constant (a repulsive force) may be needed to explain the historical expansion of the universe. Explain how Einstein was "right for the wrong reason."

18. What do astronomers mean by *dark energy*?

19. If the universe is being forced apart by dark energy, why isn't the Milky Way Galaxy, the Solar System, or the planet Earth being torn apart?

20. In Chapter 21, we said we could estimate the age of the universe with Hubble time ($1/H_0$). Why does that method not give the best answer?

21. What is the flatness problem, and why has it created difficulties for cosmologists?

22. During the period of inflation, the universe may have briefly expanded at 10^{30} (a million trillion trillion) or more times the speed of light. Why did this ultra-rapid expansion not violate Einstein's special theory of relativity, which says that neither matter nor communication can travel faster than the speed of light?

23. Why is particle physics important for understanding the early universe?

24. The fundamental forces of the universe are generally assumed not to change.
    a. How would the fate of the universe be affected if Newton's gravitational constant changed with time?
    b. What if, instead, the electric force between charged particles changed with time?

25. The standard model cannot explain why neutrinos have mass or why electron-positron asymmetry existed in the early universe. Do these failings make it an incomplete theory? Should all of its predictions be ignored until the theory can resolve these remaining issues?

26. Explain the process of pair production.

27. Describe the Planck era.

28. What are the basic differences between a grand unified theory (GUT) and a theory of everything (TOE)?

29. Consider the term *superstring theory* in light of the discussion of scientific theory in Chapter 1. Some scientists object to using the word *theory* to describe superstring theory. Why?

30. Suggest another example of how a different value of a physical constant would affect conditions in a "parallel" universe.

Applying the Concepts

31. Study Figure 22.2.
    a. Is the vertical axis linear or logarithmic?
    b. There are two labels for the horizontal axes. The top label is measured in billions of years. Is this axis linear or logarithmic?
    c. The bottom label for the horizontal axis is measured in relative brightness. Is this axis linear or logarithmic?
    d. What is the relationship between billions of years and relative brightness?

32. Figure 22.3 shows a blue curve that indicates a model in which the universe first decelerated and then accelerated, as well as a red curve indicating continual deceleration.
    a. How are the two curves different?
    b. What is it about one of these curves that indicates deceleration? What indicates acceleration?
    c. If a straight line were plotted on this graph, what would the model that the new line represents indicate about the expansion of the universe?

33. Study Figure 22.2. On this graph, the colored lines represent various models, and the black dots represent data taken in the actual universe.
    a. Why are there no data points on the right-hand side of the graph?
    b. Which models are excluded by the data?
    c. Roughly how far back in time do the data go?
    d. What fraction of the age of the universe is the answer to part (c) (assuming an age of 13.8 billion years)?

34. Study Figure 22.8. Is the plot linear or logarithmic? How much did the universe increase in size during the time of inflation? How much smaller was the universe at the beginning in an inflationary model than in the noninflationary model?

35. Suppose that new data coming in from a new instrument give the value 0.1 for both Ω_Λ and Ω_m. How would astronomers probably respond to these new data?

36. Study Figure 22.13.
    a. Is the time axis (the vertical dimension of the figure) approximately linear or approximately logarithmic?
    b. By how many orders of magnitude (factors of 10) has the density ρ of the universe dropped since earliest time?
    c. By how many orders of magnitude has the temperature dropped since earliest time?

37. Currently, the Hubble constant has an uncertainty of about 4 percent. What are the corresponding maximum and minimum ages allowed for the universe?

38. How many hydrogen atoms need to be in 1 cubic meter (m^3) of space to equal the critical density of the universe?

39. The universe today has an average density $\rho_0 = 3 \times 10^{-28}$ kg/m³. Assuming that the average density depends on the scale factor, as $\rho = \rho_0/R_U^3$, what was the scale factor of the universe when its average density was about the same as Earth's atmosphere at sea level ($\rho = 1.2$ kg/m³)?

40. The proton and antiproton each have the same mass, $m_p = 1.67 \times 10^{-27}$ kg. What is the energy (in joules) of each of the two gamma rays created in a proton-antiproton annihilation?

41. There are about 500 million CMB photons in the universe for every hydrogen atom. Using $E = mc^2$, what is the equivalent mass of these photons? Is it large enough to factor into the overall density of the universe?

42. Suppose you brought together a gram of ordinary-matter hydrogen atoms (each composed of a proton and an electron) and a gram of antimatter hydrogen atoms (each composed of an antiproton and a positron). Keeping in mind that 2 grams is less than the mass of a dime:
    a. Calculate how much energy (in joules) would be released as the ordinary-matter and antimatter hydrogen atoms annihilated one another.
    b. Compare this amount of energy with the energy released by a 1-megaton hydrogen bomb (4.2×10^{15} J).

43. One GUT theory predicts that a proton will decay in about 10^{31} years, which means if you have 10^{31} protons, you should see one decay per year. The Super-Kamiokande observatory in Japan holds about 20 million kg of water in its main detector, and it did not see any decays in 5 years of continual operation. What limit does this observation place on proton decay and on the GUT theory described here?

44. Assume a planet's orbit is perfectly circular as it travels in the gravitational well of its star. If this were true, would the orbit's circumference be greater than, less than, or equal to 2π times the radius of the orbit? Explain.

45. To get a feeling for the emptiness of the universe, compare its density (about 9.9×10^{-27} kg/m³) with that of Earth's atmosphere at sea level (1.2 kg/m³). How much denser is our atmosphere? Write this ratio using standard notation.

USING THE WEB

46. Go to the website for the JINA Center for the Evolution of the Elements and watch the animation of Big Bang nucleosynthesis (http://www.jinaweb.org/movies/bangmovie.html). What are the major stages after the Big Bang? What elements are created? Why are heavier elements not created?

47. Go to the website for the Dark Energy Survey, an international project that began in 2013 (https://www.darkenergysurvey.org/index.shtml). What observations will be made for this project? What will it tell scientists about dark energy? Click on "News." What is the status of this project? Are there any results yet?

48. a. Go to the website of the European Organization for Nuclear Research (CERN—http://home.web.cern.ch/about/physics/early-universe) and read through the pages indexed on the right. What was the role of the Higgs boson after the Big Bang? (Note: The World Wide Web was invented at CERN.)
    b. Citizen science: Go to Higgs Hunters (http://www.higgshunters.org/) and log in with your Zooniverse account. Click on "Science" and watch the brief videos; keep going until "How you can help." Why do they expect that human eyes are more likely than computers to find exotic decays? Click on "Classify" and "Restart the Tutorial" to see examples of how to mark the images with "Off-center vertex" or "Something weird." Mark up a few images, keeping a record for your homework.

49. Scientists debate whether there ever can be scientific evidence for a multiverse. A good example is a discussion in the journal *Scientific American*. Read the article "Does the Multiverse Really Exist?" in the August 2011 issue (it is probably accessible online through your school library) and the response at http://scientificamerican.com/article/multiverse-the-case-for-parallel-universe. What are some of the arguments for and against multiverses?

50. Clips and full episodes from a four-part series of the television show *NOVA* called "The Fabric of the Cosmos" can be accessed on PBS's website (http://pbs.org/wgbh/nova/physics/fabric-of-cosmos.html). Watch at least one of the episodes. Are the arguments made in these programs compelling? Is the science explained in a way that makes sense to a general audience?

smartwork

If your instructor assigns homework in SmartWork, access your assignments at digital.wwnorton.com/astro5.

digital.wwnorton.com/astro5

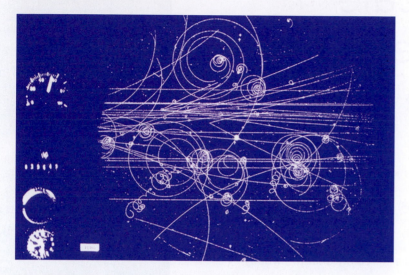

Figure 22.15 The decay of a particle in a hydrogen bubble chamber.

Just as astronomers use images to study distant clouds of dust and gas, so too do particle physicists use images to study the tiny particles. These images capture the tracks the particles make. As the particle passes through vapor in a chamber, it ionizes some of the molecules of the vapor. The vapor then condenses around the ionized molecules, leaving a thin trail of cloud. A strong magnetic field fills the chamber, so that the charged particles will turn as they travel through it.

Figure 22.15 shows tracks of particles moving through a bubble chamber. Studying this figure in more detail will give you a feel for how scientists have gained an understanding of these tiny particles.

1 Estimate the number of particle tracks that you can see in Figure 22.15.

2 Recall that charged particles spiral around magnetic field lines. From the image of Figure 22.15, determine which way the magnetic field points: perpendicular or parallel to the image.

3 Identify each of the following types of tracks in Figure 22.15:
   a. a track that looks straight
   b. a track that spirals tightly (a small spiral) and a track that spirals loosely (a large spiral)
   c. a bright track and a fainter track
   d. two tracks that form a "V" shape
   e. a track with an obvious beginning and ending
   f. a track with a kink, or abrupt change of direction

Oppositely charged particles spiral in opposite directions. For example, if the positive particles spiral clockwise, the negative ones spiral counterclockwise. More massive particles are more difficult to turn than less massive particles. Faster-moving particles are also harder to turn. The fact that both the mass and the velocity of a particle affect how easily it turns makes it difficult to separate the two properties.

4 Only charged particles interact with the vapor in the cloud chamber to produce tracks, so all the tracks should be curved in the magnetic field. But some of these tracks are very nearly straight. What can you say about the velocity of particles that leave straight tracks?

5 Consider the tight spiral and the loose spiral tracks.
   a. If both tracks were made by electrons, which particle was traveling faster?
   b. If both tracks were not made by electrons, what can you say about the total energy of each particle, including its mass?

6 The brightness of the track tells you how long the particle spent in each position: a brighter track is produced when the particle stays near the same spot longer.
   a. Consider the bright track and the faint track. Which track shows the faster particle?
   b. Now apply this reasoning to the tight and loose spiral tracks. Can you tell which particle was moving faster?
   c. If the looser track is also brighter, were both the loose and the tight tracks made by the same type of particle?

7 A track that is shaped like a "V" indicates that two charged particles were produced at the base of the "V" and then proceeded onward. This can occur, for example, if a neutral particle decays to produce a positive and a negative charge. If the two particles were a proton and an electron, but both arms of the "V" curved the same (possibly very small) amount, what can you say about their relative velocities?

8 A track with an obvious beginning and endpoint shows an event initiated by a neutral particle. Because neutral particles do not leave tracks in the chamber, this type of track must be a charged particle that suddenly begins moving. How might a neutral particle initiate one of these tracks?

9 A kink occurs when a charged particle decays, emitting a particle and in the process becoming a different charged particle.
   a. Is the particle emitted in the decay neutral or charged? How do you know?
   b. Does the remaining charged particle have more or less mass than before it decayed?
   c. Sketch the kink that you are studying, and indicate the direction the particle was traveling along the track. (Hint: Consider the tightness of the spiral!)

To learn more you would have to know more about the experimental setup—the strength of the magnetic field, the speed of the particles as they enter the chamber, and so on. By studying images like these, and "crunching the numbers," scientists determine the properties and interactions of the fundamental building blocks of the universe.

23

Large-Scale Structure in the Universe

The universe that emerged from the Big Bang was incredibly uniform, wholly unlike today's universe of galaxies, stars, and planets. In this chapter, we investigate the origin of the current structure of the universe and find that complex structure is a natural consequence of the action of physical law in an evolving universe.

LEARNING GOALS

By the conclusion of this chapter, you should be able to:

LG 1   Describe the distribution of galaxies in the universe.

LG 2   Explain how the structure that began to form shortly after the Big Bang evolved to the large-scale structure of the universe today.

LG 3   Describe the formation of the first stars and the first galaxies.

LG 4   Explain how observations of galaxies at many different redshifts illustrate the evolution of the large-scale structure of the universe.

The Hubble Ultra Deep Field 2014 image shows thousands of galaxies in a small section of space in ultraviolet, visible, and infrared light. ▶▶▶

What does this image tell us about galaxy evolution?

Figure 23.1 The Virgo Cluster of galaxies is the closest cluster to the Milky Way, at a distance of about 16.5 Mpc (54 million light-years).

23.1 Galaxies Form Groups, Clusters, and Larger Structures

Just as stars and clouds of glowing gas reveal the structure of the Milky Way, the distribution of galaxies indicates the structure of the universe. And just as it is gravity that holds galaxies together, giving them their shape, it is gravity that produces larger structures. The vast majority of galaxies are parts of gravitationally bound collections of galaxies. In this section, we will look at how these clusters of galaxies form.

Types of Galaxy Structures

Galaxy groups are the smallest and most common galaxy collections and contain up to several dozen galaxies, most of them dwarf galaxies. Recall from Chapter 20 that the Milky Way is a member of the Local Group, which consists of two giant spiral galaxies—the Milky Way Galaxy and the Andromeda Galaxy—along with more than 30 smaller dwarf galaxies in a volume of space roughly 3 megaparsecs (Mpc) in diameter. Most of the mass in the Local Group, dark and luminous, resides in the two giant galaxies.

Larger gravitationally bound systems of galaxies, called **galaxy clusters**, can consist of thousands of galaxies. Galaxy clusters are larger than groups, typically occupying a volume of space 2–10 Mpc across. Galaxy clusters also contain far more dwarf galaxies than giant galaxies. However, most of the galaxy mass in galaxy clusters resides in the giant galaxies. Most clusters contain spiral galaxies, though the more massive clusters have a much higher fraction of giant elliptical galaxies. The Virgo Cluster (**Figure 23.1**), located 16.5 Mpc from the Local Group, is an example of a cluster that contains mostly spiral galaxies. Giant elliptical and S0 galaxies dominate the more distant Coma Cluster.

Galaxy clusters and groups of galaxies bunch together to form enormous **superclusters**, which contain tens of thousands or even hundreds of thousands of galaxies and span regions of space typically more than 30 Mpc in size. Our Local Group is part of the Laniakea Supercluster, which also includes the Virgo Cluster (**Figure 23.2**).

Figure 23.2 This computer-generated visualization shows a slice of the Laniakea Supercluster. Individual galaxies are white dots, blue areas are voids, red dots indicate the Virgo Cluster, and green areas contain many galaxies. White lines indicate the movement of galaxies toward the center of the supercluster. The orange contour encloses the outer limits of those galaxies streaming to the center. The dark blue dot shows the location of the Milky Way.

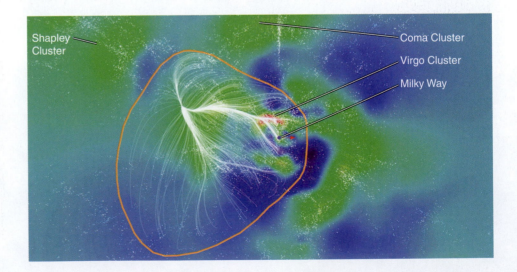

Mapping the Universe

Hubble's law for the expansion of the universe is a powerful tool for mapping the distribution of galaxies, groups, clusters, and superclusters in space. Using this law, astronomers estimate the distance to a galaxy by measuring the redshift in the galaxy's spectrum. The first redshifts were measured from spectra recorded on photographic plates, which required exposures of several hours to capture the faint signal. By 1975, astronomers had documented redshifts for only about a thousand of the several hundred billion observable galaxies. Now astronomers use larger telescopes with electronic detectors and spectrographs capable of observing many galaxies at once to measure the redshifts of galaxies. We now know the redshifts to more than 1 million galaxies, and therefore we know the approximate distances to these galaxies. From this information, we can develop a map of the structure of the universe on the largest scales.

The first large redshift survey looked only at local space, but later surveys extended much farther. These surveys were used to illustrate a "slice of the universe" as seen in **Figure 23.3a**. Observations show that clusters and superclusters of galaxies are not scattered randomly through space, but are linked in an intricate network of relatively thin structures known as filaments and walls. These concentrations of galaxies surround large regions of space with very few galaxies, known as **voids**. Voids are some of the largest structures seen in the universe. Though the voids may seem empty, we do not know that they are completely empty of matter—only that they have very few observable galaxies.

Clusters and superclusters are located within the walls and filaments. Subsequent surveys have looked at much larger volumes of space. The Sloan Digital Sky Survey (SDSS) includes the Sloan Great Wall, a string of galaxies 400 Mpc long, shown in Figure 23.3b. For as far out as observations can currently measure, the universe has a porous structure much like a sponge or a pile of soap bubbles. Together, galaxies and the larger groupings in which they are found are called **large-scale structure**.

The **peculiar velocity** of a galaxy is its motion relative to the cosmic microwave background radiation (CMB; see Chapter 21). Observations of the peculiar velocity provide information about the distribution of mass near that galaxy. For example, the peculiar velocity of the Local Group was originally attributed to the Great Attractor, which has a mass of several thousand times the mass of the Milky Way and is located about 75 Mpc away. Recently, it was found that the Great Attractor is actually the center of mass of the larger supercluster called Laniakea, with the Local Group and the Milky Way located at the outer edge, as seen in Figure 23.2.

These observations of the large-scale structure of the universe provide direct observational evidence for the cosmological principle—that the universe is homogeneous (the same everywhere) and isotropic (the same in every direction)—because they show that on this very largest scale, the structure of the universe is the same everywhere, in every direction. If, for example, the universe in the top half of the multicolored image of Figure 23.3b showed a uniform distribution of galaxies while the universe in the bottom half of the image showed walls and voids, then the universe would not be homogeneous or isotropic. All conclusions that are based on the cosmological principle would have been called into doubt. As it is, however, observations support this underlying principle of cosmology.

Figure 23.3 Redshift surveys use Hubble's law to map the universe. (a) In 1986, the Harvard-Smithsonian Center for Astrophysics redshift survey, called "A Slice of the Universe," was the first to show that clusters and superclusters of galaxies are part of even larger-scale structures. (b) The 2008 Sloan Digital Sky Survey map of the universe extends outward to a distance of about 600 Mpc. Shown here is a sample of 67,000 galaxies colored according to the ages of their stars, with the redder, more strongly clustered points being galaxies that are made of older stars.

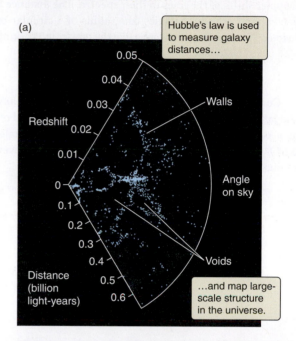

(a)

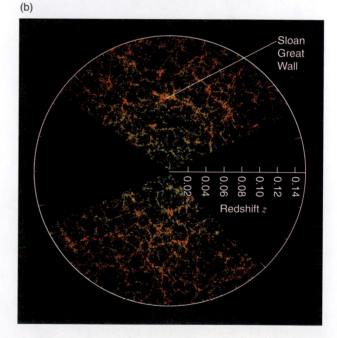

(b)

23.1 Working It Out Mass of a Cluster of Galaxies

Recall from Working It Out 20.1 that astronomers use the orbit of a star about the center of a galaxy to estimate the mass of the galaxy within the star's orbit. A similar estimation is made with groups or clusters of galaxies. These groups or clusters are often dominated by one giant galaxy at the center, with smaller ones orbiting around it. The orbital velocities of the smaller galaxies are measured from the Doppler shifts of the lines in their spectra. The distance between the central and orbiting galaxies (the radius of a circular orbit) is estimated. The equation then looks like the one from Working It Out 20.1:

$$M = \frac{r v_{\text{circ}}^2}{G}$$

Consider a typical cluster. Suppose a smaller galaxy is orbiting the large galaxy at the center of the cluster at a speed of 1,000 km/s at a distance of about 3 Mpc. The gravitational constant is $G = 6.67 \times 10^{-20} \ \text{km}^3/(\text{kg s}^2)$. As in Chapter 20, we must convert the orbital radius (3 Mpc) to kilometers:

$$(3 \times 10^6 \ \text{pc}) \times (3.09 \times 10^{13} \ \text{km/pc}) = 9.3 \times 10^{19} \ \text{km}$$

The mass of the cluster core is given by

$$M = \frac{(9.3 \times 10^{19} \ \text{km}) \times (10^3 \ \text{km/s})^2}{6.67 \times 10^{-20} \ \text{km}^3/(\text{kg s}^2)}$$

$$M = 1.4 \times 10^{45} \ \text{kg}$$

We can divide this by the mass of the Sun, 2.0×10^{30} kg, to get a cluster mass of $7.0 \times 10^{14} \ M_{\text{Sun}}$. If we divide this core cluster mass by the mass of the Milky Way Galaxy, $10^{12} \ M_{\text{Sun}}$, we see that the cluster core has a mass of about 700 Milky Way galaxies. This includes the dark matter inside of the orbit.

In 1933, astronomer Fritz Zwicky (1898–1974) measured the velocities of many galaxies within a collection of clusters, and in each case he found the entire cluster contained more mass than could reside in the stars that produce the cluster's visible light. He concluded that there must be dark matter within clusters of galaxies. Later in the 1970s, Vera Rubin (1928–) and colleagues measured rotation curves of individual galaxies and discovered dark matter in these galaxies too.

Dark Matter in Galaxy Groups and Clusters

Just as dark matter dominates galaxies, it also dominates galaxy groups and clusters. Astronomers can infer this in several ways. They can look at the motion of a small satellite galaxy orbiting the central dominant galaxy of the cluster and estimate the mass of the cluster inside of this orbit similar to the way they measure dark matter in spiral galaxies. Or they can look at the motions of all of a cluster's galaxies and calculate how strong gravity must be to hold the cluster together. And again, the conclusion is that the total mass of the clusters, including dark matter, must be about 8–10 times greater than the normal matter they contain (**Working It Out 23.1**).

Another piece of evidence for the presence of dark matter is that the space between galaxies in a cluster is filled with extremely hot gas that is 10 million to 100 million kelvins (K), making it bright in X-rays (**Figure 23.4**). Even though this gas is of extremely low density, the volume of space that it occupies is enormous: the mass of this hot gas can be up to 5 times the mass of all the stars in that cluster. X-ray spectra show that the gas contains significant amounts of massive elements that must have formed in stars. This chemically enriched gas has been either blown out of galaxies in winds driven by the energy of massive stars or stripped from galaxies during encounters with neighboring galaxies. The amount of luminous mass is not enough to keep this hot gas from escaping, thus this hot gas would have dispersed long ago, were it not for the gravity of the dark matter filling the volume of the cluster.

An additional way to look for dark matter relies on the predictions of Einstein's general theory of relativity (see Chapter 18), which states that mass distorts the

(a) (b)

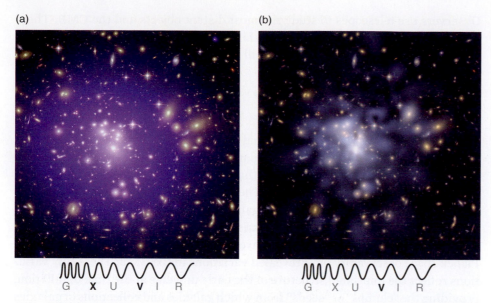

Figure 23.4 (a) Galaxy clusters such as Abell 1689 are rich in hot, X-ray-emitting gas (shown here in purple overlaying the image taken in the visible part of the spectrum). (b) This is also Abell 1689, with the inferred dark matter distribution shown in blue-white.

geometry of spacetime, causing even light to bend near a massive object. In particular, light from a distant object is bent by the gravity of a galaxy or cluster of galaxies, so that images of the distant object can be seen magnified on either side of the intervening galaxy or cluster. The result is a gravitational lens, demonstrated in **Figure 23.5a**. Recall that we mentioned gravitational lensing in the discussion of massive compact halo objects (MACHOs) in Chapter 19, where we noted that lenses can make background objects appear brighter. Lenses can also show multiple images of background objects, and these magnified images are often drawn out into arcs. The greater the gravitational lensing, the greater the mass that must be in the cluster. Figure 23.5b shows an image of a galaxy cluster that is acting as a gravitational lens for a number of background galaxies. Analysis of such images reveals the mass of the lensing cluster.

Regardless of how astronomers measure the masses of galaxy clusters—by looking at the motions of their galaxies, by measuring their hot gas, or by using them as gravitational lenses—the results are the same. Dark matter dominates the mass of galaxy clusters and superclusters.

CHECK YOUR UNDERSTANDING 23.1

Place the following types of galaxy collections in order of increasing size: (a) wall; (b) cluster; (c) group; (d) supercluster

23.2 Gravity Forms Large-Scale Structure

How did the universe evolve from a very smooth distribution of radiation and matter after the Big Bang to the large-scale structure of walls and voids seen today? Astronomers approach this question observationally and theoretically:

(a)

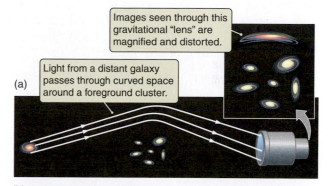

Images seen through this gravitational "lens" are magnified and distorted.

Light from a distant galaxy passes through curved space around a foreground cluster.

(b)

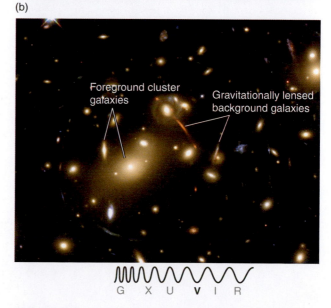

Foreground cluster galaxies

Gravitationally lensed background galaxies

Figure 23.5 (a) This illustration shows the geometry of a gravitational lens. A mass can gravitationally focus the light from a distant object, thereby magnifying and distorting the image. (b) A Hubble Space Telescope image of the cluster Abell 2218 shows many gravitationally lensed galaxies, seen as arcs.

Observers use telescopes to study the most distant objects and the CMB. Theorists use the largest supercomputers to simulate the growth of small- and large-scale structure. In this section, we will examine the role of gravity and dark matter in the formation of structure.

Gravitational Instabilities

In Chapter 15, we discussed gravitational instabilities involved in star formation. Star formation can begin in a molecular cloud with clumps inside it. Gravity causes those clumps to collapse faster than their surroundings: gravity can turn density variations of clouds into stars. The same gravitational instability can turn density variations of the universe into galaxies.

As discussed in Chapter 21, space missions such as the Cosmic Background Explorer (COBE), Wilkinson Microwave Anisotropy Probe (*WMAP*), and the Planck space observatory have revealed variations in the CMB. These tiny variations reflect the imprinted structure of the early universe at the time of inflation, providing the "clumps" or "seeds" from which galaxies and collections of galaxies grew. Over time, gravity amplified these clumps. Smaller structures such as dwarf galaxies formed first, whereas larger structures such as clusters, superclusters, filaments, and voids formed later. This process is called **hierarchical clustering** because the structure forms in a "bottom-up" hierarchy. Hierarchical clustering is supported by observations and is fundamental to how structure formed in the universe.

If the universe is not totally uniform but has small density variations from place to place, gravity will amplify those variations over time. For example, on the scale of a galaxy, the early universe may have regions that are more dense than average by perhaps 1 percent. Over time, gravity can amplify this difference in density so that it can become more than twice the average, and a galaxy can form. But the seeds are essential, as gravity cannot produce structure in a perfectly uniform universe where there is nothing to amplify. These seeds are thought to have formed from quantum fluctuations during inflation in the early universe. This model has a profound implication: the seeds leading to galaxies, clusters, and superclusters (the largest structures in the universe) arose from the same quantum physics that describes the smallest structures in the universe (atoms, nuclei, and elementary particles).

Galaxies Formed because of Dark Matter

Whereas cosmology theorists think about gravity, inflation, and dark energy and consider the balance among radiation, normal matter, and dark matter in the early universe, observational astronomers measure the constants of cosmology, such as the Hubble constant (H_0), the cosmological constant (Ω_Λ), and matter density (Ω_m). The available observations and measured constants of cosmology become inputs to theoretical models and supercomputer simulations of how the large-scale structure formed. By the beginning of the 21st century, a "standard model" of Big Bang cosmology had broad support for explaining the large-scale structure and accelerated expansion of the universe, as well as the CMB and the amounts of the light elements from Big Bang nucleosynthesis (discussed in Chapters 21 and 22). This model is called **Lambda-CDM**, from *lambda* for the cosmological constant Ω_Λ, and *CDM* for the "cold dark matter" that played a vital role in structure formation.

Astronomers measure the ratio of normal and dark matter and look for galaxies with the highest redshifts. Remember that more distant galaxies have higher measured values of redshift (z), so the observed light was emitted when the universe was *younger*, closer to the time of the Big Bang (**Table 23.1**). Observations of galaxies at many different redshifts show how the universe has changed over time. However, there is a gap of about 400 million years between the CMB maps of the universe at age 400,000 years and the highest-redshift galaxies observed. New telescopes such as the international Atacama Large Millimeter/submillimeter Array (ALMA) in Chile and the future James Webb Space Telescope (JWST) in space may be able to observe even younger galaxies closer to the time when the CMB became observable.

The details of the very early universe affected the growth of the large-scale structure seen today. The values of Ω_Λ and Ω_m are important in part because they determine how rapidly the universe expanded and therefore how difficult it is for gravity to overcome this expansion in a particular region. The more rapidly the universe is expanding, or the less mass it contains, the more difficult it will be for gravity to pull material together into galaxies and larger-scale structures.

Variations in the cosmic microwave background radiation are found at a level of about one part in 100,000. The theoretical models clearly show that such tiny variations at the time of recombination (when the universe was about 400,000 years old) are far too small to explain the structure observed in today's universe. Gravity is not strong enough to grow galaxies and clusters of galaxies from such small clumps. These models indicate that for today's galaxies to have formed, the density of those clumps must have been a few tenths of a percent greater than the average density of the universe at the time of recombination. But if normal luminous matter in the early universe had clumps with this higher density, the variations in the CMB today would be at least 30 times larger than they are. How do astronomers reconcile this problem? Dark matter holds the key.

Dark matter is an essential ingredient in the formation of the structure we observe. The density of normal matter inferred from Big Bang nucleosynthesis is about the same as the density of normal matter in today's universe and much less than the density of dark matter. This agreement in the density of normal matter provides a powerful constraint on the nature of dark matter, which dominates the mass in the universe. Dark matter cannot consist of normal matter made up of neutrons and protons. If it did, the density of neutrons and protons in the early universe would have been much higher, and the resulting amounts of light elements in the universe from Big Bang nucleosynthesis would have been very different from what we actually observe (see Chapter 21).

Dark matter must be something that interacts only weakly with normal matter and has no electric charge and therefore does not interact with electromagnetic radiation. Dark matter and normal matter behaved differently in the early universe. Because clumps of such dark matter in the early universe did not interact with radiation or normal matter, astronomers would not see them directly when looking at the CMB. Dark matter clumpiness can be large enough to form galaxies without producing too much variation in the CMB, as long as the clumpiness in the ordinary matter is much smaller. This unseen dark matter solves the problems of modeling the formation of galaxies and clusters of galaxies.

TABLE 23.1 Redshift and Age

Observed z	Age of Universe (years)
1,100	380,000 (recombination)
30	100 million
20	200 million
15	270 million
10	480 million
9	560 million
8	650 million
7	750 million
6	900 million
5	1.2 billion
4	1.6 billion
3	2.2 billion
2	3.3 billion
1	5.9 billion
0.5	8.6 billion
0.25	10.5 billion
0	13.8 billion

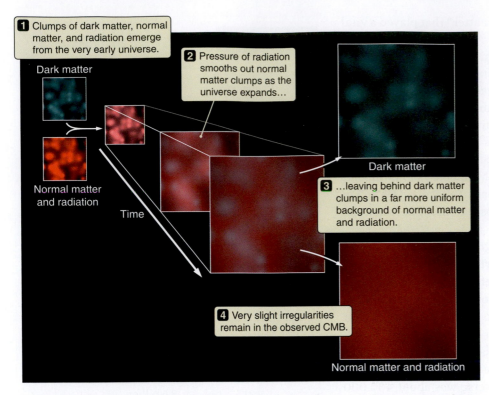

1 Clumps of dark matter, normal matter, and radiation emerge from the very early universe.

Dark matter

Normal matter and radiation

Time

2 Pressure of radiation smooths out normal matter clumps as the universe expands…

3 …leaving behind dark matter clumps in a far more uniform background of normal matter and radiation.

Dark matter

4 Very slight irregularities remain in the observed CMB.

Normal matter and radiation

Figure 23.6 Radiation pressure and other processes in the early universe smoothed out variations in normal matter, but clumps in the dark matter survived and pulled in normal matter to form galaxies.

After the Big Bang, radiation pressure smoothed out ripples in the distribution of normal matter. Feebly interacting dark matter is immune to these processes, so while normal matter smoothed out over time, dark matter remained clumpy. **Figure 23.6** shows how the distributions of normal and dark matter differ due to the difference in how they interact with radiation. In addition, the dark matter in these clumps does not glow, so astronomers do not see it in the cosmic microwave background radiation. Although these clumps of dark matter do cause slight gravitational redshifts in the light coming from normal matter, the resulting variations fit well with current observations of the CMB. The mass of dark matter controlled the growth of gravitational instabilities in the early universe, analogous to how dark matter dominates the gravitational fields of today's galaxies and clusters. Dark matter solves the problems of the formation of galaxies and clusters of galaxies by providing a stronger gravitational attraction without being smoothed out by radiation in the early universe.

Hot and Cold Dark Matter

Dark matter in the early universe was much more strongly clumped than normal matter. Within a few million years after recombination, these dark matter clumps pulled in the surrounding normal matter. Later, gravitational instabilities caused these clumps to collapse. The normal matter in the clumps went on to form visible galaxies. The details of how this happened depend greatly on the properties of the dark matter itself. Recall that in Chapter 20, we discussed the search for MACHOs and weakly interacting massive particles (WIMPs) as possible explanations of dark matter and concluded that MACHOs could not explain the unseen mass. Even though we do not yet know exactly what dark matter is made of, on the basis of how dark matter behaves we can categorize two broad classes: cold dark matter and hot dark matter.

Cold dark matter consists of feebly interacting particles that are moving about relatively slowly, like the slow-moving atoms and molecules in a cold gas. There are several candidates for the composition of cold dark matter. Most likely, cold dark matter consists of an unknown **elementary particle**. One candidate is the **axion**, a hypothetical particle first proposed to explain some observed properties of neutrons. Axions should have very low mass, and they would have been produced in great abundance in the Big Bang. Another candidate is the **photino**, an elementary particle related to the photon. Some theories of particle physics predict that the photino exists and has a mass about 10,000 times that of the proton. Physicists are looking for these types of particles using particle accelerators such as the Large Hadron Collider, and several experiments are under way to search for axions and photinos that are trapped in the dark matter halo of the Milky Way.

Hot dark matter consists of particles that are moving so rapidly that gravity cannot confine them to the same region as the luminous matter in the galaxy. Neutrinos are one example of hot dark matter. Recall from Chapter 14 that neutrinos interact with matter so weakly that they are able to flow freely outward from the center of the Sun, passing through the overlying layers of matter as if they

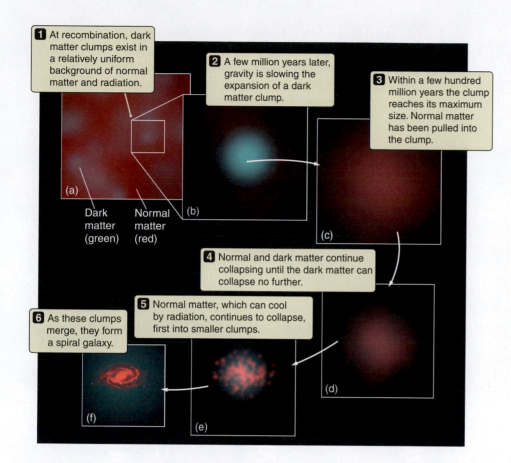

1 At recombination, dark matter clumps exist in a relatively uniform background of normal matter and radiation.

2 A few million years later, gravity is slowing the expansion of a dark matter clump.

3 Within a few hundred million years the clump reaches its maximum size. Normal matter has been pulled into the clump.

(a)

Dark matter (green)   Normal matter (red)   (b)

(c)

4 Normal and dark matter continue collapsing until the dark matter can collapse no further.

5 Normal matter, which can cool by radiation, continues to collapse, first into smaller clumps.

6 As these clumps merge, they form a spiral galaxy.

(d)

(f)   (e)

Figure 23.7 A spiral galaxy passes through roughly six stages as it forms from the collapse of a clump of cold dark matter.

were not there. The universe is filled with neutrinos, which might account for a few percent of its mass. Although this percentage is not high enough to account for all of the dark matter in the universe, it may still have had a noticeable effect on the formation of structure.

Cold and hot dark matter have different effects on structure formation because of the way they respond to a gravitational field. Slow-moving particles are more easily held by gravity than are fast-moving particles, so particles of cold dark matter clump together more easily into galaxy-sized structures than do particles of hot dark matter. As a result, theoretical models show that on the largest scales of massive superclusters, both hot dark matter and cold dark matter can form the kinds of structures observed; but on much smaller scales, only cold dark matter can clump enough to produce structures like the galaxies filling the universe. To account for the formation of today's galaxies, we need cold dark matter.

We can best see how models of galaxy formation work by following the events predicted by the models step-by-step. Consider a universe made up primarily of cold dark matter, clumped together with normal matter in a manner consistent with observations of the CMB. On the scale of an individual galaxy, the effect of the cosmological constant is so small that it can be ignored.

Figure 23.7a shows a model simulation of one clump of dark matter at the time of recombination. According to the model, the dark matter was slightly clumpier than normal matter, but overall the distribution of matter was remarkably uniform. As illustrated in Figure 23.7b, the model shows that by a few million years after recombination, the universe in the simulation had expanded severalfold; spacetime is expanding, so the clumps of dark matter are also expanding. However, the clumps of dark matter did not expand as rapidly as their surroundings because their self-gravity slowed down their expansion. The clumps of dark

Process of Science

MULTIPLE STREAMS OF EVIDENCE

Roughly 85 percent of the mass in the universe is in an unknown form or composition, called dark matter. So many observations, models, and experiments provide supporting evidence for dark matter that scientists have concluded that it exists.

DARK MATTER

Galaxy rotation curves

Motions in galaxy clusters

Cosmic microwave background

Gravitational lensing

Confined hot gas in clusters

Models of cluster formation   Models of galaxy formation

Models of the Big Bang

Scientists do not have the luxury of ignoring evidence that does not fit current theories. They must see "being wrong" as an opportunity to learn and as a challenge to try harder.

matter stood out more with respect to their surroundings. The gravity of the dark matter clumps began to pull in normal matter. Eventually, the clumps of dark matter stopped expanding when their own self-gravity slowed and then stopped their initial expansion. By the stage shown in Figure 23.7c, normal matter clumped in much the same way as dark matter. Unlike dark matter, which cannot emit radiation, the normal matter in the clumps radiated away energy and cooled, collapsing toward the center of the dark matter clumps.

These clumps did not exist in isolation: they were tugged on by the gravity of neighboring clumps and were pushed around by the pressure waves that ran through the young universe, smoothing out its structure. As a result, each clump had a little bit of rotation when it began its collapse. As normal matter fell inward toward the center of the dark matter clump, this rotation forced much of the gas to settle into a rotating disk, just as the collapsing clouds of protostars settled first into an accretion disk. Later, this rotating disk became the disk of a spiral galaxy (Figures 23.7d–f). This model has been successful because it yields the right masses for observed galaxies. For example, too low a mass system would not cool fast enough to separate the visible matter from the dark matter. The dark matter we have been talking about as a major component of galaxies turns out to be responsible for their formation as well (**Process of Science Figure**).

CHECK YOUR UNDERSTANDING 23.2

The dominant factor in the formation of galaxies is the distribution of _____ in the early universe. (a) ordinary matter; (b) dark matter; (c) energy; (d) dark energy

23.3 First Light of Stars and Galaxies

Recall from Chapter 21 that recombination occurred approximately 400,000 years after the Big Bang, when the universe cooled enough for electrons and protons to combine to form hydrogen and helium atoms. Before this time the universe was opaque; after recombination it became transparent. The CMB was emitted at this moment, and we can observe it today. This epoch in the history of the universe is called the **Dark Ages** because there was no visible "light" from astronomical objects. The only light available at this time was from the cooling—and darkening—CMB, and perhaps some amount of 21-centimeter (21-cm) radio radiation from hydrogen

The Dark Ages lasted from about 200 million to about 550 million years after the Big Bang. During this time, the first stars were forming from the elements created in the Big Bang. As these stars formed, they heated up until they emitted ultraviolet (UV) photons with enough energy to reionize neutral hydrogen in interstellar space. During this **reionization** stage, when the UV light from the first stars stripped electrons from hydrogen atoms, the ionized hydrogen became transparent to the light generated by the first stars. Reionization continued with star formation in the first low-luminosity galaxies and with radiation from the first supermassive black holes inside of the first quasars. Reionization was completed by about 750 million to 900 million years after the Big Bang (**Figure 23.8**).

Only within the past few years have astronomers been detecting objects representing light from the first billion years of the universe: these objects have redshifts greater than 6. (Table 23.1 shows how the times in the history of the universe compare with observed redshifts.) Many astronomers were surprised by the identification of galaxies, quasars, and gamma-ray bursts at such high

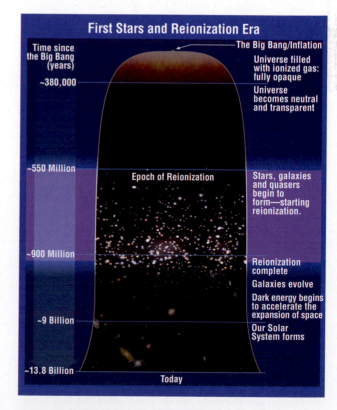

Figure 23.8 This timeline illustrates the formation of the first stars and galaxies and the epoch of reionization.

redshifts, because it had been thought that these objects did not form until after the universe was at least a billion years old. For example, recall from Chapter 18 that gamma-ray bursts (GRBs) are extremely luminous and result from the explosive deaths of massive stars. For GRBs to be detected at $z = 8$, there must have been massive stars that had already died by 650 million years after the Big Bang. Similarly, the detection of quasars at $z = 7$ indicates that supermassive black holes must have formed in less than 750 million years after the Big Bang. The study of these highest-redshift objects and what they tell astronomers about the early universe is one of the most dynamic topics in astronomy today. New telescopes and new instruments are regularly detecting objects with higher and higher redshifts. By the time you read this book, the highest known redshifts likely will be higher.

The First Stars

Astronomers use what has been learned in stellar evolution theory to study galaxy evolution and cosmology. The very first stars must have formed from the elements created in Big Bang nucleosynthesis: hydrogen, helium, and a very small amount of lithium. Observational astronomers look for stars with only these elements, but so far none have been detected. Old stars in the halo of the Milky Way contain very low abundances of heavy elements, but not quite as low as zero. Astronomers use computer simulations that combine data from these old stars with the conditions in the early universe to figure out what happened before these old stars formed.

The formation of the first stars was different from the processes we discussed in Chapter 15. Because there were no heavy elements, there was no dust, and also no molecular clouds filled with cold, dense gas for the stars to form in. Instead, these first stars formed inside of dark matter minihalos, which were about 0.5 million to 1.0 million solar masses (M_{Sun}) and 100 parsecs (pc) across. These minihalos formed a few hundred million years after the Big Bang ($z \approx 20$ to 30). Primordial gas clouds within these minihalos contained neutral hydrogen, and over time some small amounts of molecular hydrogen (H_2) formed. The molecular hydrogen cooled the gas, and as it cooled, the gas pressure dropped and the gas collapsed to the center of the minihalo. A protostar grew in this gas cloud, accreting more gas to become a star. Theoretical models and computer simulations predict that these first stars were likely to be hot and massive. **Figure 23.9** is an illustration of what such a star might have looked like. Estimates of the masses of the first stars range from 10 to more than 100 M_{Sun} for single stars and 10–40 M_{Sun} for double stars. These stars had high luminosity, peaking in the ultraviolet, which ionized the gas near the star. Large star clusters likely didn't form, and these stars were singles, doubles, or small multiples.

These first stars were much more massive than the average star observed today. Recall from Table 16.1 and Chapter 17 what happens to stars with these high masses: they have very short lifetimes, burning hydrogen in their cores for 10 million years or less. Today, massive stars use the carbon-nitrogen-oxygen (CNO) cycle for more efficient hydrogen burning, but carbon, nitrogen, and oxygen were not available to these first stars. These first stars ended their brief lives in supernova explosions, scattering some heavy elements into nearby space. If the core of such a star had rapid rotation at the time of the supernova explosion, it might have emitted a gamma-ray burst of extremely high luminosity. Astronomers have observed a few gamma-ray bursts that might be from redshift 8 or 9.

Some of these stars were massive enough to have become black holes after. Black holes with companions can become energetic X-ray binary systems, as mass falls onto the accretion disk of the black hole as the companion evolves. Because

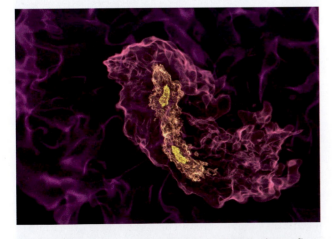

Figure 23.9 This supercomputer simulation shows first star formation from primordial gas in a dark matter minihalo a few hundred million years after the Big Bang. In this illustration, two massive stars are forming a few hundred astronomical units apart. The brighter yellow regions are of higher density than the purple regions.

these stars were all so massive, it is likely that both stars in a binary system (or all of them in a multiple system) would become black holes, and these black holes could then merge. Gravitational waves emitted during a black-hole merger might be detectable in future experiments. Some theorists think that these merged black holes might have become the seeds for the supermassive black holes found in galaxies, but other models suggest it would take too long for these stellar black holes to have built up to a mass of 1 million to 1 billion M_{Sun}. When these first generation stars died, supernova explosions scattered the heavy elements they formed through nucleosynthesis in their interiors.

Carbon, oxygen, and other elements might have mixed in and cooled nearby star-forming gas clouds. Some of these elements condensed into dust grains, which further cooled the clouds, so that the next generation of stars formed in a manner similar to the way stars form in today's cold molecular clouds. These "second-generation stars" had very low abundances of heavy elements, but measurably more than in the first stars. Because they formed in a cooler environment than the first stars, these stars could have had less mass. Any stars less massive than 0.8–0.9 M_{Sun} have such long lifetimes that they are still burning hydrogen on the main sequence today. These stars are not very luminous, but a few of them have been found in the halo of the Milky Way. Such stars have very low amounts of heavy elements, but their spectra show small amounts of many of the elements on the periodic table—including uranium. Astronomers are very interested in studying these small second-generation stars because they offer clues about the nature of the first stars and the conditions of the very young Milky Way.

The First Galaxies

How did the first galaxies form? From a theoretical perspective, the minihalos just discussed are not considered galaxies. A galaxy is supposed to have stellar systems that live a long time and are able to retain gas heated by the ultraviolet light of stars or by supernova explosions. After the first stars died, the energy from supernovae, GRBs, or X-ray binaries may have heated any remaining gas in the minihalo too much for further star formation or such gas may have escaped the minihalo because of its relatively low gravitational pull. Thus, the minihalos may have produced one generation of short-lived massive stars and may not have been able to hang on to heated or blasted gas with their small mass.

One theory is that the first galaxies were made up of the first systems of stars that were gravitationally bound in a dark matter halo. These stars may have been first stars or chemically enriched second stars. The properties of the first galaxies were shaped by the first stars: their radiation, their production of some heavier elements, and the black holes resulting from the deaths of these stars. The masses of these galaxies are thought to have been about a hundred million M_{Sun}, and they were built up hierarchically from the merging of minihalos, shown in **Figure 23.10**.

One piece of evidence for this theory comes from infrared observations, shown in **Figure 23.11**. Figure 23.11a includes the usual nearby stars and galaxies, but when these are all subtracted, a glow remains. This remaining structure, seen in Figure 23.11b, likely arose from the first stars and galaxies about 500 million years after the Big Bang.

Another piece of evidence comes from the discoveries of galaxies and quasars at higher and higher redshifts, so we see them when they were very young. These observations constrain the timeline by indicating how soon the first galaxies—and first supermassive black holes—formed after the Big Bang. The peak of the

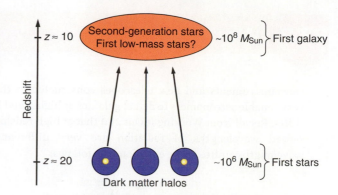

Figure 23.10 In "bottom-up" (hierarchical) growth, the first stars formed in minihalos and then came together to form larger halos, which became the first small dwarf galaxies.

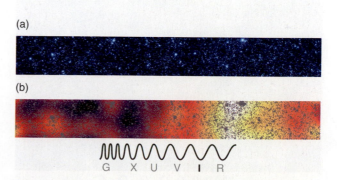

Figure 23.11 (a) A standard infrared image from the Spitzer Space Telescope shows stars and some galaxies in this strip of sky. (b) In this image, the nearby stars and galaxies have been subtracted out (gray) and the remaining glow enhanced, showing some structure from the early universe at a time when the earliest stars and galaxies were forming.

23.2 Working It Out Observing High-Redshift Objects

New instruments and new telescopes constructed in the past few years enable astronomers to detect galaxies at higher and higher redshifts. Recall from Working It Out 21.2 that at high redshifts, the observed wavelengths of radiation are very different from the wavelengths emitted. The equation is as follows:

$$1 + z = \frac{\lambda_{observed}}{\lambda_{emitted}}$$

where z is the redshift. For $z = 1$, the observed wavelength is twice that emitted; for $z = 2$, three times that emitted; and so on. For the highest-redshift galaxies discovered in the past few years, with $z = 8$ to 10, the observed wavelengths are 9–11 times that emitted.

Neutral hydrogen in gas clouds absorbs light at wavelengths shorter (bluer) than 121.6 nanometers (nm). As a result of this absorption, a galaxy spectrum is brighter at emitted wavelengths longer than 121.6 nm, and the spectrum is fainter, or "drops out," at emitted wavelengths shorter than 121.6 nm. This dropout is noticeable even in a weak spectrum from a very faint galaxy, and even if other spectral details cannot be discerned. For a nearby galaxy at $z = 0$, this dropout occurs in the far-ultraviolet, but for distant galaxies the dropout is red-shifted. For example, at $z = 9$:

$$1 + 9 = \frac{\lambda_{observed}}{121.6 \, nm}$$

Solving for the observed wavelength of the dropout: $\lambda_{observed} = 10 \times 121.6 \, nm = 1,216 \, nm = 1.2$ microns (μm).

Galaxies at the highest redshifts are detected in observations at wavelengths longer than 1.2 μm in the near-infrared, but they do not show up in optical images.

---

spectrum of these early galaxies has been cosmologically redshifted into the infrared (**Working It Out 23.2**). The near-infrared instruments installed on the Hubble Space Telescope (HST) and the infrared instruments on the Spitzer Space Telescope and the Herschel Space Observatory have provided images of these very young, highly redshifted objects (**Figure 23.12**). The images of the highest-redshift objects are small and faint, without the detail seen in the pictures of closer galaxies. Astronomers are excited by these images, because just the detection of these objects contributes to an understanding of when and how galaxies formed.

The first galaxies are thought to have formed by about $z \approx 15$ to 10, or by about 400 million to 550 million years after the Big Bang. For the second generation of stars, the physics of formation was more complicated than for the first, so astronomers need to account for heavy elements and dust from the first generation mixed into the halo, magnetic fields, and turbulence. The heavier elements carbon, oxygen, and iron cooled the gas, which then collapsed to the center of this larger dark matter halo, probably to a disk, and stars formed in the dark matter halo. These highest-redshift—that is, youngest—galaxies appear to be small, 20 times smaller than the Milky Way, which adds support for the bottom-up models of galaxy formation. Only in the past few years have astronomers been identifying galaxies at $z = 7$ to 10, so there are not so many yet to compare with somewhat older galaxies at $z = 2$ to 6.

Recall from Chapter 20 that the Local Group has small, faint dwarf galaxies orbiting the Milky Way. Streams of material from the dwarfs are falling onto the Milky Way, indicating that the Milky Way is still accreting mass. Recently, scientists have discovered many more of these galaxies. About a dozen of the faint dwarf galaxies are called **ultrafaint dwarf galaxies** because they are dim, only 1,000–100,000 times the Sun's luminosity. They contain mostly old, faint stars with low amounts of heavy elements. One of these small, ultrafaint dwarf galaxies is Segue 1. It contains only a few hundred stars and fewer heavy elements than any other observed galaxy, suggesting it may not have had much more star formation after its first stars formed. (This is unlike the additional bursts of star formation that occurred in regular dwarf galaxies.) The ultrafaint dwarfs may

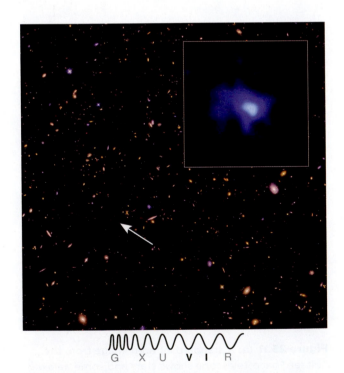

Figure 23.12 This image from the Hubble Space Telescope shows a high-redshift galaxy (arrow; enlarged in the inset). Spectroscopy with the Keck telescope in Hawaii indicates the galaxy is at redshift $z = 7.7$ and is only ~100 million years old.

have contributed to building the Milky Way's halo, and they may be the oldest galaxies around that have been involved in galaxy mergers. They may even be the fossil remains of the first galaxies or of the first minihalos. Because they are more massive than their luminosity suggests, dark matter dominates even these very old galaxies.

Parallels between Galaxy and Star Formation

Think back to the discussion of star formation in Chapter 15. Both star and galaxy formation involve the gravitational collapse of vast clouds to form denser, more concentrated structures. To help you with the comparison, this section describes some of the similarities and differences between the formation of galaxies and the formation of stars after the first generation.

Gravitational Instability In both star and galaxy formation, the collapse begins with a gravitational instability. Regions only slightly denser than their surroundings are pulled together by their own self-gravity. As the matter in these regions becomes more compact, gravity becomes stronger, and the collapse process snowballs. One key difference between galaxy and star formation is that for a galaxy to form, the dark matter clump must collapse rapidly enough to counteract the overall expansion of the universe itself.

Fragmentation The order of fragmentation differs between star and galaxy formation. In molecular clouds, first large regions begin to collapse, and then they fragment further to form individual stars. In contrast to this "top-down" star formation process, galaxy formation is "bottom-up": smaller structures collapse first and then merge to form galaxies and, eventually, assemblages of galaxies.

Compression, Heating, and Thermal Support As an interstellar molecular cloud collapses, its temperature climbs and the pressure in the cloud increases. The higher pressure would eventually be enough to prevent further collapse, except that the cloud core is able to radiate away thermal energy. That energy is the bright infrared radiation that enables astronomers to see star-forming cores. Compare this process with galaxy formation: As a dark matter clump collapses, the random velocities of its particles increase, and it too quickly reaches a point at which there is a balance between gravity and the random motions of the dark matter particles. However, dark matter is *not* able to radiate away energy, so once this balance is reached, the collapse of the dark matter is over. Only the normal matter within the cloud of dark matter is able to radiate away thermal energy and continue collapsing. That's why normal matter collapses to form galaxies, while dark matter remains in much larger dark matter halos.

As galaxies form, dark matter remains in extended halos. Dark matter may be the dominant form of matter in the universe, and it may determine the structure of galaxies; but dark matter can never collapse enough to play a role in the processes that shape stars, planets, or the interstellar medium.

Angular Momentum and the Formation of Disks Conservation of angular momentum is responsible for the formation of disk galaxies, just as it is responsible for the formation of the accretion disks around young stars and for the flatness of both the Milky Way and the Solar System. The origin of the angular momentum is different, though. Whereas turbulent motions within star-forming molecular clouds produce the net angular momentum for stellar disks, gravita-

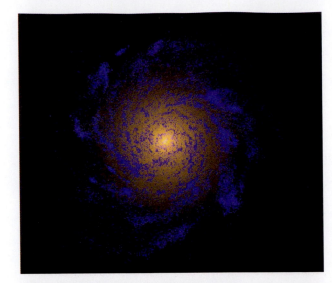

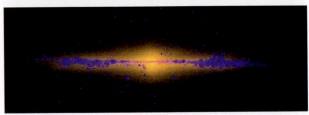

Figure 23.13 This supercomputer simulation of the formation of a Milky Way–sized spiral galaxy has been able to reproduce the small bulge and big disk using the Lambda-CDM model. At the bottom is an edge-on view of the galaxy. Blue colors indicate recent star formation, while older stars are redder.

tional interactions with nearby clumps are responsible for the angular momentum of the Milky Way.

Timescale The time a star spends as a protostar is a small fraction of the time it will spend on the main sequence. The formation of a galaxy is a slower process, taking from redshift 10–20 down to redshift 1–2, or about half of the age of the galaxy.

The End Product Once a stellar accretion disk forms, most of the matter moves inward and is collected into a star. In contrast, much of the matter in a spiral galaxy remains in the disk, as discussed in Chapter 19.

CHECK YOUR UNDERSTANDING 23.3

The first stars formed in the universe had _____ compared to the stars formed today. (a) more heavy elements and higher mass; (b) more heavy elements and lower mass; (c) fewer heavy elements and higher mass; (d) fewer heavy elements and lower mass

23.4 Galaxies Evolve

Galaxies continued to evolve hierarchically in the young universe, with smaller "protogalactic" fragments merging to form larger ones. The young universe was smaller and denser. Recall from Chapter 21 that the light from a galaxy currently at redshift z was emitted when the universe was $z + 1$ times smaller than it is today, with a volume $(1 + z)^3$ times smaller than its volume today. These early fragments and galaxies were closer together because the universe was smaller, and therefore mergers were more common. Computer simulations indicate that small concentrations of normal matter within the dark matter would have clumped and

Figure 23.14 A comparison of the Hubble classification of galaxies today (spirals on the top, barred spirals on the bottom, elliptical galaxies in the middle) with galaxies throughout the history of the universe. There were more irregular galaxies in the past, indicating that it took some time for spirals to form.

collapsed under their own gravity as they radiated and cooled, forming clumps of normal matter that ranged from the size of globular clusters to the size of dwarf galaxies.

In a large spiral galaxy such as the Milky Way, faint dwarf spheroidal galaxies (with dark matter) and the oldest globular clusters (without much dark matter) may be leftover protogalactic fragments. The gas collapsed to form a rotating disk as it cooled. A supercomputer simulation that included dark matter, gravity, star formation, and supernova explosions was able to reproduce a Milky Way–like galaxy with a large disk and a small bulge, as illustrated in **Figure 23.13**. Observationally, astronomers conduct "stellar archaeology" on the oldest parts of the Milky Way to understand better how the components are all assembled into a galaxy. For example, the oldest globular clusters may be 1 billion to 2 billion years older than the halo. Most of the stars in the Milky Way formed between 11 billion and 7 billion years ago. The disk and bulge formed at about that same time.

The Most Distant Galaxies

Figure 23.14 shows images of galaxies throughout the history of the universe. The galaxies observed in the very early universe, before about 11 billion years ago, are so faint that no structure can be seen. Observations of galaxies at about 11 billion years ago have shown visible structure that is much less regular than that of galaxies today. Even at 4 billion years ago, galaxies were much more irregular. Early irregular galaxies are merging galaxies, which fits with our ideas that in the early universe, when galaxies were closer together, there were more mergers. Most of today's galaxies conform to the Hubble classification, and only about 10 percent are irregular. However, 4 billion years ago, more than half of the galaxies were irregular, the number of elliptical and S0 galaxies was about the same, and there were many fewer spirals. This difference in galaxy types at different times suggests that it took time for spirals to form. These later mergers likely produced spiral galaxies over time.

The bottom-up hierarchical merging also may have triggered the formation of the supermassive black holes at the centers of galaxies. The first supermassive black holes, which power the distant quasars seen at $z = 6$ to 7 with masses of $10^9 M_{Sun}$, could have grown from the merging of minihalos with stellar black holes left after the first stars. Or they could have formed through the accretion of gas from the material between the galaxies during mergers of the first galaxies or through rapid collapse from hot, dense gas at the center of the first galaxies. In nearby galaxies, the mass of the supermassive black hole and the bulge properties are related, suggesting that the growth of the black hole and the bulge might have been linked when they were younger. Supermassive black holes could have grown even more massive from the mergers of large galaxies too. **Figure 23.15** shows a nearby galaxy with two supermassive black holes about 900 pc (3,000 light-years) apart, which are in the process of merging.

The hierarchical merging and growth of the supermassive black hole also affected the rates of star formation in the evolving galaxies. The tidal interactions between the galaxies and the collisions between gas clouds in the galaxies probably triggered many regions of star formation throughout the combined system. Star formation generally increased sporadically over time, including a rapid increase in the 200 million years between $z = 10$ and $z = 8$. As you can see in **Figure 23.16**, the star formation rate seems to have peaked around $z = 3$ (2.5 billion to 3 billion years after the Big Bang), before decreasing again to the current star formation ratio.

Figure 23.15 Chandra X-ray observations (red, orange, and yellow), combined with Hubble Space Telescope observations (blue and white) of NGC 6240 show that it has two black holes less than 1,000 pc apart. The black holes at the center of the white features will likely merge in about 100 million years.

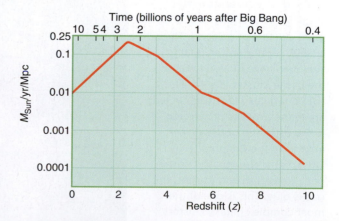

Figure 23.16 The rate of star formation has changed over time in the universe, peaking at about 2 billion to 3 billion years after the Big Bang.

(a)

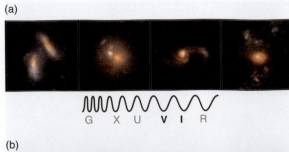

G X U **V** I R

(b)

G X U **V** I R

Figure 23.17 (a) These HST images show young, merging galaxies. From left to right, galaxies at 2.4 billion to 6.2 billion light-years from Earth merging at 11 billion to 7.5 billion years after the Big Bang. (b) These tidally interacting galaxies in the more nearby universe show severe distortions, including stars and gas drawn into long tidal tails.

Young galaxies in the process of merging are seen in HST images (**Figure 23.17a**). By observing galaxy mergers at different distances, astronomers can see how they differ at various times in the history of the universe. Large ellipticals are now thought to result from the merger of two or more spiral galaxies. The dark matter halos of the galaxies merge, and the stars eventually settle down into the blob-like shape of an elliptical galaxy. Elliptical galaxies are known to be more common in dense clusters, where mergers are likely to have been more frequent. Compare these young mergers with those of closer, older galaxies (Figure 23.17b—a computer simulation of such a merger is shown in Figure 6.29).

Just as galaxies merge, clusters of galaxies also merge. **Figure 23.18a** shows the high-speed collision and merging of two galaxy clusters in the Bullet Cluster. Images in visible light show the individual galaxies. Ordinary matter, mostly hot gas, is seen in X-rays (shown in red), and the distribution of the total mass can be found from the gravitational lensing produced by the clusters. The ordinary matter slowed down in the collision, but the dark matter (shown in blue) did not. This separation provides evidence for dark matter in galaxy clusters. The collision of four galaxy clusters is shown in Figure 23.18b. Clusters of galaxies also evolve hierarchically, growing from smaller structures to larger ones over time. As with galaxies themselves, younger, distant clusters are messier than older, nearby ones. This is additional evidence that the formation of structure in the universe was hierarchical.

Simulating Structure

Astronomers use the most powerful supercomputers available to simulate the universe. These simulations start with billions of particles of dark matter and use the most recent observations of the CMB. The simulations model the formation and evolution of dark matter clumps and halos, filaments and voids, small and large galaxies, and galaxy groups and clusters. These computations also simulate the flow of ordinary gas within these structures as stars form, and they are used to create images of what the universe should have looked like at different times (and different redshifts). These images are then compared to images of the actual universe. This comparison sets limits on the parameters of the universe: the

Figure 23.18 (a) The Bullet Cluster of galaxies, at a redshift of $z = 0.3$, represents a later stage in the merging of two giant clusters of galaxies. The smaller cluster on the right seems to have moved through the larger cluster like a bullet. (b) Four galaxy clusters (circled) are in the process of merging in the direction of the yellow arrows. In the Chandra X-ray Observatory image, the cooler gas is magenta and the hotter gas is blue. This is one of the most complex galaxy clusters observed.

(a)
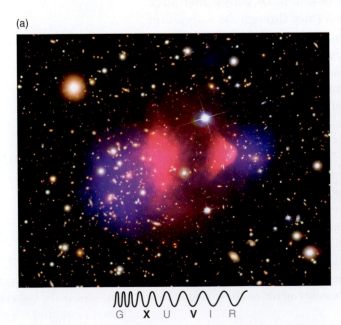
G **X** U V I R

(b)
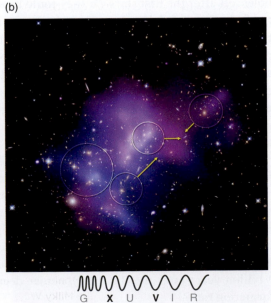
G **X** U V I R

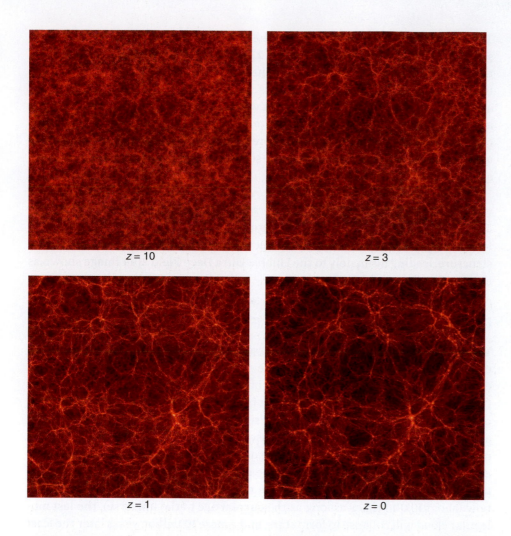

$z = 10$ $z = 3$

$z = 1$ $z = 0$

Figure 23.19 The Bolshoi supercomputer simulation of the formation of very large-scale structure in a universe filled with cold dark matter. The growth of filaments and voids is seen in these images.

amount of mass, for example, or the type of dark matter. If the inputs to the simulation are correct, the two sets of images should look very similar. If not, the two sets of images look very different.

In 2009, a Lambda-CDM simulation called the Bolshoi was run on NASA supercomputers. The simulation shows that slight variations in density after inflation led to higher-density regions that became the seeds for the growth of structure (**Figure 23.19**). During the first few billion years, dark matter fell together into structures comparable in size to today's clusters of galaxies. The spongelike filaments, walls, and voids became well defined later. Zooming in on some simulated filaments and voids shows a cluster of galaxies (**Figure 23.20**).

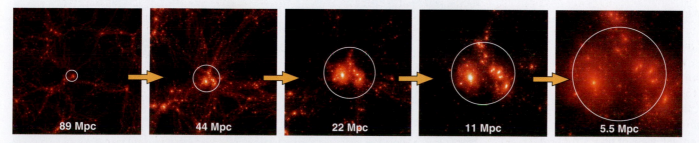

89 Mpc   44 Mpc   22 Mpc   11 Mpc   5.5 Mpc

Figure 23.20 Simulations enable astronomers to model structure at different size scales. Each image zooms in more to show smaller structures. Dark matter halos are seen as dense blobs in the images. The smallest blob in the last image could become a giant spiral galaxy like the Milky Way.

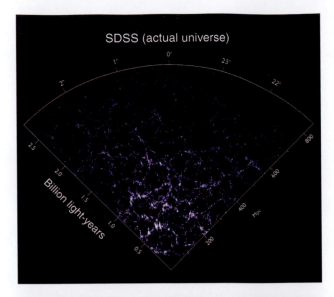

SDSS (actual universe)

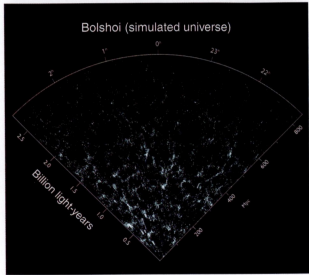

Bolshoi (simulated universe)

Figure 23.21 The large-scale structure of dark matter halos produced by the Bolshoi simulation (bottom) is remarkably similar to the distribution of galaxies observed in the Sloan Digital Sky Survey (SDSS; top).

Figure 23.22 A schematic view of how structure formed in the universe, from smaller systems to larger ones.

The similarities between the results of the models and observations of large-scale structure are quite remarkable. **Figure 23.21** compares the simulated view with the observed slice of the universe from the Sloan Digital Sky Survey. Only simulations with certain combinations of mass, CMB variations, types of dark matter, and values for the cosmological constant will produce structure similar to what is actually observed. This is a very important result. Models contain assumptions consistent with observational and theoretical knowledge of the early universe, and they predict the formation of large-scale structure that is similar to what is actually seen in today's universe.

The precision cosmology developed over the past two decades has given astronomers a detailed model of the universe in space and time, so that galaxy evolution can now be reliably sequenced. **Figure 23.22** neatly summarizes the galaxy formation process, in which smaller objects form first and merge into ever-larger structure, leading ultimately to the Hubble Ultra Deep Field 2014 image shown as the opening figure of this chapter.

Predictions about the Deep Future

What does the future hold for the universe and its structure? Scientists use well-established physics to speculate how the existing structures in the universe will evolve over a very long time. The study of galaxies and cosmology in the distant future will be much more difficult, if not impossible. In 90 billion years or so, the only visible galaxy will be the one that resulted from the merger of the Milky Way and Andromeda (see Chapter 20). Because of the acceleration of the universe, the other galaxies will be too far away to be detectable from here. In a trillion years, even the wavelength of the CMB will be greater than the size of the observable universe, and the CMB won't be visible either. In 100 trillion (10^{14}) years from now (about 10,000 times as long as the current age of the universe), the last molecular cloud will collapse to form stars, and a mere 10 trillion years later the least massive of these stars will evolve to form white dwarfs.

At that time, most of the normal matter in the universe will be locked up in degenerate stellar objects: brown dwarfs, white dwarfs, and neutron stars. Unless

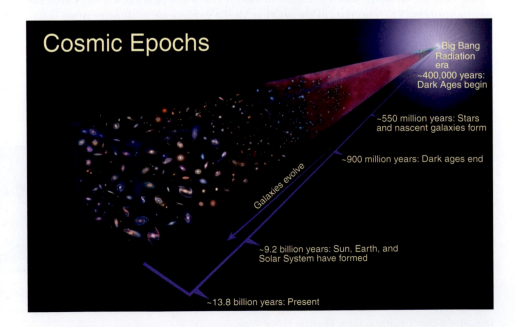

Cosmic Epochs

Big Bang
Radiation era
~400,000 years: Dark Ages begin

~550 million years: Stars and nascent galaxies form

~900 million years: Dark ages end

Galaxies evolve

~9.2 billion years: Sun, Earth, and Solar System have formed

~13.8 billion years: Present

there is a Big Rip (where everything is torn apart by expansion), every condensed object, including white dwarfs and neutron stars, will ultimately decay via quantum effects. Eventually, the only significant remaining concentrations of mass will be black holes. These will range from small ones with the masses of single stars to supermassive black holes that grew to the size of galaxy clusters. These black holes will slowly evaporate into elementary particles through the emission of Hawking radiation (see Chapter 18). A black hole with a mass of a few solar masses will evaporate into elementary particles in 10^{65} years, and galaxy-sized black holes will evaporate in about 10^{98} years. By the time the universe reaches an age of 10^{100} years, even the largest of the black holes will be gone. A universe vastly larger than our current one will contain little but photons with colossal wavelengths, neutrinos, electrons, positrons, and other waste products of black-hole evaporation. The universe will continue to expand forever—into the long, cold, dark night of eternity. Ultimately, the universe will be structureless and empty except for any residual uniform dark energy.

CHECK YOUR UNDERSTANDING 23.4

We expect the kinds of galaxies that we see at a redshift of $z = 4$ to be: (a) much like what we see today; (b) smaller and much more irregular looking than today; (c) far more numerous but with more spiral galaxies; (d) larger versions of what we see today.

Origins

We Are the 4 or 5 Percent

The very first stars initiated post–Big Bang nucleosynthesis, which chemically enriched the universe with elements heavier than lithium, including the elements necessary to form planets and life on Earth. The first stars and galaxies "lit up" the universe and brought it out of the Dark Ages, thanks to the glow of their ordinary matter. But this ordinary matter is not the primary constituent of the universe. In their study of the largest structures in the universe—the galaxies and groups and clusters of galaxies—astronomers have realized that these objects are not composed of the same stuff as stars and planets; instead, they are dominated by dark matter.

As you have seen in these past few chapters, several types of independent observations suggest that dark matter accounts for about 85 percent of all the matter in the universe. But matter is only about 32 percent of the universe, so dark matter is about 27 percent of today's universe. Dark matter dominated over normal matter at the time the first stars and galaxies formed. Dark matter dominated the evolution of the galaxies as they went through mergers to become the larger systems seen today. But despite decades of study, astronomers don't know exactly what dark matter is.

As the universe evolved, dark energy became more important to the structure of the universe. When the universe was younger and galaxies were forming, it was dominated by matter. The expansion of the universe was slowing because of the pull of gravity from all of this matter. But about 5 billion to 7 billion years ago, dark energy began to dominate over matter in the universe, and the expansion of the universe accelerated. Dark energy currently makes up about 68 percent of the mass and energy of the universe—more than 2 times as much as the dark matter—and astronomers don't know exactly what it is either.

Ordinary matter is only about 4.9 percent of the mass and energy of the present-day universe. Most of what astronomers have studied since people first looked at the sky is this 4.9 percent. The parts of the local universe that are important to life on Earth—the Sun, its planets and their moons, and the local environment in the Milky Way—are composed of this 4.9 percent. The matter that constitutes us is a surprisingly small part of the universe.

Astronomers report on the discovery that the Milky Way is part of a huge supercluster of galaxies.

Welcome to Laniakea, Your Galactic Supercluster Home

By **AMINA KHAN, Los Angeles Times**

Home, supercluster home. Astronomers studying the skies have found that our galaxy, the Milky Way, is part of an enormous supercluster of galaxies that has been named Laniakea, which comes from the Hawaiian words for "immeasurable heaven."

Laniakea, described in the journal *Nature*, stretches some 500 million light-years across and holds 100,000 galaxies with the mass of a hundred million billion Suns. This supercluster, scientists said, provides a whole new lens on the mysterious internal dynamics of such giant clusters, the biggest structures in the universe (see Figure 23.3).

The universe has a certain order, defined by gravity: Planets circle stars, which spin around in galaxies, which hang out together in large clusters, which are nestled in gargantuan superclusters. These superclusters give the cosmos its large-scale structure—matter concentrates in massive, dense nodes and thin tendrils connect those nodes across the empty spaces in between.

Even though superclusters define the nodes of this cosmic web, scientists know very little about how they work. Until now, they didn't even know where exactly our own Milky Way lives—just that we're in a local group of at least 54 galaxies, which is itself part of a larger conglomeration long called the Virgo (or Local) Supercluster.

It's a little like knowing your street address without knowing which city you live in—which sounds like a pretty basic problem, but has proved deeply challenging to solve.

"This is what I've been doing for a long time, trying to understand where we live," said lead author R. Brent Tully, an astronomer at the University of Hawaii in Honolulu.

The giveaway is gravity. Gravity holds superclusters together, so watching the movements of galaxies speeding around their neighborhoods could help define a supercluster's borders.

Here's the problem. The universe is expanding faster and faster, due to a mysterious force called dark energy. This mucks up the speed readings: Scientists don't know how much of a particular galaxy's movement is due to the inward pull from the gravity of a cluster, and how much is due to the outward pull of the universe expanding faster and faster.

So a team of cosmic cartographers used the Superflows-2 database, a catalog of the motions of galaxies, to tease apart these movements. For each of more than 8,000 galaxies, they subtracted the expansion-related movement from the overall motions—leaving only the local, gravitationally driven movement.

The researchers found that the galaxies were flowing along these long, beautiful lines, many of them toward a gravitational dense basin of galaxies known as the Great Attractor. The boundary between those flowing toward this spot and those flowing away marked the edges of the galaxy supercluster.

It's like watching a watershed where the flows divide, Tully said.

"If you're standing at a certain place at the divide, water might be going in two different directions—one going into the Mississippi basin, one going off to the Great Lakes," Tully said.

The Milky Way sits in a backwater among these flowing galaxies, near a far edge of the supercluster. In fact, the whole Virgo Supercluster is just a small fraction of the Laniakea Supercluster.

"The Local Supercluster, we now see quite clearly, is just an appendage on Laniakea," Tully said. "It's like a suburb in a city of a metropolitan area."

Elmo Tempel, an astronomer at Tartu Observatory in Estonia who was not involved in the paper, praised the findings.

"This is the first clear definition of a supercluster," Tempel wrote in a commentary. "The downside of it is that it requires dynamical information that is available only for the nearby universe."

Understanding the dynamics of our own supercluster could help shed light on the large-scale forces at work in the universe, Tully said, particularly dark matter, the mysterious unseen mass whose gravity holds such enormous structures together, and dark energy, the strange force that is tearing them inevitably apart.

1. How large is Laniakea in megaparsecs?
2. Why was it hard to figure out the shape and extent of the Milky Way's supercluster?
3. Are the Milky Way and the Local group in the center of this supercluster?
4. Why will it be difficult to map out more distant superclusters using a similar method?
5. Watch the video at http://www.nature.com/nature/videoarchive/laniakea/index.html for a 3-D view of Laniakea. How do astronomers set the boundaries of the different clusters of galaxies?

Summary

Galaxies are not distributed uniformly, but rather are clumped into groups, clusters, superclusters, and walls. Galaxies develop because of gravitational instabilities in the presence of cold dark matter. The first stars formed in minihalos of dark matter, while the first galaxies formed later, in larger dark matter halos. Over time, smaller galaxy fragments merged to form larger galaxies. Mergers still happen today. Dark matter and dark energy are responsible for the formation of structure and the future of the universe. Ordinary matter, which makes up stars, planets, and us, is only 15 percent of the total mass in the universe and, once dark energy is included, only 4 or 5 percent of all of the universe. Dark energy is playing an increasingly dominant role in the universe and will shape its future.

LG 1 **Describe the distribution of galaxies in the universe.** Galaxies are hierarchically gathered into groups, clusters, superclusters, and larger structures. The walls surround voids in which very few galaxies are present. Dark matter dominates the mass of galaxy groups and clusters.

LG 2 **Explain how the structure that began to form shortly after the Big Bang evolved to the large-scale structure of the universe today.** Structure formed in the universe as slight variations in the density of the dark matter emerging from the Big Bang grew. These "seeds" then collapsed under

the force of gravity, pulling in normal matter as well. Structure formed "bottom-up."

LG 3 **Describe the formation of the first stars and the first galaxies.** The formation of the first stars occurred in dark matter minihalos, rather than in clouds of dust and gas. Because of the absence of heavy elements in these dark matter halos, star formation and evolution were different than in the nearby universe. Gravitationally bound groupings of stars formed, which were the earliest galaxies. Radiation from the first stars and their supernovae affected the growth of the first galaxies. Radiation from the first stars, galaxies, and black holes ended the Dark Ages.

LG 4 **Explain how observations of galaxies at many different redshifts illustrate the evolution of the large-scale structure of the universe.** Larger structures form from smaller structures. The earliest galaxies merged to become larger galaxies, which in turn accumulated into clusters. The clusters grew hierarchically, through mergers, to become larger clusters, superclusters, and walls. Distant young galaxies look very different from the nearby galaxies in the present-day universe: they are smaller, fainter, and more likely to be merging.

? UNANSWERED QUESTIONS

- How do supermassive black holes form? The detection of quasars at very high redshifts indicates that supermassive black holes formed early in the Dark Ages, but it is not known exactly how or how much they may have contributed to reionization.

- How can astrophysicists gain a better understanding of dark matter? Will there be a way to detect dark matter by means other than gravity? If it is a particle, is it stable against decay? Experiments in particle accelerators on the ground and observations of Local Group halos in space have put some limits on the type of dark matter particles that might be out there, but there is still not an answer.

Questions and Problems

Test Your Understanding

1. Place the following in order of size, from smallest to largest.
   a. a galaxy
   b. star clusters
   c. the Local Group
   d. a wall
   e. Virgo Cluster
   f. Laniakea
   g. a star

2. The dominant force in the formation of galaxies is
   a. gravity.
   b. angular momentum.
   c. the electromagnetic force.
   d. the strong nuclear force.

3. Larger galaxies form from the merging of small protogalaxies. This process is similar to the formation of
   a. stars.
   b. planets.
   c. molecular clouds.
   d. asteroids.

4. Which is not a characteristic difference between cold and hot dark matter?
   a. temperature
   b. composition
   c. the way they clump under the influence of gravity
   d. mass

5. Gravitational lenses can be used to find
   a. dwarf galaxies near the Milky Way.
   b. dust and gas in the voids.
   c. the masses of galaxy clusters.
   d. the structure of Laniakea.

6. If dark energy is constant, in the far distant future the universe
   a. will be cold and dark.
   b. will be bright and hot.
   c. will collapse and re-form.
   d. will be the same as it is now, on large scales.

7. What is the primary difference between galaxy groups and galaxy clusters?
   a. how tightly they are bound by gravity
   b. the size of the largest galaxy
   c. the total mass of the galaxies
   d. there is no dark matter in galaxy groups

8. Once the redshift of a galaxy has been found, its _____ is also known.
   a. mass
   b. velocity
   c. distance
   d. both b and c

9. Galaxy formation is similar to star formation because both
   a. are the result of gravitational instabilities.
   b. are dominated by the influence of dark matter.
   c. end with the release of energy through fusion.
   d. result in the formation of a disk.

10. Dark matter clumps stop collapsing because
    a. angular momentum must be conserved.
    b. they are not affected by normal gravity.
    c. fusion begins, and radiation pressure stops the collapse.
    d. the particles are moving too fast to collapse any further.

11. Elliptical galaxies come from
    a. the gravitational collapse of clouds of normal and dark matter.
    b. the collision of smaller elliptical galaxies.
    c. the fragmentation of large clouds of normal and dark matter.
    d. the merging of two or more spiral galaxies.

12. Which of the following statements about the Dark Ages of the universe is *false*?
    a. The first stars began forming during the Dark Ages.
    b. The end of the Dark Ages coincided with reionization.
    c. The Dark Ages lasted from 200 million to 600 million years after the Big Bang.
    d. During the Dark Ages, photons could travel freely through the universe.

13. Astronomers have never observed a star that has no elements heavier than lithium. What does this imply about the first stars?
    a. They must have died before galaxies were fully formed.
    b. The first stars did not form until after galaxies formed.
    c. The first stars must have had very low masses.
    d. The first stars must have been enriched in heavy elements.

14. Reionization of the neutral gas in the universe occurred because of the
    a. decay of dark matter particles.
    b. emission of neutrinos by the first stars that formed.
    c. release of jets of charged particles from supermassive black holes.
    d. radiation from the first stars, supernovae, and black holes that formed.

15. Place the following in increasing order.
    a. the fraction of the universe that is stars, planets, dust, and gas
    b. the fraction of the universe that is dark energy
    c. the fraction of the universe that is dark matter

Thinking about the Concepts

16. Suppose you were able to view the early universe at a time when galaxies were first forming. How would it be different from today's universe?

17. Is it likely that voids are filled with dark matter? Why or why not?

18. Imagine that there are galaxies in the universe composed mostly of dark matter, with relatively few stars or other luminous normal matter. If this were true, how might you learn of the existence of such galaxies?

19. How are the processes of star formation and galaxy formation similar? How do they differ?

20. What is the origin of large-scale structure?

21. Why is dark matter essential to the galaxy formation process?

22. Which of the following is the correct evolutionary sequence? (a) Small star clusters formed first, which were bound together into galaxies, which were later bound together in clusters and superclusters; or (b) supercluster-sized regions collapsed to form clusters, which then later collapsed to form galaxies, which formed small clusters of stars. Justify your answer.

23. Why does the current model of large-scale structure require dark matter?

24. What is the difference between a galaxy cluster and a super-cluster? Is our galaxy part of either? How do we know this?

25. How does a roughly spherical cloud of gas collapse to form a disk-like, rotating spiral galaxy?

26. What are some of the observational signs that dark matter exists? Explain why this evidence challenged earlier theories and forced astronomers to change their minds about the existence of matter they could not see.

27. Using the current model of galaxy formation, describe how galaxies should appear as you look further back in time. Are the features you described observed?

28. Why do astronomers think that the dark matter in the universe must be mostly cold, rather than hot, dark matter?

29. Describe the process of structure formation in the universe, starting at recombination and ending today.

30. Why do scientists think that gravity, and not the other funda-mental forces, is responsible for large-scale structure?

Applying the Concepts

31. Figure 23.3a shows the redshifts and velocities for a large number of galaxies. Find the average recession velocity for the galaxies in the wall indicated by the line labeled "Walls."

32. The theory of cosmology assumes that on large scales, the structure in the universe is uniform no matter where you look. Maps of structure, like the ones shown in Figure 23.3, support this assumption. Does the presence of large masses like Laniakea violate this principle? Explain your answer.

33. As clumps containing cold dark matter and normal matter collapse, they heat up. When a clump collapses to about half its maximum size, the increased thermal motion of particles tends to inhibit further collapse. Whereas normal matter can overcome this effect and continue to collapse, dark matter cannot. Explain the reason for this difference.

34. In previous chapters, we painted a fairly comprehensive picture of how and why stars form. Why, then, is it difficult to model the star formation history of a young galaxy? Is this difficulty a failure of scientific theories?

35. If 300 million neutrinos fill each cubic meter of space, and if neutrinos account for only 5 percent of the mass density (including dark matter) of the universe, estimate the mass of a neutrino.

36. What are the approximate masses of (a) an average group of galaxies, (b) an average cluster, and (c) an average supercluster?

37. The lifetime of a black hole varies in direct proportion to the cube of the black hole's mass. How much longer does it take a supermassive black hole of 3 million M_{Sun} to decay compared to a stellar black hole of 3 M_{Sun}?

38. Knowing what elliptical galaxies are made of, estimate how old they must be. Knowing that ellipticals form via mergers of spirals, and knowing when galaxies first formed, estimate how long it took to complete the merging events that formed the elliptical galaxies seen today.

39. The chapter-opening figure is the Hubble Ultra Deep Field 2014.
    a. Explain how you might identify which galaxies might be at higher redshifts.
    b. The image shows about one 10-millionth of the entire sky. Estimate the number of galaxies in the entire sky. Explain how you used the cosmological principle in the calculation.

40. The Bullet Cluster image in Figure 23.18a shows the collision of two galaxy clusters. Estimate the number of galaxies you can see in each cluster.

41. It is likely that the initial fluctuations leading to large-scale structure arose from quantum fluctuations in the early universe. How would the universe look different today if those fluctuations were 10 times bigger? What about 10 times smaller?

42. Examine Figure 23.16. Is the early universe on the left or the right in this graph? (Alternatively, you could wonder if "now" was on the left or right in the graph.) Compare the star forma-tion rate today with the star formation rate at the peak. How much more star formation occurred during the peak than occurs now?

43. Figure 23.21 shows real data in the top panel and simulated data in the bottom panel. These two panels are not in exact agreement. Do these differences indicate a significant prob-lem in the simulation's ability to represent reality? Why or why not?

44. Compare galaxies of redshift $z = 0.5$, $z = 4$, and $z = 8$, respectively. About how old was the universe when the light was emitted from each of the galaxies? At what spectral wavelengths would you see the "dropout" of the spectrum? Can you observe these dropouts from the ground or do you need a telescope in space?

45. Suppose a dwarf galaxy is orbiting a giant elliptical galaxy at the center of a cluster at a distance of 4 Mpc and a speed of 800 km/s. Estimate the core mass of the cluster.

USING THE WEB

46. a. Go to the website for the Sloan Digital Sky Survey III (SDSS-III; http://sdss3.org/index.php), which has made a three-dimensional map of the sky. The 2012 video fly-through can be accessed at http://sdss3.org/press/dr9.php, and new ones may be posted as the project acquires more data. Why did the SDSS-III scientists make this map? What were the goals of the SDSS-III project? What do astronomers learn from this fly-through?

b. Go to the website for the Dark Energy Survey (http://www.darkenergysurvey.org/). What are the science goals of this project? What are the observations? Are there any results yet?

47. Go to the website for the Bolshoi simulation (http://hipacc.ucsc.edu/Bolshoi). What is this simulation? Click on "Videos" to watch the Bolshoi and some other videos that compare observed and simulated universes. What do astronomers learn from these simulations? Are there any results yet from the AGORA project?

48. Go to the website for "Galaxy Crash" (http://burro.cwru.edu/JavaLab/GalCrashWeb), a Java applet that lets you run simple models of galaxy collisions. Read the sections under "Background" and then click on "Lab." Pick an exercise from the list (or go to the one suggested by your instructor), and work through the questions.

49. Use your favorite search engine to find the highest-redshift galaxy, quasar, and GRB observed to date. Why are astronomers interested in finding objects at higher and higher redshifts? Why is it also important for astronomers to estimate the relative frequency of such objects compared to their frequency at $z = 6$ or $z = 2$?

50. Go to the website for the new ALMA telescope (http://almaobservatory.org/en). What is unique about this telescope? How will it study the Dark Ages? Why is this telescope also going to do a "deep-field" project? How will this project be different from the deep-field observations with the Hubble Space Telescope? Look at the items under "ALMA Latest News" and "Press Releases." Are there any reports about galaxy formation?

smartwork

If your instructor assigns homework in SmartWork, access your assignments at digital.wwnorton.com/astro5.

EXPLORATION

The Story of a Proton

digital.wwnorton.com/astro5

Now that you have surveyed the current astronomical understanding of the universe, you are prepared to put the pieces together to make a story of how you came to be sitting in your chair and reading these pages. It is valuable to take a moment to work your way backward through the book, starting at the Big Bang and reviewing all the intervening steps that had to occur back to the beginning of the book, which began with looking at the sky.

1 In the Big Bang, how did a proton form?

_____

_____

2 How might that proton have become part of one of the first stars?

_____

_____

3 Suppose that proton later became part of a carbon atom in a 4-M_{Sun} star. Through what type of nebula would it have passed before returning to the interstellar medium?

_____

4 Suppose that carbon atom then became part of the molecular-cloud core forming the Sun and the Solar System. What two physical processes dominated the core's collapse as the Solar System formed and that carbon atom became part of a planet?

_____

_____

5 Beginning with the Big Bang, create a timeline that traces the full history of a proton that becomes a part of the nucleus of a carbon atom in you.

_____

_____

24 Life

Throughout history, many people have wondered if we are alone in the universe. But it is only in recent times that science has been able to address the underlying questions that must be answered, such as: How common are planets? What is the range of conditions under which life can thrive? How does life begin and evolve? Even answering the question "What is life?" is surprisingly complicated. Answering these questions is part of a science called **astrobiology**— the study of the origin, evolution, distribution, and future of life in the universe. Throughout this book, the "Origins" section in each chapter has discussed how astronomers think about these questions. In this chapter, we expand on some of these topics and provide a more systematic overview of how astronomers think about life in the universe and how they search for signs of it.

LEARNING GOALS

By the conclusion of this chapter, you should be able to:

LG 1 Explain our current understanding of how and when life began on Earth and how it has evolved.

LG 2 Explain how life is a structure that has evolved through the action of the physical and chemical processes that shape the universe.

LG 3 Describe the locations in our Solar System and around other stars where astronomers think life might be possible.

LG 4 Describe some of the methods used to search for intelligent extraterrestrial life.

The Allen Telescope Array searches for signs of intelligent life. ▶ ▶ ▶

Are we alone in the universe?

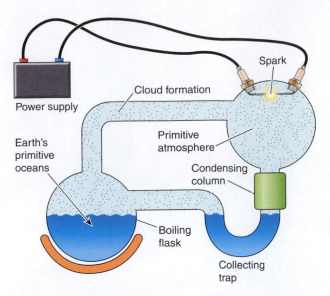

Figure 24.1 The Urey-Miller experiment was designed to simulate conditions in an early-Earth atmosphere.

(a)

(b)

Figure 24.2 (a) Life on Earth may have arisen near ocean hydrothermal vents like this one. Similar environments might exist elsewhere in the Solar System. (b) Living organisms around hydrothermal vents, such as the giant tube worms shown here, rely on hydrothermal rather than solar energy for their survival.

24.1 Life Evolves on Earth

What is **life**? Many scientists suggest that there is no single definition of life that would encompass all the different forms of life that may exist in the universe. To date, we have discovered only a single example of life: that found here on Earth. Life on Earth may be very different from life found in other places in the universe. Indeed, even comparing varied organisms on Earth leads to complications in the definition of life. Viruses, for example, meet some criteria for life, but not others. A complete definition of life may also have to take into account life-forms that have not yet been discovered. From studies of known life, we conclude that like planets, stars, and galaxies, life is a structure that has evolved in the universe. Life draws energy from the environment in order to survive and reproduce. On Earth, all life involves carbon-based chemistry and uses liquid water as its biochemical solvent, while specific biological molecules such as ribonucleic acid (RNA) and deoxyribonucleic acid (DNA) enable life to reproduce and evolve. In this section, we will briefly review what is known about the origin and evolution of life on Earth.

The Origin of Life on Earth

How did life begin on Earth? Recall from Chapter 7 that Earth's secondary atmosphere was formed in part by carbon dioxide and water vapor emitted by volcanoes. Comets and asteroids likely added large quantities of water, methane, and ammonia to the mix. Liquid water is considered essential for any terrestrial-type life to get its start and evolve because it is an effective solvent that can move other atoms and molecules around, therefore making them more accessible to cells. Early Earth had abundant sources of energy, such as lightning and ultraviolet solar radiation, that fragmented these simple molecules. These fragments subsequently reassembled into molecules of greater mass and complexity. Some of these were organic molecules; that is, molecules that contained carbon. Rain carried the heavier molecules out of the atmosphere into Earth's oceans, forming a primordial soup.

In 1952, chemists Harold Urey (1893–1981) and Stanley Miller (1930–2007) attempted to create conditions similar to these early-Earth conditions. Using equipment illustrated in **Figure 24.1**, they placed water in a sterilized laboratory jar to represent the ocean, and then added methane, ammonia, and hydrogen as a primitive atmosphere; electric sparks simulated lightning as a source of energy. Within a week, the Urey-Miller experiment yielded molecules associated with life: amino acids and components of nucleic acids. Proteins, the structural molecules of life, are made of 20 amino acids. Eleven of these were synthesized in the Urey-Miller experiment. Nucleic acids are the precursors of RNA and DNA.

Additional sealed samples from this old experiment were examined 50 years later. These samples had added hydrogen sulfide to the "primitive atmosphere." When the samples were analyzed, 23 amino acids were found. This suggests that hydrogen sulfide, which would have come from volcanic plumes in the early Earth, was important. More recent experiments with carbon dioxide and nitrogen as the primitive atmosphere have produced results similar to those of Urey and Miller. A feasible atmospheric composition with an energy source can produce significant quantities of amino acids and other substances important to life.

From laboratory experiments such as these, scientists have developed various models to explain how life might have begun in an early-Earth environment.

However, the details of how these precursor molecules evolved into the molecules of life are not yet clear. Some biologists think life began in the ocean depths, where volcanic vents provided the hydrothermal energy needed to create the highly organized molecules responsible for biochemistry (**Figure 24.2**). Others think that life originated in tide pools, where lightning and ultraviolet radiation supplied the energy (**Figure 24.3**). In either case, short strands of molecules that could replicate themselves may have formed first, later evolving into RNA, and finally into DNA, the huge molecule that serves as the biological "blueprint" for self-replicating organisms.

A few scientists have suggested that life on Earth may have been "seeded" from space in the form of microorganisms brought here by meteoroids or comets. However, there is no scientific evidence at this time to support the "seeding" hypothesis. In addition, while this hypothesis might explain how life came to Earth, it does not explain how life itself began.

When Life Began

If life did indeed get its start in Earth's oceans, when did it happen? Recall from Chapter 7 that Earth was bombarded by Solar System debris for several hundred million years after it formed roughly 4.6 billion years ago. These conditions might have been too harsh for life to form and evolve on Earth. Once the bombardment abated and oceans formed, the opportunities for life to begin greatly improved. It seems that terrestrial life quickly took advantage of this more favorable environment. Scientists debate whether carbonized material in Greenland rocks dating back 3.65 billion to 3.85 billion years provides indirect evidence of early life. Stronger and more direct evidence for early life appears in the form of fossilized **stromatolites** (masses of simple microorganisms) that date back about 3.5 billion years. Fossilized stromatolites have been found in western Australia and southern Africa, and living examples still exist today (**Figure 24.4**). This evidence suggests that the earliest life formed less than a billion years after the formation of the Solar System, and within 500 million years of the end of the late heavy-bombardment period.

All life on Earth shares a similar genetic code that originated from a common ancestor. Close comparison of DNA of different species enables biologists to trace backward to the time when different types of life first appeared on Earth and to identify the species from which these life-forms evolved. The earliest organisms were **extremophiles**—life-forms that not only survive, but thrive, under extreme environmental conditions. Extremophiles include organisms such as thermophiles, which flourish in water temperatures as high as 120°C and occur in the vicinity of deep-ocean hydrothermal vents, like the one shown in Figure 24.2a. Other extremophiles thrive in conditions of extraordinary cold, salinity, pressure, dryness, acidity, or alkalinity. Scientists today study extremophiles in boiling-hot sulfur springs in Yellowstone National Park, in salt crystals beneath the Atacama Desert in Chile, at the bottoms of glaciers, in ice fields in the Arctic, and in other extreme environments (**Figure 24.5**).

Among the early life-forms was an ancestral form of **cyanobacteria**, single-celled organisms otherwise known as *blue-green algae*. These microorganisms form extensive sheets on the

Figure 24.3 Life may have begun in tide pools, where lightning and ultraviolet light provide energy for chemical processes.

Figure 24.4 These modern-day stromatolites are growing in colonies along an Australian shore.

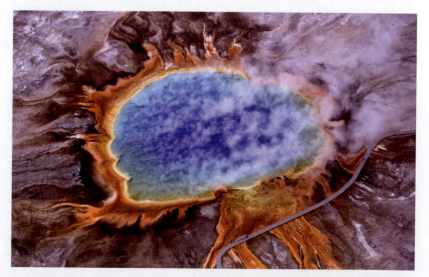

Figure 24.5 These thermophiles in the Grand Prismatic Spring in Yellowstone National Park live in temperatures of 70°C. The different colors result from different amounts of chlorophyll.

(a)

(b)

Figure 24.6 (a) Cyanobacteria today form sheets on lakes and other bodies of water. (b) Under a microscope, the individual microorganisms are visible.

surface of bodies of water, as shown in **Figure 24.6a**. Under a microscope, it becomes clear that they are colonies of individual microorganisms, as seen in Figure 24.6b. Cyanobacteria **photosynthesize**, using sunlight and carbon dioxide as food and generating oxygen as a waste product. Initially, the highly reactive oxygen produced by cyanobacteria was quickly removed from Earth's atmosphere by oxidation, or rusting, of surface minerals. Once the exposed minerals could no longer absorb oxygen, atmospheric levels of oxygen began to rise. Oxygenation of Earth's atmosphere and oceans began about 2 billion years ago, and the current level was reached only about 250 million years ago, as shown in **Figure 24.7**. Without cyanobacteria and other photosynthesizing organisms, Earth's atmosphere would be as oxygen-free as the atmospheres of Venus and Mars.

Biologists comparing DNA sequences find that terrestrial life is divided into two types: prokaryotes and eukaryotes. Prokaryotes, which include Bacteria and Archaea, are simple organisms that consist of free-floating DNA inside a cell wall; as shown in **Figure 24.8a**, they lack both cell structure and a nucleus. Eukaryotes, which form the cells in animals, plants, and fungi, have a more complex form of DNA contained within the cell's membrane-enclosed nucleus, illustrated in Figure 24.8b. The first eukaryote fossils date from about 2 billion years ago, coincident with the rise of free oxygen in the oceans and atmosphere, although the first multicellular eukaryotes did not appear until a billion years later.

Increasing Complexity

Scientists have used DNA sequencing to establish what is known as the "phylogenetic tree of life," shown in **Figure 24.9**. This complex tree describes the evolutionary interconnectivity of all species of Bacteria, Archaea, and Eukarya and has revealed some interesting relationships. For example, Archaea were initially thought to be the same as Bacteria, but genetic studies show they diverged long ago, and the Archaea have genes and metabolic pathways that are more similar to those of Eukarya than to those of Bacteria. On the macroscopic scale, the phylogenetic tree places animals closest to fungi, which branched off the evolutionary tree after slime molds and plants.

Living creatures in Earth's oceans remained much the same—a mixture of single-celled and relatively primitive multicellular organisms—for more than 3 billion years after the first appearance of terrestrial life. Between 540 million and 500 million years ago, the number and diversity of biological species increased spectacularly. Biologists call this event the **Cambrian explosion**. The trigger of this sudden surge in biodiversity remains unknown, but possibilities include rising oxygen levels, an increase in genetic complexity, major climate change, or a combination of these. The "Snowball Earth" hypothesis suggests that before the Cambrian explosion, Earth was in a period of extreme cold between about 750 million and 550 million years ago, and was covered almost entirely by ice. During this period of extreme cold, predatory animals died out, making it easier for new species to adapt and thrive. Another possibility is that the marked increase in atmospheric oxygen (O_2) would have been accompanied by a corresponding increase in stratospheric ozone (O_3), which shields Earth's surface from deadly solar ultraviolet radiation. With this protective ozone

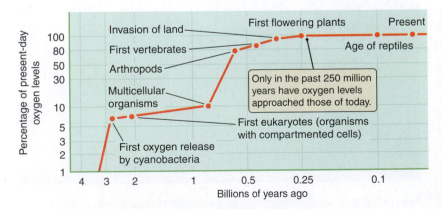

Figure 24.7 The amount of oxygen in Earth's atmosphere has built up over time as a result of cyanobacteria and plant life on the planet.

layer in place, life was free to leave the oceans and move to land. *Tiktaalik*, a fish with limblike fins and ribs, was an animal in a midevolutionary step of leaving the water for dry land, as shown in the artist's illustration in **Figure 24.10**.

The first plants appeared on land about 475 million years ago. Large forests and insects go back 360 million years. The age of dinosaurs began 230 million years ago and ended abruptly 65 million years ago, when a small asteroid or comet collided with Earth. The collision threw so much dust into the atmosphere that the sunlight was dimmed for months, causing the extinction of more than 70 percent of all existing plant and animal species. Mammals were the big winners in the aftermath. Primates evolved from the last ancestor common with other mammals about 70 million years ago. The great apes (gorillas, chimpanzees, bonobos, and orangutans) split off from the lesser apes about 20 million years ago (Figure 24.9, inset). DNA tests show that humans and chimpanzees share about 98 percent of their DNA, indicating that they evolved from a common ancestor about 6 million years ago. By comparison, all humans share 99.9 percent of their DNA. The earliest human ancestors appeared a few million years ago, and the first civilizations occurred a mere 10,000 years ago. Present-day industrial society, barely more than two centuries old, is but a moment in the history of life on Earth.

Humans are here today because of a series of events that occurred throughout the history of the universe. Some of these events are common in the universe, such as the formation of heavy elements in earlier generations of stars and the formation of planets. Other events in Earth's history may have been less likely to

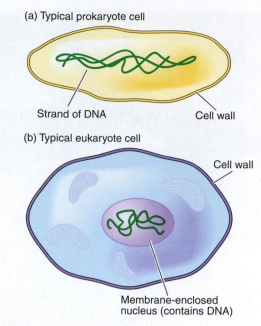

(a) Typical prokaryote cell

Strand of DNA   Cell wall

(b) Typical eukaryote cell

Cell wall

Membrane-enclosed
nucleus (contains DNA)

Figure 24.8 (a) A simple prokaryote cell contains little more than the cell's genetic material. (b) A eukaryote cell contains several membrane-enclosed structures, including a nucleus that houses the cell's genetic material.

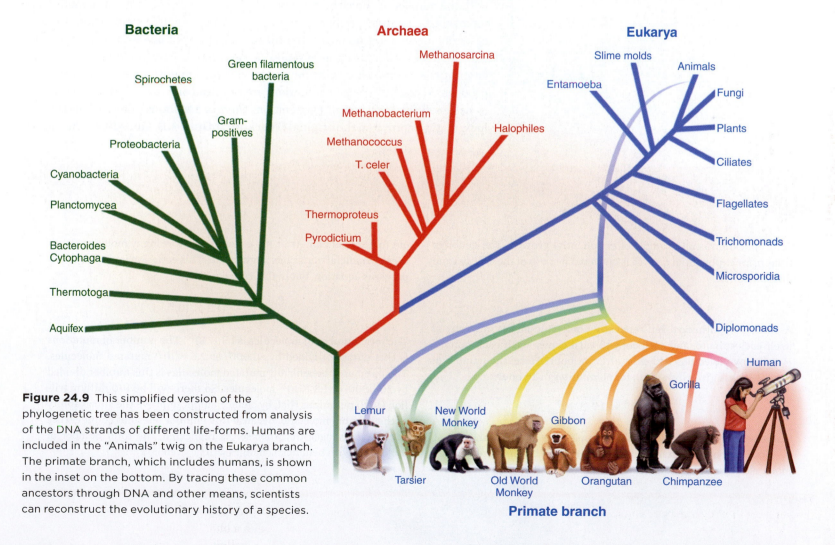

Bacteria

Green filamentous bacteria
Spirochetes
Gram-positives
Proteobacteria
Cyanobacteria
Planctomycea
Bacteroides
Cytophaga
Thermotoga
Aquifex

Archaea

Methanosarcina
Methanobacterium
Halophiles
Methanococcus
T. celer
Thermoproteus
Pyrodictium

Eukarya

Slime molds
Animals
Entamoeba
Fungi
Plants
Ciliates
Flagellates
Trichomonads
Microsporidia
Diplomonads

Human
Gorilla
Lemur
New World Monkey
Gibbon
Tarsier
Old World Monkey
Orangutan
Chimpanzee

Primate branch

Figure 24.9 This simplified version of the phylogenetic tree has been constructed from analysis of the DNA strands of different life-forms. Humans are included in the "Animals" twig on the Eukarya branch. The primate branch, which includes humans, is shown in the inset on the bottom. By tracing these common ancestors through DNA and other means, scientists can reconstruct the evolutionary history of a species.

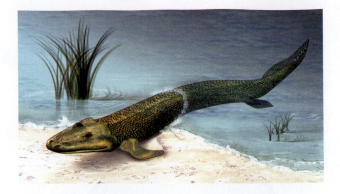

Figure 24.10 This illustration is an artist's reconstruction from a fossil of *Tiktaalik* found in the Canadian Arctic.

happen elsewhere, such as the formation of a planet with life-supporting conditions like Earth or the development of self-replicating molecules that led to Earth's earliest life. A few events stand out, such as major extinctions that allowed the evolution of mammalian life and, ultimately, human beings.

Evolution as a Mechanism of Change

Imagine that just once during the first few hundred million years after the formation of Earth, a single molecule formed somewhere in Earth's oceans. That molecule had a very special property: chemical reactions between that molecule and other molecules in the surrounding water caused the molecule to make a copy of itself. The molecule became "self-replicating." Chemical reactions then produced copies of each of these two molecules, making four molecules. Four molecules became eight, eight became 16, 16 became 32, and so on. By the time the original molecule had copied itself just 100 times, more than a million trillion trillion (10^{30}) of these molecules existed. That is about 100 million times more of these molecules than there are stars in the observable universe. The molecules of DNA that make up the chromosomes in the nuclei of the cells of all advanced life today are direct descendants of those early self-replicating molecules that flourished in the oceans of the young Earth.

Over the course of time, not all replications are exact. The likelihood that a copying variation will occur while a molecule is replicating increases significantly with the number of copies being made. For DNA, which contains the genetic code for an entire organism, a change in the genetic code is called a **mutation**. In some cases, a mutation has no effect. In others, it can prevent an organism from flourishing. In still other cases, a mutation can make an organism better suited to its environment. Organisms with these advantageous mutations will survive to reproduce successfully. Even if mutations are rare, and only a small fraction turn out to be beneficial, after just 100 generations there are trillions of mutations that, by luck, might improve on the original (**Working It Out 24.1**). **Heredity**—the ability

24.1 Working It Out Exponential Growth

Self-replication is an example of exponential growth. The doubling time, n, for exponential growth is given by the ratio of the original and final amounts:

$$\frac{P_F}{P_O} = 2^n$$

Assume a hypothetical self-replicating molecule makes one copy of itself each minute, and each copy in turn copies itself each minute. How many molecules will exist after an hour? Here, the number of generations is given by $n = 60$, for 60 minutes in an hour:

$$\frac{P_F}{P_O} = 2^{60} = 1.2 \times 10^{18}$$

There will be a billion billion of these molecules after 1 hour.

Now suppose a mutation occurs once every 50,000 times that a molecule reproduces itself, and one out of 200,000 mutations turns

out to be beneficial. After 100 generations, how many molecules with these beneficial mutations might exist? This equation is similar to the previous equation, but in this case $n = 100$:

$$\frac{P_F}{P_O} = 2^{100} = 1.3 \times 10^{30}$$

The total number of molecules is 1.3×10^{30}. The number of mutations is this number divided by 50,000, or 2.6×10^{25} mutated molecules. The number of beneficially mutated molecules is this number divided by 200,000, or 1.3×10^{20} molecules. So there will be 100 million trillion (10^{20}) mutations that, by chance, might improve the survivability of the original molecule. Because this number does not count earlier beneficial changes that themselves replicated, the total number of molecules with beneficial changes will be even larger!

of one generation to pass on its genetic code to future generations—allow beneficial mutations to persist and be incorporated into a species' genetic code.

As the organisms of the early Earth continued to interact with their surroundings and make copies of themselves, mutations caused them to diversify into many different species. In some cases, the resources they needed to reproduce became scarce. In the face of this scarcity of resources, varieties that were more successful reproducers became more numerous. Competition for resources, predation by one species on another, and cooperation between organisms became important to the survival of different varieties. Some varieties were more successful and reproduced to become more numerous, while less successful varieties became less and less common. This process, in which better-adapted organisms reproduce and thrive, while less well-adapted organisms become extinct, is called **natural selection**.

Life has existed on Earth for about 4 billion years, which is a very long time—long enough for the combined effects of heredity and natural selection to shape the descendants of that early self-copying molecule into a huge variety of complex, competitive, successful structures. Geological processes on Earth have preserved a fossil record of the history of some of these structures (**Figure 24.11**). Among these descendants are human "structures" capable of thinking about their own existence and unraveling the mysteries of the stars.

Figure 24.11 Fossils, such as this Parasaurolophys ("near crested lizard"), record the history of the evolution of life on Earth. This plant-eating dinosaur lived in North America about 75 million years ago.

CHECK YOUR UNDERSTANDING 24.1

Extremophiles are organisms that: (a) are extremely reactive; (b) are extremely rare; (c) have an extreme quality, such as mass or size; (d) live in extreme conditions.

24.2 Life Involves Complex Chemical Processes

The evolution of life on Earth cannot be separated from the narrative of astronomy: it is one of many examples of the emergence of structure in an evolving universe (**Process of Science Figure**). This leads naturally to a profound question: Has life arisen elsewhere? Unlike the study of planets, stars, and galaxies, there is only one known case for the study of life—Earth—and scientists do not know how much can be generalized to other places. To explore this question, we need to take a closer look at the processes that have led to life on Earth. In this section, we explore the chemical and physical properties of life on Earth.

The infant universe was composed basically of hydrogen and helium and very little else. After 9 billion years of stellar nucleosynthesis, all the heavier chemical elements essential to life were present and available in the molecular cloud that gave birth to the Solar System. Those heavier elements were formed by nuclear fusion in the cores of earlier generations of stars and were then dispersed into space. At times, this dispersal was passive. For example, low-mass stars such as the Sun, when they become puffed up, dying red giants, may shed their extended atmospheres, sending some newly created carbon into space. Other dispersals were more violent. High-mass stars produce even heavier elements through nucleosynthesis in their cores—up to and including iron. But some of the trace elements essential to biology on Earth are even more massive than iron. They are

Process of Science

ALL OF SCIENCE IS INTERCONNECTED

More than many other subjects, astrobiology draws on all of science and makes clear that all the fields are interconnected.

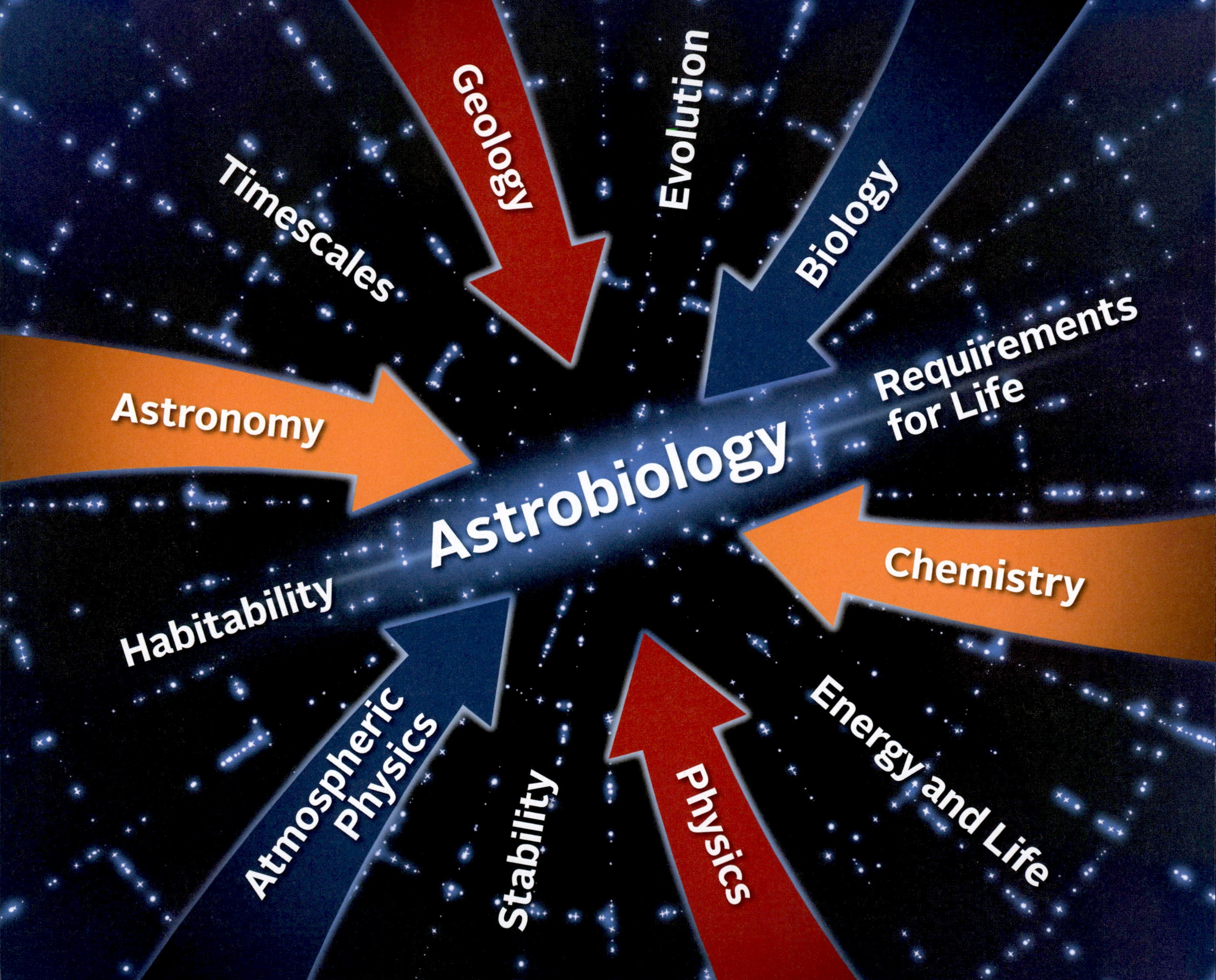

Geology

Evolution

Timescales

Biology

Astronomy

Requirements for Life

Astrobiology

Habitability

Chemistry

Atmospheric Physics

Stability

Physics

Energy and Life

No science stands alone. All are connected. Interdisciplinary fields of study like astrobiology provide opportunities for new tests for theories from many fields.

produced within a matter of minutes during the violent supernova explosions that mark the death of high-mass stars and then are thrown into the chemical mix found in molecular clouds.

All known living organisms on Earth are composed of a more or less common suite of complex chemicals. Approximately two-thirds of the atoms in the human body are hydrogen (H), about one-fourth are oxygen (O), a tenth are carbon (C), and a few hundredths are nitrogen (N). Carbon, nitrogen, and oxygen are the three most abundant products of stellar nucleosynthesis after helium; see the "astronomer's periodic table" in Figure 5.17. The several dozen remaining atomic elements in the human body make up only 0.2 percent of the total. All known living creatures are assemblages of molecules composed almost entirely of these four elements, sometimes called CHON (carbon, hydrogen, oxygen, nitrogen), along with small amounts of phosphorus and sulfur. Some of these molecules are enormous. Consider DNA, which is responsible for genetic codes, illustrated in **Figure 24.12**. DNA is made up entirely of only five atomic elements: CHON and phosphorus. But the DNA in each cell of the human body is composed of combinations of *tens of billions* of atoms of these same five elements. Proteins, the huge molecules responsible for the structure and function of living organisms, are long chains of smaller molecules called amino acids. Terrestrial life uses 20 specific amino acids, which also consist of no more than five atomic elements—in this case CHON plus sulfur instead of phosphorus.

The chemistry of life is far too complex to have only a half-dozen atomic elements. Many of the other elements, which are present in smaller amounts, are essential to the chemical processes that living organisms carry out. These elements include sodium, chlorine, potassium, calcium, magnesium, iron, manganese, and iodine. Trace elements such as copper, zinc, selenium, and cobalt also play a crucial role in biochemistry but are needed in only tiny amounts.

Notice that carbon, which can bond to four other atoms or molecules, forms the backbone of the DNA molecule shown in Figure 24.12. This is why carbon is so important to life on Earth. There could be forms of extraterrestrial life that are also carbon based but have chemistries quite different from that of life on Earth. For example, there are countless varieties of amino acids in addition to the 20 used by terrestrial life. Most other atoms are more limited than carbon in the number of bonds they can make, but silicon, like carbon, can bind to four other atoms, so that a large number of combinations is possible. As a potential life-enabling atom, silicon has both advantages and disadvantages compared to carbon. Silicon-based molecules remain stable at much higher temperatures than carbon-based molecules, perhaps enabling possible silicon-based life to thrive in high-temperature environments, such as on planets that orbit close to their parent star. But silicon is also a larger and more massive atom than carbon: it cannot form molecules as complex as those based on carbon. Any silicon-based life probably would be simpler than life-forms here on Earth, but it might exist in high-temperature niches somewhere within the universe. Although carbon's unique properties make it readily adaptable to the chemistry of life on Earth, other types of life might be found elsewhere.

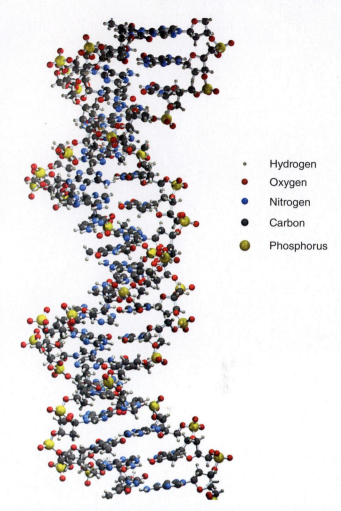

- Hydrogen
- Oxygen
- Nitrogen
- Carbon
- Phosphorus

Figure 24.12 DNA, the hereditable molecule that forms the basis for life on Earth, contains only five different atoms. Even so, these atoms are combined in billions of ways, giving rise to the diversity of life on Earth.

CHECK YOUR UNDERSTANDING 24.2

Carbon is a favorable base for life because: (a) it can bond to many other atoms in long chains; (b) it is nonreactive; (c) it forms weak bonds that can be readily reorganized as needed; (d) it is organic.

24.3 Where Do Astronomers Look for Life?

One approach to the scientific search for extraterrestrial life is to use robotic spacecraft to explore the planets and moons of the Solar System (see Chapter 6). Spacecraft have visited all of the planets and some moons and sent back at least some information about the conditions on these worlds. Another approach is to use telescopes to detect planets outside of our Solar System (see Chapter 7). Telescopes on the ground and in space have detected a few thousand planets orbiting other stars. In this section, we will survey the locations where life might be found, both in our own Solar System and around other stars.

Life within Our Solar System

Scientists start the search for evidence of extraterrestrial life here in our own Solar System. Early conjectures about life in our Solar System seem naïve, considering what we now know. Two centuries ago, the eminent astronomer Sir William Herschel, discoverer of Uranus, proclaimed, "We need not hesitate to admit that the Sun is richly stored with inhabitants." In 1877, astronomer Giovanni Schiaparelli (1835–1910) observed what appeared to be linear features on Mars and dubbed them *canali* ("channels" in Italian). The famous observer of Mars, Percival Lowell (1855–1916), misinterpreted Schiaparelli's *canali* as "canals," suggesting that they were constructed by intelligent beings.

Because Mercury and the Moon lacked atmospheres, astronomers determined they were not conducive to life. The giant planets and their moons were thought to be too remote and too cold to sustain life. The surface of Venus was far too hot, but Mars seemed more promising. During the mid-20th century, astronomers using ground-based telescopes discovered that Mars possesses an atmosphere, water ice, and carbon dioxide ice. During the 1960s, the United States and the Soviet Union sent reconnaissance spacecraft to the Moon, Venus, and Mars, but the instruments on these spacecraft probed the physical and geological properties of these astronomical bodies, rather than searching for life. Serious efforts to look for signs of life—past or present—require more advanced spacecraft with specialized instrumentation.

In the mid-1970s, two American *Viking* spacecraft were sent to Mars with detachable landers containing a suite of instruments designed to find evidence of a terrestrial type of life. When the *Viking* landers failed to find convincing evidence of life on Mars, hopes faded for finding life on any other body orbiting the Sun. Since that time, however, further exploration of the Solar System has generated renewed optimism. A better understanding of the history of Mars indicates the planet's climate has changed. Mars was once wetter and warmer than it is today.

In the 1990s, Mars missions began to map the planet's surface from the ground and from space. In 2008, NASA's *Phoenix* spacecraft landed at a far-northern latitude, inside the planet's arctic circle, where specialized instruments dug into and analyzed the martian water-ice permafrost. *Phoenix* found that the martian arctic soil has a chemistry similar to the Antarctic dry valleys on Earth, where life exists deep below the surface at the ice-soil boundary. Minerals that form in water, for example calcium carbonate, have been detected. This suggests that oceans existed in the past on Mars. However, *Phoenix* did not find evidence of life.

The *Curiosity* rover (originally known as the *Mars Science Laboratory*) landed in Gale Crater on Mars in 2012. This rover studies the rocks and soil of Mars to

Figure 24.13 The Mars *Curiosity* rover detected evidence that Mars had a watery past. The rounded gravel surrounding the bedrock suggests there was an ancient, flowing stream.

provide data for a better understanding of the history of the planet's climate and geology. Shortly after landing, *Curiosity* found evidence that a stream of liquid water had once flowed in the crater. The rover observed rounded, gravelly pebbles stuck together, which have been interpreted as coming from a stream that varied at times from ankle-deep to hip-deep, and moved at about 1 meter per second (**Figure 24.13**). It found sedimentary rocks containing clay, which suggested that at one time there had been a freshwater lake bed. Later observations found that the surface soil contained up to 2 percent water by weight—or about 1 quart per cubic foot of martian dirt. This is too dry to support plant life, which permanently wilts in soil that is about 10% water by weight.

The *Mars Atmosphere and Volatile EvolutioN* (*MAVEN*) mission arrived at Mars in September 2014. It is studying the upper atmosphere in order to learn more about the escape of carbon dioxide, hydrogen, and nitrogen from the planet's atmosphere and how the loss of those gases affected surface pressure and the existence of liquid water.

In September 2015, NASA announced that spectroscopic observations from the orbiting Mars Reconnaissance Orbiter indicate that there is liquid water on Mars today. Darkish streaks on Mars that change seasonally contain hydrated salts and minerals, indicating that liquid water is important to their formation (**Figure 24.14**). This water is briny (salty), which has a much lower freezing point than non-briny water and thus could exist in a liquid state during the martian summer. Future experiments will look for liquid water—and fossil or living microorganisms—below the martian surface.

NASA's instrumented robotic spacecraft reached the outer Solar System starting in the 1980s, and many astrobiologists were surprised by the findings. Although the outer planets themselves did not appear to be habitats for life, some of their moons became objects of special interest. Jupiter's moon Europa is covered with a layer of water ice that appears to overlie a great ocean of briny liquid water (**Figure 24.15**). The water remains liquid because of high pressure and tidal heating by Jupiter. Impacts by comet nuclei may have added a mix of organic material, another essential ingredient for life. Once thought to be a frozen, inhospitable world, Europa is now a candidate for biological exploration. Recently, scientists using the Hubble Space Telescope observed water geysers taller than Mount Everest erupting from the icy surface of Europa. Ejected material from these geysers may make it possible to search for life on Europa without drilling down through the ice.

Saturn's moon Titan has an atmosphere that is rich in organic chemicals, many of which are thought to be precursor molecules of a type that existed on Earth before life appeared here. A probe from the *Cassini* spacecraft in orbit around Saturn descended through Titan's atmosphere and found additional evidence for a variety of molecules that might be necessary for life, as well as liquid lakes of methane on the surface and probably a liquid-water ocean under the surface. As noted in Chapter 1, the *Cassini* spacecraft also detected water-ice crystals spouting from cryovolcanoes (which erupt ice crystals instead of rocks) near the south pole of Saturn's tiny moon Enceladus. Liquid water must lie beneath its icy surface, and Enceladus therefore joins Europa as a possible habitat of extremophile life, perhaps life similar to that found near hydrothermal vents deep within Earth's oceans.

Figure 24.14 This MRO image shows narrow, dark, downhill streaks, about 100 meters long, which are thought to indicate liquid water flowing on Mars today. Hydrated salts and minerals, including pyroxene (the blue color) were detected by spectroscopy.

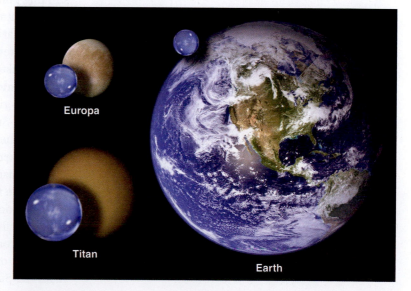

Figure 24.15 The total amount of liquid water (blue ball) on Jupiter's moon Europa and Saturn's moon Titan compared to the amount on Earth. All figures are drawn to scale and assume average ocean depths of 4 km (Earth), 100 km (Europa), and 200 km (Titan).

The discovery of life on even one Solar System body beyond Earth would be exciting. If scientists discover that life arose independently *twice* in the same planetary system, this finding could suggest that the spontaneous appearance of life is not rare at all.

CHECK YOUR UNDERSTANDING 24.3

Which of the following Solar System objects is *not* a good candidate for future searches for life? (a) Mars; (b) Jupiter's moon Europa; (c) Saturn's moon Titan; (d) Uranus

Habitable Zones

Recall from Chapter 7 that there are more than 1,000 confirmed and thousands of candidate extrasolar planets within the Milky Way Galaxy. To decide which planets to focus on for further study, astronomers are narrowing the possibilities by searching for planets with environments conducive to the formation and evolution of life, as we understand it, while eliminating planets that are clearly unsuitable. Astronomers consider issues such as each planet's orbit, its inferred temperature, its distance from its star, and its location in the galaxy. One criterion astrobiologists look for is planetary systems that are stable. As noted in Chapters 2, 3, and 4, astronomers think about the effects of a planet's rotation and orbit. Planets in stable systems have nearly circular orbits that preserve relatively uniform climatological environments. Planets in very elliptical orbits or planets with a large axial tilt can experience more intense temperature swings that could be detrimental to the survival of life. A stable temperature that maintains the existence of water in a liquid state might be important. We know that liquid water was essential for the formation and evolution of life on Earth. Of course, we don't know if liquid water is an absolute requirement for life elsewhere, but it's a good starting point.

In Chapter 7, we discussed the idea of the habitable zone, the location of a planet relative to its parent star that provides a range of temperatures in which liquid water can exist. On planets that are too close to their parent stars, water would exist only as a vapor—if at all. On planets that are too far from their stars, water would be permanently frozen as ice. Planet size is another consideration: Large gas giants retain most of their light gases during formation and do not have a surface. Small planets may be rocky or a mix of water, rock, and ice: measurement of the mass and radius enables scientists to estimate the density. Planets that are very small may have insufficient surface gravity to retain their atmospheric gases and so end up like our Moon. Calculating whether any particular planet is in the habitable zone is complicated. Recall from Chapter 7 that even if a planet is located in the habitable zone, it only means that liquid water could exist on the surface: it does not mean that astronomers have confirmed the presence of liquid water or that the planet has inhabitants.

In our own Solar System, Venus, which orbits at 0.7 times Earth's distance from the Sun, has become an inferno because of its runaway greenhouse effect (see Chapter 9). Any liquid water that might once have existed on Venus has long since evaporated and been lost to space. Mars orbits about 1.5 times farther from the Sun than the orbit of Earth, and the water that we see on Mars today is nearly always frozen. But the orbit of Mars is more elliptical and variable than Earth's, giving the planet a greater variety of climate, including long-term cycles that might occasionally permit liquid water to exist. Most astrobiologists put the habitable zone of our Solar System at about 0.9–1.4 astronomical units (AU), which includes Earth but just misses Venus and Mars. However, this range excludes the possibility of liquid water under ice, as occurs on the moons of the outer Solar System.

Astronomers must also think about the type of star they are observing in their search for planets that could have liquid water. Stars that are less massive than the Sun and thus cooler will have narrower habitable zones. A planet in the habitable zone of a cool star is close in to its star. As a result, it is more likely to be tidally locked to the star so there is no day/night cycle. Stars that are more massive than the Sun are hotter and will have a larger and more distant habitable zone (**Figure 24.16**). However, massive stars have shorter main-sequence lifetimes and might not last long enough for evolution to take place. For example, a star of 3 M_{Sun} has a lifetime of only a few hundred million years. On Earth, a billion years was long enough for bacterial life to form and cover the planet but insufficient for anything more advanced to evolve. On Earth, it took 3.5 billion years of evolution to reach the period known as the Cambrian explosion. Even though evolution might happen at a different pace elsewhere, stellar lifetime is still a sufficiently strong consideration, so astronomers focus their efforts on stars with longer lifetimes; specifically, stars of 0.6–1.4 M_{Sun}, which corresponds to spectral types F, G, K, and M.

Another factor to consider is a planet's atmosphere. The ability of a planet to keep an atmosphere depends on its mass and radius (and therefore its escape velocity) and its temperature. Planets that are very small may have insufficient surface gravity to retain their atmospheric gases. In the inner Solar System, the Moon and Mercury were too small to keep any atmosphere. Mars lost its atmosphere over time, but the larger Earth and Venus were able to keep a thick atmosphere. Another important consideration is the greenhouse effect, which traps heat underneath an atmosphere and raises the temperature on a planetary

Nebraska Simulation: Circumstellar Habitable Zone

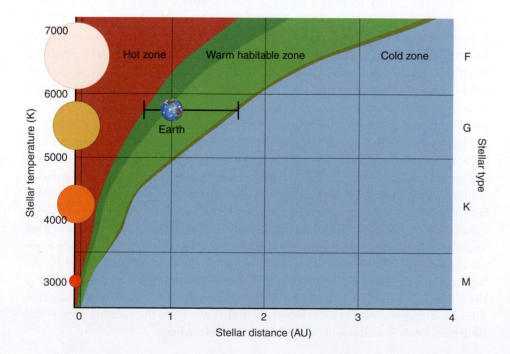

Figure 24.16 The habitable zone changes with the mass and temperature of a star. Habitable zones around hot, high-mass stars are larger and more distant than the zones around cooler, lower-mass stars.

surface. This has happened on Venus, Earth, and Mars, each of which has a higher surface temperature because of its atmosphere. The thickness and chemical content of the atmosphere affect the strength of the greenhouse effect, so that, for example, Venus is much hotter than its distance from the Sun would suggest because of its thick atmosphere of carbon dioxide (see Figure 5.24 and the Chapter 5 "Origins"). The total amount of atmosphere affects the atmospheric pressure at the surface, which, along with the temperature, determines whether water (or other molecules) can be in a liquid state on the surface. The current thin atmosphere of Mars does not permit standing liquid water on its surface.

In the next few years, there will be many more observations from planet-finding projects that identify atmospheres on extrasolar planets. Water vapor, hydrogen, and carbon monoxide have been found on a few extrasolar planets already. In particular, the discovery of oxygen in the atmosphere of an extrasolar planetary atmosphere would be exciting, but not definitive. The oxygen in Earth's atmosphere makes it stand out from the rest of the planets and moons in the Solar System, and we know that most of the terrestrial oxygen was created by photosynthetic life. However, oxygen can also come from the breakup of water molecules, so the presence of oxygen alone doesn't necessarily mean life exists there.

Astrobiologists continue to develop ways to classify planets more clearly as they try to narrow down the possibilities about where life might exist. One such classification, the Earth Similarity Index, uses the currently available data on an extrasolar planet to estimate how much it is like Earth. Factors include the radius, density, escape velocity, and surface temperature. The ESI ranges from 0 to 1. A value of 0.8–1 is used for rocky planets that can retain an atmosphere at temperatures suitable for liquid water; that is, that are Earth-like. This is an Earth-centric approach based on the experience of life on Earth. The Planetary Habitability Index aims to be less Earth-centric and to broaden the options for habitability, but it depends on factors not yet measured or measurable for most extrasolar planets. This index depends on whether the planet has a surface on which organisms can grow, as well as the right kind of chemistry, a source of energy, and the ability to hold a liquid solvent. Saturn's moon Titan or Jupiter's moon Europa might satisfy these conditions, and Mars might have done so in the past. Over the next decade, improvements in observations will likely lead to enough information that at least some of the extrasolar planets can be classified by their Planetary Habitability Index. These indices are preliminary, representing various efforts to proceed with limited data. As observations of extrasolar planets become more complete, astrobiologists will undoubtedly develop new classification schemes that are more accurate and informative.

Nebraska Simulation: Milky Way Habitability Explorer

Astronomers also consider the *galactic habitable zone*—the idea that there may be some locations within the Milky Way Galaxy where planets might have a higher probability of hosting life. Stars that are situated too far from the galactic center may be without enough heavy elements—such as oxygen, silicon (silicates), iron, and nickel—in their protoplanetary disks to form rocky planets like Earth. Conversely, regions too close to the galactic center experience less star formation and therefore fewer opportunities to gather heavy elements into planetary environments. Stars that are too close to the galactic center may be affected by the high-energy radiation environment (X-rays and gamma rays from supermassive black holes or gamma-ray bursts), which can damage RNA and DNA. Stars that migrate within the galaxy and change their distance from the galactic center over time may move in and out of any galactic habitable zone.

24.4 Scientists Are Searching for Signs of Intelligent Life

Are we alone? This question is approached by scientists from many directions. Biologists consider the origin and evolution of life and the definition of intelligence. Astronomers send messages and search for alien signals in the vast array of astronomical data. In this section, you will learn about the search for intelligent life and how scientists think about the probability of finding it.

Sending Messages

During the 1970s, messages were sent from Earth to space. The *Pioneer 11* spacecraft, which will probably spend eternity drifting through interstellar space, carries the plaque shown in **Figure 24.17**. It pictures humans and the location of Earth for any future interstellar traveler who might happen to find it and understand its content. Another message to the cosmos accompanied the two *Voyager* spacecraft on identical phonograph records that contained greetings from planet Earth in 60 languages, samples of music, animal sounds, and a message from then-President Jimmy Carter. Some politicians were concerned that scientists were dangerously advertising our location in the galaxy, even though radio signals had already been broadcast into space for nearly 80 years. Some philosophers also worried that these messages contained anthropomorphic assumptions about aliens being sufficiently like us to decode the messages. However, sending messages on spacecraft is not an efficient way to make contact with extraterrestrial life. The probability that any of these messages will actually be found by an alien species is very, very small.

In 1974, astronomers used the 300-meter-wide dish of the Arecibo radio telescope to beam a message in binary code (**Figure 24.18**) toward the star cluster M13, located 25,000 light-years away. That is sufficiently far that by the time the message arrives, the core of M13 will have moved, and the radio signal will not actually arrive there. However, the intention of the experiment was to demonstrate that such a message could be sent, not to make contact, as it would take 50,000 years for a reply to come back. The experiment confirmed that such a message could be sent. In 2008, a radio telescope in Ukraine sent a message to the exoplanet Gliese 581c. The message was composed of 501 digitized images and text messages selected by users on a social networking site and will arrive at Gliese 581c in 2029.

The Drake Equation

The first serious effort to quantify the probability of the existence of intelligent extraterrestrial life was made by astronomer Frank Drake in 1960. He developed the **Drake equation**, which estimates the likely number (N) of intelligent

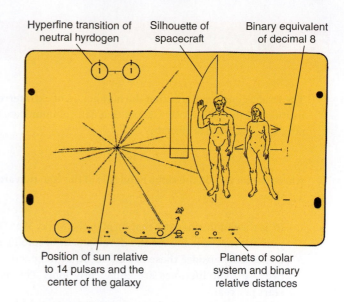

Hyperfine transition of neutral hyrdogen

Silhouette of spacecraft

Binary equivalent of decimal 8

Position of sun relative to 14 pulsars and the center of the galaxy

Planets of solar system and binary relative distances

Figure 24.17 This plaque is carried by the *Pioneer 11* probe, which was launched in 1973 and will eventually leave the Solar System to travel in interstellar space.

Figure 24.18 This message was beamed toward the star cluster M13 in 1974. This binary-encoded message contains the numbers 1–10, hydrogen and carbon atoms, some interesting molecules, DNA, a human figure and its size, the basics of the Solar System, and a depiction of the Arecibo telescope.

The Drake equation states that the number, N, of extraterrestrial civilizations in the Milky Way Galaxy that can communicate by electromagnetic radiation is given by

$$N = R^* \times f_p \times n_e \times f_l \times f_i \times f_c \times L$$

where the factors on the right-hand side of the equation are explained as follows:

1. R^* is the number of stars that form in the Milky Way Galaxy each year that are suitable for the development of intelligent life. Astronomers consider these to be F, G, K, or M spectral-type stars because their lifetimes are sufficiently long. This is about 5–7 stars per year.

2. f_p is the fraction of stars that form planetary systems. The discoveries of extrasolar planets over the past two decades have shown that planets form as a natural by-product of star formation and that many—perhaps most—stars have planets. For this calculation, astronomers assume that f_p is between 0.5 and 1.

3. n_e is the number of planets and moons in each planetary system with an environment suitable for life. In the Solar System, this number is at least 1 (for Earth), but it could be more if Mars or an outer-planet moon or two has suitability for life. Only recently have stars with multiple planets been discovered, so astronomers are just starting to get data on this factor. Generally they estimate 0–3.

4. f_l is the fraction of suitable planets and moons on which life actually arises. Remember that just a single self-replicating molecule may be enough to get the ball rolling. Some biochemists think that if the right chemical and environmental conditions are present,

then life *will* develop, but others disagree. Values of f_l range from 100 percent (life always develops) to 1 percent (life is more rare). Astronomers use a range of 0.01–1.

5. f_i is the fraction of those planets harboring life that eventually develop intelligent life. Intelligence is certainly the kind of survival trait that might often be strongly favored by natural selection. Yet on Earth, it took about 4 billion years—roughly half the expected lifetime of the Sun—to evolve tool-building intelligence. The correct value for f_i might be close to 0.01 or it might be closer to 1. The truth is, no one knows.

6. f_c is the fraction of intelligent life-forms that develop technologically advanced civilizations; that is, civilizations that send communications into space. With only one example of a technological civilization to work with, f_c also is unknown. Astronomers estimate f_c to be between 0.1 and 1.

7. L is the number of years that technologically advanced civilizations exist. This factor is certainly difficult to estimate because it depends on the long-term stability of these civilizations. On Earth, the longest-lived civilizations have existed for, at most, thousands of years. These civilizations, however, were not at the level of technology that allows interstellar communication—thus far, the first "technologically advanced civilization" on Earth is less than 100 years old. We do not know whether life that is intelligent enough to manipulate its environment technologically can maintain its planetary resources for any extended length of time. Astronomers usually put a value between 1,000 years and 1 million years in their estimates, but the average number could be much smaller or much larger.

civilizations currently existing in the Milky Way Galaxy. The Drake equation is different from the other equations in this book because the values for many of the variables are quite uncertain. However, it is a useful way to categorize some of the factors that relate to the conditions that must be met for a civilization to exist. The equation is discussed further in **Working It Out 24.2**.

As illustrated in **Figure 24.19**, the conclusions we draw using the Drake equation depend a great deal on the assumptions we make, therefore on the numbers used in the equation. For the most pessimistic estimates, the Drake equation sets the number of technological civilizations in our galaxy at about 1, in which case we are the *only* technological civilization in the Milky Way at this time. Such a universe could still be full of intelligent life. With 100 billion galaxies in the observable universe, even these pessimistic assumptions mean that there could be 100 billion technological civilizations out there somewhere. However, if the nearest neighbors are in another galaxy, they are *very* far away—millions of parsecs on average.

At the other extreme—with the most optimistic numbers, which assume that intelligent life arises and survives everywhere it gets the chance—there could be tens of millions of technological civilizations in the Milky Way alone! In this case, the nearest neighbors may be "only" 40 or 50 light-years away.

If humans did meet a technologically advanced civilization, what would it be like? The Drake equation suggests that it's highly unlikely there are neighbors nearby, unless civilizations typically live for many thousands or even millions of years (see Working It Out 24.2). If that were the case, any civilization we encountered would probably have been around for much longer than we have.

Technologically Advanced Civilizations

During lunch with colleagues, the physicist Enrico Fermi (1901–1954), a firm believer in extraterrestrial life, is reported to have asked, "If the universe is teeming with aliens . . . where is everybody?" Fermi's question—first posed in 1950 and sometimes called the *Fermi paradox*—remains unanswered. If intelligent life-forms are common but interstellar travel is difficult or impossible, would the aliens send out messages—perhaps by electromagnetic waves instead? And if they did, why haven't astronomers detected their signals?

Drake used what was then astronomy's most powerful radio telescope to listen for signals from intelligent life around two nearby stars, but he found nothing unusual. His original project has grown over the years into a much more elaborate program called the Search for Extraterrestrial Intelligence, or **SETI**. Scientists from around the world have thought carefully about what strategies might be useful for finding life in the universe. Most of these use radio telescopes to listen for signals from space that bear an unambiguous signature of an intelligent source. Some have focused on significant parts of the spectrum, assuming that a civilization will broadcast on a channel that astronomers throughout the galaxy should find interesting; for example, the 21-centimeter (21-cm) line from hydrogen gas. More recent searches have made use of advances in technology to record as broad a range of radio signals from space as possible. Analysts search these databases for regular signals that might be intelligent in origin.

The SETI Institute's Allen Telescope Array (ATA) received much of its initial financing from Microsoft cofounder Paul Allen. The ATA consists of a "farm" of small, inexpensive radio dishes like those used to capture signals from orbiting television-broadcasting satellites (see the chapter-opening photograph). One of the key projects of the ATA is to observe the planets discovered by the Kepler Mission. Each dish has a diameter of 6.1 meters, but all of the telescopes working together have a total signal-receiving area greater than that of a 100-meter radio telescope. Just as your brain can sort out sounds coming from different directions, this array of radio telescopes is able to determine the direction a signal is coming from, allowing it to listen to many stars at the same time. Over several years' time, astronomers using the ATA are expected to survey as many as a million stars, hoping to find a civilization that has sent a signal toward Earth.

As we stated earlier in this chapter, finding even one nearby civilization in the Milky Way Galaxy—that is, a *second* technological civilization in Earth's small corner of the universe—will make scientists optimistic that the universe as a whole is teeming with intelligent life. The likelihood of SETI's success is difficult to predict, but its potential payoff is enormous. Few discoveries would have a more profound impact than the certain knowledge that we on Earth are not alone.

Science fiction is filled with tales of humans who leave Earth to "seek out new life and new civilizations." Unfortunately, these scenarios are not scientifically realistic. The distances to the stars and their planets are enormous: to explore a significant sample of stars would require extending the physical search over tens

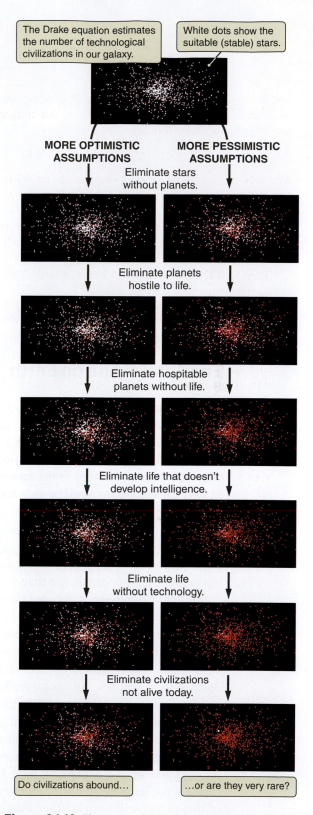

Figure 24.19 The two columns show estimates of the existence of intelligent, communicative civilizations in the Milky Way Galaxy using the Drake equation. White dots represent stars with possible civilizations. Notice how widely these estimates vary, given optimistic and pessimistic assumptions about the seven factors in the equation.

or hundreds of light-years. Special relativity limits how fast one can travel. The speed of light is the limit, and even at that rate it would take more than 4 years to reach the *nearest* star. The relativistic effect of time dilation would mean that time passes slower for astronauts traveling at very high speeds, and they would return to Earth younger than if they had stayed at home. For example, suppose astronauts visited a star 15 light-years distant. Even if they traveled at speeds close to the speed of light, by the time they returned to Earth, 30 years would have passed at home.

Some science fiction writers get around this problem by invoking "warp speed" or "hyperdrive," which enables travel faster than the speed of light, or by using wormholes as shortcuts across the galaxy—but there is absolutely no evidence that any of these options are possible. And most of these imaginative stories ignore the vast number of other complications that accompany human space travel. Humans are just beginning to learn to live in space even for short periods of time.

Origins
The Fate of Life on Earth

Astronomers have used their understanding of physics and cosmology to look back through time and watch as structure formed throughout the universe and to look forward to the future of our Sun, our galaxy, and the ultimate fate of the universe. About 5 billion years from now, the Sun will end its long period of relative stability. It will expand to become a red giant star, swelling to hundreds of times larger than it is at present and thousands of times more luminous. The giant planets, orbiting outside the extended red giant atmosphere, will probably survive. But at least some of the planets of the inner Solar System will not. Just as an artificial satellite is slowed by drag in Earth's tenuous outer atmosphere and eventually falls to the ground, so, too, will a planet caught in the Sun's atmosphere be engulfed by the expanding Sun. If this is what happens to Earth, no trace of this planet will remain other than a slight increase in the amount of massive elements in the Sun's atmosphere.

Another possibility is that the red giant Sun will lose mass in a powerful wind, its gravitational pull on the planets will weaken, and the orbits of both the inner and outer planets will spiral outward. If Earth moves out far enough, it may survive as a seared cinder, orbiting the small, hot, white dwarf star that the Sun will become. Barely larger than Earth and with its nuclear fuel exhausted, the white dwarf Sun will slowly cool, eventually becoming a cold sphere of densely packed carbon, orbited by what remains of its planets. The ultimate outcome for Earth—consumed in the heart of the Sun or left behind as a cold, burned rock orbiting a long-dead white dwarf—is not yet known.

In either case, however, Earth's status as a garden spot in the habitable zone will be at an end. If the Sun does not expand too far or Mars also migrates outward, Mars could become the habitable planet in the Solar System, at least for a while. As the dying Sun loses more and more of its atmosphere in a stellar wind, Earth's atoms might be expelled back into the reaches of interstellar space from which they came, perhaps to be recycled into new generations of stars, planets, and even life itself.

But even before the Sun's change into a red giant star, the Sun's luminosity will begin to rise. As solar luminosity increases, so will temperatures on all the planets. The inner edge of the Sun's habitable zone will slowly move out past the orbit of Earth. Eventually, Earth's temperatures will climb so high that all animal and plant life will perish. Even the extremophiles that inhabit the oceanic depths will die, as the oceans themselves boil away. Models of the Sun's evolution are still not precise enough for astronomers to predict with certainty when that fatal event will occur, but the end of all terrestrial life may be 1 billion to 4 billion years away. Additionally, the Milky Way Galaxy is headed for a collision with the Andromeda Galaxy in 4 billion to 5 billion years' time. Galaxies are mostly empty space, so the Sun is not likely to collide with another star, but one effect of the collision is that our Solar System may be gravitationally flung to a different part of our galaxy.

There is much work to be done before we can realistically contemplate even voyages within the Solar System.

Some people claim that aliens have already visited Earth: tabloid newspapers, books, and websites are filled with tales of UFO sightings, government conspiracies and cover-ups, alleged alien abductions, and UFO religious cults. However, none of these reports meet the basic standards of science. They are not falsifiable—they lack verifiable evidence and repeatability—and we must conclude that there is no scientific evidence for any alien visitations.

CHECK YOUR UNDERSTANDING 24.5

The Drake equation enables astronomers to: (a) calculate precisely the number of alien civilizations; (b) organize their thoughts about probabilities; (c) locate the stars they should study to find life; (d) find new kinds of life.

It is far from certain, however, that the descendants of today's humanity will even be around a billion years from now. Some of the threats to life come from beyond Earth. For the remainder of the Sun's life, the terrestrial planets, including Earth, will continue to be bombarded by asteroids and comets. Perhaps a hundred or more of these impacts will involve kilometer-sized objects, capable of causing the kind of devastation that eradicated the dinosaurs. Although these events may create new surface scars and may be harmful to human life as we know it, they will have little effect on the integrity of Earth itself. Earth's geological record is filled with such events, and each time they happen, life manages to recover and reorganize.

Humans might protect themselves from the fate of the dinosaurs, but in the long run humanity will either leave this world or die out. Planetary systems are common to other stars, and many other Earth-like planets may well exist throughout the Milky Way Galaxy. Colonizing other planets is currently the stuff of science fiction, but if the descendants of modern-day humans are ultimately to survive the death of the home planet, off-Earth colonization must become science fact at some point in the future.

By studying astronomy, you have learned where you come from. You have learned about the self-correcting nature of science—how it continually adapts to new information to give us the ability to make better and better predictions about the behavior of the physical world. This predictive ability makes science an extremely powerful tool. No other species in the history of Earth has been able to understand its position, predict what will happen next, and therefore adapt its behavior to seek the best possible future for its members.

Figure 24.20 shows Earth as seen by the *Cassini* spacecraft, looking near Saturn's rings. That tiny dot, which is Earth, is the only place in the entire universe where we know that life exists. Compare the size of that dot to the size of the universe. Compare the history of life on Earth to the history of the universe. Compare Earth's future to the fate of the universe. Astronomy is humbling. We occupy a tiny part of space and time. Yet we are unique, as far as we know. Think for a moment about what that means to you. This may be the most important lesson the universe has to offer.

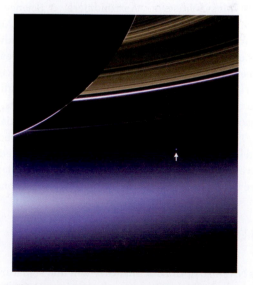

Figure 24.20 This image from the *Cassini* spacecraft shows Earth as seen from Saturn. The pale dot in the lower right (arrow) is Earth.

In this article, scientists discuss how they will search for life beyond Earth.

Finding Life Beyond Earth Is within Reach

NASA

Many scientists believe we are not alone in the universe. It's probable, they say, that life could have arisen on at least some of the billions of planets thought to exist in our galaxy alone—just as it did here on planet Earth. This basic question about our place in the universe is one that may be answered by scientific investigations. What are the next steps to finding life elsewhere?

Experts from NASA and its partner institutions addressed this question on July 14, 2014, at a public talk held at NASA headquarters in Washington. They outlined NASA's road map to the search for life in the universe, an ongoing journey that involves a number of current and future telescopes.

"Sometime in the near future, people will be able to point to a star and say, 'that star has a planet like Earth,'" says Sara Seager, professor of planetary science and physics at the Massachusetts Institute of Technology in Cambridge, Massachusetts. "Astronomers think it is very likely that every single star in our Milky Way Galaxy has at least one planet."

NASA's quest to study planetary systems around other stars started with ground-based observatories, then moved to space-based assets like the Hubble Space Telescope, the Spitzer Space Telescope, and the Kepler Space Telescope. Today's telescopes can look at many stars and tell if they have one or more orbiting planets. Even more, they can determine if the planets are the right distance away from the star to have liquid water, the key ingredient to life as we know it.

The NASA road map will continue with the launch of the Transiting Exoplanet Surveying Satellite (TESS) in 2017, the James Webb Space Telescope (Webb telescope) in 2018, and perhaps the proposed Wide Field Infrared Survey Telescope–Astrophysics Focused Telescope Assets (WFIRST-AFTA) early in the next decade. These upcoming telescopes will find and characterize a host of new exoplanets—those planets that orbit other stars—expanding our knowledge of their atmospheres and diversity. The Webb telescope and WFIRST-AFTA will lay the groundwork, and future missions will extend the search for oceans in the form of atmospheric water vapor and for life as in carbon dioxide and other atmospheric chemicals on nearby planets that are similar to Earth in size and mass, a key step in the search for life.

"This technology we are using to explore exoplanets is real," said John Grunsfeld, astronaut and associate administrator for NASA's Science Mission Directorate in Washington. "The James Webb Space Telescope and the next advances are happening now. These are not dreams—this is what we do at NASA."

Since its launch in 2009, Kepler has dramatically changed what we know about exoplanets, finding most of the more than 5,000 potential exoplanets, of which more than 1,700 have been confirmed. The Kepler observations have led to estimates of billions of planets in our galaxy and shown that most planets within 1 astronomical unit are less than 3 times the diameter of Earth. Kepler also found the first Earth-size planet to orbit in the "habitable zone" of a star, the region where liquid water can pool on the surface.

"What we didn't know five years ago is that perhaps 10 to 20 percent of stars around us have Earth-size planets in the habitable zone," says Matt Mountain, director and Webb telescope scientist at the Space Telescope Science Institute in Baltimore. "It's within our grasp to pull off a discovery that will change the world forever. It is going to take a continuing partnership between NASA, science, technology, the U.S. and international space endeavors, as exemplified by the James Webb Space Telescope, to build the next bridge to humanity's future."

This decade has seen the discovery of more and more super-Earths, which are rocky planets that are larger and heftier than Earth. Finding smaller planets, the Earth twins, is a tougher challenge because they produce fainter signals. Technology to detect and image these Earth-like planets is being developed now for use with the future space telescopes. The ability to detect alien life may still be years or more away, but the quest is under way.

Said Mountain, "Just imagine the moment, when we find potential signatures of life. Imagine the moment when the world wakes up and the human race realizes that its long loneliness in time and space may be over—the possibility we're no longer alone in the universe."

1. Why is NASA so optimistic that they will find signatures of alien life?
2. What are some potential signatures they hope to find?
3. Why are telescopes in space needed to find these signatures?
4. Do a search to see what is the status of the TESS and James Webb space telescopes.
5. How do you think the world would react to a discovery of signatures of life on another planet?

Summary

Astrobiology seeks to answer the question, Are we alone? All the sciences have a part to play in answering this question, from astronomy to zoology. Theories of how life began and evolved on Earth help to limit the number of targets astronomers study in their search for life elsewhere. Even before the Sun expands in size and evolves to a red giant star, its luminosity will increase enough to alter the location of its habitable zone. When that happens, Earth may no longer be the planet at the right location for maintaining liquid water.

LG 1 **Explain our current understanding of how and when life began on Earth and how it has evolved.** Life on Earth is a form of complex carbon-based chemistry, made possible by self-replicating molecules. Life likely formed in Earth's oceans, then evolved chemically from simple molecules into self-replicating organisms through a combination of mutation and heredity.

LG 2 **Explain how life is a structure that has evolved through the action of the physical and chemical processes that shape the universe.** All terrestrial life is composed primarily of only six elements: carbon, hydrogen, oxygen, nitrogen, sulfur, and phosphorus. Life-forms that are very different from those on Earth, including those based on silicon chemistry, cannot be ruled out.

LG 3 **Describe the locations in our Solar System and around other stars where astronomers think life might be possible.** Within the Solar System, Mars and some moons of Jupiter and Saturn are the most promising candidates for life. A habitable zone around a star is a location within which the temperature of a planet will support liquid water on its surface. Thus, astronomers look for extrasolar planets that orbit in habitable zones surrounding solar-type stars. This concept of habitable zone can be somewhat extended by special circumstances, such as tidal heating from a giant planet or atmospheres containing greenhouse gases.

LG 4 **Describe some of the methods used to search for intelligent extraterrestrial life.** The Drake equation includes different factors that astronomers consider when thinking about the possibility of life in the universe. Astronomers use radio telescopes to search for signals from extraterrestrial life, particularly in astronomically important regions of the spectrum. In recent times, this has been expanded to a search through as broad a region of the radio spectrum as possible. No intelligent extraterrestrial life has yet been detected.

? UNANSWERED QUESTIONS

- Will humans spread life into space? Some scientists have suggested that seeding from Earth may have already happened as Earth microorganisms scattered into space after giant impacts. It is also possible that humans have unintentionally sent microorganisms to space aboard our spacecraft. More intentional methods of seeding include sending microorganisms from Earth to other planets or moons to try to jump-start evolution, "terraforming" Mars or a moon to change conditions on it to make it more habitable for humanity, or sending humans in spaceships to colonize the galaxy.

- Will humans themselves spread into space? At some point, humans must leave planet Earth if the species is to survive. But space is a dangerous place, and many of the problems that humans encounter in space have not yet been solved. These range from purely physical issues such as the loss of bone density in low gravity to the societal problems that occur when a small number of people are confined together for long periods of time. We do not yet know if humanity can overcome these problems and journey to other planets.

Questions and Problems

Test Your Understanding

1. The study of life and the study of astronomy are connected because (select all that apply)
   a. life may be commonplace in the universe.
   b. studying other planets may help explain why there is life on Earth.
   c. explorations of extreme environments on Earth suggest where to look for life elsewhere.
   d. life is a structure that evolved through physical processes, and life on Earth may not be unique.
   e. life elsewhere is most likely to be found by astronomers.

2. Scientists look for water to indicate places where life might exist because
   a. water is a common molecule in interstellar space.
   b. life on Earth depends on it.
   c. no other molecules are solvents.
   d. the spectrum of water is very complex.

3. The Urey-Miller experiment produced _____ in a laboratory jar.
   a. life
   b. RNA and DNA
   c. amino acids
   d. proteins

4. A mutation is
   a. always a deadly change to DNA.
   b. always a beneficial change to DNA.
   c. a change to DNA that is sometimes beneficial and sometimes not.
   d. a change to DNA that is not inheritable.

5. Scientists think that terrestrial life probably originated in Earth's oceans, rather than on land, because (select all that apply)
   a. all the chemical pieces were in the ocean.
   b. energy was available in the ocean.
   c. the earliest evidence for life on Earth is from fossils of ocean-dwelling organisms.
   d. the deepest parts of the ocean have hydrothermal vents.

6. Any system with the processes of heredity, mutation, and natural selection will (choose all that apply)
   a. change over time.
   b. become larger over time.
   c. become more complex over time.
   d. develop intelligence.

7. The fact that no alien civilizations have yet been detected indicates that
   a. they are not there.
   b. they are rare.
   c. Earth is in a "blackout," and they are not talking to us.
   d. we don't know enough yet to draw any conclusions.

8. During the Cambrian explosion
   a. the dinosaurs were killed.
   b. all the carbon that is now here on Earth was produced.
   c. biodiversity increased significantly.
   d. a lot of carbon dioxide was released into the atmosphere.

9. The habitable zone is the place around a star where
   a. life has been found.
   b. atmospheres can contain oxygen.
   c. liquid water exists.
   d. liquid water can exist on the surface of a planet.

10. The difference between a prokaryote and a eukaryote is that prokaryotes
    a. have no DNA.
    b. have no cell wall.
    c. have no nucleus.
    d. do not exist today.

11. A thermophile is an organism that lives in extremely _____ environments.
    a. salty
    b. hot
    c. cold
    d. dry

12. The search for life elsewhere in the Solar System is carried out primarily by
    a. astronauts.
    b. robotic spacecraft.
    c. astronomers using optical telescopes.
    d. astronomers using radio telescopes.

13. Astronomers think that intelligent life is more likely to be found around stars of types F, G, K, and M because
    a. those stars are hot enough to have planets and moons with liquid water.
    b. those stars are cool enough to have planets and moons with liquid water.
    c. those stars live long enough for life to begin and evolve.
    d. those stars produce no ultraviolet radiation or X-rays.

14. Life first appeared on Earth
    a. billions of years ago.
    b. millions of years ago.
    c. hundreds of thousands of years ago.
    d. thousands of years ago.

15. In the phrase "theory of evolution," the word *theory* means that evolution
    a. is an idea that can't be tested scientifically.
    b. is an educated guess to explain natural phenomena.
    c. probably doesn't happen anymore.
    d. is a tested and corroborated scientific explanation of natural phenomena.

Thinking about the Concepts

16. How does the evolution of life on Earth depend on RNA and DNA?

17. How do scientists think that amino acids first formed on Earth?

18. Today, most organisms on Earth enjoy relatively moderate climates and temperatures. Compare this environment to some of the conditions in which early life developed.

19. In the Process of Science Figure, many areas of science are represented that all inform the science of astrobiology. Choose any two of these, and explain how one of these areas of science depends on the other.

20. How was Earth's carbon dioxide atmosphere changed into today's oxygen-rich atmosphere? How long did that transformation take?

21. What was the Cambrian explosion, and what might have caused it?

22. Why did plants and forests appear in high numbers before large animals did?

23. What is a habitable zone? What defines its boundaries?

24. Which general conditions were needed on Earth for life to arise?

25. Is evolution under way on Earth today? If so, how might humans continue to evolve?

26. Where did all the atoms in your body come from?

27. The two *Viking* spacecraft did not find convincing evidence of life on Mars when they visited that planet in the late 1970s nor did the *Phoenix* lander when it examined the martian soil in 2009. Do these results imply that life never existed on the planet? Why or why not?

28. Some scientists believe that humans may be the only advanced life in the galaxy today. If this is indeed the case, which factors in the Drake equation must be extremely small?

29. The second law of thermodynamics says that the entropy (a measure of disorder) of the universe is always increasing. Yet complex living organisms exist. Why does this not violate the second law of thermodynamics?

30. Why is it likely that life on Earth as we know it will end long before the Sun runs out of nuclear fuel?

Applying the Concepts

31. The Kepler Mission is currently searching for planets in the habitable zones of stars. Explain which factors in the Drake equation are affected by this search and how the final number *N* will be affected if Kepler finds that most stars have planets in their habitable zones.

32. Study Figure 24.18. The white blocks at the top represent the numbers 1–10. The bottom row of the set is a placeholder, and the top three rows of the set represent the numbers, in order from left to right. Explain the "rule" for the kind of counting shown here. (For example, how do three white blocks represent the number 7?)

33. Suppose that an organism replicates itself each second. If you start with a single specimen, what will the final population be after 10 seconds?

34. Suppose that an organism replicates itself each second. After how many seconds will the population increase by a factor of 1,024?

35. As a general rule, you can find the doubling time for exponential growth by dividing 70 by the rate of increase. So, if the population increases by 7 percent per year, the doubling time is 10 years (70/7 = 10). Suppose Earth's human population continues to grow by 1 percent annually. What is the doubling time? How much time will pass before there are 4 times as many humans on Earth?

36. As discussed in Working It Out 24.1, the doubling time for exponential growth is given by $P_F/P_O = 2^n$. Assume a self-replicating molecule that makes one copy of itself each second. Make a graph of the number of molecules versus time for the first 60 seconds after the molecule begins replicating.

37. The doubling time for *Escherichia coli* is 20 minutes, and you start getting sick when just 10 bacteria enter your system. How many bacteria are in your body after 12 hours?

38. If the chance that a given molecule will mutate is one in 100,000, how many generations are needed before, on average, at least one mutation has occurred?

39. Study the Drake equation in Working It Out 24.2. Make your own most optimistic and most pessimistic assumptions for each of the variables in the equation. What values do you find for *N*?

40. Study the Drake equation in Working It Out 24.2. Keep the other variables as is, but put in different values for the lifetime of a technological civilization. How does this affect your value of *N*? How does this affect the possibility of making contact with extraterrestrials?

41. Look back at Figure 5.24 in Chapter 5. Trace (or photocopy) this graph, and then add horizontal lines for the temperatures at which water freezes and boils. Which planets in the Solar System have measured temperatures that fall within those lines? Which planets have predicted temperatures (based on the equilibrium model) that fall within those lines? What does this tell you about assumptions about the habitable zone?

42. Figure 5.24 shows that Mercury's measured range of temperatures overlaps the temperatures at which water is a liquid. Is Mercury in the habitable zone? Why or why not?

43. Suppose astronomers announce the discovery of a new planet around a star with a mass equal to the Sun's. This planet has an orbital period of 87 days. Is this planet in the habitable zone for a Sun-like star?

44. Why do you think astronomers sent a coded radio signal to the globular cluster M13 in 1974—rather than, say, to a nearby star?

45. As noted in Section 24.1, some scientists suspect the early Earth was "seeded" with primitive life stored in comets and meteoroids. Knowing when and how our Solar System, galaxy, and universe formed, what timeline is required for such seeding to be possible?

USING THE WEB

46. a. Go to the online *Astrobiology Magazine* (http://astrobio.net), which covers many topics included in this chapter. Under "Origins," click on "Extreme Life." What are some new findings about extremophiles? Why is this important for astrobiologists? Under "Deep Space" click on "New Planets." What is a recent discovery?
    b. Go to the "Life, Unbounded" blog (http://blogs.scientificamerican.com/life-unbounded). What is a recent topic of interest? Is the discussion based on some new data? A new theory?

47. Solar System space missions:
    a. Go to the website for the *Cassini* mission to Saturn (http://saturn.jpl.nasa.gov). Is it still collecting data? What has been found recently on one of the moons that would be of interest to astrobiologists?
    b. Go to the website for the *Mars Science Laboratory* (http://mars.jpl.nasa.gov/msl/mission). The rover *Curiosity* landed on the surface in 2012. What are the science goals of this mission, and how do they relate to astrobiology? What are some recent results?

c. The *MAVEN* Mars mission (http://mars.nasa.gov/maven/) arrived at Mars in 2014. What are the science goals of this mission? Why are astrobiologists interested in the history of the climate and atmosphere of Mars? Are there any results?

48. Go to the "Habitable Exoplanets Catalog" website (http://phl.upr.edu/projects/habitable-exoplanets-catalog). Click on "Methods" to read about the different criteria for evaluating a planet's habitability. How many confirmed habitable exoplanets are there? How many candidates? How might the number of confirmed exoplanets affect the terms in the Drake equation?

49. a. Go to the website for the Kepler space telescope (http://kepler.nasa.gov): click on "News" and "Planet-finding News." What is a recent discovery of a planet in the habitable zone? What is a recent discovery of a planet with an interesting atmosphere?
    b. Go to the website for the European Space Agency (ESA) Gaia mission (http://sci.esa.int/gaia). What are the science objectives of this mission? Click on "Exoplanets" on the left. How will Gaia identify new planets? What properties of the planet will it be able to measure? What are some recent results?

50. Go to the website for "Super Planet Crash" (http://www.stefanom.org/spc/). Start with an "Earth" in the habitable zone and start adding in other planets. Does the system remain stable if you add in a second planet to the habitable zone? What happens if you add a third planet? What happens if you add planets inside or outside of the habitable zone?

smartwork

If your instructor assigns homework in SmartWork, access your assignments at digital.wwnorton.com/astro5.

EXPLORATION Fermi Problems and the Drake Equation

digital.wwnorton.com/astro5

The Drake equation is a way of organizing ideas about other intelligent, communicating civilizations in the galaxy. This type of thinking is very useful for estimating a value, particularly when analyzing systems for which counting is not possible. The types of problems that can be solved in this way are often called Fermi problems after Enrico Fermi, who was mentioned in this chapter. For example, we might ask, What is the circumference of Earth?

You could Google this question, or you could already "know" the answer, or you might look it up in this textbook. Alternatively, you could very carefully measure the shadow of a stick in two locations at the same time on the same day. Or you could drive around the planet.

Or you could reason this way:

How many time zones are between New York and Los Angeles?
3 time zones. You know this from traveling or from television.

How many miles is it from New York to Los Angeles?
3,000 miles. You know this from traveling.

So, how many miles per time zone?
3,000/3 = 1,000

How many time zones in the world?
24, because there are 24 hours in a day, and each time zone marks an hour.

So, what is the circumference of Earth?
24,000 miles, because there are 24 time zones, each 1,000 miles wide.

The measured circumference is 24,900 miles, which agrees with our estimate to within 4 percent.

Here we list several Fermi problems. Time yourself for an hour, and work as many of them as possible. (You don't have to do them in order!)

1 How much has the mass of the human population on Earth increased in the past year?

_____

2 How much energy does a horse consume in its lifetime?

_____

3 How many pounds of potatoes are consumed in the United States annually?

_____

4 How many cells are there in the human body?

_____

5 If your life earnings were given to you by the hour, how much is your time worth per hour?

_____

6 What is the weight of solid garbage thrown out by American families each year?

7 How fast does human hair grow (in feet per hour)?

_____

8 If all the people on Earth were crowded together, how much area would be covered?

_____

9 How many people could fit on Earth if every person occupied 1 square meter of land?

_____

10 How much carbon dioxide (CO_2) does an automobile emit each year?

_____

11 What is the mass of Earth?

_____

12 What is the average annual cost of an automobile, including overhead (maintenance, looking for parking, insurance, cleaning, and so on)?

_____

APPENDIX 1: Mathematical Tools

Mathematics helps scientists to understand the patterns they see and to communicate that understanding to others. Appendix 1 presents some tools that will be useful in our study of astronomy.

Powers of 10 and Scientific Notation

Astronomy is a science of both the very large and the very small. The mass of an electron, for example, is

0.0000000000000000000000000000009109 kilograms (kg)

whereas the distance to a galaxy far, far away might be about

100,000,000,000,000,000,000,000,000 meters

All it takes is a quick glance at these two numbers to see why astronomers need a more convenient way to express numbers.

Our number system is based on **powers of 10**. Going to the left of the decimal place,

$$10 = 10 \times 1$$
$$100 = 10 \times 10 \times 1$$
$$1{,}000 = 10 \times 10 \times 10 \times 1$$

and so on. Going to the right of the decimal place,

$$0.1 = \frac{1}{10} \times 1$$

$$0.01 = \frac{1}{10} \times \frac{1}{10} \times 1$$

$$0.001 = \frac{1}{10} \times \frac{1}{10} \times \frac{1}{10} \times 1$$

and so on. In other words, each place to the right or left of the decimal place in a number represents a power of 10. For example, 1 million can be written as

$$1 \text{ million} = 1{,}000{,}000 = 1 \times 10 \times 10 \times 10 \times 10 \times 10 \times 10$$

That is, 1 million is "1 multiplied by six factors of 10." **Scientific notation** combines these factors of 10 in convenient shorthand.

Rather than all being written out, the six factors of 10 are expressed using an exponent:

$$1 \text{ million} = 1 \times 10^6$$

which also means "1 multiplied by six factors of 10."

Moving to the right of the decimal place, each step *removes* a power of 10 from the number. One-millionth can be written as

$$1 \text{ millionth} = 1 \times \frac{1}{10} \times \frac{1}{10} \times \frac{1}{10} \times \frac{1}{10} \times \frac{1}{10} \times \frac{1}{10}$$

This divides by powers of 10, so this number can be written by use of a negative exponent

$$\frac{1}{10} = 10^{-1}$$

as

$$1 \text{ millionth} = 1 \times 10^{-1} \times 10^{-1} \times 10^{-1} \times 10^{-1} \times 10^{-1} \times 10^{-1}$$
$$= 1 \times 10^{-6}$$

Returning to our earlier examples, the mass of an electron is 9.109×10^{-31} kg, and the distant galaxy is located 1×10^{26} meters away. These are much more convenient ways of writing these values. Notice that *the exponent in scientific notation gives you a feel for the size of a number at a glance*. The exponent of 10 in the electron mass is −31, which quickly indicates that it is a very small number. The exponent of 10 in the distance to a remote galaxy, +26, quickly indicates that it is a very large number. This exponent is often called the *order of magnitude* of a number. When you see a number written in scientific notation while reading *21st Century Astronomy* (or elsewhere), just remember to look at the exponent to understand the size of the number.

Scientific notation is also convenient because it makes multiplying and dividing numbers easier. For example, 2 billion multiplied by eight-thousandths can be written as

$$2{,}000{,}000{,}000 \times 0.008$$

but it is more convenient to write these two numbers using scientific notation:

$$(2 \times 10^9) \times (8 \times 10^{-3})$$

We can regroup these expressions in the following form:

$$(2 \times 8) \times (10^9 \times 10^{-3})$$

The first part of the problem is just $2 \times 8 = 16$. The more interesting part of the problem is the multiplication in the right-hand parentheses. The first number, 10^9, is just shorthand for $10 \times 10 \times 10 \ldots$ nine times. That is, it represents nine factors of 10. The second number stands for three factors of 1/10—or removing three factors of 10 if you prefer to think of it that way. Altogether, that makes $9 - 3 = 6$ factors of 10. In other words,

$$10^9 \times 10^{-3} = 10^{9-3} = 10^6$$

Putting the problem together, we get

$$(2 \times 10^9) \times (8 \times 10^{-3}) = (2 \times 8) \times (10^9 \times 10^{-3})$$
$$= 16 \times 10^6$$

By convention, when a number is written in scientific notation, only one digit is placed to the left of the decimal point. In this case, there are two. However, 16 is 1.6×10, so we can add this additional factor of 10 to the exponent at right, making the final answer

$$1.6 \times 10^7$$

Dividing is just the inverse of multiplication. Dividing by 10^3 means removing three factors of 10 from a number. Using the previous number,

$$(1.6 \times 10^7) \div (2 \times 10^3) = (1.6 \div 2) \times (10^7 \div 10^3)$$
$$= 0.8 \times 10^{7-3}$$
$$= 0.8 \times 10^4$$

This time we have only a zero to the left of the decimal point. To get the number into proper form, we can substitute 8×10^{-1} for 0.8, giving

$$0.8 \times 10^4 = (8 \times 10^{-1}) \times 10^4 = 8 \times 10^3$$

Adding and subtracting numbers in scientific notation is somewhat more difficult because all numbers must be written as values multiplied by the *same* power of 10 before they can be added or subtracted. Therefore, you will need to use a calculator that has scientific notation. Most scientific calculators have a button that says EXP or EE. These mean "times 10 to the." So for 4×10^{12}, you would type [4][EXP][1][2] or [4][EE][1][2] into your calculator. Usually, this number shows up in the window on your calculator either just as you see it written in this book or as a 4 with a smaller 12 all the way over in the right side of the window.

Significant Figures

In the previous example, we actually broke some rules in the interest of explaining how powers of 10 are treated in scientific notation. The rules we broke involve the *precision* of the numbers. When expressing quantities in science, it is extremely important to know not only the value of a number but also how precise that value is.

The most complete way to keep track of the precision of numbers is to actually write down the uncertainty in the number. For example, suppose you know that the distance to a store (call it d) is between 0.8 and 1.2 kilometers (km); you can then write

$$d = 1.0 \pm 0.2 \text{ km}$$

where the symbol "$\pm$" is pronounced "plus or minus." In this example, d is between $1.0 - 0.2 = 0.8$ km and $1.0 + 0.2 = 1.2$ km. This is an unambiguous statement about the limitations on knowing the value of d, but carrying along the formal errors with every number written would be cumbersome at best. Instead, you keep track of the approximate precision of a number by using *significant figures*.

The convention for significant figures is this: Assume the written number has been rounded from a number that had one additional digit to the right of the decimal point. If a quantity d, which might represent the distance to the store, is "1.", then d is close to 1. It is likely not as small as "0.", and it is likely not as large as "2.". If instead it is written as

$$d = 1.0$$

then d is likely not 0.9 and is likely not 1.1. It is roughly 1.0 to the nearest tenth. The greater the number of significant figures, the more precisely the number is being specified. For example, 1.00000 is not the same number as 1.00. The first number, 1.00000, represents a value that is probably not as small as 0.99999 and probably not as large as 1.00001. The second number, 1.00, represents a value that is probably not as small as 0.99 nor as large as 1.01. The number 1.00000 is much more precise than the number 1.00.

In mathematical operations, significant figures are important. For example, $2.0 \times 1.6 = 3.2$. It does *not* equal 3.20000000000. *The product of two numbers cannot be known to any greater precision than the numbers themselves.* As a general rule, when you multiply and divide, the answer should have the same number of significant figures as the less precise of the numbers being multiplied or divided. In other words, $2.0 \times 1.602583475 = 3.2$. Because all you know is that the first factor is probably closer to 2.0 than to 1.9 or 2.1, all you know about the product is that it is between about 3.0 and 3.4. It is 3.2. It is not 3.205166950 (*even if that is the answer on your calculator*). The rest of the digits to the right of 3.2 just do not mean anything.

When two numbers are added or subtracted, if one number has a significant figure with a particular place value but another number does not, their sum or difference cannot have a significant figure in that place value. For example,

$$1,045.$$
$$+1.34567$$
$$\overline{1,046.}$$

The answer is "1,046." *not* "1,046.34567". Again, the extra digits to the right of the decimal place have no meaning because "1,045." is not known to that precision.

What is the precision of the number 1,000,000? As it is written, the answer is unclear. Are all those zeros really significant or are they placeholders? If the number is written in scientific notation, however, there is never a question. Instead of 1,000,000, you write 1.0×10^6 for a number that is known to the nearest hundred thousand or so; or you write 1.00000×10^6 for a number that is known to the nearest 10.

So the earlier example would have been more correct if written as

$$(2.0 \times 10^9) \times (8.0 \times 10^{-3}) = 1.6 \times 10^7$$

Algebra

There are many branches of mathematics. The branch that focuses on the relationships between quantities is called **algebra**. Basically, algebra begins by using symbols to represent quantities.

For example, you could write the distance you travel in a day as d. As it stands, d has no value. It might be 10,000 miles. It might be 30 feet. It does, however, have **units**; in this case, the units of distance. The average speed at which you travel is equal to the distance you travel divided by the time you take. By using the symbol v to represent your average speed and the symbol t to represent the time you take, instead of writing out "Your average speed is equal to the distance you travel divided by the time you take" you can write

$$v = \frac{d}{t}$$

The meaning of this algebraic expression is exactly the same as the sentence quoted before it, but it is much more concise. As it stands, v, d, and t still have no specific values. There are no numbers assigned to them yet. However, this expression indicates what the relationship between those numbers will be when you look at a specific example. For example, if you go 500 km ($d = 500$ km) in 10 hours ($t = 10$ hours), this expression tells you that your average speed is

$$v = \frac{d}{t} = \frac{500\,\text{km}}{10\,\text{h}} = 50\,\text{km/h}$$

Notice that the units in this expression act exactly like the numerical values. Dividing the two shows that the units of v are

kilometers divided by hours, or km/h (pronounced "kilometers per hour").

We introduced algebra as shorthand for expressing relations between quantities, but it is far more powerful than that. Algebra provides rules for manipulating the symbols used to represent quantities. We begin with a bit of notation for *powers* and *roots*. Similar to powers of ten, raising a quantity to a power means multiplying the quantity by itself some number of times. For example, if S is a symbol for something (anything), then S^2 (pronounced "S squared" or "S to the second power") means $S \times S$, and S^3 (pronounced "S cubed" or "S to the third power") means $S \times S \times S$. Suppose S represents the length of the side of a square. The area of the square is given by

$$\text{Area} = S \times S = S^2$$

If $S = 3$ meters (m), then the area of the square is

$$S^2 = 3\,\text{m} \times 3\,\text{m} = 9\,\text{m}^2$$

(pronounced "9 square meters"). This is why raising a quantity to the second power is called *squaring* the quantity. We could have done the same thing for the sides of a cube and found that the volume of the cube is

$$\text{Volume} = S \times S \times S = S^3$$

If $S = 3$ meters, then the volume of the cube is

$$S^3 = 3\,\text{m} \times 3\,\text{m} \times 3\,\text{m} = 27\,\text{m}^3$$

(pronounced "27 cubic meters"). This is why raising a quantity to the third power is called *cubing* the quantity.

Roots are the reverse of this process. The square root of a number is the value that, when squared, gives the original quantity. The square root of 4 is 2, which means that $2 \times 2 = 4$. The square root of 9 is 3, which means that $3 \times 3 = 9$. Similarly, the cube root of a quantity is the value that, when cubed, gives the original quantity. The cube root of 8 is 2, which means that $2 \times 2 \times 2 = 8$. Roots are written with the symbol $\sqrt{}$. For example, the square root of 9 is written as

$$\sqrt{9} = 3$$

and the cube root of 8 is written as

$$\sqrt[3]{8} = 2$$

If the volume of a cube is $V = S^3$, then

$$S = \sqrt[3]{V} = \sqrt[3]{S^3}$$

Roots can also be written as powers. Powers and roots behave exactly like the exponents of 10 in our discussion of scientific notation. (The exponents used in scientific notation are just powers of 10.) For example, if a, n, and m are all algebraic quantities, then

$$a^n \times a^m = a^{n+m} \quad \text{and} \quad \frac{a^n}{a^m} = a^{n-m}$$

(The square root of a can also be written $a^{1/2}$ and the cube root of a can be written $a^{1/3}$.)

The rules of arithmetic can be applied to the symbolic quantities of algebra. As long as the rules of algebra are applied properly, then the relationships among symbols arrived at through algebraic manipulation remain true for the physical quantities that those symbols represent.

Here we summarize a few algebraic rules and relationships. In this summary, a, b, c, m, n, r, x, and y are all algebraic quantities.

Associative rule:

$$a \times b \times c = (a \times b) \times c = a \times (b \times c)$$

Commutative rule:

$$a \times b = b \times a$$

Distributive rule:

$$a \times (b + c) = (a \times b) + (a \times c)$$

Cross-multiplication:

$$\text{If } \frac{a}{b} = \frac{c}{d}, \text{ then } ad = bc$$

Working with exponents:

$$\frac{1}{a^n} = a^{-n} \qquad a^n a^m = a^{n+m}$$

$$\frac{a^n}{a^m} = a^{n-m} \qquad (a^n)^m = a^{n \times m} \qquad \left(\frac{a}{b}\right)^n = \frac{a^n}{b^n}$$

Equation of a line with slope *m* and *y*-intercept *b*:

$$y = mx + b$$

Equation of a circle with radius *r* centered at *x* = 0, *y* = 0:

$$x^2 + y^2 = r^2$$

Angles and Distances

The farther away something is, the smaller it appears. This is common sense and everyday experience. Because astronomers cannot walk up to the object they are studying and measure it with a meterstick, their knowledge about the sizes of things usually depends on relating the size of an object, its distance, and the angle it covers in the sky.

One way to measure angles is to use a unit called the **radian**. As shown in **Figure A1.1a**, the size of an angle in radians is the length of the arc subtending the angle, divided by the radius of the circle. In the figure, the angle $x = S/r$ radians.

Because the circumference of a circle is 2π multiplied by the radius ($C = 2\pi r$), a complete circle has an angular measure of $(2\pi r)/r = 2\pi$ radians. In more conventional angular measure, a complete circle is 360°, so

$$360° = 2\pi \text{ radians}$$

or

$$1 \text{ radian} = \frac{360°}{2\pi} = 57.2958°$$

Often, seconds of arc (**arcseconds**) are used to measure angles for stars and galaxies. A degree is divided into 60 minutes of arc (**arcminutes**), each of which is divided into 60 seconds of arc—so there are 3,600 seconds of arc in a degree. Therefore,

$$3{,}600\frac{\text{arcseconds}}{\text{degree}} \times 57.2958\frac{\text{degree}}{\text{radian}} = 206{,}265\frac{\text{arcseconds}}{\text{radian}}$$

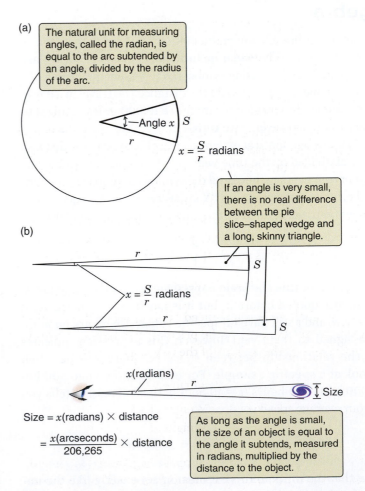

(a) The natural unit for measuring angles, called the radian, is equal to the arc subtended by an angle, divided by the radius of the arc.

—Angle x S
r
$x = \dfrac{S}{r}$ radians

If an angle is very small, there is no real difference between the pie slice–shaped wedge and a long, skinny triangle.

(b)
r S
$x = \dfrac{S}{r}$ radians
r S

x(radians) r Size

$$\text{Size} = x(\text{radians}) \times \text{distance}$$

$$= \frac{x(\text{arcseconds})}{206{,}265} \times \text{distance}$$

As long as the angle is small, the size of an object is equal to the angle it subtends, measured in radians, multiplied by the distance to the object.

Figure A1.1 Measuring angles.

If the angle is small enough (which it usually is in astronomy), there is very little difference between the pie slice just described and a long skinny triangle with a short side of length S, as Figure A1.1b illustrates. So, if you know the distance d to an object and you can measure the angular size x of the object, then the size of the object is given by

$$S = x(\text{in radians}) \times d = \frac{x(\text{in degrees})}{57.2958\,\text{degrees/radian}} \times d$$

$$S = \frac{x(\text{in arcseconds})}{206{,}265\,\text{arcseconds/radian}} \times d$$

which is how astronomers relate an object's angular size, distance, and physical size.

Circles and Spheres

To round out these mathematical tools, here are a few useful formulas for circles and spheres. The circle or sphere in each case has a radius r.

$$\text{Circumference}_{\text{circle}} = 2\pi r$$

$$\text{Area}_{\text{circle}} = \pi r^2$$

$$\text{Surface area}_{\text{sphere}} = 4\pi r^2$$

$$\text{Volume}_{\text{sphere}} = \frac{4}{3}\pi r^3$$

Working with Proportionalities

Most of the mathematics in *21st Century Astronomy* involves **proportionalities**—statements about how one physical quantity changes when another quantity changes. We began a discussion of proportionality in Working It Out 1.1; here, we offer a few examples of working with proportionalities.

To use a statement of proportionality to compare two objects, begin by turning the proportionality into a ratio. For example, the price of a bag of apples is **proportional** to the weight of the bag:

$$\text{Price} \propto \text{Weight}$$

Here, the symbol $\propto$ is pronounced "is proportional to." What this means is that the ratio of the prices of two bags of apples is equal to the ratio of the weights of the two bags:

$$\text{Price} \propto \text{Weight means}$$

$$\frac{\text{Price of A}}{\text{Price of B}} = \frac{\text{Weight of A}}{\text{Weight of B}}$$

Let's work a specific example. Suppose bag A weighs 2 pounds and bag B weighs 1 pound. That means bag A will cost twice as much as bag B. We can turn this proportionality into the following equation:

$$\frac{\text{Price of A}}{\text{Price of B}} = \frac{\text{Weight of A}}{\text{Weight of B}} = \frac{2\,\text{lb}}{1\,\text{lb}} = 2$$

In other words, the price of bag A is 2 times the price of bag B. The price per pound is an example of a **constant of proportionality**.

Now let's work another, more complicated example. In Chapter 13, we discuss how the luminosity, brightness, and distance of stars are related. The luminosity of a star—the total energy that the star radiates each second—is proportional to the star's brightness multiplied by the square of its distance:

$$\text{Luminosity} \propto \text{Brightness} \times \text{Distance}^2$$

We can turn this proportionality into a ratio for two stars, A and B:

$$\frac{\text{Luminosity of A}}{\text{Luminosity of B}} = \frac{\text{Brightness of A}}{\text{Brightness of B}} \times \left(\frac{\text{Distance of A}}{\text{Distance of B}}\right)^2$$

If we use the symbols L, b, and d to represent luminosity, brightness, and distance, respectively, this equation becomes

$$\frac{L_A}{L_B} = \frac{b_A}{b_B} \times \left(\frac{d_A}{d_B}\right)^2$$

As an example, suppose that star A appears twice as bright in the sky as star B, but star A is located 10 times as far away as star B. How do the luminosities of the two stars compare? We know that

$$\text{Luminosity} \propto \text{Brightness} \times \text{Distance}^2$$

we write

$$\frac{\text{Luminosity of A}}{\text{Luminosity of B}} = \frac{\text{Brightness of A}}{\text{Brightness of B}} \times \left(\frac{\text{Distance of A}}{\text{Distance of B}}\right)^2$$

$$\frac{\text{Luminosity of A}}{\text{Luminosity of B}} = \frac{2}{1} \times \left(\frac{10}{1}\right)^2 = 200$$

In other words, star A is 200 times as luminous as star B.

Two quantities may also be inversely proportional, such that making one of them smaller makes the other larger. For example, when you are driving to another town, if you drive twice as fast, it takes half the time to get there. The travel time is inversely proportional to the travel speed. We write this relationship as

$$\text{Time} \propto \frac{1}{\text{Speed}}$$

Proportionalities are used to compare one object to another. Constants of proportionality are used to calculate actual values. In *21st Century Astronomy*, it is usually the proportionality that is important.

APPENDIX 2: Physical Constants and Units

Fundamental Physical Constants

Constant	Symbol	Value
Speed of light in a vacuum	c	2.99792×10^8 m/s
Universal gravitational constant	G	6.6738×10^{-11} m³/(kg s²) 6.6738×10^{-20} km³/(kg s²)
Planck constant	h	6.62607×10^{-34} J-s
Boltzmann constant	k	1.38065×10^{-23} J/K
Stefan-Boltzmann constant	σ	5.67037×10^{-8} W/(m² K⁴)
Mass of electron	m_e	9.10938×10^{-31} kg
Mass of proton	m_p	1.67262×10^{-27} kg
Mass of neutron	m_n	1.67493×10^{-27} kg
Electric charge of electron or proton	e	1.60218×10^{-19} C

SOURCE: Data from the Particle Data Group (http://pdg.lbl.gov).

Unit Prefixes

Prefix*	Name	Factor†
n	nano-	10^{-9}
μ	micro-	10^{-6}
m	milli-	10^{-3}
k	kilo-	10^3
M	mega-	10^6
G	giga-	10^9
T	tera-	10^{12}

These prefixes (*), when appended to a unit, change the size of the unit by the factor (†) given. For example, 1 km is 10^3 meters.

Units and Values

Quantity	Fundamental Unit	Values
Length	meters (m)	radius of Sun (R_{Sun}) = 6.9551×10^8 m astronomical unit (AU) = 1.49598×10^{11} m 1 AU = 149,598,000 km light-year (ly) = 9.4605×10^{15} m 1 ly = 6.324×10^4 AU 1 parsec (pc) = 3.262 *ly* = 3.0857×10^{16} m 1 m = 3.281 feet
Volume	meters^3 (m^3)	1 m^3 = 1,000 liters = 264.2 gallons
Mass	kilograms (kg)	1 kg = 1,000 grams mass of Earth (M_{Earth}) = 5.9726×10^{24} kg mass of Sun (M_{Sun}) = 1.9885×10^{30} kg
Time	seconds (s)	1 hour (h) = 60 minutes (min) = 3,600 s solar day (noon to noon) = 86,400 s sidereal day (Earth rotation period) = 86,164.1 s tropical year (equinox to equinox) = 365.24219 days = 3.15569×10^7 s sidereal year (Earth orbital period) = 365.25636 days = 3.15581×10^7 s
Speed	meters/second (m/s)	1 m/s = 2.237 miles/h 1 km/s = 1,000 m/s = 3,600 km/h c = 2.99792×10^8 m/s = 299,792 km/s
Acceleration	meters/second^2 (m/s^2)	g = gravitational acceleration on Earth = 9.81 m/s^2
Energy	joules (J)	1 J = 1 kg m^2/s^2 1 megaton = 4.18×10^{15} J
Power	watts (W)	1 W = 1 J/s solar luminosity (L_{Sun}) = 3.828×10^{26} W
Force	newtons (N)	1 N = 1 kg m/s^2 1 pound (lb) = 4.448 N 1 N = 0.22481 lb
Pressure	newtons/meter^2 (N/m^2)	atmospheric pressure at sea level = 1.013×10^5 N/m^2 = 1.013 bar
Temperature	kelvins (K)	absolute zero = 0 K = -273.15°C = -459.67°F

SOURCE: Data from the Particle Data Group (http://pdg.lbl.gov).

APPENDIX 3: Periodic Table of the Elements

Legend:
- 1 — Atomic number
- H — Symbol
- Hydrogen — Name
- 1.00794 — Average atomic mass

- Metals
- Metalloids
- Nonmetals

1 (1A)	2 (2A)	3 (3B)	4 (4B)	5 (5B)	6 (6B)	7 (7B)	8 (8B)	9 (8B)	10 (8B)	11 (1B)	12 (2B)	13 (3A)	14 (4A)	15 (5A)	16 (6A)	17 (7A)	18 (8A)
1 H Hydrogen 1.00794																	2 He Helium 4.002602
3 Li Lithium 6.941	4 Be Beryllium 9.012182											5 B Boron 10.811	6 C Carbon 12.0107	7 N Nitrogen 14.0067	8 O Oxygen 15.9994	9 F Fluorine 18.9984032	10 Ne Neon 20.1797
11 Na Sodium 22.98976928	12 Mg Magnesium 24.3050											13 Al Aluminum 26.9815386	14 Si Silicon 28.0855	15 P Phosphorus 30.973762	16 S Sulfur 32.065	17 Cl Chlorine 35.453	18 Ar Argon 39.948
19 K Potassium 39.0983	20 Ca Calcium 40.078	21 Sc Scandium 44.955912	22 Ti Titanium 47.867	23 V Vanadium 50.9415	24 Cr Chromium 51.9961	25 Mn Manganese 54.938045	26 Fe Iron 55.845	27 Co Cobalt 58.933195	28 Ni Nickel 58.6934	29 Cu Copper 63.546	30 Zn Zinc 65.38	31 Ga Gallium 69.723	32 Ge Germanium 72.64	33 As Arsenic 74.92160	34 Se Selenium 78.96	35 Br Bromine 79.904	36 Kr Krypton 83.798
37 Rb Rubidium 85.4678	38 Sr Strontium 87.62	39 Y Yttrium 88.90585	40 Zr Zirconium 91.224	41 Nb Niobium 92.90638	42 Mo Molybdenum 95.96	43 Tc Technetium [98]	44 Ru Ruthenium 101.07	45 Rh Rhodium 102.90550	46 Pd Palladium 106.42	47 Ag Silver 107.8682	48 Cd Cadmium 112.411	49 In Indium 114.818	50 Sn Tin 118.710	51 Sb Antimony 121.760	52 Te Tellurium 127.60	53 I Iodine 126.90447	54 Xe Xenon 131.293
55 Cs Cesium 132.9054519	56 Ba Barium 137.327	57 La Lanthanum 138.90547	72 Hf Hafnium 178.49	73 Ta Tantalum 180.94788	74 W Tungsten 183.84	75 Re Rhenium 186.207	76 Os Osmium 190.23	77 Ir Iridium 192.217	78 Pt Platinum 195.084	79 Au Gold 196.966569	80 Hg Mercury 200.59	81 Tl Thallium 204.3833	82 Pb Lead 207.2	83 Bi Bismuth 208.98040	84 Po Polonium [209]	85 At Astatine [210]	86 Rn Radon [222]
87 Fr Francium [223]	88 Ra Radium [226]	89 Ac Actinium [227]	104 Rf Rutherfordium [261]	105 Db Dubnium [262]	106 Sg Seaborgium [266]	107 Bh Bohrium [264]	108 Hs Hassium [277]	109 Mt Meitnerium [268]	110 Ds Darmstadtium [271]	111 Rg Roentgenium [272]	112 Cn Copernicium [285]	113 Uut Ununtrium [284]	114 Fl Flerovium [289]	115 Uup Ununpentium [288]	116 Lv Livermorium [292]	117 Uus Ununseptium [294]	118 Uuo Ununoctium [294]

6 Lanthanides

58 Ce Cerium 140.116	59 Pr Praseodymium 140.90765	60 Nd Neodymium 144.242	61 Pm Promethium [145]	62 Sm Samarium 150.36	63 Eu Europium 151.964	64 Gd Gadolinium 157.25	65 Tb Terbium 158.92535	66 Dy Dysprosium 162.500	67 Ho Holmium 164.93032	68 Er Erbium 167.259	69 Tm Thulium 168.93421	70 Yb Ytterbium 173.05	71 Lu Lutetium 174.967

7 Actinides

90 Th Thorium 232.03806	91 Pa Protactinium 231.03588	92 U Uranium 238.02891	93 Np Neptunium [237]	94 Pu Plutonium [244]	95 Am Americium [243]	96 Cm Curium [247]	97 Bk Berkelium [247]	98 Cf Californium [251]	99 Es Einsteinium [252]	100 Fm Fermium [257]	101 Md Mendelevium [258]	102 No Nobelium [259]	103 Lr Lawrencium [262]

We have used the U.S. system as well as the system recommended by the International Union of Pure and Applied Chemistry (IUPAC) to label the groups in this periodic table. The system used in the United States includes a letter and a number (1A, 2A, 3B, 4B, etc.), which is close to the system developed by Mendeleev. The IUPAC system uses numbers 1–18 and has been recommended by the American Chemical Society (ACS). While we show both numbering systems here, we use the IUPAC system exclusively in the book. Elements with atomic numbers higher than 112 have been reported but not yet fully authenticated.

APPENDIX 4: Properties of Planets, Dwarf Planets, and Moons

Physical Data for Planets and Dwarf Planets

Planet	EQUATORIAL RADIUS (km)	EQUATORIAL RADIUS (R/R_{Earth})	MASS (kg)	MASS (M/M_{Earth})	Average Density (relative to water*)	Rotation Period (days)	Tilt of Rotation Axis (degrees, relative to orbit)	Equatorial Surface Gravity (relative to Earth[†])	Escape Velocity (km/s)	Average Surface Temperature (K)
Mercury	2,440	0.383	3.30×10^{23}	0.055	5.427	58.65	0.01	0.378	4.3	340 (100, 700)[§]
Venus	6,052	0.950	4.87×10^{24}	0.815	5.243	243.02[‡]	177.3	0.907	10.36	735
Earth	6,378	1.000	5.97×10^{24}	1.000	5.513	1.000	23.44	1.000	11.19	288 (185, 331)[§]
Mars	3,396	0.532	6.42×10^{23}	0.107	3.934	1.026	25.19	0.377	5.03	210 (120, 293)[§]
Ceres	481.5	0.076	9.47×10^{20}	0.0002	2.09	0.378	~3	0.28	1.86	168
Jupiter	71,492	11.209	1.90×10^{27}	317.8	1.326	0.4135	3.13	2.528	6.02	125
Saturn	60,268	9.449	5.68×10^{26}	95.16	0.687	0.4440	26.73	1.065	36.1	134
Uranus	25,559	4.007	8.68×10^{25}	14.54	1.270	0.7183[‡]	97.77	0.887	21.4	76
Neptune	24,764	3.883	1.02×10^{26}	17.148	1.638	0.6713	28.32	1.14	23.6	59
Pluto	1,195	0.187	1.30×10^{22}	0.0022	2.050	6.387[‡]	122.53	0.070	1.23	44
Haumea	~650	0.11	4.0×10^{21}	0.0007	~3	0.163	?	0.045	0.84	<50
Makemake	715	0.11	4.0×10^{21}	0.0007	~1.7	0.937	?	~0.5	~0.8	~30
Eris	1,163	0.182	1.67×10^{22}	0.0028	2.5	1.08	?	0.084	1.38	43

*The density of water is 1,000 kg/m³.
[†]The surface gravity of Earth is 9.81 m/s².
[‡]Venus, Uranus, and Pluto rotate opposite to the directions of their orbits. Their north poles are south of their orbital planes.
[§]Where provided, values in parentheses give extremes of recorded temperatures.

Orbital Data for Planets and Dwarf Planets

Planet	MEAN DISTANCE FROM SUN (A*)		Orbital Period (P) (sidereal years)	Eccentricity	Inclination (degrees, relative to ecliptic)	Average Speed (km/s)
	(10^6 km)	(AU)				
Mercury	57.9	0.387	0.241	0.2056	7.0	47.36
Venus	108.2	0.723	0.615	0.0068	3.39	35.02
Earth	149.6	1.000	1.000	0.0167	0.000	29.78
Mars	227.9	1.524	1.881	0.0934	1.85	24.08
Ceres	413.7	2.765	4.603	0.079	10.59	17.88
Jupiter	778.3	5.203	11.863	0.0484	1.30	13.06
Saturn	1,426.7	9.537	29.447	0.0539	2.49	9.64
Uranus	2,870.7	19.189	84.017	0.0473	0.77	6.80
Neptune	4,498.4	30.070	164.79	0.0086	1.769	5.43
Pluto	5,906.4	39.48	247.92	0.2488	17.14	4.67
Haumea	6,432.0	43.0	281.9	0.198	28.22	4.50
Makemake	6,783.3	45.5	305.3	0.164	29.00	4.39
Eris	10,180	68.0	561.4	0.434	43.17	3.43

*A is the semimajor axis of the planet's elliptical orbit.

Properties of Selected Moons*

Planet	Moon	ORBITAL PROPERTIES		PHYSICAL PROPERTIES		Relative Density (g/cm³) (water = 1.00)
		P (days)	A (10³ km)	R (km)	M (10²⁰ kg)	
Earth (1 moon)	Moon	27.32	384.4	1,737.5	735	3.34
Mars (2 moons)	Phobos	0.32	9.38	13.4 × 11.2 × 9.2	0.0001	1.9
	Deimos	1.26	23.46	7.5 × 6.1 × 5.2	0.00002	1.5
Jupiter (67 known moons)	Metis	0.30	128	21.5	0.00012	3
	Amalthea	0.50	181.4	131 × 73 × 67	0.0207	0.8
	Io	1.77	421.8	1,822	893	3.53
	Europa	3.55	671.1	1,561	480	3.01
	Ganymede	7.15	1,070	2,631	1,482	1.94
	Callisto	16.69	1,883	2,410	1,076	1.83
	Himalia	250.56	11,461	85	0.067	2.6
	Pasiphae	744†	23,624	30	0.0030	2.6
	Callirrhoe	759†	24,102	4.3	0.00001	2.6
Saturn (62 known moons)	Pan	0.58	133.58	14	0.00005	0.42
	Prometheus	0.61	139.38	74 × 50 × 34	0.0016	0.48
	Pandora	0.63	141.70	55 × 44 × 31	0.0014	0.49
	Mimas	0.94	185.54	198	0.38	1.15
	Enceladus	1.37	238.04	252	1.08	1.6
	Tethys	1.89	294.67	533	6.18	0.97
	Dione	2.74	377.42	562	11.0	1.48
	Rhea	4.52	527.07	764	23.1	1.23
	Titan	15.95	1,222	2,575	1,346	1.88
	Hyperion	21.28	1,501	205 × 130 × 110	0.0559	0.54
	Iapetus	79.33	3,561	736	18.1	1.08
	Phoebe	550.3†	12,948	107	0.08	1.6
	Paaliaq	687.5	15,024	11	0.0001	2.3
Uranus (27 known moons)	Cordelia	0.34	49.8	20	0.0004	1.3
	Miranda	1.41	129.9	236	0.66	1.21
	Ariel	2.52	190.9	579	12.9	1.59
	Umbriel	4.14	266.0	585	12.2	1.46

(continued)

Properties of Selected Moons*

(continued)

Planet	Moon	ORBITAL PROPERTIES		PHYSICAL PROPERTIES		Relative Density[†] (g/cm³) (water = 1.00)
		P (days)	A (10³ km)	R (km)	M (10²⁰ kg)	
Uranus (27 known moons)	Titania	8.71	436.3	789	34.2	1.66
	Oberon	13.46	583.5	761	28.8	1.56
	Setebos	2,225[†]	17,420	24	0.0009	1.5
Neptune (14 known moons)	Naiad	0.29	48.2	48 × 30 × 26	0.002	1.3
	Larissa	0.56	73.5	108 × 102 × 84	0.05	1.3
	Proteus	1.12	117.6	218 × 208 × 201	0.5	1.3
	Triton	5.88[†]	354.8	1,353	214	2.06
	Nereid	360.13	5,513.8	170	0.3	1.5
Pluto (5 moons)	Charon	6.39	17.54	604	15.5	1.68
Haumea (2 moons)	Namaka	18	25.66	85	0.018	~1
	Hi'iaka	49	49.88	170	0.179	~1
Eris	Dysnomia	15.8	37.4	100–500?	?	?

*Innermost, outermost, largest, and/or a few other moons for each planet.
[†]Irregular moon (has retrograde orbit).

APPENDIX 5: Space Missions

Selected Recent and Current Solar System Missions

Spacecraft	Sponsoring Nation(s)*	Destination	Launch Year	Type	Status (mid-2015)
Voyager 1 and 2	USA	Jupiter, Saturn, Uranus (2), Neptune (2)	1977	Flyby	Actively exploring outer edge of Solar System
Galileo	USA	Jupiter	1989	Orbiter/probe	Ended 2003
Ulysses	USA, Europe	Sun	1990	Solar polar orbiter	Ended 2008
SOHO	USA, Europe	Sun	1995	Orbiter	Active
Mars Global Surveyor	USA	Mars	1996	Orbiter	Ended 2006
Cassini-Huygens	USA, Europe, Italy	Saturn, Titan	1997	Saturn orbiter, Titan probe/lander	Orbiter active
Stardust	USA	Comets	1999	Sample return/flyby	Ended 2011
Mars Odyssey	USA	Mars	2001	Orbiter	Active
Mars Exploration Rover	USA	Mars	2003	Two landers	One rover active
Hayabusa	Japan	Asteroid	2003	Sample return	Ended 2010
Mars Express	Europe	Mars	2003	Orbiter	Active
Messenger	USA	Mercury (2011)	2004	Orbiter	Ended 2015
Venus Express	Europe	Venus	2005	Orbiter	Ended 2014
Mars Reconnaissance Orbiter (MRO)	USA	Mars	2005	Orbiter	Active
Deep Impact/EPOXI	USA	Comet Hartley (2010)	2005	Impactor/flyby	Ended 2010
STEREO	USA	Sun	2006	Two orbiters	Active
New Horizons	USA	Pluto (2015)	2006	Flyby	Active
Chang'e 1	China	Moon	2007	Orbiter	Ended 2009
Kayuga	Japan	Moon	2007	Orbiter	Ended 2009

(continued)

Selected Recent and Current Solar System Missions

(continued)

Spacecraft	Sponsoring Nation(s)*	Destination	Launch Year	Type	Status (mid-2015)
Artemis	USA	Moon, solar wind	2007		Active
Dawn	USA	Vesta (2011), Ceres (2015)	2007	Orbiter	Active
Chandrayaan	India	Moon	2008	Orbiter/impactor	Ended 2009
Lunar Reconnaissance Orbiter (LRO)	USA	Moon	2009	Orbiter	Active
Lunar Crater Observation and Sensing Satellite (LCROSS)	USA	Moon	2009	Impactor	Ended 2009
Chang'e 2	China	Moon	2010	Orbiter	Ended 2011
Juno	USA	Jupiter (2016)	2011	Orbiter	En route
Gravity Recovery and Interior Laboratory (GRAIL)	USA	Moon	2011	Two orbiters	Ended 2012
Mars Science Laboratory (Curiosity rover)	USA	Mars	2011	Lander	Active
Mars Atmosphere and Volatile EvolutioN (MAVEN) mission	USA	Mars	2013	Orbiter	Active
Chang'e 3	China	Moon	2013	Lander	On lunar surface

*Countries are represented by the following agencies: China = CNSA (China National Space Administration); Europe = ESA (European Space Agency); India = ISRO (Indian Space Research Organisation); Italy = Italian Space Agency; Japan = JAXA (Japan Aerospace Exploration Agency); USA = NASA (National Aeronautics and Space Administration).

APPENDIX 6: Nearest and Brightest Stars

Stars within 12 Light-Years of Earth

Name*	Distance (ly)	Spectral Type†	Relative Visual Luminosity‡ (Sun = 1.000)	Apparent Magnitude	Absolute Magnitude
Sun	1.55×10^{-5}	G2V	1.000	−26.74	4.83
Alpha Centauri C (Proxima Centauri)	4.24	M5.0V	0.000052	11.05	15.48
Alpha Centauri A	4.36	G2V	1.5	0.01	4.38
Alpha Centauri B	4.36	K0V	0.44	1.34	5.71
Barnard's star	5.96	M4Ve	0.00043	9.57	13.25
CN Leonis	7.78	M5.5	0.000019	13.53	16.64
BD +36-2147	8.29	M2.0V	0.0057	7.47	10.44
Sirius A	8.58	A1V	22.1	−1.43	1.47
Sirius B	8.58	DA2	0.0025	8.44	11.34
BL Ceti	8.73	M5.5V	0.000059	12.61	15.40
UV Ceti	8.73	M6.0	0.000039	13.06	15.85
V1216 Sagittarii	9.68	M3.5V	0.00050	10.44	13.08
HH Andromedae	10.32	M5.5V	0.00010	12.29	14.79
Epsilon Eridani	10.52	K2V	0.28	3.73	6.20
Lacaille 9352	10.74	M1.0V	0.011	7.34	9.76
FI Virginis	10.92	M4.0V	0.00033	11.16	13.53
EZ Aquarii A	11.26	M5.0V	0.000063	13.03	15.33
EZ Aquarii B	11.26	M5e	0.000050	13.27	15.58
EZ Aquarii C	11.26	MV	0.000010	15.07	17.37
Procyon A	11.40	F5IV-V	7.38	0.38	2.66
Procyon B	11.40	DA	0.00055	10.70	12.98

(continued)

Stars within 12 Light-Years of Earth

(continued)

Name*	Distance (ly)	Spectral Type†	Relative Visual Luminosity‡ (Sun = 1.000)	Apparent Magnitude	Absolute Magnitude
61 Cygni A	11.40	K5.0V	0.087	5.20	7.48
61 Cygni B	11.40	K7.0V	0.041	6.03	8.31
Gliese 725 A	11.52	M3.0V	0.0029	8.90	11.17
Gliese 725 B	11.52	M3.5V	0.0014	9.69	11.96
Andromedae GX	11.62	M1.5V	0.0064	8.08	10.32
Andromedae GQ	11.62	M3.5V	0.00041	11.06	13.30
Epsilon Indi A	11.82	K5Ve	0.15	4.68	6.89
Epsilon Indi B (brown dwarf)	11.82	T1.0	—	—	—
Epsilon Indi C (brown dwarf)	11.82	T6.0	—	—	—
DX Cancri	11.82	M6.0V	0.000012	14.90	17.10
Tau Ceti	11.88	G8.5V	0.46	3.49	5.68
Gliese 1061	11.99	M5.5V	0.000067	13.09	15.26

SOURCE: From the Research Consortium on Nearby Stars (http://www.recons.org).
*Stars may carry many names, including common names (such as Sirius), names based on their prominence within a constellation (such as Alpha Canis Majoris, another name for Sirius), or names based on their inclusion in a catalog (such as BD +36-2147). Addition of letters A, B, and so on, or superscripts indicates membership in a multiple-star system.
†Spectral types such as M3 are discussed in Chapter 13. Other letters or numbers provide additional information. For example, "V" after the spectral type indicates a main-sequence star, and "III" indicates a giant star. Stars of spectral type T are brown dwarfs.
‡Luminosity in this table refers only to radiation in "visual" light.

The 25 Brightest Stars in the Sky

Name	Common Name	Distance (ly)	Spectral Type	Relative Visual Luminosity* (Sun = 1.000)	Apparent Visual Magnitude	Absolute Visual Magnitude
Sun	Sun	1.58×10^{-5}	G2V	1.000	−26.74	4.83
Alpha Canis Majoris	Sirius	8.60	A1V	22.9	−1.46	1.43
Alpha Carinae	Canopus	313	F0II	14,900	−0.72	−5.60
Alpha¹ Centauri	Rigil Kentaurus A	4.36	G2V	1.51	−0.01	4.38
Alpha² Centauri	Rigil Kentaurus B	4.36	K1V	0.44	1.33	5.71
Alpha Bootis	Arcturus	36.7	K1.5III	113	−0.04	−0.30
Alpha Lyrae	Vega	25.3	A0Va	49.2	0.03	0.60
Alpha Aurigae	Capella	43	G5IIIe+G0III	137	0.08	−0.51
Beta Orionis	Rigel	860	B8Iab	54,000	0.12	−7.0
Alpha Canis Minoris	Procyon	11.5	F5IV-V	7.73	0.34	2.61
Alpha Eridani	Achernar	140	B3Vpe	1,030	0.46	−2.70
Beta Centauri	Hadar	392	B1III	7,180	0.61	−4.81
Alpha Orionis	Betelgeuse	570	M1-2Iab	13,600	0.7	−5.5
Alpha Aquilae	Altair	16.7	A7V	11.1	0.77	2.22
Alpha Crucis	Acrux	325	B0.5IV+B1V	3,100	1.3	−3.9
Alpha Tauri	Aldebaran	67	K5III	163	0.85	−0.70
Alpha Scorpii	Antares	550	M1.5Ib	16,300	0.96	−5.7
Alpha Virginis	Spica	250	B1IV+B4V	1,920	1.04	−3.38
Beta Geminorum	Pollux	34	K0III	32.2	1.14	1.06
Alpha Piscis	Fomalhaut	25	A3V	17.4	1.16	1.73
Beta Crucis	Mimosa	280	B0.5III	1,980	1.25	−3.41
Alpha Cygni	Deneb	1,425	A2Ia	58,600	1.25	−7.09
Alpha Leonis	Regulus	79	B7V	146	1.35	−0.58
Epsilon Canis Majoris	Adhara	405	B2II	3,400	1.50	−4.0
Alpha Gemini	Castor	51	A1V+A5Vm	49	1.58	0.61
Gamma Crucis	Gacrux	88	M3.5III	138	1.63	−0.52

SOURCES: Data from Jim Kaler's *STARS* page (http://stars.astro.illinois.edu/sow/bright.html); SIMBAD Astronomical Database (http://simbad.u-strasbg.fr/simbad).
*Luminosity in this table refers only to radiation in "visual" light.

APPENDIX 7: Observing the Sky

The purpose of this appendix is to provide enough information so that you can make sense of a star chart or list of astronomical objects and find a few objects in the sky.

Celestial Coordinates

In Chapter 2, we discuss the **celestial sphere**—the imaginary sphere with Earth at its center upon which celestial objects appear to lie. A number of different coordinate systems are used to specify the positions of objects on the celestial sphere. The simplest of these is the *altitude-azimuth coordinate system*. The altitude-azimuth coordinate system is based on the "map" direction to an object (the object's azimuth, with north = 0°, east = 90°, south = 180°, and west = 270°) combined with how high the object is above the horizon (the object's altitude, with the horizon at 0° and the zenith at 90°). For example, an object that is 10° above the eastern horizon has an altitude of 10° and an azimuth of 90°. An object that is 45° above the horizon in the southwest is at altitude 45°, azimuth 225°.

The altitude-azimuth coordinate system is the simplest way to tell someone where in the sky to look at the moment, but it is not a good coordinate system for cataloging the positions of objects. The altitude and azimuth of an object are different for each observer, depending on the observer's position on Earth, and they are constantly changing as Earth rotates on its axis. To specify the direction to an object in a way that is the same for everyone requires a coordinate system that is fixed relative to the celestial sphere. The most common such coordinates are called *celestial coordinates*.

Celestial coordinates are illustrated in **Figure A7.1**. Celestial coordinates are much like the traditional system of latitude and longitude used on the surface of Earth. On Earth, latitude specifies how far you are from Earth's equator, as discussed in Chapter 2. If you are on Earth's equator, your latitude is 0°. If you are at Earth's North Pole, your latitude is 90° north. If you are at Earth's South Pole, your latitude is 90° south.

The latitude-like coordinate on the celestial sphere is called **declination**, often signified with the lowercase Greek letter δ (delta). The celestial equator has $\delta = 0°$. The north celestial pole has $\delta = +90°$. The south celestial pole has $\delta = -90°$. (See Chapter 2 if you need to refresh your memory about the celestial equator

or celestial poles.) Declination is usually expressed in degrees, minutes of arc, and seconds of arc. For example, Sirius, the brightest star in the sky, has $\delta = -16°42'58''$ meaning that it is located not quite 17° south of the celestial equator.

On Earth, east–west position is specified by longitude. Lines of constant longitude run north–south from one pole to the other. Unlike latitude, for which the equator provides a natural place to call "zero," there is no natural starting point for longitude, so one was invented. By arbitrary convention, the Royal Observatory in Greenwich, England, is defined to lie at a longitude of 0°. On the celestial sphere, the longitude-like coordinate is called **right ascension**, often signified with the lowercase Greek letter α (alpha). Unlike the case with longitude, there *is* a natural point on the celestial sphere to use as the starting point for right ascension: the vernal equinox, or the point at which the ecliptic crosses the celestial equator with the Sun moving from the southern sky into the northern sky. The (Northern Hemisphere) vernal equinox defines the line of right ascension at which $\alpha = 0°$. The (Northern Hemisphere) autumnal equinox, located on the opposite side of the sky, is at $\alpha = 180°$.

Normally, right ascension is measured in units of time rather than degrees. It takes Earth 24 hours (of sidereal time) to rotate on its axis, so the celestial sphere is divided into 24 hours of right ascension, with each hour of right ascension corresponding to 15°. Hours of right ascension are then subdivided into minutes and seconds of time. Right ascension increases going to the east. The right ascension of Sirius, for example, is $\alpha = 06^\mathrm{h}45^\mathrm{m}08.9^\mathrm{s}$, meaning that Sirius is about 101° (that is, $06^\mathrm{h}45^\mathrm{m}$) east of the vernal equinox. Time is a natural unit for measuring right ascension because time naturally tracks the motion of objects due to Earth's rotation on its axis. If stars on the meridian at a certain time have $\alpha = 06^\mathrm{h}$, then an hour later the stars on the meridian will have $\alpha = 07^\mathrm{h}$, and an hour after that they will have $\alpha = 08^\mathrm{h}$. The *local sidereal time*, or *star time*, at your location right now is equal to the right ascension of the stars that are on your meridian at the moment. Because of Earth's motion around the Sun, a sidereal day is about 4 minutes shorter than a solar day, so local sidereal time constantly gains on solar time. At midnight on September 22, the local sidereal time is 0^h. By midnight on December 21, local sidereal time has advanced to 06^h. On March 20, local sidereal time at midnight is 12^h. And at midnight on June 20, local sidereal time is 18^h.

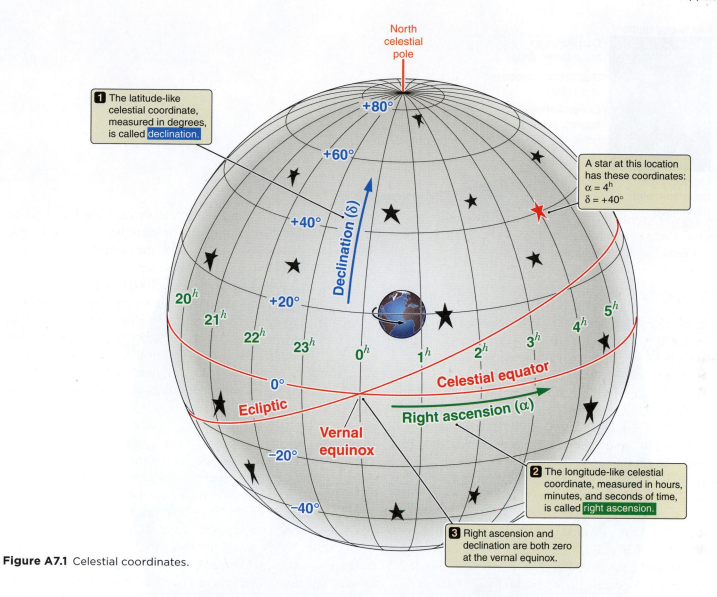

1 The latitude-like celestial coordinate, measured in degrees, is called declination.

North celestial pole

+80°

+60°

Declination (δ)

+40°

A star at this location has these coordinates:
α = 4ʰ
δ = +40°

20ʰ

21ʰ

22ʰ

23ʰ

+20°

0ʰ   1ʰ   2ʰ   3ʰ   4ʰ   5ʰ

0°

Celestial equator

Ecliptic

Right ascension (α)

Vernal equinox

−20°

2 The longitude-like celestial coordinate, measured in hours, minutes, and seconds of time, is called right ascension.

−40°

3 Right ascension and declination are both zero at the vernal equinox.

Figure A7.1 Celestial coordinates.

Putting this all together, right ascension and declination provide a convenient way to specify the location of any object on the celestial sphere. Sirius is located at $\alpha = 06^h45^m08.9^s$, $\delta = -16°42'58''$, which means that at midnight on December 21 (local sidereal time = 06^h), you will find Sirius about 45^m east of the meridian, not quite 17° south of the celestial equator.

There is just one final caveat. As we discussed in Chapter 2, the directions of the celestial equator, celestial poles, and vernal equinox are constantly changing as Earth's axis wobbles like the axis of a spinning top. In Chapter 2, we called this 26,000-year wobble the **precession of the equinoxes**, meaning that the location of the equinoxes is slowly advancing along the ecliptic. So when we specify the celestial coordinates of an object, we need to specify the date at which the positions of the vernal equinox and celestial poles were measured. By convention, coordinates are usually referred to with the position of the vernal equinox on January 1, 2000. A complete, formal specification of the coordinates of Sirius would then

be $\alpha(2000) = 06^h45^m08.9^s$, $\delta(2000) = -16°42'58''$, where the "2000" in parentheses refers to the equinox of the coordinates.

Constellations and Names

Although it is certainly possible to specify any location on the surface of Earth exactly by giving its latitude and longitude, it is usually convenient to use a more descriptive address. We might say, for example, that one of the coauthors of this book works near latitude 37° north, longitude 122° west; but it would probably mean a lot more to you if we said that George Blumenthal works in Santa Cruz, California.

Just as the surface of Earth is divided into nations and states, the celestial sphere is divided into 88 **constellations**, the names of which are often used to refer to objects within their boundaries (see the star charts in **Figure A7.2**). The brightest stars within the boundaries of a constellation are named using a Greek letter

(a) Key to star maps

Constellation boundaries (blue)

Star names (match star color)

Constellation figures, names (yellow)

Lines of right ascension and declination (green)

Meridian at midnight, September 22

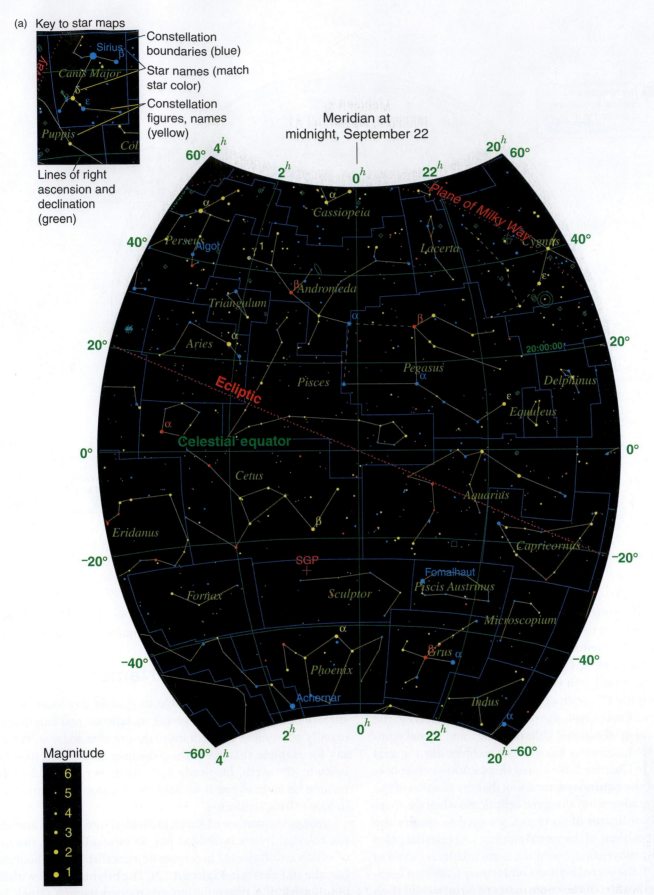

Magnitude
6
5
4
3
2
1

Figure A7.2A The sky from right ascension 20ʰ to 04ʰ and declination −60° to +60°.

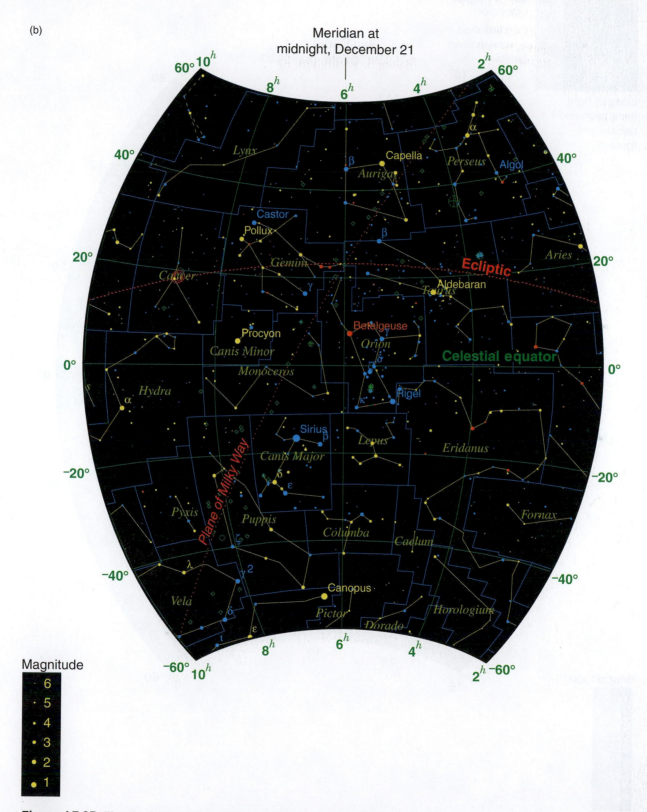

(b)

Meridian at
midnight, December 21

Figure A7.2B The sky from right ascension 02ʰ to 10ʰ and declination −60° to +60°.

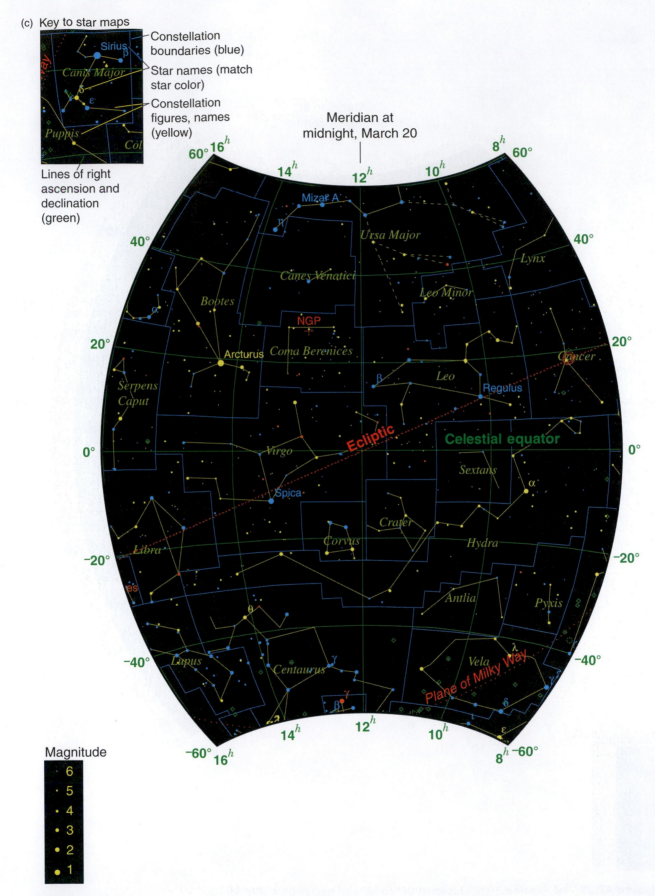

(c) Key to star maps

Constellation boundaries (blue)

Star names (match star color)

Constellation figures, names (yellow)

Lines of right ascension and declination (green)

Meridian at midnight, March 20

Magnitude

6
5
4
3
2
1

Figure A7.2C The sky from right ascension 08ʰ to 16ʰ and declination −60° to +60°.

(d)

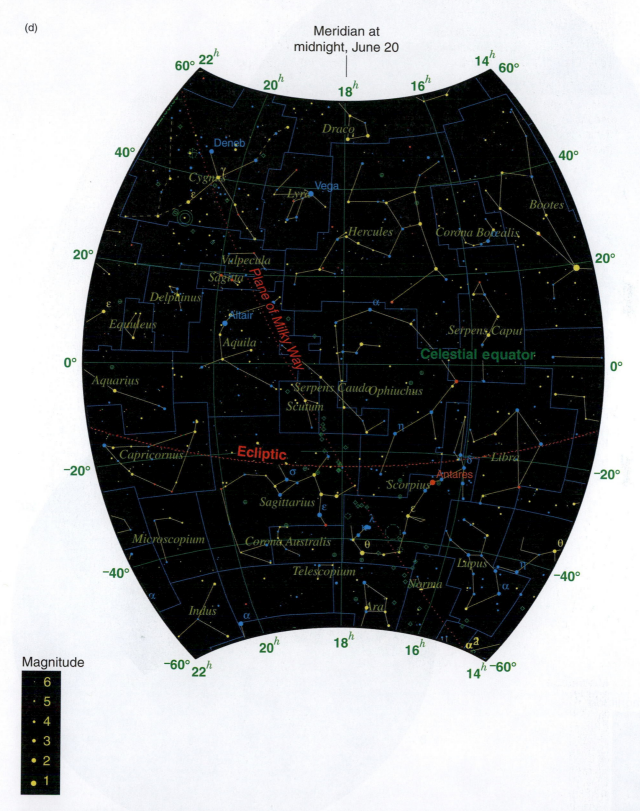

Figure A7.2D The sky from right ascension 14ʰ to 22ʰ and declination −60° to +60°.

(e)

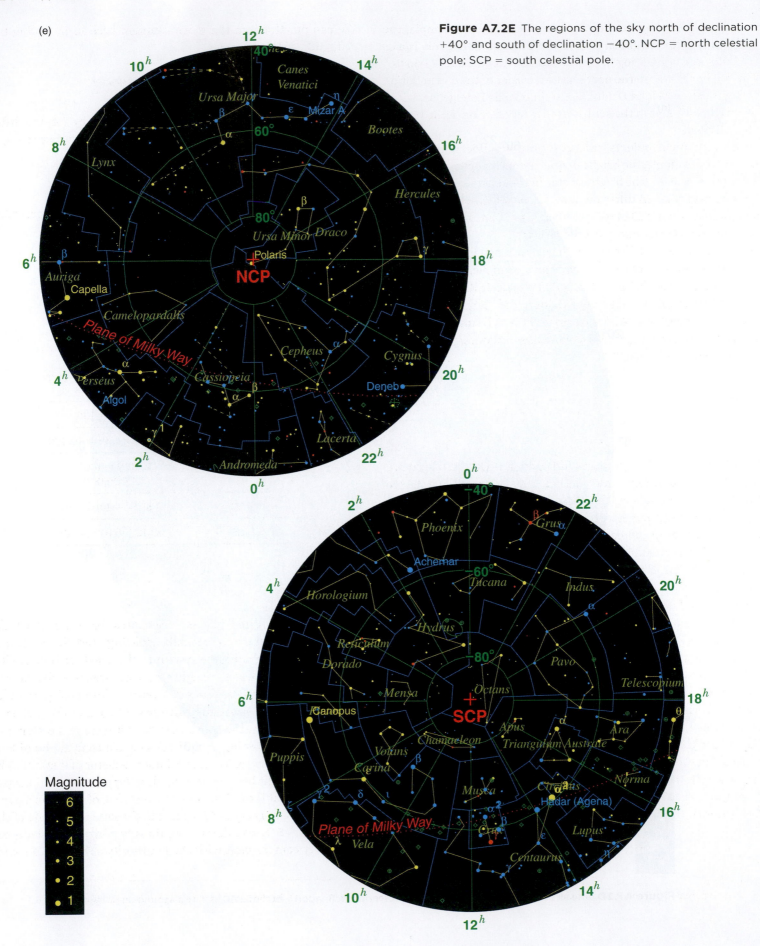

Figure A7.2E The regions of the sky north of declination +40° and south of declination −40°. NCP = north celestial pole; SCP = south celestial pole.

Magnitude

combined with the name of the constellation. For example, the star Sirius is the brightest star in the constellation Canis Major (literally, the "great dog"), so it is called "Alpha Canis Majoris." The bright red star in the northeastern corner of the constellation Orion is called "Alpha Orionis," also known as Betelgeuse. Rigel, the bright blue star in the southwest corner of Orion, is also called "Beta Orionis."

Astronomical objects can take on a bewildering range of names. For example, the bright southern star Canopus, also known as "Alpha Carinae" (the brightest star in the constellation Carina), has no fewer than 34 different names, most of which are about as memorable as "SAO 234480" (number 234,480 in the Smithsonian Astrophysical Observatory catalog of stars).

There is a slight difference in the way a constellation is spelled when it becomes part of a star's name. For example, Sirius is called "Alpha Canis Majoris," not "Alpha Canis Major"; Rigel is referred to as "Beta Orionis," not "Beta Orion"; and Canopus becomes "Alpha Carinae," not "Alpha Carina." This is because the Latin genitive, or possessive, case is used with star names; for example, *Orionis* means "of Orion."

Astronomical Magnitudes

Apparent Magnitudes

We first introduced magnitudes in Working It Out 13.2; here, we provide some additional information. You are most likely to see this system if you take a lab course in astronomy or if you use a star catalog. Astronomers use the logarithmic system of **apparent magnitudes** to compare the apparent brightness of objects in the sky. Other common systems of logarithmic measurements that you may have encountered include decibels for measuring sound levels and the Richter scale for measuring the strength of earthquakes. For example, an earthquake of magnitude 6 is not just a little stronger than an earthquake of magnitude 5; it is, in fact, 10 times stronger.

As discussed in Working It Out 13.2, a difference of five magnitudes between the apparent brightness of two stars (say, a star with $m = 6$ and a star with $m = 1$), corresponds to 100 times difference in brightness, and *the greater the magnitude, the fainter the object*. If five steps in magnitude correspond to a factor of 100 in brightness, then one step in magnitude must correspond to the fifth root of 100; that is, a factor of $100^{1/5}$ = approximately 2.512 in brightness ($100^{1/5} \times 100^{1/5} \times 100^{1/5} \times 100^{1/5} \times 100^{1/5} = 100$).

If star 1 has a brightness of b_1 and star 2 has a brightness of b_2, then the ratio of the brightness of the stars is given by

$$\frac{b_1}{b_2} = (2.512)^{m_2 - m_1} = 100^{\frac{(m_2 - m_1)}{5}}$$

We can put this into the more common base 10 by noting that $100 = 10^2$, so this becomes

$$\frac{b_1}{b_2} = 10^{2 \times \frac{(m_2 - m_1)}{5}} = 10^{0.4(m_2 - m_1)}$$

After taking the log of both sides and dividing by 0.4, the difference in magnitude ($m_2 - m_1$) between the two stars is given by

$$m_2 - m_1 = 2.5 \log_{10} \frac{b_1}{b_2}$$

The following table shows some examples using the preceding equations.

Apparent Magnitude Difference ($m_2 - m_1$)	Ratio of Apparent Brightness (b_1/b_2)
1	2.512
2	$2.512^2 = 6.3$
3	$2.512^3 = 15.8$
4	$2.512^4 = 39.8$
5	$2.512^5 = 100$
10	$2.512^{10} = 100^2 = 10,000$
15	$2.512^{15} = 100^3 = 1,000,000$
20	$2.512^{20} = 100^4 = 10^8$
25	$2.512^{25} = 100^5 = 10^{10}$

Absolute Magnitudes

Recall that stars differ in their brightness for two reasons: the amount of light they are actually emitting, and their distance from Earth. The magnitude system is also used for **luminosity**, with the same scale as for brightness: a difference of five magnitudes corresponds to 100 times difference in luminosity. Astronomers call these **absolute magnitudes** (M_{abs}), and the idea is to imagine how bright the star would be if it were at a distance of 10 parsecs (pc). Absolute magnitudes enable comparison of how luminous two stars really are, without the factor of distance. The Sun is very bright because it is so close (apparent visual magnitude = −27), but if the Sun were at a distance of 10 pc, its magnitude would be only about 5.[1] Thus, the absolute magnitude of the Sun is $M_{abs} = 5$. Recall that the luminosity of a star is usually expressed by comparing it with the luminosity of the Sun. As with

---

[1]The apparent and absolute magnitudes of the Sun are −26.74 and +4.83, respectively. We use +5 for the Sun's absolute magnitude as an approximation.

apparent magnitudes, higher magnitude numbers corresponding to lower luminosity. Thus, a star that is 100 times less luminous than the Sun will be 5 absolute magnitudes fainter, or $M_{abs} = 10$. A star that is 10,000 times more luminous than the Sun will be 10 absolute magnitudes brighter, or $M_{abs} = -5$.

Absolute magnitudes and luminosities follow the same equations as those we provided already, using L instead of b and M_{abs} instead of m:

$$\frac{L_1}{L_2} = 10^{2 \times \frac{(M_{abs(2)} - M_{abs(1)})}{5}} = 10^{0.4(M_{abs(2)} - M_{abs(1)})}$$

and

$$M_{abs(2)} - M_{abs(1)} = 2.5 \log_{10} \frac{L_1}{L_2}$$

Most often, astronomers think about the luminosity of a star compared to the luminosity of the Sun. In this case $L_1 = L_{star}$ and $L_2 = L_{Sun}$. The following table compares luminosity (where $L_{Sun} = 1$) with absolute magnitude of a star.

L_{Star}/L_{Sun}	M_{abs}
1,000,000	−10
10,000	−5
100	0
1	5
1/100	10
1/10,000	15

Distance Modulus

The difference between the apparent magnitude and the absolute magnitude depends on the star's distance. By definition, a star at a distance of exactly 10 pc will have an apparent magnitude equal to its absolute magnitude. Astronomers can always measure the brightness of a star and thus its apparent magnitude and can estimate the luminosity of a star and thus its absolute magnitude using the Hertzsprung-Russell (H-R) diagram. This is the way the distances to most stars are found.

Using the preceding equations and the definition of absolute magnitude, we can get to the following relatively simple expression:

$$m - M_{abs} = 5 \log_{10} d - 5$$

where distance d is in parsecs.

We can rewrite this equation to solve for distance as follows:

$$d = 10^{(\frac{m - M_{abs} + 5}{5})}$$

The following table shows how the difference between an object's apparent and absolute magnitudes leads to its distance in parsecs.

$m - M_{abs}$	Distance (pc)
−3	2.5
−2	4.0
−1	6.3
0	10
1	16
2	25
3	40
4	63
5	100
10	1,000
15	10,000
20	100,000

Although the system of astronomical magnitudes is convenient in many ways—which is why astronomers continue to use it—it can also be confusing to new students. Just remember three things and you will probably get by:

1. The greater the magnitude, the fainter the object.

2. One magnitude *smaller* means about two and a half times *brighter*.

3. The brightest stars in the sky have magnitudes of less than 1, and the faintest stars that can be seen with the naked eye on a dark night have magnitudes of about 6.

A final note: Astronomers sometimes use "colors" based on the ratio of the brightness of a star as seen in two different parts of the spectrum. The "b_B/b_V color," for example, is the ratio of the brightness of a star seen through a blue filter, divided by the brightness of a star seen through a yellow-green (visual) filter. Normally, astronomers instead discuss the "$B - V$ color" of a star, which is equal to the difference between a star's blue magnitude and its visual magnitude. We can use the previous expression for a magnitude difference to write

$$B - V \text{ color} = m_B - m_V = -2.5 \log_{10} (b_B/b_V)$$

Thus, a star with a b_B/b_V color of 1.0 has a $B - V$ color of 0.0, and a star with a b_B/b_V color of 1.4 has a $B - V$ color of −0.37. Notice that, as with magnitudes, $B - V$ colors are "backward": the bluer a star, the greater its b_B/b_V color but the less its $B - V$ color.

APPENDIX 8: Uniform Circular Motion and Circular Orbits

Uniform Circular Motion

In Chapter 4 (see Section 4.2 and Figure 4.7), we discuss the motion of an object moving in a circle at a constant speed. This motion, called **uniform circular motion**, is the result of the fact that centripetal force always acts toward the center of the circle. The key question when thinking about uniform circular motion is, How hard does something have to pull to keep the object moving in a circle? Part of the answer to this question is obvious: the more massive an object is, the harder it will be to keep it moving on its circular path. According to Newton's second law of motion, $F = ma$, or in this case, the centripetal force equals the mass multiplied by the centripetal acceleration. The larger the mass, the greater the force required to keep it moving in its circle.

The centripetal force needed to keep an object moving in constant circular motion also depends on two other quantities: the speed of the object and the size of the circle. The faster an object is moving, the more rapidly it has to change direction to stay on a circle of a given size. The second quantity that influences the needed acceleration is the radius of the circle. The smaller the circle, the greater the pull needed to keep it on track. You can understand this by looking at the motion. A small circle requires a continuous "hard" turn, whereas a larger circle requires a more gentle change in direction. It takes more force to keep an object moving faster in a smaller circle than it does to keep the same object moving more slowly in a larger circle. (To get a better feel for how this works, think about the difference between riding in a car that is taking a tight curve at high speed and a car that is moving slowly around a gentle curve.)

To arrive at the circular velocity and other results discussed in Chapter 4, these intuitive ideas about uniform circular motion are turned into a quantitative expression of exactly how much centripetal acceleration is needed to keep an object moving in a circle with radius r at speed v. **Figure A8.1** shows a ball moving around a circle of radius r at a constant speed v at two different times. The centripetal acceleration that is keeping the ball on the circle is a. Remember that the acceleration is always directed toward the center of the circle, whereas the velocity of the ball is always perpendicular to the acceleration. The ball's velocity and its acceleration are always at right angles to each other. As the object moves around the circle, the direction of motion and the direction of the acceleration change together in lockstep.

Figure A8.1 contains two triangles. Triangle 1 shows the velocity (speed and direction) at each of the two times. The arrow labeled "Δv" connecting the heads of the two velocity arrows shows how much the velocity changed between time 1 (t_1) and time 2 (t_2). This change is the effect of the centripetal acceleration. If you imagine that points 1 and 2 are very close together—so close that the direction of the centripetal acceleration does not change

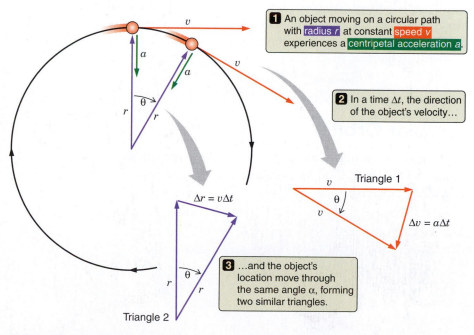

Figure A8.1 Similar triangles are used to find the centripetal force that is needed to keep an object moving at a constant speed on a circular path.

by much between the two—then the centripetal acceleration equals the change in the velocity divided by the time between the two, $\Delta t = t_2 - t_1$. So, $\Delta v = a\Delta t$.

Triangle 2 shows something similar. Here, the arrow labeled "Δr" indicates the change in the position of the ball between time 1 and time 2. Again, if you imagine that the time between the two points is very short, Δr is equal to the velocity multiplied by the time, or $\Delta r = v\Delta t$.

The line between the center of the circle and the ball is always perpendicular to the velocity of the ball. So if the direction of the ball's velocity changes by an angle θ, then the direction of the line between the ball and the center of the circle must also change by the same angle α. In other words, triangles 1 and 2 are "similar triangles." They have the same *shape*. If the triangles are the same shape, the ratio of two sides of triangle 1 must equal the ratio of the two corresponding sides of triangle 2. Then,

$$\frac{a\Delta t}{v} = \frac{v\Delta t}{r}$$

If we divide by Δt on both sides of the equation and then cross-multiply, we obtain

$$ar = v^2$$

which, after dividing both sides of the equation by r, becomes

$$a_{centripetal} = \frac{v^2}{r}$$

The subscript "centripetal" is added to a to signify that this is the centripetal acceleration needed to keep the object moving in a circle of radius r at speed v. The centripetal force required to keep an object of mass m moving on such a circle is then

$$F_{centripetal} = ma_{centripetal} = \frac{mv^2}{r}$$

Circular Orbits

In the case of an object moving in a circular orbit, there is no string to hold the ball on its circular path. Instead, this force is provided by **gravity**.

Think about an object with mass m in orbit about a much larger object with mass M. The orbit is circular, and the distance between the two objects is given by r. The force needed to keep the smaller object moving at speed v in a circle with radius r is given by the previous expression for $F_{centripetal}$. The force actually provided by gravity (see Chapter 4) is

$$F_{grav} = G\frac{Mm}{r^2}$$

If gravity is responsible for holding the mass in its circular motion, then it should be true that $F_{grav} = F_{centripetal}$. That is, if mass m is moving in a circle under the force of gravity, the force provided by gravity must equal the centripetal force needed to explain that circular motion. Setting the two expressions for $F_{centripetal}$ and F_{grav} equal to each other gives

$$\frac{mv^2}{r} = G\frac{Mm}{r^2}$$

All that remains is a bit of algebra. Dividing by m on both sides of the equation and multiplying both sides by r gives

$$v^2 = G\frac{M}{r}$$

Taking the square root of both sides then yields the desired result:

$$v_{circular} = \sqrt{\frac{GM}{r}}$$

This is the **circular velocity** we presented in Chapter 4. It is the velocity at which an object in a circular orbit must be moving. If the object were not moving at this velocity, then gravity would not be providing the needed centripetal force, and the object would not move in a circle.

APPENDIX 9*: IAU 2006 Resolutions: "Definition of a Planet in the Solar System" and "Pluto"

August 24, 2006, Prague

Resolutions

Resolution 5 is the principal definition for the IAU usage of "planet" and related terms. Resolution 6 creates, for IAU usage, a new class of objects, for which Pluto is the prototype. The IAU will set up a process to name these objects.

Resolution 5: Definition of a Planet in the Solar System

Contemporary observations are changing our understanding of planetary systems, and it is important that our nomenclature for objects reflect our current understanding. This applies, in particular, to the designation "planets." The word "planet" originally described "wanderers" that were known only as moving lights in the sky. Recent discoveries lead us to create a new definition, which we can make using currently available scientific information.

The IAU therefore resolves that planets and other bodies, except satellites, in our Solar System be defined into three distinct categories in the following way [**Figure A9.1**]:

(1) A "planet"[1] is a celestial body that

   (a) is in orbit around the Sun,

   (b) has sufficient mass for its self-gravity to overcome rigid body forces so that it assumes a hydrostatic equilibrium (nearly round) shape, and

   (c) has cleared the neighborhood around its orbit.

(2) A "dwarf planet" is a celestial body that

   (a) is in orbit around the Sun,

   (b) has sufficient mass for its self-gravity to overcome rigid body forces so that it assumes a hydrostatic equilibrium (nearly round) shape,[2]

   (c) has not cleared the neighborhood around its orbit, and

   (d) is not a satellite.

(3) All other objects,[3] except satellites, orbiting the Sun shall be referred to collectively as "Small Solar-System Bodies."

Resolution 6: Pluto

The IAU further resolves:

Pluto is a "dwarf planet" by the above definition and is recognized as the prototype of a new category of Trans-Neptunian Objects.[4]

---

*SOURCE: International Astronomical Union (IAU).
[1]The eight planets are Mercury, Venus, Earth, Mars, Jupiter, Saturn, Uranus, and Neptune.
[2]An IAU process will be established to assign borderline objects to the dwarf planet or to another category.
[3]These currently include most of the Solar System asteroids, most Trans-Neptunian Objects (TNOs), comets, and other small bodies.
[4]An IAU process will be established to select a name for this category.

Figure A9.1 Planets and dwarf planets of the Solar System.

Glossary

A

aberration of starlight The apparent displacement in the position of a star that is due to the finite speed of light and Earth's orbital motion around the Sun.

absolute magnitude A measure of the intrinsic brightness, or luminosity, of a celestial object, generally a star. Specifically, the apparent magnitude an object would have if it were located at a standard distance of 10 parsecs (pc). Compare *apparent magnitude*.

absolute zero The temperature at which thermal motions cease. The lowest possible temperature. Zero on the Kelvin temperature scale.

absorption The process by which an atom captures energy from a passing photon. Compare *emission*.

absorption line A minimum in the intensity of a spectrum that is due to the absorption of electromagnetic radiation at a specific wavelength determined by the energy levels of an atom or molecule. Compare *emission line*.

acceleration The rate at which the speed and/or direction of an object's motion is changing.

accretion disk A flat, rotating disk of gas and dust surrounding an object, such as a young stellar object, a forming planet, a collapsed star in a binary system, or a black hole.

achondrite A stony meteorite that does not contain chondrules. Compare *chondrite*.

active comet A comet nucleus that approaches close enough to the Sun to show signs of activity, such as the production of a coma and tail.

active galactic nucleus (AGN) A highly luminous, compact galactic nucleus whose luminosity may exceed that of the rest of the galaxy.

active region An area of the Sun's chromosphere anchoring bursts of intense magnetic activity.

adaptive optics Electro-optical systems that largely compensate for image distortion caused by Earth's atmosphere.

AGB See *asymptotic giant branch*.

AGN See *active galactic nucleus*.

albedo The fraction of electromagnetic radiation striking a surface that is reflected by that surface.

algebra A branch of mathematics in which numeric variables are represented by letters.

alpha particle A ^{4}He nucleus, consisting of two protons and two neutrons. Alpha particles are given off in the type of radioactive decay referred to as *alpha decay*.

altitude The location of an object above the horizon, measured by the angle formed between an imaginary line from an observer to the object and a second line from the observer to the point on the horizon directly below the object.

Amors A group of asteroids whose orbits cross the orbit of Mars but not the orbit of Earth. Compare *Apollos* and *Atens*.

amplitude In a wave, the maximum deviation from its undisturbed or relaxed position. For example, in a water wave the amplitude is the vertical distance from the wave's crest to the undisturbed water level.

angular momentum A conserved property of a rotating or revolving system whose value depends on the velocity and distribution of the system's mass.

angular resolution The ability of an imaging device such as a telescope (or the eye) to separate two objects that appear close together.

annular solar eclipse The type of solar eclipse that occurs when the apparent diameter of the Moon is less than that of the Sun, leaving a visible ring of light ("annulus") surrounding the dark disk of the Moon. Compare *partial solar eclipse* and *total solar eclipse*.

Antarctic Circle The circle on Earth with latitude 66.5° south, marking the northern limit where at least one day per year is in 24-hour daylight. Compare *Arctic Circle*.

anthropic principle The idea that this universe (or this bubble in the universe) must have physical properties that allow for the development of intelligent life.

anticyclonic motion The rotation of a weather system resulting from the Coriolis effect as air moves outward from a region of high atmospheric pressure. Compare *cyclonic motion*.

antimatter Matter made up of antiparticles.

antiparticle An elementary particle of antimatter identical in mass but opposite in charge and all other properties to its corresponding ordinary matter particle.

aperture The clear diameter of a telescope's objective lens or primary mirror.

aphelion (pl. **aphelia**) The point in a solar orbit that is farthest from the Sun. Compare *perihelion*.

Apollos A group of asteroids whose orbits cross the orbits of both Earth and Mars. Compare *Amors* and *Atens*.

apparent magnitude A measure of the apparent brightness of a celestial object, generally a star. Compare *absolute magnitude*.

arcminute (arcmin) A minute of arc ('), a unit used for measuring angles. An arcminute is 1/60 of a degree of arc.

arcsecond (arcsec) A second of arc ("), a unit used for measuring very small angles. An arcsecond is 1/60 of an arcminute, or 1/3,600 of a degree of arc.

Arctic Circle The circle on Earth with latitude 66.5° north, marking the southern limit where at least one day per year is in 24-hour daylight. Compare *Antarctic Circle*.

asteroid Also called *minor planet*. A primitive rocky or metallic body (planetesimal) that has survived planetary accretion. Asteroids are parent bodies of meteoroids.

asteroid belt The region between the orbits of Mars and Jupiter that contains most of the asteroids in our Solar System.

astrobiology An interdisciplinary science combining astronomy, biology, chemistry, geology, and physics to study life in the cosmos.

astrology The belief that the positions and aspects of stars and planets influence human affairs and characteristics, as well as terrestrial events.

astronomical seeing A measurement of the degree to which Earth's atmosphere degrades the resolution of a telescope's view of astronomical objects.

astronomical unit (AU) The average distance from the Sun to Earth: approximately 150 million kilometers (km).

astronomy The scientific study of planets, stars, galaxies, and the universe as a whole.

astrophysics The application of physical laws to the understanding of planets, stars, galaxies, and the universe as a whole.

asymptotic giant branch (AGB) The path on the H-R diagram that goes from the horizontal branch toward higher luminosities and lower temperatures, asymptotically approaching and then rising above the red giant branch.

Atens A group of asteroids whose orbits cross the orbit of Earth but not the orbit of Mars. Compare *Amors* and *Apollos*.

atmosphere The gravitationally bound, outer gaseous envelope surrounding a planet, moon, or star.

atmospheric greenhouse effect A warming of planetary surfaces produced by atmospheric gases that transmit optical solar radiation but partially trap infrared radiation. Compare *greenhouse effect*.

atmospheric probe An instrumented package designed to provide on-site measurements of the chemical and/or physical properties of a planetary atmosphere.

atmospheric window A region of the electromagnetic spectrum in which radiation is able to penetrate a planet's atmosphere.

atom The smallest unit of a chemical element that retains the properties of that element. Each atom is composed of a nucleus (neutrons and protons) surrounded by a cloud of electrons.

AU See *astronomical unit*.

aurora Emission in the upper atmosphere of a planet from atoms that have been excited by collisions with energetic particles from the planet's magnetosphere.

autumnal equinox 1. One of two points where the Sun crosses the celestial equator. 2. The day on which the Sun appears at this location, marking the first day of autumn (about September 22 in the Northern Hemisphere and March 20 in the Southern Hemisphere). Compare *vernal equinox*. See also *summer solstice* and *winter solstice*.

axion A hypothetical elementary particle first proposed to explain certain properties of the neutron and now considered a candidate for cold dark matter.

B

backlighting Illumination from behind a subject as seen by an observer. Fine material such as human hair and dust in planetary rings stands out best when viewed under backlighting conditions.

bar A unit of pressure. One bar is equivalent to 10^5 newtons per square meter—approximately equal to Earth's atmospheric pressure at sea level.

barred spiral galaxy A spiral galaxy with a bulge having an elongated, barlike shape. Compare *elliptical galaxy*, *irregular galaxy*, *S0 galaxy*, and *spiral galaxy*.

basalt Gray to black volcanic rock, rich in iron and magnesium.

beta decay 1. The decay of a neutron into a proton by emission of an electron (beta ray) and an antineutrino. 2. The decay of a proton into a neutron by emission of a positron and a neutrino.

Big Bang The event that occurred 13.8 billion years ago that marks the beginning of time and the universe.

Big Bang nucleosynthesis The formation of low-mass nuclei (H, He, Li, Be) during the first few minutes after the Big Bang.

Big Crunch A hypothetical cosmic future in which the expansion of the universe reverses and the universe collapses onto itself.

Big Rip A hypothetical cosmic future in which all matter in the universe, from stars to subatomic particles, is progressively torn apart by expansion of the universe.

binary star A system in which two stars are in gravitationally bound orbits about their common center of mass.

binding energy The minimum energy required to separate an atomic nucleus into its component protons and neutrons.

biosphere The global sum of all living organisms on Earth (or any planet or moon). Compare *hydrosphere* and *lithosphere*.

bipolar outflow Material streaming away in opposite directions from either side of the accretion disk of a young star.

black hole An object so dense that its escape velocity exceeds the speed of light; a *singularity* in spacetime.

blackbody An object that absorbs and can reemit all electromagnetic energy it receives.

blackbody spectrum See *Planck spectrum*.

blue-green algae See *cyanobacteria*.

blueshift The Doppler shift toward shorter (bluer) wavelengths of light from an approaching object. Compare *redshift*.

Bohr model A model of the atom, proposed by Niels Bohr in 1913, in which a small positively charged nucleus is surrounded by orbiting electrons, similar to a miniature solar system.

bolide A very bright, exploding meteor.

bound orbit An orbit in which an object is gravitationally bound to the body it is orbiting. A bound orbit's velocity is less than the escape velocity. Compare *unbound orbit*.

bow shock The boundary at which the speed of the solar wind abruptly drops from supersonic to subsonic in its approach to a planet's magnetosphere; the boundary between the region dominated by the solar wind and the region dominated by a planet's magnetosphere.

brightness The apparent intensity of light from a luminous object. Brightness depends on both the *luminosity* of a source and its distance. Units at the detector: watts per square meter (W/m^2).

brown dwarf A "failed" star that is not massive enough to cause hydrogen fusion in its core. An object whose mass is intermediate between that of the least massive stars and that of supermassive planets.

bulge The central region of a spiral galaxy that is similar in appearance to a small elliptical galaxy.

C

C See *Celsius*.

C-type asteroid An asteroid made of material that has remained mostly unmodified since the formation of the Solar System; the most primitive type of asteroid. Compare *M-type asteroid* and *S-type asteroid*.

caldera The summit crater of a volcano.

Cambrian explosion The spectacular rise in the number and diversity of biological species that occurred between 540 million and 500 million years ago.

carbon-nitrogen-oxygen (CNO) cycle One of the ways in which hydrogen is converted to helium (hydrogen burning) in the interiors of main-sequence stars. Compare *proton-proton chain*.

carbon star A cool red giant or asymptotic giant branch star that has an excess of carbon in its atmosphere.

carbonaceous chondrite A primitive stony meteorite that contains chondrules and is rich in carbon and volatile materials.

Cassini Division The largest gap in Saturn's rings, discovered by Jean-Dominique Cassini in 1675.

catalyst An atomic and molecular structure that permits or encourages chemical and nuclear reactions but does not change its own chemical or nuclear properties.

CCD See *charge-coupled device*.

celestial equator The imaginary great circle that is the projection of Earth's equator onto the celestial sphere.

celestial sphere An imaginary sphere with celestial objects on its inner surface and Earth at its center. The celestial sphere has no physical existence but is a convenient tool for picturing the directions in which celestial objects are seen from the surface of Earth.

Celsius (C) Also called *centigrade scale*. The arbitrary temperature scale, formulated by Anders Celsius (1701–1744), that defines 0°C as the freezing point of water and 100°C as the boiling point of water at sea level. Unit: degrees Celsius (°C). Compare *Fahrenheit* and *Kelvin scale*.

center of mass 1. The weighted average location of all the mass in a system of objects. The point in any isolated system that moves according to Newton's first law of motion. 2. In a binary star system, the point between the two stars that is the focus of both their elliptical orbits.

centigrade scale See *Celsius*.

centripetal force A force directed toward the center of curvature of an object's curved path.

Cepheid variable An evolved high-mass star with an atmosphere that is pulsating, leading to variability in the star's luminosity and color.

Chandrasekhar limit The upper limit on the mass of an object supported by electron degeneracy pressure; approximately 1.4 solar masses (M_{Sun}).

chaotic Behavior in complex, interrelated systems in which a small change in the initial state of a system can lead to a large change in the final state of the system.

charge-coupled device (CCD) A common type of solid-state detector of electromagnetic radiation that transforms the intensity of light directly into electric signals.

chondrite A stony meteorite that contains chondrules. Compare *achondrite*.

chondrule A small, crystallized, spherical inclusion of rapidly cooled molten droplets found inside some meteorites.

chromatic aberration A detrimental property of a lens in which rays of different wavelengths are brought to different focal distances from the lens.

chromosphere The region of the Sun's atmosphere located between the *photosphere* and the *corona*.

circular velocity The orbital velocity needed to keep an object moving in a circular orbit.

circumpolar Describing the part of the sky, near either celestial pole, that can always be seen above the horizon from a specific location on Earth.

circumstellar disk See *protoplanetary disk*.

climate The state of an atmosphere averaged over an extended time. Compare *weather*.

closed universe A finite universe with a curved spatial structure such that the sum of the angles of a triangle always exceeds 180 degrees. Compare *flat universe* and *open universe*.

CMB See *cosmic microwave background radiation*.

CNO cycle See *carbon-nitrogen-oxygen cycle*.

cold dark matter Particles of dark matter that move slowly enough to be gravitationally bound even in the smallest galaxies. Compare *hot dark matter*.

color index The color of a celestial object, generally a star, based on the ratio of its brightness in blue light to its brightness in "visual" (yellow-green) light. The difference between an object's blue (*B*) magnitude and visual (*V*) magnitude, *B* − *V*.

coma (pl. **comae**) The nearly spherical cloud of gas and dust surrounding the nucleus of an active comet.

comet A complex object consisting of a small, solid, icy nucleus; an atmospheric halo; and a tail of gas and dust.

comet nucleus A primitive planetesimal composed of ices and refractory materials that has survived planetary accretion. The "heart" of a comet, containing nearly the entire mass of the comet. A "dirty snowball."

comparative planetology The study of planets through comparison of their chemical and physical properties.

composite volcano A large, cone-shaped volcano formed by viscous, pasty lava flows alternating with pyroclastic (explosively generated) rock deposits. Compare *shield volcano*.

compound lens A lens made up of two or more elements of differing refractive index, the purpose of which is to minimize chromatic aberration.

concave mirror A telescope mirror with a surface that curves inward toward the incoming light.

conduction The transfer of energy in which the thermal energy of particles is transferred to adjacent particles by collisions or other interactions. Conduction is the most important way that thermal energy is transported in solid matter. Compare *convection*.

conservation law A physical law stating that the amount of a particular physical quantity (such as energy or angular momentum) of an isolated system does not change over time.

conservation of angular momentum The physical law stating that the amount of angular momentum of an isolated system does not change over time.

conservation of energy The physical law stating that the amount of energy of an isolated, closed system does not change over time.

constant of proportionality The number by which one quantity is multiplied to get another number.

constellation An imaginary image formed by patterns of stars; any of 88 defined areas on the celestial sphere used by astronomers to locate celestial objects.

constructive interference A state in which the amplitudes of two intersecting waves reinforce one another. Compare *destructive interference*.

continental drift The slow motion (centimeters per year) of Earth's continents relative to each other and to Earth's mantle. See also *plate tectonics*.

continuous radiation Electromagnetic radiation with intensity that varies smoothly over a wide range of wavelengths.

continuous spectrum A spectrum containing all wavelengths, without specific spectral lines.

convection The transport of thermal energy from the lower (hotter) to the higher (cooler) layers of a fluid by motions within the fluid driven by variations in buoyancy. Compare *conduction*.

convective zone A region within a star where energy is transported outward by convection. Compare *radiative zone*.

co-orbital moons Moons that occupy the same orbit.

core 1. The innermost region of a planetary interior. Compare *crust* and *mantle*. 2. The innermost part of a star.

core accretion–gas capture A process for forming giant planets, in which large amounts of surrounding hydrogen and helium gas are gravitationally captured onto a massive rocky core.

Coriolis effect The apparent displacement of objects in a direction perpendicular to their true motion as viewed from a rotating frame of reference. On a rotating planet, different latitudes rotating at different speeds cause this effect.

corona The hot, outermost part of the Sun's atmosphere. Compare *chromosphere* and *photosphere*.

coronal hole A low-density region in the solar corona containing "open" magnetic-field lines along which coronal material is free to stream into interplanetary space.

coronal mass ejection (CME) An eruption on the Sun that ejects hot gas and energetic particles at much higher speeds than are typical in the solar wind.

cosmic microwave background radiation (CMB) Also called simply *cosmic background radiation*. Isotropic microwave radiation from every direction in the sky having a 2.73-kelvin (K) blackbody spectrum. The CMB is residual radiation from the Big Bang.

cosmic ray A very fast-moving particle (usually an atomic nucleus) that originated in outer space; cosmic rays fill the disk of the Milky Way.

cosmological constant A constant, introduced into general relativity by Einstein, that characterizes an extra, repulsive force in the universe due to the vacuum of space itself.

cosmological principle The (testable) assumption that the same physical laws that apply here and now also apply everywhere and at all times, and that there are no special locations or directions in the universe.

cosmological redshift The redshift that results from the expansion of the universe rather than from the motions of galaxies or gravity. Compare *gravitational redshift*.

cosmology The study of the large-scale structure and evolution of the universe as a whole.

crescent Any phase of the Moon, Mercury, or Venus in which the object appears less than half illuminated by the Sun. Compare *gibbous*.

Cretaceous-Tertiary (K-T) boundary The boundary between the Cretaceous and Tertiary periods in Earth's history. This boundary corresponds to the time of the impact of an asteroid or comet and the extinction of the dinosaurs.

critical density The value of mass density of the universe that, ignoring any cosmological constant, is just barely capable of halting expansion of the universe.

crust The relatively thin, outermost, hard layer of a planet, which is chemically distinct from the interior. Compare *core* and *mantle*.

cryovolcanism Low-temperature volcanism in which the magmas are composed of molten ices rather than rocky material.

cyanobacteria Also called *blue-green algae*. Single-celled organisms that were responsible for creating oxygen in Earth's atmosphere by photosynthesizing carbon dioxide and releasing oxygen as a waste product.

cyclonic motion The rotation of a weather system resulting from the Coriolis effect as air moves toward a region of low atmospheric pressure. Compare *anticyclonic motion*.

D

Dark Ages The epoch in the history of the universe during which there was no visible "light" from astronomical objects.

dark energy A form of energy that permeates all of space (including the vacuum), producing a repulsive force that accelerates the expansion of the universe.

dark matter Matter in galaxies that does not emit or absorb electromagnetic radiation. Dark matter is thought to constitute most of the mass in the universe. Compare *luminous matter*.

dark matter halo The centrally condensed, greatly extended dark matter component of a galaxy that accounts for up to 95 percent of the galaxy's mass.

daughter product An element resulting from radioactive decay of a more massive *parent element*.

decay 1. The process of a radioactive nucleus changing into its daughter product. 2. The process of an atom or molecule dropping from a higher energy state to a lower energy state. 3. The process of a satellite's orbit losing energy.

declination A measure, analogous to *latitude*, that tells you the angular distance of a celestial body north or south of the celestial equator (from 0° to ±90°). Compare *right ascension*.

density The measure of an object's mass per unit of volume. Possible units include kilograms per cubic meter (kg/m^3).

destructive interference A state in which the amplitudes of two intersecting waves cancel one another. Compare *constructive interference*.

differential rotation Rotation of different parts of a system at different rates.

differentiation The process by which materials of higher density sink toward the center of a molten or fluid planetary interior.

diffraction The spreading of a wave after it passes through an opening or beyond the edge of an object.

diffraction grating An optical component with many narrow parallel rules lines that separates the wavelengths of light to produce a spectrum.

diffraction limit The limit of a telescope's angular resolution caused by diffraction.

diffuse ring A sparsely populated planetary ring spread out both horizontally and vertically.

dispersion The separation of rays of light into their component wavelengths.

distance ladder A sequence of techniques for measuring cosmic distances: each method is calibrated using the results from other methods that have been applied to closer objects.

Doppler effect The change in wavelength of sound or light that is due to the relative motion of the source toward or away from the observer.

Doppler shift The amount by which the wavelength of light is shifted by the Doppler effect.

Drake equation A prescription for estimating the number of intelligent civilizations existing in the Milky Way Galaxy.

dust devil A small tornado-like column of air containing dust or sand.

dust tail A type of comet tail consisting of dust particles that are pushed away from the comet's head by radiation pressure from the Sun. Compare *ion tail*.

dwarf galaxy A small galaxy with a luminosity ranging from 1 million to 1 billion solar luminosities (L_{Sun}). Compare *giant galaxy*.

dwarf planet A body with characteristics similar to those of a planet except that it has not cleared smaller bodies from the neighboring regions around its orbit. Compare *planet* (definition 2).

dynamic equilibrium A state in which a system is constantly changing but its configuration remains the same because one source of change is exactly balanced by another source of change. Compare *static equilibrium*.

dynamo theory A theory postulating that Earth's magnetic field (and those of other planets) is generated from a rotating and electrically conducting liquid core.

E

Earth Similarity Index (ESI) An index for quantifying the habitability of extrasolar planets in which currently available data on an extrasolar planet are used to estimate how much it is like Earth. Factors include the radius, density, escape velocity, and surface temperature. Compare *Planetary Habitability Index*.

eccentricity (e) The ratio of the distance between the two foci of an ellipse to the length of its major axis, which measures how noncircular the ellipse is.

eclipse 1. The total or partial obscuration of one celestial body by another. 2. The total or partial obscuration of light from one celestial body as it passes through the shadow of another celestial body.

eclipse season Any time during the year when the Moon's line of nodes is sufficiently close to the Sun for eclipses to occur.

eclipsing binary A binary system in which the orbital plane is oriented such that the two stars appear to pass in front of one another as seen from Earth. Compare *spectroscopic binary* and *visual binary*.

ecliptic 1. The apparent annual path of the Sun against the background of stars. 2. The projection of Earth's orbital plane onto the celestial sphere.

ecliptic plane The plane of Earth's orbit around the Sun. The ecliptic is the projection of this plane on the celestial sphere.

effective temperature The temperature at which a blackbody, such as a star, appears to radiate.

Einstein ring Light bent by gravitational lensing into a ring.

ejecta 1. Material thrown outward by the impact of an asteroid or comet on a planetary surface, leaving a crater behind. 2. Material thrown outward by a stellar explosion.

electric field A field that is able to exert a force on a charged object, whether at rest or moving. Compare *magnetic field*.

electric force The force exerted on electrically charged particles such as protons and electrons, arising from their electric charges. Compare *magnetic force*. See also *electromagnetic force*.

electromagnetic force The force, including both electric and magnetic forces, that acts on electrically charged particles. One of four fundamental forces of nature, along with the *strong nuclear force*, *weak nuclear force*, and *gravity* (definition 1). The force is mediated by the exchange of photons.

electromagnetic radiation A traveling disturbance in the electric and magnetic fields caused by accelerating electric charges. In quantum mechanics, a stream of photons. Light.

electromagnetic spectrum The spectrum made up of all possible frequencies or wavelengths of electromagnetic radiation, ranging from gamma rays through radio waves and including the portion our eyes can use.

electromagnetic wave A wave consisting of oscillations in the electric-field strength and the magnetic-field strength.

electron (e^-) A subatomic particle having a negative electric charge of 1.6×10^{-19} coulomb (C), a rest mass of 9.1×10^{-31} kilogram (kg), and a rest energy of 8×10^{-14} joule (J). The antiparticle of the *positron*. Compare *proton* and *neutron*.

electron-degenerate Describing matter, compressed to the point at which electron density reaches the limit imposed by the rules of quantum mechanics.

electroweak theory The quantum theory that combines descriptions of both the electromagnetic force and the weak nuclear force.

element One of 92 naturally occurring substances (such as hydrogen, oxygen, and uranium) and more than 20 human-made ones (such as plutonium). Each element is chemically defined by the specific number of protons in the nuclei of its atoms.

elementary particle One of the basic building blocks of nature that is not known to have substructure, such as the *electron* or the *quark*.

ellipse A conic section produced by the intersection of a plane with a cone when the plane is passed through the cone at an angle to the axis other than 0° or 90°. The shape that results when you attach the two ends of a piece of string to a piece of paper, stretch the string tight with the tip of a pencil, and then draw around those two points while keeping the string taut.

elliptical galaxy A galaxy of Hubble type "E" class, with a circular to elliptical outline on the sky, and containing almost no disk and a population of old stars. Compare *barred spiral galaxy*, *irregular galaxy*, *S0 galaxy*, and *spiral galaxy*.

emission The production of a photon when an atom decays to a lower energy state. Compare *absorption*.

emission line A peak in the intensity of a spectrum that is due to the emission of electromagnetic radiation at a specific wavelength determined by the energy levels of an atom or molecule. Compare *absorption line*.

empirical science Scientific investigation that is based primarily on observations and experimental data. It is descriptive rather than based on theoretical inference.

energy The conserved quantity that gives objects and systems the ability to do work. Possible units include: joules (J).

energy transport The transfer of energy from one location to another. In stars, energy transport is carried out mainly by radiation or convection.

entropy A measure of the disorder of a system related to the number of ways a system can be rearranged without its appearance being affected.

equator The imaginary great circle on the surface of a body midway between its poles that divides the body into northern and southern hemispheres. The equatorial plane passes through the center of the body and is perpendicular to its rotation axis. Compare *meridian*.

equilibrium The state of an object in which physical processes balance each other so that its properties or conditions remain constant.

equinox Literally, "equal night." 1. One of two positions on the ecliptic where it intersects the celestial equator. 2. Either of the two times of year (the *autumnal equinox* and *vernal equinox*) when the Sun is at one of these two positions. At this time, night and day are of the same length everywhere on Earth. Compare *solstice*.

equivalence principle The principle stating that there is no difference between a frame of reference that is freely floating through space and one that is freely falling within a gravitational field.

erosion The degradation of a planet's surface topography by the mechanical action of wind and/or water.

escape velocity The minimum velocity needed for an object to achieve a parabolic trajectory and thus permanently leave the gravitational grasp of another mass.

ESI See *Earth Similarity Index*.

eternal inflation The idea that a universe might inflate forever. In such a universe, quantum effects could randomly cause regions to slow their expansion, eventually stop inflating, and experience an explosion resembling the Big Bang.

event A particular location in spacetime.

event horizon The effective "surface" of a black hole. Nothing inside this surface—not even light—can escape from a black hole.

evolutionary track The path that a star follows across the H-R diagram as it evolves through its lifetime.

excited state Any energy level of a system or part of a system, such as an atom, molecule, or particle, that is higher than its ground state. Compare *ground state*.

exoplanet See *extrasolar planet*.

exosphere A very thin atmosphere or layer of atmosphere, where the molecules are bound by gravity to the moon or planet but their density is too low to behave like a gas of colliding particles

extrasolar planet Also called *exoplanet*. A planet orbiting a star other than the Sun.

extremophile A life-form that thrives under extreme environmental conditions.

eyepiece A lens that is closest to the eye in a telescope. Changing the eyepiece will change the magnification of the image in the telescope.

F

F See *Fahrenheit*.

Fahrenheit (F) The arbitrary temperature scale, formulated by Daniel Gabriel Fahrenheit (1686–1736), that defines 32°F as the melting point of water and 212°F as the boiling point of water at sea level. Unit: degrees Fahrenheit (°F). Compare *Celsius* and *Kelvin scale*.

falsified A hypothesis that is shown to be false.

fault A fracture in the crust of a planet or moon along which blocks of material can slide.

filter An instrument element that transmits a limited wavelength range of electromagnetic radiation. For the optical range, such elements are typically made of different kinds of glass and take on the hue of the light they transmit.

first quarter Moon The phase of the Moon in which only the western half of the Moon, as viewed from Earth, is illuminated by the Sun. It occurs about a week after a new Moon. Compare *third quarter Moon*. See also *full Moon* and *new Moon*.

fissure A fracture in the planetary lithosphere from which magma emerges.

flat rotation curve A rotation curve of a spiral galaxy in which rotation rates do not decline in the outer part of the galaxy but remain relatively constant to the outermost points.

flat universe An infinite universe whose spatial structure obeys Euclidean geometry, such that the sum of the angles of a triangle always equals 180 degrees. Compare *closed universe* and *open universe*.

flatness problem The surprising result that the sum of Ω_m plus Ω_Λ is extremely close to 1 in the present-day universe; equivalent to saying that it is surprising the universe is so close to being exactly flat.

flux The total amount of energy passing through each square meter of a surface each second. Unit: watts per square meter (W/m²).

flux tube A strong magnetic field contained within a tubelike structure. Flux tubes are found in the solar atmosphere and connecting the space between Jupiter and its moon Io.

flyby A spacecraft that first approaches and then continues flying past a planet or moon. Flybys can visit multiple objects, but they remain in the vicinity of their targets only briefly. Compare *orbiter*.

focal length The optical distance between a telescope's objective lens or primary mirror and the plane (called the focal plane) on which the light from a distant object is focused.

focal plane The plane, perpendicular to the optical axis of a lens or mirror, on which an image is formed.

focus (pl. **foci**) 1. One of two points that define an ellipse. 2. A point in the focal plane of a telescope.

force A push or a pull on an object.

frame of reference A coordinate system within which an observer measures positions and motions.

free fall The motion of an object when the only force acting on it is gravity.

frequency The number of times per second that a periodic process occurs. Unit: hertz (Hz), or cycles per second (1/s).

full Moon The phase of the Moon in which the near side of the Moon, as viewed from Earth, is fully illuminated by the Sun. It occurs about two weeks after a *new Moon*. See also *first quarter Moon* and *third quarter Moon*.

G

galaxy A gravitationally bound system that consists of stars and star clusters, gas, dust, and dark matter; typically greater than 1,000 light-years across and recognizable as a discrete, single object.

galaxy cluster A large, gravitationally bound collection of galaxies containing hundreds to thousands of members; typically 3–5 megaparsecs (Mpc) across. Compare *galaxy group* and *supercluster*.

galaxy group A small, gravitationally bound collection of galaxies containing from several to a hundred members; typically 1–2 megaparsecs (Mpc) across. Compare *galaxy cluster* and *supercluster*.

gamma ray Also called *gamma radiation*. Electromagnetic radiation with higher frequency, higher photon energy, and shorter wavelength than all other types of electromagnetic radiation.

gamma-ray burst (GRB) A brief, intense burst of gamma rays from a distant energetic explosion.

gas giant A giant planet formed mostly of hydrogen and helium. In the Solar System, Jupiter and Saturn are the gas giants. Compare *ice giant*.

gauss A basic unit measuring the strength of a magnetic field.

general relativistic time dilation The verified prediction that time passes more slowly in a gravitational field than in the absence of a gravitational field. Compare *time dilation*.

general relativity See *general theory of relativity*.

general theory of relativity Sometimes referred to as simply *general relativity*. Einstein's theory explaining gravity as the distortion of spacetime by massive objects, such that particles travel on the shortest path between two events in spacetime. This theory deals with all types of motion. Compare *special theory of relativity*.

geocentric model A historical cosmological model with Earth at its center, and all of the other objects in the universe in orbit around Earth. Compare *heliocentric model*.

geodesic The path an object will follow through spacetime in the absence of external forces.

geometry A branch of mathematics that deals with points, lines, angles, and shapes.

giant galaxy A galaxy with luminosity greater than about 1 billion solar luminosities (L_{Sun}). Compare *dwarf galaxy*.

giant molecular cloud An interstellar cloud composed primarily of molecular gas and dust, having hundreds of thousands of solar masses.

giant planet Also called *Jovian planet*. One of the largest planets in the Solar System (Saturn, Jupiter, Uranus, or Neptune), typically 10 times the size and many times the mass of any *terrestrial planet* and lacking a solid surface.

gibbous Any phase of the Moon, Mercury, or Venus in which the object appears more than half illuminated by the Sun. Compare *crescent*.

global circulation The overall, planetwide circulation pattern of a planet's atmosphere.

globular cluster A spherically symmetric, highly condensed group of stars, containing tens of thousands to a million members. Compare *open cluster*.

gluon The particle that carries (or, equivalently, mediates) interactions due to the strong nuclear force.

grand unified theory (GUT) A unified quantum theory that combines the strong nuclear, weak nuclear, and electromagnetic forces but does not include gravity.

granite Rock that is cooled from magma and is relatively rich in silicon and oxygen.

grating An optical surface containing many narrow, closely and equally spaced parallel grooves or slits that spectrally disperse reflected or transmitted light.

gravitational lens A massive object that gravitationally focuses the light of a more distant object to produce multiple brighter, magnified, possibly distorted images.

gravitational lensing The bending of light by gravity.

gravitational potential energy The stored energy in an object that is due solely to its position within a gravitational field.

gravitational redshift The shifting to longer wavelengths of radiation from an object deep within a gravitational well. Compare *cosmological redshift*.

gravitational wave A wave in the fabric of spacetime emitted by accelerating masses.

gravity 1. The mutually attractive force between massive objects. One of four fundamental forces of nature, along with the *electromagnetic force*, the *strong nuclear force*, and the *weak nuclear force*. 2. An effect arising from the bending of spacetime by massive objects.

GRB See *gamma-ray burst*.

great circle Any circle on a sphere that has as its center the center of the sphere. The celestial equator, the meridian, and the ecliptic are all great circles on the sphere of the sky, as is any circle drawn through the zenith.

Great Red Spot The giant, oval, brick red anticyclone seen in Jupiter's southern hemisphere.

greenhouse effect The solar heating of air in an enclosed space, such as a closed building or car, resulting primarily from the inability of the hot air to escape. Compare *atmospheric greenhouse effect*.

greenhouse gas One of a group of atmospheric gases such as carbon dioxide that are transparent to visible radiation but absorb infrared radiation.

greenhouse molecule A molecule such as water vapor or carbon dioxide that transmits visible radiation but absorbs infrared radiation.

Gregorian calendar The modern calendar. A modification of the Julian calendar decreed by Pope Gregory XIII in 1582. By that time, the less accurate Julian calendar had developed an error of 10 days over the 13 centuries since its inception.

ground state The lowest possible energy state for a system or part of a system, such as an atom, molecule, or particle. Compare *excited state*.

GUT See *grand unified theory*.

H

H II region A region of interstellar gas that has been ionized by ultraviolet radiation from nearby hot, massive stars.

H-R diagram The Hertzsprung-Russell diagram, a plot of the luminosities versus the surface temperatures of stars. The evolving properties of stars are plotted as tracks across the H-R diagram.

habitable zone The distance from its star at which a planet must be located in order to have a temperature suitable for water to exist in a liquid state.

Hadley circulation A simplified, and therefore uncommon, atmospheric global circulation that carries thermal energy directly from the equator to the polar regions of a planet.

half-life The time it takes for half a sample of a particular radioactive parent element to decay into a daughter product.

halo The spherically symmetric, low-density distribution of stars and dark matter that defines the outermost regions of a galaxy.

harmonic law See *Kepler's third law*.

Hawking radiation Radiation from a black hole.

Hayashi track The path that a protostar follows on the H-R diagram as it contracts toward the main sequence.

head The part of a comet that includes both the nucleus and the inner part of the coma.

heat death The possible eventual fate of an open universe, in which entropy has triumphed and all energy- and structure-producing processes have come to an end.

heavy element Also called *massive element*. Any element more massive than helium.

Heisenberg uncertainty principle The physical limitation that the product of the position and the momentum of a particle cannot be smaller than a well-defined value, Planck's constant (h).

heliocentric model A model of the Solar System, with the Sun at its center, and the planets, including Earth, in orbit around the Sun. Compare *geocentric model*.

helioseismology The use of solar oscillations to study the interior of the Sun.

heliosphere A region surrounding the Solar System in which the solar wind blows against the interstellar medium and clears out an area like the inside of a bubble. The heliosphere protects the Solar System from cosmic rays.

helium flash The runaway explosive burning of helium in the degenerate helium core of a red giant star.

Herbig-Haro (HH) object A glowing, rapidly moving knot of gas and dust that is excited by bipolar outflows in very young stars.

heredity The process by which one generation passes on its characteristics to future generations.

hertz (Hz) A unit of frequency equivalent to cycles per second.

Hertzsprung-Russell diagram See *H-R diagram*.

HH object See *Herbig-Haro object*.

hierarchical clustering The "bottom-up" process of forming large-scale structure. Small-scale structure first produces groups of galaxies, which in turn form clusters, which then form superclusters.

high-mass star A star with a main-sequence mass of greater than about 8 solar masses (M_{Sun}). Compare *low-mass star*.

homogeneous In cosmology, describing a universe in which observers in any location would observe the same properties. Compare *isotropic*.

horizon The boundary that separates the sky from the ground.

horizon problem The puzzling observation that the cosmic background radiation is so uniform in all directions, even though widely separated regions should have been "over the horizon" from each other in the early universe.

horizontal branch A region on the H-R diagram defined by stars burning helium to carbon in a stable core.

hot dark matter Particles of dark matter that move so fast that gravity cannot confine them to the volume occupied by a galaxy's normal luminous matter. Compare *cold dark matter*.

hot Jupiter A large, Jupiter-type extrasolar planet located very close to its parent star.

hot spot A place where hot plumes of mantle material rise near the surface of a planet.

Hubble constant (H_0) The constant of proportionality relating the recession velocities of galaxies to their distances. Compare *Hubble time*.

Hubble time An estimate of the age of the universe from the inverse of the *Hubble constant*, $1/H_0$.

Hubble's law The law stating that the speed at which a galaxy is moving away from Earth is proportional to the distance of that galaxy.

hurricane A large tropical cyclonic system circulating counterclockwise in the Northern Hemisphere and clockwise in the Southern Hemisphere. Hurricanes can extend outward from their center to more than 600 kilometers (km) and generate winds in excess of 300 kilometers per hour (km/h).

hydrogen burning The release of energy from the nuclear fusion of four hydrogen atoms into a single helium atom.

hydrogen shell burning The fusion of hydrogen in a shell surrounding a stellar core that may be either degenerate or fusing more massive elements.

hydrosphere The portion of Earth that is largely liquid water. Compare *biosphere* and *lithosphere*.

hydrostatic equilibrium The condition in which the weight bearing down at a particular point within an object is balanced by the pressure within the object.

hypernova (pl. **hypernovae**) A very energetic supernova from a very high-mass star.

hypothesis A well-considered idea, based on scientific principles and knowledge, that leads to testable predictions. Compare *theory*.

Hz See *hertz*.

I

ice The solid form of a volatile material; sometimes the *volatile material* itself, regardless of its physical form.

ice giant A giant planet formed mostly of the liquid form of volatile substances (ices). In the Solar System, Uranus and Neptune are the ice giants. Compare *gas giant*.

ideal gas law The relationship of pressure (P) to density of particles (n) and temperature (T) expressed as $P = nkT$, where k is Boltzmann's constant.

igneous activity The formation and action of molten rock (magma).

impact crater The scar of the impact left on a solid planetary or moon surface by collision with another object. Compare *secondary crater*.

impact cratering The process in which solid planetary objects collide with each other, leaving distinctive scars.

index of refraction (*n*) The ratio of the speed of light in a vacuum (*c*) to the speed of light in an optical medium (*v*).

inert gas A gaseous element that combines with other elements only under conditions of extreme temperature and pressure. Examples include helium, neon, and argon.

inertia The tendency for objects to retain their state of motion.

inertial frame of reference 1. A frame of reference that is moving in a straight line at constant speed; that is, not accelerating. 2. In general relativity, a frame of reference that is falling freely in a gravitational field.

inertial reference frame See *inertial frame of reference*.

inferior planet A Solar System planet that orbits the Sun at a closer distance than Earth's orbit. *See superior planet.*

inflation An extremely brief phase of ultra-rapid expansion of the very early universe. After inflation, the standard Big Bang models of expansion apply.

infrared (IR) radiation Electromagnetic radiation with frequencies, photon energies, and wave-lengths between those of visible light and microwaves.

instability strip A region of the H-R diagram containing stars that pulsate with a periodic variation in luminosity.

integration time The time interval during which photons are collected and added up in a detecting device.

intensity Of light, the amount of radiant energy emitted per second per unit area. Units for electromagnetic radiation: watts per square meter (W/m^2).

intercloud gas A low-density region of the interstellar medium that fills the space between interstellar clouds.

interference The interaction of two sets of waves producing high and low intensity, depending on whether their amplitudes reinforce (*constructive interference*) or cancel (*destructive interference*).

interferometer Linked optical or radio telescopes whose overall separation determines the angular resolution of the system.

interferometric array An interferometer that is made up of several telescopes arranged in an array.

interstellar cloud A discrete, high-density region of the interstellar medium made up mostly of atomic or molecular hydrogen and dust.

interstellar dust Small particles or grains (0.01–10 microns [μm] in diameter) of matter, primarily carbon and silicates, distributed throughout interstellar space.

interstellar extinction The dimming of visible and ultraviolet light by interstellar dust.

interstellar gas The tenuous gas, far less dense than air, composing 99 percent of the matter in the interstellar medium.

interstellar medium The gas and dust that fill the space between the stars within a galaxy.

inverse square law The rule stating that a quantity or effect diminishes with the square of the distance from the source.

ion An atom or molecule that has lost or gained one or more electrons.

ionize see *ionization*

ionization The process by which electrons are stripped free from an atom or molecule, resulting in free electrons and a positively charged atom or molecule.

ionosphere A layer high in Earth's atmosphere in which most of the atoms are ionized by solar radiation.

ion tail A type of comet tail consisting of ionized gas. Particles in the ion tail are pushed directly away from the comet's head in the antisolar direction at high speeds by the solar wind. Compare *dust tail*.

IR Infrared. See *infrared radiation*.

iron meteorite A metallic meteorite composed mostly of iron-nickel alloys. Compare *stony-iron meteorite* and *stony meteorite*.

irregular galaxy A galaxy without regular or symmetric appearance. Compare *barred spiral galaxy*, *elliptical galaxy*, *S0 galaxy*, and *spiral galaxy*.

irregular moon A moon that has been captured by a planet rather than having formed along with that planet. Some irregular moons revolve in a direction opposite to the rotation of the planet, and many are in distant, unstable orbits. Compare *regular moon*.

isotopes Forms of the same chemical element that have the same number of protons but a different number of neutrons.

isotropic In cosmology, having the same appear-ance to an observer in all directions. Compare *homogeneous*.

J

J See *joule*.

jansky (Jy) The basic unit of flux density. Unit: watts per square meter per hertz (W/m^2/Hz).

jet 1. A stream of gas and dust ejected from a comet nucleus by solar heating. 2. A stream of material that moves away from a protostar or active galactic nucleus at hundreds of kilometers per second.

joule (J) A unit of energy or work. 1 J = 1 newton meter.

Jovian planet See *giant planet*.

Jy See *jansky*.

K

K See *kelvin*.

K-T boundary See *Cretaceous-Tertiary boundary*.

KBO See *Kuiper Belt object*.

kelvin (K) The basic unit of the Kelvin scale of temperature.

Kelvin scale The temperature scale, formulated by William Thomson, better known as Lord Kelvin (1824–1907), that uses Celsius-sized degrees, but defines 0 K as absolute zero instead of as the melting point of water. Unit: kelvins (K). Compare *Celsius* and *Fahrenheit*.

Kepler's first law A rule of planetary motion, inferred by Johannes Kepler, stating that planets move in elliptical orbits with the Sun at one focus.

Kepler's laws The three rules of planetary motion inferred by Johannes Kepler from the data collected by Tycho Brahe.

Kepler's second law Also called *law of equal areas*. A rule of planetary motion, inferred by Johannes Kepler, stating that a line drawn from the Sun to a planet sweeps out equal areas in equal times as the planet orbits the Sun.

Kepler's third law Also called *harmonic law*. A rule of planetary motion, inferred by Johannes Kepler, that describes the relationship between the period of a planet's orbit and its distance from the Sun. The law states that the square of the period of a planet's orbit, measured in years, is equal to the cube of the semimajor axis of the planet's orbit, measured in astronomical units: $(P_{years})^2 = (A_{AU})^3$.

kiloparsec A unit of distance equal to 1 thousand parsecs, or 3.26 thousand light-years

kinetic energy (*E*$_K$) The energy of an object due to its motions. $E_K = \frac{1}{2}mv^2$. Possible units include joules (J).

Kirkwood gap A gap in the main asteroid belt related to orbital resonances with Jupiter.

Kuiper Belt A disk-shaped population of comet nuclei extending from Neptune's orbit to perhaps several thousand astronomical units (AU) from the Sun. The highly populated innermost part of the Kuiper Belt has an outer edge approximately 50 AU from the Sun.

Kuiper Belt object (KBO) Also called *trans-Neptunian object*. An icy planetesimal (comet nucleus) that orbits within the Kuiper Belt beyond the orbit of Neptune.

L

Lagrangian equilibrium point Also called simply *Lagrangian point*. One of five points of equilibrium in a system consisting of two massive objects in nearly circular orbit around a common center of mass. Only two Lagrangian points (L_4 and L_5) rep-resent stable equilibrium. A third, smaller body located at one of the five points will move in lock-step with the center of mass of the larger bodies.

Lambda-CDM The standard model of the Big Bang universe in which most of the energy density of the universe is dark energy (similar to Einstein's cosmological constant), and most of the mass in the universe is cold dark matter.

lander An instrumented spacecraft designed to land on a planet or moon. Compare *rover*.

large-scale structure Observable aggregates on the largest scales in the universe, including galaxy groups, clusters, and superclusters.

latitude The angular distance north (+) or south (−) from the equatorial plane of a nearly spherical body. Compare *longitude*.

lava Molten rock flowing out of a volcano during an eruption; also the rock that solidifies and cools from this liquid.

law of equal areas See *Kepler's second law.*

law of gravitation See *universal law of gravitation.*

leap year A year that contains 366 days. Leap years occur every 4 years when the year is divisible by 4, correcting for the accumulated excess time in a normal year, which is approximately 365¼ days long.

length contraction The relativistic compression of moving objects in the direction of their motion.

Leonids A November meteor shower associated with the dust debris left by comet Tempel-Tuttle.

life A biochemical process in which living organisms can reproduce, evolve, and sustain themselves by drawing energy from their environment. All terrestrial life involves carbon-based chemistry, assisted by the self-replicating molecules ribonucleic acid (RNA) and deoxyribonucleic acid (DNA).

light All electromagnetic radiation, which composes the entire electromagnetic spectrum.

light-year (**ly**) The distance that light travels in 1 year—about 9.5 trillion kilometers (km).

limb The outer edge of the visible disk of a planet, moon, or the Sun.

limb darkening The darker appearance caused by increased atmospheric absorption near the limb of a planet or star.

limestone A common sedimentary rock composed of calcium carbonate.

line of nodes 1. A line defined by the intersection of two orbital planes. 2. The line defined by the intersection of Earth's equatorial plane and the plane of the ecliptic.

lithosphere The solid, brittle part of Earth (or any planet or moon), including the crust and the upper part of the mantle. Compare *biosphere* and *hydrosphere.*

lithospheric plate A separate piece of Earth's lithosphere capable of moving independently. See also *continental drift* and *plate tectonics.*

Local Group The group of galaxies that includes the Milky Way and Andromeda galaxies as members.

long-period comet A comet with an orbital period of greater than 200 years. Compare *short-period comet.*

longitude The angular distance east (+) or west (−) from the prime meridian at Greenwich, England. Compare *latitude.*

longitudinal wave A wave that oscillates parallel to the direction of the wave's propagation. Compare *transverse wave.*

look-back time The amount of time that the light from an astronomical object has taken to reach Earth.

low-mass star A star with a main-sequence mass of less than about 3 solar masses (M_{Sun}). Compare *high-mass star.*

luminosity The total amount of light emitted by an object. Unit: watts (W). Compare *brightness.*

luminosity class A spectral classification based on stellar size, ranging from supergiants at the large end to white dwarfs at the small end.

luminosity-temperature-radius relationship A relationship among these three properties of stars indicating that if any two are known, the third can be calculated.

luminous matter Also called *normal matter.* Matter in galaxies—including stars, gas, and dust—that emits electromagnetic radiation. Compare *dark matter.*

lunar eclipse An eclipse that occurs when the Moon is partially or entirely in Earth's shadow. Compare *solar eclipse.*

lunar tide A tide on Earth that is due to the differential gravitational pull of the Moon. Compare *solar tide.*

ly See *light-year.*

M

μm See *micron.*

M-type asteroid An asteroid made of material that was once part of the metallic core of a larger, differentiated body that has since broken into pieces; made mostly of iron and nickel. Compare *C-type asteroid* and *S-type asteroid.*

MACHO Short for *massive compact halo object.* MACHOs include brown dwarfs, white dwarfs, and black holes and are candidates for dark matter. Compare *WIMP.*

magma Molten rock, often containing dissolved gases and solid minerals.

magnetic field A field that is able to exert a force on a moving electric charge. Compare *electric field.*

magnetic force The force exerted on electrically charged particles such as protons and electrons, arising from their motion. Compare *electric force.* See also *electromagnetic force.*

magnetometer A device that measures magnetic fields.

magnetosphere The region surrounding a planet that is filled with relatively intense magnetic fields and plasmas.

magnitude A system used by astronomers to describe the brightness or luminosity of stars. The brighter the star, the lower its magnitude.

main asteroid belt See *asteroid belt.*

main sequence The strip on the H-R diagram where most stars are found. Main-sequence stars are fusing hydrogen to helium in their cores.

main-sequence lifetime The amount of time a star spends on the main sequence, fusing hydrogen into helium in its core.

main-sequence turnoff The location on the main sequence of an H-R diagram made from a population of stars of the same age (such as a star cluster) where stars are just evolving off the main sequence. This location is determined by the age of the population of stars.

mantle The solid portion of a rocky planet that lies between the *crust* and the *core.*

mare (pl. **maria**) A dark region on the Moon composed of basaltic lava flows.

mass 1. Inertial mass: the property of matter that determines its resistance to changes in motion. Compare *weight.* 2. Gravitational mass: the property of matter defined by its attractive force on other objects. According to general relativity, the two are equivalent.

mass-luminosity relationship An empirical relationship between the luminosity (L) and mass (M) of main-sequence stars; for example, $L \propto M^{3.5}$.

mass transfer The transfer of mass from one member of a binary star system to its companion. Mass transfer occurs when one of the stars evolves to the point that it overfills its Roche lobe, so that its outer layers are pulled toward its binary companion.

massive element Also called *heavy element.* Any element more massive than helium.

matter 1. Objects made of particles that have mass, such as protons, neutrons, and electrons. 2. Anything that occupies space and has mass.

Maunder Minimum The period from 1645 to 1715, when very few sunspots were observed.

medium The substance that light travels through, such as air or glass. Compare *vacuum.*

megabar A unit of pressure equal to 1 million bars.

megaparsec (**Mpc**) A unit of distance equal to 1 million parsecs, or 3.26 million light-years.

meridian The imaginary arc in the sky running from the horizon at due north through the zenith to the horizon at due south. The meridian divides the observer's sky into eastern and western hemispheres. Compare *equator.*

mesosphere The layer of Earth's atmosphere immediately above the stratosphere, extending from an altitude of 50 kilometers (km) to about 90 km. Compare *troposphere, stratosphere,* and *thermosphere.*

meteor The incandescent trail produced by a small piece of interplanetary debris as it travels through the atmosphere at very high speeds. Compare *meteorite* and *meteoroid.*

meteor shower A larger-than-normal display of meteors, occurring when Earth passes through the orbit of a disintegrating comet, sweeping up its debris. Compare *sporadic meteor.*

meteorite A piece of rock or other fragment of material (a meteoroid) that survives to reach a planet's surface. Compare *meteor* and *meteoroid.*

meteoroid A small cometary or asteroidal fragment, ranging in size from 100 microns (μm) to 100 meters. When entering a planetary atmosphere, the meteoroid creates a *meteor.* Compare *meteor* and *meteorite;* also *planetesimal* and *zodiacal dust.*

micrometer (μm) See *micron.*

micron (μm) One-millionth (10^{-6}) of a meter; a unit of length used for the wavelength of infrared light.

microwave radiation Electromagnetic radiation with frequencies, photon energies, and wavelengths between those of infrared radiation and radio waves.

Milky Way Galaxy The galaxy in which the Sun and Solar System reside.

minor planet See *asteroid.*

minute of arc See *arcminute.*

modern physics Usually, the physical principles, including relativity and quantum mechanics, developed after 1900.

molecular cloud An interstellar cloud composed primarily of molecular hydrogen.

molecular-cloud core A dense clump within a molecular cloud that forms as the cloud collapses and fragments. Protostars form from molecular-cloud cores.

molecule Generally, the smallest particle of a substance that retains its chemical properties and is composed of two or more atoms.

momentum The product of the mass and velocity of a particle. Possible units include: kilograms times meters per second (kg m/s).

moon A less massive satellite orbiting a more massive object. Moons are found around planets, dwarf planets, asteroids, and Kuiper Belt objects. The term is usually capitalized when referring to Earth's Moon.

Mpc See *megaparsec.*

multiverse A collection of parallel universes that together comprise all that is.

mutation In biology, an imperfect reproduction of self-replicating material.

N

N See *newton.*

nadir The point on the celestial sphere located directly below an observer, opposite the *zenith.*

nanometer (nm) One-billionth (10^{-9}) of a meter; a unit of length used for the wavelength of visible light.

natural selection The process by which forms of structure, ranging from molecules to whole organisms, that are best adapted to their environment become more common than less well-adapted forms.

NCP See *north celestial pole.*

neap tide An especially weak tide that occurs around the time of the first or third quarter Moon, when the gravitational forces of the Moon and the Sun on Earth are at right angles to each other. Compare *spring tide.*

near-Earth asteroid An asteroid whose orbit brings it close to the orbit of Earth. See also *near-Earth object.*

near-Earth object (NEO) An asteroid, comet, or large meteoroid whose orbit intersects Earth's orbit.

nebula (pl. **nebulae**) A cloud of interstellar gas and dust, either illuminated by stars (bright nebula) or seen in silhouette against a brighter background (dark nebula).

nebular hypothesis The first plausible theory of the formation of the Solar System, proposed by Immanuel Kant in 1734, which stated that the Solar System formed from the collapse of an interstellar cloud of rotating gas.

NEO See *near-Earth object.*

neutrino A very low-mass, electrically neutral particle emitted during beta decay. Neutrinos interact with matter only very feebly and so can penetrate through great quantities of matter.

neutrino cooling The process in which thermal energy is carried out of the center of a star by neutrinos rather than by electromagnetic radiation or convection.

neutron A subatomic particle having no net electric charge and a rest mass and rest energy nearly equal to that of the proton. Compare *electron* and *proton.*

neutron star The neutron-degenerate stellar core left behind by a Type II supernova.

new Moon The phase of the Moon in which the Moon is between Earth and the Sun, and from Earth we see only the side of the Moon not being illuminated by the Sun. Compare *full Moon.* See also *first quarter Moon* and *third quarter Moon.*

newton (N) The force required to accelerate a 1-kilogram (kg) mass at a rate of 1 meter per second per second (m/s^2). Unit: kilograms multiplied by meters per second squared ($kg\ m/s^2$).

Newton's first law of motion The law, formulated by Isaac Newton, stating that an object will remain at rest or will continue moving along a straight line at a constant speed until an unbalanced force acts on it.

Newton's laws The three physical laws of motion formulated by Isaac Newton.

Newton's second law of motion The law, formulated by Isaac Newton, stating that if an unbalanced force acts on a body, the body will have an acceleration proportional to the unbalanced force and inversely proportional to the object's mass: $a = F/m$. The acceleration will be in the direction of the unbalanced force.

Newton's third law of motion The law, formulated by Isaac Newton, stating that for every force there is an equal force in the opposite direction.

nm See *nanometer.*

normal matter See *luminous matter.*

north celestial pole (NCP) The northward projection of Earth's rotation axis onto the celestial sphere. Compare *south celestial pole.*

North Pole The location in the Northern Hemisphere where Earth's rotation axis intersects the surface of Earth. Compare *South Pole.*

nova (pl. **novae**) A stellar explosion that results from runaway nuclear fusion in a layer of material on the surface of a white dwarf in a binary system.

nuclear burning Release of energy by nuclear fusion of low-mass elements.

nuclear fusion The combination of two less massive atomic nuclei into a single more massive atomic nucleus.

nucleosynthesis The formation of more massive atomic nuclei from less massive nuclei, either in the Big Bang (Big Bang nucleosynthesis) or in the interiors of stars (stellar nucleosynthesis).

nucleus (pl. **nuclei**) 1. The dense, central part of an atom. 2. The central core of a galaxy, comet, or other diffuse object.

O

objective lens The primary optical element in a telescope or camera that produces an image of an object.

oblateness The flattening of an otherwise spherical planet or star caused by its rapid rotation.

obliquity The inclination of a celestial body's equator to its orbital plane.

observational uncertainty The fact that real measurements are never perfect; all observations are uncertain by some amount.

Occam's razor The principle that the simplest hypothesis is the most likely, named after William of Occam (circa 1285–1349), the medieval English cleric to whom the idea is attributed.

Oort Cloud A spherical distribution of comet nuclei stretching from beyond the Kuiper Belt to more than 50,000 astronomical units (AU) from the Sun.

opacity A measure of how effectively a material blocks the radiation going through it.

open cluster A loosely bound group of a few dozen to a few thousand stars that formed together in the disk of a spiral galaxy. Compare *globular cluster.*

open universe An infinite universe with a negatively curved spatial structure (much like the surface of a saddle) such that the sum of the angles of a triangle is always less than 180 degrees. Compare *closed universe* and *flat universe.*

orbit The path taken by one object moving around another object under the influence of their mutual gravitational or electric attraction.

orbital resonance A situation in which the orbital periods of two objects are related by a ratio of small integers.

orbiter A spacecraft that is placed in orbit around a planet or moon. Compare *flyby.*

organic Containing the element carbon.

P

P wave See *primary wave.*

pair production The creation of a particle-antiparticle pair from a source of electromagnetic energy.

paleoclimatology The study of changes in Earth's climate throughout its history.

paleomagnetism The record of Earth's magnetic field as preserved in rocks.

palimpsest The flat circular patch of bright terrain that remains after a crater has been deformed over time.

parallax Also called *parallactic angle.* The displacement in the apparent position of a nearby star caused by the changing location of Earth in its orbit.

parent element A radioactive element that decays to form more stable *daughter products.*

parsec (pc) Short for *parallax second.* The distance to a star with a parallax of 1 arcsecond (arcsec) using a base of 1 astronomical unit (AU). One parsec is approximately 3.26 light-years.

partial lunar eclipse An eclipse that occurs when the Moon is partially in Earth's shadow.

partial solar eclipse The type of eclipse that occurs when Earth passes through the penumbra of the Moon's shadow, so that the Moon blocks only a portion of the Sun's disk. Compare *annular solar eclipse* and *total solar eclipse*.

pc See *parsec*.

peculiar velocity The motion of a galaxy relative to the overall expansion of the universe.

penumbra (pl. **penumbrae**) 1. The outer part of a shadow, where the source of light is only partially blocked. Compare *umbra* (definition 1). 2. The region surrounding the umbra of a sunspot. The penumbra is cooler and darker than the surrounding surface of the Sun but not as cool or dark as the umbra. Compare *umbra* (definition 2).

penumbral lunar eclipse A lunar eclipse in which the Moon passes through the penumbra of Earth's shadow. Compare *total lunar eclipse*.

perihelion (pl. **perihelia**) The point in a solar orbit that is closest to the Sun. Compare *aphelion*.

period The time it takes for a regularly repetitive process to complete one cycle.

period-luminosity relationship The relationship between the period of variability of a pulsating variable star, such as a Cepheid or RR Lyrae variable, and the luminosity of the star. Longer-period pulsating variable stars are more luminous than shorter-period ones.

Perseids A prominent August meteor shower associated with the dust debris left by comet Swift-Tuttle.

phase One of the various appearances of the sunlit surface of the Moon or a planet caused by the change in viewing location of Earth relative to both the Sun and the object. Examples include crescent phase and gibbous phase.

PHI See *Planetary Habitability Index*.

photino An elementary particle related to the photon. One of the candidates for cold dark matter.

photochemical Resulting from the action of light on chemical systems.

photodissociation The breaking apart of molecules into smaller fragments or individual atoms by the action of photons. Compare *recombination* (definition 1).

photoelectric effect The emission of electrons from a substance that is illuminated by electromagnetic radiation greater than a certain critical frequency.

photometry The process of measuring the brightness of a source of light, generally over a specific range of wavelength.

photino A hypothetical subatomic particle.

photon Also called *quantum of light*. A discrete unit or particle of electromagnetic radiation. The energy of a photon is equal to Planck's constant (h) multiplied by the frequency (f) of its electromagnetic radiation: $E_{photon} = h \times f$. The photon is the carrier of the electromagnetic force.

photosphere The apparent surface of the Sun as seen in visible light. Compare *chromosphere* and *corona*.

photosynthesize The process by which plants and algae convert energy from sunlight into chemical energy.

physical law A broad statement that predicts a particular aspect of how the physical universe behaves and that is supported by many empirical tests. See also *theory*.

pixel The smallest picture element in a digital image array.

Planck era The early time, just after the Big Bang, when the universe as a whole must be described with quantum mechanics.

Planck spectrum Also called *blackbody spectrum*. The spectrum of electromagnetic energy emitted by a blackbody per unit area per second, which is determined only by the temperature of the object.

Planck's constant (h) The constant of proportionality between the energy and the frequency of a photon. This constant defines how much energy a single photon of a given frequency or wavelength has. Value: 6.63×10^{-34} joule-second.

planet 1. A large body that orbits the Sun or other star that shines only by light reflected from the Sun or star. 2. In the Solar System, a body that orbits the Sun, has sufficient mass for self-gravity to overcome rigid body forces so that it assumes a spherical shape, and has cleared smaller bodies from the neighborhood around its orbit. Compare *dwarf planet*.

planet migration The theory that a planet can move to a location away from where it formed, through gravitational interactions with other bodies or loss of orbital energy from interaction with gas in the protoplanetary disk.

Planetary Habitability Index (**PHI**) An index for quantifying the habitability of extrasolar planets that aims to be less Earth-centric than the *Earth Similarity Index* and to broaden the options for habitability. It depends on factors not yet measured (or measurable) for most extrasolar planets, including the availability of energy, the presence of some kind of liquid, the type of surface, and the chemical makeup.

planetary nebula The expanding shell of material ejected by a dying asymptotic giant branch star. A planetary nebula glows from fluorescence caused by intense ultraviolet light coming from the hot, stellar remnant at its center.

planetary system A system of planets and other smaller objects in orbit around a star.

planetesimal A primitive body of rock and ice, 100 meters or more in diameter, that combines with others to form a planet. Compare *meteoroid* and *zodiacal dust*.

plasma A gas that is composed largely of charged particles but also may include some neutral atoms.

plate tectonics The geological theory concerning the motions of lithospheric plates, which in turn provides the theoretical basis for *continental drift*.

positron A positively charged subatomic particle; the antiparticle of the *electron*.

power The rate at which work is done or at which energy is delivered. Possible units include watts (W) and joules per second (J/s).

precession of the equinoxes The slow change in orientation between the ecliptic plane and the celestial equator caused by the wobbling of Earth's axis.

pressure Force per unit area. Possible units include newtons per square meter (N/m^2) and bars.

primary atmosphere An atmosphere, composed mostly of hydrogen and helium, that forms at the same time as its host planet. Compare *secondary atmosphere*.

primary mirror The principal optical mirror in a reflecting telescope. The primary mirror determines the telescope's light-gathering power and resolution. Compare *secondary mirror*.

primary wave Also called *P wave*. A longitudinal seismic wave, in which the oscillations involve compression and decompression parallel to the direction of travel. Compare *secondary wave*.

principle A general idea or sense about how the universe is that guides us in constructing new scientific theories. Principles can be testable theories.

prograde motion 1. Rotational or orbital motion of a moon that is in the same direction as the planet it orbits. 2. The counterclockwise orbital motion of Solar System objects as seen from above Earth's orbital plane. Compare *retrograde motion*.

prominence An archlike projection above the solar photosphere often associated with a sunspot.

proportional See *proportionality*

proportionality A relationship between two things whose ratio is a constant.

proton (***p*** or ***p⁺***) A subatomic particle having a positive electric charge of 1.6×10^{-19} coulomb (C), a rest mass of 1.67×10^{-27} kilogram (kg), and a rest energy of 1.5×10^{-10} joule (J). Compare *electron* and *neutron*.

proton-proton chain One of the ways in which hydrogen burning can take place. This is the most important path for hydrogen burning in low-mass stars such as the Sun. Compare *carbon-nitrogen-oxygen cycle*.

protoplanetary disk The remains of the accretion disk around a young star from which a planetary system may form. Sometimes called *circumstellar disk*.

protostar A young stellar object that derives its luminosity from the conversion of gravitational energy to thermal energy, rather than from nuclear reactions in its core.

pulsar A rapidly rotating neutron star that beams radiation into space in two searchlight-like beams. To a distant observer, the star appears to flash on and off.

pulsating variable star A variable star that undergoes periodic radial pulsations.

Q

QCD See *quantum chromodynamics*.

QED See *quantum electrodynamics*.

quantized Existing as discrete, irreducible units.

quantum chromodynamics (**QCD**) The quantum theory describing the strong nuclear force and its mediation by gluons. Compare *quantum electrodynamics*.

quantum efficiency The likelihood that a particular photon falling on a detector will actually produce a response in the detector.

quantum electrodynamics (QED) The quantum theory describing the electromagnetic force and its mediation by photons. Compare *quantum chromodynamics*.

quantum mechanics The branch of physics that deals with the quantized and probabilistic behavior of atoms and subatomic particles.

quantum of light See *photon*.

quark The building block of protons and neutrons.

quasar Short for *quasi-stellar radio source*. The most luminous of the active galactic nuclei, seen only at great distances from the Milky Way.

R

radial velocity The component of velocity that is directed toward or away from the observer.

radian The angle at the center of a circle subtended by an arc equal to the length of the circle's radius; 2π radians equals 360°, and 1 radian equals approximately 57.3°.

radiant The direction in the sky from which the meteors in a meteor shower seem to come.

radiation Waves or particles of energy traveling through space or a medium.

radiation belt A toroidal ring of high-energy particles surrounding a planet.

radiative transfer The transport of energy from one location to another by electromagnetic radiation.

radiative zone A region within a star where energy is transported outward by radiation. Compare *convective zone*.

radio galaxy A type of elliptical galaxy that has an active galactic nucleus at its center and very strong emission (10^{35} to 10^{38} watts [W]) in the radio part of the electromagnetic spectrum. Compare *Seyfert galaxy*.

radio telescope An instrument for detecting and measuring radio frequency emissions from celestial sources.

radio wave Electromagnetic radiation in the extreme long-wavelength region of the spectrum, beyond the region of microwaves.

radioisotope A radioactive element.

radiometric dating Use of the radioactive decay of elements to measure the ages of materials such as minerals.

ratio The relationship in quantity or size between two or more things.

ray 1. A beam of electromagnetic radiation. 2. A bright streak emanating from a young impact crater.

recombination 1. The combining of ions and electrons to form neutral atoms. Compare *photodissociation*. 2. An event early in the evolution of the universe in which hydrogen and helium nuclei combined with electrons to form neutral atoms. The removal of electrons caused the universe to become transparent to electromagnetic radiation.

red giant A low-mass star that has evolved beyond the main sequence and is now fusing hydrogen in a shell surrounding a degenerate helium core.

red giant branch A region on the H-R diagram defined by low-mass stars evolving from the main sequence toward the horizontal branch.

reddening The effect by which stars and other objects, when viewed through interstellar dust, appear redder than they actually are. Reddening is caused by the fact that blue light is more strongly absorbed and scattered than red light.

redshift The Doppler shift toward longer (redder) wavelengths of light from an approaching object. Compare *blueshift*.

reflecting telescope A telescope that uses mirrors for collecting and focusing incoming electromagnetic radiation to form an image in their focal planes. The size of a reflecting telescope is defined by the diameter of the primary mirror. Compare *refracting telescope*.

reflection The redirection of a beam of light that strikes, but does not cross, the surface between two media having different refractive indices. If the surface is flat and smooth, the angle of incidence equals the angle of reflection. Compare *refraction*.

refracting telescope A telescope that uses objective lenses for collecting and focusing incoming electromagnetic radiation to form an image. Compare *reflecting telescope*.

refraction The redirection or bending of a beam of light when it crosses the boundary between two media having different refractive indices. Compare *reflection*.

refractory material Material that remains solid at high temperatures. Compare *volatile material*.

regular moon A moon that formed together with the planet it orbits. Compare *irregular moon*.

reionization A period after the Dark Ages during which objects formed that radiated enough energy to ionize neutral hydrogen, at redshift $6 < z < 20$.

relative humidity The amount of water vapor held by a volume of air at a given temperature compared (stated as a percentage) to the total amount of water that could be held by the same volume of air at the same temperature.

relative motion The difference in motion between two individual frames of reference.

relativistic Describing systems that travel at nearly the speed of light or are located in the vicinity of very strong gravitational fields.

relativistic beaming The effect created when material moving at nearly the speed of light beams the radiation it emits in the direction of its motion.

relativistic speed A speed high enough that special relativity, rather than Newtonian physics, is needed to describe the motion. Speeds greater than about 10% the speed of light are relativistic.

remote sensing The use of images, spectra, radar, or other techniques to measure the properties of an object from a distance.

resolution The ability of a telescope to separate two point sources of light. Resolution is determined by the telescope's aperture and the wavelength of light it receives.

rest wavelength The wavelength of light that is seen coming from an object at rest with respect to the observer.

retrograde motion 1. Rotation or orbital motion of a moon that is in the opposite direction to the rotation of the planet it orbits. 2. The clockwise orbital motion of Solar System objects as seen from above Earth's orbital plane. Compare *prograde motion*. 3. Apparent retrograde motion is a motion of the planets with respect to the "fixed stars," in which the planets appear to move westward for a period of time before resuming their normal eastward motion.

revolve Motion of one object in orbit around another.

rift zone A region created by a geological fault, in which mantle material rises up, cools, and slowly spreads out, forming new crust.

right ascension A measure, analogous to *longitude*, that tells you the angular distance of a celestial body eastward along the celestial equator from the vernal equinox. Compare *declination*.

ring An aggregation of small particles orbiting a planet or star. The rings of the four giant planets of the Solar System are composed variously of silicates, organic materials, and ices.

ring arc A discontinuous, higher-density region within an otherwise continuous, narrow ring.

ringlet A narrowly confined concentration of ring particles.

Roche limit The distance at which a planet's tidal forces exceed the self-gravity of a smaller object—such as a moon, asteroid, or comet—causing the object to break apart.

Roche lobes The hourglass-shaped or figure eight–shaped volume of space surrounding two stars, which constrains material that is gravitationally bound by one or the other.

rotation curve A plot showing how the orbital velocity of stars and gas in a galaxy changes with radial distance from the galaxy's center.

rover A remotely controlled instrumented vehicle designed to move and explore the surface of a terrestrial planet or moon. Compare *lander*.

RR Lyrae variable A variable giant star whose regularly timed pulsations are good predictors of its luminosity. RR Lyrae variables are used for distance measurements to globular clusters.

S

S-type asteroid An asteroid made of material that was once part of the outer layer of a larger, differentiated body that has since broken into pieces. Compare *C-type asteroid* and *M-type asteroid*.

S wave See *secondary wave*.

S0 galaxy A galaxy with a bulge and a disk-like spiral, but smooth in appearance like ellipticals. Compare *barred spiral galaxy*, *elliptical galaxy*, *irregular galaxy*, and *spiral galaxy*.

satellite An object in orbit around a more massive body; for example, a human-made satellite or a moon of any planet.

scale factor (R_U) A dimensionless number proportional to the distance between two points in space. The scale factor increases as the universe expands.

scattering The random change in the direction of travel of photons, caused by their interactions with molecules or dust particles.

Schwarzschild radius The distance from the center of a nonrotating, spherical black hole at which the escape velocity equals the speed of light.

scientific method The formal procedure—including hypothesis, prediction, and experiment or observation—used to test (attempt to falsify) the validity of scientific hypotheses and theories.

scientific notation The standard expression of numbers with one digit (which can be zero) to the left of the decimal point and multiplied by 10 to the exponent required to give the number its correct value. Example: $2.99 \times 10^8 = 299,000,000$.

SCP See *south celestial pole*.

second law of thermodynamics The law stating that the entropy or disorder of an isolated system always increases as the system evolves.

second of arc See *arcsecond*.

secondary atmosphere An atmosphere that forms—as a result of volcanism, comet impacts, or another process—sometime after its host planet has formed. Compare *primary atmosphere*.

secondary crater A crater formed from ejecta thrown from an *impact crater*.

secondary mirror A small mirror placed on the optical axis of a reflecting telescope that returns the beam back through a small hole in the *primary mirror*, thereby shortening the mechanical length of the telescope.

secondary wave Also called *S wave*. A transverse seismic wave, which involves the sideways motion of material. Compare *primary wave*.

sedimentation A process in which material carried by water or wind deposits layers of material and buries what lies below.

seismic wave A vibration due to an earthquake, a large explosion, or an impact on the surface that travels through a planet's interior.

seismometer An instrument that measures the amplitude and frequency of seismic waves.

self-gravity The gravitational attraction among all parts of the same object.

semimajor axis Half of the longer axis of an ellipse.

SETI The Search for Extraterrestrial Intelligence project, which uses advanced technology combined with radio telescopes to search for evidence of intelligent life elsewhere in the universe.

Seyfert galaxy A type of spiral galaxy with an active galactic nucleus at its center; first discovered in 1943 by Carl Seyfert. Compare *radio galaxy*.

shepherd moon A moon that orbits close to rings and gravitationally confines the orbits of the ring particles.

shield volcano A volcano formed by very fluid lava flowing from a single source and spreading out from that source. Compare *composite volcano*.

short-period comet A comet with an orbital period of less than 200 years. Compare *long-period comet*.

sidereal day Earth's period of rotation with respect to the stars—about 23 hours 56 minutes—which is the time it takes for Earth to make one rotation and face the exact same star on the meridian. It differs from the *solar day* because of Earth's motion around the Sun.

sidereal period An object's orbital or rotational period measured with respect to the stars. Compare *synodic period*.

silicate One of the family of minerals composed of silicon and oxygen in combination with other elements.

singularity The point where a mathematical expression or equation becomes meaningless, such as a fraction whose denominator approaches zero See also *black hole*.

solar abundance The relative amount of an element detected in the atmosphere of the Sun, expressed as the ratio of the number of atoms of that element to the number of hydrogen atoms.

solar day The day in common use—24 hours, which is Earth's period of rotation that brings the Sun back to the same local meridian where the rotation started. Compare *sidereal day*.

solar eclipse An eclipse that occurs when the Sun is partially or entirely blocked by the Moon. Compare *lunar eclipse*.

solar flare An explosion on the Sun's surface associated with a complex sunspot group and a strong magnetic field.

solar maximum (pl. **maxima**) The time, occurring about every 11 years, when the Sun is at its peak activity, meaning that sunspot activity and related phenomena (such as prominences, flares, and coronal mass ejections) are at their peak.

solar neutrino problem The historical observation that only about a third as many neutrinos as predicted by theory seemed to be coming from the Sun.

Solar System The gravitationally bound system made up of the Sun, planets, dwarf planets, moons, asteroids, comets, and Kuiper Belt objects, along with their associated gas and dust.

solar tide A tide on Earth that is due to the differential gravitational pull of the Sun. Compare *lunar tide*.

solar wind The stream of charged particles emitted by the Sun that flows at high speeds through interplanetary space.

solstice Literally, "Sun standing still." 1. One of the two most northerly and southerly points on the ecliptic. 2. Either of the two times of year (the *summer solstice* and *winter solstice*) when the Sun is at one of these two positions. Compare *equinox*.

south celestial pole (SCP) The southward projection of Earth's rotation axis onto the celestial sphere. Compare *north celestial pole*.

South Pole The location in the Southern Hemisphere where Earth's rotation axis intersects the surface of Earth. Compare *North Pole*.

spacetime A concept that combines space and time into a four-dimensional continuum with three spatial dimensions plus time.

special relativity See *special theory of relativity*.

special theory of relativity Sometimes referred to as simply *special relativity*. Einstein's theory explaining how the fact that the speed of light is a constant affects nonaccelerating frames of reference. Compare *general theory of relativity*.

spectral type A classification system for stars based on the presence and relative strength of absorption lines in their spectra. Spectral type is related to the surface temperature of a star.

spectrograph Also called *spectrometer*. A device that spreads out the light from an object into its component wavelengths.

spectrometer See *spectrograph*.

spectroscopic binary A binary star system whose existence and properties are revealed to astronomers only by the Doppler shift of its spectral lines. Most spectroscopic binaries are close pairs. Compare *eclipsing binary* and *visual binary*.

spectroscopic parallax Use of the spectroscopically determined luminosity and the observed brightness of a star to determine the star's distance.

spectroscopy The study of an object's electromagnetic radiation in terms of its component wavelengths.

spectrum (pl. **spectra**) Waves sorted by wavelength. See also *electromagnetic spectrum*.

speed The rate of change of an object's position with time, without regard to the direction of movement. Possible units include meters per second (m/s) and kilometers per hour (km/h). Compare *velocity*.

spherically symmetric Describing an object whose properties depend only on distance from the object's center, so that the object has the same form viewed from any direction.

spin-orbit resonance A relationship between the orbital and rotation periods of an object such that the ratio of their periods can be expressed by simple integers.

spiral density wave A stable, spiral-shaped change in the local gravity of a galactic disk that can be produced by periodic gravitational kicks from neighboring galaxies or from nonspherical bulges and bars in spiral galaxies.

spiral galaxy A galaxy of Hubble type "S" class, with a discernible disk in which large spiral patterns exist. Compare *barred spiral galaxy*, *elliptical galaxy*, *irregular galaxy*, and *S0 galaxy*.

spoke One of several narrow radial features seen occasionally in Saturn's B Ring. Spokes appear dark in backscattered light and bright in forward, scattering light, indicating that they are composed of tiny particles. Their origin is not well understood.

sporadic meteor A meteor that is not associated with a specific *meteor shower*.

spreading center A zone from which two tectonic plates diverge.

spring tide An especially strong tide that occurs around the time of a new or full Moon, when lunar tides and solar tides reinforce each other. Compare *neap tide*.

stable equilibrium An equilibrium state in which the system returns to its former condition after a small disturbance. Compare *unstable equilibrium*.

standard candle An object whose luminosity either is known or can be predicted in a distance-independent way, so its brightness can be used to determine its distance via the inverse square law of radiation.

standard model The theory of particle physics that combines electroweak theory with quantum chromodynamics to describe the structure of known forms of matter.

star A luminous ball of gas that is held together by gravity. A normal star is powered by nuclear reactions in its interior.

star cluster A group of stars that all formed at the same time and in the same general location.

static equilibrium A state in which the forces within a system are all in balance so that the system does not change. Compare *dynamic equilibrium*.

Stefan-Boltzmann constant (σ) The constant of proportionality that relates the flux emitted by an object to the fourth power of its absolute temperature. Value: 5.67×10^{-8} W/(m^2 K^4) (W = watts, m = meters, K = kelvins).

Stefan-Boltzmann law The law, formulated by Josef Stefan and Ludwig Boltzmann, stating that the amount of electromagnetic energy emitted from the surface of a body (flux), summed over the energies of all photons of all wavelengths emitted, is proportional to the fourth power of the temperature of the body: $\mathcal{F} = \sigma T^4$.

stellar-mass loss The loss of mass from the outermost parts of a star's atmosphere during the course of its evolution.

stellar occultation An event in which a planet or other Solar System body moves between the observer and a star, eclipsing the light emitted by that star.

stellar population A group of stars with similar ages, chemical compositions, and dynamic properties.

stereoscopic vision The way an animal's brain combines the different information from its two eyes to perceive the distances to objects around it.

stony-iron meteorite A meteorite consisting of a mixture of silicate minerals and iron-nickel alloys. Compare *iron meteorite* and *stony meteorite*.

stony meteorite A meteorite composed primarily of silicate minerals, similar to those found on Earth. Compare *iron meteorite* and *stony-iron meteorite*.

stratosphere The atmospheric layer immediately above the troposphere. On Earth, it extends upward to an altitude of 50 kilometers (km). Compare *troposphere*, *mesosphere*, and *thermosphere*.

string theory See *superstring theory*.

stromatolite A structure created by living or fossilized cyanobacteria..

strong nuclear force The attractive short-range force between protons and neutrons that holds atomic nuclei together. One of the four fundamental forces of nature, along with the *electromagnetic force*, the *weak nuclear force*, and *gravity* (definition 1). The force is mediated by the exchange of gluons.

subduction zone A region where two tectonic plates converge, with one plate sliding under the other and being drawn downward into the interior.

subgiant A giant star that is smaller and lower in luminosity than normal giant stars of the same spectral type. Subgiants evolve to become giants.

subgiant branch A region of the H-R diagram defined by stars that have left the main sequence but have not yet reached the red giant branch.

sublimation The process by which a solid becomes a gas without first becoming a liquid.

subsonic Moving within a medium at a speed slower than the speed of sound in that medium. Compare *supersonic*.

summer solstice 1. One of two points where the Sun is at its greatest distance from the celestial equator. 2. The day on which the Sun appears at this location, marking the first day of summer (about June 20 in the Northern Hemisphere and December 21 in the Southern Hemisphere). Compare *winter solstice*. See also *autumnal equinox* and *vernal equinox*.

Sun The star at the center of the Solar System.

sungrazer A comet whose perihelion is within a few solar diameters of the surface of the Sun.

sunspot A cooler, transitory region on the solar surface produced when loops of magnetic flux break through the surface of the Sun.

sunspot cycle The approximate 11-year cycle during which sunspot activity increases and then decreases. This is one-half of a full 22-year cycle, in which the magnetic polarity of the Sun first reverses and then returns to its original configuration.

supercluster A large conglomeration of galaxy clusters and galaxy groups; typically, more than 100–300 megaparsecs (Mpc) in size and containing tens of thousands to hundreds of thousands of galaxies. Compare *galaxy cluster* and *galaxy group*.

super-Earth An extrasolar planet with about 2–10 times the mass of Earth.

superior planet A Solar System planet that orbits the Sun at a greater distance than Earth's orbit.

superluminal motion The appearance (though not the reality) that a jet is moving faster than the speed of light.

supermassive black hole A black hole of 1,000 solar masses (M_{Sun}) or more that resides in the center of a galaxy, and whose gravity powers active galactic nuclei.

supernova (pl. **supernovae**) A stellar explosion resulting in the release of tremendous amounts of energy, including the high-speed ejection of matter into the interstellar medium. See also *Type Ia supernova* and *Type II supernova*.

supernova remnant The material ejected from the outer layers of a star following a supernova explosion.

supersonic Moving within a medium at a speed faster than the speed of sound in that medium. Compare *subsonic*.

superstring theory The theory that conceives of particles as strings in 10 dimensions of space and time; the current contender for a theory of everything.

surface brightness The amount of electromagnetic radiation emitted or reflected per unit area.

surface wave A seismic wave that travels on the surface of a planet or moon.

symmetry In theoretical physics, the properties of physical laws that remain constant when certain things change, such as the symmetry between matter and antimatter even though their charges may be different.

synchronous rotation The case that occurs when a body's rotation period equals its orbital period around another body. A special type of spin-orbit resonance.

synchrotron radiation Radiation from electrons moving at close to the speed of light as they spiral in a strong magnetic field; named because this kind of radiation was first identified on Earth in particle accelerators called synchrotrons.

synodic period An object's orbital or rotational period measured with respect to the Sun. Compare *sidereal period*.

T

T Tauri star A young stellar object that has dispersed enough of the material surrounding it to be seen in visible light.

tail A stream of gas and dust swept away from the coma of a comet by the solar wind and by radiation pressure from the Sun.

tectonism Deformation of the lithosphere of a planet.

telescope The basic tool of astronomers. Working over the entire range from gamma rays to radio waves, astronomical telescopes collect and concentrate electromagnetic radiation from celestial objects.

temperature A measure of the average kinetic energy of the atoms or molecules in a gas, solid, or liquid.

terrestrial planet An Earth-like planet, made of rock and metal and having a solid surface. In the Solar System, the terrestrial planets are Mercury, Venus, Earth, and Mars. Compare *giant planet*.

theoretical model A detailed description of the properties of a particular object or system in terms of known physical laws or theories; often, a computer calculation of predicted properties based on such a description.

theory A well-developed idea or group of ideas that are tied solidly to known physical laws and make testable predictions about the world. A very well-tested theory may be called a *physical law*, or simply a fact. Compare *hypothesis*.

theory of everything (TOE) A theory that unifies all four fundamental forces of nature: strong nuclear, weak nuclear, electromagnetic, and gravitational forces.

thermal conduction See *conduction*.

thermal energy The energy that resides in the random motion of atoms, molecules, and particles, by which we measure their temperature.

thermal equilibrium The state in which the rate of thermal-energy emission by an object is equal to the rate of thermal-energy absorption.

thermal motion The random motion of atoms, molecules, and particles that gives rise to thermal radiation.

thermal radiation Electromagnetic radiation resulting from the random motion of the charged particles in every substance.

thermosphere The layer of Earth's atmosphere at altitudes greater than 90 kilometers (km), above the mesosphere. Near its top, at an altitude of 600 km, the temperature can reach 1000 K. Compare *troposphere*, *stratosphere*, and *mesosphere*.

third quarter Moon The phase of the Moon in which only the eastern half of the Moon, as viewed from Earth, is illuminated by the Sun. It occurs about one week after the full Moon. Compare *first quarter Moon*. See also *full Moon* and *new Moon*.

tidal bulge Distortion of a body resulting from tidal stresses.

tidal force A force caused by the change in the strength of gravity across an object.

tidal locking Synchronous rotation of an object caused by internal friction as the object rotates through its tidal bulge.

tide On Earth, the rise and fall of the oceans as Earth rotates through a tidal bulge caused by the Moon and the Sun. See also *lunar tide*, *neap tide*, *solar tide*, and *spring tide*.

time dilation The relativistic "stretching" of time. Compare *general relativistic time dilation*.

TOE See *theory of everything*.

topographic relief The differences in elevation from point to point on a planetary surface.

tornado A violent rotating column of air, typically 75 meters across with 200-kilometer-per-hour (km/h) winds. Some tornadoes can be more than 3 km across, and winds up to 500 km/h have been observed.

torus (pl. **tori**) A three-dimensional, doughnut-shaped ring.

total lunar eclipse A lunar eclipse in which the Moon passes through the umbra of Earth's shadow. Compare *penumbral lunar eclipse*.

total solar eclipse The type of eclipse that occurs when Earth passes through the umbra of the Moon's shadow, so that the Moon completely blocks the disk of the Sun. Compare *annular solar eclipse* and *partial solar eclipse*.

transform fault The actively slipping segment of a fracture zone between lithospheric plates.

transit method A method of detecting extrasolar planets by measuring the decrease in light from a star as its orbiting planet passes in front of the star as viewed from Earth.

trans-Neptunian object See *Kuiper Belt object*.

transverse wave A wave that oscillates perpendicular to the direction of the wave's propagation. Compare *longitudinal wave*.

triple-alpha process The nuclear fusion reaction that combines three helium nuclei (alpha particles) together into a single nucleus of carbon.

Trojans A group of asteroids orbiting in the L_4 and L_5 Lagrangian points of Jupiter's orbit.

tropical year The time between one crossing of the vernal equinox and the next. Because of the precession of the equinoxes, a tropical year is slightly shorter than the time that it takes for Earth to orbit once around the Sun. Compare *year*.

Tropics The region on Earth between latitudes 23.5° south and 23.5° north, where the Sun appears directly overhead twice during the year.

tropopause The top of a planet's troposphere.

troposphere The convection-dominated layer of a planet's atmosphere. On Earth, the atmospheric region closest to the ground within which most weather phenomena take place. Compare *stratosphere*, *mesosphere*, and *thermosphere*.

tuning fork diagram The two-pronged diagram showing Hubble's classification of galaxies into ellipticals, S0s, spirals, barred spirals, and irregular galaxies.

turbulence The random motion of blobs of gas within a larger cloud of gas.

Type Ia supernova A supernova explosion with a calibrated peak luminosity that occurs as a result of runaway carbon burning in a white dwarf star that accretes mass from a companion and approaches the Chandrasekhar mass limit of $1.4\ M_{\text{Sun}}$.

Type II supernova A supernova explosion in which the degenerate core of an evolved massive star suddenly collapses and rebounds.

U

ultrafaint dwarf galaxy A dim dwarf galaxy with only 1,000–100,000 times the Sun's luminosity. Ultrafaint dwarf galaxies differ from globular clusters in that they are composed of large amounts of dark matter.

ultraviolet (UV) radiation Electromagnetic radiation having frequencies and photon energies greater than those of visible light but less than those of X-rays and having wavelengths shorter than those of visible light but longer than those of X-rays.

umbra (pl. **umbrae**) 1. The darkest part of a shadow, where the source of light is completely blocked. Compare *penumbra* (definition 1). 2. The darkest, innermost part of a sunspot. Compare *penumbra* (definition 2).

unbalanced force The nonzero net force acting on a body.

unbound orbit An orbit in which an object is no longer gravitationally bound to the body it was orbiting. An unbound orbit's velocity is greater than the escape velocity. Compare *bound orbit*.

uncertainty principle See *Heisenberg uncertainty principle*.

unified model of AGN A model in which many different types of activity in the nuclei of galaxies are all explained by accretion of matter around a supermassive black hole.

uniform circular motion Motion in a circular path at a constant speed.

unit A fundamental quantity of measurement. The meter is an example of a metric unit; the foot is an example of an English unit.

universal gravitational constant (G) The constant of proportionality in the universal law of gravitation. Value: 6.67×10^{-11} meters cubed per kilogram second squared $[\text{m}^3/\text{kg s}^2 = \text{N m}^2/\text{kg}^2]$.

universal law of gravitation The law, formulated by Isaac Newton, stating that the gravitational force between any two objects is proportional to the product of their masses and inversely proportional to the square of the distance between them:

$$F_{\text{grav}} = G \times \frac{m_1 \times m_2}{r^2}$$

universe 1. All of space and everything contained therein. 2. Our own universe in a collection of parallel universes that together comprise all that is.

unstable equilibrium An equilibrium state in which a small disturbance will cause a system to move away from equilibrium. Compare *stable equilibrium*.

UV Ultraviolet. See *ultraviolet radiation*.

V

vacuum A region of space that contains very little matter. In quantum mechanics and general relativity, however, even a perfect vacuum has physical properties.

variable star A star with varying luminosity. Many periodic variables are found within the instability strip on the H-R diagram.

velocity The rate and direction of change of an object's position with time. Possible units include: meters per second (m/s) and kilometers per hour (km/h). Compare *speed*.

vernal equinox 1. One of two points where the Sun crosses the celestial equator. 2. The day on which the Sun appears at this location, marking the first day of spring (about March 20 in the Northern Hemisphere and September 22 in the Southern Hemisphere). Compare *autumnal equinox*. See also *summer solstice* and *winter solstice*.

virtual particle A particle that, according to quantum mechanics, comes into existence only momentarily. According to theory, fundamental forces are mediated by the exchange of virtual particles.

visual binary A binary system in which the two stars can be seen individually from Earth. Compare *eclipsing binary* and *spectroscopic binary*.

void A region in space containing little or no matter. Examples include regions in cosmological space that are largely empty of galaxies.

volatile material Sometimes called *ice*. Material that remains gaseous at moderate temperature. Compare *refractory material*.

volcanism A form of geological activity on a planet or moon in which molten rock (magma) erupts at the surface.

W

W See *watt*.

waning Describing the changing phases of the Moon as it becomes less fully illuminated between full Moon and new Moon as seen from Earth. Compare *waxing*.

waning gibbous Moon The phases of the moon between gibbous and 3rd quarter

water cycle The flow of water on, above, and through Earth's surface.

watt (W) A measure of *power*. Unit: joules per second (J/s).

wave A disturbance moving along a surface or passing through a space or a medium.

wavefront The imaginary surface of an electromagnetic wave, either plane or spherical, oriented perpendicular to the direction of travel.

wavelength The distance on a wave between two adjacent points having identical characteristics. The distance a wave travels in one period. Possible units include meters (m).

waxing Describing the changing phases of the Moon as it becomes more fully illuminated between new Moon and full Moon as seen from Earth. Compare *waning*.

waxing crescent Moon The phases of the Moon between new and 1st quarter.

waxing gibbous Moon The phases of the Moon between 1st quarter and full.

weak nuclear force The force underlying some forms of radioactivity and certain interactions between subatomic particles. It is responsible for radioactive beta decay and for the initial proton-proton interactions that lead to nuclear fusion in the Sun and other stars. One of the four fundamental forces of nature, along with the *electromagnetic force*, the *strong nuclear force*, and *gravity* (definition 1). The force is mediated by the exchange of *W* and *Z* particles.

weather The state of an atmosphere at any given time and place. Compare *climate*.

weight The gravitational force acting on an object; that is, the force equal to the mass of an object multiplied by the local acceleration due to gravity. In general relativity, the force equal to the mass of an object multiplied by the acceleration of the frame of reference in which the object is observed. Compare *mass*.

white dwarf The stellar remnant left at the end of the evolution of a low-mass star. A typical white dwarf has a mass of 0.6 solar mass (M_{Sun}) and a size about equal to that of Earth; it is made of nonburning, electron-degenerate carbon.

Wien's law A law, named for Wilhelm Wien, stating that location of the peak wavelength in the electromagnetic spectrum of an object is inversely proportional to the temperature of the object.

WIMP Short for *weakly interacting massive particle*. A hypothetical massive particle that interacts through the weak nuclear force and gravity but not with electromagnetic radiation. WIMPs are candidates for dark matter. Compare *MACHO*.

winter solstice 1. One of two points where the Sun is at its greatest distance from the celestial equator. 2. The day on which the Sun appears at this location, marking the first day of winter (about December 21 in the Northern Hemisphere and June 20 in the Southern Hemisphere). Compare *summer solstice*. See also *autumnal equinox* and *vernal equinox*.

X

X-ray Electromagnetic radiation having frequencies and photon energies greater than those of ultraviolet (UV) light but less than those of gamma rays and having wavelengths shorter than those of UV light but longer than those of gamma rays.

X-ray binary A binary system in which mass from an evolving star spills over onto a collapsed companion, such as a neutron star or black hole. The material falling in is heated to such high temperatures that it glows brightly in X-rays.

Y

year The time it takes Earth to make one revolution around the Sun. A solar year is measured from equinox to equinox. A sidereal year, Earth's true orbital period, is measured relative to the stars. Compare *tropical year*.

Z

zenith The point on the celestial sphere located directly overhead from an observer. Compare *nadir*.

zero-age main sequence The strip on the H-R diagram plotting where stars of all masses in a cluster begin their lives.

zodiac The 12 constellations lying along the plane of the ecliptic.

zodiacal dust Particles of cometary and asteroidal debris less than 100 microns (μm) in size that orbit the inner Solar System close to the plane of the ecliptic. Compare *meteoroid* and *planetesimal*.

zodiacal light A band of light in the night sky caused by sunlight reflected by zodiacal dust.

zonal wind The east–west component of a wind.

Selected Answers

These pages contain selected answers for all 24 chapters in the complete volume of *21st Century Astronomy*, Fifth Edition. The Solar System Edition does not include Chapters 15–23. The Stars and Galaxies Edition does not include Chapters 8–12.

Chapter 1

Check Your Understanding
1. Radius of Earth–light-minute–distance from Earth to Sun–light-hour–radius of Solar System–light-year.
2. b
3. c

Thinking about the Concepts
19. 2.5 million years.
22. *Falsifiable* means that something can be tested and shown to be false/incorrect through an experiment or observation.
29. The use of mathematics is not the hallmark of good science. Rather, it is following the scientific method, which astrology does not employ.

Applying the Concepts
35. 20 days
39. 1/16th
44. The 18-inch pizza is more economical.

Chapter 2

Check Your Understanding
1. b
2. d
3. d
4. Full
5. If the lunar cycle was 30 days and the year was exactly 12 cycles, holidays like these based on a lunar calendar would stay on the same date.
6. b

Thinking about the Concepts
17. Zenith.
24. Midnight, noon.
29. Longer and more severe seasonal differences.

Applying the Concepts
35. (a) 23.5°, (b) 23.5°
37. 78.5°
44. 3.67 times larger

Chapter 3

Check Your Understanding
1. Planets wander through the sky.
2. c
3. b
4. c-b-d-a
5. d
6. b

Thinking about the Concepts
17. *Empirical* means basing conclusions on what one has found in observations and data
21. A planet is always moving the fastest when it is closest to the Sun, and slowest when furthest.
24. Yes, the laws of physics are universal.

Applying the Concepts
34. 30.1 AU, 367.5 days.
41. (a) 200 km/h. (b) 200 km/h.
44. 5 m/s^2

Chapter 4

Check Your Understanding
1. d
2. c, a
3. a
4. c-a-d-b
5. a

Thinking about the Concepts
18. *Mass* is a measurement of an object's inertia. *Weight* is how strongly gravity acts on the object.
21. The Earth rotates the fastest at the equator, thus launching satellites from here gives them the highest initial speed. Launched toward the east because Earth rotates to the east.
25. Near sunrise or sunset during the new and full Moon phases.

Applying the Concepts
34. 1.12 m/s
39. 7.7 km/s
44. (a) 0.71 yr. (b) 42.1 km/s

Chapter 5

Check Your Understanding
1. e-c-b-d-a
2. c
3. Spectral features correspond to elements present.
4. d
5. e
6. c

Thinking about the Concepts
19. The red beam would have to be more intense.
24. The object is moving toward you.
26. The blue star is hotter than the yellow one.

Applying the Concepts
33. 10,000 km/s, moving away.
38. 9 times more luminous.
42. 9.35 microns. 131 W

Chapter 6

Check Your Understanding
1. a, b, c, d
2. b
3. c
4. d
5. c
6. a, b, d

Thinking about the Concepts
16. Simple lenses suffer from chromatic aberration, that is, they bend different colors by different amounts, so the resulting image is not nearly as crisp as in telescopes with multiple lenses.
23. An adaptive-optic system will use a star and sophisticated computer programs to control a deformable mirror in an attempt to keep the star's image steady and as small as possible, removing the effects of atmospheric blurring and providing higher-resolution images.
25. The nearest star is about 4 light-years away, while low-Earth orbit is about 400 kilometers above the ground. Putting a telescope in orbit gets the telescope closer to that closest star by such an infinitesimally small amount that there is no suitable analogy that immediately comes to mind.

Applying the Concepts
33. 16 times.
40. (a) 35 milliarcsec. (b) 1 millarcsec.
45. 1.06 mm, microwave.

Chapter 7

Check Your Understanding
1. a, b, d
2. a
3. d
4. a
5. c

Thinking about the Concepts
22. An accretion disk is the thin, rotating disk that forms as a gas cloud collapses on itself.
28. Stars are very bright and far away while planets appear very close to their stars. Thus it is difficult to mask out the light of the star and still see close enough to it to see the reflected light of a planet.
30. The Kepler satellite uses planetary transits to search for exoplanets. By "Earth-like," we first mean planets that are terrestrial and have roughly the same mass.

Applying the Concepts
31. 80 mph, or 838 times smaller than Earth's orbital speed.
41. 1%
45. (a) 4.43×10^{27} kg. (b) 1.02×10^8 m. (c) 4.45×10^{24}. (d) 996 kg/m^3, slightly less than that of water. The planet is gaseous.

Chapter 8

Check Your Understanding
1. b
2. d
3. d
4. c
5. c
6. b

Thinking about the Concepts
18. We can select igneous rocks from various levels of the Grand Canyon and measure their ages using radiometric dating.
20. The Moon is smaller than the Earth, so it contained less thermal energy in its core to begin with, and a smaller core cools faster than a larger one.
29. Mars shows dry river beds, canyons, and teardrop-shaped erosions around craters,

Applying the Concepts
34. Figure 8.23 shows a very smooth surface, with fewer small craters than Figure 8.18, suggesting this region is younger
38. The moon cools off 3.7 times faster.
41. (a) 6 half lives. (b) 34,200 y. (c) overestimate.

Chapter 9

Check Your Understanding
1. a, b, c, d
2. c
3. c-a-e-b-d
4. a
5. d, c, b, a
6. a-b-c

Thinking about the Concepts
19. Impacts from icy comets, and outgassing from volcanoes.
24. (a) Artic ice floats, so it adds no extra mass to the ocean when it melts. (b) Glaciers reside on land, adding to the mass of water in the oceans.
25. Venus has perpetual, opaque cloud cover due to its thick atmosphere.

Applying the Concepts
35. (a) 2000 N. (b) 200 kg. (c) 3. (d) Our bones and muscle hold us up.
42. (a) 0.85 times. (b) No.
45. 920 m

Chapter 10

Check Your Understanding
1. a
2. b
3. b
4. d
5. a-d-b-c

Thinking about the Concepts
18. The size of a planet, and study the presence and properties of the planet's atmosphere and rings.
19. Coriolis forces
28. Synchrotron radiation.

Applying the Concepts
35. 632 AU; 21 times further than Neptune.
36. (a) 51,100 km. (b) a lower limit to the size.
43. 60 K

Chapter 11

Check Your Understanding
1. a, b, c
2. a
3. a-d-b-c
4. a
5. b

Thinking about the Concepts
16. Tidal stresses from Jupiter continuously flex Io by many tens of meters per orbit and keep the interior mantle molten.
24. Breakup of moons, asteroids, or comets; volcanic processes; impact events; atmospheric escape.
25. Gaps can be created by small moons; larger gaps are created by orbital resonances of shepherd moons.

Applying the Concepts
31. 2.6 km/s
36. Increase; increase; increase
44. 0.69; separate particles.

Chapter 12

Check Your Understanding
1. Because they do not orbit another body.
2. c
3. c
4. a, b, c, e
5. Pre-collision orbit, structure, and chemical make-up.

Thinking about the Concepts
18. Asteroids are loose agglomerations of refractory rock and metal held together only by self-gravity. Comet nuclei are "dirty snowballs" that are composed primarily of ices and organic materials held together by a loose rocky matrix
24. The meteorite probably came from Mars, or another object that was geologically active 1 billion years ago.
30. Comets and asteroids might have brought water and organic molecules to Earth, or their impacts might have created enough energy to cause chemical reactions that formed self-replicating proteins in our "primordial soup."

Applying the Concepts
35. 3.68×10^{20} J, or 87,600 H-bombs.
39. 2.5 AU.
43. 5×10^{16} cm³.

Chapter 13

Check Your Understanding
1. b
2. d
3. a, b, d
4. b, d

Thinking about the Concepts
18. (a) The sapphire blue star has the higher temperature. (b) The golden star is larger.
19. A star with a 2,500 K blackbody emits mostly in the red and infrared but still gives off a measurable amount of blue light.
28. Eclipsing and spectroscopic binaries; visual binaries.

Applying the Concepts
32. 1.5 times further
38. (a) 0.85 M_{Sun}. (b) 19 AU.
42. 4×10^6 AU. 6.32 ly.

Chapter 14

Check Your Understanding
1. c
2. b
3. c
4. b
5. d

Thinking about the Concepts
18. Neutrinos are very hard to detect because they carry no charge, travel at nearly the speed of light, and interact very weakly with matter.
21. The core would heat and expand, cool down, fuse less, contract, and return to equilibrium. The Sun would brighten for a short time.
28. Sunspots tell us the Sun's overall rotation period, but more important, that the Sun has differential rotation, whereby the equator rotates faster than the poles

Applying the Concepts
34. A photon scatters around in the Sun, traveling about 1 cm before hitting another atom. Thus, the distance required for a photon to escape the Sun is much longer than the straight path out.
37. (a) 6×10^{26} kg. (b) 0.03%.
38. 1/6 as long as the Sun, i.e., 1.7 billion years.

Chapter 15

Check Your Understanding
1. d
2. c
3. a
4. a

Thinking about the Concepts
16. Dust grains are comparable in size to visible-light wavelengths, so they act as a barrier to visible light. Gas molecules are much smaller than dust grains, so they do not obstruct light at visible wavelengths.
22. Tenuous gases have very few collisions with other atoms or dust; thus the atoms have a hard time getting rid of their energy, and as a result, they stay hot. On the other hand, dense gases have frequent collisions; thus they can cool easily.
27. As a protostar collapses, more mass implies higher gravity, but that higher gravity compresses the gas further. This compression raises its temperature and thus its pressure, so the star maintains hydrostatic equilibrium at all times.

Applying the Concepts
33. 10^{-12} grains/m³.
38. 43 times.
43. 39 times.

Chapter 16

Check Your Understanding
1. a
2. c
3. a
4. a
5. d

Thinking about the Concepts
16. To burn helium, a star must have a massive amount of it in its core and the core must be hotter than about 100 million K. Neither are possible with a newly-born star.
19. Close binary stars share mass, shifting stars higher and lower on the main sequence.
27. A degenerate gas does not expand when you heat it, so there is no safety valve to slow down the nuclear reactions.

Applying the Concepts
35. (a) 0.1 L_{Sun}, (b) 529 L_{Sun}, and (c) 1.7×10^6 L_{Sun}.
39. 3.16×10^{13} km (about 1 pc)
40. 800 million years

Chapter 17

Check Your Understanding
1. b, d
2. b
3. d
4. b

Thinking about the Concepts
24. The neutrinos from SN 1987A escaped directly from the core, while the shock wave from core collapse took a few hours to break out of the envelope.
26. The ejecta from a supernova loses energy as it blows into low-density interstellar gas and dust, and thus the ejecta slow down.
27. Under the assumption that all stars in a cluster formed at approximately the same time, the "turnoff point" where main sequence stars are beginning to evolve into supergiants gives the age, since these stars have just finished their main-sequence lifetime.

Applying the Concepts
35. Approximately 3,000 L_{Sun}. Observed (or apparent) brightness.
37. 3.7×10^9 km/s^2 (380 trillion times stronger than on Earth)
40. 3.8×10^{23} kg. (b) 5.2 times the mass of the moon.

Chapter 18

Check Your Understanding
1. d
2. a
3. d
4. c

Thinking about the Concepts
17. *c, c*
24. It must be elongated (i.e., an oval or rugby-shaped ball) on the ship.
26. It will not affect us any differently than the stars of similar mass that are already at that distance.

Applying the Concepts
38. You would be 87% of your twin's age.
40. 10^{-25} m.
44. at 99% C, 3.5 yrs, 25.3 yrs, 25.3 yrs

Chapter 19

Check Your Understanding
1. d
2. c
3. c
4. a
5. c

Thinking about the Concepts
17. Cepheids found in the Andromeda Galaxy (M31) by Hubble showed that M31 was far too large and too far away to be part of our own Milky Way.
22. The bulge of a spiral has an older population of stars and the stars are on randomized orbits, which mimics on a small scale the structure of an elliptical galaxy.
26. AGNs are about the size of our Solar System. We know because the time variability of an object constrains its size, that is, an object can't be much larger than the timescales of variability, otherwise that signal would be washed out.

Applying the Concepts
33. (a) 0.143. (b) 4.3×10^4 km/s. 613 Mpc.
37. 512 AU
41. 5,540 Mpc

Chapter 20

Check Your Understanding
1. The Milky Way contains gas and dust as well as ongoing star formation. Observations of 21-cm radiation emitted from neutral hydrogen clouds show the existence of spiral arms. The galaxy has a well-defined rotation curve that flattens toward the visible edge. Finally, most stars in our galactic neighborhood share the same relative motion around the galactic center.
2. The disk contains old and young stars; the halo contains old stars.
3. c
4. Andromeda is moving toward us because of our mutual gravitational attraction.

Thinking about the Concepts
17. Globulars are distributed roughly spheroidally, showing that there is a large, spheroidal halo that encloses the disk of the galaxy. The center of that spheroid is roughly 8,000 pc from the Sun, showing that we are roughly that distance from the center of the galaxy.
21. Like most other spiral galaxies, ours has a flat rotation curve, which implies that up to 90 percent of the mass of our galaxy must be dark matter.
24. There is far too much dust along the path to the galactic center to see optical light from that region. Whereas dust is opaque to optical and UV light, it is transparent to radio, X-ray, and infrared, hence we can (and do) observe this region in these wavelengths.

Applying the Concepts
33. 1.8 billion years
36. (a) 92 kpc. (b) The halo is at least 3 times larger than the disk.
44. 4.5 million M_{Sun}

Chapter 21

Check Your Understanding
1. b, a
2. d
3. c
4. b
5. a

Thinking about the Concepts
19. Quite the opposite, Hubble's law implies that there is no center of the universe, since galaxies are moving away from each other.
23. Hubble's redshift-distance relationship, the chemical abundances of stars and the interstellar medium, the presence of the cosmic-microwave background, inhomogeneities and fluctuations within the CMB.
27. The tiny variations in the CMB are the microscopic fluctuations that grew over time into the large-scale structure we observe and live in today.

Applying the Concepts
35. 1.96×10^{10} yr, 9.79×10^9 yr.
37. 9 times.
42. 0.24 atoms/m^3

Chapter 22

Check Your Understanding

1. c
2. c
3. flatness problem and horizon problem
4. d
5. b

Thinking about the Concepts

19. Gravity holds them together.
20. The Hubble time does not take into account the changing dynamics that result from having different cosmological densities, a cosmological constant, or dark energy.
26. If a photon contains more energy than the combined mass energy of a particle and its antiparticle, then the photon can become a particle-antiparticle pair that flies off in different directions, conserving mass-energy and momentum.

Applying the Concepts

38. 5
40. 1.5×10^{-10} J.
41. 2×10^{-31} kg. We can ignore it.

Chapter 23

Check Your Understanding

1. c-b-d-a
2. b
3. c
4. b

Thinking about the Concepts

18. In the absence of luminous matter, we could detect a dark matter galaxy only from the gravitational influence it exerts on its surroundings or by gravitational lensing.
20. Quantum-sized (e.g., ultramicroscopic) fluctuations in density after the Big Bang
28. Hot dark matter could not be gravitationally bound enough to form the halos needed to contain galaxies and to explain galactic rotation curves.

Applying the Concepts

31. 1.05×10^{7} m/s.
37. 10^{18} times more time.
45. $6 \times 10^{14}\ M_{Sun}$.

Chapter 24

Check Your Understanding

1. d
2. a
3. d
4. a
5. b

Thinking about the Concepts

17. Amino acids were probably built from simpler elements and molecules, using energy from the Sun or lightning as a catalyst.
22. Plants and forests use photosynthesis, which relies on CO_2, and which was abundant far earlier than oxygen. Large animals required oxygen.
27. Absence of evidence is not evidence of absence. Conditions on Mars have changed over time, and these spacecraft did not look deep under the surface of Mars.

Applying the Concepts

34. 10 sec
38. 17 generations.
43. This is at the location of Mercury, outside the habitable zone.

Credits

These pages contain credits for all 24 chapters in the complete volume of *21st Century Astronomy,* Fifth Edition. The Solar System Edition does not include Chapters 15–23. The Stars and Galaxies Edition does not include Chapters 8–12.

Photos

Chapter 1

2-3 NASA **5 top** NASA **6** NASA and STScI **6** NASA **7 bottom** Sebastian Kaulitzki / Alamy **7 center bottom** Subaru Telescope (NAOJ), Hubble Space Telescope; Processing & Copyright: Roberto Colombari & Robert Gendler **11** Courtesy of the Archives, California Institute of Technology **12** Matheisl / Getty Images **15** ESA/NASA **16 top** NASA/JPL-Caltech **16 bottom** NASA/JPL-Caltech/Space Science Institute

Chapter 2

22-23 Raymond Patrick/National Geographic Creative/Corbis **24 bottom** Henry Westheim Photography / Alamy **27 left** Pekka Parviainen / Science Source **27 right** D. Nunuk / Science Source **38** Arnulf Husmo/Getty Images **44** akg-images / Rabatti - Dominigie / The Image Works **46** Nick Quinn **48** GSFC/NASA **49 top** Johannes Schedler / Panther Observatory **49 bottom** Anthony Ayiomamitis (TWAN)

Chapter 3

58-59 © Damian Peach **60 top** Royal Astronomical Society / Science Source **60 bottom** Tunc Tezel **61 center** Nicolaus Copernicus Museum, Frombork, Poland / Bridgeman Images **61 bottom** Bettmann/ Corbis **64** The Granger Collection, NYC **65 top** De Mundi Aetherei Recentioribus Phaenomenis, 1588. The Tycho Brahe Museum, Sweden. www.tychobrahe.com **65 bottom** SSPL / The Image Works **69** Galleria Palatina, Palazzo Pitti, Florence, Italy / The Bridgeman Art Library **70 top** The Granger Collection, NYC **70 bottom** SSPL/ Jamie Cooper / The Image Works **71** Lebrecht Music and Arts Photo Library / Alamy

Chapter 4

82-83 NASA **97 both** © Christopher Mackay **101** NASA

Chapter 5

108-109 Calvin_Bradshaw/Wikimedia Commons **114** Yva Momatiuk & John Eastcott/ Minden

Pictures/National Geographic Stock **123** Nigel Sharp, NOAO/NSO/Kitt Peak FTS/AURA/NSF **135 (all)** NASA

Chapter 6

142-143 © Laurie Hatch **144** ASU Physics Instructional Resource Team **145** Vik Dhillon/ University of Sheffield **147 top** ASU Physics Instructional Resource Team **147 bottom** Jim Sugar/Corbis **149 top** © Laurie Hatch **149 center** The Space Telescope Science Institute **149 bottom** ASU Physics Instructional Resource Team **151 center** Courtesy TMT Observatory Corporation **151 bottom** NASA Earth Observatory/NOAA NGDC **151 top** NASA, ESA, HEIC, and The Hubble Heritage Team (STScI/AURA) **153 top** Science Photo Library **153 bottom** Jean-Charles Cuillandre (CFHT) **155** NASA **155a** © Laurie Hatch **155b** NASA **155c** NASA **155d** NASA **155e** East Asian Observatory **155f** Photo by Dave Finley, courtesy National Radio Astronomy Observatory and Associated Universities, Inc. **155g** NRAO/AUI, James J. Condon, John J. Broderick, and George A. Seielstad **155** David Parker/Science Source **156 top** Roger Ressmeyer/Corbis **156 center** David Parker/ Science Source **156 bottom** Photo by Dave Finley, courtesy National Radio Astronomy Observatory and Associated Universities, Inc. **157 top** ALMA (ESO/NAOJ/NRAO)/W. Garnier (ALMA) **157 center** NASA Photo / Carla Thomas, (bottom right): NASA/Tom Tschida **157 bottom** NASA Photo / Carla Thomas, (bottom right): NASA/Tom Tschida **159** NASA/JPL-Caltech/MSSS **160** NASA/ JPL-Caltech/MSSS **162 top** CERN **162 bottom** Jim Haugen/NSF **163 top** University of Maryland **163 center** UPPA/Photoshot **164** Patrik Jonsson, Greg Novak & Joel Primack, UC Santa Cruz, 2008

Chapter 7

172-173 Caltech/NASA/JPL **175 top** NASA **175 bottom** Photograph by Pelisson, SaharaMet **178** Reuters/Corbis **185** NASA/Johns Hopkins University Applied Physics Laboratory/Carnegie Institution of Washington **189** NASA **191 bottom** NASA **193** NASA Ames/SETI Institute/JPL-Caltech

Chapter 8

200-201 NASA/JPL-Caltech/Univ. of Arizona **202** NASA **203** Montes De Oca & Associates **204 top** Photograph by D.J. Roddy and K.A. Zeller, USGS, Flagstaff, AZ. **204 bottom** NASA/Goddard/ MIT/Brown **205** NASA/JPL/Caltech **209** NASA **213 top** Grant Heilman Photography **213 bottom** Art Directors & TRIP / Alamy **213 bottom** Art Directors & TRIP / Alamy **214** Donald Duckson/ Visuals Unlimited/Corbis **218 bottom right** NASA **218 top** NASA/JSC **218 bottom left** NSSDC/NASA **219** NASA/Magellan Image/JPL **221 top** NASA/ JSC **221 center** K.C. Pau **221 bottom** NASA/Johns Hopkins University Applied Physics Laboratory/ Arizona State University/Carnegie Institution of Washington. Image reproduced courtesy of Science/AAAS. **222** NASA/MOLA Science Team **223** NASA **224** NASA/JPL/University of Arizona **225 top** NASA **225 bottom** NASA/JPL-Caltech/ University of Arizona/Texas A&M University **225 top right** NASA/JPL **227** © Don Davis **228** NASA/GSFC/Arizona State University

Chapter 9

234-235 Stockli, Nelson, Hasler, Goddard Space Flight Center/NASA. **244** NOAA **249 top** NASA **249 center** Design Pics Inc / Alamy **249 bottom** NASA **252 top** Dmitri Titov/ESA **252 center** © Ted Stryk 2007 **252 bottom** Dmitri Titov/ESA **253** NASA/JPL/Caltech **254 top** NASA/JPL-Caltech/MSSS **254 bottom** NASA/STScI

Chapter 10

268-269 NASA **273** NASA/JPL/University of Arizona, **274** NASA **275** NASA, JPL; Digital processing: Bjorn Jonsson (IAAA) **276 top** NASA/ JPL/Caltech. **276 bottom left** Lawrence Sromocsky, University of Wisconsin-Madison W. M. Keck Obervatory **276 bottom right** NASA, L. Sromovsky, and P. Fry (University of Wisconsin-Madison) **276 bottom right** NASA **276 bottom right** NASA **280** NASA, ESA, L. Sromovsky and P. Fry (University of Wisconsin), H. Hammel (Space Science Institute), and K. Rages (SETI Institute) **285** http://www.freenaturepictures.com/ **286** NASA/STScI. **288** NASA's Goddard Space

Flight Center/S. Wiessinger **290 (all)** NASA **295** NASA/JPL

Chapter 11

296-297 NASA/JPL/Space Science Institute **298** NASA/JPL/Caltech **299** NASA/JPL/University of Arizona, (right): ESA/DLR/FU Berlin (G. Neukum) **301(all)** NASA **302 top** NASA/JPL/Caltech **302 center** NASA/JPL/Caltech **302 bottom** NASA/JPL-CalTech **303 top** NASA **303 bottom** NASA/JPL/Space Science Institute **304 top** NASA/JPL/Space Science Institute **304 bottom** NASA/JPL **305 top** NASA **305 center** A. D. Fortes/UCL/STFC/NASA **305 bottom** NASA/JPL/USGS **306 bottom** NASA, ESA and A. Feild/STScI **306 top** NASA, ESA and A. Feild/STScI **307** NASA/JPL/Space Science Institute **308 top** NASA/JPL/Ted Stryk **308 bottom** NASA/JPL/Space Science Institute **310** NASA/JPL/Caltech **312** NASA/JPL/Space Science Institute **313 top** NASA/JPL/Space Science Institute **313 center** Stephanie Swartz/Photolia **313 bottom** NASA/JPL/Space Science Institute **314** NASA **316 top** NASA/JPL/Space Science Institute **316 bottom** NASA/JPL/Space Science Institute **317 top** NASA/JPL/Caltech **317 bottom right** NASA, ESA, and M. Showalter (SETI Institute) **318 top** NASA/JPL/Caltech. **318 bottom** NASA **320** NASA/JPL **325** NASA/JPL/Caltech

Chapter 12

326-327 ESA/Rosetta/NAVCAM, CC BY-SA IGO 3.0 **329** NASA **331** Images courtesy of NASA, ESA, JPL, and A. Feild (STScI) **332 top** NASA/JPL-Caltech/UCLA/MPS/DLR/IDA **332 bottom** NASA/JPL-Caltech/UCLA/MPS/DLR/IDA **336 top** NASA/JPL/USGS **336 bottom** NASA/JPL-Caltech/UCLA/MPS/DLR/IDA **337** NASA/JPL-Caltech/UCLA/MPS/DLR/IDA **342 top** Dr Robert McNaught **342 bottom** Courtesy of Terry Acomb **343 top** NASA/JPL/Caltech **343 bottom (both)** NASA/JPL/Caltech/UMD. **344** NASA/JPL-Caltech/UMD **345 left** Tony Hallas/Science Faction/Corbis **345 right** Barrie Rokeach/Getty Images **346 (all)** Courtesy of Ron Greeley **347** NASA/JPL-Caltech/LANL/CNES/IRAP/LPGNantes/CNRS/IAS/MSSS **348** ESO/Y. Beletsky **349 top** NASA **349 bottom left** Camera Press/Ria Novosti/Redux **349 bottom right** Lunin Gleb/ITAR-TASS Photo/Corbis **357** Larry Marschall and Christy Zuidema, Gettysburg College

Chapter 13

358-359 Tyler Nordgren **373** ESA/Hubble **378 center** Bettmann/Corbis **382 top** Norman Lockyer Observatory, Sidmouth **383** NASA/ESA, A.Feild/STScI

Chapter 14

390-391 NASA/SDO/AIA **399** Hinode JAXA/NASA/PPARC **404 top** Photo by Brocken Inaglory, 2006; http://creativecommons.org/licenses/by-sa/3.0/deed.en **404 bottom** Nigel Sharp, NOAO/NSO/Kitt Peak FTS/AURA/NSF **405 top (all)** Hinode JAXA/NASA **405 bottom** NASA/Science Source **407 (both)** NASA/SDO/Solar Dynamics Observatory **408 top** SOHO/ESA/NASA **409** NASA/SDO **409 top center** © 2001 by Fred Espenak, courtesy of www.MrEclipse.com. **409 top right** 2001 by Fred Espenak, courtesy of www.MrEclipse.com. **410 top** NASA SDO **410 bottom** SOHO/ESA and NASA **412** NASA **416** JPL/NASA

Chapter 15

420-421 ESO **422 top** Axel Mellinger **422 bottom** NASA/JPL-Caltech/IRAS/2MASS/COBE **423** AP Photos **425 top** NASA/JPL-Caltech/UCLA **425 bottom** IRAS/COBE (NASA) **426 top** Masashi Kimura/MAXI/SSC **426 bottom** Dr. Douglas Finkbeiner **427 bottom** Image courtesy of NRAO/AUI **427 top left** Courtesy of Stefan Seip (**427 top center** Courtesy of Emmanuel Mallart www.astrophoto-mallart.com & www.axisinstruments.com, **427 top right** ESO **428 (both)** ESO **430** ESA/Planck Collaboration **433 left** ESA/Herschel/PACS/SPIRE/Hill, Motte, HOBYS Key Programme Consortium **433 center** MPG/ESO **433 right** VLT/ISAAC/McCaughrean & Andersen/AIP/ESO **439** NASA/JPL-Caltech/ALMA **440 top** Anglo-Australian Observatory /Royal Observatory, Edinburgh, Photograph UK Schmidt, plates by David Malin. **440 bottom** Anglo-Australian Observatory /Royal Observatory, Edinburgh, Photograph UK Schmidt, plates by David Malin. **441** NASA/JPL-Caltech/UIUC

Chapter 16

448-449 NASA and ESA **463 left** NASA/JPL-Caltech/Univ.of Ariz. **463 right center** Minnesota Astronomical Society **463 left center** NASA, ESA, and the Hubble SM4 ERO Team **463 right center** NASA, ESA and The Hubble Heritage Team **470 top (all)** NASA/GSFC/D.Berry **470 bottom (all)** NASA

Chapter 17

478-479 NASA/CXC/SAO **484** NASA/CXC/GSFC/M.Corcoran et al.; Optical: NASA/STScI **488** X-ray: NASA/CXC/SAO; Optical: NASA/STScI; Infrared: NASA/JPL-Caltech/Steward/O.Krause et al. **489 top, center** Anglo-Australian Observatory, photographs by David Malin **489 bottom** Alexandra Angelich (NRAO/AUI/NSF); NASA

Hubble; NASA Chandra **489 top, center** Anglo-Australian Observatory, photographs by David Malin **495 both** NASA **495** NASA **496 top left** The Electronic Universe Project **496 top right** (NOAO / AURA / NSF / Science Source **496 bottom left** European Space Agency & NASA **496 bottom right** Barbara Mochejska Hedberg, J. Kaluzny (Copernicus Astronomical Center)

Chapter 18

506-507 NASA/ESA **521 top** NASA/ESA **521 bottom** ESA/Hubble & NASA **526** NASA/CXC/M.Weiss **527** NASA/CXC/Cal; Chandra X-ray Observatory Center

Chapter 19

534-545 NASA, ESA, and The Hubble Heritage Team (STScI/AURA) **536 top** NASA, ESA, and the Hubble Heritage Team (STScI/AURA) **536 bottom** NASA **539** NOAO/AURA/NSF **540** NASA and The Hubble Heritage Team (STScI/AURA) **541 top** European Southern Observatory **541 bottom** Johannes Schedler / Panther Observatory **543** BJ Fulton (LCOGT) / PTF/Byrne Observatory at Sedgwick Reserve **543 top** BJ Fulton/Bryne Observatory at Sedgewick Reserve and the Palomar Transient Factory **543 bottom** NASA/Swift/Peter Brown, Univ. of Utah **545** NASA **548** NASA/CXC/Penn State/G. Garmire; Optical: NASA/ESA/STScI/M. West **550 top** HST **550 bottom** NASA/CXC/CfA/R.Kraft et al; Radio: NSF/VLA/Univ. Hertfordshire/M.Hardcastle; Optical: ESO/WFI/M.Rejkuba et al. **556** NASA / Swift / NOAO / Michael Koss and Richard Mushotzky (Univ. Maryland) **558** NASA, ESA, and D. Coe and G. Bacon (STScI) **563** NASA, ESA, and the Hubble Heritage Team (STScI/AURA)

Chapter 20

564-565 Brad Goldpaint / Aurora Open / Alamy **566 top** © Axel Mellinger **566 bottom** C. Howk and B. Savage (University of Wisconsin); N. Sharp (NOAO)/WIYN, Inc. **567 top** © Lynette Cook, all rights reserved **567 center** (bottom left): NOAO/AURA/NSF **567 bottom left** NASA/JPL-Caltech **567 bottom right** © Robert Gendler / robgendlerastropics.com **568 top** Todd Boroson/NOAO/AURA/NSF **568 bottom** Richard Rand, University of New Mexico **569** IMG100 / Media Bakery **578** NASA **580** David A. Aguilar (CFA) **581 top** Anglo-Australian Observatory, David Malin **581 bottom** Amanda Smith/University of Cambridge **582 top** NASA **582 bottom** NASA, ESA, Z. Levay and R. van der Marel (STScI), T. Hallas, and A. Mellinger

Chapter 21

590-591 ESA and the Planck Collaboration **594** NASA, ESA, H. Teplitz and M. Rafelski (IPAC/Caltech), et al. **605 top** Lucent Technologies, Bell Labs **605 center** Ann and Rob Simpson Nature Photography **605 bottom** Ann and Rob Simpson Nature Photography **607** ESA and the Planck Collaboration **615** Courtesy of Stacy Palen

Chapter 22

616-617 NASA/JPL/Space Science Institute **622** NASA, ESA, H. Teplitz and M. Rafelski (IPAC/Caltech), A. Koekemoer (STScI), R. Windhorst (Arizona State University), and Z. Levay (STScI) **645** CERN PhotoLab

Chapter 23

646-647 NASA, ESA, H. Teplitz and M. Rafelski (IPAC/Caltech), A. Koekemoer (STScI), R. Windhorst (Arizona State University), and Z. Levay (STScI) **648 top** NASA/ESA/ESO/NAOJ/G. Paglioli **648 bottom** R. Brent Tully (U. Hawaii) et al., SDvision, DP, CEA/Saclay **649** Max Tegmark/SDSS Collaboration **651 top** NASA/CXC/MIT/E.-H Peng et al; Optical: NASA/STScI **651 bottom** NASA, A. Fruchter and the ERO team (STScI, ST-ECF). NASA, ESA **657** NASA/WMAP Science Team **658** Matthew Turk, Tom Abel, Brian O'Shea Visualization: Matthew Turk, Samuel Skillman **659** NASA/JPL-Caltech/GSFC **660** NASA, ESA, P. Oesch and I. Momcheva (Yale University), and the 3D-HST and HUDF09/XDF Teams **662 bottom** NASA/Hubble Space Telescope Cosmic Assembly Near-infrared Deep Extragalactic Legacy Survey **662 top** Guedes, Javiera, et al. The Astrophysical Journal, Volume 742, Issue 2, article id. 76 (2011). **663** NASA/CXC/SAO/P **664 top** NASA, ESA, and N. Pirzkal (STScI/ESA) **664 bottom** X-ray (NASA/CXC/IfA/C. Ma et al.); Optical (NASA/STScI/IfA/C. Ma et al. **666 top** Courtesy Joel Primack and George Blumenthal **666 bottom** NASA, ESA, and A. Feild (STScI)

Chapter 24

674-675 SETI Institute **676 bottom** Woods Hole Oceanographic Institution, Deep Submergence Operations Group, Dan Fornari **676 top** Dr. Michael Perfit, University of Florida, Robert Embley/NOAA **677 top** Gaertner / Alamy **677 center** Chris Boydell/Australian Picture Library/Corbis **677 bottom** National Park Service Photo by Jim Peaco **678 bottom** Peter Essick/Aurora Photos **678 top** Dr. Robert Calentine/Visuals Unlimited/Corbis **680** Courtesy of National Science Foundation **681** With permission of the Royal Ontario Museum © ROM **685** NASA/JPL-Caltech/MSSS **689** NASA **691** F. Drake (UCSC) et al., Arecibo Observatory (Cornell, NAIC) **693** NASA/JPL-Caltech/Space Science Institute

Text

Appendix 9 text excerpt: Resolutions 5A and 6A from "Press Release - IAU 2006 General Assembly: Result of the IAU Resolution Votes." Reprinted by permission of IAU.

Line Art

188 (Figure 7.17) Graph: "Radial Velocity/ Year" from Exoplantets.org. Reprinted by permission of the Department of Terrestrial Magnetism, Carnegie Institution of Washington.
659 (Figure 23.10) Figure from "The First Galaxies," by V. Bromm & N. Yoshida. Reproduced with permission of Annual Review of Astronomy & Astrophysics, Volume 49 (373-407) © by Annual Reviews, http://www.annualreviews.org
687 (Figure 24.15) Figure 1 from "Summary of the Limits of the New Habitable Zone." PHL.UPR.edu, March 23, 2013. Reprinted by permission of PHL@UPR Arecibo.

Index

This index contains page references for all 24 chapters in the complete volume of *21st Century Astronomy,* Fifth Edition. The Solar System Edition does not include Chapters 15–23 (pp. 420–673). The Stars and Galaxies Edition does not include Chapters 8–12 (pp. 200–357).